Baltic Human-Animal Histories

STUDIES IN LITERATURE, CULTURE, AND THE ENVIRONMENT

Edited by
Hannes Bergthaller, Gabriele Dürbeck,
Robert Emmett, Serenella Iovino, Ulrike Plath

Editorial Board:
Stefania Barca (University of Coimbra, Portugal)
Axel Goodbody (University of Bath, UK)
Isabel Hoving (Leiden University, The Netherlands)
Dolly Jørgensen (Luleå University of Technology, Sweden)
Timo Maran (University of Tartu, Estonia)
Serpil Oppermann (Cappadocia University, Ürgüp / Nevşehir, Turkey)
Dana Phillips (Towson University, Baltimore, USA)
Stephanie Posthumus (McGill University, Montreal, Canada)
Christiane Solte-Gresser (Saarland University, Saarbrücken, Germany)
Keijiro Suga (Meiji University, Tokyo, Japan)
Pasquale Verdicchio (University of California, San Diego, USA)
Berbeli Wanning (University of Siegen, Germany)
Sabine Wilke (University of Washington, Seattle, USA)
Hubert Zapf (University of Augsburg, Germany)
Evi Zemanek (University of Freiburg, Germany)

VOLUME 12

Linda Kaljundi, Anu Mänd, Ulrike Plath, Kadri Tüür (eds.)

Baltic Human-Animal Histories

Relations, Trading, and Representations

Lausanne - Berlin - Bruxelles - Chennai - New York - Oxford

Bibliographic Information published by the Deutsche Nationalbibliothek
The Deutsche Nationalbibliothek lists this publication in the Deutsche Nationalbibliografie; detailed bibliographic data is available in the internet at http://dnb.d-nb.de.

Library of Congress Cataloging-in-Publication Data A CIP catalog record for this book has been applied for at the Library of Congress.

PAUL KAEGBEIN STIFTUNG

The publication of this book was supported by the Book Publication Subvention issued by the Association for the Advancement of Baltic Studies and by Paul-Kaegbein-Stiftung.

Cover illustration: Johann Conrad Susemihl. Goldfinch and wagtail. Coloured copper engraving Illustration for *Getreue Abbildungen und naturhistorische Beschreibung des Thierreichs aus den nördlichen Provinzen Russlands, vorzüglich Liefland, Ehstland und Kurland betreffend* by Ernst Wilhelm Drümpelmann and Wilhelm Christian Friebe. Riga, 1807. The image is in the public domain.

ISSN 2365-645X
ISBN 978-3-631-87992-4 (Print)
E-ISBN 978-3-631-87993-1 (E-PDF)
E-ISBN 978-3-631-87994-8 (EPUB)
DOI 10.3726/b20592

© 2024 Peter Lang Group AG, Lausanne
Published by Peter Lang GmbH, Berlin, Deutschland

info@peterlang.com - www.peterlang.com

All rights reserved.
All parts of this publication are protected by copyright.
Any utilization outside the strict limits of the copyright law, without the permission of the publisher, is forbidden and liable to prosecution.
This applies in particular to reproductions, translations, microfilming, and storage and processing in electronic retrieval systems.

This publication has been peer reviewed.

Contents

List of Contributors 7

Linda Kaljundi, Ulrike Plath, Kadri Tüür
Introduction. Entangled Human–Animal Histories: Tracing Multispecies
Relations in the Eastern Baltic Region 11

I. HUMAN–ANIMAL RELATIONS

Inna Põltsam-Jürjo
Animals in Medieval Livonian Laws 43

Ken Ird
When Men Stayed in the Barn for Too Long: Cases of Bestiality in the
Early Modern Baltic Provinces 59

Kaarina Rein
Depiction of Animals in the Medical Works of the Seventeenth Century
University of Tartu 75

Ulrike Plath
Animal Abolitionism and Early Environmentalism in Nineteenth Century
Riga 95

Anita Zariņa / Dārta Treija / Ivo Vinogradovs
Bison in the Latvian Ethnoscape: Contingency of (Not) Becoming 121

Eve Rannamäe / Anneli Ärmpalu-Idvand
Estonian Aboriginal Sheep in Modern History: Description, Importance,
and the Story of Becoming the Kihnu Native Breed 141

II. TRADING

Juhan Kreem
Horses of the Teutonic Order in Livonia: An Attempt to Map This Vital
Resource for Knights 165

Ivar Leimus
The Horse in Livonia as a Strategic Commodity in the Middle Ages 183

Lembi Lõugas
Fish and Fish Trade by the Archaeological Sources 201

Kadri Tüür
Trawling for Atlantic Herring in Estonian Literature 217

III. REPRESENTATIONS

Tõnno Jonuks
Griffins in the Eastern Baltic Late Iron Age 253

Anu Mänd
Visual Representation of Animals in Livonian Urban Space,
ca. 1400–1550 275

Stefan Donecker (+)
Dogs of War. Wolves and Warfare in Early Modern Livonia
(ca. 1555–1605) 305

Meelis Friedenthal
Cats, Allergy and Occult Powers in Early Modern Disputations 321

Jaanika Anderson / Hilkka Hiiop
Fashion or Conceptual Choice: The Motifs of Animals, Birds, and
Semi-Animals in Pompeian-Style Interiors in Estonia 339

Index 365

List of Contributors

Jaanika Anderson is the Director of Research in the University of Tartu Museum and curator of several exhibitions. She has researched and published on museum education and reception of ancient art. Her research interest focuses on the history of university collections and methods for implementation the scientific collections in teaching processes during 19th and 20th centuries. Her special interest is interpretation and reception of ancient art in Estonian art scene. She received the PhD in classical philology in 2015.

Stefan Donecker (†) was a historian specializing on Baltic, Scandinavian and European early modern cultural history and history of ideas. He received his PhD from the European University Institute, and held fellowships at different institutions. His research focused primarily on the construction of identities and stereotypes, cultural transfers, and migration. In addition to numerous articles, his publications include the monograph *Origines Livonorum. Frühneuzeitliche Hypothesen zur Herkunft der Esten und Letten* (2017).

Meelis Friedenthal is an Associate Professor of Intellectual History at the University Library and in the University of Tartu, where his main research topic is the intellectual history of the Baltic Sea region in the 17th century. He has published articles on early modern intellectual history and recently edited a volume with Robert Seidel and Hanspeter Marti on *Early Modern Disputations and Dissertations in an Interdisciplinary and European Context* (2020).

Hilkka Hiiop is Professor at the Estonian Academy of Art, Department of Cultural Heritage and Conservation and Dean of the Faculty of Art and Culture. Her PhD thesis treated the conservation management of contemporary art. She has studied and worked as a conservator in Berlin, Amsterdam and Rome, supervised a number of conservation and technical investigation projects in Estonia, curated exhibitions, and conducted scientific research on conservation and technical art history.

Ken Ird is a trained historian working as a curator in the University of Tartu museum. His fields of study comprise social history, history of sexuality, as well as educational history, with main focus on 18th- and 19th-centuries Estonia.

Tõnno Jonuks, PhD in archaeology, has been working mostly on prehistoric religion, focusing on the materiality of religion. He has conducted studies about Estonian religion-related artefacts since the Mesolithic to Modern Age and recently investigated the contemporary materiality of religion. He is working at the department of folkloristics at Estonian Literary Museum and part-time at Tallinn University.

Inna Jürjo studied history at the University of Tartu, she received her PhD at Tallinn University. She is working on medieval and early modern urban and social history. Her research interests include history of culture, everyday life and food culture in Estonia. She is a senior researcher at Tallinn University, School of Humanities, Institute of History, Archaeology and Art History.

Linda Kaljundi is a Professor of cultural history at the Estonian Academy of Arts, and a research fellow at Tallinn University. Her research interests focus on Baltic medieval and early modern history, memory studies, and environmental humanities. She has published on history writing, historical novels and images, heritage, nature and nationalism.

Juhan Kreem studied history in Tartu and medieval studies in Budapest. Since 1996 he is working in Tallinn City Archives. His main field of interest is Teutonic Order, about what he has published extensively. Above that he has done research also on urban history, history of history writing and maritime history. He is a board member of International Commission for the History of the Teutonic Order and member of the International Commission for the History of Towns.

Ivar Leimus is a senior research fellow at the Estonian History Museum. Holding a PhD in numismatics, he has published widely on Estonian economic history from the Viking period to the present. His main research focus has been on medieval economic history and numismatics. Next to research, he has also been active as a translator of early modern chronicles into Estonian.

Lembi Lõugas is a research track Associate Professor at the Tallinn University and the head of the Archaeological Research Collection department. Her areas of specialisation are archaeozoology and environmental archaeology. Her main research interests are palaeozoology, fauna history (fish and sea mammals), archaeo-ichthyology, archaeogenetics, digital archaeozoology (data management).

Anu Mänd is Professor of Art History at the University of Tartu, and Associate Professor at Tallinn University. She has specialized in medieval art, animal history and social history in the Baltic Sea region. She was the head of the Institute of History, Archaeology and Art History at Tallinn University from 2015 to 2021.

Ulrike Plath is a Professor for Baltic German studies and environmental history at Tallinn University. She has been dealing with Baltic food, animal and climate history, but also with Baltic German cultural and literature history. She is a founder of the Estonian Center for Environmental History (KAJAK) and its first head in 2011 and has been serving the European Society for Environmental History (ESEH) in different committees since 2013. She is also working as a researcher at the Estonian Academy of Sciences.

Eve Rannamäe, PhD, is an Associate Professor of Archaeology at the University of Tartu, Estonia. Her main field of research is zooarchaeology, with a focus on livestock husbandry in the past, development of native breeds, and the use of biomolecular methods in zooarchaeological studies. Her works include general overviews on animal remains from archaeological sites, including discussions on the consumption and use of different species in the past; and analyses on ancient DNA, where affinities between the past and present-day populations are studied. In collaboration with present-day Kihnu native sheep breeders, one of her main interests has been the development of sheep populations in Estonia from the Prehistory to present-day.

Kaarina Rein is a research fellow at the Research Centre of the University of Tartu Library. She has studied classical philology at the University of Tartu and defended her PhD thesis at the Faculty of Theology of the same university. Her works include articles about early modern disputations and orations in the Baltic region, reception of ancient literature in Estonia and translation studies, as well as several translations from Ancient and Modern Greek and Latin into Estonian.

Dārta Treija is an environmental geographer, she has studied at Vidzeme University College and University of Latvia. Her research focuses on landscape values and spatial aspects of nature conservation, blending in communication theories. She has been involved in practical nature conservation, working for both nongovernmental and governmental institutions.

Kadri Tüür is a researcher in the project *Estonian Environmentalism in the 20th century: ideology, discourses, practices* at Tallinn University. Her PhD thesis applied semiotic methodology in the analysis of written nature representations. She served as the head of Estonian Center for Environmental History KAJAK from 2018 to 2021.

Ivo Vinogradovs is an environmental geographer who currently works as a leading researcher at the Department of Geography, University of Latvia. Ivo does research in Landscape Ecology, Ecosystem Services and Geoinformatics (GIS).

Anneli Ärmpalu-Idvand got her veterinary diploma in 1991 from the Estonian Agricultural Academy and has since actively worked as a veterinarian. Since 2001 she has been involved in several research projects on the investigation, conservation, and acknowledgement of the Kihnu native sheep, and since 2007 she is one of the founders and the head of the Kihnu Native Sheep Breeders Society. In her everyday work in breeding and conserving the endangered Kihnu native sheep, she also contributes to preserving the skills and knowledge related to traditional sheep keeping.

Anita Zariņa is an environmental geographer, associate professor at University of Latvia. In her past projects she has studied path dependent landscape change and modernist transformation of wetlands, as well as discursive aspects of Latvia's landscapes. Currently her work is related to nature discourses, wildlife conservation that includes her interest in rewilding and human-animal relations.

Linda Kaljundi, Ulrike Plath, Kadri Tüür

Introduction
Entangled Human–Animal Histories: Tracing Multispecies Relations in the Eastern Baltic Region

> *Humans and animals exist in entangled agential networks, and their interrelations have consequences on the world they co-inhabit and co-create.*[1]

Entanglement as an Approach

Animals are newcomers in Estonian and Latvian history writing. Their bioinvasion of the habitat of Baltic historians started at the very end of the last millennium following a global trend.[2] In the present edited volume we aim to show how the study of Baltic history and other branches of historical humanities are entangled with the field of interdisciplinary human–animal studies[3] and how historical entanglements between the human and the more-than-non-human world can be explained and narrated.

The possibilities of writing multi-species histories have been tested recently[4], inspired by the fascinating outcomes from the field of multispecies

---

1 Villanueva, Gonzalo: "Animals Are Their Best Advocates: Interspecies Relations, Embodied Actions, and Entangled Activism". *Animal Studies Journal* 8(1), 2019, pp. 190–217, p. 190.
2 For a broad overview over methodological approaches in historical animal studies see Roscher, Mieke / Krebber, André / Mizelle, Brett (eds.): *Handbook of Historical Animal Studies*. Walter de Gruyter: Berlin / Boston 2021.
3 For entangled approaches in environmental history see: Jørgensen, Finn Arne / Karlsdottir, Unnur B. /Mårald, Erland et al.: "Entangled Environments: Historians and Nature in the Nordic Countries". *Norsk Historisk Tidsskrift* 1, 2013, pp. 9–34. DOI:10.18261/ ISSN1504-2944-2013-01-02; Mullin, Molly H.: "Mirrors and Windows: Sociocultural Studies of Human–Animal Relationships". *Annual Review of Anthropology* 28(1), 1999, pp. 201–224. DOI:10.1146/annurev.anthro.28.1.201; Kohn, Eduardo: "Anthropology of Ontologies". *Annual Review of Anthropology* 44, 2015, pp. 311–327. DOI:10.1146/ annurev-anthro-102214-014127; Bargheer, Stefan: *Moral Entanglements. Conserving Birds in Britain and Germany*. The University of Chicago Press: Chicago / London 2018.
4 Tsing, Anna Lowenhaupt / Mathews, Andrew S. / Bubandt, Nils: "Patchy Anthropocene: Landscape Structure, Multispecies History, and the Retooling of

ethnography[5] and human–animal studies.[6] Although historians have not been so eager in taking the holobiont turn and therefore the wilderness inside us seriously,[7] they increasingly raise the question of whether or not we can aim for a "multispecies knowledge of the past"[8] and by doing this agree that history has become a "record of many trajectories of world-making, human and not human."[9] Leaving aside the philosophical question of if and to what extent we can narrate the history of another species from a human perspective, we will concentrate here on entanglements as a historical method that can be used not only to bridge the human and the non-human spheres, but also to overcome the

---

Anthropology. An Introduction to Supplement 20". *Current Anthropology* 60, 2019, pp. 186–197. DOI:10.1086/703391; Schoenbrun, David L. / Johnson, Jennifer L.: "Toward multi-species histories: Introduction: Ethnic Formation with Other-than-Human-Beings". *History in Africa* 45, 2018, pp. 307–345. DOI:10.1017/hia.2018.11; Tamm, Marek / Simon, Zoltán Boldizsár: "More-than-Human History: Philosophy of History at the Time of the Anthropocene". In: Kuukkanen, Jouni-Matti (ed.): *Philosophy of History: Twenty-First-Century Perspectives*. Bloomsbury: London 2020, pp. 198–215, Domanska, Ewa: "Animal History". *History and Theory* 56(2), 2017, pp. 267–287. DOI:10.1111/hith.12018.

5   Kirksey, Eben / Helmreich, Stefan: "The Emergence of Multispecies Ethnography". *Cultural Anthropology* 25(4), 2010, pp. 545–576. DOI:10.1111/j.1548-1360.2010.01069.x; Livingston, Julie / Puar, Jasbir K.: "Interspecies". *Social Text* 29(1), 2011, pp. 3–14. DOI:10.1215/01642472-1210237; Ogden, Laura A. / Hall, Billy / Tanita, Kimiko: "Animals, Plants, People, and Things: A Review of Multi-species Ethnography". *Environment and Society: Advances in Research* 4, 2013, pp. 5–24. DOI:10.3167/ares.2013.040102; Country, Bawaka / Wright, Sarah / Suchet-Pearson, Sandie et al.: "Co-becoming Bawaka: toward a Relational Understanding of Place / Space". *Progress in Human Geography* 40(4), 2016, pp. 455–475. DOI:10.1177/0309132515589437.

6   Shapiro, Kenneth / DeMello, Margo: "The State of Human–Animal Studies". *Society and Animals* 18, 2010, pp. 307–318. DOI:10.1163/156853010X510807; Métraux, Alexandre: "On Some Issues of Human–Animal Studies: An Introduction". *Science in Context*, 29(1), 2016, pp. 1–10. DOI:10.1017/S0269889715000368.

7   Simon, Jean-Christophe / Marchesi, Julian R. / Mougel, Christophe et al. "Host-microbiota Interactions: From Holobiont Theory to analysis". *Microbiome* 7(5), 2019. DOI:10.1186/s40168-019-0619-4; Haraway, Donna J.: *Staying with the Trouble: Making Kin in the Chthulucene*. Duke University Press: New York 2016, pp. 58–98.

8   Domanska, Ewa: "Posthumanist History". In: Tamm Marek / Burke Peter (eds.): *Debating New Approaches to History*. London: Bloomsbury 2018, pp. 327–338, p. 337.

9   Tsing, Anna Lowenhaupt: *The Mushroom at the End of the World. On the Possibility of Life in Capitalist Ruins*. Princeton University Press: Oxford and Princeton 2015, p. 168.

gap between human–animal and multi-species studies and classical approaches as cultural history or the history of religion.[10]

We strongly agree with the feminist philosopher Lori Gruen and her latest book "Entangled Empathy" that entanglements between species are given by nature and do not need to be constructed: "Our relationships with human and animal others co-constitute who we are and how we configure our identities and agency, even our thoughts and desires. We can't make sense of living without others, and that includes other animals. We are entangled in complex relationships and rather than trying to accomplish the impossible by pretending we can disentangle, we would do better to think about how to be more perceptive and more responsive to the deeply entangled relationships we are in."[11] Or, as Winfried Speitkamp put this in 2017: "Separation and unrelatedness between humans and animals are neither historically provable nor theoretically thinkable."[12] This holds true especially in our anthropocenic world where human influence and 'bad entanglements' have grown to an unavoidable factor of natural and even planetary life. Entanglements thus can help to unfold deep mutual dependency of actors[13] and can evoke a feeling of existential spookiness or uncanniness as Doug Jackson put it referring to Albert Einstein.[14] Although the analysis of the deep causality between different bodies across time and space might be reserved for quantum physicists, a new deep understanding of interspecies entanglements has entered the realm of the humanities.

In this volume we are staying within the scope of humanities and not delving into physical and bodily entanglements between humans and their animal co-mates – a research field that has been exploding since the beginning

---

10 McInerney, Jeremy: *The "Entanglement" of Gods, Humans, and Animals in Ancient Greek Religion*. Routledge: London 2020. The six-volume *A Cultural History of Animals*, ed. by Linda Kalof and Brigitte Resl, Oxford and Berg: New York 2007, followed a multidisciplinary approach, combining history, philosophy, art history, etc.

11 Gruen, Lori: *Entangled Empathy: An Alternative Ethic for Our Relationships with Animals*. Lantern books: Brooklyn 2015.

12 "Trennung und Beziehungslosigkeit zwischen Menschen und Tieren sind freilich weder historisch nachweisbar noch theoretisch denkbar." Speitkamp, Winfried: "Vielfältig verflochten? Zugänge zur Tier-Mensch Relationalität". In: Forschungsschwerpunkt «Tier–Mensch–Gesellschaft» (ed.): *Vielfältig verflochten. Interdisziplinäre Beiträge zur Tier-Mensch-Relationalität*. (Human–animal Studies 13). Transcript: Bielefeld 2017, pp. 9–32, p. 10. DOI:10.1515/9783839436851-001.

13 Hodder, Ian: "The Entanglements of Humans and Things: A Long-Term View". *New Literary History* 45(1), 2014, pp. 19–36, 20. DOI:10.1353 /nlh.2014.0005.

14 Jackson 2017.

of the pandemic.¹⁵ Nevertheless we have had larger and more complex scales of entanglements and their dimensions in mind while editing this volume.¹⁶ Taking interspecies relatedness and entanglements as a given, we skip classifying animals according to their relation with their human co-mates (wild animals, livestock, pets), but will concentrate on the different forms of entanglements and how they were discussed in Baltic historical texts, ongoing history writing, and within historically oriented Baltic humanities. With Speitkamp we want to ask what forms of human–animal entanglements can we find in regional history records from the medieval times up to the twentieth century and how their interpretations can be framed within environmental humanities and human–animal studies.

In doing so, we are using the competence of Baltic humanities in entangled research, which during the last decades has taught us to analyse local history as a multiperspective phenomenon, bringing together Estonian, Latvian, Danish, German, Swedish, Polish and Russian cultural and history writing.¹⁷ Coming from a highly transnational form of writing and interpreting cultural history, we wanted to open this approach to a closer analysis of interspecies' entanglements. We define entanglements as main elements of human–animal relations that may be conceptualised on at least three different scales. On the emotional scale entanglements can vary between emotional and rational forms of relatedness; on the subjective scale they range from personal to impersonal involvement; on a practical scale, they can be active or passive, defining thence the agency not only for the human, but also for the non-human actors. Although running the risk of over-simplification, we can easily describe certain forms of relations based on the three scales of entanglements: the relation between butchers in the slaughterhouses and the animals there can be described as an active, personal, rational relation, while from the animal perspective it would be passive, personal, emotional. A poet's relation toward animals might be definitively very personal

---

15 Nading, Alex M.: "Humans, Animals, and Health. From Ecology to Entanglement". *Environment and Society: Advances in Research* 4, 2013, pp. 60–78. DOI:10.3167/ares.2013.040105.
16 Chakrabarty, Dipesh: *The Climate of History in a Planetary Age*. The University of Chicago Press: Chicago, London 2021.
17 Andrejevs, Nikita / Braedder, Anne / Dabrowska, Malgorzata, et al. *Baltic Sea History: New Perspectives on the History of the Baltic Sea Region*. Academia Baltica: Lübeck / Oeversee 2019.

and emotional, but still passive. Animal activism on the contrary would be personal, emotional, and active.[18]

In the case of diseases, the care of a certain animal can also develop a closer individualised relationship, which, however, can turn into a mass problem in times of pandemics such as swine flu or other diseases that are carried forth through human–animal entanglements. Also from the human perspective, we can witness constant switches from one category to the other. So, for example, professional vivisectors such as Nicolai Ivanovich Pirogov could change their mind concerning human–animal relations and prefer certain individualised relations due to the private–ethical approach. Changes in food habits that are personal choices, but are driven by ethical understandings of whole generations, mark the importance of historical change in human–animal relations. While the problem of mass relations is mainly a problem of the long twentieth century, marked by urbanisation, overpopulation and industrial agriculture of meat, we can still also find unpersonalised relations toward animals in previous times.

In general, the majority of material and written sources on human–animal relations in the Baltics from prehistoric time onwards speak about unpersonalised relationships. While personalised animals or individualised relations are difficult to find from historical sources, these few stories are, of course, of great value. We take historical developments that might have changed the character of human–animal entanglements seriously. Alienation of former close entanglements in the course of urbanisation and mass production has to be taken into account, although this never means that single persons and groups involved in the process were unable to have very emotional feelings and relations with each other. Even more, we want to claim that interactions, emotions, and discourses often counterbalance each other. The deeper the entanglements were between humans and animals in everyday life, the more intensive was the need for othering animals from the human sphere.[19] Alienation from the animal world evokes the need for new affective discourses that emphasise our close relations. Even the whole variety of disentanglement and othering can be seen as an offspring of the primarily given entanglements.

---

18 Villanueva 2019.
19 Borkfelt, Sune: "Non-Human Otherness: Animals as Others and Devices for Othering". In: Sencindiver, Susan Yi / Beville, Maria / Lauritzen, Marie (eds.): *Otherness. A Multilateral Perspective*. Peter Lang: Frankfurt a. M. et al. 2011, pp. 137–154.

Multispecies Relations in Baltic History

The Baltic region can be defined in a number of ways – it can designate the broader area neighbouring the Baltic Sea, the Eastern Baltic territories, or the three Baltic countries that exist today: Estonia, Latvia and Lithuania. In this volume, the Eastern Baltic region is defined in a narrow way, meaning the territory of today's Estonia and Latvia, which have had a shared history since medieval times. Their history is shaped by features that are typical to border areas: multiculturalism and ethnicity, numerous changes of power and domination, different and complicated forms of colonialism, various cultural influences and transfers. Lithuania became a part of this narrow definition of the Baltic region only at the turn of the eighteenth and nineteenth centuries, when it started to share some very comparable features with its history, as well as the identities of Latvia and Estonia.[20]

Today, one of the most prominent shared features of the three Baltic countries concerns the myths about these peoples as nature's nations.[21] While this identity narrative emphasises the uniqueness of the Baltic relationship with nature, it nevertheless relies on the transnational importance of nature in nation-building and close entanglements with the German, Russian Imperial and Soviet traditions of thinking about the relations between nature and national identity. Moreover, the idea of a "natural" close relationship to nature is also closely related to the postcolonial character of Baltic identity, as the idea that Baltic peoples are closer to nature than to culture was first constructed among the colonial, German-speaking and Russian imperial elites, and only later adapted by the young Estonian, Latvian and Lithuanian nations in the nineteenth and twentieth centuries.[22] While today, the myths about the Baltic nations' close relations with the natural world are mostly constructed around the forest[23] (another indicator

---

20 Kasekamp, Andres: *A History of the Baltic State*s. Palgrave: London 2010; Angermann, Norbert / Brüggemann, Karsten: *Geschichte der baltischen Länder*. Reclam: Stuttgart 2018.
21 Schwartz, Katrina Z. S.: *Nature and National Identity After Communism: Globalizing the Ethnoscape*. University of Pittsburgh Press: Pittsburgh, PA 2006; Annus, Epp: "A Post-Soviet Eco-Digital Nation? Metonymic Processes of Nation-Building and Estonia's High-Tech Dreams in the 2010s." *East European Politics and Societies and Cultures* 36(2), 2020. DOI:10.1177 /0888325420958138.
22 Plath, Ulrike: *Esten und Deutsche in den baltischen Provinzen Russlands. Fremdheitskonstruktionen, Lebenswelten, Kolonialphantasien 1750–1850*. Harrassowitz: Wiesbaden 2011.
23 Jonuks, Tõnno / Remmel, Atko: "From Nature Romanticism to Eco-Nationalism: The Development of the Concept of Estonians as a Forest Nation". *Folklore: Electronic*

of close entanglements with German and Russian national identity where forest has been highly significant[24]), earlier periods have also witnessed the prominence of various ideas about the importance of diverse close human–animal relations and animism in the culture and religion of the Baltic peoples.[25]

This short introduction does not enable us to give any deeper overview of human–animal history in the Baltics, but it allows us to rationalise and periodise some of the most crucial developments, topics and characteristics. In the Eastern Baltic region, the beginning of entangled human–animal relations can be marked with the end of the Ice Age when these territories were first populated by human hunters and gatherers. While the idealisation of the pre-Christian and pre-colonial period is of paramount importance to the nineteenth-century and later constructions of Baltic peoples as nature's nations, human–animal histories of the ancient period inevitably rely only on material sources, as there are no written sources produced in the Baltics prior to the Crusades in the late twelfth and early

---

Journal of Folklore 81, 2021, pp. 33–62. DOI:10.7592 /FEJF2021.81.remmel_jonuks; Kaljundi, Linda: "Eestlus – loodusrahvamüüt keskkonnakriisi ajastul." [Estonianness – The Myth of Nature's Nation in the Age of Environmental Crisis.] *Vikerkaar* 9, 2019, pp. 114–117; Kaljundi, Linda: "Uusmetsik Eesti." [New-wild Estonia]. *Vikerkaar*, 7–8, 2018, pp. 68–80.

24 Zechner, Johannes: "From Poetry to Politics. The Romantic Roots of the 'German Forest'". In: Beinart, William / Middleton, Karen / Pooley, Simon (eds.): *Wild Things. Nature and the Social Imagination*. White Horse Press: Winwick 2013, pp. 185–210. Costlow, Jane T.: *Heart-Pine Russia: Walking and Writing the Nineteenth-Century Forest*. Ithaca: Cornell University Press: Ithaca 2012; Wilson, Jeffrey K.: *The German Forest: Nature, Identity, and the Contestation of a National Symbol, 1871–1914*. University of Toronto Press: Toronto 2012; Imort, Michael: "A Sylvan People: Wilhelmine Forestry and the Forest as a Symbol of Germandom". In: Lekan, Thomas / Zeller, Thomas (eds.): *Germany's Nature: Cultural Landscapes and Environmental History*. Rutgers University Press: New Brunswick 2005, pp. 55–80; Lehmann, Albrecht: *Von Menschen und Bäumen. Die Deutschen und ihr Wald*. Rowohlt: Reinbek bei Hamburg 1999.

25 In the first half of the twentieth century, the significance of animism in the ancient religion, folklore and culture of Estonians was emphasised by several leading folklorists, such as Matthias Johann Eisen and Oskar Loorits. Jonuks, Tõnno: "Nationalism and Prehistoric Religion: Religion in The Creation of Estonian Identity". In: Peedu, Indrek (ed.): *Estonian Study of Religion. A Reader*. University of Tartu Press: Tartu 2019, pp. 241–260. Selart, Anti: "Gehören die Esten zum Westen oder Osten? Oskar Loorits zwischen friedvollen Finnen und aggressiven Germanen." In: Drost, Alexander / North, Michael (eds.): *Die Neuerfindung des Raumes. Grenzüberschreitungen und Neuordnungen*. Böhlau: Cologne et al. 2013, pp. 143–159. See also the chapter by Tõnno Jonuks in this volume.

thirteenth centuries. Concerning the ancient period, archaeological evidence enables us to map the traces of the agricultural revolution in the Neolithic period (around 3000–2500 BC), the domestication of animals, and other ways of consuming animals for food, such as fishing and hunting.[26] The representation of animals on different material objects suggests their diverse cultural and religious significance, but the interpretation of these figures nevertheless poses a challenge, as also explained in the chapter by Tõnno Jonuks. As we are focusing on direct entanglements between humans and animals and leaving aside the burning question of long-term developments, we are not delving here into the question of animal and insect extinction, which has been studied for example by analysing Baltic amber.[27]

The history of medieval Livonia began when the territories of today's Estonia and Latvia were conquered, converted and colonised as a result of the German and Danish crusades.[28] Due to diverse contacts, Christianity had already been known in the area before, but prior to the Crusades, these lands had remained outside of the Christian structures and networks of power. The German-speaking elites started to become dominant in the territory early on. No central power developed in Livonia, but the land remained divided between the bishoprics and the Livonian branch of the Teutonic Order, and was affected by frequent internal power struggles. Frontier identity remained central for medieval Livonia, as its leaders conceptualised these lands as the last bulwark of Western Christendom.

Next to reorganising society and building up new networks and institutions of power, conquest also meant the construction of a new, Christian identity for Livonia. In addition, the Crusades and colonisation had a deep impact on the local environment, transforming the landscape and exploiting natural resources

---

26 Courel, Blandine / Robson, Henry K. / Lucquin, Alexandre; et al. "Organic Residue Analysis Shows Sub- Regional Patterns in the Use of Pottery by Northern European Hunter- Gatherers". *Royal Society Open Science* 7(4), 2020. DOI:10.1098 /rsos.192016; Gunnarssone, Alise / Oras, Ester / Talbot, Helen M. et al. "Cooking for the Living and the Dead: Lipid Analyses of Rauši Settlement and Cemetery Pottery from the 11th–13th Century". *Estonian Journal of Archaeology* 24(1), 2020, 45–69. DOI:10.3176 / arch.2020.1.02.

27 Labandeira, Conrad C.: "The Fossil Record of Insect Extinction: New Approaches and Future Directions". *American Entomologist* 51(1), 2005, pp. 14–29, DOI:10.1093/ae/ 51.1.14.

28 For the medieval history of the region, see Murray Alan V. (ed.): *Crusade and Conversion on the Baltic Frontier, 1150–1500*. Ashgate: Aldershot 2001; Mänd, Anu / Tamm, Marek (eds.): *Making Livonia: Actors and Networks in the Medieval and Early Modern Baltic Sea Region*. Routledge: London and New York 2020.

for building new centres of power, developing agricultural production, etc.[29] The main animal force behind building up the new rule was horses. They became symbols of the supremacy of the new German-speaking elites in wars, as well as in trading, as explained by Juhan Kreem and Ivar Leimus in this volume. The social hierarchies influenced the patterns of consumption, especially of wild game, which from then on was predominantly reserved for the tables of the nobility and the city governments.[30] Christianisation also markedly changed the relations to fish. Around the monasteries and manors fish ponds were established, and new fish species spread. Christian religious beliefs and practices, especially fasting, increased fish consumption.[31] Baltic fish specialties such as lamprey were exported to Central Europe using the Hanseatic trading companies and connections that were emblematic to medieval Livonia and its new urban culture.[32]

Christianity transformed the ontological status and symbolic meaning of animals. According to the medieval Christian worldview, animals were of a lower status than humans.[33] At the same time, animals were associated with

---

29 As recently demonstrated in connection to the broader Eastern Baltic region, including Lithuania and Prussia. Pluskowski, Aleksander (ed.): *Environment, Colonization, and the Baltic Crusader States: Terra Sacra I*. Brepols: Turnhout 2019. Pluskowski Aleksander (ed.): *Ecologies of Crusading, Colonization, and Religious Conversion in the Medieval Baltic: Terra Sacra II*. Brepols: Turnhout 2019.

30 Põltsam, Inna: "Essen und Trinken in den livländischen Städten im Spätmittelalter". In: Kivimäe, Jüri / Kreem, Juhan (eds.): *Quotidianum Estonicum* (Medium Aevum Quotidianum, Sonderband 5). Gesellschaft zur Erforschung der Materiellen Kultur des Mittelalters: Krems 1996, pp. 118–128; Mänd, Anu: "Festive Food in Medieval Riga and Reval", *Medium Aevum Quotidianum* 41, 1999, pp. 64–65; Mänd, Anu: *Urban Carnival: Festive Culture in the Hanseatic Cities of the Eastern Baltic, 1350–1550*. Brepols: Turnhout 2005, pp. 195, 217, 224–228.

31 Malve, Martin: "Tartu kesk- ja varauusaegsete kalmistute osteoloogilised uuringud 2008–2017." [Osteological Studies in Medieval and Early Modern Period Cemeteries in Tartu between 2008 and 2017.] *Tartu Linnamuuseumi Aastaraamat* 21, 2018, pp. 67–84.

32 Põltsam-Jürjo, Inna: "Kala tähtsusest kaubanduses, majanduses ja toidumenüüs 13.–16. sajandi Eestis." [On the Importance of Fish to Trade, Economy and Daily Menu in Estonia in the 13th to 16th Centuries.] *Acta Historica Tallinnensia* 24, 2018, pp. 3–23. DOI.ORG/10.3176/HIST.2018.1.01; Põltsam-Jürjo, Inna: "Fisch, Fischerei, Fischhandel und -konsum im estnischen Gebiet vom 13.–16. Jahrhundert". *Forschungen zur Baltischen Geschichte* 16, 2021, pp. 22–50.

33 See, e.g., Page, Sophie: "Good Creation and Demonic Illusions: The Medieval Universe of Creatures". In: Resl, Brigitte (ed.): *A Cultural History of Animals*, vol. 2. Berg: Oxford and New York 2007, pp. 30–31.

rich symbolic meanings, both negative and positive. Christian culture brought along the spread of animal imaginaries in clerical and secular visual culture, coats of arms and other symbols of the nobility, etc. On the one hand, medieval Christianity was characterised by a fascination toward fantastic animals. This is also reflected in animal representations depicted on the visual and material culture that has been preserved from medieval Livonia, discussed by Anu Mänd in this collection. On the other hand, wild beasts of the local forests, such as bears and wolves, became symbols of the pagan or the newly converted and potentially perfidious countryside and its inhabitants. Hence the association of the local peoples, Estonians, Latvians, and Livs, with nature (rather than culture), and the representations of their cult of nature – all of which became widely popular from the nineteenth century onwards – date back to the medieval period and originally often served the need to legitimise the Christianisation of these lands.[34]

The medieval organisation of power was put to an end with the Livonian War (1558–1583), as the increasingly stronger early modern states, Sweden, Poland and Russia started to fight over the fragmented Livonia. As a result of the war, the Livonian territory was split between Poland and Sweden. Struggles between Protestantism and Catholicism resulted in religious and cultural competition (Lutheran Reformation had already been introduced to Livonia in the 1520s). The development of scholarly culture, and the establishment of the University of Tartu (1632) introduced to Livonia the early modern discourse of animals, discussed in the articles by Meelis Friedenthal and Kaarina Rein. Next to this, military conflicts and confessional rivalry resulted in numerous descriptions of the disbelief of the local peasantry[35], which was first and foremost associated with the cult of nature and animals. These early modern narratives formed the other significant layer of text that later fed into the idea of the Baltic peoples' exceptionally close relations to nature. At the same time, the traffic moving in the opposite direction, the early modern and also later transmissions of local knowledge into the German-speaking learned culture have been largely forgotten (for example, the Baltic-German adaptations of Estonian or Latvian habits to forecast weather).

---

34 Kaljundi, Linda: "The Workings of Cultural Memory and Colonialism: Friedrich Ludwig von Maydell's Baltic History in Images." *Proceedings of the Art Museum of Estonia* 5(10), 2015, pp. 233–267.
35 Raik, Katri: "Talupojapilt Eesti- ja Liivimaa kohalikus humanistlikus ajalookirjutuses." [The Image of Peasants in the Local Humanist Historiography of Estonia and Livonia.] In: Raik, Katri: *Eesti- ja Liivimaa kroonikakirjutuse kõrgaeg 16. sajandi teisel poolel ja 17. sajandi algul*. PhD Dissertation. Tartu Ülikooli Kirjastus: Tartu 2004, pp. 221–244.

In the late medieval and early modern period, the socio-ethnic segregation between Germans and non-Germans grew bigger.[36] As the Estonian and Latvian peasants were gradually forced into serfdom,[37] comparisons with domestic animals and the local peasantry started to spread, creating a transnational discourse about the misery of the Livonian peasantry.[38] Abundant stories about werewolves reflected the cultural and social crisis, as explained in Stefan Donecker's article. Numerous wars that had been accompanied by plague and other diseases, bad harvest, and hunger, had also left their mark on the local population and seriously decreased the number of inhabitants. At the same time, the tumultuous early modern centuries also saw German migration toward the Baltic provinces, one of the pull-factors being the propagation of the richness of wild game and foal in these regions.

After the Great Nordic War (1700–1721), and the annexation of Curonia (1795), the territories of today's Latvia and Estonia became part of the Russian Empire. During all these major changes, the Baltic-German elites managed to ensure their political and social position. The eighteenth and nineteenth centuries saw the recovery of the manor economy, which relied on grain export and increasingly on the production of spirit for the Russian empire. From the perspective of environmental history, (pre)industrialisation put heavy pressure on local forests, as well as increased cattle farming. Keeping livestock was challenging due to diseases and wolves, but it was also crucial for farm economies,

---

36 Johansen, Paul / von zur Mühlen, Heinz: *Deutsch und Undeutsch im mittelalterlichen und frühneuzeitlichen Reval*. Böhlau: Cologne, Vienna 1973. Kala, Tiina: "Gab es eine „nationale Frage" im mittelalterlichen Reval?" *Forschungen zur baltischen Geschichte* 7, 2012, pp. 11–34.

37 Seppel, Marten: "The Growth of the State and its Consequences on the Structure of Serfdom in the Baltic Provinces, 1550–1750." In: Simonetta Cavaciocchi (ed.): *Schiavitù e servaggio nell'economia europea. Secc. XI-XVIII = Serfdom and Slavery in the European Economy. 11th–18th Centuries*. Florence University Press: Florence 2014, pp. 291–305. For the conceptual history of Livonian serfdom, see Seppel, Marten: "The Semiotics of Serfdom: How Serfdom Was Perceived in the Swedish Conglomerate State, 1561–1806". *Scandinavian Journal of History*, 45(1), 2020, pp. 48–70. DOI:10.1080/ 03468755.2019.1612466.

38 Kaljundi, Linda / Plath, Ulrike: "Serfdom as Entanglement: Narratives of a Social Phenomenon in Baltic History Writing". *Journal of Baltic Studies* 51(3), 2020, pp. 349–372. DOI:10.1080 /01629778.2020.1776349; Kaljundi, Linda: "Pagans into Peasants: Ethnic and Social Boundaries in Early Modern Livonia." In: Lehtonen, Tuomas M.S. / Kaljundi, Linda (eds.): *Re-forming Texts, Music, and Church Art in the Early Modern North*. Amsterdam University Press: Amsterdam 2016, pp. 357–392.

which strongly depended on the workforce of animals for agricultural labour and transportation, as well as needed their manure. From the mid-nineteenth century onwards, various agricultural innovations (cultivation of potato and clover, cattle breeding, development of amelioration and technology) increased animal husbandry.[39] The growing demand for milk products enlarged the number of dairy cattle, as well as the overall economic significance of husbandry. In parallel, the socio-economic organisation behind agricultural production changed profoundly, as the early nineteenth century emancipation laws had put an end to serfdom, and the reforms of the 1850s–1860s finally gave the peasants the right to buy farmland, as well as improved their legal status and thereby started the social mobility of the peasantry, as well as the Estonian and Latvian national movement that began in the 1860s.

The roots of the emancipation of non-German peasantry were, however, much earlier, dating back to the Enlightenment. In order to criticise serfdom and to argue against the civilising narrative legitimising the German colonisation, the Enlightenment authors had constructed a positive image of the local pre-Christian religion, culture and society. At the end of the nineteenth century, their texts fed into the national romanticist conceptualisations of Estonian and Latvian culture. In parallel, popular Enlightenment began to increase, as well as shape and control the knowledge about the natural world.[40] Calendars, journals, newspapers, and text books in all local languages introduced not only local nature, but typically to the age of empires and colonies, tended to focus on the more exotic flora and fauna. Attitudes toward local farm animals were more and more strongly influenced by agricultural societies, which were first established among the Baltic-Germans and thereafter also among the Estonian and Latvian peasantry. On the academic level, the Baltic nobility participated in the race for developing life sciences in the Russian Empire, and were thereby closely related to exploration and colonisation of the empire's vast territories in the far north, Siberia, Caucasus, and Far East. Although the scholars of the

---

39 A first glimpse into Latvian and Estonian breeds and their history is given in Porter Valerie / Alderson Lawrence / Hall Stephen J. G. et al. (eds.): *Mason's World Encyclopedia of Livestock Breeds and Breeding* Vol. 1–2, CABI: Oxfordshire 2016; Strautmanis, D. / Lidaks, M: "Possibilities of Maintenance and Reproduction of Local Latvian Brown Dairy Breed". In: Olev Saveli (ed.): *Animal Breeding in the Baltics*. Institute of Animal Science of Estonian Agricultural University Estonian Animal Breeding Association: Tartu 2004, pp. 91–95.
40 Daija, Pauls: *Literary History and Popular Enlightenment in Latvian Culture*. Cambridge Scholars Publishing: Newcastle upon Tyne 2017.

Baltic Provinces were particularly known for their work in plant systematics, their legacy also includes various studies of animals. A particular feature of the Russian exploration was its fascination with the polar regions, which was also motivated by the considerable economic value of the animals living in the cold habitat. Many of the Baltic explorers also became known for their discoveries near the poles.[41] In parallel to this, the well-known Baltic naturalist Karl Ernst von Baer also initiated novel research on fisheries and overexploitation.[42]

The manorial lifestyle cultivated other kinds of human–animal relations, and ways of cohabitation with animals. Hunting game became a privilege of the upper classes, and the number and quality of game, wild birds and fish in the Baltics was praised throughout the early modern times. Salmon was the everyday food for servants in Riga even in the nineteenth century, and in the seventeenth century craftsmen preferred to eat sausages and ham to the ordinary game meat. Taming wild animals and birds became popular during the Enlightenment. About the same time, exotic animals were brought to the Baltics by menageries, and from then on parrots, monkeys and guinea pigs spread among the upper classes in towns and the countryside. The first zoo in the Baltics was established in Riga in 1911. During the harsh time of the war in 1915 it is said that the animals of the zoo were eaten (at least this has been used as an argument in local propaganda).[43] In Estonia the first zoo was founded in Tallinn in 1939.

By the end of the nineteenth century, animals in the cities became a serious problem. While wild animals in the cities were killed, the slaughter of fatstock got centralised. Smaller animals were killed in large numbers for the sake of science. Baltic-German scholars and Russian legislative authorities helped to initiate nature conservation, which in the case of today's Estonia and Latvia first concerned the protection of migratory birds. Next to these well-known overwhelmingly male endeavours, however, the role of women and their

---

41 Tammiksaar, Erki: "The Russian Antarctic Expedition under the Command of Fabian Gottlieb von Bellingshausen and its Reception in Russia and the World." *Polar Record* 52(5), 2016, pp. 578–600. DOI:10.1017 /S0032247416000449.
42 Lajus, Julia / Ojaveer, Henn / Tammiksaar, Erki: "Fisheries on the Northeast Coast of the Baltic Sea in the First Half of the 19th Century: What Can Be Learned from the Archives of Karl Ernst von Baer." *Fisheries Research* 87, 2007, pp. 126–136. DOI:10.1016 /j.fishres.2007.07.004.
43 Hatlie, Mark R.: *Riga at War 1914–1919. War and Wartime Experience in a Multi-ethnic Metropolis.* (Studien zur Ostmitteleuropaforschung 30). Herder Institut: Marburg 2014, p. 79.

societies in propagating animal rights has been largely forgotten, as shown by Ulrike Plath.

The early twentieth century saw many fundamental changes in the Baltic societies. As a result of the First World War and the Russian Revolution of 1917, Estonia, Latvia and Lithuania[44] became independent nation states. This profoundly re-organised land ownership, as the lands previously owned by the nobility were largely given to the local peasantry, making the lands more agrarian again. As elsewhere, it was difficult for family farms to succeed in the market for agricultural goods. The situation changed somewhat in the 1930s, after the founding of the authoritarian regime in Estonia and Latvia (1934; in Lithuania the authoritarian *coup d'état* had already occurred in 1926) when the state started to support the export of agricultural goods, most notably dairy products (butter) and pork. The founding of the nation state also affected the symbolic meaning of animals, which nevertheless showed signs of colonial entanglements. Among other things, the nation state supported the promotion of native breeds (also discussed in the article by Eve Rannamäe): the Estonian and Latvian horse, the Estonian hound, etc. The history of the Estonian coat of arms shows an equally complex relationship to colonial legacy: although initially many had argued for depicting local animals (that could not easily be associated with the symbols of the nobility), eventually lions from the medieval coat of arms of the Duchy of Estonia were chosen for the insignia.

The Second World War and the following Soviet occupation of all three Baltic countries resulted in another wave of major socio-economic changes. Land was nationalised and farms, along with their livestock, were forced to merge into kolkhozes. This led to the industrialisation of agricultural production, including cattle farming, poultry and fishing. The Soviet Union also offered a large market for meat, fish, poultry and dairy products. The industrialisation of agriculture was idealised well into the 1960s and 1970s, reflected well by visual culture, media, etc. As elsewhere, this changed from these days onwards, when environmentalism became more and more prominent, along with the increasing popularity of nature conservation societies, the visibility of environmental topics in culture and media, etc. The Baltic republics in the 1970s–1980s were also characterised by the so-called folklore, or rustic turn, which brought along a new valorisation and interpretation of folk culture, including also the idealisation of family farms. While the urbanised elites behind this movement did not return

---

44 Lithuania had been annexed by the Russian Empire during the partitions of Poland in the late eighteenth century.

to live in the countryside and contribute to the more traditional cattle-keeping (instead many old farms turned into summer cottages), animals were present in the everyday life of late Socialism. These decades witnessed the boom of nature writing and films. Focusing on endangered and often also attractive species (e.g. eagles), these books and films criticised the harms of industrial modernity on wildlife and often idealised the past "natural" relations between animals and traditional peasants, hunters, etc. These developments also prepared the way for the large environmental protests of the Perestroika period.[45] Often interpreted as a surrogate for the fight for political and national sovereignty[46], the decades-long development of environmental citizen activism and environmentalist thought demonstrates that environmental thinking was far from arbitrary in the massive protests of the 1980s. Across Eastern Europe, environmental activism reached its peak in this decade, eventually feeding into the collapse of the Soviet Union and the Eastern Bloc. While environmental activism was mainly targeted against extensive extractivism, infrastructures, and energy politics, it also created negative attitudes toward industrial agriculture. In the Baltics, cattle farming gained an especially negative meaning, often being blamed for polluting the local environment and exporting all the meat to the rest of the Soviet Union, leaving the local food stores empty.

After regaining independence in 1991, the Baltic States experienced the disintegration of kolkhozes and the restitution of family farms to their pre-war owners. Although the Soviet regime had collectivised farming, keeping farm animals was still widespread in the countryside. In small towns, animals were frequently kept in individual households as well: chicken and ducks, as well as rabbits and coypu (nutria). This changed in the 1990s and 2000s. This development was furthered by new European Union regulations, as well as the spread of new global diseases. After the Baltic countries joined the EU in 2004, the EU regulations and support programmes have made large farms again the dominant form of agricultural production in the Baltic States from the 2000s onwards. Today, the overwhelming majority of meat and dairy is again produced in large farms.

The 2010s have witnessed the new rise of environmental activism in diverse forms. The new millennium has also brought along a new rise of animal rights activism. Activists have protested fur, poultry and dairy farms, as well as

---

45 See *Methis* 30, 2022, special issue on Estonian environmentalism. DOI: 0.7592/methis.v24i30
46 Dawson, Jane I.: *Eco-Nationalism: Anti-Nuclear Activism and National Identity in Russia, Lithuania, and Ukraine*. Duke University Press: Durham, 1996.

problematised pet keeping and breeding. In comparison to the earlier periods, this has moved the focus of animal conservation away from focusing on just some charismatic species, such as the Siberian flying squirrel (who is also depicted in the logo of the Estonian Fund for Nature). Yet, some native animal species have remained in the focus of conservation. In Estonia, the best example of this is the European mink, which has been in the focus of Estonian nature conservation for decades as an example of a native species that has come under threat due to the invasion of a new species, the American mink (a newcomer originating from the fur farms). As shown by Anita Zariņa, Dārta Treija, Ivo Vinogradovs, in Latvia, the veneration of native fauna has also led to reintroducing native species, such as the European bison. Thus the twentieth century has seen many changes in the fauna, including the return of some local species such as wild boar and beaver. New invasive species have also arrived, and they are perceived increasingly negatively, as signs of bioinvasion. In the twenty-first century, this is particularly true of the golden jackal, which arrived in the early 2000s, and the Portuguese slug, arriving in the 2010s.

Although animals are a part of contemporary environmental activism in the Baltic countries, they have never been at the very core of it. Even today, the public resonance of animal rights activism is considerably smaller in comparison to the forestry protests. This is also reflected in culture and identity. As explained above, today, myths about the Baltic peoples as nature's nations are mainly constructed around the forest, even if in earlier times ideas about indigenous animism and close "natural" relations with animals have also been relevant in identity building and environmentalism. Stereotypes about forest people spread in popular essayist literature, as well as in nation-branding. Images of Estonians or Latvians with a laptop in the middle of a pristine forest and bog landscapes (that do not include any animals) are widely used for promoting these countries for IT companies and specialists. Just like the earlier identity narratives, such images are deeply entangled and shaped by transnational transfers and shared by a number of countries, including the neighbouring Scandinavian countries. The prominence of forests both in cultural imaginaries and environmental activism also stems from large public protests of major infrastructure and industrial projects: such as the plans to build a cellulose factory near Tartu (2017–2018), the prospective construction of the Rail Baltic (the high-speed railway that is planned to unite the three Baltic States with the EU). In addition, national and international forestry companies alike are the subject of broad public activism. All this has encouraged academic research into the cultural imaginaries of forests, as well as the forms of pro-forest activism. In this context, animal studies have had to make way to these more pressing environmental issues and related research agenda.

Research Trends on Humans and Animals in the Eastern Baltic Region

The fact that animals are currently mostly excluded from the identity narratives, as well as mainstream activism, does not mean that they are missing from Baltic history and culture. The following overview focuses mainly on recent research trends, and does not aim to give any detailed insight into the no doubt diverse scholarly traditions that have touched upon the history of animals in the Baltic countries. While in large European countries cultural history has been at the turning point of animal history since the 1980s,[47] in Estonia the animal turn in humanities was driven throughout the 1990s mainly by bio- and ecosemiotics and ecocriticism, where human–animal interaction was modelled in a semiotic way.[48] The development of this line of research was also encouraged by the much earlier local research tradition, the studies of the Baltic-German biologist Jacob von Uexküll. According to the definition derived on the basis Uexküll's works,[49] an *Umwelt* is composed of all the meaning relations in the perception-based and in the action-based functional circles of an animal. A human observer must have at least partly the same perceptual capacities that the animal has, as well as some knowledge of the underlying biology that helps us to create a contact zone between different species, in order to be able to relate to other species. It also means that we are prone to better understanding the animals with whom our *Umwelts* overlap to a greater degree.

These special regional forms of interdisciplinary entanglements had a certain influence on the field. Cross-fertilisation within environmental humanities is beneficial for all the disciplines involved. Zoosemiotics,[50] for example, can well

---

47   Ritvo, Harriet: "On the Animal Turn." *Daedalus* 136(4), 2007, pp. 118–122.
48   A recent more general overview of Baltic ecocriticism also briefly touches upon the state of animal studies in the Baltics. Tüür, Kadri / Soovik, Ene-Reet: "Among Forests, Wetlands and Animals: Ecocriticism in the Baltics". *Ecozon@ European Journal of Literature Culture and Environment* 11: 2, 2020, pp. 42–51. DOI:10.37536/ECOZONA.2020.11.2.3498.
49   See Uexküll, Jakob von: *A Foray into the Worlds of Animals and Humans*. University of Minnesota Press: Minneapolis 2010 [1921].
50   For a general overview, see Maran, Timo / Martinelli, Dario / Turovski, Aleksei: "Readings in Zoosemiotics. Introduction." In: Maran, Timo / Martinelli, Dario / Turovski, Aleksei (eds.): *Readings in Zoosemiotics*. DeGruyter: Berlin 2011, pp. 1–18; Maran, Timo /

be complemented with a historical perspective that helps to understand how human–animal relations have changed over time. It is interesting to observe how the emphasis of a relationship between one and the same species with humans has been guided by different lead functions (resource / trade, partnership / relationship, symbol / representation) during different historical periods. The connecting point that ties together humans' and other animals' *Umwelts* is indeed the life cycle: each individual is born (or hatched), it matures, moves about, mates, produces some offspring, and finally dies. All of these stages of life are observable in most vertebrates, thus providing a nice common ground for drawing effective parallels between the lives of different species. Kinship relations and life cycles are universally present across species. This fosters sympathy toward other species.

Biosemiotics analyses signs in living systems, emphasising the connection between chemical and biological processes and sign systems. Biosemiotics does not differentiate between human and non-human animals sign systems, but emphasises their mutual interdependence, influence and hybridisation. Examples of this would be the impact of the changes in zoos to the animals living there[51] or communication between different species[52]. Characteristic to the development of bio- and ecosemiotics in Estonia is their close ties with literary studies, which have widely used and further developed the same conceptual tools.[53]

Tønnessen, Morten / Magnus, Riin et al.: "Methodology of Zoosemiotics: Concepts, Categorisations, Models." In: Maran, Timo / Tønnessen, Morten / Rattassepp, Silver (eds.): *Animal Umwelten in a Changing World: Zoosemiotic Perspectives*. University of Tartu Press: Tartu 2016, pp. 29–50.

51 Mäekivi, Nelly / Maran, Timo: "Semiotic Dimensions of Human Attitudes toward Other Animals: A Case of Zoological Gardens". Sign Systems Studies 44 (1–2), 2016, pp. 209–230. DOI:10.12697/SSS.2016.44.1-2.12.

52 Magnus, Riin: *The Semiotic Grounds of Animal Assistance: Sign use of Guide Dogs and Their Visually Impaired Handlers*. PhD Dissertation. University of Tartu. University of Tartu Press: Tartu 2015.

53 Maran, Timo: "Biosemiotic Criticism: Modelling the Environment in Literature". *Green Letters: Studies in Ecocriticism* 18(3), 2014, pp. 297–311. DOI:10.1080/14688417.2014.901898; Tüür, Kadri: "Bird Sounds in Nature Writing: Human Perspective on Animal Communication." Sign Systems Studies 37 (3/4), 2009, pp. 226–255. DOI:10.12697/SSS.2009.37.3-4.11; Timo Maran / Kadri Tüür (eds.): *Eesti looduskultuur*. [Estonian Natureculture.] Eesti Kirjandusmuuseum: Tartu 2005; Tüür, Kadri: *Semiotics of Nature Representations: On the Example of Nature Writing*. PhD Dissertation. University of Tartu. University of Tartu Press: Tartu 2017.

In Latvia, human–animal studies have been mainly developed in association with ecocriticism,[54] as well as with geography, as also explained by Anita Zarina et al. in this collection. Most notably the field has been influenced by the philosopher and journalist Artis Svece who has addressed human–animal relations from a philosophical perspective.[55] In Estonia, similarly to Latvia, humanistic geography was one of the first streams of early environmental humanities, developed by cultural geographers such as Hannes Palang, and art historians Kaia Lehari and Virve Sarapik.[56] Focusing on reconceptualisations of landscape, space and place, most of these studies have yet not concentrated on human–animal relations. In contrast, animals have been very central for environmental ethics, which have also started to develop in the Baltic countries. In Estonia, this research has been developed in the directions of animal rights and the rights of nature in general both in semiotics and in philosophy.[57] In Lithuania, an emerging centre for human–animal research strives to embrace the Baltic area in a wider sense, led by Kristupas Sabolius and Audronė Žukauskaitė. Their focus is on biopolitics, the philosophical dimension of human–animal relations and on the possible ways of re-conceptualising them in the Anthropocene.[58] There are also some nice overviews of the history of fisheries research in Latvia from the perspective of the history of science.[59]

In Baltic history and art history writing, the mapping of human–animal entanglements began in the mid-1990s. In particular, researchers have been focusing on human–animal relations in the Middle Ages. This phenomenon can at least be partly explained by the close relations between Baltic medievalists and

---

54 See Svece, Artis: „Meža tiesības." [Forest Rights.] *Rīgas Laiks*, December 2019. https://www.rigaslaiks.lv/zurnals/decembris-2019. Accessed 10.07.2023.
55 Svece has also been lecturing on these topics at the University of Latvia and Latvian Academy of Art.
56 See the series of article collections *Place and Location: Studies in Environmental Aesthetics and Semiotics*, published in 2000–2008. http://www.eki.ee/km/place/place_about.htm. Accessed 10.07.2023.
57 One of the results of this is a collection of translated key texts of environmental ethics Vaher, Aire / Keskpaik, Riste / Keerus, Külli (eds.): *Keskkonnaeetika võtmetekste.* [Key Texts of Environmental Ethics.] Eesti Keele Sihtasutus: Tartu 2008.
58 The centre is based at the Department of philosophy of Vilnius University. In 2020, Vilnius University organised an online conference, "Reimagining the Human That is Available", on YouTube.
59 Vitins, Maris / Gaumiga, Ritma / Mitans, Andis: "History of Latvian Fisheries Research". *Proceedings of the Estonian Academy of Sciences Biol. Ecol.*, 50-2, 2001, pp. 85–109. DOI:10.3176/biol.ecol.2001.2.03.

the Medieval Studies Department of the Central European University (CEU), where, at the initiative of Professors Gerhard Jaritz and Alice M. Choyke, the Medieval Animal Data-Network (MAD) was founded in 2005.[60] The Baltic alumni of the CEU and their colleagues have studied medieval human–animal relations from various perspectives, ranging from animal symbolism[61] to animals in trade, food consumption, and diplomacy.[62] Benefitting from the rapid development of technological and chemical analyses, another rising field of studies has been zoo-archaeology, with its focus on animal bones and other animal remains, which has notably contributed to our understanding of the use of animals and their products in material culture, food, trade, and religion.[63] Although modern

---

60 See https://mad.hypotheses.org/; http://mad.imareal.sbg.ac.at/. Accessed 10.07.2023.
61 Kivimaa, Katrin: "Dualistlik maailmavaade eesti keskaegses loomasümboolikas". [Dualistic Worldview in Estonian Medieval Animal Symbolism.] In: Alttoa, Kaur (ed.): *Ars Estoniae Medii Aevi Grates Villem Raam*. Eesti Muinsuskaitse Selts: Tallinn 1995, pp. 157–170; Mänd, Anu: "Kass voodi all. Ühest motiivist Hermen Rode ja Bernt Notke Tallinna retaablitel". *Kunstiteaduslikke Uurimusi = Studies on Art and Architecture* 21(1–2), 2012, pp. 231–246; Mänd, Anu: "A Mouser in the Bedroom". Berggren, Lars / Landen, Annette (eds.): *Ecce leones! Om djur och odjur i bildkonsten*. Artifex, Lund 2018, pp. 191–205.
62 Mänd, Anu: "Beaver Tails and Roasted Herring Heads: Fast as Feast in Late-Medieval Livonia". *Medium Aevum Quotidianum* 50, 2004, pp. 5–12; Põltsam-Jürjo, Inna: "Siga, seapidamine ja sealiha toiduks tarvitamine Eestis 13.–16. sajandil". [Pig, Pig Farming and Consumption of Pork in 13th–16th Century Estonia.] *Tuna: Ajalookultuuri ajakiri* 4, 2017, pp. 8–24; Mänd, Anu: "Horses, Stags and Beavers: Animals as Presents in Late-Medieval Livonia". *Acta Historica Tallinnensia* 22, 2016, pp. 3–17. DOI:10.3176/hist.2016.1.01; Mänd, Anu: "Animals as Presents in Late-Medieval Livonia". In: Choyke, Alice M. / Jaritz, Gerhard (eds.): *Animaltown: Beasts in Medieval Urban Space*: BAR Publishing: Oxford 2017, pp. 59–65.
63 Lõugas, Lembi / Bläuer, Auli: "Detecting Medieval Foodways in the North-eastern Baltic: Fish Consumption and Trade in Towns and Monasteries of Finland and Estonia", *Environmental Archaeology*, 2020, pp. 1–12. DOI:10.1080/14614103.2020.1758993; Lõugas, Lembi / Rannamäe, Eve: "Investigating Animal Remains in Estonia". *Archaeologia Lituana* 21, 2020, pp. 132–141. DOI:10.15388/ArchLit.2019.21.8; Luik, Heidi / Maldre, Liina: "Animal Bones and Bone Artefacts from the Viking Age Site of Tornimäe in Saaremaa". *Archaeologia Lituana* 21, 2020, pp. 41–58. DOI:10.15388/ArchLit.2019.21.3; Pluskowski, Aleksander / Rannamäe, Eve / Lõugas, Lembi et al.: "The Baltic Crusades and Ecological Transformation: The Zooarchaeology of Conquest and Cultural Change in the Eastern Baltic in the Second Millennium AD". *Quaternary International* 50, 2019, pp. 28–43. DOI:10.1016/j.quaint.2018.11.039; Lõugas, Lembi / Maldre, Liina / Tomek, Teresa et al.: "Archaeozoological Evidence from Padise Monastery". *Archeological Fieldwork in Estonia 2011*, 2012, pp. 83–92; Maldre, Liina: "Big and Small Bovids from

times offer significantly richer sources for the study of human–animal history, these sources have been investigated less. Among the topics studied so far, the history of slaughterhouses[64] and pet-keeping[65] could be mentioned. Concerning the modern period, research on animal history has been largely shaped by the favouring of the history of the local peasantry in the nation-states of the interwar period, as well as in the Soviet period. This means that animals have been mostly discussed in village contexts, and not urban. More recent studies on human–animal relations also depart from peasant history, such as a study of violence against animals and related regulations, which also well points to the importance of legal and court materials as sources of human–animal history.[66] In cultural history, focusing on local peasantry has resulted in the long-time favouring of folklore and ethnology studies. This has likewise contributed to the Baltic tradition of addressing human–animal histories mainly in the context of the village and folk culture. Animals have been researched in connection to various aspects of peasant history and culture: for instance, there are studies about husbandry-related architecture and material culture, transportation, and tools,[67]

Mediaeval Towns in Estonia". *Anthropozoologica*, 1997, pp. 707–714; Glykou, Aikaterini / Lõugas, Lembi / Piličiauskienė, et. al.: "Reconstructing the Ecological History of the Extinct Harp Seal Population of the Baltic Sea". *Quaternary Science Reviews* 251, 2021. DOI:10.1016 /j.quascirev.2020.106701.

64 Gustavson, Heino: *Lihatööstus Eestis XIX sajandi teisest poolest kuni 1917. aastani.* [The Meat Industry in Estonia from the Second Half of the 19th Century until 1917.] Valgus: Tallinn 1988; Gustavson, Heino: *Linna- ja alevitapamajad Eestis, 1918–1940.* [City and Town Slaughterhouses in Estonia, 1918–1940.] Valgus: Tallinn 1989.

65 Marju Kõivupuu has studied pet cemeteries in Estonia. Torp-Kõivupuu, Marju: "Risti peale kirjutas: ühel papil oli peni… Eesti loomakalmistukultuurist". [On the Cross He Wrote: A Man Had a Dog… From Estonian Animal Cemetery Culture.] *Mäetagused* 25, 2004, pp. 47–76.

66 Kaaristo, Maarja: "Vägivald loomade vastu: inimene ja koduloom Lõuna-Eesti külas 19. sajandi II poolel vallakohtute protokollide näitel". [Violence against Animals: Man and Domestic Animals in a South Estonian Village in the Second Half of the 19th Century Exploring Municipal Court Records.] *Mäetagused* 31, 2006, pp. 49–62.

67 Paramonov, Riho: *Eesti ühiskonna moderniseerumine avaliku privaattranspordi (voorimees ja takso) näitel, 1900–1940.* [The Cabman and the Taxi Driver: The Modernization of Estonian Society During 1900–1940 on the Example of Public Transport.]. PhD Dissertation. Tallinn University Press: Tallinn, 2019. Viires, Ants: *Talurahva veovahendid. Baltimaade rahvapäraste põllumajanduslike veokite ajalugu.* [Peasants' Vehicles.The History of Agricultural Vehicles in the Baltic Countries.] Valgus: Tallinn 1980. Viires, Ants: "Ratsutamine – vana rahvapärane liikumisviis." [Horse Riding – an Old Popular Way of Movement.] *Eesti Rahva Muuseumi aastaraamat*

as well as about herders.[68] The animal that has the most folklore materials related to it is the wolf,[69] but folkloristic sources cover all kinds of different animals from oxen to hedgehogs, herring, insects, etc. Estonian folklore about birds, mammals, fish and insects has been studied most notably by Mall Hiiemäe[70] and Mart Mäger.[71] Among other similar topics in Latvian and Lithuanian folklore studies, there is some remarkable research on house snakes.[72]

Baltic museums are also showing increasing interest toward animals in history and visual culture. In 2020–2021, Estonian museums have organised several interdisciplinary exhibitions on the topic.[73] This resonates with a broader change

---

1(15). Eesti Rahva Muuseum: Tartu 1947, pp. 36–80. Viires, Ants: "Baltikumi rahvaste põllumajanduslikus transpordis avalduvaist ajaloolis-kultuurilistest suhetest." [About the Historical-Cultural Relations Manifested in Agricultural Transport of the Baltic Nations.] *Etnograafiamuuseumi aastaraamat XX*. Valgus: Tallinn 1965, pp. 29–29.

68 Liiv, Ildike: "Ühiskarjatamisest Saaremaal." [About Communal Grazing in Saaremaa.] *Etnograafiamuuseumi aastaraamat XVIII*. Valgus: Tallinn 1962, pp. 38–79; Jaagosild, Ildike: "Loomade karjatamine Nõukogude Eesti külas". [Animal Grazing in the Soviet Estonian Village.] *Etnograafiamuuseumi aastaraamat XXI*. Valgus: Tallinn 1966, pp. 272–295; Eisen, Matthias Johann: "Karjane". [Shepherd.] *Eesti Rahva Muuseumi aastaraamat I*. Eesti Rahva Muuseum: Tartu 1925, pp. 10–34.

69 Rootsi, Ilmar: *Hunt ja inimene: suhted Eestis XVIII sajandi keskpaigast XIX sajandi lõpuni*. [Wolf and Man: Relations in Estonia from the Middle of the 18th Century to the End of the 19th Century.] PhD Dissertation. Tartu Ülikool: Tartu 2011.

70 Hiiemäe, Mall: *Väike linnuraamat rahvapärimusest: looduse sõnumitoojad*. [Little Bird Book from Folklore: Nature's Messengers.] Eesti kirjandusmuuseumi Teaduskirjastus: Tartu 2017, 2nd ed. 2020; Hiiemäe, Mall: *Väike loomaraamat rahvapärimusest* [Little Animal Book from Folklore.] Eesti Kirjandusmuuseumi Teaduskirjastus: Tartu 2019. Hiiemäe, Mall: "Geschichten von Waldtieren als Tatsachenberichte." In: Beyer, Jürgen / Hiiemäe, Reet (eds.): *Folklore als Tatsachenbericht*. Eesti Kirjandusmuuseum, Tartu 2001, pp. 37–53, Hiiemäe, Mall: "Kakskümmend kaks kala eesti rahvausundis I-III." [Twenty-two Fish in Estonian Folk Religion I-III.] *Mäetagused* 11–12, 1999, 13, 2000, pp. 7–33, 7–29, 7–23.

71 Mäger, Mart: *Eesti linnunimetused*. [Estonian Bird Names.] Eesti NSV Teaduste Akadeemia: Tallinn 1967; Mäger, Mart: *Linnud rahva keeles ja meeles*. [Birds in the Language and Mind of the People.] Eesti Raamat: Tallinn 1969.

72 Luven, Yvonne: *Der Kult der Hausschlange: eine Studie zur Religionsgeschichte der Letten und Litauer*. Köln et al.: Böhlau 2001.

73 *Always by Our Side: Cats and Dogs in 16th–19th-Century Art* at the Kadriorg Art Museum (2020–2021, curated by Anu Allikvee, Tiina-Mall Kreem, Anu Mänd). Kreem, Tiina-Mall Kreem / Murre, Aleksandra (ed.): *Alati meie kõrval: kassid ja koerad 16.–19. sajandi kunstis = Always by Our Side: Cats and Dogs in 16th–19th-Century Art = Всегда рядом: кошки и собаки в искусстве XVI–XIX веков*. Eesti Kunstimuuseum: Tallinn

of approach in curatorial practices, and moving away from the previous tradition that focused mainly on animals in art.[74] In Estonia, earlier research and curatorial practices related to more traditional animal paintings have also previously been shaped by long-term collaboration between the artists, and Tallinn Zoo.[75] In recent years, animals have also moved into the focus of contemporary art in the works of Edith Karslon, Katja Novitskova, and others.

Since the establishment of the position of Baltic regional representative (first Ulrike Plath, then Kati Lindström) at the European Society for Environmental History in 2013, the joint initiatives have grown and strengthened. Therefore, the collection of papers presented in this edited volume already have their roots stretching beyond several years. As new topics are constantly emerging in Baltic environmental history, the current collection is definitely already a snapshot of a situation that has already changed during the publishing process. Initiatives behind the volume include a series of international graduate seminars dedicated to animals between 2015 and 2018, organised by the Estonian Centre for Environmental History (KAJAK).[76] In cooperation with the Centre for Medieval Studies at Tallinn University, the topic of animal studies in the humanities was brought to Estonian historians. In 2016, Linda Kaljundi, Kadri Tüür and Ulrike Plath organised this first conference of the series with the title, "Animals in medieval and early modern Estonia and Livonia. Stories, sources, research perspective".[77] It was an attempt to deviate from the dominant human-based narratives and to define the new field of animal history for Estonian history

---

2020; *From Caves to Cuddles: The Story of Dogs and Humans* at the Estonian History Museum (2021–2022, curated by Krista Sarv, Mari-Leen Tammela); *Urban Animal: Cow, Bedbug and Dragon in the History of Tallinn* at the Tallinn City Museum (2021–2022).

74 As already indicated by the exhibition *From a Lion to a Bullfinch. Animals in Art from the Stern Family Collection* at the Mikkel Museum (2013, curated by Tiina-Mall Kreem, Mai Levin). An example of the earlier, more conventionally art-focused approach would be, for instance, *Animals and Humans: Animal Theme in Art* at the Art Museum of Estonia (2003, curated by Mai Levin).

75 Stern, A. (ed.): *Loomad kunstis: animalistliku kunsti näitus Karoly Sterni erakogust.* [Animals in Art: An Exhibition of Animalistic Art from the Private Collection of Karoly Stern.] S.n.: Tallinn 1989.

76 "Animals in transdisciplinary environmental history" organised by Ulrike Plath and Kadri Tüür, co-organised by the Department of Semiotics at Tartu University, Rachel Carson Centre for Environment and Society, and financially supported by the European Society for Environmental History.

77 Kaljundi, Linda / Tüür, Kadri / Plath, Ulrike, et al.: "Loomad ajaloos. Kus nad on?" [Animals in History. Where are they?] *Sirp*, 6.5.2016.

research, to bring new necessary approaches into the discussions, and to broaden the scope of Baltic history. Several leading Estonian historians gladly joined the endeavour. In 2017, a second conference on the topic followed, this time organised by the Under and Tuglas Literature Museum. In 2018, at a third meeting, it was decided to go for an edited volume. From here, the Baltic cooperation has been advanced in the framework of the BALTEHUMS[78] conference series initiated by KAJAK.[79]

Thus the animal turn in Baltic history writing has been diverse; the influences and impulses are received from a wide array of other disciplines, and it is gradually shifting toward the mainstream of history-writing. The specifically characteristic features of human–animal history in the Baltics still need to be mapped and linked to the entangled history of this border area, which would analyse together different ethnic and social groups, as well as different species.

This Book

Putting animals more vigorously in the centre of Baltic history writing is what this volume intends to do. The chronological scope of the articles stretches from the Middle Ages to the twentieth century. The volume brings together authors who were already with us during the first graduate seminar (Ird, Rannamäe, Tüür) or the first conferences (Kõivupuu, Mänd, Kreem, Leimus, Jürjo, Kaljundi, Hiiop, Rein). There are also a couple of new authors who were added during the editing phase (Jonuks, Zarina, Donecker, Friedenthal) in order to add new approaches to the discussion and to broaden the scope of Baltic history. The contributions in this volume are organised on a thematic basis, featuring studies of animals as related to human activities, animals as trade objects, and the visual representations of animals. These three forms of entanglement fit best with our articles, which have a strong focus on medieval and early modern times.

The first group of animals associated with trade are domesticated animals – most notably horses, as well as livestock, sheep, dogs, etc. Sooner or later in the course of their lifespans they are considered a resource, an object of trade and exchange, a source of income to their breeders. At the same time, it is not unusual that personal relations (or even trans-species bonds) develop in the process of raising domesticated animals. Fish (cod and herring discussed in the present

---

78 Acronym for Baltic Conference on the Environmental Humanities and Social Sciences. The first conference was held in 2018 in Riga; the second one online in Tallinn in 2021.
79 Estonian Centre for Environmental History (KAJAK) at Tallinn University, established in 2011.

volume by Lembi Lõugas and Kadri Tüür) usually appear as an anonymous mass object of trade. Establishing special bonds between humans and individual fish is unusual. Wild animals, such as bison or wolves, are integrated into human societies predominantly as a resource – something to be hunted, traded, eaten, skinned.

In the case of wild animals, representations play a considerable role. As most members of human society never have close first-hand contact with wild animals living beyond human habitats, most people gain access and develop their ideas about wild animals, based on secondary representations, such as stories, pictures, songs, or performances. Then there are purely "fantastic" animals, such as griffins or werewolves. It is difficult to trade an animal that only exists in the form of a representation. The relations to such animals can also only be virtual. The collection concludes with a couple of treatments of purely fictional beings, such as paintings of creatures on the wall inspired by Pompeian mural decorations, analysed by Jaanika Anderson and Hilkka Hiiop. Those animal representations already reach to the realm of fantasy, far beyond the animal kingdom, as well as into the temporal and spatial depths of human history.

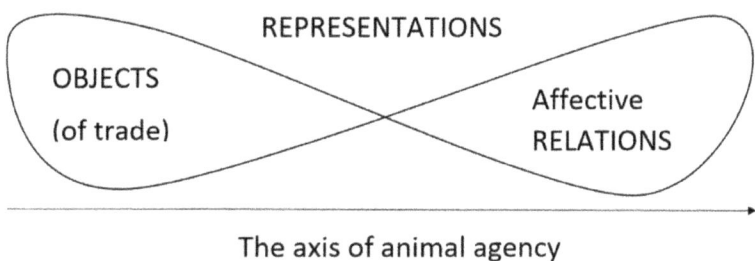

Figure 1. The dynamics between objects, representations, and relations on the axis of agency. Drawing by Kadri Tüür.

The categories of objects, relations and representations may easily slide and merge into one another: a horse who has a personal relation to his breeder may be traded as a "strategic resource" or depicted in a painting as an "animal hero". However, those three categories serve as good heuristic devices in imagining the multitude of animals and human–animal relations that exist around us today and that have existed in historical times.

So in the case of trade, the agency of the individual animal is almost non-existent, but the traded animals may be perceived either as anonymous mass or as valuable individuals, depending on the nature of the deal (see the contributions of Juhan Kreem and Ivar Leimus in the current volume). Animals also come in handy when it is necessary to demonstrate metaphorically, via representations, a sovereign's power over his subordinates (see Anu Mänd in the volume). Attributes of power, such as lions, leopards, dogs, horses, unicorns and other representatives of charismatic megafauna, be they real or fictional, are indeed ubiquitous, spectacular, and perhaps also most often mentioned in historical accounts of the past societies where no special emphasis on human–animal interactions has been employed. Those instances fall into the category of animal representations. The agency, or the power of the represented animals themselves to somehow influence the cultural images of their species, is little here, although perhaps not entirely non-existent.

Human–animal relations on a personal basis involve the greatest degree of agency as the development of a mutually strong inter-species' bond requires willingness and motivation from both parties. Such personal relations are nowadays often experienced between humans and their pets. Historically, domesticated agricultural animals and peasants must have had personal relations. Such personal relations could, in some extreme cases, also lead to sodomy, as Ken Ird's contribution to the present volume eloquently demonstrates.

The first block of the volume on quite different – but often even very personal forms of entanglement – is opened by historian Inna Põltsam-Jürjo with an overview of animals in medieval Livonian laws. Jürjo remarks that written sources about animals in human history are scarce. Laws provide some material for insights. She records the frequency of mentions of animals in medieval laws, and comes to the conclusion that mainly the species judged to be of value for humans are mentioned in those texts. Animals generally appear in juridical texts as objects of possession, so that the relation from the human perspective can be described according to the above-named scheme as an impersonal, rational, active / passive one.

Ken Ird in his chapter also looks at juridical written sources (eighteenth century), but his material yields a different relation between humans and domestic animals – one that is intimate, even overly so. The cases of bestiality taken to a county court in a Livonian province allow the author to discuss the notions of "natural" vs "unnatural", the question of demarcating the border between humans and animals, and the division of labour between men and women in relation to the domestic animals that were engaged in the acts of bestiality. Punishment was assigned not only to the human counterpart, but also to the animal involved – most often such animals were slaughtered.

Kaarina Rein, a classical philologist with a focus on medicinal texts, contributes to the volume with an insight into the depiction of animals in the seventeenth century treatises by scholars of the University of Tartu (Academia Gustaviana). She points out that the understanding of animals at that time was based on the doxas of antiquity. As the discussed texts demonstrate, animals are discussed either as capable of curing themselves, as metaphors, or as indications that humans appear in close affinity to (other) animals. In the subsequent period, the discussion of animals gradually becomes more technically medicine-centred.

Historian and literary scholar Ulrike Plath, the figure behind the initiative for this volume, writes about animal abolitionism and the animal protection movement that spread in Riga, the capital of Livonia, in the nineteenth century. She demonstrates that the then widely discussed question of freedom vs slavery was also extrapolated to work animals. The Baltic animal welfare societies developed from advancing animal protection to advocating animal rights. There was also a strong educational dimension involved in the promotion of animal welfare. It can be concluded that this movement contributed to the development of entangled human–animal ethics in an exemplary manner.

Latvian geographers Anita Zariņa, Dārta Treija and Ivo Vinogradovs discuss the re-creation of past landscapes by means of introducing a once-extinct big herbivore – the bison – in Latvia. Their example comes from the twentieth century, and it examines the surprising juxtaposition of a mythical animal with the contemporary landscape. The authors propose a three-fold categorisation of animals according to Deleuze and Guattari, namely an animal as an individual, an animal as a representative of one's species, and an animal as a source of conflict. Re-wilding gone wild provides us with some puzzling questions about the impact of wild animals on our contemporary societies – whether in the form of ideas or in their physical presence in one's backyard.

Animal histories of the twentieth century continue in the chapter by archaeologist Eve Rannamäe and native breed promoter Anneli Ärmpalu-Idvand. Their chapter is about the past, present and future of Kihnu native sheep, and it combines archaeological data with genetic research. They also discuss the history of breeding with regard to native breeds, which often just evolved in relative isolation, just as the sheep on Kihnu Island. Why is it important that a local group of domestic animals with archaic traits should be juridically recognised as a breed? The authors poetically say that native sheep are a gift from the past.

The second part of the book discusses animals as trade objects and concentrates on rational and active forms of entanglements. First, there are two contributions on horses. Juhan Kreem offers an attempt to conduct a census of the horses of the Teutonic Order in Livonia. The horse makes the knight, as the saying goes. Kreem introduces the written sources that contain information about the horses of the Teutonic Order. He summarises the numeric data, as well as the information that can be found in these documents about the categories of horses, their welfare, and breeding. Of native breeds, the horses of Saaremaa were especially valued. This reassures Rannamäe's and Ärmpalu-Idvand's observation that native breeds have valuable traits that deserve more attention. The value of horses can be illustrated with the fact that the Order possessed large amounts of horses that could be sent to other locations over considerable distances as gifts.

Ivar Leimus, an economic historian and numismatist, in his contribution discusses horses as an important resource in trading, especially between Livonia and Russia. Leimus presents a summarising table that compares known data about horse prices in medieval Livonia. He concludes that horses were a strategic commodity whose trade the authorities attempted to regulate.

Lembi Lõugas provides us with an insight into the archaeology of Baltic fish – fish bones in archaeological findings, as well as tools related to fishing. In the context of fish populations in the Baltic Sea, the question of cod is discussed: has it been an imported trade object, or could it have been caught by local fishermen? Studying archaeological fish is important from the perspective of historical fish populations, as well as for the advancement of food history.

Literary scholar Kadri Tüür discusses Atlantic herring, asking whether it has been a trade object or a mythological figure on the coasts of the Eastern Baltic. In order to find a possible answer, she turns to runic songs recorded in the nineteenth century and to literary fiction written during the twentieth century where herring appears – either as a side character or a protagonist. As Atlantic herring does not live in the Baltic Sea, it has been imported to the Baltic region, but as a number of other goods that had to be purchased for money, it has had a tendency to be regarded as a "luxury item" with a symbolic gloss about it.

The third part of the volume is dedicated to human (artistic) representations of animals and therefore concentrates on impersonal entanglements. More often than not such representations contain cultural prejudices toward animals, symbolic meanings attributed to them, and other human-specific ideas that are playfully projected onto the various representatives of the animal kingdom.

Archaeologist and folklorist Tõnno Jonuks takes us to the Late Iron Age in order to ask where eagles with ears come from. Griffin-shaped objects in archaeological material spark questions about the categorisation of animals and the representations of mythological creatures over the course of time. The ideas about griffins may well be living on even in nineteenth and twentieth-century folklore about eagle stones that were believed to carry mythical knowledge and wealth. The mythological layer is definitely stronger in the case of griffins than biological knowledge or imaginaries.

Historian and art historian Anu Mänd discusses the visual representation of animals in late medieval Livonian cities, focusing on secular art in public spaces. Images of animals often adorned the houses of wealthy city dwellers, as the façades of the houses acted as important channels of communication or "social media". Many animals were associated with saints, but they could also appear as heraldic elements with a certain position within this particular complex of meanings. Animal images could offer symbolic protection or convey didactic-moralistic messages.

Historian Stefan Donecker focuses on the living environment in the sixteenth century, with his research objects appearing mostly beyond the city walls. Narratives about wolves and werewolves tend to multiply at the times of wars and troubles. Such crises have increased the probability of wolf attacks, but they have also fuelled stories about werewolves. Donecker discusses three narratives about werewolves from the territory of present-day Estonia: wolves at the war camp of Toolse, "wolves of hell" in Tartu, and werewolves in Rakvere. He concludes that the wolf is an animal whose manifested presence is associated with times of crisis.

Writer and theologian Meelis Friedenthal takes us to the world of cats in the seventeenth century when those animals were often associated with occult powers. As cat allergies were unknown, some other explanation had to be found for the phenomenon that some people started to feel bad in the presence of cats. Some of the reasons for this reaction in humans, offered in contemporary learned texts, include pregnancy, temperament, and magic. The seventeenth century in general was a time period where new theories for explaining the surrounding world were actively sought, the question about the nature of cats being no exception.

The final chapter of the volume by art historians Hilkka Hiiop and Jaanika Anderson discusses the bizarre world that opens up in the murals from the nineteenth century in three Estonian manor houses. They tell the story of the discovery of the Pompeian style paintings depicting centaurs, domestic animals and birds, but also insects, such as cicadas, and offer the interpretations of those figures and scenes. (Neo-)classical culture was cherished among the local nobility of that time, but the décor of their private and public halls was decidedly different. Much more fantasy was allowed in decorating the private spaces with scenes from the "animal world".

In general, we can state that animals are in Baltic history and they come to the fore as soon as we start looking for them, in all the different forms – as historical subjects relating to humans, as objects of trade, or as symbolic representations. Putting all those diverse appearances and encounters into a larger context is a work that largely remains to be done. The present collection hopefully lays a foundation for a number of further studies in Baltic animal history.

Acknowledgements

We wish to thank all the contributors to the volume who have patiently been with us through the whole process of publication. Our special gratitude goes to Tõnno Jonuks, who's sensible advice helped us to finalise the volume, and to Tambet Muide, who helped us to format the papers. The preparation of this volume has been carried out with the support of numerous research projects of Tallinn University as well as the Under and Tuglas Literature Centre: PRG908 Estonian Environmentalism in the 20th century: ideology, discourses, practices; PRG1276 Digital Livonia: For a Digitally Enhanced Study of Medieval Livonia (ca. 1200–1550); TF1220, Developing KAJAK's (TLU's Centre for Environmental History) competences in Applied Environmental Humanities; TF 1620 Digital Livonia: Visual Sources; IUT18-8, The Making of Livonia: Actors, Institutions and Networks in the Medieval and Early Modern Baltic Sea Region; IUT28-1, Entangled Literatures: Discursive History of Literary Culture in Estonia. We are also very grateful for the Association for the Advancement of Baltic Studies (AABS) for their Book Publication Subvention and the Paul-Kaegbein-Stiftung.

We hope this book will initiate new entanglements between research groups, disciplines, humans, and animals!

I. HUMAN-ANIMAL RELATIONS

Inna Põltsam-Jürjo
Animals in Medieval Livonian Laws

Abstract: In the Middle Ages, animals constituted a natural part of people's domestic environment. A useful animal had to be subjected to humans, and people were convinced that they could control the animals that were part of their daily life. Medieval attitudes toward animals become visible in different legal orders, regulations, and rulings. This article gives an overview of the animals mentioned in different legal texts of Medieval Livonia, and looks into the different contexts in which these animals appear. The study is based on an analysis of three town, five peasant and three feudal-land laws.

Keywords: Medieval Livonia, Livonian laws, status symbol, home environment, living property, wild animals, domestic animals

Introduction

In the Middle Ages, people dealt with animals daily. Most people owned an animal or two, took care of them, and made use of what they had to offer. Back then, daily life without domestic animals would have been unthinkable. Animals were a natural part of people's home environment, and often they even lived under the same roof. For instance, in winter peasants would bring their animals into the only heated room they had for warmth and shelter. Even in the cities, poor people shared their modest households with domestic animals.

There are scarce sources at the disposal of historians that shed light on the medieval man's attitudes toward and his knowledge of animals. Such information mainly comes from the writings of different scholars, and the clergy. In the Middle Ages, attitudes toward and perceptions of animals were strongly influenced and shaped by the Christian faith. According to the teachings of the church an animal had no soul; therefore, studying them would in no way help people in their journey toward salvation.[1] The church emphasised the dominant

---

1 Fossier, Robert: *Das Leben im Mittelalter*. Piper: München, Zürich 2008, p. 242. Nevertheless, in philosophical debates on the soul and discussions concerning moral and virtuous behaviour, animals were important to some extent, cf. Prietzel, Kathrin: "Animals in Religious, and Non-Religious Anglo-Saxon Writings". In: Obermaier, Sabine (ed.): *Tiere und Fabelwesen im Mittelalter*. Walter de Gruyter: Berlin / New York 2009, pp. 235–260, p. 257. DOI:https://doi.org/10.1515/9783110213591.4.235. Research for this chapter was supported by the Estonian Research Council grant PRG1276. The

position of man and did not favour the scientific research or study of animals.[2] Medieval scholars viewed animals either as useful or harmful to man. The danger stemming from an animal and the fear it generated could be both real – the fear of an angry dog or a kicking horse, and abstract – a cat as a symbol of black magic, the devil, and witchcraft. A useful animal had to be subjected to man. It is unclear whether people then indeed believed they could fully control the animals that were part of their everyday environment. In any case, they did try to subject animals to their will, and not without success as is demonstrated by animal taming, procreation management, castration, and "grading up" animal breeding.[3] In medieval literature, animals are seen as having symbolic meanings. Their socio-economic importance, however, is less expanded upon.[4] Whether the writings of scholars reflect the actual attitudes of the common folk in their everyday lives is another question. To what extent – and if at all – a peasant's knowledge of domestic and wild animals matched that of what has been described in the bestiaries is unclear.

Medieval attitudes toward animals also become visible through much more practical texts, such as different legal orders, regulations, and rulings. While for the most part the law established the rules of interpersonal communication, it also touched upon the relationship between man and animal.[5] For instance, animals receive much attention in one of the earliest so-called barbarian laws of medieval Europe, the Salic law, written in the 6th century.[6] When dealing

---

research results in Estonian have been published as "Loomad ja õigus [Animals and the Law]". In: Kreem, Juhan / Leimus, Ivar / Mänd, Anu / Põltsam-Jürjo, Inna: *Loomad keskaegse Liivimaa ajaloos ja Kunstis*. [Animals in Medieval Livonian Society and Art.]. Tallinn University Press: Tallinn, 2022, pp. 13–34.

2   However, Dominic Alexander, for example, has pointed out that all the stories of the animal and saint tradition can be classified into two categories: those demonstrating power over nature, and those exemplifying empathy with animals, cf. Alexander, Dominic: *Saints and Animals in the Middle Ages*. The Boydell Press: Woodbridge 2008, p. 169.

3   Fossier 2008, pp. 246–251.

4   Le Goff, Jacques: *Medieval Civilization, 400–1500*. Blackwell: Oxford UK, Cambridge US 1988, pp. 376–377.

5   Cf. Deutsch, Andres / König, Peter (eds.): *Das Tier in der Rechtsgeschichte*. (Schriftenreihe des Deutschen Rechtswörterbuchs. Akademiekonferenzen 27). Universitätsverlag Winter: Heidelberg 2017.

6   Clement, Knut Jungbohn / Zoepfl, Heinrich (eds.): *Forschungen über das Recht der Salischen Franken vor und in der Königszeit. Lex Salica und Malberische Glossen*. Verlag von Theobald Grieben: Berlin 1879, pp. 89–104.

with animals in medieval legal texts, historians have been most interested in the so-called animal trials.[7] As a phenomenon, the latter are no doubt unique and extraordinary, and they convey and reflect the attitudes toward animals that were common at that period of time.

In terms of written sources from medieval Livonia, which roughly encompassed the areas of today's Estonia and Latvia, a number of city, peasant, and feudal law documents have survived. In all of them there are provisions that deal with animals. This paper will give an overview of the animals mentioned in different Livonian legal texts, and it will also look into the different contexts in which these animals appear and describe the attitudes toward animals that these legal documents convey.

Legal Texts

Medieval Livonia was not politically unified; it was divided into several small territories. The internal subdivisions of Livonia were determined during the thirteenth century, with the exception of the northern Estonian lands belonging to the king of Denmark – these areas were sold to the Teutonic Order in 1346. The division of the area between different lords – the Livonian branch of the Teutonic Order (67,000 km^2), the Archbishopric of Riga (18,000 km^2), the Bishoprics of Tartu (9600 km^2), Saaremaa (7600 km^2) and Courland (4500 km^2) – shaped and established the local legal system. The research at hand is based on a selection of important legal texts of medieval Livonia.

We know of several feudal and land laws of medieval Livonia.[8] The oldest of these is thought to be the Valdemar-Erik law for the vassals of Harju–Viru. It was written at the behest of Erik VI Menved of Denmark in 1315, but its preface contains references to Valdemar II (1170–1241). The second oldest is a collection of legal customs known as the oldest knight law of Livonia. The so-called Land and Feudal Law Mirror is especially significant: it is basically the Livonian version of the most important German-language legal text, the *Sachsenspiegel* (Mirror of the Saxons), written in 1220–1235. Based on the needs of the area, a number of articles of the *Sachsenspiegel* were left out of the Livonian version, several articles were adapted, and, in fact, only articles of practical value were transferred.[9]

---

7   Dinzelbacher, Peter: *Das fremde Mittelalter. Gottesurteil und Tierprozess*. Magnus Verlag: Essen 2006.
8   Cf. Bunge, Friedrich Georg v. (ed.): *Altlivlands Rechtsbücher*. Verlag von Breitkopf und Härtel: Leipzig 1879, pp. 1–4.
9   Bunge 1879, pp. 18–21.

Throughout its existence Medieval Livonia was open to migration from elsewhere in Europe. This migration mainly contributed to the upper and middle classes of society, while at the same time, local indigenous people mostly belonged to the lower class. As there was no colonisation of the peasantry, it remained ethnically homogenous. On the basis of local peasant laws and German land laws, the medieval Livonian Peasant Laws were developed, which for the most part entailed criminal rather than civil provisions. We know of five Livonian Peasant Law versions written in the fourteenth and fifteenth centuries: from the Archbishopric of Riga, from the Bishopric of Courland, the Livonian/Estonian version, the so-called village law originated probably from Bishopric of Tartu and the peasant law of Läänemaa.[10]

Towns had separate legal systems. Livonian towns mostly adhered to either the Riga Law or the Lübeck Law. Taking into account the influence and traditions of merchants from Gotland, in 1211 Bishop Albert gave Riga the so-called Gotland Law, which by around 1300 had been replaced by a town charter based on the Hamburg Law. More than half of the articles of the Riga Law are derived from the Hamburg Law, while the rest were specifically designed by the Riga town council to fit the city's needs; there were also some articles taken from the Riga–Haapsalu Statute, earlier laws of Riga, from the Lübeck Law and the Second *Skraa* of Novgorod.[11] The Riga Law applied in most of the towns in Livonia, and the Lübeck Law applied only in the towns of northern Estonia – Tallinn, Rakvere and Narva. Tallinn received the Lübeck Law Code from the king of Denmark both in 1257 and in 1282, and the research at hand is based on the latter.[12] The town law of Haapsalu dates back to the year 1294 and it is basically an early adaptation of the Riga Law to the specificities of Haapsalu.[13]

In all the laws of Livonia the influence of German law is dominant. This also pertains to the laws of peasantry, even though the peasantry was indigenous. The overview at hand is based on the legal texts and codes mentioned above, although there are a number of other contemporary rulings and regulations that

---

10 Arbusow, Leonid: "Die altlivländischen Bauerrechte". In: *Mitteilungen aus der livländischen Geschichte* 23, Kymmel: Riga 1924–1926, pp. 5–6.
11 Napiersky, Jacob G. L. (ed.): *Die Quellen des Rigischen Stadtrechts bis zum Jahr 1673*. Deubner: Riga 1876, pp. LIII–LXXXVIII; Zühlke, Raoul: *Bremen und Riga. Zwei mittelalterliche Metropolen im Vergleich. Stadt– Land–Fluß*. (Arbeiten zur Geschichte Osteuropas 12). LIT: Münster et al. 2002, pp. 150–164.
12 Cf. Kala, Tiina (ed.): *Lübecki õiguse Tallinna koodeks 1282 = Der Revaler Kodex des Lübischen Rechts 1282*. Ilo: Tallinn 1998.
13 Napiersky 1876, pp. XXIII–XXXI.

also deal with animals to some extent. These documents constitute a collection of sources that enable historians to monitor the changes that took place over time and the issues that needed additional regulation; however, these topics will not be discussed in this paper.

Statistics

Simple statistics provide us with the best general overview of animals in Livonian laws. The following table (Table 1) demonstrates the type and number of animals in different town charters.

Table 1. Animals in town laws, 1282–1300

	Tallinn Code of Lübeck Law, 1282	Haapsalu Town Law, 1294	Law of Riga, ca. 1300
Horse	4	6	4
Dog	1	-	-
Chicken	-	-	1
Goose	-	-	1
Cow	1	-	-
Domestic animal / farm animal / beast	3	2	4

Pigsties have been mentioned in the 1282 Tallinn Code of Lübeck Law, but as pigs have not been noted, they have not been listed in the table above.[14] When comparing the different town charters, the picture is rather similar: there are no real differences in the frequency of animals being mentioned – they appear in eight to ten articles of the laws. Taking into account the livelihoods and lifestyles of citizens, it comes as no surprise that most articles pertain to horses, as they were the main means of transportation.

---

14 The 1282 Code forbade the building of pigsties closer than 5 feet (ca. 1.5 metres) to the street or church yard, cf. Kala 1998, p. 162 § 140.

Table 2. Animals in peasant laws, 1300–1500

	Archbishopric of Riga	Bishopric of Courland	Livonian / Estonian version	Village law	Peasant law of Läänemaa
Horse	2	-	-	-	-
Domestic animal / farm animal / beast	2	1	3	-	2

The scarcity or even lack of animals in peasant laws is surprising (Table 2). Peasants were close to nature and dealt with animals in their everyday activities, but this close relationship is barely reflected in legal texts. It is likely, however, that different animal-related incidents that needed some legal regulation occurred quite often. We can assume that in many such cases the unwritten common law was referred to.

When looking at feudal laws (Table 3) and the number of animal-related articles in them, the *Sachsenspiegel*-based Land and Feudal Law Mirror of Livonia stands out. In other feudal law documents, animals are barely mentioned.

Table 3. Animals in land and feudal laws, 1315–1500

	Valdemar-Erik law	Livonia's oldest knight law	Land and Feudal Law Mirror
Horse	1	1	11
Dog	-	-	6
Hunting dog / hound / bloodhound	-	-	1/1/1
Boar	-	-	4
Ox	-	-	2
Wolf / tame wolf	-	-	1/1
Songbird	-	-	1
Tame bird	-	-	1
Fox	-	-	1
Pig	-	-	1
Goose	-	-	2
Domestic animal / farm animal / beast	-	-	11

When compared to town and peasantry laws, the Land and Feudal Law Mirror is somewhat distinctive as it deals with wild animals such as the fox, wolf and boar. The nobility lived in the countryside and they had the right to hunt; this explains the presence both of wild animals and several dog breeds in the law, such as hunting dogs and hounds, the *bracke*.[15] The bracke is one of the oldest hound breeds in Western Europe.[16] The mentioning of *bracke* in the law does not in itself prove that these types of hounds were in fact kept in medieval Livonia. Taking into account the fact, however, that in the Livonian version of the *Sachsenspiegel* articles not seen as pertinent to the Livonian context were cast aside, it seems likely that there indeed were *bracke*-type hounds in Livonia. The birds mentioned in the Land and Feudal Law Mirror are also noteworthy: songbirds and so-called tame birds (*tame vögel*),[17] the latter being most likely falconry birds – hawks trained and used for hunting.

To conclude, animals are very much present in the legal texts of medieval Livonia – articles pertaining to them are not many but that is expected. Most texts deal with domestic animals, such as horses, dogs and cows. Horses have received by far the most attention. Often the type of animal has not been specified and instead a more general term such as "domestic animal" (*queck*) or "beast" (*ve*) has been used. For instance, cats had no place in Livonian law, while in everyday life they were seen as a useful remedy against vermin. Still, cats' economic importance remained insignificant. In general, the "animal park" of each legal text – town, feudal or peasantry – reflected the life environment of its respective societal group. Animal-related articles of the legal texts all deal with very practical issues and situations, and any magical essence or meaning animals may have had has not been mentioned.

Animals as Property

According to Livonian legal texts animals were property: they were bought, sold, pawned, inherited, lent, and given for safe-keeping, as well as stolen, lost, and found. At times, the notion of an animal as property has been clearly stated: "[…] a domestic animal or whichever kind of property it is […];"[18] If from

---

15 Bunge 1879, p. 150 § 40: "Singende vögel edder tame vögel, unde winde- unde haszhunde unde braken [...]."
16 Fossier 2008, p. 252; Wendt, Ulrich: *Kultur und Jagd. Ein Birschgang durch die Geschichte* 1. Georg Reimer: Berlin 1907, pp. 34, 242–243.
17 Bunge 1879, p. 150 § 40.
18 Bunge 1879, p. 127 § 25.

the servant (*knechte*) his horse or his other property is stolen, robbed or taken [...]".[19] As the aforementioned examples demonstrate, animal-related articles of medieval Livonian law mainly dealt with issues of compensation. While animals themselves may have been the ones to sustain injuries or other damages, they also could be the ones causing harm, both to people and their property. The law determined the cases in which an owner had to be compensated for the damages he incurred through the injury or killing of his animal. In all probability many such disputes were solved outside the court: the city sheriff (*Stadtvogt*; Low German *voget*) became involved if the sides involved in the dispute could not reach an agreement amongst themselves. This is exactly what was prescribed in the Tallinn Code of Lübeck Law: "If someone hurts someone else's horse or cow or whichever other animal, if he wants, he can compensate the owner for it, so the sheriff will not get involved. If the sheriff has received a complaint or if the bailiff has come then the sheriff has to decide on the accusation in order to solve it".[20] A distinction was made between an intentional (*dankes edder ane not*) and an unintentional (*ane sine willen*) act.[21] In case of the latter, the person had to swear in the name of saints[22] that he had no guilt in the death or the injury of the animal.[23] No compensation had to be paid if an act was committed in self-defence (*dorch notwere*).[24]

In Livonian legal texts, it is mostly people who are seen as potentially imperilling animals, but in the Livonian Land and Feudal Law Mirror there are also articles in which another animal can be the one causing harm: for instance, a dog biting a cow or a domestic animal hurting another.[25] A person killing or injuring an animal in a situation in which it could have potentially attacked a

---

19 Bunge 1879, p. 141 § 5.
20 Kala 1998, p. 157 § 125.
21 Cf. Bunge 1879, pp. 150–151 § 42.
22 Napiersky 1876, p. 149 § 20: "[...] wil he dat sweren, dat et ane sine scult doot si ghebleuen".
23 Napiersky, p. 149 § 20: "Ofte en man deme andere sin gut doyt tho holdende. [...] Unde iset ve ofte quik, unde storve dat: deme dat tho holdende was ghedan, de ne sal dar nene noot umme liden, wil he dat sweren, dat et ane sine scult doot si ghebleven"; Bunge 1879, p. 151 § 42: "[...] he blift es ane wandel, sweret he dat up den hilligen, dat he dem nicht anders stüren konde".
24 Bunge 1879, p. 135 § 52: "Sleit ein man einen hunt edder einen beren [...] dot, [...], he blift des ane wandel, ift dat swere, dat he dat dorch notwere gedan heft".
25 Cf. Bunge 1879, p. 133 § 44, p. 135 § 51, p. 150–151 § 42.

person or another animal did not bear any responsibility for the consequences.[26] Additionally, according to the Livonian Land and Feudal Law Mirror no compensation was to be paid if pigs or geese trespassing on another lord's field were killed by dogs meant to chase them away.[27] In other situations, for injuring or killing another person's domestic animal damages had to be paid.[28]

According to medieval Livonian laws it was punishable to use someone else's personal living property, such as a horse "to ride or harness a horse to a wagon or a plough".[29] If such an act was committed within the boundaries of a certain legal space, for instance, within town borders (*stades markede*), the owner of the horse was to receive three marks as compensation and one mark was to be paid to the town. However, if the animal was taken outside the town, the act was considered to be theft and this meant that the perpetrator's life was forfeit. Punishment for animal theft depended on the value of the specific animal. According to the Riga Law, stealing a chicken or a goose was considered to be a minor theft, for which the thief was to sit in the stocks.[30] A horse thief, on the other hand, was considered to be a thief like any other and, as such, faced the death penalty – this becomes evident through contemporary case law, as codified legal texts of the time do not specifically address the issue of horse theft.

Cities were artisanal and commercial centres, and as such city laws stood for the best practices of trading and doing business. Trading animals was part of this. Livonian city laws established procedures that were to be followed if a bona fide purchaser happened to buy a stolen horse, for instance. According to the Haapsalu town charter, a buyer had to be able to prove the validity of his purchase with the help of witnesses. If he failed to do so, the animal had to be returned to its rightful owner, yet no compensation was to be paid, however. The rightful owner, on the other hand, had to be able to prove that the animal in question was indeed his.[31] According to the 1282 Tallinn Code of Lübeck Law, witnesses were to be heard in matters pertaining to the ownership of horses or

---

26 Bunge 1879, pp. 150–151 § 42: "Belemet överst ein man einen hunt, edder sleit en, dorch dat he en biten wolde, edder dat he sin vee bitt up der straten edder up einem velde [...]".
27 Bunge 1879, p. 129 § 31.
28 Bunge, Friedrich Georg v. (ed.): *Die Quellen des Revaler Stadtrechts 1*. Franz Kluge: Dorpat 1844, p. 170; Napiersky 1876, p. 28 § 26.
29 Napiersky 1876, p. 40 § 59.
30 Napiersky 1876, p. 193 § 3.
31 Napiersky 1876, p. 36 § 49.

other animals, which suggests that relatives and close acquaintances were also well informed of each other's animals.

Animals being living creatures were seen as living property – his property was capable of acting independently on its own, without the will or control of its owner. For example, an animal could run away from home and get lost. Livonian laws had provisions for dealing with cases of runaway animals. In rural areas, runaway animals were often found on and sometimes causing damage to another farmer's or landlord's field or meadow.[32] In such cases the victim was allowed to keep the animal in his custody until the damages the animal had caused had been compensated for. On the other hand, if the animal's guilt could not be proven, then a fine had to be paid for holding the animal in custody.[33] Naturally, such provisions were not established out of consideration for the rights of the animal, but of its owner. In any case, according to medieval laws, an owner was responsible for any damages his animal caused to another man's property: these damages had to be compensated for and, in some cases an additional fine was to be paid as well.[34]

The law prescribed how to handle a runaway animal so that the one finding it would later not be accused of stealing or appropriation. The 1282 Tallinn Code of Lübeck Law stipulated that a person, having found such an animal, had to make this publicly known in church during religious festivals. If no one claimed the animal as theirs, the animal was to be sold and the revenue made from the sale "kept in the house of god for a year and a day", and if by that time the owner still had not appeared, the money was to "stay in the house of god for the soul of whosoever's animal it was".[35] In this way, aided by law, an animal could help his owner on his path to salvation. The Haapsalu town charter made a clear distinction in this matter between animals: "a horse [...] or some other living item" and lifeless items: "an item of clothing or some other thing", namely, the finding of a runaway animal had to be declared publicly at the marketplace, while the finding of a lifeless thing was to be publicly announced from the chancel three consecutive times.[36] If the owner of a "horse or some other thing" was not found, it was to be given to the town council.

---

32 Arbusow 1824–1826, p. 39 § 17: "Findet ener queck up seinem acker adder heuschlegen dat vehe sall he en sien huss driwen, beth dat em sien schade betalt werdt".
33 Ibid., p. 83 § 7: "Schuttet er queck und kan keinen schaden beweisen, ist 4 mr".
34 Bunge 1879, p. 137 § 36: "We sin vee drift up eines andern mannes korn edder up sinen hoislach wedende, he schal em sinen schaden gelden mit recht, unde böten mit dren marken langudes".
35 Kala 1998, p. 166 § 151.
36 Napiersky 1876, pp. 36–37 § 49.

Attitudes Toward Animals

Legal texts not only regulate and stipulate, they also reflect the attitudes and values of the time. In the Middle Ages, people's attitudes toward (domestic) animals was first and foremost practical, and this is also reflected in Livonian legal texts, where animals are basically seen as things: "Also, when someone has had the use of his item of clothing or horse or some other thing for a day and a year [...]";[37] "Also, if someone buys a horse or some other thing in this town ..."[38]; "If someone has borrowed something from someone else, be it a horse, clothing or some other thing [...]".[39] "If someone finds a horse or a domestic animal or whatever thing it is [...]".[40] In some cases the law attributes animals certain qualities. For example, in the Riga city law there is the phrase "senseless animal" (*unvernünfftig thier*).[41] An article of the Livonian Land and Feudal Law Mirror touches upon the issue of domestic animals falling prey to wolves and in this article wolves and their actions are seen as something common for robbers: "What a robber (*röver*) or a wolf [...]".[42] Medieval Livonian legal texts mainly deal with domestic animals, and wild animals are only mentioned in the Land and Feudal Law Mirror. The tame birds and tame animals mentioned in the latter are especially interesting.[43] The tame animals mentioned are most likely falconry birds. However, the purpose or benefits of a tame wolf in the household of a nobleman remain a mystery. In any case, the desire to proclaim the superiority and dominant position of man over animals as well as the desire to subject animals to man's will was evident even in medieval legal texts.

Medieval Livonian legal texts also represent animals as symbols of status. For example, in the Land and Feudal Law Mirror a horse was part of the criteria for a nobleman to prove his legal competency: "[...] therefore he is (legally) competent, if he is able [...] to mount and ride a horse [...]".[44] This clearly demonstrates that a nobleman's horse was a symbol of his status. The inseparable bond of knight and horse is also referred to in an article that is present in all three of the feudal laws under view, which states that a knight always inherits the best horse of the bequest. The dog breeds mentioned in the Land and Feudal Law

---

37 Ibid., p. 34 § 41.
38 Ibid., p. 34 § 43.
39 Ibid., p. 46 § 27.
40 Ibid., p. 159 § 25.
41 Ibid., p. 191 § 23.
42 Bunge 1879, p. 132 § 43.
43 Ibid., p. 135 § 51, p. 150 § 40.
44 Ibid., p. 103 § 25.

Mirror – hounds, the *bracke,* other hunting dogs – were also important symbols of the nobility.

A considerable amount of articles pertaining to animals in medieval Livonian laws deal with violence: hitting, injuring, hurting someone's leg or eye, beating (an animal) to death are mentioned among other things.[45] To a great extent, in medieval Livonian legal texts, violence was the thing that characterised the relationship between man and animal the best. It was not a one-sided relationship: animals were a danger to man as well. From medieval Europe there are several accounts of people having been killed or injured by domestic animals. Medieval Livonian laws also touch upon this danger. The 1282 Tallinn Code of Lübeck Law established that if a domestic animal or a beast (*ve*) injured someone, then the owner of the animal was not liable – both in case the incident took place at his home or on the street.[46] There is a similar article in the Livonian Land and Feudal Law Mirror, where among other domestic animals boars (*bere*) have also been mentioned.[47] Such an approach was based on the notion of an animal as basically a thing.[48] If the court was unable to ascertain any intention of causing harm on behalf of the owner, or if the owner was unable to anticipate his animal hurting someone, then according to the law, the owner had no guilt. It was common to give the victim the animal that attacked them.[49] The 1586 Code of Lübeck Law states that such an animal (*das schadhaffige Viehe*) is to be divided equally between the victim and the court.[50]

There is a certain discord in medieval law; therefore, it is not so extraordinary that in Europe trials were held against domestic animals that had killed people: these animals were dressed up as people and judged as common murderers and criminals. There have been attempts to place the phenomenon of animal trials into the wider context of mental and social developments in European society[51]; however, for the most part there are good grounds to agree with the opinion of Markus Feigl according to whom animal trials mainly addressed a purely legal issue: punishing animals in this way enabled people to

---

45 Ibid., p. 150 § 42: "We överst dödet edder belemet in einem vote [...]; Lemet he it överst in eine ouge [...]".
46 Kala 1998, p. 142 § 68, § 69.
47 Bunge 1879, p. 129 § 30.
48 E.g. 643 In the Lombard king's edict (*Edictum Rothari*), animals are referred to as *muta res* – silent thing, cf. Dinzelbacher, p. 111, 115.
49 Ibid., p. 114.
50 Bunge 1844, p. 184.
51 Dinzelbacher, pp. 103–156.

avoid situations wherein, because of disproportionate human guilt or liability, justice would not be served at all.[52]

The question of the liability of the animal's owner deserves scrutiny in its own right. As mentioned earlier, according to medieval law, an owner was not responsible for his animal's actions if these actions did not reflect or took place independently of the owner's intentions. Nor was a person responsible when he caused harm to animals for reasons justifiable in the eyes of the law, for example, when driving someone else's animals away from his land. According to medieval law, being an owner of an animal did not automatically mean that the owner was liable for his animal's actions. For instance, if a servant was walking his lord's dog and the dog bit someone, then, in the eyes of the law, the one responsible was not the noble owner but the servant.[53] The same applied to cases with other animals: in case of any damages, the person responsible was the one under whose care the animal was at the time.[54] Only if an animal hurt another in a communal herd under the care of a shepherd, and the shepherd swore to it under oath, did the owner of the culprit have to nurture the injured animal back to health at his own cost, or if the animal died, compensate its owner.[55]

Conclusion

Among other issues, legal texts of medieval Livonia deal with animals. The analysis of the three town, five peasant and three feudal-land laws that this study is based on showed that the number of animal-related articles in them is rather low. The Livonian Land and Feudal Law Mirror has by far the most articles pertaining to animals. In general, the "animal park" of each legal text – town, feudal or peasantry – reflected the life environment of its respective societal group: for instance, wild animals are only mentioned in the Land and Feudal Law Mirror. The Mirror also shed light on the fact that horses and hunting dogs served as symbols of status for the nobility. The fact that animals were barely mentioned in peasant laws came as a surprise, and it highly likely suggests that the unwritten common law played an important role in village society.

Medieval Livonian laws mostly mention animals that were useful, important or valuable to man in some way or another. Understandably, both in terms of importance and value, horses stood out. In the eyes of the law, animals were

---

52 Ibid., p. 134.
53 Bunge 1879, p. 151 § 43.
54 Bunge 1879, p. 129 § 31.
55 Ibid., p. 133 § 44.

property and because of this, most of the animal-related articles of Livonian laws deal with issues of compensation in case an animal-caused or incurred damages. The law also reflected the largely practical attitude toward animals at the time – animals were seen as things. At least somewhat, the objectification of animals also affected the nature and extent of the owner's liability; in general, in the eyes of the law an owner was not automatically liable for the actions of his animals.

Medieval Livonian laws mainly deal with domestic animals. However, the tame wild animals and birds mentioned in some legal texts evoke an image of a different kind of relationship with animals. Among other things, it illustrates the desire of man to dominate animals. In a few cases the law attributed some animals certain qualities, for example, a wolf was seen as a robber. To sum up, medieval sources offer information both about the practical and material aspects of the relationship between man and animal and the general values of the time. Hence, legal texts not only regulated and stipulated, but also reflected the attitudes common in the Middle Ages.

References

Alexander, Dominic: *Saints and Animals in the Middle Ages*. The Boydell Press: Woodbridge 2008.

Arbusow, Leonid: "Die altlivländischen Bauerrechte". In: *Mitteilungen aus der livländischen Geschichte* 23, Kymmel: Riga 1924–1926, pp. 5–6.

Bunge, Friedrich Georg v. (ed.): *Altlivlands Rechtsbücher*. Breitkopf und Härtel: Leipzig 1879.

Bunge, Friedrich Georg v. (ed.): *Die Quellen des Revaler Stadtrechts 1*. Franz Kluge: Dorpat 1844.

Clement, Knut Jungbohn / Zoepfl, Heinrich (eds.): *Forschungen über das Recht der Salischen Franken vor und in der Königszeit. Lex Salica und Malberische Glossen*. Theobald Grieben: Berlin 1879.

Dinzelbacher, Peter: *Das fremde Mittelalter. Gottesurteil und Tierprozess*. Magnus Verlag: Essen 2006.

Fossier, Robert: *Das Leben im Mittelalter*. Piper: München, Zürich 2008.

Kala, Tiina (ed.): *Lübecki õiguse Tallinna koodeks 1282 = Der Revaler Kodex des Lübischen Rechts 1282*. Ilo: Tallinn 1998.

Le Goff, Jacques: *Medieval Civilization, 400–1500*. Blackwell: Oxford UK, Cambridge US 1988.

Napiersky, Jacob G. L. (ed.): *Die Quellen des Rigischen Stadtrechts bis zum Jahr 1673*. Deubner: Riga 1876.

Prietzel, Kathrin: "Animals in religious, and Non-Religious Anglo-Saxon Writings". In: Obermaier, Sabine (ed.): *Tiere und Fabelwesen im Mittelalter*. Walter de Gruyter: Berlin, New York 2009, pp. 235–260.

Wendt, Ulrich: *Kultur und Jagd. Ein Birschgang durch die Geschichte* 1. Verlag von Georg Reimer: Berlin 1907.

Zühlke, Raoul: *Bremen und Riga. Zwei mittelalterliche Metropolen im Vergleich. Stadt–Land–Fluß*. (Arbeiten zur Geschichte Osteuropas 12). LIT: Münster et al. 2002.

Ken Ird

When Men Stayed in the Barn for Too Long: Cases of Bestiality in the Early Modern Baltic Provinces

Abstract: Before the arrival of modern society in Estonia in the second half of the nineteenth century hardly any ego documents were left behind by the local Estonian peasantry. Therefore, one of the few possibilities to give voice to a local commoner emerges in the rich amount of county court materials from the early modern period of Estonian history. This article is based upon court cases of bestiality in the Baltic provinces from the seventeenth until the early nineteenth century, with the focus on the communication between the accused, the witnesses and the animals concerned in the cases. It demonstrates how easily "natural" understanding of cohabitation with domestic animals and their everyday exploitation became "unnatural," and eventually brought about their stigmatisation and abuse.

Keywords: Baltic provinces, legal history, social marginality, bestiality, treatment of animals, history of mentality

Introduction

[...] the accused have been sent to the barn to fetch the horse, but [he] stayed away for too long [...], where at that moment cattle were still standing tethered.[1]

This statement of a farmer's wife, Maret, was made in 1755 at one of the county courts of the Baltic provinces[2] during a trial against a young farmhand named

---

1   "[...] Inquisit in dem Stall das Pferd zu holen gesand worden, aber gar zu lange weggeblieben [...], wo das Vieh der Zeit annoch angebunden gestanden." *Protocollum Inquisitionis in Sachen actor officious contra des Pollenhoffschen Hofes Böttigers Puisa Hans seinen Jungen Peter in puncto Sodomiae insimulatae* (1755–1756). Estonian National Archives (EAA), fond 915, inventory 1, file 1285, p. 6. All translations of the source material from German to English have been made by the author.

2   The term "Baltic provinces" is used to indicate an area that was formed approximately by nowadays Estonia and Northern Latvia. Since the seventeenth century it has been called *Östersjöprovinserna* in Swedish and *Ostseeprovinzen* in German ("Baltic Sea Provinces" or literally "East Sea Provinces"). In the given article the word "Sea" is omitted in order to connect the area more directly to the contemporary geopolitical name of the region: "the Baltic States". The area was divided into two governorates – Estonia

Peeter who was accused of bestiality. In the quoted eyewitness's testimony, it is demonstrated how in the early modern Estonian rural community an everyday routine with domestic animals could easily become suspicious and possibly even criminal.

Sociologist Piers Beirne has referred in his book "Confronting Animal Abuse: Law, Criminology, and Human–Animal Relationships" that the clerical, criminal, psychiatric and academic approaches toward bestiality over time have paid almost no attention to the harm it inflicts on animals. When taking such a viewpoint into account, Beirne stresses the incapacity of animals to resist sexual advances, thus renaming bestiality in its essence an act of animal sexual assault.[3] As seen below, the given argument is well backed by the data offered by historical sources.

The given article is based upon court cases of bestiality from the seventeenth until the beginning of the nineteenth centuries in the Baltic provinces with a focus on the communication between the accused, witnesses of, and animals concerned in cases. More specifically, the analysed cases are mostly taken from archival funds of local county courts in Estonian National Archives and Latvian State Historical Archives. These courts processed criminal cases, i.e. cases of bestiality concerning local rural, low-rank populations as well.[4]

Bestiality as a Phenomenon

Evidently bestiality was not just an agricultural crime in early modern European society, there is clear evidence of such behaviour among people from different social strata (cabmen, merchants, gardeners, artisans, chemists, soldiers, civil

---

(*Ehstland, Estland*) and Livonia (*Liefland, Livland*). In 1795 Courland (*Kurland*), i.e. nowadays Western and Southern Latvia, was incorporated into the Russian empire and thenceforth also considered as a part of the Baltic provinces. However, as the main focus of the given article lies in the eighteenth century, Courland is not included.

3   Beirne, Piers: *Confronting Animal Abuse. Law, Criminology, and Human–Animal Relationships*. Rowman & Littlefield Publishers: Lanham 2009, pp. 113–115.

4   Contemporary court officials have indexed the analysed court cases under the term "sodomy" (*Sodomie, Sodomia*). Throughout the early modern period this conception held a much broader meaning, ranging from non-reproductive heterosexual practices and masturbation to homosexual intercourse and bestiality. In the given article all the analysed sodomitical cases feature solely the matter of bestiality. Schwerhoff, Gerd: *Historische Kriminalitätsforschung*. Campus Verlag: Frankfurt / New York 2011, p. 161.

servants, etc.).[5] However, as the jurisdiction of the county courts in the Baltic provinces expanded only to the common people living in rural areas, the lion's share of the studied cases concern the local peasantry. In that way, juridical material from the early modern period of Estonian history offers a unique possibility to give a voice to the local common man, as there are hardly any ego documents left behind by the local Estonian peasantry before the arrival of modern society in Estonia in the second half of the nineteenth century.

In Western European and North American historiography bestiality in the early modern period has been mostly viewed within the context of social and criminal history.[6] In Scandinavia academic research on the subject has been published since the 1990s.[7] However, the latter approaches analyse bestiality on the viewpoint of masculinity, which is just one possible way to handle the topic.

---

5 Muravyeva, Marianna: "Sexual Variations". In: Peakman, Julie (ed.): *A Cultural History of Sexuality in the Enlightenment*. Bloomsbury: London / New York 2015, p. 93.
6 Some examples of more recent studies on the subject: Ben-Atar, Doron S. / Brown, Richard D.: *Taming Lust. Crimes against Nature in the Early Republic*. University of Pennsylvania Press: Philadelphia 2014; Canup, John: *"The Cry of Sodom Enquired Into": Bestiality and the Wilderness of Human Nature in Seventeenth-Century New England*. American Antiquarian Society: Worcester 1988; Fudge, Erica: "Monstrous Acts. Bestiality in Early Modern England". *History Today* 50(8), 2000, pp. 20–25; Jordan, William Chester / Creager, Angela N. H.: *The animal/human boundary. historical perspectives*. University of Rochester Press: Rochester / New York 2002; Lang, Dominik: *Sodomie und Strafrecht. Geschichte der Strafbarkeit des Geschlechtsverkehrs mit Tieren*. (Europäische Hochschulschriften / European University Studies / Publications Universitaires Européennes. Reihe II. Rechtswissenschaft, 4750). Peter Lang: Frankfurt a. M. et al. 2009; Thomas, Courtney: "'Not Having God Before his Eyes'. Bestiality in Early Modern England". *Seventeenth Century* 26(1), 2011, pp. 149–173. DOI:10.1080/0268117X.2011.10555663.
7 Liliequist, Jonas: *Brott, synd och straff. Tidelagsbrottet i Sverige under 1600- och 1700-talet*. [Crime, Sin and Punishment. The Breach of Time Law in Sweden during the 17th and 18th Centuries.] (Doctoral thesis). Umeå University: Umeå 1991; Liliequist, Jonas: "Peasants Against Nature: Crossing the Boundaries between Man and Animal in Seventeenth- and Eighteenth-Century Sweden". *Journal of the History of Sexuality* 3(1), 1991, pp. 393–423; Keskisarja, Teemu: *"Secoituxesta järjettömäin luondocappalden canssa". Perversiot, oikeuselämä ja kansankulttuuri 1700-luvun Suomessa*. [Perversions, Legal Life and Folk Culture in 18th-Century Finland.] (Doctoral thesis). University of Helsinki: Helsinki 2006; Sjödin Lindenskoug, Susanna: *Manlighetens bortre gräns. Tidelagsrättegånar i Livland åren 1685–1709*. [The Outer Limit of Manhood. Tidal Law Trials in Livonia in the Years 1685–1709.] (Doctoral thesis). Stockholm University: Stockholm 2011.

Cases of bestiality can also be seen as rare historical evidence of the interaction between humans and animals and as sources for finding notions concerning mentality toward domestic animals.

Human–animal communication is closely linked to the understanding of the natural and the unnatural. In the recent decades the former concept has not been seen to be limited just to the methods and theories of the natural sciences. In fact, both of these concepts are determined by the social, moral, and political norms and practices. Understanding of something being unnatural implicates that it has shifted from its original or essential state. Such alteration is most commonly associated with some sort of human interference, thus allowing the defenders of "the natural" to stigmatise either concrete individuals (e.g. accused at court) or whole social groups (e.g. different ethnic, racial, religious or sexual minorities).[8] All this can also be well observed in the way how early modern society dealt with cases of bestiality. On the basis of that the main issue in the given article is the question of how a "natural" understanding of cohabitation with domestic animals and their everyday exploitation in the early modern Estonian rural community became "unnatural", which eventually brought about their stigmatisation and abuse.

Livonian Peasants at the Pärnu County Court

The historical social structure in the Baltic provinces had already been determined in the Middle Ages with a feudal system in which the indigenous majority of Estonian, Livonian and Latvian people formed the farming serfs, whereas the German-speaking minority of settlers embodied the clerical, administrative and manufacturing elite (gentry, clergy, scholars, professional artisans, etc.). Accordingly, German was constituted as an official administrative language.[9]

---

8 Kasemets, Kadri et al.: *The Unnatural and Cultural Theory. Policing Boundaries, Articulating Claims and Positioning the Human.* retrieved 10.07.2023, from https://www.eurozine.com/the-unnatural-and-cultural-theory/.

9 Andresen, Andres: "Unifying the Periphery of a Conglomerate State. Ecclesiastical Legislation in Estland and Livland (1686–1832)". In: Luts-Sootak, Marju / Osipova, Sanita / Schäfer, Frank L. (eds.): *Einheit und Vielfalt in der Rechtsgeschichte im Ostseeraum. Unity and Plurality in the Legal History of the Baltic Sea Area.* (Rechtshistorische Reihe 428). Peter Lang: Frankfurt a. M. et al. 2012, pp. 7–14, here pp. 8–9; Loit, Aleksander: "Reformation und Konfessionalisierung in den ländlichen Gebieten der baltischen Lande von ca. 1500 bis zum Ende der schwedischen Herrschaft". In: Asche, Matthias / Buchholz, Werner / Schindling, Anton (eds.): *Die baltischen Lande im Zeitalter der Reformation und Konfessionalisierung. Livland, Estland, Ösel, Ingermanland, Kurland und Lettgallen. Stadt, Land und Konfession 1500–1721,*

By the beginning of the seventeenth century the Baltic provinces had become part of the kingdom of Sweden. As Lutheran orthodoxy was the sole accepted state religion in the Swedish empire, it thus resulted in active cooperation with clerical, juridical and state authorities in the Baltic provinces in order to secure better official control over the morals and behaviour of the people.[10] Above all this meant that county courts in the Baltic provinces, exclusively formed by the local German gentry, were rather eager to take up cases connected with sexuality (including bestiality). Although the Baltic provinces were conquered by Russia in the beginning of the eighteenth century, the trials against sexual offences remained practically unaltered, because the ruling Russian czar Peter the First granted extensive privileges to the local German elite, which cemented a status quo polity, juridical system, and Lutheran faith in the region.[11]

From the seventeenth century until the end of the nineteenth century all criminal cases of local peasantry in the Baltic provinces were tried in the county courts.[12] Using materials of one of these courts, the Pärnu county court (*Pernauer Landgericht*) in the governorate of Livonia, it is possible to point out some general aspects about bestiality that are easily extendable to the whole region with similar faith, social structure, juridical system and common customs.[13]

In the Pärnu county court the first independent court case file with an accusation of bestiality was opened in 1669.[14] Until the abolition of serfdom in the governorate of Livonia in 1819, altogether 78 independent files about cases

---

1. (Katholisches Leben und Kirchenreform im Zeitalter der Glaubensspaltung 69). Aschendorff: Münster 2009, pp. 49–216, here pp. 82–83.
10. Ird, Ken: "Sozialdisziplinierung im frühneuzeitlichen Livland: Fälle von Sodomie vor dem Landgericht Pernau im 17. bis 19. Jahrhundert". *Forschungen zur baltischen Geschichte* 9, 2014, pp. 67–82; Liliequist 1991a, pp. 175–176.
11. Luts-Sootak, Marju: "Die baltischen Kapitulationen von 1710 und die Gesetzbücher des 19. Jahrhunderts". In: Brüggemann, Karsten / Laur, Mati / Piirimäe, Pärtel (eds.). *Die baltischen Kapitulationen von 1710*. (Quellen und Studien zur baltischen Geschichte 23). Böhlau: Köln, Weimar, Wien 2014, pp. 153–182, here p. 153.
12. These courts were called "land courts" (*Landgericht*) in the governorate of Livonia and "vassal courts" (*Manngericht*) in the governorate of Estonia. Laur, Mati: *Eesti ala valitsemine (1710–1783)*. [Administration of the Estonian Area (1710–1783).] Eesti Ajalooarhiiv: Tartu 2000, p. 106–107.
13. Material derives from archival fonds of Estonian National Archives (EAA, fond 915), and Latvian State Historical Archives (LVVA, fond 109).
14. *Akte in Untersuchungssachen wider den Tackerortschen Bauern Andres Kempi Willems Sohn wegen sodomie* (1669–1670). Estonian National Archives (EAA), fond 915, inventory 1, file 9443.

of bestiality can be determined in the Pärnu county court.[15] From the same period cases concerning the sexual sphere in the Pärnu county court were as follows: 28 about adultery and prenuptial intercourse, 27 about incest, and five about polygamy. Therefore, in comparison to those numbers the ratio of accusations of bestiality becomes more than evident.

Because in the Pärnu county court only the cases of the local low-rank rural population were processed, the accused belonged almost exclusively to the ranks of Estonian serf peasants. The peasantry formed around 95 % of the population of eighteenth century Estonia, whereas 0.5–1 % belonged to the gentry. Between the minority of the German gentry and majority of Estonian serfs, a distinguished and diverse rank of common freemen living in the countryside was 4–5 %.[16] Therefore, it is understandable that out of 78 studied cases there was only one case concerning a freeman, a German miller named Gotthard Friedrich Schack.[17]

Around 40 % of the accused belonged to the lower strata of the local peasantry (farmhands, cottagers), but there were also many men who were farmers and rural craftsmen or their descendants. The lion's share of the accused consisted of single men under 25 years who were traditionally occupied as herders and/or farmhands in the Estonian rural community.[18] In the historiography of bestiality

---

15 The abolition of serfdom has been selected as the final year of selection rather arbitrarily, in order to secure a wider time frame of 150 years for the applied research. Although the abolition itself eventually did not change the local criminal jurisprudence, it is a threshold for the overall social change in the local traditional peasant community. Cf. Lust, Kersti: "The Impact of the Baltic Emancipation Reforms on Peasant–Landlord Relations: A Historiographical Survey". *Journal of Baltic Studies* 44(1), 2013, pp. 1–18.
16 Hupel, August Wilhelm: *Topographische Nachrichten von Lief- und Ehstland*, 1. Johann Friedrich Hartknoch: Riga 1774, pp. 140–141.
17 The eyewitness in this case, German miller Jürgen Johann Normann confessed to the court that after he had caught Schack red handed at the deed, he had shouted, "Was machst du da Unverschämter, und vermischt dich mit meiner Kuh, gehe zum Teufel, du soltest dich schämen, um so mehr, da du ein Teutscher bist." In English: "What are you doing there, you shameless, mingling with my cow, go to hell, you should feel ashamed all the more so as you are a German." Evidently Normann was disturbed by the fact that a person belonging to the same social strata of the society had done something which to his mind was done only by uneducated and half barbarian peasants. *Acta in Inquisitions-Sachen contra den Müller Gotthard Friedrich Schack in puncto Sodomie* (1771–1772). EAA, fond 915, inventory 1, file 1509, p. 4.
18 Here it is also important to note that there was no accusation against a woman. This was first and foremost determined by the general understanding of the European early modern jurisprudence that in order to convict anyone for an accusation of bestiality, it

it is stressed that such behaviour is considered to be chiefly an issue of young uneducated boys/adolescents who were commonly herders.[19] Jonas Liliequist, researcher of the cases of bestiality in seventeenth and eighteenth-century Sweden, has come to a similar conclusion, adding that the fact of a large amount of herders being accused in bestiality can also be nevertheless explained by their giving in to a simple sexual temptation, not just their being less cultivated. The latter is also backed by the matter that in the early modern period it was ordinary for the common people to marry rather late.[20]

Baltic-German Pastors and Their View on Bestiality

Similarly, one contemporary local pastor, Johann Georg Eisen (1717–1779), offered an explanation of the same kind about the prevalence of bestiality among the local Estonian peasantry, written around 1773. The argument in his piece of writing on the subject is quite straightforward: due to the fact that young peasant men cannot afford lavish wedding ceremonies, they demeaned themselves with acts of bestiality. Eisen also refers that the ordinances since the times of the Swedish rule had tried to constrain the lavishness of peasant weddings, but in fact had only a little effect.[21] Eisen hinted here at a widespread tendency in early modern Europe – a social constraint of young people from lower social spheres to postpone their marriage until they were able to maintain an independent household. Marrying in the mid to late 20s was not common only in Western Europe and Scandinavia, but also in Estonia.[22]

Therefore, in the early modern Baltic provinces bestiality was first and foremost a practice of juveniles and single young men who satisfied their unrestrained lust in an exceptional manner. The majority of the accused in the investigated cases were young boys under 20 years old. This corresponds to the studies of Jonas Liliequist on bestiality in seventeenth and eighteenth-century Sweden, as well as

---

was necessary to prove penetration, which therefore made women unlikely offenders. Cf. Thomas, p. 158.
19　Schubert, Ernst: Räuber, *Henker, arme Sünder. Verbrechen und Strafe im Mittelalter.* WBG: Darmstadt 2007, p. 217.
20　Liliequist 1991b, p. 412.
21　Bartlett, Roger / Donnert, Erich (eds.): *Johann Georg Eisen (1717–1779) Ausgewählte Schriften. Deutsche Volksaufklärung und Leibeigenschaft im Russischen Reich.* Verlag Herder-Institut: Marburg 1998, p. 548.
22　Crawford, Katherine: *European Sexualities, 1400–1800.* Cambridge University Press: Cambridge 2007, p. 21; Palli, Heldur: *Eesti rahvastiku ajalugu 1712–1799.* (Academia 7). Teaduste Akadeemia Kirjastus: Tallinn 1997, p. 81.

to the results from the doctoral thesis of Teemu Keskisarja about bestiality cases in the eighteenth century in Finland. Both of them emphasise the importance of pastoral culture of young Scandinavian boys and its probable effect on the practice of bestiality.[23]

In addition to that a contemporary pastor and literate in Livonia, August Wilhelm Hupel (1737–1819), wrote in his scrupulous description of the Baltic provinces, "Of course from the early years on they see mating of their domestic animals; since all [otherwise] not worksome village children spend the whole summer at pasture."[24] Hupel associates the phenomenon of bestiality with an inevitable everyday interaction between peasantry and their domestic animals.

Children started working in farms when they were six or seven years old or even younger. The most common chores for children were babysitting younger siblings, and herding.[25] All this meant that almost every Estonian peasant child had worked as a herder. Thus young farm children undoubtedly formed a close bond with domestic animals.[26] In addition to that, during wintertime the peasants in the Baltic provinces dwelled with their domestic animals, fowl and cattle alike, in order to minimise expenditure of resources and to benefit from their body heat. This certainly intensified the bond between domestic animals and peasantry even more.

Other Views on Livonian Bestiality

At the same time anthropologist Maarja Kaaristo has noticed in her works about relations between the Estonian peasantry and domestic animals that in fact there is a lot of ambivalence. From one side they were full members of the farm (household) whose welfare was taken care of in every way, but on the other side the animals were often still seen as hierarchically below "objects", and only with agricultural value.[27]

---

23 Liliequist 1991a, pp. 174, 176.
24 "Freylich sehen sie von Jugend auf das Begatten ihres Viehes; denn alle nicht arbeitsame Dorfskinder bringen den ganzen Sommer auf der Wiehweide zu." Hupel, p. 517.
25 Berg, Eiki et al: *Eesti rahvakultuur*. Eesti Entsüklopeediakirjastus: Tallinn 2008, p. 306.
26 Torp-Kõivupuu, Marju: "Risti peale kirjutas: Ühel papil oli peni... . Eesti loomakalmistukultuurist." *Mäetagused* 25, 2004, pp. 47–76, here p. 49.
27 Kaaristo, Maarja: "Vägivald loomade vastu: inimene ja koduloom Lõuna-Eesti külas 19. sajandi II poolel vallakohtute protokollide näitel." *Mäetagused* 32, 2006, pp. 49–62, here pp. 53–54; Kaaristo, Maarja: "Märkmeid maarahva koduloomade kohta 19. sajandil." *Vikerkaar* 7–8, 2012, pp. 123–134.

The domestication of animals for beneficial purposes was an extremely important starting point, because people started to functionally communicate with a domestic animal in a totally different manner than with a wild, i.e. "strange" animal.[28] Animal husbandry was an organic part of the agricultural life of early modern Estonians and it stayed rather invariant up until the middle of the nineteenth century – domestic animals were kept in order to gain driving power (horses, oxen), as well as to get a sufficient supply of manure (cows). In addition to that sheep, goats, pigs, chicken, and geese were also kept.[29]

Therefore, when taking into account the conviction that domestic animals had first and foremost a practical value for early modern people, one can also agree with an argument of Courtney Thomas, who stated that in the early modern period bestiality was in essence a selfish crime. Sodomitical acts at that time were not initiated by emotional and physical affection toward some concrete animal.[30] Such a view is diametrically opposed to the understandings of a group of "zoophiles" of the twenty-first century who were studied by Colin J. Williams and Martin S. Weinberg. Members of that group straightforwardly expressed their deep love, devotion, and passion toward animals.[31] The court protocols used for this article clearly indicate that for the accused an animal was a simple instrument, which helped them to satisfy their carnal lust. No emotional bond between the two counterparts is to be detected from the given sources.

This is well illustrated by the manner of behaviour toward molested animals. In order to copulate with them, the accused usually tied an animal down or approached it when it was harnessed. For example, in the case of Vana Hinriku Jüri's son Hinrik, the latter explains his actions, "(…) for the given purpose he bound a cow with a whipcord that he had in his pocket".[32] Such behaviour was very typical, and what is more, many culprits had to take into account that an animal would not stay calm. For instance, a church beggar named Jaak testified, "(…) only because the mare did not stay still, he said that he could not finish the

---

28 Kaaristo 2012, p. 123.
29 Berg, p. 93.
30 Thomas, pp. 159–160.
31 Williams, Colin J. / Weinberg, Martin S.: "Zoophilia in Men: A Study of Sexual Interest in Animals." *Archives of Sexual Behavior* 32(6), 2003, pp. 523–535, here pp. 527–529.
32 „[...] er hätte auch zu dem Ende eine Kuh im Viehstalle mit einer Peitschen Schnur, welche er in der Tasche gehabt angebunden." *Protocollum Inquisitionis Actor officiosus Fiskal Bruno contra des Wanna Hinricko Jürri seinen Sohn Hinrick aus dem Testamaschen Gebiete in puncto Sodomiae insimulatae* (1757). EAA, fond 915, inventory 1, file 1303, p. 2p.

deed".[33] Later Jaak also testified that he had even hit the animal.[34] Many other accused also described how they were unable to keep an animal still. In the given sources there is no indication that the accused showed the slightest regard toward molested animals.

When taking a closer look at the inflicted animals, one must self-evidently take just domestic animals into account, because only they belong to the Estonian peasant's "domestic circle" – a physical and mental space that in a broader sense occupied all farmed land with fields, meadows and pastures, in a narrow sense the farm, its yard, buildings, etc.[35] In the majority, i.e. two-thirds of the cases, those accused of bestiality in the Pärnu county court copulated or tried to do that with a mare. Susanna Sjödin Lindenskoug, using the data from Livonia at the turn of the eighteenth century, has come to the same conclusion and points out a clear distinction in comparison with early modern Sweden, where sexual assault on cows prevails. Lindenskoug rightfully explains such a difference with the fact that in the Baltic provinces horses belonged to the masculine sphere of Estonian peasant society.[36] In addition to that notion it is also presumably that all the studied animal molesters decided to choose a mare to copulate with due to the reason that as they had close daily contact with horses, they knew their behaviour better.

In Estonian agrarian society, the chores were divided between men and women, and this had symbolic value within the understanding of masculinity and femininity. The division of labour of course had its regional differences, but usually the chores of men were sowing, ploughing, caring for horses, slaughtering animals, hunting, fishing, forestry, and hauling.[37] Therefore, it is quite understandable why only less than one-third of the accused chose cows as their "partner": tending to cows, milking, and everything concerning processing milk was explicitly a women's field. The "femininity" of dealing with a cow was determined by giving birth and suckling, both of which were closely connected

---

33 „[...] allein weil die Stuhte nicht stille gestanden, wäre es dazu nicht gekommen." *Protocollum Inquisitionis in Denunciations-Sachen des Paistelschen Pastoris Zimermann contra den Kirchen Bettler Saint Jaack aus dem Euseküllschen in puncto Sodomiae* (1741–1742). EAA, fond 915, inventory 1, file 1148, L 8p.
34 loc. cit.
35 Paulson, Ivar: *Vana Eesti rahvausk*. Ilmamaa: Tartu 1997, p. 104.
36 Sjödin Lindenskoug, p. 142.
37 Kalkun, Andreas: "Work First and Love Will Follow. Division of Labor and Male Beauty in Estonian Folk Culture." In: Samma, Jaanus / Paistik, Alo (eds.): *AAFAGC. Applied Art for a Gay Club*. Gepard: Tallinn, Paris 2010, pp. 51–62, here p. 54.

with milk. It was considered odd or even dangerous when a man milked a cow, and it was also thought that milk milked by men and cream skinned by them was filthy and sour in its essence.[38] All these things considered, it is clear that cows were molested less in the early modern Baltic provinces because Estonian peasant men had little if any connection with cattle.

Conclusion

Contemporary understanding of breaking borders between humans and animals was seen as a grave danger to the whole of society and thus according to the early modern law those convicted by the court were usually executed along with the molested animal(s), although it was evident that the animal itself was not to be blamed for the deed. The juridical system of the early modern period thought such a measure necessary in order not to let the animal remind the community of the wickedness of this kind of crime. Lawyers and lawmakers also believed that an animal was "spoiled" because of intercourse with a human.[39]

The effective law in the early modern Baltic provinces condemned both the convicted and the molested animal to be burnt at the stake. However, usually both parties were at first decapitated and then their bodies burnt. During the reign of the Empress Elizabeth Petrovna (1741–1762) the death penalty was suspended throughout Russia, including the Baltic provinces. Thus persons condemned to death from thereon were subdued to mutilation, flogging, and lifelong hard labour.[40] However, imperial suspensions did not apply to the molested animals, and their decapitation and burning was not ceased. The turn of the eighteenth and the nineteenth centuries brought about some changes in the legal process. In order not to wind up the local peasantry, it was decided not to execute the molested animal, but rather to expel them from their original location.[41]

---

38 Ibid., p. 54–55. The same notions about the Swedish peasantry have been made by Jonas Liliequist. Cf. Liliequist 1991b, pp. 415–417.
39 Crawford, p. 162.
40 Ird, p. 79–80; Marasinova, Elena: "The Prayer of an Empress and the Death Penalty Moratorium in Eighteenth-Century Russia." *The Journal of Religious History, Literature and Culture* 3:2, 2017, pp. 36–55.
41 "[...] die Stute mit der er das delictum vollziehen wollen, zur Vermeidung alles Aergers aus dem Gebiethe zu entfernen seÿ." In English: "[...] the mare, with which he wanted the delict to be consummated, is to be dislodged from the area in order to avoid any trouble." *Acta in inquisitions Sachen gegen den zum publique Gute Enge gehörigen Jungen Mursep Mardi Sohn Mert wegen versuchter Sodomiterey mit einer Stute* (1810). EAA, fond 915, inventory 1, file 2464, L 12.

The cases of bestiality in the Pärnu county court from the early modern period are rich material, offering some insight to the everyday lives of the local Estonian peasantry. More specifically, these enable us to have a closer look at interspecies communication and rural mentality toward domestic animals. All in all, the relation between early modern Estonian peasants and their domestic animals tends to be ambivalent – on one side the peasantry took care of their domestic animals and treated them as members of their household, but at the same time the relationship between them was rather selfish. The abundance of cases of bestiality in the Pärnu county court from the seventeenth to the nineteenth centuries demonstrates how in some cases such selfishness transformed the "natural" everyday exploitation of domestic animals into "unnatural" sexual assault on them and could eventually bring about unpleasant or even fatal consequences to both parties.

References

Source Materials

Acta in inquisitions Sachen gegen den zum publique Gute Enge gehörigen Jungen Mursep Mardi Sohn Mert wegen versuchter Sodomiterey mit einer Stute (1810). EAA, fond 915, inventory 1, file 2464.

Acta in Inquisitions-Sachen contra den Müller Gotthard Friedrich Schack in puncto Sodomie (1771–1772). EAA, fond 915, inventory 1, file 1509.

Akte in Untersuchungssachen wider den Tackerortschen Bauern Andres Kempi Willems Sohn wegen sodomie (1669–1670). EAA, fond 915, inventory 1, file 9443.

Protocollum Inquisitionis Actor officiosus Fiskal Bruno contra des Wanna Hinricko Jürri seinen Sohn Hinrick aus dem Testamaschen Gebiete in puncto Sodomiae insimulatae (1757). EAA, fond 915, inventory 1, file 1303.

Protocollum Inquisitionis in Denunciations-Sachen des Paistelschen Pastoris Zimermann contra den Kirchen Bettler Saint Jaack aus dem Eusekullschen in puncto Sodomiae (1741–1742). EAA, fond 915, inventory 1, file 1148.

Protocollum Inquisitionis in Sachen actor officious contra des Pollenhoffschen Hofes Böttigers Puisa Hans seinen Jungen Peter in puncto Sodomiae insimulatae (1755–1756). EAA, fond 915, inventory 1, file 1285.

Literature

Andresen, Andres: "Unifying the Periphery of a Conglomerate State. Ecclesiastical Legislation in Estland and Livland (1686–1832)". In: Luts-Sootak, Marju

/ Osipova, Sanita / Schäfer, Frank L. (eds.): *Einheit und Vielfalt in der Rechtsgeschichte im Ostseeraum. Unity and Plurality in the Legal History of the Baltic Sea Area.* (Rechtshistorische Reihe 428). Peter Lang: Frankfurt a. M. et al. 2012, pp. 7–14.

Bartlett, Roger / Donnert, Erich (eds.): *Johann Georg Eisen (1717–1779) Ausgewählte Schriften. Deutsche Volksaufklärung und Leibeigenschaft im Russischen Reich.* Verlag Herder-Institut: Marburg 1998.

Beirne, Piers: *Confronting Animal Abuse. Law, Criminology, and Human–Animal Relationships.* Rowman & Littlefield Publishers: Lanham 2009.

Ben-Atar, Doron S. / Brown, Richard D.: *Taming Lust. Crimes against Nature in the Early Republic.* University of Pennsylvania Press: Philadelphia 2014.

Berg, Eiki et al: *Eesti rahvakultuur.* Eesti Entsüklopeediakirjastus: Tallinn 2008.

Canup, John: *"The Cry of Sodom Enquired Into". Bestiality and the Wilderness of Human Nature in Seventeenth-Century New England.* American Antiquarian Society: Worcester 1988.

Crawford, Katherine: *European Sexualities, 1400–1800.* Cambridge University Press: Cambridge 2007.

Fudge, Erica: "Monstrous Acts. Bestiality in Early Modern England". *History Today* 50(8), 2000, pp. 20–25. DOI:10.1080/0268117X.2011.10555663.

Hupel, August Wilhelm: *Topographische Nachrichten von Lief- und Ehstland,* 1. Johann Friedrich Hartknoch: Riga 1774.

Ird, Ken: "Sozialdisziplinierung im frühneuzeitlichen Livland: Fälle von Sodomie vor dem Landgericht Pernau im 17. bis 19. Jahrhundert". *Forschungen zur baltischen Geschichte* 9, 2014, pp. 67–82.

Jordan, William Chester / Creager, Angela N. H.: *The animal/human boundary. historical perspectives.* University of Rochester Press: Rochester / New York 2002.

Kaaristo, Maarja: "Märkmeid maarahva koduloomade kohta 19. sajandil." *Vikerkaar* 7–8, 2012, pp. 123–134.

Kaaristo, Maarja: "Vägivald loomade vastu: inimene ja koduloom Lõuna-Eesti külas 19. sajandi II poolel vallakohtute protokollide näitel." *Mäetagused* 32, 2006, pp. 49–62.

Kalkun, Andreas: "Work First and Love Will Follow. Division of Labor and Male Beauty in Estonian Folk Culture." In: Samma, Jaanus / Paistik, Alo (eds.): *AAFAGC. Applied Art for a Gay Club.* Gepard: Tallinn, Paris 2010, pp. 51–62.

Kasemets, Kadri et al.: *The unnatural and cultural theory. Policing boundaries, articulating claims and positioning the human,* retrieved 10.07.2023, from https://www.eurozine.com/the-unnatural-and-cultural-theory/.

Keskisarja, Teemu: *"Secoituxesta järjettömäin luondocappalden canssa".* *Perversiot, oikeuselämä ja kansankulttuuri 1700-luvun Suomessa.* [Perversions, Legal Life and Folk Culture in the 18th-Century Finland.] (Doctoral thesis). University of Helsinki: Helsinki 2006.

Lang, Dominik: *Sodomie und Strafrecht. Geschichte der Strafbarkeit des Geschlechtsverkehrs mit Tieren.* (Europäische Hochschulschriften / European University Studies / Publications Universitaires Européennes. Reihe II. Rechtswissenschaft, 4750). Peter Lang: Frankfurt a. M. et al. 2009.

Laur, Mati: *Eesti ala valitsemine (1710–1783).* Eesti Ajalooarhiiv: Tartu 2000.

Liliequist, Jonas: *Brott, synd och straff. Tidelagsbrottet i Sverige under 1600- och 1700-talet.* (Doctoral thesis). Umeå University: Umeå 1991a.

Liliequist, Jonas: "Peasants Against Nature: Crossing the Boundaries between Man and Animal in Seventeenth- and Eighteenth-Century Sweden". *Journal of the History of Sexuality* 3(1), 1991b, pp. 393–423.

Loit, Aleksander: "Reformation und Konfessionalisierung in den ländlichen Gebieten der baltischen Lande von ca. 1500 bis zum Ende der schwedischen Herrschaft". In: Asche, Matthias / Buchholz, Werner / Schindling, Anton (eds.): *Die baltischen Lande im Zeitalter der Reformation und Konfessionalisierung. Livland, Estland, Ösel, Ingermanland, Kurland und Lettgallen. Stadt, Land und Konfession 1500–1721,* 1. (Katholisches Leben und Kirchenreform im Zeitalter der Glaubensspaltung 69). Aschendorff: Münster 2009, pp. 49–216.

Lust, Kersti: "The Impact of the Baltic Emancipation Reforms on Peasant-Landlord Relations: A Historiographical Survey". *Journal of Baltic Studies* 44(1), 2013, pp. 1–18.

Luts-Sootak, Marju: "Die baltischen Kapitulationen von 1710 und die Gesetzbücher des 19. Jahrhunderts". In: Brüggemann, Karsten / Laur, Mati / Piirimäe, Pärtel (eds.). *Die baltischen Kapitulationen von 1710.* (Quellen und Studien zur baltischen Geschichte 23). Böhlau: Cologne et al. 2014, pp. 153–182.

Marasinova, Elena: "The Prayer of an Empress and the Death Penalty Moratorium in Eighteenth-Century Russia." *The Journal of Religious History, Literature and Culture* 3(2), 2017, pp. 36–55. DOI:10.16922/jrhlc.3.2.3.

Muravyeva, Marianna: "Sexual Variations". In: Peakman, Julie (ed.): *A Cultural History of Sexuality in the Enlightenment.* Bloomsbury: London / New York 2015, pp. 85–106.

Palli, Heldur: *Eesti rahvastiku ajalugu 1712–1799.* (Academia 7). Teaduste Akadeemia Kirjastus: Tallinn 1997.

Paulson, Ivar: *Vana Eesti rahvausk.* Ilmamaa: Tartu 1997.

Schubert, Ernst: *Räuber, Henker, arme Sünder. Verbrechen und Strafe im Mittelalter*. WBG: Darmstadt 2007.

Schwerhoff, Gerd: *Historische Kriminalitätsforschung*. Campus Verlag: Frankfurt / New York 2011.

Sjödin Lindenskoug, Susanna: *Manlighetens bortre gräns. Tidelagsrättegånar i Livland åren 1685-1709*. [The Outer Limit of Manhood. Tidal Law Trials in Livonia in the Years 1685-1709.] Doctoral Dissertation. Stockholm University: Stockholm 2011.

Thomas, Courtney: ""Not Having God Before his Eyes". Bestiality in Early Modern England". *Seventeenth Century* 26:1, 2011, pp. 149-173. DOI:10.1080/0268117X.2011.10555663.

Torp-Kõivupuu, Marju: "Risti peale kirjutas: Ühel papil oli peni… . Eesti loomakalmistukultuurist". *Mäetagused* 25, 2004, pp. 47-76.

Williams, Colin J. / Weinberg, Martin S.: "Zoophilia in Men: A Study of Sexual sInterest in Animals." *Archives of Sexual Behavior* 32(6), 2003, pp. 523-535. DOI:10.1023/a:1026085410617.

Kaarina Rein

Depiction of Animals in the Medical Works of the Seventeenth Century University of Tartu[1]

Abstract: The foundation of the University of Tartu as well as the beginning of Estonian scientific research occurred in the seventeenth century. At the time four medical works were defended in the University of Tartu in which animals were described or mentioned in some connection. In these works animals were often used as metaphors in order to draw parallels with humans. In this regard, animals rather tended to be considered inferior to humans, as was typical of the early modern period. The medical works of the seventeenth-century University of Tartu mention veterinary medicine as well as the use of animals as a source of remedies or even animals as their own doctors, teaching humans how to heal certain diseases. There are hints to animals in scientific experiments. In the medical work *Dissertatio prima de oeconomia corporis animalis* originating from 1698, the development of early modern physiology can be observed. There is a clear difference between the works that were compiled in the first half of the seventeenth century and the dissertation written at the end of the seventeenth century. The works written earlier were mostly based on authors of classical antiquity, whereas the dissertation from the end of the seventeenth century forwards empirical knowledge. Animals made an impact on the medicine of the seventeenth-century University of Tartu as well as on the ways of speaking and writing about medicine.

Keywords: Early Modern disputations, Early Modern medicine, natural sciences in the Early Modern era, orations and dissertations, the Swedish period of the University of Tartu

Introduction

The University of Tartu was founded in 1632 by King Gustavus Adolphus as the second university in the Swedish Empire. The institution was called *Academia Gustaviana* during its first period of activity from 1632 to 1656. During its second

---

1 The article has been written in the framework of the grant POST 108, the First Scientific Works at the University of Tartu. Disputation as a Literary Genre in the Academia Gustaviana Period. The study has been supported by the fellowship programme of Wolfenbüttel Herzog August Bibliothek in 2019. The research results in Estonian have been published as Rein, Kaarina: "Loomad Rootsi-aegse Tartu ülikooli meditsiinitöödes" [Animals in the Medical Works of the Swedish-period University of Tartu.] *Mäetagused* 78, 2020, pp. 155–172. DOI:10.7592/mt2020.78.rein.

period of activity between 1690 and 1710 the university was called *Academia Gustavo-Carolina*. One of the three higher faculties at the Swedish University of Tartu was the Faculty of Medicine. The first constitution of the University declared that the future physicians had to study natural history on the basis of Johannes Magirus' textbook *Physiologiae Peripateticae libri sex* (1597).[2] In this book there is a special chapter (Book V, chapter 14) devoted to animals and their division according to Aristotelian philosophy with the commentaries of Johannes Magirus (ca 1560–1596).[3] The beginning of modern zoology was laid even earlier by Conrad Gessner (1516–1565) and others in the sixteenth century, and the invention of zoology was one of the fascinating features that the redefinition of the notion of animal and animal species brought about.[4] However, there are no purely zoological works from the University of Tartu originating from the seventeenth century. There are just a few medical works from that time, but we can find several references to animals in them.

The aim of this article is to study how animals were depicted in the medical disputations, dissertations and orations of the Swedish University of Tartu from 1632 to 1710. One can presume that the impact of animals in medicine can be distinguished, e.g. in the field of pharmacy as well as in the medical vocabulary. The literary tradition of classical antiquity was dominant in the zoological discourses of the early modern period. The authorisation of zoological knowledge by Aristotle, Pliny, Athenaus and others, even in cases in which empirical knowledge was available, forms one of the highly fascinating features of early modern zoology.[5]

Research about zoology in early modern cultures has recently been published in article collections by Karl Enenkel and Paul Smith,[6] which have had the intention of investigating early modern zoology in connection with the use of animals

---

2   Lepajõe, Marju (ed.): *Constitutiones Academiae Dorpatensis (Academia Gustaviana). Tartu Akadeemia (Academia Gustaviana) Põhikiri*. Tartu Ülikooli Kirjastus: Tartu 1997, pp. 56–57.
3   Magirus, Joannes: *In physiologiam suam Peripateticam commentarius*. Nebenius: Lichae 1601, pp. 540–560.
4   Enenkel, Karl A. E. / Smith, Paul J. (eds.): *Early Modern Zoology. The Construction of Animals in Science, Literature and the Visual Arts*, vol. 1. (Intersections 7). Brill: Leiden / Boston 2007, pp. 1–3.
5   Ibid., p. 7.
6   Enenkel / Smith 2007, p. 7; Enenkel Karl A. E. / Smith Paul J. (eds.): *Zoology in Early Modern Culture. Intersections of Science, Theology, Philology and Political and Religious Education*. (Intersections 32). Brill: Leiden / Boston 2014.

in various fields, theory and practice alike, such as hunting, horsemanship, veterinary medicine, courtly life, jurisdiction, literature, and the visual arts.[7] As the editors state in the introduction of their first collection, botany has received considerably more attention than zoology in studies on the history of biology.[8] In the edited volume by Christian Hoffstadt, Franz Peschke, Michael Nagenborg and Sabine Müller the relations between animals and humans in medicine is described from antiquity to modern times.[9] The use of animals in the treatment of illnesses and injuries has been analysed observing medieval bestiaries,[10] as well as in Estonian folk medicine.[11] One can presume that it was the practice in early modern times as well. There is a treatise about allergies to cats described in early modern texts in this collection of articles by Meelis Friedenthal. Zoology in early modern academic culture is also one of the themes in Maija Kallinen's monograph about the disputations in natural history at the Academia of Turku. The author mentions descriptions of a whale, doves and elephants, and states that most of the information was taken either directly from the classical or from the seventeenth-century Aristotelian authors' texts.[12]

The present article can be regarded as a part of the similar research on natural history and medicine in early modern Tartu.[13] The aim of the article is to find the sources used by the authors and their attitude toward animals – thus a descriptive-comparative method is applied. The main question is whether their knowledge about animals is empirical or based on ancient authors.

---

7   Enenkel / Smith 2007, p. 4.
8   Ibid., p. 3.
9   Hoffstadt, Christian et al. (eds.): *Humana – Animalia. Mensch und Tier in Medizin, Philosophie und Kultur.* (Aspekte der Medizinphilosophie 13). Projektverlag: Bochum 2012.
10 White, Cynthia: "Potiones ad sanandum. Text as Remedy in a Medieval Latin Bestiary". In: Classen, Albrecht (ed.): *Bodily and Spiritual Hygiene in Medieval and Early Modern Literature. Explorations of Textual Presentations of Filth and Water.* De Gruyter: Berlin / Boston 2017, pp. 221–274, p. 222. DOI:10.1515/9783110523799-008.
11 E.g. Sõukand, Renata: "Koera keele otsas rohi, kassi keele otsas tõbi" [Dog Tongue Carries Medicine, Cat Tongue Carries Illness.] *Mäetagused* 31, 2006, pp. 87–105; Kõivupuu, Marju: *Loomad eestlaste elus ja folklooris.* Tänapäev: Tallinn 2017, pp. 261–302.
12 Kallinen, Maija: *Change and Stability. Natural Philosophy at the Academy of Turku (1640–1713).* (Studia Historica 51). Finnish Historical Society: Helsinki 1995, pp. 228–230.
13 See also Rein 2020.

Animals in Medicine from Antiquity to Early Modern Period

Early modern zoology departs from the dogma that man has the full right to exploit animals.[14] In European medicine animals have been a source of remedies since antiquity. It can easily be proved, as due to the use of serpents and their venom in the process of healing in Ancient Greece,[15] the serpent has become the symbol of medicine. Of course, a serpent can be dangerous as well and one should remember its deceitful role in Adam's Fall. In early modern France, however, every drug had to have at least one ingredient from the animal realm.[16]

Animals function in many roles in ancient Greek literature, but basically they seem to figure as metaphors, objects of comparison both in factual literature as well as in "fiction".[17] In the context of the early modern period it has been mentioned that animals have influenced theology and philology a lot.[18] In medicine there can also be found several phrases that include animals even in terms of anatomy and diagnostics: *cochlea, cor bovinum, dens caninus* (dogtooth), *facies leontina* (leonine facies), *fames canina* or *bulimia* or *lycorexia* (wolfish appetite), *faux lupina* (cleft palate), *labium leporinum* (cleft lip), *pes anserinus* (goose foot), *pes equinus* (equinus deformity), *vermis cerebelli* (cerebellar worm).[19] Thus animals have been productive metaphors while describing human conditions.

The first literary work in European literature, the Greek epic *The Iliad*, begins with a description of a plague sent by the god Apollo to the Greek camp as punishment for insulting a priest. The first victims of the plague in the first song of *The Iliad* were mules and dogs.[20] Animals have also been victims of scientific experiments in antiquity as well as during the early modern period. The most renowned physician in Ancient Rome, Galen of Pergamon (129–ca. 200 / 216 AD), performed anatomical dissections on living and dead mammals,

---

14 Enenkel / Smith 2007, p. 2.
15 Leven, Karl-Heinz (ed.): *Antike Medizin. Ein Lexikon*. C. H. Beck: München 2005, p. 777.
16 Brockliss, Laurence / Jones, Colin: *The Medical World of Early Modern France*. Clarendon Press: Oxford 1997, p. 161.
17 Korhonen, Tua / Ruonakoski, Erika: *Human and Animal in Ancient Greece. Empathy and Encounter in Classical Literature*. I. B. Tauris: London / New York 2017, p. 42.
18 Enenkel / Smith 2014, p. 1.
19 Valdes, Albert / Veski, Johann Voldemar: *Ladina-eesti-vene meditsiinisõnaraamat* [Latin-Estonian-Russian medical dictionary.]. Vol. 1–2. Valgus: Tallinn 1982–1983.
20 Murray, August Taber (ed.): *The Iliad with an English translation*. William Heinemann LTD: London / G. P. Putnam's Sons: New York 1928, Song I, Verse 50, retrieved 14.10.2019, from https://ryanfb.github.io/loebolus-data/L170N.pdf.

mostly focusing on pigs and primates.²¹ Galen's principal interest was in human anatomy, but Roman law had prohibited the dissection of human cadavers. However, Galen believed that the anatomical structures of these animals closely mirrored those of humans and the conclusions he drew thereby influenced the perception and approach to human anatomy and physiology until the sixteenth century, as humans were thought to be analogous to animals.²² In the early modern period animal bodies were even used in order to understand the depths of human character.²³ René Descartes's (1596–1650) description of animals as "machine-like" was heavily criticised by many of his contemporaries, but nevertheless provided scientists with a way to justify what would now be considered extremely gruesome experiments, in a time when anaesthesia, for humans and animals alike, was not available.²⁴

Animals occur in early modern medicine not only as objects of medical experiments, but as clients as well. Veterinary medicine as a branch of medicine can be found in ancient literature (Aristotle, Cato, Varro, Columella, etc.). The main interest of veterinary medicine was directed toward horses and cattle, i.e. economically important animals.²⁵ One of the first preserved collections of texts on veterinary medicine is *Mulomedicina Chironis* (ca. 400 A.D.), which proves that the level of ancient veterinary medicine did not lag behind human medicine in the field of diagnostics, aetiology and therapeutics.²⁶ In the early modern period institutionalised veterinary medicine did not exist: it was completely the province of the itinerant quack and local uncertificated healers, especially the village blacksmith.²⁷

The abovementioned examples expose that animals can have connections with medicine in different ways. It is the hypothesis of the present article that

---

21 Gerabek, Werner E. et al. (eds.): *Enzyklopädie Medizingeschichte*. Walter de Gruyter: Berlin / New York 2007, p. 450.
22 Garrison, Fielding H.: *An Introduction to the History of Medicine*. W. B. Saunders Company: Philadelphia / London 1960, pp. 113–114.
23 Muratori, Cecilia: "From Animal Bodies to Human Souls. (Pseudo-)Aristotelian Animals in Della Porta's Physiognomics". *Early Science and Medicine* 22, 2017, pp. 1–23, p. 2. DOI: 10.1163/15733823-00221p01.
24 Franco, Nuno Henrique: "Animal Experiments in Biomedical Research. A Historical Perspective." *Animals* 3(1), 2013, pp. 238–273, p. 241. DOI:10.3390/ani3010238.
25 Achner, Heike: *Ärzte in der Antike*. Verlag Philipp von Zabern: Mainz 2009, pp. 146–152.
26 Leven, p. 864.
27 Brockliss / Jones, pp. 455–456.

these relations can be traced in the early modern academic texts written and published in Tartu as well.

Animals in the Medical Texts of *Academia Gustaviana Dorpatensis*

From the first period of the existence of the University of Tartu, the *Academia Gustaviana* period, there are three texts, two orations and one disputation, which mention or describe animals in some way: 1) The *Oratio de medicina* by Friedrich Hein (1637)[28]; 2) the *Oratio de homine* by Sequardus Wallander (1640)[29]; and 3) the *Disputatio medica de natura et constitutione medicinae* by Sebastian Wirdig and Andreas Arvidi (1648)[30]. A closer examination of these works should reveal the way of approaching animals in them.[31]

Oratio de medicina

Friedrich Hein (1533–1604), the supposed author of *Oratio de medicina*, tells us that humans should take care of animals as animals are created for the use of men and thus medicine of animals, i.e. veterinary medicine is appreciated by many. A list of ancient authors is presented, who have written about diseases of animals: Vergil describes diseases of horses and bees in his *Georgica*, Columella writes about diseases of sheep and Varro of goats.[32] There is no mention of famous Greek scientific authors like Aristotle or Galen in the speech. Hein also

---

28 Hein, Friedrich: *Oratio de medicina*. Lit. acad: Dorpati Livonorum 1637.
29 Wallander, Sigvardius Olai: *Oratio de homine*. Lit. Acad.: Dorpati Livonorum 1640.
30 Wirdig, Sebastian / Arvidi Stregnensis, Andreas: *Disputatio medica de natura et constitutione medicinae*. J. Vogelius: Dorpati Livonorum 1648.
31 In order to get an overview of these medical texts, see the following works: Rein, Kaarina: *Arstiteadus rootsiaegses Tartu gümnaasiumis ja ülikoolis aastatel 1630–1656. Meditsiinialased disputatsioonid ja oratsioonid ning nende autorid*. [Medicine at the Gymnasium and University of Tartu from 1630 to 1656. Medical Disputations, Orations and their Authors.] University of Tartu Press: Tartu 2011; Rein, Kaarina: "Tartu as the Eastern Outpost of European Medicine in the First Half of the 17th Century". *Acta Baltica Historiae et Philosophiae Scientiarum* 2(1), 2014, pp. 37–52. DOI: 10.11590/abhps.2014.1.03.
32 "In vita communi receptum est, ut Medicinam brutorum animalium magnifaciant multi. ... Etenim, cum in usum hominum Bruta animantia sint creata; quando male habere inceperunt, cura indigent & sollicitudine. ... Sic de Equorum morbis, agit Virgil. 3 Georg., De ovium cura, Columella lib. 8. De Caprarum medela, Varro de re Rustica lib. 2. c. 2. De cura apum & morbis, Virg. 4. Georg." Hein, p. A4 v.

brings an example from the Old Testament Book of Tobit (6:5), where Archangel Raphael tells Tobit to keep the heart, bile and liver of a fish in order to use them as remedies, and the bile of a fish is especially told to be a useful remedy for eyes.[33] On the other hand, the author of the oration further mentions that fish can be either harmful or useful to health.[34]

Probably the most interesting parts of the speech are the passages that tell us about the use of animals in medicine. Hein calls his audience to witness the dissections performed in Tartu.[35] These must have been dissections of animals, as no autopsies of human cadavers are known from that time in Tartu. He also claims that animals act as their own doctors. Hereby he stresses that intelligent physicians should learn from animals. Amongst the examples food and herbs are shown, which animals use in certain cases, either against poisons or in order to improve their health, e.g. sight. Thus it is claimed that the Egyptian bird ibis demonstrates therapy with emetics, the deer uses dittany in order to make extraction of arrows easier, the wild boar uses ivy to cure venomous bites, the turtle uses oregano against snakes, and the weasel rue against venom.[36] The list about the wisdom of animals is even longer in the oration; however, the description of the connection of animals with certain remedies is taken from Pliny the Elder's *Natural History*.[37] Thus the animals mentioned in the speech are not necessarily the same that one could meet in the northern or eastern part of Europe. In fact the described passage about clever and caring animals must have been commonly known in the 17th century, as one can find it in the popular novel by Hans Jacob Christoffel Grimmelshausen (ca. 1622–1676) "Simplicius Simplicissimus" (1669) as well.[38]

---

33 "Sic Raphael Angelus monuit Tobiam, ut capti piscis cor, fel & jecur reponeret: ea enim esse necessaria ad medicamenta utiliter, praesertim fel valere ad oculos sanandos". Hein, p. B3 v.
34 Hein, p. C v.
35 "Etenim, versamur in sancto hoc Auditorio Academico-Medico, in quo publice Docentes identidem audimus Medicos ingeniosissimos. Vos Auditores inclyti, testes estis Anatomicae sectionis in corporibus soepius institutae ...". Hein, p. B.
36 "Etenim, monstravit Ibis, Aegyptica avis, vomendi modum. Cervus, dictamnum pro extrahendis sagittis; ... ut & apri ijsdem contra ederam esam utuntur: ... testudo cunilae contra serpentis usum: mustela, contra venenum rutam ...". Hein, p. B4 v.
37 Pliny the Elder: *Natural History*. Book VIII, chapter 41. Retrieved 15.1.2019, from http://www.perseus.tufts.edu/hopper/text?doc=Plin.+Nat.+7&fromdoc=Perseus%3At ext%3A1999.02.0138.
38 Grimmelshausen, Hans Jakob Christoph: *Der abenteuerliche Simplicissimus*. Eduard Kaiser Verlag: Klagenfurt 1950, pp. 80–81.

In the *Oratio de medicina* from 1637 the attitude toward animals is clearly positive, i.e. he describes them as clever and useful creatures. There is just one hint from the negative side, describing the story of Hippolytus, who was torn into pieces by his own horses.[39] However, typically for the early modern period, the author states that animals are created for the use of men, although he stresses the importance of veterinary medicine and taking care of animals. The empirical knowledge about animals is mentioned by hints of dissections. Otherwise, ancient authors and the Old Testament are the main sources, while mentioning animals and exposing information about remedies, treatment of diseases or veterinary medicine.

Oratio de homine

Sequardus Wallander's speech *Oratio de homine* was presented three years later in 1640 and it can be noticed that animals are mentioned here only metaphorically and rather as negative metaphors. It is stressed in the introduction of the speech that if a man does not behave properly, he rather belongs among animals.[40] There is an example of King Nebuchadnezzar from the Book of Daniel (4:30) of the Old Testament, who ate hay like an ox and his hair became similar to the feathers of an eagle and his claws like those of birds.[41] The phrases *beluina voluptas* (lust of a beast) and *Est Homo Homini soepius Lupus* (a man is usually a wolf to another man) used in the oration reflect the negative side of animals as well.[42] The last sentence is to be found in the comedy of Plautus.[43] In the final part of the speech *Oratio de homine* there is a quotation from Ovid, which stresses the man's authority over animals.[44]

---

39 "Postea, praenominatus ille Apollo excelluit, & ejus filius Aesculapius, qui hujus Artis peritus, Dianae precibus, Hippolytum ab equis discerptum vitae restituit". Hein, p. B4.
40 "Vita Hominum media inter Vitam & Angelorum, & Vitam pecorum. Si vixerit Homo secundum carnem & ad delictorum illecebras feretur, pecoribus conjungetur". Wallander, p. A2.
41 "Erat & Rex superbissimus Nabuchdonosor, qui abjectus, foenum ut bos comedit, & rore coeli corpus ejus infectum est, donec capilli ejus in similitudinem aquilarum crescerent, & ungues ejus quasi ungues avium". Wallander, p. A3 v.
42 Wallander, pp. C–C2.
43 Plautus: *Asinaria*. p. 495.
44 Ovid: *Metamorphoses*, Book I, verses 76–88, retrieved 15.01.2019, from http://www.thelatinlibrary.com/ovid/ovid.met1.shtml.

Except for the hints to Plautus and Ovid, the main source of examples with animals is the Old Testament in Sequardus Wallander's speech *Oratio de homine*. There is no mention of empirical experience with animals.

Disputatio medica de natura et constitutione medicinae

The third medical work at the Swedish University of Tartu, where there were hints at animals, *Disputatio medica de natura et constitutione medicinae* was printed in 1648 by Andreas Arvidi (around 1620–1673). The first part of it is devoted to defining medicine and while speaking about homonymy, the author tells it to be a producer of errors, and thus one has to be careful when using the word "medicine". Namely, in the broader sense veterinary medicine also belongs to this discipline.[45] Thus human medicine and veterinary medicine were distinguished at the seventeenth-century University of Tartu. Another context, where animals are mentioned in the disputation of Andreas Arvidi, is in the passage where it is explained who is capable of studying medicine. There is a quotation from Hippocrates' treatise *The Law*,[46] where it is said that if someone who does not have a natural talent for medicine tries to learn this art, it is like uniting wolves with lambs or a serpent with a monkey.[47]

Thus there are not many mentions of animals in the disputation of Andreas Arvidi, and the attitude to animals in the work *Disputatio medica de natura et constitutione medicinae* seems to be neutral. Animals are mentioned rather in a metaphorical way, and there is just one hint of the practical part, i.e. the existence of veterinary medicine.

In conclusion it can be said that the three medical works from *Academia Gustaviana* reflect the Aristotelian idea about *scala naturae* (the great chain of being), which places humans at the top of the hierarchy in complexity and value.

---

45 "Circa *Homonomiam*, quae Errorum genetrix a Philosophis salutari solet, tenendum est, Vocabulum Medicinae accipi hoc loco, *non Improprie & Abusive*. Sic enim obtinet Significationem I. *Latiorem*, & ambitu suo, veterinariam & Mulomedicinam includit". Wirdig / Arvidi, p. A3 v.

46 *Hippocrates with an English translation by W. H. S. Jones*, vol. II. (The Loeb Classical Library). Harvard University Press: Cambridge, Massachusetts / William Heinemann LTD: London 1981, pp. 262–263.

47 "*Causa Efficiens Secunda* Medicinae *addiscendae* est I. *Naturalis aptitudo*; quippe ut ait Medicorum ille Coryphaeus Hippocrates Lib. de Lege, Natura reluctante irrita sunt omnia; & vere, qui enim natura ineptus natus ad Medicinam, artem illam nihilominus addiscere tentat, is aut lupos agnis, aut serpentem simiae conciliare, aut certe sigillum lapidi imprimere conatur". Wirdig / Arvidi, pp. B3 v.–B4.

At the same time animals are used both as positive and negative examples in the text of orations. The themes of veterinary medicine, dissections, animals as remedies, damage done by animals, and of animals as their own doctors are treated. Animals are often used as metaphors for comparison. It is to be noticed that in the two orations there are several examples from the Old Testament, but none from the New Testament. The speeches are presented by the students of theology, thus the biblical examples are not surprising. One can guess that animals in the medical context occur more often in the Old Testament, whereas in the New Testament the human factor is more important. There are also many hints to Greek and Roman authors in the medical works, whereas the empirical part in connection with animals is almost non-existent.

Animals in a Medical Text of *Academia Gustavo-Carolina*

The Medical Dissertation by Jacob Friedrich Below and Salomon Matthiae

The first three medical works from the Swedish University of Tartu, treating or mentioning animals, were relatively within the same time period. The fourth work was compiled much later, in 1698, when the university was called *Academia Gustavo-Carolina* (1690–1710).[48] It was presided by Professor of Medicine Jacob Friedrich Below (1669–1716), and was presented by a student of theology named Salomon Matthiae (1609–1655); it was entitled *Circuli anatomico-physiologici minoris, segmentum primum; seu dissertatio prima de oeconomia corporis animalis, cogitata, functionum animalium potissimarum Formalitatem & causas concernantia.*[49] One can presume from the title that it should be a dissertation relating to the composition of a living being and its functions and causes. It can also be seen that the dissertation is entitled as the first on this theme, thus probably a continuation was expected from it.

---

48 In order to get an overview of the medical works written at that time, cf. Rauch, Georg von: *Die Universität Dorpat und das Eindringen der Frühen Aufklärung in Livland 1690–1710*. Essener Verlagsanstalt: Essen 1943, pp. 276–277; Piirimäe, Helmut (ed.): *Tartu Ülikooli ajalugu I 1632–1798*. Valgus: Tallinn 1982, pp. 232–238.

49 Below, Jacob Friedrich / Matthiae, Salomon: *I. N. J. Circuli Anatomico-physiologici Minoris, Segmentum Primum; Seu Dissertatio Prima De Oeconomia Corporis Animalis*. Johannes Brendeken, Acad.Typographus: Dorpat 1698.

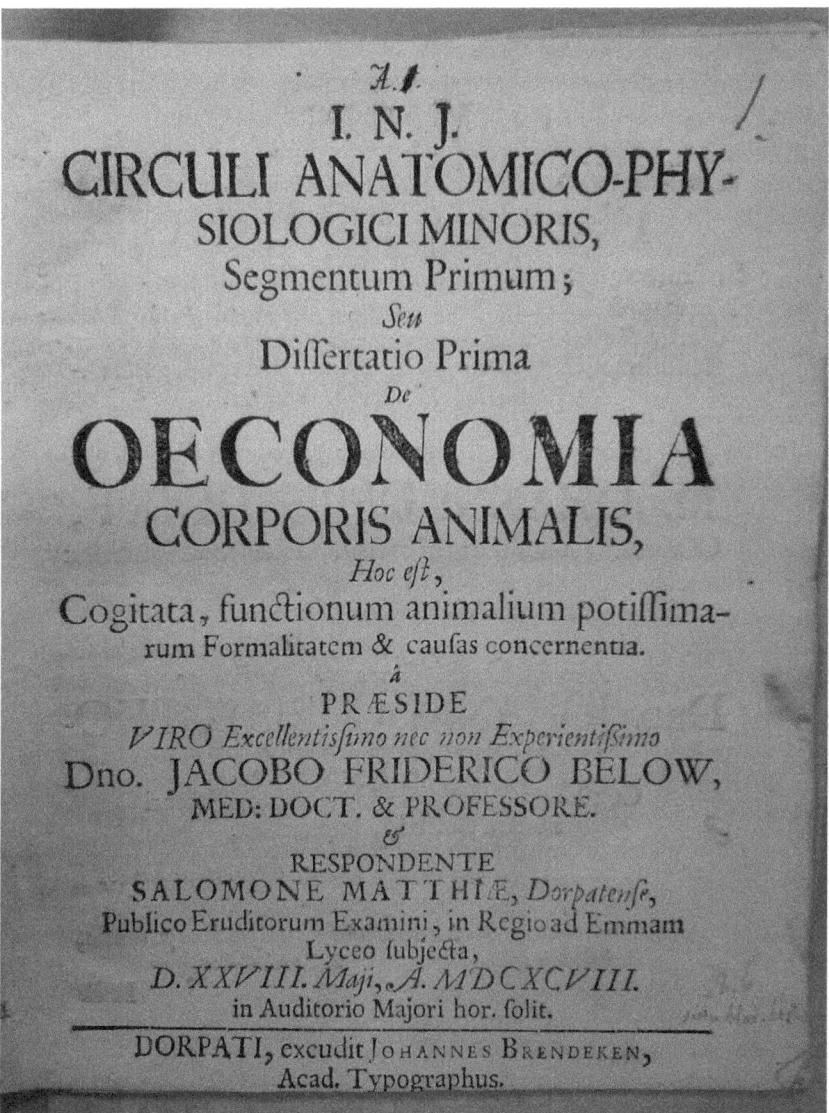

Picture 1. Title page of the dissertation by Jacob Friedrich Below and Salomon Matthiae *De oeconomia corporis animalis* ("About the constitution of animal's (living being's) body"). Title page of the exemplar at the Uppsala University Library.

It was a common practice from antiquity to the beginning of the early modern period that on the basis of dissections of animal cadavers conclusions were drawn about human anatomy,[50] and there was a short description of the practice of Galen in the second chapter of this article. In the work *Dissertatio prima de oeconomia corporis animalis* it occurs to be the opposite, as in the introduction of his work the author states that he has received inspiration for his work from the dissection of a human corpse, which had recently taken place, and the author's aim was to continue with the dissections in order to satisfy public curiosity.[51]

And indeed, when Jacob Friedrich Below became a professor of medicine at the University of Tartu in 1695, he considered it important to illustrate his lectures of medicine by dissections. Consequently, in the winter of the academic year 1697/98 Below dissected two human corpses.[52] It was a historical event, as it happened for the first time at the University of Tartu.

On the 28th of May 1698, the abovementioned dissertation about the constitution of a living being was presented at the University of Tartu, where the author claimed that the dissertation has a direct connection with the previous dissections of human corpses.[53] There is no concrete living being or animal described in this dissertation, and the theme of the work is the physiology of digestion in general. The explanation of the title of the work is presented in its introduction – the dissertation is claimed to be based on the work of Johannes Bohn (1640–1718) from 1697, which has a similar title.[54] The latter was a professor of medicine at the University of Leipzig.

Johannes Bohn was influenced by Cartesian ideas and his impact in Germany as well as in Scandinavia was significant.[55] His book *Circulus anatomico-physiologicus* is dedicated to his friend Marcello Malpighi (1628–1694), discoverer of capillaries.[56] Johannes Bohn's monograph covers 500 pages and has

---

50 Wright, Thomas: *William Harvey. A Life in Circulation.* Oxford University Press: New York / Oxford 2013, pp. 30–39.
51 Below / Matthiae, p. A1.
52 Rauch, p. 276.
53 Below / Matthiae, p. A1.
54 Bohn, Johannes: *Circulus anatomico-physiologicus seu oeconomia corporis animalis, seu Oeconomia corporis animalis, hoc est, cogitata, functionum animalium potissimarum formalitatem et causas concernantia.* Apud Thomam Fritsch: Lipsiae 1697. The former edition of this work was published in 1686.
55 Tering, Arvo: "René Descartes'i ideede jõudmisest Eesti- ja Liivimaale XVII sajandil ja XVIII sajandi algul". [Die Verbereitung der cartesianischen Ideen im 17. und zu Beginn des 18. Jahrhundert in Est- und Livland.] *Keel ja Kirjandus* 3, 1996, p. 187.
56 Bohn, p. a2.

30 chapters, where the themes begin with generation and conception and give an account of themes in physiology, like respiration, circulation of blood, digestion, function of different inner organs, senses, etc.; the final chapter is about the importance of sleep.[57]

The theme of the dissertation compiled in Tartu by Below and Salomon Matthiae on the other hand, is just the physiology of digestion, thus forming a small part of the themes treated in Johannes Bohn's monograph. According to its author(s), its aim is to discreetly correct the controversies in Johannes Bohn's treatise.[58]

The Depiction of Digestion in the Dissertation by Jacob Friedrich Below and Salomon Matthiae

It seems however that Johannes Bohn and Below and Matthiae interpreted the Latin word *animal* differently. For Johannes Bohn *animal* is a living being, someone, who has got *anima* (soul), whereas in the dissertation presented in Tartu there are much more hints at animals and their physiology.

In Johannes Bohn's work there is a separate chapter devoted to the description of the circulation of blood, which was discovered by William Harvey in 1628,[59] whereas the dissertation by Below and Matthiae just mentions the system of blood circulation as an important discovery in the 17th century.[60] The work compiled in Tartu concentrates on the physiology of digestion, explaining hunger and thirst, the function of teeth and the tongue, e.g. the author highlights the long tongue of frogs. The author of the dissertation also explains the constitution of saliva and describes all the organs that are connected with digestion. In Johannes Bohn's work there are several chapters that have a connection with digestion, talking, e.g. about nutrition, appetite, chylification, and excretion.[61] Animals are seldom mentioned there, and one of the few examples is the comparison of the teeth of fish, hare, squirrel, mouse, dog, cattle, and horse with those of a man.[62]

The dissertation by Below and Matthiae has got five chapters,[63] which describe the process of digestion of animals in more detail. The author tells us that there is always liquid in the maw of animals, which is simple and pure in case of

---

57 Bohn, pp. 1–478.
58 Below / Matthiae, p. A1 v.
59 Bohn, pp. 97–114.
60 Below / Matthiae, p. A2.
61 Bohn, pp. 114–205.
62 Bohn, p. 131.
63 For the description of these chapters see Rein 2020, pp. 163–164.

those animals that eat at longer intervals, and thick and not homogeneous in case of those animals that eat every day and often.[64] The authors of both works agree that nutrition is connected with *spiritus animales* (animal spirits).[65] In the dissertation by Below and Matthiae a metaphorical parallel with animals can be seen, where the author compares the movement of intestines with that of worms, and explains that one part of the intestines is stretching out like passing through a snail shell.[66]

At the end of the dissertation *Dissertatio prima de oeconomia corporis animalis* there are 36 theses, where animals are mentioned in some cases. Generally, these theses reflect the philosophy of René Descartes on iatrophysics and mechanics and treat animals like machines. However, there are some statements about concrete animals, as well as the use of their body liquids in pharmacies. Thus there is a description of the anatomy of the beaver and the explanation of the origin of castor, an exudation secreted by beavers, used in folk medicine in antiquity in order to heal various illnesses. The explanation presented in the dissertation to this phenomenon is that this exudation is not from the testicles of the male beaver as it has been thought, but it is secreted from a special gland of both male and female species, it is of the size of a chicken egg, and it produces an oily strong-smelling liquid, which the beaver licks.[67] It is said to be an idea of Thomas Bartholin (1616–1680), who published a description of the lymphatic system in 1652. The latter is mentioned as the discoverer of the salivary duct in the monograph of Johannes Bohn as well.[68]

---

64 "Reperitur nempe semper in quorumcunque animalium ventriculis fluidum quoddam simplex & sincerum magis in illis, quae aliquam mediam passasunt, turbidum vero crassius & heterogeneum magis, si victu quotidiano ac frequenti fruatur animal". Below / Matthiae, p. B3 v.
65 Bohn, pp. 123; Below / Matthiae, p. A2 v.
66 "Hic intestinorum motus … vulgo peristalticus lumbricalis atque vermicularis dicitur, quod pariter ac vermis flexuosis gyris per totam sui longitudinem modo constringantur, modo dilatentur intestina, & horum una pars quasi per cochleam explicetur, proxima constricta, nunc subsideat haec, attollatur altera". Below / Matthiae, p. C4.
67 "XXXIII. Castorei testiculi magnitudine vix superant gallorum gallinaceorum testes, falsum igitur est, illud medicamen quod in pharmacopoliis castorei nomine venditur, hujus animalis esse testiculos. Sed, sunt tantum binae glandulae quae tam in maribus quam in foeminis horum animalium reperiuntur: & quidem utrinque in inguinibus, statim sub cute, ovorum gallinaceorum magnitudine & saepe majores conspiciuntur, ex quibus liquor oleosus & gravi odore emanat, quem animal lingua lambit. Bartholin". Below / Matthiae, p. D4 v.
68 Bohn, pp. 135.

The author of the dissertation also explains the origin of spermaceti. It is written there that spermaceti is not sperm, but a liquid that is created in the brains of whales, which the author claims to be named *orca* (killer whales).[69] As a comment it can be added that this waxy substance is in fact created in the head of cachalots or sperm whales (*Physeter macrocephalus*) and not killer whales. The last thesis of the work *Dissertatio prima de oeconomia corporis animalis* is written in German instead of Latin. It tells us that the greatest difference between a man and stupid livestock is the rational soul, which a man possesses and livestock does not, which sets an obligation to a man according to his understanding to deal with fine arts in order to be considered a proper human. Those, on the other hand, who feel contempt for studies should be considered as non-rational men, which means – monsters.[70]

Thus, in the dissertation by Johannes Below and Salomon Matthiae animals tend to be considered negative metaphors as is seen from the last thesis of the treatise. There are no quotations from the ancient authors or the Bible in this work – on the contrary, empirical knowledge is prevalent in it.

Conclusion

At the Swedish University of Tartu there are four medical works in which animals are described or mentioned in some connection. Animals were often used as metaphors in these works in order to draw parallels with humans, as it was typical to antiquity as well. According to Aristotle's *scala naturae* (the great chain of being) animals were considered to be inferior to humans, and according to early modern ideas animals existed for the use of man.

There is a clear difference between the works that were compiled in the first half of the seventeenth century and the dissertation written at the end of the seventeenth century. The works that were written earlier were mostly based on ancient authors and the Bible, whereas the dissertation from the end of the

---

69 "XXXIV. Sperma ceti non est sperma, sed liquor arte paratus, saepe maris summitati innatans, ex cerebro, non cujusvis ceti, sed hujus generis qui orca vocatur, oriundus". Below / Matthiae, p. D4 v.
70 "XXXVI. Weil der grösseste Unterscheid/ so zwischen einem Menschen und einem tummen Viehe ist/ darin bestehet/ daß der Mensch eine Vernünftige Seele hat/ das Vieh aber nicht: so wird ein jeder gestehen müssen/ daß nur die jenige/ so ihren Verstand durch schöne Wissenschafften exoliren, für rechte Menschen zu halten: alle andere aber/ so die Studien verachten/ und ihr Gespött damit treiben/ für unvernünftige Menschen/ das ist für Unmenschen zu achten sind". Below / Matthiae, p. D4 v.

seventeenth century delivered empirical knowledge. We can see that the medical works of the seventeenth-century University of Tartu mention veterinary medicine and the use of animals as a source of remedies or even animals as their own doctors, teaching humans how to heal certain diseases. There are hints at animals in scientific experiments.

We could not see from the medical works of the Swedish University of Tartu how much contact the authors of these works had with the animals described by them, as their texts were usually based on previous authors from antiquity or the early modern period. However, it is probable that they had experience with animal autopsies and the use of animals in pharmacies. In the medical work *Dissertatio prima de oeconomia corporis animalis* originating from 1698 the development of early modern physiology can be observed. Thus animals have their impact on the medicine of the seventeenth-century University of Tartu as well as the way of talking and writing about medicine.

References

Source Materials

Below, Jacob Friedrich / Matthiae, Salomon: *I. N. J. Circuli anatomico-physiologici minoris, Segmentum Primum; Seu Dissertatio Prima De oeconomia corporis animalis*. Johannes Brendeken, Acad. Typographus: Dorpat 1698.

Bohn, Johannes: *Circulus anatomico-physiologicus seu oeconomia corporis animalis*. Sumptibus Joh. Friedrich Gleditsch, Typis Christophori Fleischeri: Lipsiae 1686.

Bohn, Johannes: *Circulus anatomico-physiologicus seu oeconomia corporis animalis, seu Oeconomia corporis animalis, hoc est, cogitata, functionum animalium potissimarum formalitatem et causas concernantia*. Apud Thomam Fritsch: Lipsiae 1697.

Book of Daniel, Vulgate. Biblia Sacra Vulgata. Retrieved 15.01.2019, from https://www.biblegateway.com/passage/?search=Daniel+4&version=VULGATE.

Grimmelshausen, Hans Jakob Christoph: *Der abenteuerliche Simplicissimus*. Eduard Kaiser Verlag: Klagenfurt 1950.

Hein, Friedrich: *Oratio de medicina*. Lit. acad: Dorpati Livonorum 1637.

Hippocrates with an English translation by W. H. S. Jones, vol. II. (The Loeb Classical Library). Harvard University Press: Cambridge, Massachusetts / William Heinemann LTD: London 1981.

Lepajõe, Marju (ed.): *Constitutiones Academiae Dorpatensis (Academia Gustaviana). Tartu Akadeemia (Academia Gustaviana) Põhikiri*. Tartu Ülikooli Kirjastus: Tartu 1997, pp. 56–57.

Magirus, Joannes: *In physiologiam suam Peripateticam commentarius*. Nebenius: Lichae 1601.

Murray, August Taber (ed.): *The Iliad with an English translation*. William Heinemann LTD: London / G. P. Putnam's Sons: New York 1928, retrieved 14.10.2019, from https://ryanfb.github.io/loebolus-data/L170N.pdf.

Ovid: *Metamorphoses*. Retrieved 15.01.2019, from http://www.thelatinlibrary.com/ovid/ovid.met1.shtml.

Pliny the Elder: *Natural History*. Retrieved 15.1.2019, from http://www.perseus.tufts.edu/hopper/text?doc=Plin.+Nat.+7&fromdoc=Perseus%3Atext%3A1999.02.0138.

Wallander, Sigvardius Olai: *Oratio de homine*. Lit. Acad.: Dorpati Livonorum 1640.

Wirdig, Sebastian / Andreas Arvidi Stregnensis: *Disputatio medica de natura et constitutione medicinae*. J. Vogelius: Dorpati Livonorum 1648.

Literature

Achner, Heike: *Ärzte in der Antike*. Philipp von Zabern: Mainz 2009.

Brockliss, Laurence / Jones, Colin: *The Medical World of Early Modern France*. Clarendon Press: Oxford 1997.

Enenkel, Karl A.E./ Smith, Paul J. (eds.): *Zoology in Early Modern Culture. Intersections of Science, Theology, Philology and Political and Religious Education*. (Intersections 32). Brill: Leiden / Boston 2014.

Enenkel, Karl A. E. / Smith, Paul J. (eds.): *Early Modern Zoology. The Construction of Animals in Science, Literature and the Visual Arts*, vol. 1. (Intersections 7). Brill: Leiden / Boston 2007.

Franco, Nuno Henrique: "Animal Experiments in Biomedical Research. A Historical Perspective." *Animals* 3(1), 2013, pp. 238–273. DOI:10.3390/ani3010238.

Garrison, Fielding H.: *An Introduction to the History of Medicine*. W. B. Saunders Company: Philadelphia / London 1960.

Gerabek, Werner E. et al. (eds.): *Enzyklopädie Medizingeschichte*. Walter de Gruyter: Berlin / New York 2007.

Hoffstadt, Christian et al. (eds.): *Humana – Animalia. Mensch und Tier in Medizin, Philosophie und Kultur.* (Aspekte der Medizinphilosophie 13). Projektverlag: Bochum 2012.

Kallinen, Maija: *Change and Stability. Natural Philosophy at the Academy of Turku (1640–1713).* (Studia Historica 51). Finnish Historical Society: Helsinki 1995.

Kõivupuu, Marju: *Loomad eestlaste elus ja folklooris.* Tänapäev: Tallinn 2017.

Korhonen, Tua / Ruonakoski, Erika: *Human and Animal in Ancient Greece. Empathy and Encounter in Classical Literature.* I. B. Tauris: London / New York 2017.

Leven, Karl-Heinz (ed.): *Antike Medizin. Ein Lexikon.* C. H. Beck: München 2005.

Muratori, Cecilia: "From Animal Bodies to Human Souls. (Pseudo-)Aristotelian Animals in Della Porta's Physiognomics". *Early Science and Medicine* 22, 2017, pp. 1–23. DOI:10.1163/15733823-00221p01.

Piirimäe, Helmut (ed.): *Tartu Ülikooli ajalugu I 1632–1798.* [History of the University of Tartu I 1632–1798.] Valgus: Tallinn 1982.

Rauch, Georg von: *Die Universität Dorpat und das Eindringen der Frühen Aufklärung in Livland 1690–1710.* Essener Verlagsanstalt: Essen 1943.

Rein, Kaarina: *Arstiteadus rootsiaegses Tartu gümnaasiumis ja ülikoolis aastatel 1630–1656. Meditsiinialased disputatsioonid ja oratsioonid ning nende autorid.* [Medicine at the Gymnasium and University of Tartu from 1630 to 1656. Medical Disputations, Orations and their Authors.] Tartu University Press: Tartu 2011.

Rein, Kaarina: "Tartu as the Eastern Outpost of European Medicine in the First Half of the 17th Century". *Acta Baltica Historiae et Philosophiae Scientiarum* 2(1), 2014, pp. 37–52. DOI:10.11590/abhps.2014.1.03.

Rein, Kaarina: "Loomad Rootsi-aegse Tartu ülikooli meditsiinitöödes". [Animals in the medical works of the Swedish-period University of Tartu.] *Mäetagused* 78, 2020, pp. 155–172. DOI:10.7592/mt2020.78.rein.

Sõukand, Renata: "Koera keele otsas rohi, kassi keele otsas tõbi". [Dog Tongue Carries Medicine, Cat Tongue Carries Illness.] *Mäetagused* 31, 2006, pp. 261–302. DOI:10.7592/MT2005.31.RENATA.

Tering, Arvo: "René Descartes'i ideede jõudmisest Eesti- ja Liivimaale XVII sajandil ja XVIII sajandi algul". [Die Verbereitung der cartesianischen Ideen im 17. und zu Beginn des 18. Jahrhundert in Est- und Livland.] *Keel ja Kirjandus* 3, 1996, pp. 179–188.

Valdes, Albert / Veski, Johann Voldemar: *Ladina-eesti-vene meditsiinisõnaraamat.* [Latin-Estonian-Russian Medical Dictionary.] Vol. 1–2. Valgus: Tallinn 1982–1983.

White, Cynthia: "Potiones ad sanandum. Text as Remedy in a Medieval Latin Bestiary". In: Classen, Albrecht (ed.): *Bodily and Spiritual Hygiene in Medieval and Early Modern Literature. Explorations of Textual Presentations of Filth and Water.* De Gruyter: Berlin / Boston 2017, pp. 221–274. DOI:10.1515/9783110523799-008.

Wright, Thomas: *William Harvey. A Life in Circulation.* Oxford University Press: New York / Oxford 2013.

Ulrike Plath

Animal Abolitionism and Early Environmentalism in Nineteenth Century Riga[1]

Abstract: Freedom has been the main narrative of history writing since the eighteenth century, but only in the nineteenth century was this concept applied to animals as well. The first Animal Welfare Society in the Russian Empire was established in Riga in 1861 – the year when serfdom was abolished and personal freedom given to the local peasantry all over the Empire. In the Baltic provinces, however, this step had been taken nearly half a century earlier. All adult persons over 18 years of age could join the newly founded animal welfare society irrespective of their gender or class. Unhappy with the animal welfare programme and following examples in England and Germany, it was local upper-class women who, in the 1870s, initiated an animal rights movement within the animal welfare society, which culminated in the foundation of another society. Despite the different goals of the local animal welfare and animal rights movements, both had a goal in common as well: the education of the next generation. The idea to protect "good" animals quickly spread in German, Latvian and Hebrew schools and the first wave of animal protection movement can be detected to have been led not only by learned societies, but also by schools and students, while women played an important role in this movement.

Keywords: Baltic history, animal welfare, societies, women, environmentalism

"For the sake of the existence of our own souls, we also have to concede those [souls] to polyps."[2]
"The eighteenth century can be proud to have promoted human rights if not to their full application but at least to their theoretical recognition; the nineteenth century deserves the credit for doing the same for animals."[3]

---

1 This article has been supported by the Estonian Research Grant PRG908 "Estonian Environmentalism in the 20th century: ideology, discourses, practices".
2 Hupel, August Wilhelm: *Anmerkungen und Zweifel über die gewöhnlichen Lehrsätze vom Wesen der menschlichen und thierischen Seele.* Hartknoch: Riga 1774, p. 354.
3 Glasenapp, Carl Friedrich: "Menschenrechte und Thierrechte." *Der Anwalt der Thiere* (1–2), 1885, pp. 18–20, p. 18. Glasenapp (1847–1915) is citing Ernst Georg Friedrich Grysanowski (1824–1888) here, the most prominent leader of the anti-vivisection

Introduction

Freedom has been the main narrative of history writing since the eighteenth century.[4] This holds especially true for Baltic history writing due to its complex layers of colonial domination.[5] In the medieval crusades Old Livonia (the territory of present-day Estonia and Latvia) was brought under German rule. German knights, monks, tradesmen and citizens formed a small upper class over Finno-Ugric and Baltic tribes, who during late medieval and early modern times became peasants and serfs. Despite assimilation processes and despite the fact that the territory was split several times between the Polish, Swedish and finally Russian Empire, the Germans or later on so-called Baltic-Germans kept their upper class position up to the early 20th century. This formed the old narrative of "700 years of serfdom under German rule," which dominated Estonian and Latvian history writing up to the nineteenth century and survived as a master narrative even after the abolition of serfdom.

It was only in the twentieth century when the history of slavery was finally replaced by a history of freedom, which was based on three major political events: 1. The abolition of serfdom in 1816/19 that guaranteed personal freedom from Baltic-German dominance; 2. Gaining national independence from the Russian Empire in 1918; and 3. The restoration of national independence in 1991, which was lost during the period of German and Soviet occupations. But freedom is not only a key concept in the identity story of the colonised. Even in Baltic-German history writing the social and cultural German domination lasting seven centuries was used as an argument for the importance of regional autonomy from the Swedish and Russian Empires, and this again was later claimed to have prepared the way to national independence.[6] Voluntary abolition of serfdom and concessions of personal freedom were interpreted in this context as highlights of positive Baltic-German cultural domination.[7] The switch from

---

movement and his Article "Rechte der Thiere", published in "Die Taube", the journal of the Animal Welfare Society in Berlin.
4   Chakrabarty, Dipesh: *The Climate of History in a Planetary Age*. The University of Chicago Press: Chicago / London 2021, p. 32.
5   Kaljundi, Linda / Plath, Ulrike: "Serfdom as entanglement. Narratives of a social phenomenon in Baltic history writing." *Journal of Baltic Studies* 51(3), 2020, pp. 349–372. DOI:10.1080/01629778.2020.1776349.
6   Undusk, Jaan: "Ajalootõde ja metahistoorilised žestid. Eesti ajaloo mitmest moraalist." [Historical Truth and Metahistorical Gestures. About Several Morals of Estonian History.] *Tuna. Ajalookultuuri ajakiri* 2, 2000, pp. 114–130.
7   Kaljundi / Plath 2020.

oppression to freedom so typical for many colonial contexts has been narrated in key Baltic texts from different perspectives and with different goals.[8]

While this narrative of freedom in history writing has been exclusively used for speaking about humans, there are also early traces of public discourse claiming freedom even for animals. In the Baltic and Russian contexts it started in nineteenth-century Riga – the oldest city of Old Livonia, founded in 1201 by the first German colonisers. At the end of the nineteenth century Riga had developed into the cultural and mercantile metropolis of the Baltic provinces and was a melting pot of Baltic-German, Estonian, Latvian, Russian and Jewish culture.[9] As in other metropolises, human and non-human interaction became an unignorable issue even during the early time of urbanisation.[10] It was not only the coach horses that often worked under miserable circumstances on the streets,[11] but also the myriad of rats and mice that flooded the cities flanked by an exploding number of stray dogs and cats, microbes and pests, and all the fatstock that was kept in and transported to the cities.[12] Urbanisation led to the co-evolution of new (eco)systems with unforeseen challenges.[13] These new entanglements questioned the ruling understanding of human dominance and led to new forms of activism for a more-than-human world. In this context the first animal welfare society was founded in 1860–61.

---

8   In 2019 the Museum of Occupation was renamed the Museum of Occupations and Freedom.
9   Hirschhausen, Ulrike von: *Die Grenzen der Gemeinsamkeit. Deutsche, Letten, Russen und Juden in Riga 1860–1914*. Vandenhoeck & Ruprecht: Göttingen 2006.
10 McShane, Clay / Tarr, Joe A.: *The Horse in the City: Living Machines in the Nineteenth Century*. Johns Hopkins University Press: Baltimore 2007; Atkins, Peter (ed.): *Animal Cities. Beastly Urban Histories*. Routledge: London 2012; Wischermann, Clemens / Steinbrecher, Aline / Howell, Philip (eds.): *Animal History in the Modern City. Exploring Liminality*. Bloomsbury Academic: London et al. 2019.
11 About coach horses in Estonia, see the PhD thesis on the history of public transport, Paramonov, Riho: *Eesti ühiskonna moderniseerumine avaliku privaattranspordi (voorimees ja takso) näitel, 1900–1940*. [The Cabman and the Taxi Driver: The Modernization of Estonian Society During 1900–1940 on the Example of Public Transport.] PhD Dissertation. Tallinn University Press: Tallinn 2019.
12 Wischermann / Steinbrecher / Howell 2019.
13 Modern slaughterhouses might be interpreted as an example of these new envirotechnical systems, Leiderer, Annette: "History of Animal Slaughter". In: Roscher, Mieke / Krebber, André / Mizelle, Brett (eds.): *Handbook of Historical Animal Studies*. Walter de Gruyter: Berlin / Boston 2021, 539–554. DOI:10.1515/9783110536553-040.

I claim that Livonia was in the forefront of this new movement in Russia due to the fact that the Baltic provinces already had 50-plus years of experience in thinking about personal freedom. The discussions about and the final abolition of serfdom were the matrix for new thinking concerning non-humans. There is a clear and direct connection between animal abolitionism and the Enlightenment idea of freedom,[14] that still has to be discovered in Baltic history.

By telling the story of animal welfare and the animal rights movements in the Baltic provinces, I hope to be able to add new facts about the spread of animal welfare ideas beyond the well-known centres in central Europe, America, and central Russia.[15] The Baltic provinces as a highly modernised border zone between different cultures and empires might be a good place to reflect on transnational forms of entanglements. Being aware of the deep philosophical and legal questions at hand when speaking about non-humans' personal rights,[16] and also being aware of the growing criticisms from environmentalists who stress the importance of thinking holistically instead of defending the rights of certain species,[17] I will not delve deeper into the arguments of Gary Lawrence Francione's "abolitionist approach to animal rights" from 1954, which radicalised the gap between animal welfarists and animal rightists.[18] I also don't aim to deepen the existing gender gap between male welfarists' and "emotional" female

---

14 Spencer, Jane: *Writing about Animals in the Age of Revolution*. Oxford University Press: Oxford, 2020; Wolloch, Nathaniel: *The Enlightenment's Animals: Changing Conceptions of Animals in the Long Eighteenth Century*. Amsterdam University Press: Amsterdam 2019; Festa, Lynn: *Fiction Without Humanity. Person, Animal, Thing in Early Enlightenment Literature and Culture*. University of Pennsylvania Press: Philadelphia 2019.
15 Obenchain, Theodore G: *The Victorian Vivisection Debate. Frances Power Cobbe, Experimental Science and "Claims of Brute"*. McFarland & Company Inc.: Jefferson et al. 2012.
16 Beauchamp, Tom L. / Frey, Raymond G. (eds.): *The Oxford Handbook of Animal Ethics*. Oxford Press: Oxford / New York 2011.
17 Wrenn, Corey: "Abolitionist Animal Rights. Critical Comparisons and Challenges within the Animal Rights Movement". *Interface: A Journal for and about Social Movements* 4(2), 2012, pp. 438–458.
18 Francione, Gary L. / Garner, Robert: *The Animal Rights Debate. Abolition or Regulation?* Columbia University Press: New York 2010; Cordeiro-Rodrigues, Luis: "Animal Abolitionism and 'Racism without Racists'". *Journal of Agricultural and Environmental Ethics* 30, 2017, pp. 745–764. DOI:10.1007/s10806-017-9697-0.

rightists' attitudes toward non-human freedom.[19] Instead I will try to discover the common ground of the new forms of activism concerning more-than-human or environmental issues as such. Many of the biased debates about the schism within animal activism in the late nineteenth century seem to be over-exaggerated in the late twentieth and twenty-first-century debates that try to make us forget about the common goals of all these different social movements – rising care for the non-humans in a more-than-human world.

The sources for this article are German reports and periodicals of the animal welfare and rights societies and public reflections and discussions in local German-speaking newspapers. Based on that I will analyse the members, goals, and importance of these societies in shaping a new public understanding of animal ethics and the concept of freedom.[20] This will bring me to the final discussion of whether the nineteenth century's movements for animal welfare and animal rights can be interpreted as forms of early environmentalism in the Baltic provinces. If so, this might add some local aspects to the exploding field of human–animal studies and studies on the rise and forms of nineteenth-century environmentalism.[21]

The Birth of Baltic Animal Welfare Societies

"What we call animal welfare should not be a disclaimer for our domination over animals, their work power, their flesh, their gains; no disclaimer for the elimination of those animals, which are harmful for us; its goal is foremost care for those animals that serve us (…)."[22]

"Cruelty to animals is the beginning, cruelty against humans the process, and crime the outcome of this state of mind."[23]

---

19 These issues were discussed thoroughly in Donovan, Josephine: "Animal Rights and Feminists Theory". *Signs* 15(2), 1990, pp. 350–375; Singer, Peter: *Animal Liberation*. Avon: New York 1975.
20 So far, I have not been able to find any archival material on that matter. I was also unable to discover the journals of the Couronian society to study, as they are not stored in the Latvian national library or elsewhere.
21 Fitzgerald, Amy J.: *Animal Advocacy and Environmentalism. Understanding and Bridging the Divide*. Polity Press: Cambridge / Medford 2019; Hadley, John: "Nonhuman Animal Property. Reconciling Environmentalism and Animal Rights". *Journal of Social Philosophy* 36(3), 2005, pp. 305–315.
22 *Kurländische Gouvernements Zeitung* (101), 18.12.1868.
23 *Rigasche Zeitung* (27), 3.2.1860.

It was in 1861, the very year the abolition of serfdom was declared all over Russia, when the first animal welfare society of the Empire was founded in Riga. The founding members in Riga were following international examples. In Great Britain the first society of this kind was founded in 1824, called The Royal Society for the Prevention of Cruelty to Animals (RSPCA), in the German countries similar societies were founded in the 1830s and 1840s, and in America the American Society for the Prevention of Cruelty to Animals (ASPCA) followed a couple of years later in 1866.[24] It was in this international context that the early animal welfarists in Riga were acting, copying the way societies elsewhere were set up and acting, and being closely connected with them.

The foundation of an animal welfare society was discussed for the first time in 1859, and since then slowly prepared by members of the Literary Practical Civic Society,[25] and two years later the Riga Society Against Cruelty to Animals or as it was called later the Riga Animal Welfare Society, was finally founded.[26] Every person over 18 years could become a member, notwithstanding gender, estate or nationality – which was remarkable in the otherwise nationally divided Baltic society of the late nineteenth century. It was the aim of the society to spread the message widely all over the provinces, to found new branches making new people involved, to support humane education for children, youth and the lower social classes, and to advance new laws concerning animals in some of the hot topics of the modernising world: transport and technology, food supply, public health, and science.[27]

As the society was founded by members of the Baltic-German upper class it is no wonder that the language and general style of the society and its branches were thoroughly German – as were most of the societies in the Baltic provinces of that time. The vast majority of the members of the new society belonged to the middle class and were civil servants, teachers, physicians, and pharmacists. Only

---

24 Similar societies were founded 1837 in Stuttgart, 1838 in Württemberg, 1841 in Berlin, 1842 in München, 1846 in Wien etc. Zerbel, Miriam: *Tierschutz im Kaiserreich. Ein Beitrag zur Geschichte des Vereinswesens*. (Münchner Studien zur neueren und neuesten Geschichte 4). Peter Lang: Frankfurt a.M. 1993; Duncan I.J.H.: "Animal Welfare: A Brief History". In: Hild, Sophie / Schweitzer, Louis (eds.), *Animal Welfare: From Science to Law*. La Fondation Droit Animal, Éthique et Sciences (LFDA): Paris 2019, pp. 13–19.
25 *Rigasche Stadtblätter* (7), 16.2.1861.
26 The idea of founding a society was discussed in late 1859, but it took some time to set up the goals and to prepare international cooperation. *Rigasche Zeitung* (27), 3.2.1860; *Rigasche Stadtblätter* (20), 19.5.1860.
27 *Rigasche Zeitung* (138), 17.6.1861; *Rigasche Zeitung* (265), 14.11.1861.

a few members were from the Baltic-German nobility, or were Estonian and Latvian social climbers.[28] Societies were most attractive for the first generation of educated Estonians and Latvians, who got their education in German, were partly Germanised or at least German-speaking.[29] The German character of the society was stressed by the close cooperation with animal welfare societies in Germany and Switzerland, which supported the development in the Baltic provinces by sharing not only ideas and structures, but also print media, medals, and other materials.[30] The main idea of all these societies, however, was not to stress local national sentiment, but to spread the ideas of transnational humanity through compassion with animals. For this task, newly published literature on the topic was translated into local languages and spread all over Europe in schools and within partner societies. So for example the Curonian Animal Welfare Society translated German books on animal welfare to Russian and Latvian and helped to spread them on a local level. The general goal of teaching ordinary people followed the goals of popular enlightenment, which aimed to educate children and lower classes.[31] In the Baltic context with its inherent problem of German cultural domination this feature can be interpreted as a form of cultural transfer with some obvious colonial elements.[32]

---

28 *Rigasche Stadtblätter* (7), 16.2.1861.
29 Hackmann, Jörg: *Vereinskultur und Zivilgesellschaft in Nordosteuropa. Regionale Spezifik und europäische Zusammenhänge. – Associational Culture and Civil Society in North-Eastern Europe. Regional Features and the European Context.* (Quellen und Studien zur baltischen Geschichte 20). Böhlau: Wien / Köln / Weimar 2012.
30 There was also cooperation with France, but very little with England. *Rigasche Stadtblätter* (15), 12.4.1862; *Kurländische Gouvernementszeitung* (30), 13.4.1863.
31 For a general background and the English case, see the old classic work of Ritvo, Harriet: *The Animal Estate. The English and Other Creatures in the Victorian Age.* Harvard University Press: Cambridge, Massachusetts 1987. So we can find passages on how to be kind and gentle not only to other children, serfs and animals in the first local children book "Kleines Buch für kleine Kinder, welche lesen lernen", written by the most important Baltic-German enlightener Sonntag, Karl Gottlob: *Das kleine Buch für kleine Kinder in Riga, welche lesen lernen*. S.n.: Riga, 1802.; Plath, Ulrike: "Baltisaksa laste- ja noorsookirjandusest kuni 1840. aastateni". [Baltic German Children's and Youth Literature until the 1840s.] *Keel ja Kirjandus* 8–9, 2011, pp. 698–715.
32 Daija, Pauls: *Literary History and Popular Enlightenment in Latvian Culture.* Cambridge Scholars Publishing: Newcastle upon Tyne 2017; Plath, Ulrike: *Esten und Deutsche in den baltischen Provinzen Russlands. Fremdheitskonstruktionen, Lebenswelten, Kolonialphantasien 1750–1850.* Harrassowitz: Wiesbaden 2012.

In Riga animal welfare had to face very practical questions. First of all hierarchical and spatial structures had to be established to make the society work. For that goal the whole city was divided into districts that were watched by officers of the society, who were obliged to trace misbehaviour against animals and bring these to the police. But all members of the society – men and women – were asked to constantly trace their urban environment and to be vigilant concerning any form of cruelty against animals.[33] The structure of the Baltic societies itself was hierarchical, but still integrative and for their time modern. While the presidency and the main positions were in the hand of the academic and social elite,[34] district officers could be local gardeners and craftsmen and their assistants, even women.[35] Their task was to interact with the local police and to motivate local society members to track cruelty against animals in every respect. Although German by structure, animal welfare societies as other agricultural and gardening societies brought together Baltic-German manor owners, Estonian and Latvian peasants, and all the national indifferent groups and social climbers in between.[36] It seems that societies that dealt with environmental and animal questions had in general a broader transcultural and national focus than other societies of the nineteenth century.[37] Thus the understanding that societies as such helped to raise nationalism might need some revisions.

The idea of animal society fitted perfectly with the spirit of the time and region. Soon branches of the society were founded in Curonia (1861) and Estonia (1869), copying the statutes of the Riga society, but still developing local particularities.[38] At the beginning of the 1870s the former city-based branches

---

33 New FSB at LMU-Munich "Cultures of Vigilance", retrieved 10.07.2023 from https://www.en.fnz.geschichte.uni-muenchen.de/research/research_projects/vigilanz/index.html.
34 The Curonia nobility seemed to be more active in the board of the societies than in Livonia. *Rigasche Stadtblätter* (31), 3.8.1861.
35 *Rigasche Zeitung* (30), 6.2.1869.
36 Zahra, Tara. "Imagined Noncommunities: National Indifference as a Category of Analysis". *Slavic Review* 69(1), 2010, pp. 93–119. DOI:10.1017/S0037677900016715.
37 Hackmann, Jörg: *Zivilgesellschaft im östlichen und südöstlichen Europa in Geschichte und Gegenwart.* (Völker, Staaten und Kulturen im östlichen Mitteleuropa 5). Oldenbourg: München 2011.
38 The three Baltic provinces of Russia were Livonia with its capital Riga, Couronia with its capital Jelgava, and Estonia with the capital Tallinn. While the peasantry in Couronia was mostly Latvian and in Estonia Estonian, Livonia had a mixed peasant population, as the governmental border was running north of Tartu, including the whole of Southern Estonia.

grew into governmental societies. Thus in 1871 the Curonian Animal Welfare Society was founded and in 1876 the Livonian Animal Welfare Society followed.[39] By the late 1870s both societies had many branches in the smaller cities and towns all over the governorates, and to promote animal welfare among the Latvian peasantry the statutes of the societies were translated into Latvian.[40] In Estonia the first animal welfare society in Reval / Tallinn was founded in 1869,[41] followed by a branch of the Russian animal welfare society that was founded in Narva in 1874.[42]

The initiative in the Baltic governments was a remarkable first step for the further development throughout the whole Russian Empire. The first Russian society (*Rossiiskoe obshchestvo pokrovitel'stva zhivotnym* or ROPZh) was founded in St. Petersburg in 1865 as an elite project following the English example.[43] All 50 founding members of the society were from the nobility or had high positions in the state governments or at universities. This society also started to spread animal welfare ideas among the local population by using public exhibitions, as within the Agricultural Exhibition in Retovo in 1881.[44] In 1882 the first Congress of Animal Welfare Societies in Russia was organised in Moscow,[45] followed by the first Russian Congress on Animal Welfare in St. Petersburg in 1890.[46] Right after the foundation of ROPZh, its rules were also implemented in the Baltic provinces replacing earlier statutes.[47] It also worked on pushing animal welfare

---

39 *Rigasche Zeitung* (109), 14.5.1877.
40 In Couronia by the end of the 1870s there were already 22 smaller societies in every bigger town. *Rigasche Zeitung* (104), 7.5.1879; *Libausche Zeitung* (188), 2.12.1876. In 1877 there were similar societies in Viljandi, Tartu, Daugavpils, Võru, Kuressaare. An explanation of the slower spread of animal welfare societies in Estonia might be that the issue was politicised and became a weapon of the early anti-German russification propaganda. *Rigasche Zeitung* (187), 14.8.1868; *Rigasche Zeitung* (257), 5.11.1877.
41 *Rigasche Zeitung* (266), 15.11.1869.
42 *Der Anwalt der Thiere*, 1889, S. 383.
43 Bonhomme, Brian: "Russian Compassion. The Russian Society for the Protection of Animals – Founding and Contexts, 1865–75". *Canadian Journal of History* 45(2), 2010, pp. 259–297. DOI:10.3138/cjh.45.2.259; *Rigasche Zeitung* (286), 8.12.1867.
44 The idea for this exhibition, planned to be held in Mitau in June 1881, and to integrate agriculture and animal welfare and should have been supported from Germany, and the animal welfare society in Retovo has been originally expressed by the Couronian Animal Welfare Society. Due to unknown problems, the exhibition was organised in the end in Retovo. *Libausche Zeitung* (21), 27.1.1881; *Rigasche Zeitung* (196); 26.8.1881.
45 *Libausche Zeitung* (86), 16.4.1882.
46 *Düna Zeitung* (238), 19.10.1890.
47 *Kurländische Gouvernements-Zeitung* (49), 29.6.1866.

questions in Russian law. So, for example, in 1868 a new law was launched that prohibited cruelty against animals and allowed the local police to punish delinquents.[48]

Baltic animal welfare societies had in general the same goals as similar societies elsewhere: First of all they wanted to "humanise" the local population and lower violence and cruelty against humans through compassion with animals. While in central Europe the main object of this new humanisation campaign were the workers, in Eastern Europe it was the peasantry who had just come out of serfdom and needed to be integrated into the new civic society. All parts of society were called to follow the model of their former landlords who followed the moral call of abolition to overcome the logic of violence and slavery: violence against unfree humans was compared with violence against animals as we will explain more closely later on. The education toward animal welfare was understood as education toward a peaceful modern society.

Baltic animal welfare societies had a clear masterplan for how to integrate the lower classes, although there were clear local differences between societies in the countryside and smaller cities and those in the big cities. While in 1863 the local animal welfare society in Mitau/Jelgava had seven peasant members and one local craftsman and actively translated its messages into Latvian,[49] the society in Riga included the lower classes, servants and domestic workers more actively. From 1865 onward the Riga society decorated one servant who had gained extra praise from their landlord every January. In 1865 this honour was given to Latvian cook Helene Lange for her humane handling of slaughter animals and overall smart service. She got a silver brooch, some money, a silk scarf, and life-long membership in the Riga Animal Welfare Society. The ceremony was accompanied by church songs and speeches based on animals in the Holy Bible.[50]

Although it might seem that the early animal welfare societies were more concerned about the human soul than about the animals, all of them were also struggling for new animal laws that would prohibit overexploitation and cruelty against animals. In the Baltic provinces the first regulation concerning animal welfare already existed in the first half of the 19th century. Thus in 1833 the police in Reval freed songbirds on the local market place.[51] However, it was only thanks to the societies that animal laws became a subfield of regional and

---

48 *Rigasche Zeitung* (165), 19.7.1868.
49 *Rigasche Stadtblätter* (31), 3.8.1861.
50 *Rigasche Zeitung* (14), 19.1.1865.
51 *Rigasche Zeitung* (87), 16.04. 1863.

international law,⁵² and violation against it was consequently prosecuted on the local and later national levels.⁵³ Vigilance was the new tonus of the members of the societies, who were asked to actively intervene whenever violence against animals was detected, and to call the police.⁵⁴ Here the beginning of active citizenship can be seen, but it also opened the doors to false accusations. All animal welfare societies were concerned about the use of animals that should be treated as "humanely" as possible; however, not a single one was against the use of animals for human food. On the contrary, they all promoted for example the use of horse meat in order to make use of many different species and to benefit from emergency slaughter of injured horses.⁵⁵ This human-centred approach changed within the second wave of animal welfare, which made the final step towards animal ethics and animal rights.

From Animal Welfare to Animal Rights

> "Children love their toys, Turks their female slaves and vivisectors their dogs. The greatest vivisectors were members of animal welfare societies."⁵⁶
>
> "The day may come when the rest of animal creation may acquire those rights that never could have been withheld from them but by the hand of tyranny. The French have already discovered that blackness of the skin is no reason why a human being should be abandoned without redress to the caprice of a tormentor. It may come one day to be recognised that the number of the legs, the villosity of the skin, or the termination of the os sacrum, are reasons equally insufficient for abandoning a sensitive being to the same fate. (...) the question is not, Can they reason? nor, Can they talk? but, Can they suffer."⁵⁷

As Harriet Rivo and others have pointed out, animals lost their subjectivity throughout the nineteenth century and became objects of human interest.⁵⁸ This

---

52 *Kurländische Gouvernementszeitung* (30), 30.4.1863.
53 Five years after the foundation of the Jelgava animal welfare society for the prevention of cruelty against animals was persecuted all over Couronia by the local police. *Kurländische Gouvernementszeitung* (49), 29.6.1866.
54 *Rigasche Zeitung* (230), 3.10.1868.
55 *Rigasche Zeitung* (78), 5.4.1863.
56 *Anwalt der Thiere* (5), 1885, p. 79.
57 Bentham, Jeremy: *An Introduction to the Principles of Morals and Legislation*. Oxford Press: London / New York / Toronto 1823, chapter XVII, footnote 122. Retrieved 10.07.2023 from https://oll.libertyfund.org/titles/bentham-an-introduction-to-the-principles-of-morals-and-legislation.
58 There are struggles about how and when this shift occurred. So for example Gary L. Francione suggests that animals have been objects throughout early modern times and got subjectives only in the nineteenth century. Francione, Gary L. et al. (eds.).

new distance from the non-human world, however, also allowed animals not only to be exploited, but to be encountered with new care and affection that finally even re-subjectified them.[59] This split outcome of the Enlightenment movement is typical for the Janus-faced movement itself, as ongoing discussions about the colonial and national layers of the Enlightenment show.[60] In the debates about animals, freedom as the main core of the Enlightenment was transferred to the more-than-human world by women who in advocating animal rights broadened the boundaries of their own social engagement. Dominik Ohrem therefore states, that "Feminism has been and continues to be a guiding force in the field of animal studies."[61] In England, Germany and America female activists stepped out against the science-based, rational and objectifying understanding of animals in the male-dominated post-Enlightenment society.[62] In Victorian England in the 1870s Frances Power Cobbe was fighting against vivisection getting into conflict with the leading male scientists of her time, but still being able to push the Cruelty against Animal Act from 1876 that imposed clear limits to the harm done in the name of science to non-humans.[63] In Germany a similar gendered public fight in the name of non-human co-mates took place between the late 1870s and early 1880s. In comparison to England, however, only half-hearted compromises were achieved, which in the end established the power of real sciences and ridiculed the emotionalised debates of the ladies. Still public debates about the borders of sciences were fought. In less than ten years 50 brochures from 22 enemies of vivisection and more than 20 brochures from advocates of the practice were published, leading to a remarkable politicisation of non-human interest in the public sphere and in 1879 to a schism between the

---

Exploring the Animal Turn. Human–animal relations in Science, Society and Culture. (Pufendorfinstitutets skriftserie 4). Lund University: Lund 2014, pp. 57–72.

59 For the English and broader context see Ritvo: *The Animal Estate*. For local Baltic developments have a look at Inna Jürjo's and Ken Ird's articles in this collected volume.

60 Daija 2017; Plath 2012.

61 Ohrem, Dominik: "Feminist Intersectionality Studies". In: Roscher, Mieke / Krebber, André / Mizelle, Brett (eds.): *Handbook of Historical Animal Studies*. Walter de Gruyter: Berlin / Boston 2021, pp. 341–355, p. 241. DOI:10.1515/9783110536553-027.

62 Sachse, Carola: "Von Männern, Frauen und Hunden. Der Streit um die Vivisektion im Deutschland des 19. Jahrhunderts". *Feministische Studien* 24(1), 2006, pp. 9–28. DOI:10.1515/fs-2006-0103; Gaarder, Emily: *Women and the Animal Rights Movement*. Rutgers University Press: New Brunswick 2011.

63 Obenchain 2012.

first German animal welfare movement and the new animal rights movement.[64] While in England this movement was led by women, in Germany only one-third of the members of the new animal rights movement (founded by Richard Wagner) were women, among them 40 % from the nobility, and not very active in the public sphere. In America women played a more active part. The first Humane Societies and animal shelters were founded, nonprofit organisations that aimed to ensure the welfare of working animals and slaughter cattle, also combining its goals quickly with child welfare creating new room for feminised emotions of care and affection toward the non-adult and non-dominating world around.[65]

Women in Riga were very well aware of what was going on and knew about the possibilities animal advocacy would bring to the women's liberation movement[66] by using the old gender stereotypes about the special addiction of women to animals.[67] The position of women in Riga and their level of organisation was remarkably high. There were not only structures for old, unmarried and widowed women that helped them to compensate for missing family support,[68] but also possibilities for young women to take an active part in social life. After the Napoleonic Wars the first women's aid committees (*Damen Comite*) were founded to help the poor in the cities, to support military hospitals[69] or prisoners.[70] By the end of the century women's aid committees could have quite

---

64 Sachse, pp. 13–14, 25; Roscher, Mieke: "Tierschutz- und Tierrechtsbewegung – ein historischer Abriss". *Aus Politik und Zeitgeschichte* 8–9, 2012, pp. 34–40.
65 In 1877 The American Humane Association (AHA) was founded uniting 27 existing associations under a national umbrella. The first humane animal shelter was developed in Philadelphia by the Women's Auxiliary of the Pennsylvania SPCA, followed by the "City Refuge for Lost and Suffering Animals" in 1874.
66 Adams, Carol J. / Donovan, Josephine: *The Feminist Care Tradition in Animal Ethics. A Reader.* Columbia University Press: New York 2007; Adams, Carol J.: *The Sexual Politics of Meat. A Feminist-Vegetarian Critical Theory.* Continuum: New York 1990; Donovan 1990, pp. 350–375; Singer 1975.
67 For the debate in Riga, see S.[chelling], M.[ary]: "Die weibliche Betheiligung an den Thierschutzbestrebungen. Ein Aufruf an die Frauen". *Der Anwalt der Thiere* (1–2), 1885, pp. 23–26.
68 Anepaio, Lembi: *Baltisaksa aadlidaamid ja vaesus.* [Baltic-German Noble Women and Poverty.] MA thesis, University of Tartu, Faculty of Arts and Humanities, Institute of History and Archaeology: Tartu 2019.
69 *Goldingenscher Anzeiger* (28), 9.7.1877.
70 *Rigasche Zeitung* (75), 31.3.1862.

a short life, as some of them were closed down directly after their goals were fulfilled.[71]

In 1872 the female members of the Riga Animal Welfare Society decided to develop this branch of practical animal care and to found an animal hospital in Riga.[72] In 1876 they had already founded a provisional animal hospital in Daugavpils that aimed to control the health of imported horses that were brought to Riga via the Daugava River.[73] In the same year a women's circle (*Damenkreis*) of the society was founded and started to develop their own structures with a separate library.[74] The idea of shelters was not new. On the contrary it was quite popular at that time in all parts of society, but they were meant exclusively for humans. Although the existing animal welfare society dealt also with suffering and sick animals and had a free a charge veterinarian for low-income animal owners,[75] the female members of the society used the apparent problems of the old structure to rebel openly against the president of the society, who was accused of corruption and mismanagement of the society.[76] In August 1877 the animal shelter in Riga became a reality – it was the first animal hospital led by women and supported by the Women's Committee of the Animal Hospital, a sub-structure of the Livonian Animal Welfare Society.[77]

In the very first years of its existence there were only women (44 women, 16 unmarried, ten members of the nobility) who were engaged in the Committee led by Mary Julie Elise v. Schilling (1824–1900). In the very first year, the animal hospital took care of 224 dogs, 43 horses, 41 geese, fowl and birds. New buildings for the animals, their servants, veterinarians, and the matron of the hospital were financed with donations and revenues.[78] With this turn toward practical care and medical help, the ladies in Riga also became a model for the animal welfare societies in and outside the Empire. In 1878 a women's committee was also founded within the Russian Animal Welfare society, and in 1888 a similar

---

71 The women's committee for the acquisition of items for military hospitals in Windau was closed after half a year of existence as the goals were fulfilled, see *Goldingenscher Anzeiger* (28), 9.7.1877.
72 *Rigasche Zeitung* (250), 27.10.1872.
73 *Rigasche Zeitung* (245), 21.10.1876.
74 *Zeitung für Stadt und Land* (220), 22.09.1876; *Zeitung für Stadt und Land* (245), 21.10.1876.
75 *Rigasche Zeitung* (30), 6.2.1869.
76 *Zeitung für Stadt und Land* (117), 22.5.1875.
77 *Rigasche Zeitung* (16), 20.01.1879; *Rigasche Zeitung* (241), 18.10.1885.
78 *Rigasche Zeitung* (16), 20.1.1878.

animal shelter was established in Berlin led by the local women's committee.[79] Although the inner development shows clear parallels to the developments in America or England, the women's committee was also closely connected to the Russian Animal Welfare Society. The women's committee was directly supported and protected by Aleksandr Vasilyevich Suvorov (1729–1800), one of the greatest military commanders and generals of its times in Russia.[80] They also played an important role as cultural mediators between Germany and Russia. So, for example, they translated Richard Knoche's (1822–1892) booklet "*Die wissenschaftliche Thierfolter, eine Reihe von Thatsachen*"[81] from German to Russian, spread translations of Frances Power Cobbe's works, and were actively involved when Cobbe became an honorary member of the Russian Animal Welfare Society.[82] The close connection to Russia can also be seen in the fact that in winter 1891 when there was a severe famine in the Russian governorate of Saratov, the Riga Committee donated 300 rubles in order to support the suffering animals in the region.[83]

With its practical care of animals, the women's committee was directed against some main understandings and actions of the former animal welfare society. While the first promoted culls of stray dogs in the city and stray cats in the countryside, this was unacceptable for the women.[84] While the women in England with Frances Power Cobbe and others were successful mainly in their fight against vivisection, the women in Riga had their main focus on the development of veterinary medicine and on ending culls on the streets. From 1876 they were thinking about splitting from the Riga Welfare Society and moving under the umbrella of the Livonian or the Central Animal Welfare Society in St. Petersburg. At request of the first the ladies remained within the local society but gained more structural independence.[85] In 1886, however, the final schism between the women's committee and the Riga Animal Welfare Society took place, and the women's committee was established as an independent society. There were

---

79 *Der Anwalt der Thiere* (1–2), 1885, p. 21.
80 *Düna Zeitung* (108), 13.5.1902.
81 65,000 copies of this booklet were published and it was translated into many languages, https://www.deutsche-biographie.de/sfz43291.html, retrieved 10.07.2023.
82 *Der Anwalt der Thiere* (1–2), 1885, p. 30.
83 *Düna Zeitung* (294), 30.12.1891.
84 *Der Anwalt der Thiere* (6), 1885, pp. 101–104; *Der Anwalt der Thiere* (1), 1886, pp. 43–56.
85 Schilling, Mary v.: "Jahresbericht des Damen-Comités des Rigaer Thierasyls pro 1884". *Der Anwalt der Thiere* (3), 1885, pp. 43–56, p. 45; *Der Anwalt der Thiere* (6–7), 1886, pp. 89–97.

many reasons for this step, most of them ideological ones: while the older welfare society was in the first place a local society with international cooperation, the women's society started to act in the time of Russian centralisation embodied by the governmental and the Central Animal Welfare Society in St. Petersburg. The women used this change in the main structure as well as the ethical development from human-centred animal welfare to animal rights to make their interests stronger. While working for practical care for animals, humanely education, anti-vivisection, anti-hunting, anti-fur, and anti-meat rhetoric, they were not alone fighting for the liberation of animals but also for the liberation of women by empowering them within the societies. This was a clear provocation of the animal welfare society and led to a clear split between the both.

After this step in 1886 the women's committee started to blossom. From 1885 they were editing their own journal "The Lawyer of Animals", edited by the former editor of the *"Vierteljahresschrift des Kurländischen Thierschutzvereins"*, Baron Edmund von Lüdinghausen-Wolff (1842–1907).[86] The society was from the beginning a mixed society whose members were mostly women, but whose president by statute had to be a male. Thanks to this gender equality, the Committee soon got a remarkable number of male members (by 1890 half of the society's members were already men). While this society grew to an integrative society that was an umbrella for all individuals interested in animal welfare, the former animal welfare society in Riga with its old-fashioned structure died in 1899. Up to 1918 when after the declaration of independence of Latvia the first Latvian Animal Welfare Society was founded, the women's committee was the only animal welfare society in Riga. Only in 1935 did the Latvian Animal Welfare Society take over the women's committee, but this was already right at the end of Baltic-German history in the Baltic provinces.

Parallels and comparisons between the slavery or serfdom of humans and animals have been the very basics of the animal rights and animal abolitionism movements all around the world. It is precisely this basic ethic understanding that caused the main split between the human-centred ideology of the animal welfare society and the women's committee, whose new ideological understanding was made clear in the very first sentences of its journal "The Lawyer of Animals" in 1885. The editors started with a comparison between the negative emotions and accusations the abolition of serfdom provoked within Couronian society when it was first discussed, and the negative public emotions concerning the split of the women's committee and its idea of animal abolitionism. And true enough, the

---

86 Schilling 1885, p. 46.

critique against the new more-than-human ethic that came from the conservative and religious sides was indeed harsh and prompt. So the Christian newspapers in Riga polemicised the "anti-Christian and Darwinistic philosophy" of the women's committee and of its "destructive organ", meaning the "Lawyer of Animals",[87] while the animal welfare society accused the ladies of reacting emotionally toward animals, not being therefore reasonable partners in the dialogue. Parallels between both abolitionist movements were highlighted several times in the "Lawyer". And even more: the forerunners and supporters of both movements in Couronia were members of the very same Lüdinghausen-Wolff family, who belonged to the most noble ones in the government.[88] So the actors working for human and more-than-human abolition in the Baltic provinces of Russia were described as being from within the same social class – the higher nobility. Knowing that the role of the Baltic knighthoods in the abolition of serfdom was not by any means only supportive,[89] this statement needs explanation. It seems to be a political gesture to link the women's committee to the nobility in order to invent for the new committee a long and glorious history of freedom.

While we are waiting for further comparative studies to be written,[90] we can find further proof of our hypothesis that the abolition of serfdom was a matrix for speaking about animal abolitionism. Though the "Lawyer" had a clear ethical agenda, it was nevertheless open for all kinds of articles and we can find the main editors even being inconsistent with their agenda. So Edmund Lüdinghausen-Wolff wrote within the same first year of existence of the journal an article stressing that "Animals are the natural slaves of humans", ending his discussions with the words: "It is time to define the conception position of the animals

---

87 Lüdinghausen-Wolff, Edmund von: "Der 'Bote aus dem Mitauer Diakonissenhause' über den 'Anwalt der Thiere'". *Der Anwalt der Thiere* (3), 1885, pp. 35–37; *Der Anwalt der Thiere* (4), 1885, pp. 57–64; *Der Anwalt der Thiere* (5), 1885, pp. 73–77.
88 Baron Georg Christoph von Lüdinghausen-Wolff (1751–1807) was a politician and the last Curonian chancellor, Baron Edmund von Lüdinghausen-Wolff (1842–1907), who was the main editor of the "Lawyer of Animals", and former editor of the "Vierteljahresschriften des Kurländischen Thierschutzvereins" and author of "Ideen zu einer Metaphysik der Materie" from 1870.
89 Plath, Ulrike: "Politik als Erlösung. Überlegungen zu Karl Gottlob Sonntags Aufklärungstheologie". In: Lukas, Liina / Schwidtal, Michael / Undusk, Jaan (eds.): *Politische Dimensionen der baltischen literarischen Kultur.* (Schriften der Baltischen Historischen Kommission 22). LIT: Berlin 2018, pp. 123–161.
90 Kruusmaa, Andre. "Baltic-German pro-serfdom thought in the Russian Baltic provinces from a comparative perspective". *Slavery and Abolition* 42(2), 2021, pp. 345–362. DOI: 10.1080/0144039X.2020.1816101.

juridically more precisely as well: the ancient ones debased their slaves to the notion of animals, we are lifting animals up to the notion of slaves."[91]

Enlightenment, Education and Early Animal Activism

Although the differences between the agenda and ethical dimensions of the earlier animal welfare and the later animal rights movement are clear, they had a common ground in the earlier Enlightenment traditions.[92] In Baltic cultural history the importance and legacy of the Enlightenment has been discussed very differently. While the innovative power of the high and popular enlightenment is often interpreted as the main driver that enabled the abolition of serfdom, the emancipation of the educated middle classes, and the development of a civic society, there are also voices that kept reminding of its patriarchal, colonial, nationalistic, and racist structures.[93] Rationalism as the main philosophy of the eighteenth century also shed a new light on the perception of animals. One of the earliest thinkers of the Enlightenment in the late eighteenth century who dealt with animal questions in the Baltic region was pastor August Wilhelm Hupel (1737–1819), known in local history for his three volumes of topographic descriptions of the Baltic provinces.[94] Less known is his book from 1774, where he concedes the existence of souls not only to humans, but also to insects, polyps and even stones.[95] Although this seems to be a founding moment of local Baltic environmental thinking, he is not developing any kind of environmental ethic out of it. Even less, it is clear that in his search for the place of the soul in the body, he – as many other enlightened men of that time – conducted vivisection on quite a few small co-mates. Hupel embodies in a way the rational, scientific and "dark" side of the Enlightenment that was presented later by the animal welfare society.[96] We should not, however, forget the power of the "light" Enlightenment with its overwhelming passion for social improvement and engagement, that was

---

91 Lüdinghausen-Wolff, Edmund von: "Der Thierschutz vom juristischen Standpunkte aus betrachtet". [Animal Protection from a Legal Point of View.] *Der Anwalt der Thiere* (11–12), 1885, pp. 16–164, p. 164.
92 Bargheer, Stefan: *Moral Entanglements. Conserving Birds in Britain and Germany.* The University of Chicago Press: Chicago / London 2018, p. 59.
93 Plath 2012; Daija 2017; Kronauer, Ulrich (ed.): *Aufklärer im Baltikum. Europäischer Kontext und regionale Besonderheiten.* Winter: Heidelberg 2011; Undusk 2000.
94 Hupel, August Wilhelm: *Topographische Nachrichten von Liv- und Ehstland.* Vol I-IV. Hartknoch: Riga 1774–1789.
95 Hupel 1774, p. 334.
96 Horkheimer, Max / Adorno, Theodor: *Dialektik der Aufklärung.* S. Fischer: Frankfurt 1969.

dominant in Baltic enlightened rhetoric and public enlightenment throughout the nineteenth century. Highly emotional language, striking metaphors, and clear vision on how to improve particular levels of society were common for that much milder face of the Enlightenment. The common ground for both branches of the Enlightenment was its educational vision to change the world through learning and teaching. Let's look into how animal issues and animal ethics were taught at schools.

The first wave of environmental education was activated by books written for children.[97] In 1864, just three years after the establishment of the animal welfare society in Riga, the first book meant for children called "The Animal Lover" (*"Der Thierfreund"*) written by Ferdinand Müller,[98] a well-known person in Latvian popular enlightenment, was published.[99] Müller, born in Germany and working as a teacher in the Russian Empire from 1822 was a productive author of school and house books for children. In many of them nature played a central role. In 1834 he published in Riga the book "Nature's Friend", which was extended and reprinted in 1835.[100] The love for local nature was part of the general new love of the near home and its surroundings (*Heimatbewegung*), that started in the 1830s.[101] The attention to animals in this context arose significantly in the second part of the century, supported by the new societies. The animal welfare society imported and spread books on animal welfare in schools[102] and translated them to Latvian, such as Katterfeld's book *"Adama valihgi un zitas mazas pasazinas"*, published in 1895.[103] All societies dealing with animal questions at that time, also including the Naturalist Society in Riga, understood that their message had to be spread among pupils to gain ground.[104]

---

97 For the general background see Ritvo, Harriet: "Learning from Animals: Natural History for Children in the Eighteenth and Nineteenth Centuries". *Children's Literature* 13, 1985, pp. 72–93. DOI 10.1353/chl.0.0198.
98 Müller, Ferdinand: *Der Thierfreund, oder: Andeutungen aus dem Seelenleben der Thiere*. Ernst Plates: Riga 1864. Under the same name, a Russian journal was published from 1869 in Tallinn.
99 Daija 2017.
100 Müller, Ferdinand: *Der Naturfreund, oder, erster Unterricht in der Sternkunde und Naturlehre. Ein Schul- und Hausbuch in katechetischer Form*. Wilhelm Ferdinand Haecker: Riga 1832.
101 Plath, Ulrike: "Heimat. Rethinking Baltic-German Spaces of Belonging". *Kunstiteaduslikke Uurimusi / Studies on art and architecture* 23(3–4), 2014, pp. 55–78.
102 *Kurländische Gouvernements-Zeitung* (22), 7.3.1862.
103 *Düna Zeitung* (128), 9.6.1895.
104 *Rigasche Zeitung* (134), 13.6.1872.

In the late 1870s and 1880s a new wave of practical environmentalism reached the local schools when pupils' associations for animal welfare (*Schülervereinigungen für Thierschutz*) were founded in large numbers, directly supported by the local animal welfare societies, which again followed international models.[105] In the late 1870s and 1880s all schools of Curonia, primary and secondary schools, and Latvian, German, and Hebrew schools became members of this association and started a program on animal activism.[106] In Windau/Ventspils schoolchildren collected money in order to buy and free captured songbirds, to buy food for bird feeding stations around their school, and to support all other activities connected with animal protection.[107] In St. Petersburg, pupils' associations for the protection of songbirds were also founded by the local animal welfare society, but this happened only in 1899. Every member of a pupils' association in the Baltics had to sign a contract that they would never destroy a bird's nest or young birds, or allow that this would be done by others.[108] The children were trained to monitor all activities around and actively interfere. This new vigilance mobilised an "army" of around 30,000 "juvenile comrades" in Curonia alone, who became active parts of the society.[109]

The new form of activism of the youth did not, however, only protect animals. It also defined who was part of the "good nature" they aimed to protect and who belonged to the "bad nature" that had to be destroyed. In Germany children were trained to kill all kinds of worms and caterpillars such as the cock chafer and its grubs, and hamsters were also as ill-fated as they were all a danger to the agricultural production of men.[110] In the Baltic provinces we can find an impressive number of birds and mammals against which the children started their "war of extinction" (*Vernichtungskrieg*): domesticated and savage cats were in first place, followed by their wild relatives, squirrels, foxes, martens, weasels, most raptors (except kestrels and buzzards), eagle owls, common ravens, crows, jays and magpies.[111] In this war against "bad nature" everything seemed to be

---

105 *Rigasche Zeitung* (296), 20.12.1878.
106 1879 in Curonia, 1884 in Dorpat. *Rigasche Zeitung* (262), 9.11.1879; *Rigasche Zeitung* (36), 13.2.1884; *Libausche Zeitung* (48), 27.3.1879.
107 *Goldigenscher Anzeiger* (46), 17.1.1879.
108 *Libausche Zeitung* (206), 13.9.1899; *Libausche Zeitung* (87), 15.4.1899.
109 *Goldigenscher Anzeiger* (46), 17.1.1879.
110 *Rigasche Zeitung* (296), 20.12.1878.
111 "Wars of extinction" against different species were described and declared in local newspapers since the mid nineteenth century; Lothar, Leo: "Unschuldig Verfolgte". *Land- und forstwirtschaftliche Zeitung für praktische und wissenschaftliche Pflege der Land-, Forst- und Volkswirtschaft* (22), 30.5.1894, 130–131.

allowed from destroying nests, and killing the specimens in all possible ways from hunting, catching to poisoning.[112]

At this point it seems to be relevant to ask if the discussed forms of actions to protect animals can be interpreted as early forms of early "bad environmentalism" involving step by step not only adults from all classes, but also children.[113] Although some pre-forms of environmentalism have been analysed dating back to the Enlightenment, environmentalism has mostly been interpreted as a phenomenon of the long twentieth century. Speaking about earlier times, interpretations of the main works of exceptional thinkers such as Humboldt and George Perkins Marsh prevail.[114] However, studies on their impact on broader society are still missing.

Conclusion

In the second part of the nineteenth century Riga, the metropolis of the Baltic provinces of Russia, became a hotspot for the animal welfare and animal rights movement. As the urbanised and modern city was closely connected with similar societies all over Europe, the information about the new societies in England and Germany spread quickly. Both societies passionately engaged in bringing a transhuman dimension to the societies, public discussions and everyday practices, even though they were concentrating on very different goals. In the end, local animals might have gained as much from the animal welfare agenda to change and introduce new laws and innovate animal transportation and slaughtering practices, as from the animal rights approach to support veterinarian medicine and strengthen emotional ties. Both of them had a very similar understanding about the importance of spreading the word among as many social groups and countries as possible and to educate the next generations and lower classes. In this respect, both movements were predecessors of popular enlightenment that translated and popularised the ideas of their intellectual forerunners.

Looking at the Latvian cases from Livonia and Couronia, I can see a broadening of interest in animal questions throughout the nineteenth century involving all parts of society, and taking more and more forms of social activism. While caring

---

112 Poisoning crows was allowed for example during the winter. "Mannigfaltige Nachrichten aus dem Sportleben. I. Jagdsport". *Dünazeitung* (208), 13.9.1900.
113 Seymour, Nicole: *Bad Environmentalism. Irony and Irreverence in the Ecological Age.* University of Minnesota Press: Minneapolis 2018.
114 Sachs, Aaron: *The Humboldt Current. Nineteenth-Century Explorations and the Roots of American Environmentalism.* Penguin Books: London 2006.

about landscapes and "wilderness" was a highly romanticised project, taking care of animals was a hands-on thing to do, a practical need in the growing cities, and provided reachable and understandable goals for everybody. This very practical dimension of animal activism might differ from the environmental activism of the later decades and centuries, but it helped to develop a sense for living in a more-than-human world.

References

Source Materials: Periodicals

Der Anwalt der Thiere 1885–1886

Dünazeitung 1890–1900

Goldigenscher Anzeiger 1877–1879

Kurländische Gouvernements Zeitung 1862–1868

Land- und forstwirtschaftliche Zeitung für praktische und wissenschaftliche Pflege der Land-, Forst- und Volkswirtschaft 1895

Libausche Zeitung 1876–1899

Rigasche Stadtblätter 1860–1862

Rigasche Zeitung 1860–1885

Zeitung für Stadt und Land 1875–1876

Literature

Adams, Carol J. / Donovan, Josephine: *The Feminist Care Tradition in Animal Ethics. A Reader.* Columbia University Press: New York 2007.

Adams, Carol J.: *The Sexual Politics of Meat. A Feminist-Vegetarian Critical Theory.* Continuum: New York 1990.

Anepaio, Lembi: *Baltisaksa aadlidaamid ja* vaesus. [Baltic-German Noble Women and Poverty.] MA thesis, University of Tartu, Faculty of Arts and Humanities, Institute of History and Archaeology, Tartu 2019.

Atkins, Peter (ed.): *Animal Cities. Beastly urban histories.* Routledge: London 2012.

Bargheer, Stefan: *Moral Entanglements. Conserving Birds in Britain and Germany.* University of Chicago Press: Chicago / London 2018

Beauchamp, Tom L. / Frey, Raymond G. (eds.): *The Oxford Handbook of Animal Ethics.* Oxford Press: Oxford / New York 2011.

Bentham, Jeremy: *An Introduction to the Principles of Morals and Legislation.* Oxford Press: London / New York / Toronto 1823. Retrieved 10.07.2023 from https://oll.libertyfund.org/titles/bentham-an-introduction-to-the-principles-of-morals-and-legislation

Bonhomme, Brian: "Russian Compassion. The Russian Society for the Protection of Animals – Founding and Contexts, 1865–75". *Canadian Journal of History* 45(2), 2010, pp. 259–297.

Cordeiro-Rodrigues, Luis: "Animal Abolitionism and 'Racism without Racists'". *Journal of Agricultural and Environmental Ethics* 30, 2017, pp. 745–764.

Chakrabarty, Dipesh: *The Climate of History in a Planetary Age*. Chicago, The University of Chicago Press: Chicago / London 2021.

Daija, Pauls: *Literary History and Popular Enlightenment in Latvian Culture*. Cambridge Scholars Publishing: Newcastle upon Tyne 2017.

Donovan, Josephine: "Animal Rights and Feminists Theory". *Signs* 15(2), 1990, pp. 350–375.

Duncan I.J.H.: "Animal Welfare: A Brief History". In: Hild, Sophie / Schweitzer, Louis (eds.), *Animal Welfare: From Science to Law*. La Fondation Droit Animal, Éthique et Sciences (LFDA): Paris 2019, pp. 13–19.

Festa, Lynn: *Fiction without humanity. Person, Animal, Thing in Early Enlightenment Literature and Culture*. University of Pennsylvania Press: Philadelphia 2019.

Fitzgerald, Amy J.: *Animal Advocacy and Environmentalism. Understanding and Bridging the Divide*. Polity Press: Cambridge / Medford 2019.

Francione, Gary L. / Garner, Robert: *The Animal Rights Debate. Abolition or Regulation?* Columbia University Press: New York 2010.

Francione, Gary L. et al. (eds.): *Exploring the Animal Turn. Human–animal relations in Science, Society and Culture*. (Pufendorfinstitutets skriftserie 4). Lund University: Lund 2014.

Gaarder, Emily: *Women and the Animal Rights Movement*. Rutgers University Press: New Brunswick 2011.

Hackmann, Jörg: *Vereinskultur und Zivilgesellschaft in Nordosteuropa. Regionale Spezifik und europäische Zusammenhänge. – Associational Culture and Civil Society in North Eastern Europe. Regional Features and the European Context*. (Quellen und Studien zur baltischen Geschichte 20). Böhlau: Wien / Köln / Weimar 2012.

Hackmann, Jörg: *Zivilgesellschaft im östlichen und südöstlichen Europa in Geschichte und Gegenwart*. (Völker, Staaten und Kulturen im östlichen Mitteleuropa 5). Oldenbourg: München 2011.

Hadley, John: "Nonhuman Animal Property. Reconciling Environmentalism and Animal Rights". *Journal of Social Philosophy* 36(3), 2005, pp. 305–315.

Hirschhausen, Ulrike von: *Die Grenzen der Gemeinsamkeit. Deutsche, Letten, Russen und Juden in Riga 1860–1914.* Vandenhoeck & Ruprecht: Göttingen 2006.

Horkheimer, Max / Adorno, Theodor: *Dialektik der Aufklärung.* S. Fischer: Frankfurt 1969.

Hupel, August Wilhelm: *Anmerkungen und Zweifel über die gewöhnlichen Lehrsätze vom Wesen der menschlichen und thierischen Seele.* Hartknoch: Riga 1774.

Hupel, August Wilhelm: *Topographische Nachrichten von Liv- und Ehstland.* Vol. I-IV. Hartknoch: Riga 1774–1789.

Kaljundi, Linda / Plath, Ulrike: "Serfdom as entanglement. Narratives of a Social Phenomenon in Baltic History Writing". *Journal of Baltic Studies* 51:3, 2020, pp. 349–372. DOI:10.1080/01629778.2020.1776349.

Kronauer, Ulrich (ed.): *Aufklärer im Baltikum. Europäischer Kontext und regionale Besonderheiten.* Winter: Heidelberg 2011.

Kruusmaa, Andre: "Baltic-German Pro-Serfdom Thought in the Russian Baltic Provinces from a Comparative Perspective". *Slavery and Abolition* 42(2), 2021, pp. 345–362. DOI:10.1080/0144039X.2020.1816101.

Leiderer, Annette: "History of Animal Slaughter". In: Roscher, Mieke / Krebber, André / Mizelle, Brett (eds.): *Handbook of Historical Animal Studies.* Walter de Gruyter: Berlin / Boston 2021, 539–553. DOI:10.1515/9783110536553-040.

McShane, Clay / Tarr, Joe A.: *The Horse in the City: Living Machines in the Nineteenth Century.* Johns Hopkins University Press: Baltimore 2007.

Müller, Ferdinand: *Der Naturfreund, oder, erster Unterricht in der Sternkunde und Naturlehre. Ein Schul- und Hausbuch in katechetischer Form.* Wilhelm Ferdinand Haecker: Riga 1832.

Müller, Ferdinand: *Der Thierfreund, oder: Andeutungen aus dem Seelenleben der Thiere.* Ernst Plates: Riga 1864.

Obenchain, Theodore G: *The Victorian Vivisection Debate. Frances Power Cobbe, Experimental Science and "Claims of Brute".* McFarland & Company Inc.: Jefferson / North Carolina / London 2012.

Ohrem, Dominik: "Feminist Intersectionality Studies".In: Roscher, Mieke / Krebber, André / Mizelle, Brett (eds.): *Handbook of Historical Animal Studies.* Walter de Gruyter: Berlin / Boston 2021, pp.341–355. DOI:10.1515/9783110536553-027.

Paramonov, Riho: *Eesti ühiskonna moderniseerumine avaliku privaattranspordi (voorimees ja takso) näitel, 1900–1940.* [The Cabman and the Taxi Driver: The Modernization of Estonian Society During 1900–1940 on the Example of Public Transport.] PhD Dissertation. Tallinn University Press: Tallinn 2019.

Plath, Ulrike: "Baltisaksa laste- ja noorsookirjandusest kuni 1840. aastateni". [Baltic German Children's and Youth Literature until the 1840s.] *Keel ja Kirjandus* 8–9, 2011, pp. 698–715.

Plath, Ulrike: *Esten und Deutsche in den baltischen Provinzen Russlands. Fremdheitskonstruktionen, Lebenswelten, Kolonialphantasien 1750–1850*. Harrassowitz: Wiesbaden 2012.

Plath, Ulrike: "Heimat. Rethinking Baltic-German Spaces of Belonging". *Kunstiteaduslikke Uurimusi / Studies on art and architecture* 23(3–4), 2014, pp. 55–78.

Plath, Ulrike. "Politik als Erlösung. Überlegungen zu Karl Gottlob Sonntags Aufklärungstheologie". Lukas, Liina / Schwidthal, Michael / Undusk, Jaan (eds.): *Politische Dimensionen der baltischen literarischen Kultur*. (Schriften der Baltischen Historischen Kommission 22). LIT: Berlin 2018, pp. 123–161.

Ritvo, Harriet: "Learning from Animals: Natural History for Children in the Eighteenth and Nineteenth Centuries". *Children's Literature* 13, 1985, pp. 72–93. DOI:10.1353/chl.0.0198.

Ritvo, Harriet: *The Animal Estate. The English and Other Creatures in the Victorian Age*. Harvard University Press: Cambridge, Massachusetts 1987.

Roscher, Mieke: "Tierschutz- und Tierrechtsbewegung – ein historischer Abriss." *Aus Politik und Zeitgeschichte* 8–9, 2012, pp. 34–40.

Sachs, Aaron: *The Humboldt Current. Nineteenth-Century Explorations and the Roots of American Environmentalism*. Penguin Books: London 2006.

Sachse, Carola: "Von Männern, Frauen und Hunden. Der Streit um die Vivisektion im Deutschland des 19. Jahrhunderts". *Feministische Studien* 24(1), 2006, pp. 9–28.

Seymour, Nicole: *Bad Environmentalism. Irony and Irreverence in the Ecological Age*. University of Minnesota Press: Minneapolis 2018.

Singer, Peter: *Animal Liberation*. Avon: New York 1975.

Sonntag, Karl Gottlob: *Das kleine Buch für kleine Kinder in Riga, welche lesen lernen*. S.n.: Riga, 1802.

Spencer, Jane: *Writing about Animals in the Age of Revolution*. Oxford University Press: Oxford, 2020.

Undusk, Jaan: "Ajalootõde ja metahistoorilised žestid. Eesti ajaloo mitmest moraalist". [Historical Truth and Metahistorical Gestures. About Several Morals of Estonian History.] *Tuna. Ajalookultuuri ajakiri* 2, 2000, pp. 114–130.

Wischermann, Clemens / Steinbrecher, Aline / Howell, Philip (eds.): *Animal history in the modern city. Exploring liminality*. Bloomsbury Academic: London et al. 2019.

Wolloch, Nathaniel: *The Enlightenment's Animals: Changing Conceptions of Animals in the Long Eighteenth Century*. Amsterdam University Press: Amsterdam 2019.

Wrenn, Corey: "Abolitionist animal rights. critical comparisons and challenges within the animal rights movement". *Interface: a journal for and about social movements* 4(2), 2012, pp. 438–458.

Zahra, Tara: "Imagined Noncommunities: National Indifference as a Category of Analysis". *Slavic Review* 69(1), 2010, pp. 93–119. DOI:10.1017/S0037677900016715.

Zerbel, Miriam: *Tierschutz im Kaiserreich. Ein Beitrag zur Geschichte des Vereinswesens*. (Münchner Studien zur neueren und neuesten Geschichte 4). Peter Lang: Frankfurt a.M. et al. 1993.

Anita Zariņa / Dārta Treija / Ivo Vinogradovs

Bison in the Latvian Ethnoscape: Contingency of (Not) Becoming

Abstract: Bisons roaming in the Latvian landscape today are part of a large herbivores' (re)introduction project that started at the beginning of the 2000s and is aimed at rewilding the area of marginal farmland around Lake Pape in the South-Western part of the country. Due to a contingency, the bison left their fenced enclosure, gradually becoming a wild pack which is now roaming the area. On the one hand, the article focuses on the bison as part of diverse place-based topologies brought about by nation-wide changes in the socio-economic situation, nature discourses, land and nature management conflicts and the locals' attitude. On the other hand, it exposes yet unarticulated complexities relating to a general understanding of the bison's place in this particular landscape, along with its social implications and the broader meaning of the bison's becoming part of the national ethnoscape. In particular, we explore the ways in which the attitudes towards the bison in Pape are formed, using Deleuze and Guattari's notion of "three kinds of animals", and the role of the local human and non-human actors in shaping the bison milieu in Pape.

Keywords: rewilding, reintroduction, large herbivores, human-animal relations, bison

Introduction

In his seminal work "Landscapes and Memory" Simon Schama[1] starts the chapter about the European bison (the Royal Beasts of Białowieża) with a taste of their meat, "[…] it tasted like nothing I had eaten before: a strange sweetness lurking beneath its cheesy pungency". Indeed, hunting for meat has long defined an important aspect of human relations with this animal. Another aspect that penetrates and in a way defines the human–bison relationship is the acknowledgment of their grace, majesty, special treatment and protection, for example, by royalty in Białowieża Forest,[2] or currently by international

---

1   Schama, Simon: *Landscape and Memory*. Vintage books: New York 1995, p. 37.
    The research results of the current study have also been published as Zariņa, Anita / Treija, Dārta / Vinogradovs, Ivo: "Bison in the Latvian Ethnoscape: A Contingency of (not)Becoming". *Letonica* 44, 2022, 153–168. DOI: 10.35539/LTNC.2022.0044.A.Z.D.T.I.V.0009.

2   Baerselman, Fred / Vera, Frans W.M.: *Nature Development. An Exploratory study for the construction of ecological networks*. Ministry of Agriculture, Nature and Fisheries: Den Haag 1995, p. 12; Schama 1995, pp. 37–53.

regulations. Moreover, they are also subject to awe of some kind, as this long forgotten Latvian tale suggests, "Far away a king ruled. His son was a great hunter [...] Once, accompanied, he went to hunt and arrived at a great forest abundant with game. Carried away by the heated hunt from other hunters, the prince spotted a bison and shot an arrow at him. The bison vanished instantly, a huge wizard appeared in its place and cursed him by pointing his hand [...]".[3]

On a grey and windy day in December 2014, along with the Director of Pasaules Dabas Fonds[4], Jānis Rozītis, we were out in a field in search of bison in the wild surroundings of Lake Pape (south-western Latvia). The landscape at this time of the year in the coastal lowlands is damp and muddy; ditches are full of water, limiting the possibility to traverse these young, deciduous forests on foot. Within eyesight, some elk passed over the road and ditches, leaving numerous footprints, similar to the ones we were looking for. Confused by the plethora of animal footprints, we were carefully looking for other hints: noises, smells, gnawed-off pieces of tree bark, etc. Yet we knew that we were wandering in a bison landscape, imagining them living there rather than truly expecting to encounter them at once. They are animals of the dusk and dawn; fast and silent, masters at remaining unseen among trees.

Although bison are large mammals, they have quite a narrow body shape enabling them to hide easily behind tree trunks. Some analogies of this can also be found in language. The word in Lithuanian for European bison is *stumbras*, which in kindred Latvian means tree trunk (*stumbrs*), but the name for bison in Latvian is *sumbrs* (ancient, *sūbrs*). However, the bison is also a meadow animal. Some excrement that we found in their former home of fenced meadows confirmed their presence in the area (as it did in the spring of 2013 during a similar quest) where they have been roaming in the wild for the past ten years.

The 'experiment' of bison (re)introduction in Pape, followed by a contingent bison escape from the enclosure, exposes yet unarticulated complexities relating to general understandings of the bison's place in this particular landscape, their 'becoming' wild animals, along with social implications and broader meaning of bison becoming part of the national landscape. This research traces the bison movements since their arrival in Pape until today, giving insight into bison cultural history in the region, as well as the ideas and meanings of rewilding

---

3   Šmits, Pēteris (ed.): *Latviešu pasakas un teikas 4* [Latvian Tales and Legends 4]. Valters un Rapa: Rīga 1927, p. 253.
4   An NGO, formerly WWF-Latvia, now an associate partner of WWF in Latvia (World Wide Fund for Nature).

itself. In particular, we wish to explore the ways in which attitudes toward bison in Pape are formed, using Deleuze and Guattari's[5] notion of "three kinds of animals", and the role of local human and non-human actors in shaping the bison milieu in Pape.

We used field observations, interviews with locals, park managers and nature protection experts, media content analysis, analyses of literature and documents to understand the geography and time-line of the events, as well as to interpret the bison project in larger socio-political contexts. Our fieldwork was conducted in 2013–2015, when we visited the surroundings of Pape, including visits to the former bison enclosure and farmsteads in the local neighbourhood.

An Insight Into the History of Bison

In 1854, The Latvian Newspaper (*Latweeschu Awises*) wrote that bison were the largest forest bull, which still can be seen in Lithuanian and Białowieża forests, while elsewhere it has been exterminated. This is presumably the oldest record in Latvian, published while Courland was a part of the Russian empire, describing in detail the management of bison in Białowieża that at the time belonged to "our emperor". Could it be so that bison have roamed Courland's forests hundreds of years before, the author rhetorically asks? Yes, they had sometimes entered from Lithuanian lands, they had also been spotted in Prussia until they were all shot. And, after all, there is a Latvian name for the beast, concludes the author.[6]

Formerly widespread throughout Europe, the European bison *Bison bonasus* became nearly extinct in the beginning of the twentieth century.[7] In the Holocene and early historical times the range of European bison covered western, central and south-eastern Europe, extending up to the Volga River and the Caucasus. Since then the population of bison has been subjected to gradual shrinkage and fragmentation: the decreasing number of individuals and isolation of subpopulations lead to extinction.[8]

---

5 Deleuze, Gilles / Guattari, Félix: *A Thousand Plateaus. Capitalism and Schizophrenia*. Bloomsbury Academic: New Delhi 2014 [1987], pp. 279–284.
6 "Sumbrs". *Latweeschu Awises* Nr. 44, 28.10.1854, retrieved 10.7.2023, from http://periodika.lv/periodika2-viewer/view/index-dev.html?lang=fr#panel:pa|issue:/p_001_laav1854n44|article:DIVL30|issueType:P.
7 Balcčiauskas, Linas: "European Bison (Bison Bonasus) in Lithuania. Status and Possibilities of Range Extension". *Acta Zoologica Lituanica. Biodiversity* 9(3), 1999, pp. 3–18, p. 3.
8 Pucek, Zdzisław (ed.): *European Bison. Status Survey and Conservation Action Plan*. Gland, Switzerland and Cambridge: IUCN, 2004, pp. 1–15.

In Latvia bison were extant at the end of the Iron Age and through the Middle Ages (fifteenth–sixteenth centuries), yet archaeological findings indicate that bison hunting remains are particularly characteristic of the Late Iron Age period (tenth–twelfth centuries) in the southern parts of Latvia's territory.[9] Afterwards its presence in the territory of Latvia is hardly evidenced; while in captivity bison were registered in the so-called manorial deer garden (*Hirschgarten*) of Valmiermuiža (*Wolmarshof*), where a pair of bison from Białowieża was brought in the end of the nineteenth century to enrich the animal diversity of a park created for royal hunting pleasures.

By the eighteenth–nineteenth centuries, bigger herds of European bison were found only in Białowieża – large sparsely populated forest, which was preserved by monarchs for over 400 years.[10] Due to the protection measures of the royal game the number of bison grew up there to almost two thousand in the 1850s, but by the end of the century only 380 animals were left there,[11] then increasing again in 1915 to approximately 785 animals.[12] War and post-war chaos lead to uncontrolled poaching of bison: during World War I, around 600 bison were killed, mainly for meat, leather and horns, but after the retreat of German soldiers only nine bison had survived.[13] Eventually, in 1919 a poacher shot the last European bison. In 1929 efforts were initiated to reintroduce bison on the basis of a few individuals, which were saved in few European zoological gardens.[14] Nowadays Białowieża is the only lowland primeval forest left in Europe – around 900 bison roam freely there.[15]

The history of bison in Latvia, which has been absent from its landscape for a long time, can thus be disclosed only through scrupulous archaeological research or folk studies, whereas in the neighbouring country of Lithuania bison

---

9   Bīrons, Anatolijs, Mugurēvičs, Ēvalds, Stubavs, Ādolfs, Šņore, Elvīra (eds.): *Latvijas PSR arheoloģija*. Zinātne: Rīga 1974, p. 245.
10 Samojlik, Tomasz: *Bialoweza Forest in the Time of Kings. Conservation and Hunting.* Mammal Research Institute, Polish Academy of Science: Bialoweza 2005, p. 87.
11 Morris, Desmond: *Bison*. Reaktion Books: London 2015.
12 Ricciuti, Edward R.: *To the Brink of Extinction*. Holt, Rinehart &Winston: New York / Chicago / San Francisco 1973.
13 Ibid.
14 Samojlik, p. 89.
15 Deinet, Stefanie et al.: *Wildlife comeback in Europe: The Recovery of Selected Mammal and Bird Species. Final Report to Rewilding Europe by ZSL, BirdLife International and the European Bird Census Council.* ZSL: London 2013, p. 26.

became extinct much later – in the seventeenth century.[16] The reintroduction of bison in Lithuania – bison has been known to be the animal of the ancient Lithuanian state[17] – started in 1969, and the first bison were set free in the wild in 1973.[18] Since 2003 the population size has been rapidly increasing and currently there are nearly 200 individuals.[19]

The Idea of a Wilderness with Large Herbivores

In Latvia, the new wilderness idea that precedes our bison story stems from WWF-Latvia's initiative in the late 1990s aimed at restoring the pre-agricultural landscape with its shifting mosaic of open land and forests, continuously reshaped through the natural disturbances of fire, wind, grazing and predation. This wilderness was established within the Pape Nature Park, comprising the landscapes of wetlands and migrating birds, natural meadows grazed by Konic horses and auroxen, and European bison that also inhabited the park's forests. However, acceptance of the idea in the beginning of the 2000s proved to be controversial here, because according to spokesman of the Latvian Fund for Nature Jānis Priednieks, "The introduction of the Konik horses signified that 'an enormous territory [had] been taken out' of the Latvian landscape; that the nationally iconic Latvian farmer had been driven off the land, to no good purpose, by 'animals of fairly bizarre genetic origins'".[20]

Indeed, wilderness, in the sense it is used in Anglo-American understandings, is not a word found in the Latvian vocabulary. Latvians, the ancient Balts, have always fought for the land, slashed and burned 'dark forests' and reclaimed impassable wetlands, transforming and maintaining the land to render it usable for living. This even applied to lands with poor and sandy soils and to areas with articulated topography or marshes. The whole national narrative of an iconic ethnoscape was built upon the ideals of peasantry[21] – the Latvian as ploughman

---

16 Kibiša, Artūras et al.: "Impact of Free-Ranging European Bison to Ecosystems in Fragmented Landscape, Lithuania". *Balkan Journal of Wildlife Research* 4(2), 2017, pp. 18–25. DOI: 10.15679/bjwr.v4i2.55.
17 Bonda, Moreno: *History of Lithuanian Historiography*. Vytautas Magnus University: Kaunas 2013, pp. 24–33.
18 Balčiauskas, p. 3.
19 Kibiša et al., p. 20.
20 Schwartz, Katrina Z.S.: "Wild Horses in a 'European Wilderness'. Imagining Sustainable Development in the Post-Communist Countryside". *Cultural Geographies* 12(3), 2005, pp. 292–320, p. 293. DOI: 10.1191/1474474005eu331oa.
21 E.g. Skultans, Vieda: "Narratives of Landscape in Latvian History and Memory". *Landscape Review* 7(2), 2001, pp. 25–39.

in folk culture, 'masters of our own land' in the years of national awakening from the mid-nineteenth century and onwards, and productivist agriculture as national business during the years of the country' first national independence. As to a social construct, the wilderness occurred perhaps as an idea of untouched nature preserved for scientific purposes during the Soviet period when strictly off-limits nature reserves were established. They were small and discrete, open to scientists, but not accessible or of interest at all to the general public.

But this wilderness was something new. According to former WWF-Latvia Director Uģis Rotbergs, this idea was based on new values regarding grazing animals as landscape inhabitants and the main caretakers of land, with the whole area functioning as a demonstration site for economic, ecological and social benefits that would come from the reintroduction of large herbivores. Pape village along with the neighbouring village of Rucava and the whole of Rucava County were always considered to be historically unique and rich with material and immaterial ethnographic values. In Rucava, locals would have had the idea of basing tourism on local ethnographic traditions and everyday life. As Schwartz observed, in the minds of many, this vision of heritage tourism directly conflicted with the western vision of wilderness-based ecotourism. For locals, "the unrestrained nature was nothing more than worthless 'jungle', unfit for human enjoyment unless redeemed through cultivation".[22]

Taking Place: Bison Arrival and Exodus

The landscape of Lake Pape's surroundings was redeemed, in fact, through cultivation during the productivism era of the mid-twentieth century.[23] On the dump lands with naturally limited agricultural potential the Pape polder (515 ha) was engineered at the end of the 1960s. The polder was seen as a solution for redeeming the kolkhoz's weak subsistence, and, indeed, both then and now providing the locals with extra, albeit still poor, agricultural lands. The natural habitats in and around Lake Pape (a Ramsar site since 2004) were considerably transformed by various forms of drainage over the course of the twentieth century.[24]

---

22 Schwartz 2005, p. 310.
23 Zariņa, Anita / Vinogradovs, Ivo / Šķiņķis, Pēteris: "Towards (Dis)continuity of Agricultural Wetlands. Latvia's Polder Landscapes after Soviet Productivism". *Landscape Research* 43(3), 2018, pp. 455–469, p. 463. DOI: 10.1080/01426397.2017.1316367.
24 Vlasakker, Joep W.G. van de: *Evaluation of the Hydrological Regime Lake Pape. FNC Report No. 2006.001.* 2006, retrieved 10.7.2023, from http://ec.europa.eu/environment/life/project/Projects/index.cfm?fuseaction=home.showFile&rep=file&fil=LAKE_PAPE_Hidrology_study_final.pdf.

As a relatively marginal location, Pape attracted WWF-Latvia because of its nature potential (lagoonal lake and migratory bird site, high peat bog, wet meadows and vast forest areas) and its depopulated landscape – in Lake Pape's immediate surroundings 15 to 20 residents lived on farmsteads, while a small Pape village consisted of a few permanent inhabitants. A regional newspaper imagined WWF-Latvia's vision of the future landscape as the one from thousands of years long past, "[…] it will be as a view from yore – in a lush meadow wild horses will roam, from a thicket bison would leer, while sky would traverse a skein of cranes. Would there be a place for man? […]".[25]

19 Konik horses arrived at Pape in 1999. In the next project phase in 2004, WWF-Latvia brought the first five European bison (Figure 1) and 24 auroxen. Another 12 bison from Poland and the Netherlands arrived in 2006. The bison were released in an enclosure built on the previously mentioned polder lands; soon after the observation tower and information stand for tourists were installed there. All things considered, the arrival of bison, in particular, was an event of national importance: up to 100 people attended the great opening of the enclosure (among them were representatives of the media, stakeholders, and locals). Back then the local newspaper wrote that "[…] the reintroduction of herbivores will be the most challenging and best investment in the landscape; both environmentally friendly and absolutely sustainable".[26] Before the reintroduction, the project's inquiry data showed that more than a half of Pape's population supported the project. Their expectations were related to tourism development and new job opportunities, etc. Yet this attitude partly changed as soon as the bison agency intervened in the private lives of locals. Bison left their grazing area twice (in 2006 and 2009); the second time no longer returning to their enclosure. The media declared quite clearly at that time[27] that the bison project in Pape had been unsuccessful.

---

25 Tišheizere, Edīte: "Savvaļas zirgiem pievienosies citas dzīvas radības". [Other God's Creatures Will Join the Wild Horses.] *Kursas Laiks*, 4.7.2002, p. 7.
26 Cepleniece, Inese: "Latvijā atgriezīsies arī sumbri". [Also Bison Will Return to Latvia.] *Jaunā Avīze*, 30.1.2003, pp. 16–17.
27 Vīksne, Imants: "Ministrija sumbrus neganīs". [Ministry Will not Shepherd the Bison.] *Neatkarīgā Rīta Avīze Latvijai*, 24.11.2009, p. 6; Vīksne, Imants: "Sumbri Papē uzvedas kā huligāni". [Bison Behave as Hooligans in Pape.] *Neatkarīgā Rīta Avīze Latvijai*, 14.11.2009, p. 6.

Figure 1. Bison arrived at their enclosure in Pape polder lands in 2004. Photo: Pasaules Dabas Fonds.

What caused the bison to leave this place? In human-related terms we would characterise this "escape" as an evacuation or exodus, because the bison were, in fact, fleeing the wet and sometimes even flooded grounds caused by the malfunctioning of the polder. As park managers noted, the second time the escape itself was more of an effort – they had crawled under the bottom of a broken fence, and then swam across the polder canal. Various experts, before and after the escape, acknowledged the unsuitability of the place for bison. Jānis Ozoliņš, a zoologist at the Latvian Silava State Forest Research Institute in 2003 commented on the forthcoming project: "There are strict rules for breeding bison in captivity [...]. Particularly the safety precautions regarding the enclosures must be considered, because these animals can be aggressive. The release of bison into the wild would be a crazy experiment. Ecological expertise is definitely needed here. [...]". On the same page was Jānis Priednieks, the director of the Department of Zoology and Ecology at the Faculty of Biology, saying, "In Latvia there are no suitable biotopes, necessary food base or living space for bison, especially in the poor surroundings of Lake Pape. They need broadleaf forests. [...] These animals need a proper fence, because they cannot be allowed to go into the wild".[28]

Essentially, the malfunctioning polder played a key role in the whole process of bison eventually returning to Latvia's landscape, which was set by a chain of relational contingencies between human and non-human actors. Another question is the suitability of the place itself. In the socio-ecological assessment of the potential European bison habitat Kuemmerle et al. have acknowledged Latvia as suitable for bison populations.[29] However, in Pape no ecological assessment had been done prior to the bison's arrival, there is also no monitoring of their movements, no particular laws or guidelines for their conservation or management have been issued so far. From the point of view of the state, it's almost as if they are non-existent. Yet there they are – known to be last seen on October 13 (2018) by early morning travellers. The ecological conditions of the place itself, apparently, is not a paramount to measure the suitability for bison in the contemporary landscape, rather the relational formations between human and non-human actors and events taking place, as well as human and society's

---

28 Zemberga, Kaija: "Sumbri atgriežas: par un pret". [Bison Return: For and against] *Lauku avīze*, 14.03.2003, p. 37.

29 Kuemmerle, Tobias et al.: "Predicting Potential European Bison Habitat Across its Former Range". *Ecological Applications* 21(3) 2011, pp. 840–841. DOI: 10.1890/10-0073.1.

attitudes (e.g. the state, locals, stakeholders) are the ones that may determine bison's potentiality in Latvia's landscape in general. As Kuemmerle et al. state, in contemporary conservation planning, moving from bioclimatic niche models toward broadscale habitat models is a necessary step to restore the ecological roles of large herbivores in human-dominated landscapes.[30] With this in mind, we must have a closer look at people's modes of relating to these animals.

Three Kinds of Animals

Essentially, this bison project involves much more than (re)introduction, a local developmental asset or locals' disapproval and eventual acceptance. Figuratively speaking, it is a precedent of an animal becoming a bison, that is, a tamed animal becoming a wild animal, likely a place becoming a place with bison in it. In human geography "becoming" designates a process-based movement, in which the world is conceived of as a dynamic and open-ended set of relational transformations.[31] Deleuze uses the term "becoming" to describe the continual production of difference intrinsic to the constitution of events, whether they are physical or other.[32] If we consider becoming as "pure movement" between particular events, as Deleuze and Guattari suggest,[33] we can trace "bison becoming" as relational to the events that triggered and changed bison and their movements in this particular story. However, we will not speak here of bison becoming as such or some inherent characteristics of bison themselves. Instead we will discuss people's modes of relating to these animals. To illustrate this, using the examples from Pape and elsewhere, we will employ Deleuze and Guattari's idea of there being three kinds of animals of which only the third kind is capable of becoming.[34]

1. First, there are individuated animals, family pets, sentimental ones that invite us to regress and draw us into narcissistic contemplation.[35] These were bison during the phase of their transfer to Latvia, the ones carefully selected and transported, the ones people were willing to see and feed in their enclosure. They occupied approximately 70 ha of territory covered partly by meadows and

---

30 Kuemmerle et al., p. 842.
31 McCormack, Derek P.: "Becoming". In: Kitchin, Rob / Thrift, Nigel (eds.): *International Encyclopaedia of Human Geography*. Elsevier: Amsterdam 2009, p. 277.
32 Stagoll, Cliff: "Becoming". In: Parr, Adrian (ed.): *The Deleuze Dictionary*. Edinburgh University Press: Edinburgh 2005, p. 21.
33 Ibid.
34 Deleuze / Guattari, pp. 279–284.
35 Deleuze / Guattari, p. 281.

partly by forest, and because of the latter, they weren't always visible to observers. Gradually adapting to the place and people, bison became more secure with coming out of the wooded area, yet still keeping their distance from visitors. "The bison, although they arrived only this year, slept calmly in the distance. It was a rainy October day, but our group watched those unusual animals with great pleasure and admiration", wrote a local newspaper.[36] Eventually, becoming unalarmed by people, the bison even approached the fence with visitors behind it, "sometimes we even tried to feed them bread from our hands…" recalls a local woman, who years later came face to face with them in her backyard. The local and even national media followed the implementation of the project, scrupulously describing the bison's adaptation to people and place. Even when they were already outside the fence, roaming quietly in the fields, sleeping for hours in the backyards of some farmsteads, some locals regarded them with compassion, as disoriented animals that were the part of a project that failed because of human negligence.

Currently bison in captivity in Latvia are encountered only in two privately owned parks in theVidzeme region: the Zemitāni deer park (a deer farm and safari park), which in its 300 ha area holds a large number of animals, among which there are more than 1200 deer, but also mouflons, wild boar, and European bison. However, of 17 bison in the beginning of the 2010s only one is left there for various reasons, among which, as park manager K. Vasiļonoks notes, prevailed territorial management and breeding issues. This free-range safari park on the one hand is as a fairy forest – that's how the official site of the Latvia Travel Agency (2019) describes it; on the other – it is a hunting park (*Hirschgarten*), similar to the ones there were in the manorial times: the hunter's club Hiršenhofa (a word play with German *Hirsch* and *Hof* would mean a manor of deer) is also established there. Three bison from Zemitāni are now owned by a private estate, perhaps of similar purposes to the ones during the manorial time. While on Līgatne Nature Trail (an open-plan zoo of indigenous fauna) of Gauja National Park a pair of bison were brought there from Białowieża during the Soviet period in 1984, they eventually died of natural causes (the last one died in 2012). Now, the park represents only indigenous Latvian fauna and, regardless of requests from visitors, they have abandoned the idea of reintroducing bison there again. This represents the official position of the state, that is, the Nature Conservation Agency.

---

36 Jansone, Zigrīda: "Sen izmirušo dzīvnieku pēcteči". [Descendants of Long Extinct Animals.] *Saldus Zeme*, 30.10.2004, p. 5.

2. The second kind of animal is the one with characteristics or attributes; genus, classification; or animals as they are treated in the great divine myths.[37] For example, they might serve a purpose for science or ideas. In this sense, bison (as well as tarpans and auroxen) represent the idea of the restoration of the pre-agricultural landscape, as well as a particular trademark for WWF-Latvia as an institution understanding and supporting the Western narratives of reintroduction, biodiversity and wilderness. But for Latvia it meant even more than that. It was an attempt to question and dispute landscape developmental paths, based on conventional yet contestable ethnoscape values, particularly in depopulated areas. As Schwartz described, it was "the first step in teaching Latvians to look beyond the limits of the agrarian paradigm and to see relatively 'wild' nature as a new kind of developmental asset". Indeed, since 1999 such wilderness territories have increased in ranging from one to eight in various places in Latvia[38]. However, the predominant attitude of the public and experts have remained the same, favouring the icons of the ethnoscape, as a result of which even now no constructive dialogue is possible. And this is not only about eco-tourism or western nature values, but also about changes in the whole system of the function and meaning of marginalised landscapes in Latvia.

As the Deleuzian second kind of animal, the bison is also the state animal, and here we must speak particularly about Lithuania, Poland and the former Russian Empire and nowadays also Belarus with bison as the national state animal of Belarus. After all, as Samojlik has put it, bison, due to the monarchs' protection, has become "a living relic of the time of Kings".[39] Schama made an apt account on the bison's historical meaning for the Lithuanian–Polish realm at the time, writing, "the bison was as important to the Lithuanian–Polish cult of knighthood as the bull was to the Spanish warrior caste".[40] In the significant poem of Hussovianus (also Hussowski – a poet and catholic priest of the Grand Duchy of Lithuania), *A Song about the Appearance, Savagery and Hunting of the Bison*, dedicated to Pope Leo X to satisfy his desire to know more about European bison,[41] the animal, writes Schama,[42] "was depicted as a miraculous

---

37 Deleuze / Guattari, p. 281.
38 Reķe, Agnese / Zariņa, Anita / Vinogradovs, Ivo: "Management of Semi-Wild Large Herbivores' Grazing Sites in Latvia". *Environment. Technology. Resources. Proceedings of the 12th International Scientific and Practical Conference*, Vol. 1, 2019, pp. 241–244.
39 Samojlik, p. 87.
40 Schama, pp. 37–53, p. 41.
41 Bonda, Moreno: *History of Lithuanian Historiography*. Vytautas Magnus University: Kaunas 2013, p. 28.
42 Schama, p. 41.

relic of a presocial, even prehistoric past – a tribal, arboreal world of hunters and gatherers, at the same time frightening and admirable". But even more than that, the whole poem was, regardless of its title, in fact, a historical and social account about the ancient Lithuanian state and its regents.[43] After all, another common name for European bison is Lithuanian bison.[44]

European bison is also a species on the IUCN Red List – declared as vulnerable species with the currently increasing population trend[45] – and for human–bison relationships such institutional status of bison in the wild in the surroundings of Pape settles tense situations, particularly in relation to the local hunter community. Bison that were used to people roamed freely from one farmstead to another, from a wildlife-feeding place to a field with fodder for livestock, creating unprecedented dissatisfaction for locals, and the overall project's critique from nature protection experts. Nevertheless, this set of relational transformations of the human attitude toward bison led to what many now acknowledge as the proper "bison return to Latvia's landscape".

3. The behaviour of bison in the yards of locals for about a year after their escape created numerous conflict situations that eventually ceased along with the bison retreat into the forest – the wild. For some locals, bison were then perceived as giants or devils that had escaped from fairy tales. "I have not grown up with them [...] I go out in the yard and there's huge livestock... a huge cow with hair," that's how contact with bison was described by a local woman during our first field visit in 2013. The first encounters by locals were all symptomatic of the fear and danger of something foreign and unfamiliar. Fear for children walking on their own, fear of a sudden encounter, fear for their properties, etc. Indeed, there are lots of stories about fearful, destructive and somewhat comical encounters. A boy running away from a bison on the way from his school, a man passing by an enraged bison with recently delivered offspring, trampling down lawns and croplands, gnawing at apple tree bark, licking the windows of a house, creating minor disorder in someone's barn, and other such incidents.

Yet some human kindness toward bison showed through the majority of these stories. It seemed that, in a way, they understood the world of this perplexed animal at the time, perhaps because at the time of these inquiries the bison were already out of their everyday landscapes. However, during the most heated

---

43 Bonda, p. 29.
44 e.g. Schama, pp. 37–53.
45 Olech, Wanda: Bison bonasus. *The IUCN Red List of Threatened Species*. T2814A9484719, 2018.

discussions amidst the first conflicts, local men were ready to "hunt them down immediately".[46] More than that, as it turned out, the most dissatisfied section of locals in Pape are hunter communities, because the bison partly feed at domestic wildlife feeding places, but local hunters cannot hunt them down because of their legal status, and more precisely, due to its uncertainty.

Years later quite a similar story took place in Latgale (south-eastern region of Latvia), when a solitary bison crossed the border with Belarus and roamed quietly in the local farmsteads, apparently used to people and their assistance. It sometimes followed some livestock herds, sometimes slept in gardens, and, of course, also created some disorders in the yards. People tried to keep the bison off their properties by throwing stones, brandishing sticks or driving toward it on their tractors. One of the farmers decided to take a bison in his care with the future intentions to settle this bison in an enclosure constructed in one of his forest properties.[47]

However, we can talk about Pape's bison as a third kind of animal, the "more demonic animals, pack or affect animals that form a multiplicity, a becoming, a population, a tale…",[48] only after their retreat to forest, that is, when they faced the deadly encounters with man. Lorimer et al. use the term "animals' affective atmospheres", which can be shaped by changing wider political ecological dynamics: for instance, landscapes torn by poaching can generate "traumatic" circumstances.[49] For the bison, the real trigger for becoming wild animals – avoiding people and living without their assistance – was the precedent

---

46 Pujēna, Sarmīte: "Vīri gatavi pat šaut". [Men are Even Ready to Shoot.] *Latvijas Avīze*, 21.2.2007, p. 5.
47 Jonāne, Egita T.: "Andris Mizāns paplašina saimniecību un glābj sumbru". [Andris Mizāns Enlarges the Farm and Saves a Bison] *Latgales laiks*, 3.1.2018, retrieved 10.7.2023, from http://latgaleslaiks.lv/raksti/2018-01-03-andris-mizans-paplasina-saimniecibu-un-glabj-sumbru; Blass, Roberts: "Apsver domu nošaut pierobežā klīstošo baltkrievu sumbru. Vai izdosies dumpinieku paglābt?" [It is Considered to Shoot down the Roaming Belarusian Bison at the Border Area. Will we Succeed in Saving the Rebel?]. *Bez Tabu*, 3.11.2017, retrieved 10.7.2023, from https://skaties.lv/beztabu/zverudarzs/apsver-domu-nosaut-pierobeza-klistoso-baltkrievu-sumbru-vai-izdosies-dumpinieku-paglabt/.
48 Deleuze / Guattari, p. 281.
49 Lorimer, Jamie / Hodgetts, Timothy / Barua, Maan: "Animals' atmospheres". *Progress in Human Geography* 43(1), 2019, p.31. DOI: 10.1177/0309132517731254; Bradshaw, Gay A.: *Elephants on the Edge. What Animals Teach Us about Humanity*. Yale University Press: New Haven 2009, pp. 119–156.

of poaching, and perhaps not the only one, as Lake Pape Project Manager Ints Mednis, as well as local rumours, suggest.

Bison as a third kind of animal, from the perspective of the human relation to it, is an animal that can be also hunted. The tales of bison hunting were particularly vividly described in the media during the end of nineteenth and the beginning of the twentieth centuries, when territorial links with Białowieża were sustained through the bounds of the Russian Empire. For example, a hunting magazine[50] published an account of a royal hunt in Białowieża in 1860, which was organised by the court of Alexander II involving a lot of preparations (this includes about 2000 peasants to encircle the hunting grounds); 28 bison among other animals were hunted down. This is how F. A. Glinsky depicted the bison hunt:

> [...] The howling and barking of dogs that resemble a hellish music, indicates the nearing of bison. The beaters stop at a designated spot and dead silence befalls the forest once again. But soon one can hear as if trees are cracking and branches breaking, and soon enough a huge shaggy head with two horns turns up from the thicket, then another one till the whole pack of bison appears. As if realising that the silence that came after this hellish noise does not mean anything good – the bison stop. Their bloodshot eyes inflamed with wildness. The leader of the pack lifts his head high in the air, all of his limbs and muscles are strained, and each posture shows the forest king's commitment to sell his skin at the highest cost possible. But when the bison notice the crowd around them, they suddenly decide to retreat trying to break through the row of beaters. Hundreds of shots that are charged only with powder fall upon them. Bison then rush forward again, where they receive the hunters' bullets. Bison rarely fall from one bullet [...].

Culling of bison nowadays, for example, in Poland takes place in order to regulate local bison populations and for the sake of bison habitat, mainly because in most free-living populations the European bison has no natural predators.[51] The authors also admit that the regulation of bison by selling hunting licences for the shooting of a protected species is legally and ethically ambiguous. However, as a journalist for *The Geographical* sharply remarks, although commercial hunting of bison is not currently allowed by Polish law, a quick Internet search for 'Bison hunting Europe' brings up several outfits offering hunters the chance to shoot

---

50 [No author] Zumbrs (Bos bonasus). *Mednieks un makšķernieks*, Nr. 8, 01.08.1928. retrieved 10.07.2023, from http://periodika.lv/periodika2-viewer/view/index-dev. html?lang=fr#panel:pa|issue:/p_001_mema1928n08|article:DIVL77|query:zumb ris%20zumbri%20zumbris|issueType:P.
51 Pucek, Zdzisław (ed.): *European Bison. Status Survey and Conservation Action Plan.* Gland, Switzerland and Cambridge: IUCN, 2004, pp. 1–15.

wild bison in Poland (the same can be said about Belarus).⁵² The hunting sites also display a variety of photographs with hunted down bison during these so-called trophy hunts.

In Latvia trophy-hunting of bison certainly is not on the agenda for now, nor for the near future. However, in a situation when an animal, regardless of its protective status, would become a lethal danger for locals, it can be shot immediately, without consulting with the Nature Conservation Agency. And we can only speculate what has happened to the rest of the bison from Pape's enclosure.

We will not observe Pape's bison roaming near human yards and gardens during daytime anymore, not even on adjacent roads and fields. Lodgers, for instance, may see bison only occasionally, during the early mornings at the edge of the deep woods (Figure 2), while hunters would trace their footprints around feeding places and some locals would observe them here and there by accident.

Figure 2. The pack of eight bison in the wild surroundings of Pape on an early morning of October, 2015. Photo: Gaston Lacombe, courtesy of Pasaules Dabas Fonds.

---

52 Stacey, Katie: "Europe's 'Black Rhino'. The culling of Poland's wild bison". *The Geographical*, 17.3.2017, retrieved 10.7.2023, from http://geographical.co.uk/nature/wildlife/item/2172-europe-s-black-rhino-the-culling-of-poland-s-wild-bison.

The remaining pack of five bison now roam an area of approximately 500 ha, following familiar paths, of which one is their former enclosure.

Conclusion

> *In this country the animals have the faces of animals.*
> *Their eyes flash once in car headlights and are gone.*
> *Their deaths are not elegant.*
> *They have the faces of no one.*
>
> *(Margaret Atwood)*

A landscape that is full of potential of encountering the unknown and the wild is the one we visited that grey December day. To humans, Pape's landscape inhabited by bison is not something wilder than it was before, when in particular instances human domains met with those of wild boars, wolves or elk. This, however, signifies a precedent that has changed the perspective for the potentiality of the wilderness in the landscape. Pape's landscape is still in the process of becoming a landscape with bison, with no units of measurement for understanding its success or failure. Nevertheless, the bison landscape of Pape can be regarded rather as an epiphenomenon of this wilderness project, brought into existence due to the set of relational interactions among nature potential in Pape, the Soviet landscape alterations (the polder) and current post-productivist landscape management, land owners and managers, wider politics (including the conflicting interests between nature protection and agricultural subsidies) and local interests, local residents, various interest groups (families with small children, small-scale farmers, hunters, tourists, nature protection institutions), the media and national discourses. However, since Latvia's territory is also located in such a geographical proximity to the heart of bison rewilding projects in Lithuania, Poland, and Belarus, the Latvian state must reconsider their position toward this majestic animal that once was a part of Latvia's history.

The bison precedent in Pape, in a way, rendered possible a space to be shared with, and occupied by an animal long absent in the national landscape. It also becomes a symbol of the multiple possibilities of the wilderness, changing inevitably the institutional thought embedded within the system of conventional landscape values. Although the constructed and desired wilderness as the face or idea of a new landscape in Latvia today is still inside the fence, the openings in its walls are there to be found.

We would like to close this bison story with a remark heard on Latvia's Radio 4 program Noev Kovcheg, where zoologist and nature protection expert Vilnis

Skuja, talking about the moose in Latvia that was declared the animal of the year 2019, mentioned that the moose was actually the second largest animal here. "But who is the first then?", asks the surprised journalist. "The bison," Skuja quickly answers with confidence and some kind of satisfaction in his voice.[53]

References

Balcčiauskas, Linas: "European Bison (*Bison bonasus*) in Lithuania. Status and Possibilities of Range Extension". *Acta Zoologica Lituanica. Biodiversity* 9(3), 1999, pp. 3–18.

Baerselman, Fred / Vera, Frans W.M.: *Nature Development. An Exploratory Study for the Construction of Ecological Networks*. Ministry of Agriculture, Nature and Fisheries: Den Haag 1995.

Bīrons, Anatolijs, Mugurēvičs, Ēvalds, Stubavs, Ādolfs, Šņore, Elvīra (eds.): *Latvijas PSR arheoloģija*. [Archaeology of Latvian SSR.] Zinātne: Riga 1974.

Blass, Roberts: "Apsver domu nošaut pierobežā klīstošo baltkrievu sumbru. Vai izdosies dumpinieku paglābt?" [It is Considered to Shoot Down the Roaming Belarusian Bison at the Border Area. Will We Succeed in Saving the Rebel?] *Bez Tabu*, 3.11.2017, retrieved 10.07.2023, from https://skaties.lv/beztabu/zverudarzs/apsver-domu-nosaut-pierobeza-klistoso-baltkrievu-sumbru-vai-izdosies-dumpinieku-paglabt/.

Bonda, Moreno: *History of Lithuanian Historiography*. Vytautas Magnus University: Kaunas 2013.

Bradshaw, Gay A.: *Elephants on the Edge. What Animals Teach Us about Humanity*. Yale University Press: New Haven 2009.

Cepleniece, Inese: "Latvijā atgriezīsies arī sumbri". [Also Bison Will Return to Latvia.] *Jaunā Avīze*, 30.1.2003.

Čera, Lidija: "Pogremok i drugije zeljonyje neproshennyje gosti v Latvii". [Rattle and Other Uninvited Guests in Latvia.] *Noev Kovcheg. LR4*, 26.1.2019, retrieved 10.07.2023, from https://lr4.lsm.lv/lv/lr4/peredachi/noev-kovcheg/.

Deinet, Stefanie et al.: *Wildlife Comeback in Europe: The Recovery of Selected Mammal and Bird Species. Final Report to Rewilding Europe by ZSL, BirdLife International and the European Bird Census Council*. ZSL: London 2013.

Deleuze, Gilles / Guattari, Félix: *A Thousand Plateaus. Capitalism and Schizophrenia*. Bloomsbury Academic: New Delhi 2014 [1987].

---

53 Čera, Lidija: "Pogremok i drugije zeljonyje neproshennyje gosti v Latvii" [Rattle and Other Uninvited Guests in Latvia]. *Noev Kovcheg. LR4*, 26.1.2019, retrieved 10.07.2023, from https://lr4.lsm.lv/lv/lr4/peredachi/noev-kovcheg/.

Jansone, Zigrīda: "Sen izmirušo dzīvnieku pēcteči". [Descendants of Long Extinct Animals.] *Saldus Zeme*, 30.10.2004.

Jonāne, Egita T.: "Andris Mizāns paplašina saimniecību un glābj sumbru". [Andris Mizāns Enlarges the Farm and Saves a Bison.] *Latgales laiks*, 3.1.2018, retrieved 10.07.2023, from http://latgaleslaiks.lv/raksti/2018-01-03-andris-mizans-paplasina-saimniecibu-un-glabj-sumbru.

Kibiša, Artūras et al.: "Impact of Free-Ranging European Bison to Ecosystems in Fragmented Landscape, Lithuania". *Balkan Journal of Wildlife Research* 4(2), 2017, pp. 18–25. DOI:10.15679/bjwr.v4i2.55.

Kuemmerle, Tobias et al.: "Predicting Potential European Bison Habitat across its Former Range". *Ecological Applications* 21(3) 2011, pp. 830–843. DOI:10.1890/10-0073.1.

Lorimer, Jamie / Hodgetts, Timothy / Barua, Maan: "Animals' Atmospheres". *Progress in Human Geography* 43(1), 2019, pp. 26–45. DOI:10.1177/0309132517731254.

McCormack, Derek P.: "Becoming". In: Kitchin, Rob / Thrift, Nigel (eds.): *International Encyclopaedia of Human Geography*, Elsevier: Amsterdam 2009, pp. 277–281.

Morris, Desmond: *Bison*. Reaktion Books: London 2015.

Olech, Wanda: "Bison bonasus". *The IUCN Red List of Threatened Species*. T2814A9484719, 2018.

Pucek, Zdzisław (ed.): *European Bison. Status Survey and Conservation Action Plan*. Gland, Switzerland and Cambridge: IUCN, 2004.

Pujēna, Sarmīte: "Vīri gatavi pat šaut". [Men are Even Ready to Shoot.] *Latvijas Avīze*, 21.2.2007.

Reķe, Agnese / Zariņa, Anita / Vinogradovs, Ivo: "Management of Semi-Wild Large 'Herbivores' Grazing Sites in Latvia". *Environment. Technology. Resources. Proceedings of the 12th International Scientific and Practical Conference*, Vol. 1, 2019, pp. 241–244.

Ricciuti, Edward R.: *To the Brink of Extinction*. Holt, Rinehart &Winston: New York / Chicago / San Francisco 1973.

Samojlik, Tomasz: *Bialoweza Forest in the Time of Kings. Conservation and Hunting*. Mammal Research Institute, Polish Academy of Science: Bialoweza 2005.

Schama, Simon: *Landscape and Memory*. Vintage books: New York 1995.

Schwartz, Katrina Z.S.: "Wild Horses in a 'European Wilderness': Imagining Sustainable Development in the Post-Communist Countryside". *Cultural Geographies* 12(3), 2005, pp. 292–320. DOI:10.1191/1474474005eu331oa.

Skultans, Vieda: "Narratives of Landscape in Latvian History and Memory". *Landscape Review* 7(2), 2001, pp. 25–39.

Šmits, Pēteris (ed.): *Latviešu pasakas un teikas 4*. [Latvian Tales and Legends 4.] Valters un Rapa: Riga 1927.

Stacey, Katie: "Europe's 'Black Rhino'. The Culling of Poland's Wild Bison". *The Geographical*, 17.3.2017, retrieved 10.07.2023, from http://geographical.co.uk/nature/wildlife/item/2172-europe-s-black-rhino-the-culling-of-poland-s-wild-bison.

Stagoll, Cliff: "Becoming". In: Parr, Adrian (ed.): *The Deleuze Dictionary*. Edinburgh University Press: Edinburgh 2005, pp. 21–23.

"Sumbrs". *Latweeschu Awises* No. 44, 28.10.1854, retrieved 10.07.2023, from http://periodika.lv/periodika2-viewer/view/index-dev.html?lang=fr#panel:pa|issue:/p_001_laav1854n44|article:DIVL30|issueType:P.

Tišheizere, Edīte: "Savvaļas zirgiem pievienosies citas dzīvas radības". [Other God's Creatures Will Join the Wild Horses.] *Kursas Laiks,* 4.7.2002.

Vlasakker, Joep W.G. van de: *Evaluation of the Hydrological Regime Lake Pape. FNC Report No. 2006.001*. 2006, retrieved 10.07.2023, from http://ec.europa.eu/environment/life/project/Projects/index.cfm?fuseaction=home.showFile&rep=file&fil=LAKE_PAPE_Hidrology_study_final.pdf.

Vīksne, Imants: *"Ministrija sumbrus neganīs"*. [Ministry Will not Shepherd the Bison.] *Neatkarīgā Rīta Avīze Latvijai*, 24.11.2009.

Vīksne, Imants: *"Sumbri Papē uzvedas kā huligāni"*. [Bison Behave as Hooligans in Pape.] *Neatkarīgā Rīta Avīze Latvijai*. 14.11.2009.

Zariņa, Anita / Treija, Dārta / Vinogradovs, Ivo: "Bison in the Latvian Ethnoscape: A Contingency of (not)Becoming". *Letonica* 44, 2022, 153–168. DOI:10.35539/LTNC.2022.0044.A.Z.D.T.I.V.0009.

Zariņa, Anita / Vinogradovs, Ivo / Šķiņķis, Pēteris: "Towards (Dis)Continuity of Agricultural Wetlands. Latvia's Polder Llandscapes After Soviet Productivism". *Landscape Research* 43(3), 2018, pp. 455–469. DOI:10.1080/01426397.2017.1316367.

Zemberga, Kaija: "Sumbri atgriežas: par un pret". [Bison Return: For and against.] *Lauku avīze*, 14.03.2003, p. 37.

Zumbrs (Bos bonasus). *Mednieks un makšķernieks*, Nr. 8, 01.08.1928. retrieved 10.07.2023, from http://periodika.lv/periodika2-viewe-r/view/index-.

Eve Rannamäe / Anneli Ärmpalu-Idvand

Estonian Aboriginal Sheep in Modern History: Description, Importance, and the Story of Becoming the Kihnu Native Breed

Abstract: The history of sheep husbandry in Estonia can be traced back for at least three thousand years. While in the past sheep were one of the main sources of subsistence, sheep farming today is rather small-scale and of minor importance economically. However, there is a growing interest in native sheep breeds and their products, such as meat, wool and skin. Furthermore, traditional ways of managing sheep and their products induce interest in the history of local types of sheep and their husbandry. In Estonia, there is an acknowledged native sheep breed – the Kihnu native sheep, who originates from the small island of Kihnu in the Gulf of Riga. In this paper we explore the recent history of the aboriginal sheep in Estonia, starting with the first descriptions deriving from the eighteenth century, moving on to the challenging times of extensive breed improvement in the twentieth century, and finally coming to the recognition and conservation of the Kihnu native sheep breed. We focus on the animals themselves and their exterior, but also on the genetic significance, ecological efficiency, and cultural value they carry.

Keywords: *Ovis aries*, traditional animal husbandry, Northern European short-tailed sheep, Kihnu Native Sheep Breeders Society

Introduction

Sheep are animals that everyone knows, but few think of in detail. They are usually seen grazing in the distance as some uniform set of wool and mutton resources, referred to in religious contexts, featured in bedtime accessories, or associated with foolishness. Yet, the meaning of sheep is much wider. They occur in countless parts of human cultures and societies, and their products form important everyday consumables. It is inarguable that sheep husbandry has influenced human history to a great extent during the last eleven thousand years, or in other words, since the domestication of wild Asiatic mouflon, *Ovis orientalis*, in the rocky outcrops of southwest Asia.[1] In the following few thousand

---

1   We are grateful to Anti Lillak, Kristi Umalas, and Kairi Kaelep (Estonian National Museum) for their help with the photo archive; and Aivar Kriiska (University of Tartu) for the information on the earliest settlement on Kihnu. This project has received funding from the European Union's Horizon 2020 research and innovation programme

years, humans transformed those great beasts with large horns and rough fur coats to obedient wool-producing livestock. Shortly after domestication, people started to spread sheep all over the world: Europe, Asia, Africa, and later on to the Americas and Australia. We are not certain what kind of sheep were those initially dispersed, but probably they resembled more their wild ancestors than the woolly animals we know today. Their initial dispersal across Europe has been referred to as the first wave.[2] Despite the very different environmental conditions that the animals faced they adapted everywhere – better in some places, less so in others. It took a long time, around five to six thousand years since the initial domestication, for the sheep to arrive in northern parts of Europe. Around the same time in south-western Asia, they were already being selected for wool, and a new wave of wool sheep started to disperse across Europe and replace the initial "primitive" populations.[3] Researchers have suggested that the second wave did not reach the peripheries as strongly – the consequences of which we see today in two groups that are considered the descendants of the first migration.[4] One of these are the feral Mouflons on the Mediterranean islands, which have shown to be successors of the once domesticated sheep, but transformed back to wild. The other are northern European short-tailed sheep, which are a group of domestic breeds.

Northern European short-tailed sheep are rather distinct from other breeds, both phenotypically and genetically. They prevail in the northern fringe of Europe, from Russia to Iceland across the Baltic and North Atlantic Sea regions.[5]

---

under Marie Skłodowska-Curie grant agreement No. 749226, and Estonian Research Council grants IUT20-7 and PRG29.
Zeder, Melinda A.: "Out of the Fertile Crescent: The dispersal of domestic livestock through Europe and Africa". In: Boivin, Nicole / Crassard, Rémy / Petraglia, Michael (eds.): *Human Dispersal and Species Movement From Prehistory to the Present*. Cambridge University Press: Cambridge 2017, pp. 261–303, p. 266.

2   Chessa, Bernardo / Pereira, Filipe / Arnaud, Frederick et al.: "Revealing the History of Sheep Domestication Using Retrovirus Integrations". *Science* 324(5926), 2009, pp. 532–536.
3   Bender Jørgensen, Lise / Rast-Eicher, Antoinette: "Fibres for Bronze Age Textiles". In: Bender Jørgensen, Lise / Sofaer, Joanna / Stig Sørensen, Marie Louise (eds.): *Creativity in the Bronze Age: Understanding Innovation in Pottery, Textile, and Metalwork Production*. Cambridge University Press: Cambridge 2018, pp. 25–36, p. 26; Chessa et al. 2009, p. 532.
4   Chessa et al. 2009, p. 532.
5   Dýrmundsson, Ólafur R. / Niżnikowski, Roman: "North European Short-Tailed Breeds of Sheep: A Review". *Animal* 4(8), 2010, pp. 1275–1282, p. 1276.

They are considered "primitive" – a term related to the inheritance of the first migration wave and not being significantly affected by the later improvement. The Soay breed in Scotland is believed to be a true relic from as early as the Bronze Age, and clearly linked to the Mediterranean and Asiatic mouflon.[6] Another example, the Icelandic sheep, is believed to have survived rather intact for the last eleven hundred years, from the time it was introduced to the island by the Viking settlers.[7] There are more than thirty short-tailed breeds, all characterised by small hardy body build, short tail, double-layered coat, primitive fleece structure, moderately fine wool, natural moult in the spring, range of colour from black and brown to grey and white, prolificacy, and strong maternal instinct.

Estonian aboriginal sheep have often been considered being part of the northern European group, because they share the same characteristics and geographical range, and they have shown to be genetically distinct from modern breeds. Today, the Kihnu native sheep breed represents the aboriginal population in Estonia. Like other "primitive" breeds, their importance lies in their ancestry, genetic resource, ecological efficiency, and socio-cultural value. In this paper, we take a closer look at Estonian aboriginal sheep and the Kihnu native breed, and discuss their origin and importance. We see that this is a story with missing facts and controversial opinions. However, the matter has been and is continuously investigated with several methodological approaches, including scientific analyses, historical observations, traditional knowledge, and personal experience by the breeders. So far, the evidence points to a direction of Kihnu native sheep being the true descendant of the aboriginal population, distinct from modern breeds, and having affinities to the northern European short-tailed group.

Archaeological Evidence

To start from the beginning, we turn back to early evidence. The first traces of domestic sheep in northern Europe occurred around four to six thousand years ago.[8] These findings are often problematic due to contextual and preservation

---

6   Ryder, Michael L.: *Sheep and Man*. Duckworth: London 1983, p. 522; Chessa et al. 2009.
7   Dýrmundsson, Ólafur R.: "Leadersheep: the Unique Strain of Iceland Sheep". *Agricultural Science and Technology* 32, 2002, pp. 45–48, p. 45.
8   Bläuer, Auli / Kantanen, Juha: "Transition from Hunting to Animal Husbandry in Southern, Western and Eastern Finland: New Dated Osteological Evidence". *Journal of Archaeological Science* 40, 2013, pp. 1646–1666, p. 1646.

issues. In Estonia, scarce findings from around 2700 BCE in the Late Neolithic seem to show only brief contact – perhaps related to imported bone items or failed attempts to rear livestock.[9] Not until two thousand years later, from the Late Bronze Age onward, comes a clear indication of laying the main subsistence strategy on livestock and agriculture: the predominance of domestic animal remains in archaeological material.[10]

Faunal remains provide primary information about past animal husbandry. However, the material possesses many problems and deficiencies, leaving the interpretations almost always open. In Estonia, sheep are usually the second most abundant domestic animal in osteological assemblages after cattle. Sheep were used for meat (evidenced by butchering marks on the bones and the presence of food and cooking waste) and wool (the presence of archaeological textile finds and the dominance of adult individuals in the bone assemblage), as well as for other products (like horn and bone items, skins, furs, and threads). Regarding the age structures and kill patterns, most of the flock usually comprised females, while males were for breeding purposes and thus slaughtered at an early age. We know little about the appearance of the past sheep. According to skeletal material, sheep were horned, polled, or scurred, and around 50–70 cm tall in withers.[11] Medieval textiles have shown that they had rather coarse wool and great variety in fibres.[12]

Although skeletal evidence shows extensive use for meat, the most important product from sheep was wool. Elsewhere in Europe, from around the twelfth and thirteenth centuries, Britain and Spain were building upon an outstanding wool market, influencing the development of finewool breeds (including the best-known Merino), and basing their whole economy on the wool and textile

---

9   Lõugas, Lembi / Kriiska, Aivar / Maldre, Liina: "New Dates for the Late Neolithic Corded Ware Culture Burials and Early Animal Husbandry in the East Baltic Region". *Archaeofauna* 16, 2007, pp. 21–31.
10 Ibid.
11 Rannamäe, Eve / Lõugas, Lembi: "Animal Exploitation in Karksi and Viljandi (Estonia) in the Late Iron Age and Medieval Period". In: Pluskowski, Aleksander G. (ed.): *The Ecology of Crusading, Colonisation and Religious Conversion in the Medieval Eastern Baltic: Terra Sacra II*. Brepols: Turnhout 2019.
12 Rammo, Riina: *Tekstiilileiud Tartu keskaegsetest jäätmekastidest: tehnoloogia, kaubandus ja tarbimine / Textile Finds from Medieval Cesspits in Tartu: Technology, Trade and Consumption*. (Dissertationes Archaeologiae Universitatis Tartuensis 4). University of Tartu Press: Tartu 2015, pp. 133–134.

industries – processes that continued up until the eighteenth century.¹³ In Estonia, local sheep seem to have been the main wool source. In prehistoric times, the occasional exchange of wool or textiles might have occurred, for example during Viking Age trade and migrations, but the first evidence comes only from the Middle Ages. In medieval Livonia (present-day Estonia and Latvia), the processes of the western European wool market were expressed in imported fabrics (not wool or live animals) within a trade network of the Hanseatic League.¹⁴ However, while western textiles were favoured among urban inhabitants, local wool prevailed in the rural areas.¹⁵ Therefore, it seems that innovations and large-scale wool production in the west did not reach the easternmost part of medieval Europe to the same extent. Even more, after the thirteenth-century Baltic crusades, despite the degree of social polarisation between the colonists and the colonised, the economic basis in Livonia seems to have remained consistent and new power centres became dependent on the indigenous population.¹⁶ This means that most of the local farmers remained intact, continued animal husbandry based on the existing animals, and there was no (extensive) livestock movement or trade. This consistency has been supported by ancient DNA studies, showing some of the maternal sheep lineages to have continued from before the crusades to the Middle Ages and Early Modern Period.¹⁷

---

13 Albarella, Umberto: "Size, Power, Wool and Veal: Zooarchaeological Evidence for Late Medieval Innovations". In: De Boe, Guy / Verhaeghe, Frans (eds.): *Environment and Subsistence in Medieval Europe*. Institute for the Archaeological Heritage of Flanders: Brugge 1997, pp. 19–30; Ryder 1983, pp. 426, 447, 456.
14 Rammo 2015, pp. 135–136.
15 Ibid., p. 64, 70.
16 Pluskowski, Aleksander / Valk, Heiki: "Conquest and Europeanisation: the Archaeology of the Crusades in Livonia, Prussia and Lithuania". In: Boas, Adrian (ed.): *The Crusader World*. Routledge: London / New York 2016, pp. 568–592; Pluskowski, Aleksander / Makowiecki, Daniel / Maltby, Mark et al.: "The Baltic Crusades and Ecological Transformation: The Zooarchaeology of Conquest and Cultural Change in the Eastern Baltic in the Second Millennium AD". *Quaternary International* 510, 2019, pp. 28–43.
17 Rannamäe, Eve et al.: "Maternal and Paternal Genetic Diversity of Ancient Sheep in Estonia from the Bronze Age to the Post-Medieval Period, and Comparison with Other Regions in Eurasia". *Animal Genetics* 47(2), 2016a, pp. 208–218; Rannamäe, Eve et al.: "Three Thousand Years of Continuity in the Maternal Lineages of Ancient Sheep in Estonia". *PLoS ONE* 11(10), 2016b, e0163676.

First Descriptions and the Beginning of Breed Improvement

Starting from the Early Modern Period, zooarchaeological material becomes less informative, mostly due to a lack of interest in collecting and studying such recent bones. Although documentary evidence becomes available from the seventeenth and eighteenth centuries, this is scarce as well. To better study those few hundred years of rich and complicated history of livestock husbandry, both sources would need more attention and systematic research.

Early documentation about imported animals of British and Spanish origin comes mainly from seventeenth and eighteenth-century manors, while any reports about trading with local sheep are almost non-existent.[18] The first descriptions of local sheep come from the end of the eighteenth century. According to a Baltic-German official and historian Wilhelm Christian Friebe,[19] there were two kinds of sheep present in Estonia at that time: the "German" and the "peasant" sheep. The German sheep in manors were slightly bigger and had shorter, finer, and whiter wool. Occasionally they were evaluated to as high standards as English finewool. The more abundant aboriginal sheep were farmed by peasants, and had black and coarse wool to produce dark-coloured clothes and furs. Interestingly, Friebe noted that more sheep were kept on Saaremaa Island where they provided better wool than on the mainland – possibly because of grazing by the sea and consuming salty seawater. Whether scientific observations would support the latter statement remains open in the framework of this paper.

As merino wool was much valued on the global market in the nineteenth century, manor holders in Estonia, starting from 1824, also put more focus on importing merinos.[20] However, despite the efforts over several decades, a local finewool breed was never developed. Merinos just did not adapt in the local climate and environment, and did not develop resistance against invasion and infection diseases. At the end of the century, the main direction was instead

---

18 Rannamäe, Eve: *Development of Sheel Populations in Estonia as Indicated by Archaeofaunal Evidence and Ancient Mitochondrial DNA Lineages from the Bronze Age to the Modern Period*. (Dissertationes Archaeologiae Universitatis Tartuensis 6). University of Tartu Press: Tartu 2016, pp. 25–26.
19 Friebe, Wilhelm C.: *Physisch-ökonomische und statistische Bemerkungen von Lief- und Ehstland oder von den beiden Statthalterschaften Riga und Reval*. Hartknoch: Riga 1794, pp. 157–159, 298–299.
20 Viinalass, Haldja / Värv, Sirje / Piirsalu, Peep et al.: "Teadmata päritoluga lammastest". *Tõuloomakasvatus* 4, 2006, pp. 20–23, p. 20.

taken toward meat productive sheep breeds and more profitable dairy cattle husbandry.[21]

Challenges of the Twentieth Century

In the 1930s, Estonian Whitehead and Blackhead sheep were developed by crossing the aboriginal animals with imported meat breeds.[22] However, despite the overall direction of new and more profitable sheep husbandry, aboriginal animals were still valued. They were described as hardy, adapted, resistant to diseases and parasites, prolific, with good wool and skin and presenting characteristics like horns, shorter tail, and double-layered wool.[23] In 1937, a survey to count the remaining aboriginal sheep showed their numbers to still be sufficient.[24] However, they were not regarded as profitable compared to the more productive crossbreds, they were unattractive for the farmers, and gained no support from the state.[25] At the time when breed books were established for other native breeds (e.g. 1914 Estonian native cattle, 1921 Estonian native horse, 1932 Estonian landrace pig, 1942 Estonian Blackhead and Whitehead sheep),[26] it was never established for the aboriginal sheep.

The Second World War and collectivisation were devastating for sheep husbandry in general: large numbers of sheep perished, and the remaining were not properly bred and were suffering from inbreeding and admixture.[27] Sheep populations continued to decrease in the following decades, mostly because of more productive dairy cattle husbandry.[28] The number of aboriginal sheep was decreasing rapidly: in the 1950s, many of them were still noted in western Estonia, the islands, and south-eastern Estonia, but by the end of the 1990s they

---

21 Jaama, Kristjan: *Eesti tumedapealine lambatõug*. Eesti Riiklik Kirjastus: Tallinn 1959, p. 24.
22 Piirsalu, Peep: *Lambakasvatus I*. Tartumaa Põllumeeste Liit: Tartu 2012, pp. 18–19.
23 Jaama 1959, p. 19; Kallit, Otilie: *Lambakasvatusest*. Eesti Põllumeeste Keskselts: Tallinn 1924, p. 6.
24 Viinalass / Värv / Piirsalu et al. 2006, p. 20.
25 loc. cit.
26 *Republic of Estonia Country Report on the State of Farm Animal Genetic Resources, Ministry of Agriculture 2004*, retrieved 17.1.2019, from http://www.fao.org/docrep/pdf/010/a1250e/annexes/CountryReports/Estonia.pdf.
27 Viinalass / Värv / Piirsalu et al. 2006, p. 20.
28 Porga, M.: *Lambakasvatus. Konspekt*. Eesti NSV Põllumajanduse Ministeeriumi Informatsiooni ja Juurutamise Valitsus: Tallinn 1979, p. 3, 8.

had only survived in peripheries like the islands and coastal regions.[29] By the 2000s, aboriginal sheep were best preserved only on Kihnu Island thanks to the geographic separateness and cultural conservatism.

Kihnu Island as a Cultural Periphery

Kihnu is a small (16.9 km^2) island in the Gulf of Riga. The first evidence of human activity on the island comes from around three thousand years ago, when it seems to have served as an occasional stopping point for seal hunters and fishermen.[30] In written history, the island was first mentioned as *Kyne* in 1386 and occasionally as *Kyna* in the 1500s.[31] Now there are four villages and around 700 people on the island.[32] In 2003, Kihnu's cultural space (along with the neighbouring island Manija) was listed on the UNESCO Representative List of the Intangible Cultural Heritage of Humanity for its songs, games, dances, wedding ceremonies, and handicraft.[33] An important feature of the culture is that while Kihnu men were fishing out on the sea, women took responsibility for all household work, including farming and animal husbandry.

Kihnu culture is very much defined by its separateness, solidarity, and conservatism. Life on the island made innovations very slow, because everything new turned out to be unsuitable in local conditions and not able to compete with traditions. Kihnu inhabitants kept to their own things, values, and people, that is, to a community that had developed over hundreds of years and had proven trustworthy. Anything unknown was regarded as a danger or failure. These traditions reach into the present day. Despite intertwining with the modern world, old customs and practices have remained important for the local inhabitants and are followed on a daily basis (Figure 1).

---

29 Roosve, Hardi-Erik / Teinberg, Rein (eds.): *Eesti põllumajandusentsüklopeedia, I köide*. Eesti Entsüklopeediakirjastus: Tallinn 1998, entry *"eesti maalammas"*.
30 Kriiska, Aivar: "Noppeid Eesti väikesaarte esi- ja varaajaloost". In: Vunk, Aldur / Martsik, Katrin (eds.): *Artiklite kogumik 3. Pärnumaa ajalugu. Vihik 5.* Pärnu Maavalitsus, Pärnu Muuseum: Pärnu 2002, pp. 12–15; Kriiska, Aivar / Lõhmus, Mari: "Archaeological excavations on Mõisaküla settlement site in Kihnu". *Arheoloogilised välitööd Eestis 2003*, 2004, pp. 132–136, p. 135.
31 Päll, Peeter / Kallasmaa, Marja (eds.): *Eesti kohanimeraamat*. Eesti Keele Sihtasutus: Tallinn 2016, pp. 199–200.
32 *Kihnu parish*, 10.07.2023, from http://kihnu.kovtp.ee/.
33 *UNESCO: Intangible cultural heritage*, retrieved 10.07.2023, from https://ich.unesco.org/en/RL/kihnu-cultural-space-00042.

Figure 1. Even today, the sight of someone driving a motorcycle in a striped skirt (*kört*) is not unusual on the islands of Kihnu and Manija. Here, one of the authors, Anneli Ärmpalu-Idvand, who first visited Manija in 1998, adopted the local customs and moved there shortly after, is ready to give a ride to her family members and guests, 2005. Photo: Herki Idvand.

Traditional Sheep-Keeping on Kihnu

The conservatism and closeness of the Kihnu culture is also reflected in sheep-keeping. In the last few decades, during her visits to the islands of Kihnu and Manija, Ärmpalu-Idvand has noted some insights to this mentality.

In the past, when sending sheep to spring grasslands, people put an iron item on the barn doorstep and sprinkled salt on the sheep's backs for their protection. Yet, most important and still relevant today has been protection from the evil eye. This is one reason why strangers, both humans and animals, have been unwelcome in the barn. Nevertheless, occasional exchange or trade with animals did happen, usually between relatives and neighbours. This was not to avoid inbreeding – local people did not care for pedigrees. Instead, sheep were gifted on occasions like weddings or exchanged for a certain colour of wool. People respected their animals. Therefore, sheep were not to be bargained off and monetary value was difficult to assign to them. Occasionally, interested buyers had to state the price for the animal themselves, so the current owner could see if the animal was truly valued. Sometimes the new owner even had to agree not to sell or give the animal away. In any case, the new owner had to make sure the animal would not miss its previous home. Exchanged or traded animals were selected mainly for their age, not for any specific property (except occasionally for the wool colour). This practice was one of the reasons why the variety within the island population was preserved. There was no intentional selection, and improvement did not affect the sheep to the extent on the mainland. Locals thought that crossing with imported breeds would ruin the sheep, although at the same time they were also not wealthy enough to afford a thoroughbred animal.[34]

Adaptation to the local climate and landscape has been very important on Kihnu. Both people and animals were dependent on the island's nature and weather, and everything had to be self-maintained. Also today, sheep seldom get feed supplements and medicines, except for the minerals. In the past, mineral sources were seawater in the summer and (salted) fish remains in the winter. Individual attempts to bring modern breeds to the island often ended with the animals' death, because they were not durable enough, especially against inner parasites. Local sheep had long lived on semi-natural landscapes rich in intermediate host animals for sheep inner parasites and had thus gained good resistance. To keep off outer parasites, wool was shorn twice or three times a year.[35] Before shearing, sheep were washed in the sea. Shearing was done during the waxing moon to ensure that the wool would grow back quickly and with high quality. People thought sheep could die because of too thick wool or the wool would go to waste if not shorn often enough. Moreover, unshorn and ill-treated sheep were disapproved of by other community members. For example, Kihnu people have talked about a widowed man who brought home a young mainland maiden, but she could not shear sheep without abrasions and cuts, and this caused great shame for that household for years.

Establishing the Native Breed

The cognition in the 1990s that the aboriginal sheep had gone extinct was due to lack of interest and its low profitability. Without any scientific grounds, small

---

34 Saarma, Urmas: "Eesti ja Euroopa põlislammaste lugu kahe teadusuuringu valguses". *Eesti Loodus* 10, 2009, pp. 509–513.
35 On Kihnu there is a riddle about shearing the sheep twice: "Eenämua, kaks korda aastass niidetässe?" ("A Meadow, Mowed Twice a Year?"). Laos, Külli (ed.): *Aabets*. Kihnu Kultuuri Instituut: Kihnu 2009, p. 63.

sheep with various colours were automatically referred to as inbred animals kept in bad conditions. A small group of sheep enthusiasts did not accept this opinion. They started searching for the remaining aboriginal sheep and saving them from extinction.

The first genetic-morphologic project "The origin and genetic diversity of Nordic sheep breeds" (*"Põhjamaade lambatõugude päritolu ja geneetiline mitmekesisus"*) took place in 1999–2001 within the Nordic Gene Bank for Farm Animals (NorthSheD) project. This was the first time sheep in Estonia were genetically investigated. Samples from the islands of Ruhnu, Saaremaa, and Kihnu differed from modern breeds.[36] Along with historical evidence, it seemed that aboriginal sheep had best preserved on Kihnu Island, inducing the idea of saving this population from extinction. Because the sheep were owned by elderly people who were not capable of keeping them any longer, the animals were assembled to a single conservation herd from ten different herds (by Ärmpalu-Idvand). Other farmers followed, additional preservation herds were created, pedigree documentation started, and each individual described.

The second genetic-morphologic project "Conserving native sheep in Estonia" (*"Maalamba kui põlisväärtuse säilitamine Eestis"*) in 2005–2006 was led by the Estonian Fund for Nature and funded by UNESCO. The project aimed to study the genetic diversity of native sheep with comparison to other northern European breeds. Of the studied samples, some turned out to be Estonian Darkheads and Whiteheads, but most of the individuals belonged to a distinct group of fifteen smaller populations that clearly diverged from the modern breeds.[37] Among those fifteen, ten groups were the most distant compared to other studied populations – these were the sheep from Kihnu Island. Morphological observations supported these results, as the animals highly resembled the descriptions and photos from

---

36 Grigaliūnaitė, Ilma et al.: "Microsatellite Variation in the Baltic Sheep Breeds". *Veterinarija ir zootechnika* 21(43), 2003, pp. 66–73; Tapio, Ilma et al.: "Unfolding of Population Structure in Baltic Sheep Breeds Using Microsatellite Analysis". *Heredity* 94(4), 2005, pp. 448–456; Tapio, Miika et al.: "Native Breeds Demonstrate High Contributions to the Molecular Variation in Northern European Sheep". *Molecular Ecology* 14, 2005, pp. 3951–3963.

37 Tapio, Ilma et al.: *Unfolding of Genetic Structure of Estonian Unknown Sheep based on microsatellite analysis. Final report, 11th of September 2006.* (Unpublished report to the Estonian Fund for Nature and the Integrative zoology work group of University of Tartu). Biotechnology and Food Research MTT Agrifood Research Finland: Jokioinen 2006; *Eestimaa Looduse Fond 2006.a. majandusaasta aruanne*, retrieved 10.07.2023, from http://elfond.ee/elf/aastaaruanded.

the last few hundred years. The remaining five groups from Ruhnu, Saaremaa, and Hiiumaa were smaller in numbers and their aboriginal characteristics were not as distinct. Therefore, the consensus agreement and recommendation was to protect and manage the more numerous and most diverse Kihnu population separately, as this was the one with the greatest prospects of conserving the local aboriginal sheep. In 2007, the Kihnu Native Sheep Breeders Society (hereafter KNSBS; Est. *Kihnu Maalambakasvatajate Selts*) was created. The aims of the society have been to restore the Kihnu native sheep as a breed that descends from the aboriginal sheep in Estonia and to deal with official breeding and conservation.

The third genetic-morphologic project "I phase of the conservation of Kihnu native sheep" (*"Kihnu maalamba säilitamise I etapp"*) in 2007–2008 was led by the KNSBS and funded by Enterprise Estonia[38] (Ärmpalu-Idvand 2008). The results verified that the descendants and ancestors of the Kihnu population form a native breed with uniform characteristics: genetic analysis proved the differentiation from other contemporary breeds, and morphological analysis confirmed the similarity to historical local sheep.

In the following years, the KNSBS established a breeding and conservation program and systemised the pedigree data. In 2016, the long, hard work paid off and the breed was finally officially acknowledged. Three years later, in 2019, Kihnu native sheep were listed as an endangered breed by the Estonian Ministry of Rural Affairs.

Archaeological and genetic studies contribute to the ongoing research as well. Archaeological material has shown similarities in size of the individuals and wool characteristics between the ancient and present populations.[39] Genetic research, including the studies on ancient DNA, has taken the existing maternal lineages to as far as three thousand years and shown ancestral connections to the northern European short-tailed breeds.[40] To interpret the development of populations even further and investigate the affinities between ancient and extant animals, several national and international studies are currently implemented.

---

38 Ärmpalu-Idvand, Anneli: *Kihnu maalamba geneetiliselt ja morfoloogiliselt uuritud algpopulatsioon*. (Unpublished report). MTÜ Kihnu Maalambakasvatajate Selts: Tõhela 2008.
39 Rannamäe / Lõugas 2019; Rammo 2015.
40 Rannamäe et al. 2016a; Rannamäe et al. 2016b; Rannamäe, Eve et al.: "Retroviral Analysis Reveals the Ancient Origin of Kihnu Native Sheep in Estonia: Implications for Breed Conservation". *Scientific Reports* 10(17340), 2020.

Describing the Breed

Old concepts from Kihnu that have been preserved and noted by the KNSBS have helped to interpret and confirm the continuity, distribution, and origin of the aboriginal sheep. These terms mostly relate to their appearance: *saarik* (ewe with horns), *tillu* (small sheep), *tilbad* (wattles), and *kasukanaha lammas* (fur-coat sheep).

The word *saarik*, used only for ewes with horns (not rams), is currently in use on Kihnu. Elsewhere in Estonia the word in this specific meaning is not known or remembered. Perhaps because in breed improvement the horns, troublesome and dangerous features, were not favoured in selection. On Kihnu, however, the fact that a separate word – *saarik* – came into use to describe a horned ewe must indicate an abundance of such animals. Since the word has been preserved well in today's vocabulary of the Kihnu people, horned ewes must have been common fairly recently. Also today, horns represent a rather frequent feature on the Kihnu sheep breed, occurring on around a tenth of the females.[41]

On Kihnu, sheep usually had no names. All sheep, adults and lambs alike, were just called *tillu* (tiny), referring to their small size.[42] Even today, sheep are called home "*tillu-tillu!*".

Wattles, appendages of skin at the side of the neck, are a feature associated with aboriginal populations, including other northern European native breeds. In Estonia, wattles on sheep occurred more frequently in the past, and not only on the islands, but on the mainland as well. However, mainland people had no specific word for them (or at least, it is not remembered). On Kihnu, on the other hand, the words *tilp* (singular) and *tilbad* (plural) are still strongly in use.

Fur-coat sheep – *kasukanaha lammas* – already features in nineteenth and twentieth-century literature[43] and refers to a double-layered coat. The term has been used both for description and typology, pointing out the fleece characteristics of the aboriginal sheep and the way to utilise it. A fur winter coat, tanned and sewn by women, was a regular clothing item on Kihnu. As fur

---

41 Ärmpalu-Idvand, Anneli / Loit, Urve: *KMKS tõuraamatu statistika*. Tõukomisjon 01.12.18. (Unpublished report). MTÜ Kihnu Maalambakasvatajate Selts: Tõhela 2018.
42 „Tillul ond tallõd" („*tillu* has lambs") refers clearly to *tillu* being an adult ewe. Pajusalu, Karl / Viikberg, Jüri (eds.): *Kihnu sõnaraamat*. Eesti Keele Sihtasutus: Kihnu / Pärnu / Tallinn / Tartu 2016, p. 486.
43 For example, Ivanov, Mihhail: *Lambakasvatuse kursus*. Eesti Riiklik Kirjastus: Tallinn 1952, p. 22.

coats were widely worn and sheep with suitable fleece abundant, the term was well known.

Reused concepts are the words that were neglected during the twentieth-century breed improvement, but were brought back to people's awareness by the KNSBS.[44] These include *maalammas* (native/aboriginal sheep) and *maatõug* (native breed), *kahekihiline vill* (double-layered coat, the presence of two different wool fibres, an adaptation to the environment), *algupärane kari* (original herd that is genetically and morphologically distinct from other investigated herds, maintained and utilised with traditional methods), *säilituskari* (conservation herd based on the original herds, only pure breeding is used), and *algpopulatsioon* (primary population on which the official breed was established, all purebred Kihnu sheep originate there).

New concepts were created by the KNSBS, including *kihnukas* (short for Kihnu sheep), *talle värvijoonis* (lamb's colour pattern, recorded for new-borns, specific for each individual), *värvustüüp* (there are six colour types: black with white markings, white, so-called dalmatian (black and white patches), black, white with black markings, and so-called blackneck (Est. *vaskkael*)), *ealine värvimuutus* (age related changes in wool colour, caused by individual and breed specific characteristics), and *ajutine värvimuutus* (temporary changes in wool colour, fading of the top most fibres in black sheep, a common feature for many breeds) (Figure 2).

---

44 For example, see Jaama.

Figure 2. Up: aboriginal sheep in Vorbuse village, southern Estonia, 1898 (photo: ERM Fk 354:11, Redlin). Below: Kihnu native sheep on a coastal meadow on Manija Island, 2017 (photo: Anneli Ärmpalu-Idvand). Note the similarities in the colour variety and face patterns. In the lower photo, notice the blackneck sheep in the foreground, and a dalmatian in the centre behind.

The Importance of Native Breeds

The importance of native breeds lies foremost in their genetics, ecological efficiency, high quality products, and cultural heritage.

Genetic significance is defined by their uniqueness, resistance, and adaptation. While the start of large-scale breed development around two hundred years ago led to fragmented populations and reduced genetic variability,[45] peripheral populations preserved their distinctness and are now considered important reservoirs for genetic diversity. Although rare breeds with small population size can suffer from inbreeding, they still harbour important adaptations and phenotypic variation.[46] Native breeds have been developed both by humans and nature to be most adapted to their environment: they are resistant to diseases, climate, and importantly – parasites.[47] They are also prolific and have good motherly instincts. All these qualities are increasingly being valued in modern sheep husbandry, where long-term sustainable strategies are sought. Unfortunately, many of those invaluable breeds are endangered or have already become extinct; and because of being small populations and with low productivity, they cannot compete with large profit-oriented livestock husbandry. Therefore, measures for promoting sustainable management of genetic resources prioritise endangered breeds.[48] The endangered status of the Kihnu native sheep is of high importance – it ensures state support for the breeders and for the conservation program, which emanates from the sustainable development strategy, ecological value, and precautionary principle in the conservation of biodiversity.[49]

Ecological efficiency means that native breeds are economical and environmentally friendly. They have an important role in sustainable animal farming – one of the key issues in global perspectives. Sheep husbandry has affected the landscape and shaped the environment to a great extent over history. In many cases, livestock husbandry and overgrazing have been devastating for natural habitats. However, these vast effects usually relate to mass production and modern-day husbandry strategies.[50] Smaller populations of rare and native breeds, on the contrary, help to preserve semi-natural environments

---

45 Taberlet, Pierre et al.: "Are Cattle, Sheep, and Goats Endangered Species?". *Molecular Ecology* 17(1), 2008, pp. 275–284.
46 Ibid., p. 280; Tapio, Miika et al. 2005, pp. 3961–3962.
47 Piedrafita, David et al.: "Increased Production Through Parasite Control: Can Ancient Breeds of Sheep Teach Us New Lessons?". *Trends in Parasitology* 26(12), 2010, pp. 568–573.
48 Taberlet et al. 2008, p. 282.
49 *Säästva arengu seadus*: § 9(2), retrieved 15.1.2019, from https://www.riigiteataja.ee/akt/874359; *Convention on Biological Diversity: Preamble*, retrieved 10.07.2023, from https://www.cbd.int/convention/; *Consolidated version of the Treaty on the Functioning of the European Union: Article 191 (ex Article 174 TEC) 2016*, OJ C 202, 7.6.2016, pp. 132–133, retrieved 10.07.2023, from https://eur-lex.europa.eu/legal-content/EN/TXT/?uri=CELEX:12016E191.
50 Ross, Louise C. et al.: "Sheep Grazing in The North Atlantic Region: A Long-Term Perspective on Environmental Sustainability". *Ambio* 45 (5), 2016, pp. 551–566.

Figure 3. Variety of fleece and wool colours of Kihnu native sheep and traditional spinning equipment. Photos: Anneli Ärmpalu- Idvand.

and biodiversity. Aboriginal animals have adapted to use vegetation most economically and since they do not depend on feed supplements, their waste is free of chemical residues.

Products from native breeds are considered environmentally friendly and have high quality and cultural value. Their use intertwines with natural and sustainable lifestyle. Native animals are universal – not bred for any specific purpose. In the past, people utilised the whole animal, and nothing was wasted. The KNSBS promotes the same principle by teaching people to use sheep products both in traditional and innovative ways. Meat and organs are consumed by humans and animals; horns, bones and fat are made into jewellery, tools, and household utensils; and wool and fur are irreplaceable raw materials in clothing and furnishing textiles (Figure 3). It is interesting to note that in Estonia in general, using sheep milk (unlike that of goats) has not been traditional. Already Friebe in the eighteenth century noted that.[51] There has been one exception,

---

51 Friebe, p. 159.

though: cheese from sheep's milk was made on Pakri Island at the northern coast of Estonia, but this tradition can be attributed to another cultural space – the Coastal Swedes.[52]

Cultural value lies in the history and heritage of the native sheep, implicitly linked to the UNESCO scope of safeguarding the intangible cultural heritage. KNSBS stands good for remembering old practices and spreads the mentality of traditional and natural human–animal coexistence; the "peasant mentality" that has proven to be economical, practical, and functional. Moreover, living animals, present-day husbandry, and existing traditions have an important role in understanding the archaeological and historical evidence of the populations, societies, and cultures in the past.

Conclusion

We have come a long way from sheep domestication and their first appearance in the Baltics to the extant native breeds of northern Europe and the Kihnu sheep. Clearly, the story has many facets and is missing several chapters, but based on scientific results, historical texts and photos, traditional knowledge, and personal experience by the breeders, Estonian aboriginal sheep are preserved and Kihnu native sheep are their descendants. As they have developed and adapted for hundreds of years, their valuable genetic resource and sustainability could be seen as a gift from the past – something we would lose forever if they went extinct. Along with local importance, Kihnu sheep contribute to the wider diversity and history of the northern European short-tailed breeds. While the breeding program and rising interest in traditional products continue conserving the native heritage, multidisciplinary research continues to study the past and bridge it with the present.

References

Albarella, Umberto: "Size, Power, Pool and Veal: Zooarchaeological Evidence For Late Medieval Innovations". In: De Boe, Guy / Verhaeghe, Frans (eds.): *Environment and Subsistence in Medieval Europe*. Institute for the Archaeological Heritage of Flanders: Brugge 1997, pp. 19–30.

Bender Jørgensen, Lise / Rast-Eicher, Antoinette: "Fibres for Bronze Age Textiles". In: Bender Jørgensen, Lise / Sofaer, Joanna / Stig Sørensen, Marie Louise

---

52 Ränk, Gustav: "Ühe pisirahva kurbmäng. Mälestusi ja elamusi Pakri evakueerimisepäevilt 1940. aasta kevadel". *Eesti Sõna* (77), 4.4.1942, p. 5.

(eds.): *Creativity in the Bronze Age: Understanding Innovation in Pottery, Textile, and Metalwork Production.* Cambridge University Press: Cambridge 2018, pp. 25–36.

Bläuer, Auli / Kantanen, Juha: "Transition from Hunting to Animal Husbandry in Southern, Western and Eastern Finland: New Dated Osteological Evidence". *Journal of Archaeological Science* 40, 2013, pp. 1646–1666. DOI:10.1016/j.jas.2012.10.033.

Chessa, Bernardo et al.: "Revealing the History of Sheep Domestication Using Retrovirus Integrations". *Science* 324 (5926), 2009, pp. 532–536. DOI:10.1126/science.1170587.

Dýrmundsson, Ólafur R.: "Leadersheep: the Unique Strain of Iceland Sheep". *Agricultural Science and Technology* 32, 2002, pp. 45–48. DOI:10.1017/S1014233900001541.

Dýrmundsson, Ólafur R. / Niżnikowski, Roman: "North European Short- Tailed Breeds of Sheep: A Review". *Animal* 4(8), 2010, pp. 1275–1282. DOI:10.1017/S175173110999156X.

Friebe, Wilhelm C.: *Physisch-ökonomische und statistische Bemerkungen von Lief- und Ehstland oder von den beiden Statthalterschaften Riga und Reval.* Hartknoch: Riga 1794.

Grigaliūnaitė, Ilma et al.: "Microsatellite Variation in The Baltic Sheep Breeds". *Veterinarija ir zootechnika* 21(43), 2003, pp. 66–73.

Ivanov, Mihhail: *Lambakasvatuse kursus.* Eesti Riiklik Kirjastus: Tallinn 1952.

Jaama, Kristjan: *Eesti tumedapealine lambatõug.* Eesti Riiklik Kirjastus: Tallinn 1959.

Kallit, Otilie: *Lambakasvatusest.* Eesti Põllumeeste Keskselts: Tallinn 1924.

Kriiska, Aivar: "Noppeid Eesti väikesaarte esi- ja varaajaloost". In: Vunk, Aldur / Martsik, Katrin (eds.): *Artiklite kogumik 3. Pärnumaa ajalugu. Vihik 5.* Pärnu Maavalitsus, Pärnu Muuseum: Pärnu 2002, pp. 12–15.

Kriiska, Aivar / Lõhmus, Mari: "Archaeological Excavations on Mõisaküla Settlement Site in Kihnu". *Arheoloogilised välitööd Eestis / Archaeological Fieldwork in Estonia 2003*, 2004, pp. 132–136.

Laos, Külli (ed.): *Aabets.* Kihnu Kultuuri Instituut: Kihnu 2009.

Lõugas, Lembi / Kriiska, Aivar / Maldre, Liina: "New Dates For The Late Neolithic Corded Ware Culture Burials and Early Animal Husbandry in The East Baltic Region". *Archaeofauna* 16, 2007, pp. 21–31.

Pajusalu, Karl / Viikberg, Jüri (eds.): *Kihnu sõnaraamat.* Eesti Keele Sihtasutus: Kihnu / Pärnu / Tallinn / Tartu 2016.

Piedrafita, David et al.: "Increased Production Through Parasite Control: Can Ancient Breeds of Sheep Teach Us New Lessons?". *Trends in Parasitology* 26(12), 2010, pp. 568–573. DOI:10.1016/j.pt.2010.08.002.

Piirsalu, Peep: *Lambakasvatus I*. Tartumaa Põllumeeste Liit: Tartu 2012.

Pluskowski, Aleksander / Valk, Heiki: "Conquest and Europeanisation: the Archaeology of The Crusades in Livonia, Prussia and Lithuania". In: Boas, Adrian (ed.): *The Crusader World*. Routledge: London, New York 2016, pp. 568–592.

Pluskowski, Aleksander et al.: "The Baltic Crusades and Ecological Transformation: The Zooarchaeology of Conquest and Cultural Change in The Eastern Baltic in The Second Millennium AD". *Quaternary International* 510, 2019, pp. 28–43. DOI:10.1016/j.quaint.2018.11.039.

Porga, M.: *Lambakasvatus. Konspekt*. Eesti NSV Põllumajanduse Ministeeriumi Informatsiooni ja Juurutamise Valitsus: Tallinn 1979.

Päll, Peeter / Kallasmaa, Marja (eds.): *Eesti kohanimeraamat*. Eesti Keele Sihtasutus: Tallinn 2016.

Rammo, Riina: *Tekstiilileiud Tartu keskaegsetest jäätmekastidest: tehnoloogia, kaubandus ja tarbimine = Textile Finds From Medieval Cesspits in Tartu: Technology, Trade and Consumption*. (Dissertationes Archaeologiae Universitatis Tartuensis 4). University of Tartu Press: Tartu 2015.

Rannamäe, Eve: *Development of Sheep Populations in Estonia as Indicated by Archaeofaunal Evidence and Ancient Mitochondrial DNA Lineages from the Bronze Age to the Modern Period*. (Dissertationes Archaeologiae Universitatis Tartuensis 6). University of Tartu Press: Tartu 2016.

Rannamäe, Eve et al.: "Maternal and Paternal Genetic Diversity of Ancient Sheep in Estonia From the Bronze Age to The Post- Medieval Period, and Comparison With Other Regions in Eurasia". *Animal Genetics* 47(2), 2016a, pp. 208–218. DOI:10.1111/age.12407.

Rannamäe, Eve et al.: "Three Thousand Years of Continuity in The Maternal Lineages of Ancient Sheep in Estonia". *PLoS ONE* 11(10), 2016b, e0163676. DOI:10.1371/journal.pone.0163676.

Rannamäe, Eve et al.: "Retroviral Analysis Reveals The Ancient Origin of Kihnu Native Sheep in Estonia: Implications For Breed Conservation". *Scientific Reports* 10(17340), 2020. DOI:10.1038/s41598-020-74415-z.

Rannamäe, Eve / Lõugas, Lembi: "Animal Exploitation in Karksi and Viljandi (Estonia) in the Late Iron Age and Medieval Period". In: Pluskowski, Aleksander G. (ed.): *The Ecology of Crusading, Colonisation and Religious Conversion in the Medieval Eastern Baltic: Terra Sacra II*. Brepols: Turnhout 2019, pp. 61–76.

Roosve, Hardi-Erik / Teinberg, Rein (eds.): *Eesti põllumajandusentsüklopeedia, I köide*. Eesti Entsüklopeediakirjastus: Tallinn 1998.

Ross, Louise C. et al.: "Sheep Grazing in The North Atlantic Region: A Long-Term Perspective on Environmental Sustainability". *Ambio* 45(5), 2016, pp. 551–566. DOI:10.1007/s13280-016-0771-z.

Känk, Gustav: "Ühe pisirahva kurbmäng. Mälestusi ja elamusi Pakri evakueerimisepäevilt 1940. aasta kevadel". *Eesti Sõna* (77), 4.4.1942, p. 5.

Ryder, Michael L.: *Sheep and Man*. Duckworth: London 1983.

Saarma, Urmas: "Eesti ja Euroopa põlislammaste lugu kahe teadusuuringu valguses". *Eesti Loodus* 10, 2009, pp. 509–513.

Zeder, Melinda A.: "Out of the Fertile Crescent: The Dispersal of Domestic Livestock through Europe and Africa". In: Boivin, Nicole / Crassard, Rémy / Petraglia, Michael (eds.): *Human Dispersal and Species Movement From Prehistory to the Present*. Cambridge University Press: Cambridge 2017, pp. 261–303.

Taberlet, Pierre et al.: "Are Cattle, Sheep and Goats Endangered Species?". *Molecular Ecology* 17(1), 2008, pp. 275–284. DOI:10.1111/j.1365-294X.2007.03475.x.

Tapio, Ilma et al.: "Unfolding of Population Structure in Baltic Sheep Breeds Using Microsatellite Analysis". *Heredity* 94(4), 2005, pp. 448–456. DOI:10.1038/sj.hdy.6800640.

Tapio, Ilma et al.: *Unfolding of Genetic Structure of Estonian Unknown Sheep Based on Microsatellite Analysis. Final Report, 11th of September 2006*. (Unpublished report to the Estonian Fund for Nature and the Integrative zoology work group of University of Tartu). Biotechnology and Food Research MTT Agrifood Research Finland: Jokioinen 2006.

Tapio, Miika et al.: "Native Breeds Demonstrate High Contributions To The Molecular Variation in Northern European Sheep". *Molecular Ecology* 14, 2005, pp. 3951–3963. DOI:10.1111/j.1365-294X.2005.02727.x.

Viinalass, Haldja / Värv, Sirje / Piirsalu, Peep et al.: "Teadmata päritoluga lammastest". *Tõuloomakasvatus* 4, 2006, pp. 20–23.

Ärmpalu-Idvand, Anneli: *Kihnu maalamba geneetiliselt ja morfoloogiliselt uuritud algpopulatsioon*. (Unpublished report). MTÜ Kihnu Maalambakasvatajate Selts: Tõhela 2008.

Ärmpalu-Idvand, Anneli / Loit, Urve: *KMKS tõuraamatu statistika Tõukomisjon 01.12.18*. (Unpublished report). MTÜ Kihnu Maalambakasvatajate Selts: Tõhela 2018.

Consolidated version of the Treaty on the Functioning of the European Union: Article 191 (ex Article 174 TEC) 2016, OJ C 202, 7.6.2016, retrieved 15.1.2019, from https://eur-lex.europa.eu/legal-content/EN/TXT/?uri=CELEX:12016E191.

Convention on Biological Diversity: Preamble, retrieved 10.07.2023, from https://www.cbd.int/convention/.

Eestimaa Looduse Fond 2006. a. majandusaasta aruanne, retrieved 10.07.2023, from http://elfond.ee/elf/aastaaruanded.

Kihnu parish, retrieved 10.07.2023, from http://kihnu.kovtp.ee/.

Republic of Estonia country report on the state of farm animal genetic resources, Ministry of Agriculture 2004, retrieved 10.07.2023, from http://www.fao.org/docrep/pdf/010/a1250e/annexes/CountryReports/Estonia.pdf.

Säästva arengu seadus: §9(2), retrieved 10.07.2023, from https://www.riigiteataja.ee/akt/874359.

UNESCO: Intangible cultural heritage, retrieved 10.07.2023, from https://ich.unesco.org/en/RL/kihnu-cultural-space-00042.

II. TRADING

Juhan Kreem

Horses of the Teutonic Order in Livonia: An Attempt to Map This Vital Resource for Knights[1]

Abstract: This article surveys the possibilities of studying the horse pool of the Teutonic Order in Livonia. In comparison with Prussia, the sources concerning this topic are scanty. Nevertheless, some visitation lists and intelligence reports mention horses. A rough estimation of the field cavalry of the Order from the mid-sixteenth century is around 3,000 horses. Some rare sources also make it possible to describe the horses according to their function (destriers, plough horses), gait (ambler) or appearance (coat colour). The horses were both imported as well as bred locally, although no special names for the local breed (such as *sweik* in Prussia) appear in Livonia. Among the Livonian regions it is especially the island of Saaremaa that stands out as a place where horses could be obtained.

Keywords: Teutonic Order, Livonia, administration, horses

Introduction

There is no knight without a horse. In most of the major European languages even the etymology of the word "knight" underlines this relation: *Ritter*, *chevalier*, *caballero* all denote a rider. Mounted warriors[2] were not unknown in the Baltics before the crusader conquest,[3] but the introduction of a professional heavy cavalry in the region should, however, be connected with the crusades of the early thirteenth century.[4] Cavalry remained to play an important role in

---

1 Research for this chapter was supported by the Estonian Research Council grants PRG1276 and IUT18-8. The research results in Estonian have been published as Juhan Kreem: "Saksa ordu hobused". In: Kreem, Juhan / Leimus, Ivar / Mänd, Anu / Põltsam-Jürjo, Inna: *Loomad keskaegse Liivimaa ühiskonnas ja kunstis*. Tallinn University Press: Tallinn 2022, pp. 35–47.
2 Contamine, Philipp: *War in the Middle Ages*. Blackwell: Oxford, Cambridge 1996, pp. 126, 179–184.
3 Stirrups appear in archaeological finds in Estonia as well. Further in the south of the Baltic countries horse culture was even more developed. See, e.g. Bliujien, Audronė (ed.) *The Horse and Man in European Antiquity: Worldview, Burial Rites, and Military and Everyday Life.* (Archaeologia Baltica 11). Klaipėda University: Klaipėda 2009.
4 Jensen, Kurt Villads: "Bigger and Better: Arms Race and Change in War Technology in the Baltics in the Early Thirteenth Century". In: Tamm, Marek / Kaljundi, Linda /

the warfare in the region.⁵ Taking this in consideration, it emerges as surprising how rarely the focus of research concerning medieval Livonia has been on the animals that literally made the knight. This contribution uses written evidence to map the horse pool of the Teutonic Order in Livonia, mainly in the fifteenth and sixteenth centuries, as regards both its quantitative side as well as its geographic reach. The main intention is to find out what possibilities the written records offer for this kind of study.

Due to the extensive administrative literacy of the Teutonic Order, a large body of written evidence, above all inventories and accounts, has been preserved from Prussia, which has made it possible to reconstruct in detail the local circumstances of horse-breeding, to calculate the number of horses and to establish the variety of breeds.⁶ In Livonia, virtually no comparable sources have been preserved. Consequently, the amount of historiography on horses in medieval Livonia is also much more limited. Ethnographers were the first to express their interest in the topic; their focus was, however, on draught animals and the rural, peasant context.⁷ Archaeologists have attempted to interpret bone findings, yet have reached no conclusive results because of the low number of horse bones in medieval cultural layers.⁸ The question of the location of horses in the castles of Livonia has been raised by historians of architecture, again without delivering answers.⁹ Transport workers have been a focus of interest in studying

---

Jensen, Carsten Selch (eds.): *Crusading and Chronicle Writing on the Medieval Baltic Frontier. A Companion to the Chronicle of Henry of Livonia*. Ashgate: Farnham / Burlington 2011, pp. 245–264, here pp. 256–258.

5   Ekdahl, Sven: "Horses and Crossbows: Two Important Warfare Advantages of the Teutonic Order in Prussia". In: Nicholson, Helen (ed.): *The Military Orders vol. 2, Welfare and Warfare*. Ashgate: Farnham / Surrey 1998, pp. 119–151.

6   Rünger, Fritz: "Herkunft, Rassenzugehörigkeit, Züchtung und Haltung der Ritterpferde des Deutschen Ordens". *Zeitschrift für Tierzüchtung und Züchtungsbiologie* 2, 1925, pp. 211–308; Toeppen, Max: "Über Pferdezucht in Preußen zur Zeit des Deutschen Ordens". *Altpreußische Monatsschrift* 4, 1867, pp. 681–702.

7   Viires, Ants: "Draught Oxen and Horses in the Baltic Countries". In: Fenton, Alexander / Podolák, Ján / Rasmussen, Holger (eds.): *Land Transport in Europe*. (Folkelivs studier 4). Nationalmuseet: Copenhagen 1973, pp. 428–454.

8   Maldre, Liina: "Hobune Eestis muinas- ja keskajal". In: Lang, Valter (ed.): *Loodus, inimene ja tehnoloogia. Interdistsiplinaarseid uurimusi arheoloogias*. Eesti TA Ajaloo Instituut: Tallinn 1998, pp. 203–220.

9   Kaljundi, Jevgeni: "Linnuste argielu ehk kus hoida hobuseid". In: Lass, A. (ed.): *Arhitektuuripärandi uurimisest Eesti NSV-s. 17. mail 1985. a. Tallinnas toimunud teoreetilis-praktilise konverentsi materjalid*. Valgus: Tallinn 1987, pp. 110–113.

Hanseatic history for a long time, and waggoners certainly were a vital link in merchant communication, yet horses themselves have never been under special scrutiny in this historiography.[10] In recent years various aspects of the medieval Livonian horse culture, such as the horse trade,[11] the role of horses in diplomacy and gift-giving,[12] and the medical care of horses[13] have been investigated. The horses of the Teutonic Order in Livonia, however, have not been a subject of special study until now.[14]

Numeric Data

Let us start with the establishment of some general figures. There is no complete inventory of the horses in the castles of the Teutonic Order preserved from Livonia; thus, we are confined to indirect data. The most comprehensive picture can be gained from a secret report on Livonia compiled for the Duke of Prussia in 1556 (see Map 1).[15] According to the report, the territories of the Order in Livonia were able to lead 3230 horses to the battlefield. This figure, however, also includes the horses of the local nobility, i.e. men who had obligations to

---

10 Mühlen, Heinz von zur: "Transportwesen und Reisetätigkeit der Revaler Fuhrleute". In: Pelc, Ortwin / Pickhan Gertrud (eds.): *Zwischen Lübeck und Nowgorod. Wirtschaft, Politik und Kultur im Ostseeraum vom frühen Mittelalter bis ins 20. Jahrhundert. Norbert Angermann zum 60. Geburtstag*. Institut Nordostdeutsches Kulturwerk: Lüneburg 1996, pp. 203–221; Kivimäe, Jüri: "Voorimehed suurkaupmehe teenistuses". In: Tarvel, Enn (ed.): *Eesti ajaloo probleeme. NSV TA korrespondentliikme Artur Vassara 70. sünniaastapäevale pühendatud teaduskonverentsi ettekannete teesid (18. nov. 1981)*. Eesti NSV Teaduste Akadeemia: Tallinn 1981, pp. 54–65.

11 Leimus, Ivar: "Eesti hobune – strateegiline kaup keskajal". *Tuna* 2, 2017, pp. 10–19; Leimus, Ivar: "Perdeteken – Prägezeichen? Über eine angebliche Quelle zur Dorpater Münzgeschichte". *Numismatisches Nachrichtenblatt* 6, 2014, pp. 220–221.

12 Mänd, Anu: "Horses, Stags and Beavers: Animals as Presents in Late-Medieval Livonia". *Acta Historica Tallinnensia* 22, 2016, pp. 3–17.

13 Rannamäe, Eve et al.: "A Month in A Horse's Life: Trauma Fracture and Healing Process of A Horse Metatarsus From Medieval Viljandi, Estonia". *International Journal of Paleopathology* 24, 2019, pp. 286–292. DOI: 10.1016/j.ijpp.2018.07.003.

14 Maltby, Mark et al.: "Farming, Hunting, and Fishing in Medieval Livonia: The Zooarchaeological Data". In: Pluskowski, Aleksander (ed.): *Environment, Colonization, and the Baltic Crusader States. Terra Sacra vol. I*. Brepols: Turnhout 2019, pp. 137–173, here pp. 150–154.

15 Benninghoven, Friedrich: "Probleme der Zahl und Standortverteilung der livländischen Streitkräfte im ausgehenden Mittelalter". *Zeitschrift für Ostforschung* 12, 1963, pp. 601–622, here pp. 605, 620–621.

the Order, while, on the other hand, it excludes the 90 horses that the deposed marshal of Livonia (*Landmarschall*) Jasper von Munster had taken to Lithuania.[16] Furthermore, the figure reflects field forces, and it is hardly likely that the castles were left completely without horses when field cavalry moved out. Thus we can regard these approximately 3000 horses as no more than a rough indicator of the situation at the time.

Actually, the information on the number of the knights is not much more definite. The only source that includes an almost complete number of the Teutonic brothers in Livonia is the visitation report from 1451.[17] According to it, at that time Livonian castles housed 197 knight-brothers, 27 sergeant-brothers and 43 priests. Five smaller castles do not appear on the visitation list, so it is clear that the complete number was somewhat higher, but plausibly by no more than 10 %.[18] The codified "customs" (*Gewohnheiten*) of the Order initially prescribed two horses for the ordinary knight, three for major officials, and four for the master, the marshal and the commanders;[19] in the fourteenth century the limit for an ordinary knight was set on three horses,[20] leaving a possibility for the higher officials to have more. When we take this as the basis of calculation, in 1451 the minimum number of horses would have been 600–700 to satisfy the needs of the knights only.

It is worth noticing that the visitation report of 1451 still lists some horses, altogether 73 animals in Livonia. Yet when we read that out of the 12 knights in the commandery of Kuldīga (Ger. Goldingen) only the commander himself had a horse,[21] it is clear that this cannot be the whole truth. One explanation could

---

16 Kreem, Juhan: "Netzwerke um Jasper von Munster. Der Deutsche Orden während der Livländischen Koadjutorfehde im Jahre 1556". *Ordines militares* 19, 2014, pp. 73–86, here p. 83. DOI: 10.12775/OM.2014.005

17 Biskup, Marian / Janosz-Biskupova, Irena (eds.): *Visitationen im Deutschen Orden im Mittelalter, Teil II, 1450–1519*. (Quellen und Studien zur Geschichte des Deutschen Ordens 50). N. G. Elwert: Marburg 2004, no. 147, pp. 19–40.

18 Jähnig, Bernhart: *Verfassung und Verwaltung des Deutschen Ordens und seiner Herrschaft in Livland*. (Schriften der Baltischen Historischen Kommission 16). Lit: Berlin 2011, pp. 132–133.

19 Perlbach, Max (ed.): *Die Statuten des Deutschen Ordens nach den ältesten Handschriften*. M. Niemeyer: Halle 1890, Gw 11, 42, 45.

20 Perlbach, Gw i V 4: The brothers in the convent cannot have more than three horses, the fourth should be taken away by superiors and not sold. Perlbach, A III 5: The Visitation statutes for Livonia in 1334 stipulate that the brothers may have up to three horses, the surplus must be taken away by the commander.

21 Biskup / Janosz-Biskupova 2004, no. 147, p. 20.

be that only destriers were counted. This is supported by the fact that during this visitation horses were counted along with personal armour. The number might seem too low for destriers as only less than a half of the brethren were reported to have them, but for the sake of comparison there had also been only two or three destriers in Kuldīga in 1341.[22] And in any case it should be borne in mind that the 1451 visitation report contains inconsistencies in registration; for example, there are castles in which armour has not been recorded for some reason, etc.

Another feature regarding the horses in the 1451 visitation report is that they are explicitly connected with individual knights just like private property. According to the statutes, the Teutonic brothers were meant to live without private property, but this remained a problem and was certainly so in the fifteenth century. The statutes and additional rules set limits on the number of horses that the brethren could own; furthermore, these horses could only be sold with the permission of a superior.[23] In the 1450s it was also stipulated that one should only accept those new members who already had armour and horses of their own.[24] It is thus possible that the visitation report may have listed such "private" horses. The number of the horses recorded per knight in 1451 remains below three, only some officials have more: the *Vogt* of Narva four, the *Vogt* of Cēsis nine, and the *Vogt* of Rēzekne as many as 12.[25]

In sum, the registration of the horses in the visitation reports does not provide a good basis for calculating the number of all horses in the castles and farms of the Teutonic Order. There are also very few examples that allow us to come closer to the total number of horses on a local level. The commandery of Kuldīga was inventoried twice in 1341; 118 horses were counted on the first occasion, and 146 on the second.[26] Karksi was said to have horses ready for 60 men in 1442.[27] Ten stallions and 100 mares were counted in the manors of the castle Maasilinn

---

22 Bauer, Albert (ed.): "Die Wartgutsteuerliste der Komturei Goldingen". *Mitteilungen aus der livländischen Geschichte* 25, 1933, pp. 113–194, here pp. 182, 184.
23 Perlbach, A III 8 (1334); Biskup / Janosz-Biskupova 2004, no. 145, p. 16. (1451).
24 Biskup / Janosz-Biskupova 2004, no. 138, p. 6. This particular regulation is issued for the Order in the Empire. Cf. no. 145, p. 16.
25 Biskup / Janosz-Biskupova 2004, no. 147, pp. 31, 26, 25.
26 Bauer, pp. 182, 184.
27 Biskup, Marian / Janosz-Biskupova, Irena (eds.): *Visitationen im Deutschen Orden im Mittelalter, Teil I, 1236–1449*. (Quellen und Studien zur Geschichte des Deutschen Ordens 50). N. G. Elwert: Marburg 2002, no. 110, p. 185.

in 1561.[28] With some caution, later data could also be taken into account to suggest the number of horses in particular castles. When the Polish authorities inspected the castle of Viljandi in 1599, they mentioned several stables with an overall capacity for 200 horses.[29] The same inspection recorded a stable for 100 horses near the castle of Tarvastu,[30] and even the castle of Helme had two stables for 100 horses each[31]. In comparison, the overall capacity of the stables in Malbork (Ger. Marienburg) in Prussia is estimated to have been 1000 in the fifteenth century.[32] It goes without saying that the circumstances certainly also varied over the centuries.

On Terminology

The resources of the Order were to be used commonly, and there were frequent reminders that the horses should be kept together and that fodder should also be provided for the animals of travelling brothers. The person overlooking the horse pool and in charge of distributing them for personal use was the marshal.[33] The original importance of the marshal in the Order is underlined by the fact that he came next to the Livonian master in the hierarchy. While the duties of the Livonian marshal were extensive and included acting on behalf of the master, the function of the marshal in each castle came closer to the original Germanic meaning of the word (*marahscalc*, "horse servant"). In the thirteenth to fifteenth centuries this office was held by one of the knight brothers,[34] while the situation in the sixteenth century is not that clear. In 1513 a Teutonic knight held the office of the marshal in Karksi,[35] but on a list of the personnel of the castle in

---

28 Blumfeldt, Evald: "Keskaja agraarajalugu". In: Sepp, Hendrik / Liiv, Otto / Vasar, Juhan (eds.): *Eesti majandusajalugu I*. Akadeemilise kooperatiivi kirjastus: Tartu 1937, pp. 71–72.
29 Jakubowski, Jan / Kordzikowski, Józef (eds.): *Polska XVI wieku pod wzgledem geograficzno-statystycznym. Tom XIII, Inflanty. Czesc I*. Gebethner i Wolff: Warszawa 1915, pp. 162–171.
30 Jakubowski / Kordzikowski, p. 19.
31 Jakubowski / Kordzikowski, p. 241.
32 Jóźwiak, Sławomir / Trupinda, Janusz: *Organizacja życia na zamku krzyżackim w Malborku w czasach wielkich mistrzów (1309–1457)*. Muzeum Zamkowe: Malbork 2011, pp. 488–492, here p. 492.
33 Perlbach, Gw 19, 27.
34 Fenske, Lutz / Militzer, Klaus (eds.): *Ritterbrüder im livländischen Zweig des Deutschen Ordens*. (Quellen und Studien zur Baltischen Geschichte 12). Böhlau: Köln / Weimar / Wien 1993, pp. 750–787.
35 Fenske / Militzer, no. 650, Hermann Overlacker von Wischlingen.

Viljandi dating from 1554, various marshals are named among other servants who were not members of the Order.[36] At that time, there was a field marshal (*veltmarschalck*, Johann van der Oste), a marshal of the house (*huismarschalck*, Gert van Luneborch), and a marshal of the geldings (*runenmarschalck*, Hans vann Munchen) in the main castle of Viljandi. In the other castles of the commandery a marshal of the house and a marshal of the stallions (*hengstmarschalck*) were recorded in Põltsamaa and a marshal of the foals (*volenmarschalck*) in Tarvastu. Some sixteenth-century sources also mention the marshal of the letters (*briefmarschalck*),[37] a person most likely occupied with the postal service of the Order and the horses used for this purpose. The personnel of the lower ranks in the castles were united into the Brotherhood of the Black Heads or the Stable Brotherhood (*Schwarzhäupter, Stallbrüder*) governed by their own *Vogt*.[38] Although the name "Stable Brothers" contains a hint at horses, the corporation was sooner a guild of all servants, and it is difficult to say how many of them were directly involved in taking care of the horses in the stables.

At first glance, the horse terminology is rather meagre in Livonian sources. Neglecting the rich possibilities of expressing the function, origin, breed, gender, size, age or colour of the animals, the majority of sources record the price as the only characteristic of a horse.[39] When, for example, the Tallinn town council presented a horse to the Grand Master of the Teutonic Order in 1497, it cost 80 Riga marks, whereas the price of an average riding horse would have remained between three and five marks.[40] The fact that only the price was given may have been influenced by the type of that particular source, the municipal account book. Yet also in court cases the value of the horse, i.e. the financial loss or gain, appears as the most important measurable, even when it was volatile at the same time. Thus, in 1556 Hermann von der Recke from Karksi and the councillor of

---

36 Johansen, Paul: "Ein Verzeichnis der Ordensbeamten und Diener im Gebiet Fellin 1554". *Sitzungsberichte der Altertumsforschenden Gesellschaft zu Pernau* 9, 1926–1929, pp. 121–132.
37 Svensk Riksarkivet [SRA] Livonica I: 11. The officials in Viljandi, Tallinn, Järva and Rakvere recommend Johann Duitscher as the marshal of the letters in Cēsis, Kareda, 13 Dec. 1536.
38 Spliet, Herbert: "Die Schwarzhäupter in ihrem Verhältnis zur deutschen kolonialen Ständegeschichte in Livland". *Zeitschrift für Orstforschung* 3, 1954, pp. 233–247; Russwurm, Carl: "Die Schwarzenhäupter auf den Schlössern Livlands". *Beiträge zur Kunde Ehst-, Liv- und Kurlands* 2, Reval 1874, pp. 360–392.
39 See the article by Ivar Leimus in this volume.
40 Mänd, p. 7.

the Teutonic Order Franz von Stitten were arguing in the Lübeck court about the value of a horse Franz had received from Hermann, and although Hermann initially demanded 80 talers for it, he had to be satisfied with 36 in the end.[41] Furthermore, when in the face of military peril in 1495 the Livonian estates prohibited leaving the country with horses priced higher than 9 Riga marks, the eventual military value was first and foremost expressed by the price.[42] It is in only a few letters of the Livonian master that the issue is elaborated further and, besides the price, it is stressed that export of horses for riding in armour is prohibited.[43]

In medieval terminology the most basic differentiation of horses was based on their function. The abovementioned inventories of the commandery of Kuldīga in 1341 mention destriers (*dextrarios*), riding horses (*equos equitales*), ploughing horses (*equos uncales*), stud animals (*equos equirreales*), and colts (*polledros*).[44] Although we may assume that the functions were even more varied, this list remains the only example from Livonia in which they cover such a broad range. In most of the preserved correspondence just the word "horse" (Latin *equus*, Middle Low German *pert*) is used. The lack of more comprehensive characterisation of horses in documents is probably due to the circumstance that most of the accounts on the economic activities of the Order in Livonia have perished.

There is, however, an exceptional source, a fragment of an account book of the commander of Tallinn from 1526–1532,[45] which deserves special attention. It is a badly damaged volume on the manors in the commandery of Tallinn, while the accounts of the central castle seem to be missing. Because of the damages it is not possible to tell exactly how many horses there were in the manors in these particular years, but the figures appearing in the text remain below twenty per manor.[46] Neither is the source very specific as to the function of the horses. Once, in 1529, when the animals of the Harku manor were listed, the horses even appeared next to oxen, which may indicate their function as draught animals.[47]

---

41 Archiv der Hansestadt Lübeck, Niederstadtbuch (Reinschrift) 1556, fol. 146v.
42 Arbusow, Leonid (ed.): *Akten und Rezesse der Livländischen Ständetage*, vol. 3. Deubner: Riga 1910, no. 2: Landtag in Walk, 29–31 March 1495, § 2 *van den perden*.
43 Arbusow, Leonid (ed.): *Liv-, Est- und Kurländisches Urkundenbuch*, 2. Abt. Vol. 1. Deubner: Riga / Moskau 1900, no. 413; no. 1020: *perden im harnische to ryden denlich*.
44 Maltby et al., p. 142.
45 SRA, Livonica I: 35. A photocopy of this volume is also available in the library of the Tallinn City Archives (XVI-756).
46 SRA, Livonica I: 35, fol. 56r, 60r.
47 SRA, Livonica I: 35, fol. 56r.

No indicators of breeds have been given either, but the horses are described by age, gender, and colour. The variety of colours is of special interest. On the one hand, the common words for colours "brown" (*brun*), "red" (*rode*), "grey" (*grawe*) and "black" (*swarte*) are used, but on the other hand, occasionally also specialised vocabulary for describing equine coat colours emerges. The most common such words is the German word *Schimmel* (*schimlich*),[48] which is used for grey (and white) horses. White (*witte*) as a horse colour is sometimes mentioned in Livonian sources[49], but not in this account book. *Schimmel*, however, is also combined with other colours, such as grey (*schemmelich graw*)[50] and red (*rotschimmelich*)[51]. Furthermore, there also appear to have been dun horses (with the mane darker than the body and a dark stripe on the back), as well as those whose colour was a specific "fox dun" (*rothfaele*) and "mouse dun" (*muszvael*). When a horse is characterised as *flaszman*, a flaxen mane could be meant.[52] Finally, out of horse markings "blaze" (*blase*)[53] is mentioned. Still, encountering a vocabulary of such sophistication does not necessarily mean that it was implied with modern rigour.

The account book of the commander in Tallinn gives us an idea how complex the horse terminology in the manors of the Teutonic Order in Livonia in the sixteenth century could be, yet it still leaves us without any information on their breeds and training. Even the size of the animal (small or large) is mentioned but rarely, and other qualities occur even more seldom in the sources. In 1439 the grand master of the Teutonic Order Paul von Rusdorf recommended that the archbishop elect of Riga Silvester Stodewescher make presents in Rome, and for that purpose acquire in Livonia one or two white ambler horses, which have a smooth gait and do not jump and scream.[54] In the same letter, the grand master also claimed that he had no such horses at his disposal. Amblers were valued and expensive horses in the Middle Ages. In 1488 the visitation indeed noted two

---

48 Schiller, Karl / Lübben, August: *Mittelniederdeutsches Wörterbuch*, vol. 4. Kühtmann: Bremen 1878, *schimelink*.
49 Mänd, p. 8, 9.
50 SRA, Livonica I: 35, fol. 10r–11v (1526 Keila).
51 SRA, Livonica I: 35, fol. 17v.
52 SRA, Livonica I: 35, fol. 10v.
53 Schiller, Karl / Lübben, August: *Mittelniederdeutsches Wörterbuch*, vol. 1. Kühtmann: Bremen 1875, *blasenhengst*, a horse with a white forehead.
54 Schwartz, Philipp (ed.): *Liv-, Est- und Kurländisches Urkundenbuch*, vol. 10. Deubner: Riga / Moskau 1896, no. 624: *czwe weisze adir weiszelichte gutte czeldende pferde, die do wol gingen und gemachsam und nicht springende adir schreyende weren*.

amblers, one red and another dun, in addition to 20 nice stallions[55] in the castle of Dobele, but these were certainly not the only amblers in Livonia at the time, for the particular visitation report mentioned horses only in two castles (Dobele and Kandava), which signals serious under-registration elsewhere.

The origin of the breeds in Livonia is an intriguing question. Most authors seem to take it for granted that the horses in medieval Livonia were local, although the sources rarely contain any hints that might support this. A local breed certainly existed in pre-conquest Livonia. Such horses are mentioned as loot and riding animals of the indigenous population even in the Chronicle of Henry of Livonia,[56] but, on the other hand, horses were imported as well. Most of the crusading knights would have arrived with their own horses in the thirteenth century, even when it meant travelling overseas with the animals.[57] In 1226, crusaders also purchased extra horses on Gotland to cross over to Livonia with them.[58] It largely remains unknown how many foreign horses came to Livonia and remained there to breed, as does the geography of their origin. In the sixteenth century some references were made to foreign horses, e.g. in 1547 there was a Frisian stallion worth 200 marks in Alūksne.[59] Frisian horses were also brought in by the last master of the Order in Livonia, Gotthard Kettler.[60] In 1556 there was a Spanish stallion in the possession of the marshal of Livonia.[61] However, at the present stage of the research it is difficult to say how representative these items of knowledge are.

---

55 Biskup / Janosz-Biskupowa 2004, no. 202, p. 190: *hoebischir hengiste 20 upp seynem stalle und 2 zelder, 1 rothir, der ander vall.*
56 Brundage, James A. (transl.): *The Chronicle of Henry of Livonia / Henricus Lettus.* Columbia University Press: New York 2003, XXI 3, 7 (loot); XV, 3 (Estonian cavalry).
57 Jensen, p. 257.
58 Brundage, XXX, 1.
59 SRA, Livonica I: 30, accusations against Kort Klingenberg.
60 Deutschordenszentralarchiv [DOZA] in Vienna, Liv 7, fol. 530r–531r, ca 1560: *etzlichen frisischen hengsten.* Regest in Wieser, Klemens: *Nordosteuropa und der Deutsche Orden. Kurzregesten 1.* (Quellen und Studien zur Geschichte des Deutschen Ordens 17). Wissenschaftliches Archiv: Bad Godesberg 1969, no. 2216.
61 Hartmann, Stefan (ed.): *Herzog Albrecht von Preußen und Livland (1551–1557). Regesten aus dem Herzoglichen Briefarchiv und de Ostpreußischen Folianten.* (Veröffentlichungen aus den Archiven Preussischer Kulturbesitz 57). Böhlau: Köln / Weimar / Wien 2005, no. 1829.

Livonian Horses

Horse farming of the Teutonic Order undoubtedly started early in Livonia. The necessities of war demanded that the horses be resupplied quickly if necessary. In the thirteenth century Livonians also operated on their large destriers in Prussia, as can be read in the Chronicle of Peter of Dusburg.[62] Still, in addition to large war-horses the Livonian stud farms must have kept other breeds as well. The status of the local Livonian horse breeds is not clear. In Prussian sources the local small-sized horse is called *sweik*[63], but there is no special term for the local breed in medieval Livonian sources. When horses called Livonians are occasionally mentioned as found in the Prussian stables,[64] this first of all indicates their place of origin and does not necessarily imply a special breed. Only when a Livonian official calls his horse a Livonian stallion (*lifflendische hengst*), as the *Vogt* of Maasilinn did in 1546,[65] is it relatively certain that a special local breed is meant.

Livonia was regarded as a possible source for the purchasing of horses. The restrictions on horse export were a heated issue throughout the Middle Ages.[66] The Livonian authorities tried to stop the flow of this strategic merchandise out of the country, while the Livonian master had to allow travelling with horses for those for whom it was a matter of status.[67] Furthermore, the Teutonic Order in Livonia was expected to support their co-brothers in Prussia in their need, and vice versa. In 1429 the horse pool in Prussia suffered from severe epidemics, and the grand master tried to stock up the supply in Livonia.[68] In 1431 the servant of the grand master again visited Livonia, but the *Vogt* of Järva reported that because of the lack of horses in his region the man had moved on to Harju and Virumaa.[69] In 1516 Virumaa was again regarded by the grand master as a region

---

62   Hirsch, Theodor / Töppen, Max / Strehlke, Ernst (eds.): *Scriptores rerum Prussicarum*. Vol. 1. Hirzel: Leipzig 1861, p. 108: *cum multis et magnis dextrariis*.
63   See Ekdahl, pp. 122–126.
64   Ziesemer, Walther (ed.): *Das grosse Ämterbuch des Deutschen Ordens*. A.W. Kafemann: Danzig 1921, p. 589, *Lyfflender*; Ziesemer, Walther (ed.): *Das Marienburger Ämterbuch*. A.W. Kafemann: Danzig 1916, pp. 154–155, (1445).
65   SRA, Livonica I: 13, the *Vogt* of Maasilinn to the Livonian master, 5 Feb. 1546.
66   Leimus in this volume.
67   Arbusow 1900, no. 1020: *ritterbortigenn guiden mannen, de uns und unsem orden […] hir int land to denste kommen, in iren uitreisen ein pert ader dre boven gesette […] uittovorgonnen […]*.
68   Hildebrand, Hermann (ed.): *Liv-, Est- und Kurländisches Urkundenbuch*, vol. 8. Deubner: Riga / Moskau 1884, no. 96.
69   Ibid., no. 424.

where one could send a horse dealer to find suitable animals.[70] In the sources there are also indications that the Bishopric of Saaremaa was a possible supplier of horses.[71] More records on the delivery of unspecified Livonian horses to Prussia can be found as well.

One of the regions that appear in the sources more often than others in connection to their horses is the largest Estonian island, Saaremaa. The information on horse trade is recorded mainly in case of conflicts. A certain Pflawme, citizen of Königsberg, complained to the grand master that his commissioner had been deprived of two horses in Courland in 1500.[72] The Livonian master explained that the man had been acquiring large war horses in Saaremaa, although their export was prohibited.[73] In 1515 the *Vogt* of Grobiņa confiscated a mare with a foal and a riding horse worth 20 guilders that a citizen of Königsberg named Lorentz Eberhartt had acquired on Saaremaa as a part of a hops deal.[74] The same pattern was repeated again in 1532, when the commander of Kuldīga confiscated the horses of a certain Lennarth, a merchant from Klaipėda. A vassal of the Order in Courland, Klaus Korff, claimed that the man was in his service and that the horses belonged to him, but the commander reported to his superiors that Lennart had repeatedly been buying eight to ten horses on Saaremaa and smuggling them to Lithuania.[75] Occasionally, Saaremaa was also a source for horses for the Livonian master. In 1546 the *Vogt* of Maasilinn replied to the request of the master that he had been looking for a young Livonian stallion for him, but there was the problem that the horses on the island on the Order's as well as on the Bishop's side were only small or medium-sized.[76]

All in all, Livonian and especially Saaremaa horses were valued and desired objects. Livonia could produce genuinely expensive horses that cost up to 120

---

70 *Virtuelles Preußisches Urkundenbuch, Dh 232,* retrieved 18.1.2019, from http://www.spaetmittelalter.uni-hamburg.de/Urkundenbuch/pub/dh/dh232.htm.
71 Schwartz, Philipp (ed.): *Liv-, Est- und Kurländisches Urkundenbuch,* vol. 11. Deubner: Riga, Moskau 1905, no. 133. The Bishop of Saaremaa Johann Kreul writes on 11 May 1451 from Haapsalu to the Grand Master Ludwig von Erlichshausen that he cannot send any horses.
72 Arbusow 1900, no. 1009.
73 Arbusow 1900, no. 1020, *grove krygbare henxste up Oeszel und dar umbtrent upkopen.*
74 *Virtuelles Preußisches Urkundenbuch, Dh 194,* retrieved 10.07.2023, from http://www.spaetmittelalter.uni-hamburg.de/Urkundenbuch/pub/dh/dh194.htm: *mutterpferd mit eijnem folen und ain redtpferd vor 20 gulden Reynisch.*
75 SRA, Livonica I: 10, the vice-commander of Kuldīga to the Livonian master, 15 Sept. 1532.
76 SRA, Livonica I: 13, the *Vogt* of Maasilinn to the Livonian master, 5 Feb. 1546.

Riga marks, which are documented as presents for foreign dignitaries.[77] What kind of qualities – such as breed, colour or training – these horses had largely remains unknown. Only in 1431, when the *Vogt* of Maasilinn sent a grey Saaremaa stallion to the grand master, the specification that it had a half cross on its side indicates that the animal had a special (maybe even symbolic) value.[78] Sending horses across long distances must have helped to diversify and improve the horse pools for breeding, but there were certainly other considerations, too. In 1548, the commander of Viljandi and the commander of Tallinn sent one horse each from Livonia to the acting Grand Master in Megentheim, Germany, but while one of the horses was a pinto stallion (*buntte hengst*), the other was a brown gelding (*brawne rune*).[79] Obviously, the latter must have been intended for purposes other than breeding.

Conclusion

In conclusion, written evidence on the horses of the Teutonic Order in Livonia is scarce. There is very little data that would allow us to reconstruct the size of the horse pool. Visitation reports pay more attention to the horses owned by the knights, and leave out all others. On a few occasions, all horses of one or another commandery have been counted, yet this still does not make an overall assessment possible. Also, it should be borne in mind that horses were a mobile resource and their number was subject to considerable volatility. Although quantitative estimates are difficult to make, some rare pieces of evidence show a remarkable variety of horses, as well as some regional foci of the horse pool. Above all, Saaremaa seems to stand out as a region specialising in horse-breeding in medieval Livonia; in addition to local breeds there were also imported horses. Collecting isolated pieces of information and interpreting these against a larger framework certainly promises to bring about rewarding results in the future as well.

---

77 Mänd, pp. 7–8. See also the article by Leimus in this volume.
78 Hildebrand, no. 419; Mänd, p. 9.
79 DOZA, Preu 407/1, pag. 123–126, Johann von der Recke to Wolfgang Schutzbar, Viljandi, 16 March 1548. Regest in Wieser, no. 1445.

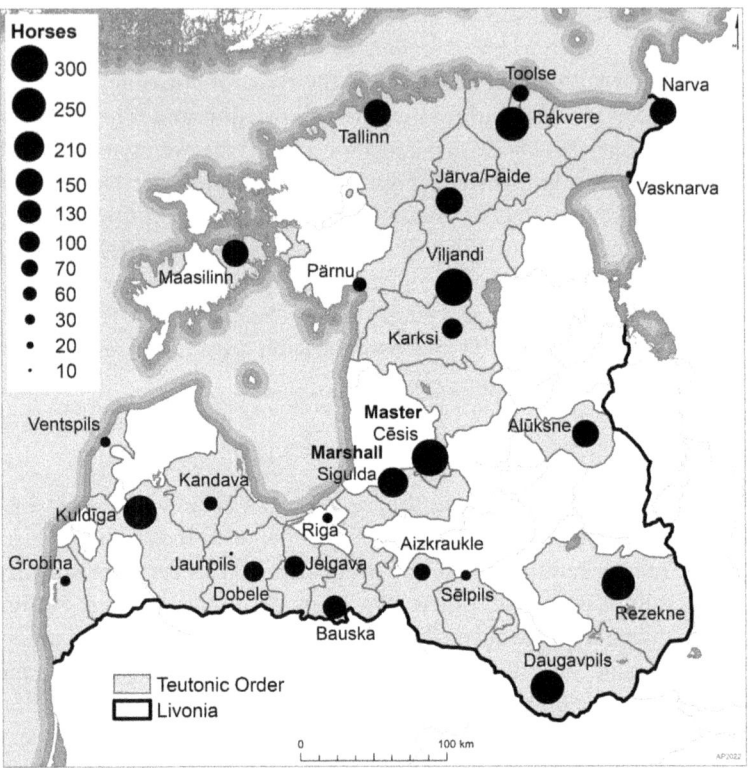

Map 1. The field cavalry of the Teutonic Order in spring 1556. Source: Benninghoven, Friedrich: "Probleme der Zahl und Standortverteilung der livländischen Streitkräfte im ausgehenden Mittelalter". Zeitschrift für Ostforschung 12, 1963, pp. 601–622, here pp. 605, 620–621.

References

Arbusow, Leonid (ed.): *Akten und Rezesse der Livländischen Ständetage*, vol. 3. Deubner: Riga 1910.

Arbusow, Leonid (ed.): *Liv-, Est- und Kurländisches Urkundenbuch*, 2. Abt. Vol. 1. Deubner: Riga / Moskau 1900.

Bauer, Albert (ed.): "Die Wartgutsteuerliste der Komturei Goldingen". *Mitteilungen aus der livländischen Geschichte* 25, 1933, pp. 113–194.

Benninghoven, Friedrich: "Probleme der Zahl und Standortverteilung der livländischen Streitkräfte im ausgehenden Mittelalter". *Zeitschrift für Ostforschung* 12, 1963, pp. 601–622.

Biskup, Marian / Janosz-Biskupova, Irena (eds.): *Visitationen im Deutschen Orden im Mittelalter, Teil II, 1450–1519.* (Quellen und Studien zur Geschichte des Deutschen Ordens 50). N. G. Elwert: Marburg 2004.

Biskup, Marian / Janosz-Biskupova, Irena (eds.): *Visitationen im Deutschen Orden im Mittelalter, Teil I, 1236–1449.* (Quellen und Studien zur Geschichte des Deutschen Ordens 50). N. G. Elwert: Marburg 2002.

Bliujien, Audronė (ed.) *The Horse and Man in European Antiquity: Worldview, Burial Rites, and Military and Everyday Life.* (Archaeologia Baltica 11). Klaipėda University: Klaipėda 2009.

Blumfeldt, Evald: "Keskaja agraarajalugu". In: Sepp, Hendrik / Liiv, Otto / Vasar, Juhan (eds.): *Eesti majandusajalugu I.* Akadeemilise kooperatiivi kirjastus: Tartu 1937, pp. 71–72.

Brundage, James A. (transl.): *The Chronicle of Henry of Livonia / Henricus Lettus.* Columbia University Press: New York 2003.

Contamine, Philipp: *War in the Middle Ages.* Blackwell: Oxford, Cambridge 1996.

Ekdahl, Sven: "Horses and Crossbows: Two Important Warfare Advantages of the Teutonic Order in Prussia". In: Nicholson, Helen (ed.): *The Military Orders vol. 2, Welfare and Warfare.* Ashgate: Farnham / Surrey 1998, pp. 119–151.

Fenske, Lutz / Militzer, Klaus (eds.): *Ritterbrüder im livländischen Zweig des Deutschen Ordens.* (Quellen und Studien zur Baltischen Geschichte 12). Böhlau: Köln et al. 1993.

Hartmann, Stefan (ed.): *Herzog Albrecht von Preußen und Livland (1551–1557). Regesten aus dem Herzoglichen Briefarchiv und den Ostpreußischen Folianten.* (Veröffentlichungen aus den Archiven Preußischer Kulturbesitz 57). Böhlau: Köln / Weimar / Wien 2005.

Hildebrand, Hermann (ed.): *Liv-, Est- und Kurländisches Urkundenbuch,* vol. 8. Deubner: Riga / Moskau 1884.

Hirsch, Theodor / Töppen, Max / Strehlke, Ernst (eds.): *Scriptores rerum Prussicarum.* Vol. 1. Hirzel: Leipzig 1861.

Jähnig, Bernhart: *Verfassung und Verwaltung des Deutschen Ordens und seiner Herrschaft in Livland.* (Schriften der Baltischen Historischen Kommission 16). Lit: Berlin 2011.

Jakubowski, Jan / Kordzikowski, Józef (eds.): *Polska XVI wieku pod wzgledem geograficzno-statystycznym. Tom XIII, Inflanty. Czesc I.* Gebethner i Wolff: Warszawa 1915.

Jensen, Kurt Villads: "Bigger and Better: Arms Race and Change in War Technology in the Baltic in the Early thirteenth century." In: Tamm, Marek / Kaljundi, Linda / Jensen, Carsten Selch (eds.): *Crusading and Chronicle Writing on the Medieval Baltic Frontier. A Companion to the Chronicle of Henry of Livonia.* Ashgate: Farnham / Burlington 2011, pp. 245–264.

Johansen, Paul: "Ein Verzeichnis der Ordensbeamten und Diener im Gebiet Fellin 1554". *Sitzungsberichte der Altertumsforschenden Gesellschaft zu Pernau* 9, 1926–1929, pp. 121–132.

Jóźwiak, Sławomir / Trupinda, Janusz: *Organizacja życia na zamku krzyżackim w Malborku w czasach wielkich mistrzów (1309–1457)*. Muzeum Zamkowe: Malbork 2011.

Kaljundi, Jevgeni: "Linnuste argielu ehk kus hoida hobuseid". In: Lass, A. (ed.): *Arhitektuuripärandi uurimisest Eesti NSV-s. 17. mail 1985. a. Tallinnas toimunud teoreetilis-praktilise konverentsi materjalid*. Valgus: Tallinn 1987, pp. 110–113.

Kivimäe, Jüri: "Voorimehed suurkaupmehe teenistuses". In: Tarvel, Enn (ed.): *Eesti ajaloo probleeme. NSV TA korrespondentliikme Artur Vassara 70. sünniaastapäevale pühendatud teaduskonverentsi ettekannete teesid (18. nov. 1981)*. Eesti NSV Teaduste Akadeemia: Tallinn 1981, pp. 54–65.

Kreem, Juhan: "Netzwerke um Jasper von Munster. Der Deutsche Orden während der Livländischen Koadjutorfehde im Jahre 1556". *Ordines militares* 19, 2014, 73–86. DOI: 10.12775/OM.2014.005.

Kreem Juhan: "Saksa ordu hobused". In: Kreem, Juhan / Leimus, Ivar / Mänd, Anu / Põltsam-Jürjo, Inna: *Loomad keskaegse Liivimaa ühiskonnas ja kunstis*. Tallinn University Press: Tallinn 2022, pp. 35–47.

Leimus, Ivar: "Eesti hobune – strateegiline kaup keskajal". *Tuna: Ajalookultuuri ajakiri* 2, 2017, pp. 10–19.

Leimus, Ivar: "Perdeteken – Prägezeichen? Über eine angebliche Quelle zur Dorpater Münzgeschichte". *Numismatisches Nachrichtenblatt* 6, 2014, pp. 220–221.

Maldre, Liina: "Hobune Eestis muinas- ja keskajal". In: Lang, Valter (ed.): *Loodus, inimene ja tehnoloogia. Interdistsiplinaarseid uurimusi arheoloogias*. Eesti TA Ajaloo Instituut: Tallinn 1998, pp. 203–220.

Maltby, Mark et al.: "Farming, Hunting, and Fishing in Medieval Livonia: The Zooarchaeological Data". In: Pluskowski, Aleksander (ed.): *Environment, Colonization, and the Baltic Crusader States. Terra Sacra 1*. Brepols: Turnhout 2019, pp. 137–173.

Mänd, Anu: "Horses, Stags and Beavers: Animals as Presents in Late-Medieval Livonia". *Acta Historica Tallinnensia* 22, 2016, pp. 3–17. DOI: 10.3176/hist.2016.1.01.

Mühlen, Heinz von zur: "Transportwesen und Reisetätigkeit der Revaler Fuhrleute". In: Pelc, Ortwin / Pickhan Gertrud (eds.): *Zwischen Lübeck und Nowgorod. Wirtschaft, Politik und Kultur im Ostseeraum vom frühen Mittelalter bis ins 20. Jahrhundert. Norbert Angermann zum 60. Geburtstag*. Institut Nordostdeutsches Kulturwerk: Lüneburg 1996, pp. 203–221.

Perlbach, Max (ed.): *Die Statuten des Deutschen Ordens nach den ältesten Handschriften*. M. Niemeyer: Halle 1890.

Rannamäe, Eve et al.: "A Month in A Horse's Life: Trauma Fracture and Healing Process of A Horse Metatarsus From Medieval Viljandi, Estonia". *International Journal of Paleopathology* 24, 2019, pp. 286–292. DOI: 10.1016/j.ijpp.2018.07.003.

Rünger, Fritz: "Herkunft, Rassenzugehörigkeit, Züchtung und Haltung der Ritterpferde des Deutschen Ordens". *Zeitschrift für Tierzüchtung und Züchtungsbiologie* 2, 1925, pp. 211–308.

Russwurm, Carl: "Die Schwarzenhäupter auf den Schlössern Livlands". *Beiträge zur Kunde Ehst-, Liv- und Kurlands* 2, Reval 1874, pp. 360–392.

Schiller, Karl / Lübben, August: *Mittelniederdeutsches Wörterbuch*, vol. 4. Kühtmann: Bremen 1878.

Schiller, Karl / Lübben, August: *Mittelniederdeutsches Wörterbuch*, vol. 1. Kühtmann: Bremen 1875.

Schwartz, Philipp (ed.): *Liv-, Est- und Kurländisches Urkundenbuch*, vol. 10. Deubner: Riga / Moskau 1896.

Schwartz, Philipp (ed.): *Liv-, Est- und Kurländisches Urkundenbuch*, vol. 11. Deubner: Riga, Moskau 1905.

Spliet, Herbert: "Die Schwarzhäupter in ihrem Verhältnis zur deutschen kolonialen Ständegeschichte in Livland". *Zeitschrift für Ostforschung* 3, 1954, pp. 233–247.

Toeppen, Max: "Über Pferdezucht in Preußen zur Zeit des Deutschen Ordens". *Altpreußische Monatsschrift* 4, 1867, pp. 681–702.

Viires, Ants: "Draught Oxen and Horses in the Baltic Countries". In: Fenton, Alexander / Podolák, Ján / Rasmussen, Holger (eds.): *Land Transport in Europe*. (Folkelivs studier 4). Nationalmuseet: Copenhagen 1973, pp. 428–454.

Wieser, Klemens: *Nordosteuropa und der Deutsche Orden. Kurzregesten 1.* (Quellen und Studien zur Geschichte des Deutschen Ordens 17). Wissenschaftliches Archiv: Bad Godesberg 1969.

Ziesemer, Walther (ed.): *Das grosse Ämterbuch des Deutschen Ordens*. A.W. Kafemann: Danzig 1921.

Ziesemer, Walther (ed.): *Das Marienburger Ämterbuch*. A.W. Kafemann: Danzig 1916.

Archiv der Hansestadt Lübeck, Niederstadtbuch (Reinschrift) 1556.

Deutschordenszentralarchiv [DOZA] in Vienna, Liv 7; Preu 407/1.

Svensk Riksarkivet [SRA], Livonica I.

Virtuelles Preußisches Urkundenbuch, retrieved 10.07.2023, from http://www.spaetmittelalter.uni-hamburg.de/Urkundenbuch/.

Ivar Leimus

The Horse in Livonia as a Strategic Commodity in the Middle Ages

Abstract: In the Middle Ages, the horse was the primary means of transport and labour. This article examines horse trade in medieval Livonia, also discussing the division of different horses into different price groups (work and draught animals, carriage and travel horses, more expensive steeds used by the nobility and in connection to various political events). Livonia's location at the boundary where the western and eastern cultural spaces met made Livonian horses a strategic commodity. For this reason, the authorities tried in every possible way to restrict and regulate their export: only the export of the cheaper horses was allowed. In troubled times, exporting horses across the Eastern border was completely prohibited. Based on surviving sources, the article reconstructs the procedure of exporting horses from Livonia to Russia.

Keywords: Livonia, Russia, horses, Hanseatic trade

Introduction

It is no exaggeration to claim that in the Middle Ages a motor consuming oats was as necessary and valuable as one operating on petrol or electricity today. The horse did everything: it transported you and fought for you, it carried your loads and ploughed your fields. The town of Tallinn even boasted a horse-drawn mill. A particular horse's specialisation correlated with its value that in its turn found expression in the horse's price. War horses were the most expensive, those meant for daily drudgery the cheapest. The rest could be found between these extremes.[1] Thus the price of a horse cannot be treated similarly with the price of other, relatively homogeneous goods (such as various foodstuffs, but also beeswax, salt, and even some sorts of woollen broadcloth), but there are several

---

1 E.g. Dyer, Christopher: *Standards of Living in the Later Middle Ages: Social Change in England c. 1200–1520*, revised edn. Cambridge University Press 1998, pp. 71–72; Hyland, Ann: *The Horse in the Middle Ages*. Sutton Publishing: Stroud 1999, pp. 75–6, 150. Research for this chapter was supported by the Estonian Research Council grants PRG1276 and IUT18-8. The research results in Estonian have been published as: Leimus, Ivar: "Eesti hobune – strateegiline kaup keskajal". *Tuna* 2, 2017, pp. 10–19; Kreem, Juhan / Leimus, Ivar / Mänd, Anu / Põltsam-Jürjo, Inna: *Loomad keskaegse Liivimaa ühiskonnas ja kunstis*. Tallinn University Press: Tallinn 2022.

other factors that have to be taken into account, such as the horse's breed, size, health, age, and so on.

Horses in Livonian Sources

Speaking of medieval Livonia, it has to be admitted that the horse prices that have reached us in historical sources are fairly numerous (see Table 1).[2] The table shows that the horses of the era can be broadly divided into three or four groups according to their value – the cheap ones, those moderately priced, the expensive, and the very expensive ones. The division is certainly somewhat arbitrary and provisional as there are always some animals positioned on the borders of categories that are difficult to classify either as one or the other. In addition, the inflation that occurred during the period observed also has to be taken into account, or rather, the diminishing of the silver content of coins and the concomitant drop in their value has to be considered. The prices took a particularly remarkable jump as a result of the monetary reform of 1422–1426. While before the reform the price of the cheapest horses used to range from one and a quarter to four Riga marks, the post-reform prices were in the range of four to ten marks. The value of medium-priced horses initially oscillated in the range of five to seven marks, but rose to 10–25 marks after the coin reform. Horses that were even more expensive and cost up to 40, and later, when the purchasing power of the currency was diminishing, up to 60 marks, are mentioned less

---

[2] Main sources: Vogelsang, Reinhard (ed.): *Kämmereibuch der Stadt Reval 1432–1463* [hereafter KBR 1432–1463]. Böhlau: Cologne 1976; Vogelsang, Reinhard (ed.): *Kämmereibuch der Stadt Reval 1463–1507* [hereafter KBR 1463–1507]. Böhlau: Cologne 1983; Bulmerincq, August von: *Kämmerei-Register der Stadt Riga 1348–1361 und 1405–1474* [hereafter KR Riga], 2 vols. Duncker & Humblot: Leipzig 1909, 1913; Bunge, Friedrich Georg von et al. (eds.): *Liv-, Est- und Kurländisches Urkundenbuch* [hereafter LEKUB], 1st series, 12 vols; 2nd series, 3 vols. Reval / Riga / Moscow 1853–1914; Stavenhagen, Oskar / Arbusow, Leonid (eds.): *Akten und Rezesse der livländischen Ständetage* [hereafter AR], 3 vols. Riga 1907–1914; Bruiningk, Hermann von / Busch, Nikolai (eds.): *Livländische Güterurkunden aus den Jahren 1207 bis 1500* [hereafter LGU 1]. Jonck & Poliewsky: Riga 1908; Bruiningk, Hermann von (ed.): *Livländische Güterurkunden aus den Jahren 1501 bis 1545* [hereafter LGU 2]. A. Gulbis: Riga 1923; Bunge, Friedrich Georg von / Toll, Robert von (eds.): *Est- und Livländische Brieflade. Eine Sammlung von Urkunden zur Adels- und Gütergeschichte Est- und Livlands*. Part 1, vol. 1: *Dänische und Ordenszeit* [hereafter Brieflade 1.1]. Kluge und Stroem: Reval 1856; Tallinn City Archives [hereafter TLA], 230_1_Ad 32.

frequently in the sources, while very expensive mounts, those that cost 80–120 marks, were very rare indeed.

The documents reveal considerably less about the kind of horse behind any given price, its purpose, and the tasks it was to carry out. As regards the low-priced horses, the account books of the Riga treasury (*Kämmerei*) from 1464–65 make note of two horses that were purchased for the brickworks and whose joint price was 16 marks.[3] Judging by the price, these may have been strong working animals. The accounts of the Tallinn town council also repeatedly note brickyard horses whose prices ranged from one mark and 25 shillings to seven marks.[4] Exactly which duties the animals were meant to perform has not been recorded. However, as the times saw vigorous building activity and several thousands of roof tiles and the like were being taken to the town from the Kopli brickyard, they could, for example, well have been draught horses. Some time in the second half of the fifteenth century, Marquard Bretholt, a Tallinn town councillor, hired five horses with sledges and complete equipment from someone called Lorentz Karman (i.e. a waggoner) – these must have been draught horses as well.[5] They were very poorly fed and one of them perished. The owner priced it at nine marks, which might, yet need not, have been a somewhat exaggerated price, considering the circumstances. However, when the town council received four marks for two horses from a certain Marten Vorman in 1490, the waggoner must have purchased two nags.[6]

Still, a horse of the cheaper kind could prove effective for other tasks as well. In 1466–67 a servant of the Riga town council bought a horse for nine and a half marks in order for him to be able to travel to Polotsk.[7] In 1469–70 the Riga town council bought a horse for the Archbishop's scribe that cost nine marks,[8] and in the following year a trip from Riga to Trakai in Lithuania was undertaken on a horse that cost six and a half marks.[9] In 1471–72 a servant of the Riga town council was sent to Lübeck and given a mount worth nine marks for this purpose.[10] In 1490 Johan Rotert, a Tallinn town councillor, bought a horse when

---

3 KR Riga 2, p. 232.
4 KBR 1463–1507, no. 2023, 2052, 2057, 2062, 2087, 2145, 2168, 2169, 2207, 2208, 2212, 2250, 2267, 2286, 2386, 2462–2464.
5 Heckmann, Dieter (ed.): *Revaler Urkunden und Briefe*. (Veröffentlichungen aus den Archiven Preussischer Kulturbesitz 25). Böhlau: Cologne 1995, no. 172.
6 KBR 1463–1507, no. 2083.
7 KR Riga 2, p. 257.
8 KR Riga 2, p. 258.
9 KR Riga 2, p. 259.
10 KR Riga 2, p. 259.

on his way to a meeting in Riga that only cost three marks and 20 shillings.[11] In 1502 the Tallinn town council sent a servant to Tartu on a Russian gelding (*Rusche rone*) whose price was four and a half marks.[12] In 1509 the horse of Tallinn town councillor Heinrich Wideman fell ill in Pärnu during a diet (*Städtetag*) and he had to purchase a new horse that he left in the town stables in his home town for nine and a quarter marks.[13] There were workhorses as well as draught horses and even riding horses among the less expensive animals; nevertheless, the latter category was meant not so much for the town councillors themselves, but rather their servants.

In Tallinn, horses from the middle price range were often bought for the town stables (called the marstall), which also accommodated the town's workyard. Quite often these animals were meant for travelling to the Hanseatic Diets, as well as to negotiations with neighbouring rulers.[14] In 1449, Gerhardus (Gerd), lector of the Tallinn Dominican friary, even rode a horse to Rome.[15] Some decades earlier, in 1431–32 and 1432–33 the scribe of the Riga town council had also been to Rome twice.[16] The source does not specify the goal of his trips but it could have been related to the inquest concerning the murder of the envoys of Livonian bishops by the Teutonic Order. Only once, in 1495, did the town council of Tallinn purchase horses for a military campaign, apparently spurred on by the topical threat from Russia.[17] All in all, 35.5 marks were paid for three horses, plus 18 marks for a single additional horse. So these could not have been warhorses, yet may have served the purpose of reaching the battlefield. However, the horse that had cost 37 marks, whom Frederic Depenebeke took along with him to Narva in 1448 in order to proceed further to a campaign against Novgorod could well have been a warhorse.[18] Yet such a horse would already belong to the next price category.

Actually, relatively more is known about the more expensive horses as the fact of purchasing them was worth recording in any case. Often these were used to ride to meetings or to attend negotiations.[19] In 1473 in Tallinn, a knight

---

11 KBR 1463–1507, no. 2113.
12 LEKUB 2.2, no. 295.
13 LEKUB 2.3, no. 652.
14 Mänd, Anu: "Horses, Stags and Beavers: Animals as Presents in Late-Medieval Livonia". *Acta Historica Tallinnensia* 22, 2016, pp. 3–17, here p. 6.
15 KBR 1432–1463, no. 810.
16 KR Riga 1, pp. 115, 195.
17 KBR 1463–1507, no. 2278, 2279.
18 KBR 1432–1463, no. 739.
19 KBR 1432–1463, no. 189, 191, 193, 602; TLA, 230_1_Ad 32, fol. 255v.

arriving from Novgorod who happened to be a relative of the Grand Master of the Teutonic Order Heinrich von Plauen, was given a 33-mark horse that had been purchased for the Tallinn town stables a little earlier, obviously for this particular purpose.[20] In 1488 the Tallinn town council gave the inspector of the Teutonic Order from Prussia a 38-mark horse as a gift.[21] In 1494, town councillor Gottschalk Remmlinckrode travelled to Moscow to conduct trade negotiations, and bought a 40-mark horse in Narva.[22] Unfortunately, his negotiations with the Russians bore no results. The kontor of the Hanseatic League in Novgorod was closed down, the goods were arrested, and the merchants who happened to be on the premises imprisoned, as was Remmlinckrode himself.[23] With winter pending, the town council of Tallinn sent Johan Hildorpe to Moscow in an attempt to free the prisoners, and he gave Ivan III a 40-mark horse as a present. Such a gift could by no means have been flattering to the Grand Duke. Hildorpe's efforts had no results and in the following summer he yet again travelled to Moscow, this time taking along a 100-mark stallion as a gift.[24] Now we know that the horse, as well as other presents, failed to have any effect, and the merchants would remain incarcerated for several years to come. 40 marks was also the price of a horse that Hermen Menning, a burgher of Tallinn, had pawned to the Commander (*Komtur*) of Tallinn, and that the town council required it to be returned via the Livonian Master of the Teutonic Order in 1497.[25] In 1507 the town council of Tallinn sold a horse in Pärnu and received 30 marks for it.[26] In 1516, Jacob Richerdes, another member of the town council, bought a roan horse for the town stables for 40 marks.[27]

It was not only members of town councils who made use of horses. In 1484, the owner of Pühajärve (Ger. Wollust) manor Bartholomeus von Tisenhusen (Tiesenhausen) put it down in his will that the bishop of Tartu owed him 100 marks for three horses, complete with saddles and bridles, who thus must have been mounts.[28] This renders the mean price of a horse at 33⅓ marks. The price of

---

20 KBR 1463–1507, no. 1495, 1496.
21 KBR 1463–1507, no. 2016; Münd, p. 8.
22 LEKUB 2.1, no. 31.
23 Goetz, Leopold Karl: *Deutsch-Russische Handelsgeschichte des Mittelalters*. Otto Waelde: Lübeck 1922, pp. 178–186.
24 LEKUB 2.1, no. 630; Münd, p. 8.
25 LEKUB 2.1, no. 497.
26 TLA, 230_1_A.d. 32, fol. 5v.
27 TLA, 230_1_A.d. 32, fol. 83r.
28 LGU 1, no. 559.

the horse that Heinrich Wrangel, a nobleman from north Estonia, owed Andreas Lode was set a mite lower, at 30 marks.[29] In 1518 Rytzhard Gutsleff sold a part of his manor to Johan Koskull and, as a result of the deal, he received 1600 marks in currency, a horse that cost 50 marks, and two so-called *Kleppers,* who appear to have been riding horses of a less expensive kind, at ten marks each.[30] In 1545 the nobleman Lorenz Versen noted in his will that he owed Dietrich Mulsdorp a stallion for a *Last* of rye that had cost 60 marks.[31] In the same year, noblemen from the Tisenhusen family in Livonia sold half of a village in the Ērgļi parish and, in addition to money, received horses for it as well.[32] Engelbrecht van Tisenhusen for instance was satisfied with a horse that cost 50 marks. Thus the pricier horses mostly changed hands in the circles of the nobility, but some were also left for the town elite, to serve either as gifts to dignitaries, to travel to diets, or just to be used by the townspeople themselves.

The 100-mark stallion given to Ivan III that was mentioned above ranked among the most expensive horses. Gifts to high-status individuals indeed fell into that approximate price category. In 1470–71 the town council of Riga sent a delegation to Trakai in Lithuania to meet the king of Poland. A grey stallion that cost 120 marks was taken along as a present for the royal.[33] In 1479 the town council of Tallinn sent the Grand Master of the Teutonic Order a horse as a gift that had been bought from the town councillor Johan Grest for 80 marks.[34] When the brothers Detloff and Johan van Tisenhusen sold the aforementioned village of the Ērgļi parish, each of them received a good stallion worth 100 marks in addition to 100 marks in currency.[35]

To sum it up, the most expensive horses in Livonia would cost approximately 100 Riga marks. The representatives of cities customarily gave them as presents to the powers that be. In addition to the above, the vogt of Maasilinn sent a grey stallion from Saaremaa to the Grand Master of the Teutonic Order in 1431; in 1499 the Livonian Master presented the Grand Master with a white stallion; and in the following year, two noblemen gave the Commander of Tallinn two white stallions.[36] Yet manor owners could also boast horses of a similar value. What is

---

29 Brieflade 1.1, no. 137 (557).
30 LGU 2, no. 277.
31 Brieflade 1.1, no. 1267. A *Last* equals about 2000 kilograms.
32 LGU 2, no. 1165.
33 KR Riga 2, p. 259.
34 KBR 1463–1507, no. 1672; Mänd, p. 7.
35 LGU 2, no. 1165.
36 Mänd, p. 9.

worth noting is that if the sources usually speak of just horses (*perde*) then in the case of the most expensive animals it is almost always specified that they were stallions. That stallions were worth more than just any horses is also indicated by a case from 1500 when a citizen of Königsberg called Pflaume disregarded the price limit established in Livonia and let his servant buy large (that means, more expensive) stallions from Saaremaa and the vicinity in order to smuggle them out of the country. Vigilant peasants, however, reported the matter to the authorities, and the horses were intercepted.[37]

Medieval Livonia bred its own horses. In the early thirteenth century both the christianising conquerors of Estonia and their allies would seize horses as loot from here.[38] The Estonian troops seem to have been mostly mounted.[39] Even in the eighteenth century the observers would praise the local *Kleppers*, which were reported to have been strong and sturdy draught and work animals due to their rough breeding conditions.[40] Data are scarce on the topic of medieval horse breeding but what little is known clearly demonstrates that horses were abundant. For example, the Teutonic Order's six manors on the islands of Saaremaa, Muhu and Hiiumaa owned ten breeding stallions and a hundred mares,[41] but numerous horses could also be found on farmsteads[42] – in the conditions of the agriculture of the time it could not possibly have been otherwise.

Trading Horses to Russia

There are very few data on horses being imported,[43] but the geographical location of Livonia gave rise to exporting them, mostly, although not exclusively, to

---

37 LEKUB 2.1, no. 1009, 1020.
38 Arbusow, Leonid / Bauer, Albert (eds.): *Heinrichs Livländische Chronik* [hereafter HCL], 2nd ed. (Scriptores rerum Germanicarum in unsum scholarum ex Monumentis Gaermaniae Historicis separatem editi 31). Hahn: Hannover 1955, *passim*.
39 E.g., HCL XV:3, XXI:3, 7.
40 [Wilde, Peter Ernst]: *Von der liefländischen Pferdezucht, und einigen bewährten Pferdecuren. Anno 1770*. Oberpahlen 1770; Hupel, August Wilhelm: *Topographische Nachrichten*. Vol. 2. Hartknoch: Riga 1777, pp. 247–250.
41 Blumfeldt, Evald: *Eesti keskaja agraarajalugu*. Akadeemiline Kooperatiiv: Tartu 1937, pp. 71–72; See also the chapter by Juhan Kreem in this volume.
42 Blumfeldt, Evald: "Talurahvas keskaja teisel poolel". Kruus, Hans (ed.): *Eesti ajalugu*, vol. 2: *Eesti keskaeg*. Eesti Kirjanduse Selts: Tartu 1937, pp. 213–240, here p. 234; Põltsam-Jürjo, Inna / Selart, Anti / Leimus, Ivar: "Maamajandus ja külarahvas". In: Selart, Anti (ed.): *Eesti ajalugu*, vol. 2: *Eesti keskaeg*. Tartu Ülikooli ajaloo ja arheoloogia instituut: Tartu 2012, pp. 185–202, here p. 186.
43 See the chapter by Juhan Kreem in this volume.

Russia. At the same time the position on the border of the Eastern and Western cultural spaces turned horses into strategic goods. For this reason, the authorities attempted to limit and regulate their export from Livonia.

The first decree of the kind was issued by Pope Innocent IV in 1248, prohibiting the residents of Livonia to provide pagans with horses, as well as weapons, iron, timber, victuals and so on, at the request of Heinrich, Bishop of Saaremaa.[44] The political background of the Pope's letter remains unclear, but Anti Selart has suggested that his intervention may have been sparked by the struggle between Livonian landowners for influence in Lithuania.[45] The Pope, this time Clemens VI, was also behind the next prohibition, as he did not allow the Bishops of Saaremaa and Tartu and the Provost of Riga to sell horses, weapons, victuals and so on, to the Russians in 1351.[46] The background for this prohibition probably lay in the plans of Magnus Eriksson, King of Sweden, to undertake a crusade in Karelia and Ingria.[47] In 1405 the Valga Diet decided that neither the people of Riga nor Tallinn could sell to the Russians any horses that were being taken to Tartu and through the bishopric.[48] The undated *Bursprake* of Tallinn from around the turn of the fourteenth and fifteenth centuries that put a ban on selling horses to Russia must have been a result of the same decision.[49] It is possible that the ban was related to the Teutonic Order's and Lithuania's joint preparations for a war against Pskov and Novgorod that took place in 1406–1409. Although peace was concluded with the Russians in 1409, trade relations with Pskov were not restored before 1411.[50]

The same year, exports of horses to Russia via Narva became particularly intensive. Within a couple of months nearly 70 horses were taken out of the country, their prices being three to five marks per animal.[51] This was a breach of

---

44 LEKUB 1, no. 201.
45 Selart, Anti: "Waffenembargo in den nordischen Kreuzzügen im 13. Jahrhundert". In: Birli, Sonja et al. (eds.): *ene vruntlike tohopesate. Beiträge zur Geschichte Pommerns, des Ostseeraums und der Hanse. Festschrift für Horst Wernicke zum 65. Geburtstag*. Verlag Dr. Kovač: Hamburg 2016, pp. 549–558, here p. 556.
46 LEKUB 6, no. 2846.
47 Selart, Anti (Селарт, Анти): "Vod´ v zapadnoevropejskix istočnikax XIII–XIV veka". *Stratum plus* 5, 2009, pp. 529–538, here pp. 532, 537.
48 LEKUB 4, no. 1656, § 6.
49 LEKUB 4, no. 1516, § 45.
50 Kazakova, Natal'ja (Казакова, Наталья): *Russko-livonskie i russko-ganzejskie otnošenija XIV – načalo XVI v*. Nauka: Leningrad 1975, pp. 88–92.
51 Horoshkevich, Anna L. (Хорошкевич, Анна Л.): *Torgovlja Velikogo Novgoroda s Pribaltikoj i Zapadnoj Evropoj v XIV–XV vekax*. Izdatel'stvo Akademij Nauk SSSR: Moskva 1963, pp. 322–323.

law, for the Master of the Order had established two marks as the highest price limit for any horses taken to Russia.[52] A glance at the Table convinces us that it was only the cheapest horses that could be exported. The Vogt of Narva informed the Livonian Master of the matter, and the latter promptly intervened. In his letter to the town council of Tallinn he prohibited the export of more expensive horses and ordered the Vogt of Narva to intercept these. Those cunning fellows who had planned to ride to Russia on their expensive horses and sell them there were obliged to leave collateral to the vogt.[53]

The trade relations and quarrels with Russians were of a somewhat local nature at the time. Thus, in 1414 the Livonian Master once again prohibited the people of Tallinn to sell horses in Pskov and asked the same from the town and the chapter of Tartu.[54] It is possible that the reason for the restriction was due to a quarrel with Pskov on grounds of checking the quality of wax.[55] In the following year, the ban on sales was also extended to the horses that were roaming freely in Livonia (*lose perde*).[56] Apparently, the prohibition must have been in force for quite a few years, as in 1420 the Vogt of Narva reported to the council of Tallinn about the imprisonment of some Swedes who had been selling horses to Russians.[57] News on horse trading would next emerge only in 1440. By then, the circumstances had become more lenient, and horses that cost up to a rouble could be sold out of the country.[58] In 1411 the price would have equaled four and a half to five marks, but in the period observed the exchange rate of the rouble was already seven marks.[59] So in principle considerably more expensive horses could be sold in comparison with 1411. This alleviation, which was apparently based on the trade treaty between the Hanseatic League and Novgorod that had been reached in 1436,[60] would remain in force for quite a lengthy period.

---

52 LEKUB 4, no. 1887.
53 Ibid.
54 LEKUB 5, no. 1955.
55 Goetz 1922, p. 106.
56 AR 1, no. 7.
57 LEKUB 5, no. 2424.
58 LEKUB 9, no. 613.
59 Leimus, Ivar (Леймус, Ивар): "Grivna serebra (stucke sulvers) v livonskix istočnikax do načala 16 v." In: Gajdukov, Petr Grigor'evič (ed.): *The Russian Rouble: 700 years of the history. Proceedings of the International Conference*, Velikiy Novgorod, April 25–27, 2016. Velikiy Novgorod 2017, pp. 36–39.
60 Kazakova 1975, pp. 114–116.

Yet towards the end of the fifteenth century the relations between Livonia and the Grand Duchy of Muscovy, which was asserting its power ever more expansively, became very tense. In 1495 the Diet in Valga established six marks as the maximum price limit for horses sold to people from Russia.[61] At first glance, the difference from the earlier price does not seem large, yet the mark had become so weak by that time that the exchange rate of the rouble had risen to 14–15 marks.[62] This meant a return to the harsh measures of a century earlier. In addition, restrictions concerning Germans were introduced for the first time. These were not as excruciating as the ones exercised against the neighbours in the east – the maximum price of a horse was established at nine marks for fellow Germans. However, in the following year the Livonian Master lowered it to eight marks.[63] That he meant business is demonstrated by his correspondence with the Grand Master whose subjects repeatedly attempted to acquire horses from Livonia that became confiscated.[64] In the reigning political situation the country itself was in need of its horses. In 1500, a *burspake* of the town council of Tartu announced a total ban on selling horses to Russians.[65] In 1505 the Bishop of Saaremaa issued a statute that declared people selling weapons and horses to Russians and unbelievers to be criminals.[66] In 1513 the Diet of Valmiera repeated the prohibition on exporting horses. Even travelling abroad was not allowed on mounts that were more expensive than 11 marks – a price that barely met the lower limit of the medium price range at the time.[67] Even in 1522 the Diet of Valmiera did not allow Russians or Votes who were acting on behalf of Russians to take any horses out of the country.[68]

The rigid prohibitions appear to have been lifted by the Diet of Valmiera of 1532. Now the upper price limit for horses exported to Russia, Lithuania and Germany was set at eight to nine marks.[69] Taking into account the weakening of the mark meanwhile, this would again mean only low-priced work horses. The masters' servants could travel to Germany on horses that cost up to 20 marks, but in order to prevent selling the horses abroad this could be done only upon

---

61   AR 3, no. 2:2.
62   Leimus 2017.
63   LEKUB 2.1, no. 413.
64   LEKUB 2.1, no. 413, 781, 998, 1009, 1020, 1067.
65   LEKUB 2.3, no. 837:34.
66   LEKUB 2.3, no. 781:27.
67   AR 3, no. 54:83.
68   AR 3, no. 136:40.
69   AR 3, no. 304:6, 7.

displaying the respective passports. The following year, the Livonian Master raised the permitted upper limit of the export price of horses to 14 marks for German or Lithuanian merchants; the horses that peasants were allowed to take out of the country could cost up to ten marks.[70]

On the one hand, repeating the prohibitions year after year testifies to the seriousness of the problem for the people of Livonia, yet on the other hand it also shows that more often than not the regulations remained on paper only. What is curious is that the "prohibition years", as it were, saw a number of decisions and treaties with Russians regulating and specifying the order of purchasing and exporting horses.[71] The earliest document of the kind that has reached us derives from 1411, the last one from 1509. On the basis of these, the procedure of complying with the required formalities, that lasted for at least a century, could be reconstructed as follows.

After purchasing a horse in Livonia, the Russian buyer had to appear before the Commander of the Order with the animal when in Tallinn, or before the bailiff of the bishop and the town vogt when in Tartu. The official would issue a certificate to the Russian, stating the origin of the horse and its price, as well as the name of the Russian in case he was planning to exchange the horse for another on the way. The certificate was marked with a sign *(teken)* that employed a special stamp. In Tartu, it apparently represented the local coat of arms – a crossed key and sword that was accompanied by the family emblem of the bishop. The design of the sign of Tallinn is unknown, yet it is fairly probable that it showed the cross of the Order. This procedure that, with certain reservations, could also be labelled as customs cost a ferding, i.e. ¼ mark.

Yet this was not everything. In Narva, the Russian had to appear before the vogt or the commander of the castle and display both the horse and the certificate. Only the Narva officials had the authority to allow horses to be taken out of the country, and not, e.g. the Commandery of Tallinn. After this procedure, an exporter or transporter (Rus. *вывoдной*) took the Russian with the horse across the border into Russia. The person had to be paid a Russian silver coin – a denga.

---

70 AR 3, no. 326.
71 LEKUB 4, no. 1887; LEKUB 5, no. 1952, 1964; LEKUB 7, no. 723; LEKUB 9, no. 613; LEKUB 10, no. 470; LEKUB 11, no. 409; *409; Akty otnosjaščiesja k istorii Zpadnoj Rossii, sobrannye i izdannye Arxeografičeskoju komissieju..* Vol. 1: 1340–1560. Sankt Peterburg 1861, no. 75, 112; AR 3, no. 2:2; LEKUB. 2.1, no. 872; LEKUB 2.3, no. 583:10; Leimus, Ivar: "Perdeteken – Prägezeichen? Über eine angebliche Quelle zur Dorpater Münzgeschichte". *Numismatisches Nachrichtenblatt* 5, 2014, pp. 220–221.

The *вывoднoй* apparently came from east of the border or else he need not have been paid in Russian currency.

If there was no sign of the commandery (of Tallinn) on the certificate, the vogt was to confiscate the horse and repay the Russian in the sum he had given for it. Also in cases when the horse's price exceeded the set price limit, the horse was to be confiscated and the Russian reimbursed. In 1495 an amendment was made in the provision. The horses were not to be confiscated and the Russian buyers remunerated anymore, but the punishment was targeted at the person who had sold the horse in question – they were to pay a fine equal to the cost of the horse.

It could also happen that a horse fell ill or died on the way while its Russian buyer was still in Livonia. In this case he could purchase a new horse for which he naturally could not have had an export certificate. So the Russian had to pay an additional fare when displaying the horse to the vogt in Narva. If, however, the Russian attempted to sneak around the city without taking the horse to the vogt to be checked, it would be intercepted from him (in case he was caught, of course).

There were also cases of abuse of power on the part of officials. For instance, a commander of Tallinn charged half a mark or a mark for a procedure that cost a ferding, i.e. it took two to four times more than the established price. He also overstepped the limits of his authority when he himself started to issue permissions for horses to be taken out of the country. In 1440 the situation was brought to the attention of superiors and the activities of the commander were banned.[72] The Vogt of Narva also used to confiscate horses from Russians forcibly and reimburse the owners in sums of money (which are likely to have been inconsiderable). Such activities were put an end to with the truce of 1493.[73] The wealth of Livonians derived from their trade with Russia, and the local merchants wished to keep the relations with Russians smooth at all cost. This attitude deriving from the geographical position of the country has survived till the present day.

Conclusion

In conclusion one can say that there were a lot of horses of diverse breeds and prices in medieval Livonia. The cheapest ones were mostly used as work and draught horses, yet sometimes trips were undertaken on them. Horses from

---

72 LEKUB 9, no. 613.
73 *Akty*, no. 112.

the middle price range normally were meant for travelling, sometimes as far as Rome. As to the expensive horses these were used to ride to meetings or to attend negotiations, but also as presents to honourable individuals. Also the noblemen's and war horses belonged to this category. The most expensive horses, stallions as a rule, were presented as gifts to high-status people.

The location of Livonia between the two worlds – western and eastern – turned horses into a strategic commodity. For this reason, the authorities attempted to limit and regulate their export from Livonia. The local merchants, on the other hand, were vitaly interested in keeping good relations with their Russian counterparts.

Table 1. Horse prices in medieval Livonia 1378–1545

Year	Cheap	Moderate	Expensive	Very expensive
1378		7		
1383	3½; 4	5; 6		
1405/06	1¼; 2; 3¾			
1406/07		7		
1409/10	1¼; 3¼			
1411	2			
1411/12	2½; 2⅜			
1412/13	c. 2½			
1415/16	3¼			
1416/17	2½	5		
1422/23		c. 6–7		
1423/24		c. 11–12		
1424/25		c. 10–11		
1425/26	c. 5–6			
1426/27		10		
1431/32		20		
1432	5; 8			
1432/33	4	c. 18–19		
1433	4; 4¼; 5; 5¼			
1433/34	7; 8			
1434	6			
1435	8 M 5 s; c. 9–10		35	

(Continued)

Table 1. Continued

Year	Cheap	Moderate	Expensive	Very expensive
1435/36	7			
1436	8	**20**	**28; 32**	
1438		13		
1439	8¾	12	**28**	
1440/41	5½			
1442	8			
1443		16		
1443/44	5½			
1444			38	
1446/47	7			
1447/48		13		
1448			37	
1448/49		15		
1449		16; 17; **26**		
1450		20		
1451		**16; 27**		
1451/52	5; 5½	8; **9**; 12; 14¾		
1452		**15; 16;** 20		
1452/53	6; 9			
1453/54		11		
1454	8½	11		
1454/55		12; 13¼		
1455		15		
1455/56		c. 13–14		
1457	7	12; **25**		
1461/62	6¼			
1462		23		
1463	8	10; **12**		
1463/64		c. 20–21		
1464	c. 6			
1464/65	c. 8			
1466/67	**9½**			
1467	9	11; 20		
1467/68	c. 6–7			
1468	7¾			
1468/69	8			
1469	10; 10 M 7 s	11–12; 13; 23		

Table 1. Continued

Year	Cheap	Moderate	Expensive	Very expensive
1469/70	9			
1470	5			
1470/71	6½	c. 23–24		120
1471	3; 4½			
1471/72	9			
1473			33	
1476		16		
1478		20½		
1479	10	13½; 14½		80
1480	10	12; 13		
1481	7; 8 M 7 s	18; 22; 24		
1482	2½	12		
1484	9	16; 20	c. 33⅓	
1485	10			
1487	7; c. 8½			
1488	c. 3; 4	26	38; 40	
1489	4; 4⅓			
1490	2; 3½; 5½; 3 M 20 s			
1491	1–2; 8			
1492	3–4			
1493	4½; 7; 9	20; 22		
1494	5–6; 6 M 15 s	**20; 26; 28**		
1495	1 M 25 s	12; 18		100
1497	5	20	30; 40	
1499	2 M 14 s; 3; 4			
1500	4; 5	**24**		
1502	4½			
1506	4			
1507		11	30	
1508		16; 20		
1509	6; **9¼**	12		
1510	9			
1511	7¾	12		
1512		**12**; 16; 20		

(*Continued*)

Table 1. Continued

Year	Cheap	Moderate	Expensive	Very expensive
1513	10	12		
1514	7; 8; 9	16		
1515		19; 22		
1516		16; **24**	40	
1517	5½; 7			
1518	10	20; **23½**	50	
1521		12		
1522		15; **18**; 19; **22**; **24**; 25		
1525	3	**20**		
1529		25		
1531	6 M 10 s			
1532	6⅓; 9		**40**	
1533	7½; 7¾	18		
1538		25		
1545			60	100

The prices are in Riga marks. Numbers in bold indicate the prices of travel horses. A line on a darker background indicates the monetary reform of 1422. Abbreviations: M – mark; s – schilling.

References

Source Materials

Tallinn City Archives [TLA], 230_1_Ad 32.

Literature

Arbusow, Leonid / Bauer, Albert (eds.): Heinrichs Livländische Chronik [HCL], 2nd ed.(Scriptores rerum Germanicarum in usum scholarum ex Monumentis Germaniae Historicis separatem editi 31). Hahn: Hannover 1955.

Blumfeldt, Evald: "Talurahvas keskaja teisel poolel". Kruus, Hans (ed.): *Eesti ajalugu*, vol. 2: *Eesti keskaeg*. Eesti Kirjanduse Selts: Tartu 1937, pp. 213–240.

Blumfeldt, Evald: *Eesti keskaja agraarajalugu*. Akadeemiline Kooperatiiv: Tartu 1937.

Bruiningk, Hermann von (ed.): *Livländische Güterurkunden aus den Jahren 1501 bis 1545* [LGU 2]. A. Gulbis: Riga 1923.

Bruiningk, Hermann von / Busch, Nikolai (eds.): *Livländische Güterurkunden aus den Jahren 1207 bis 1500* [LGU 1]. Jonck & Poliewsky: Riga 1908.

Bulmerincq, August von: *Kämmerei-Register der Stadt Riga 1348–1361 und 1405–1474* [KR Riga], 2 vols. Duncker & Humblot: Leipzig 1909, 1913.

Bunge, Friedrich Georg von / Toll, Robert von (eds.): *Est- und Livländische Brieflade. Eine Sammlung von Urkunden zur Adels- und Gütergeschichte Est- und Livlands.* Part 1, vol. 1: *Dänische und Ordenszeit* [Brieflade 1.1]. Kluge und Stroem: Reval 1856.

Bunge, Friedrich Georg von et al. (eds.): *Liv-, Est- und Kurländisches Urkundenbuch* [LEKUB], 1st series, 12 vols; 2nd series, 3 vols. J. Deubner: Reval et al. 1853–1914.

Dyer, Christopher: *Standards of Living in the Later Middle Ages: Social Change in England c. 1200–1520*, revised edn. Cambridge University Press: Cambridge 1998.

Goetz, Leopold Karl: *Deutsch-Russische Handelsgeschichte des Mittelalters.* Otto Waelde: Lübeck 1922.

Heckmann, Dieter (ed.): *Revaler Urkunden und Briefe.* (Veröffentlichungen aus den Archiven Preussischer Kulturbesitz 25). Böhlau: Cologne 1995.

Horoshkevich, Anna L. (Хорошкевич, Анна Л.): Torgovlja Velikogo Novgoroda s Pribaltikoj i Zapadnoj Evropoj v XIV–XV vekax. Izdatel'stvo Akademij Nauk SSSR: Moskva 1963.

Hupel, August Wilhelm: *Topographische Nachrichten.* Vol. 2. Hartknoch: Riga 1777.

Hyland, Ann: *The Horse in the Middle Ages.* Sutton Publishing: Stroud 1999.

Kazakova, Natal'ja (Казакова, Наталья): Russko-livonskie i russko-ganzejskie otnošenija XIV – načalo XVI v. Nauka: Leningrad 1975.

Leimus, Ivar (Леймус, Ивар): "Grivna serebra (stucke sulvers) v livonskix istočnikax do načala 16 v." In: Gajdukov, Petr Grigor'evič (Ed.): *The Russian Rouble: 700 years of the history. Proceedings of the International Conference, Velikiy Novgorod, April 25–27, 2016.* Velikiy Novgorod 2017, pp. 36–39.

Leimus, Ivar: "Perdeteken – Prägezeichen? Über eine angebliche Quelle zur Dorpater Münzgeschichte". *Numismatisches Nachrichtenblatt* 5, 2014, pp. 220–221.

Leimus, Ivar: "Eesti hobune – strateegiline kaup keskajal". *Tuna* 2, 2017, pp. 10–19.

Mänd, Anu: "Horses, Stags and Beavers: Animals as Presents in Late-Medieval Livonia". *Acta Historica Tallinnensia* 22, 2016, pp. 3–17. DOI: 10.3176/hist.2016.1.01.

Põltsam-Jürjo, Inna / Selart, Anti / Leimus, Ivar: "Maamajandus ja külarahvas". In: Selart, Anti (ed.): *Eesti ajalugu,* vol. 2: *Eesti keskaeg.* Tartu Ülikooli ajaloo ja arheoloogia instituut: Tartu 2012, pp. 185-202.

Selart, Anti: "Waffenembargo in den nordischen Kreuzzügen im 13. Jahrhundert". In: Birli, Sonja et al. (eds.): *ene vruntlike tohopesate. Beiträge zur Geschichte Pommerns, des Ostseeraums und der Hanse. Festschrift für Horst Wernicke zum 65. Geburtstag.* Verlag Dr. Kovač: Hamburg 2016, pp. 549-558.

Selart, Anti (Селарт, Анти): "Vod´ v zapadnoevropejskix istočnikax XIII-XIV veka". *Stratum plus* 5, 2009, pp. 529-538.

Stavenhagen, Oskar / Arbusow, Leonid (eds.): *Akten und Rezesse der livländischen Ständetage* [AR], 3 vols. Riga 1907-1914.

Vogelsang, Reinhard (ed.): *Kämmereibuch der Stadt Reval 1432-1463* [KBR 1432-1463]. Böhlau: Cologne 1976.

Vogelsang, Reinhard (ed.): *Kämmereibuch der Stadt Reval 1463-1507* [KBR 1463-1507]. Böhlau: Cologne 1983.

[Wilde, Peter Ernst]: *Von der liefländischen Pferdezucht, und einigen bewährten Pferdecuren. Anno 1770.* Oberpahlen 1770.

Акты относящиеся к истории Западной России, собранные и изданные Археографическою комиссиею. Vol. 1: 1340-1560. Санкт-Петербург 1861.

Lembi Lõugas

Fish and Fish Trade by the Archaeological Sources[1]

Abstract: Fish has been serving as a very important food resource for people throughout the history of humankind. Archaeological data show that fish appeared to be an essential component of food in the Eastern Baltic especially during the Stone Age, while with the appearance of agricultural activities fishing began to gradually lose its weight. However, next to the farming products fish never lost its importance. It is only that the frequency and focus of fishery became more limited: e.g. we do not find bones of marine fish at the Iron-Age archaeological sites in the Eastern Baltic. The archaeological finds of fish show that marine fishing of herring and cod started to evolve only in the fourteenth century and onwards, but seafood had been traded to the Eastern Baltic since the thirteenth century. Dried cod first reached Estonia probably with the Danes in the thirteenth century, but was imported to a greater extent by Hanseatic merchants in the fourteenth and fifteenth centuries. According to written sources, herring was being fished already at the beginning of thirteenth century; however, this was done mainly by the settlers of Swedish origin on Estonia's western and northern coasts. Still, herring fishing is not well reflected in the archaeological material of the time, but emerges somewhat later. Opposite to the trade in cod, pike was exported from the Eastern Baltic to the Western countries. It is well described in historical sources, but we have also archaeological evidence of special processing of pikes, interpreted as a preparation for sale.

Keywords: Eastern Baltic, fish trade, Hanseatic, cod, pike, herring

Introduction

Fish has been a very important food resource for people throughout the history of humankind. According to archaeological records, fish appeared to be an essential component of food especially during the Mesolithic and Neolithic in the eastern Baltic (i.e. from ca 9000 to 1800 BCE–Before Current Era), while fishing began to gradually lose its weight when agricultural activities emerged around 2200 BCE. However, fish never lost its importance compared to farming

---

[1] Research for this chapter was supported by the Estonian Research Council grant PRG29 "Foreign *vs* local in medieval and modern age foodways in the eastern Baltic: tracing the changing food consumption through provenance analyses" and core facility NATARC TT14.

products, but was a substantial source of protein in the human diet. Only the frequency and focus of fishery changed and became more local in sense of fishing grounds, e.g. prehistoric farmers did not practice fishing of marine fish or deep sea fishing and therefore we do not find bones of marine fish from archaeological sites dated to the Late Prehistoric in the eastern Baltic. Thus, fishery was limited by freshwater bodies and / or brackish coastal waters where the same species of freshwater fish could live. The archaeological finds of fish show that only in the fourteenth century and onwards, the marine fishing of herring and cod started to evolve. Nevertheless, seafood had been traded to the eastern Baltic since the thirteenth century.[2] The dried cod first reached Estonia probably with Danes in the thirteenth century, but more intensively by Hanseatic merchants in the fourteenth to fifteenth centuries. According to written sources, herring was fished even at the beginning of the thirteenth century, but this was mainly by the western and northern coastal settlers of Swedish origin in northern Estonia.[3] However, fishing for herring at that time is not well reflected in the archaeological material, but appears somewhat later. Opposite the trade of cod, pike was exported from the eastern Baltic to western countries.[4] It is well described in historical sources, but we also have archaeological evidence of special processing of pike, interpreted as preparation for sale.

In this chapter, the focus is laid on the archaeological sources of fishery, fish trade and the post-glacial formation of fish fauna in the eastern Baltic. As examples, few case studies are presented that demonstrate original study material and conclusions based on the analyses of archaeological fish remains.

Archaeological Sources in Detection of Fish and Fish Trade

Fish remains are the most commonly found archaeological evidence of fishery, fish consumption and trade in human settlement sites. Bones, scales and otoliths

---

2   Holm, Poul: "Commercial Sea Fisheries in the Baltic Region c. AD 1000–1600". In: Barrett, James H. / Orton, David (eds.): *Cod and Herring: The Archaeology and History of Medieval Sea Fishing*. Oxbow Books: Oxford 2016, pp. 13–21.
3   Kahk, Juhan / Tarvel, Enn (eds.): *Eesti talurahva ajalugu* 1. Olion: Tallinn 1992; Põltsam-Jürjo, Inna: "Kala tähtsusest kaubanduses, majanduses ning toidumenüüs 13.–16. sajandi Eestis". *Acta Historica Tallinnensia* 24, 2018, pp. 3–23. DOI: 10.3176/hist.2018.1.01.
4   Verliin, Aare et al.: "Quantification of the Early Small-Scale Fishery in the North-Eastern Baltic Sea in the Late 17th Century". *PLoS ONE* 8(7), 2013, pp. 1–11. DOI: 10.1371/journal.pone.0068513.

(ear bones) are what survive for the archaeological record, although fish otoliths are usually less well preserved than other skeletal parts. Recovery of fish bones can be made more difficult by the small size of many of them, often lost unless careful sieving and flotation procedures are used. Fish remains found from the archaeological layers of human habitation sites give us information about the fish species caught or transported in a certain time period, and the anatomical composition of these remains show how the fish was processed and whether fish was eaten within the site or prepared for transport and / or for sale.

The archaeological record of fish as a food resource is sometimes hampered by the later settlement over earlier in many coastal sites, especially towns, where evidence of extensive fishing might be found. Furthermore, archaeologists have tended to neglect fishing because it very often does not leave behind those "nice items", which attract archaeologists to collect fishery evidence (except from the Stone Age). From the settlement sites of the late prehistoric and historical periods, fish remains are the main direct evidence of fishery, fish consumption, and trade.

Fishing gear is not often found from the late prehistoric and medieval sites. Different from that of the Stone Age, of which many fish hooks, spears and even net remains[5] have been found, the later periods are known by the modest amount of finds of fishing gear.[6] The most common items among fishing gear are iron hooks since they preserve much better in soil than nets and / or basket traps. The reason why we find less fishing gear from the late prehistoric and historical sites is that this equipment was probably kept separately from the other everyday stuff and / or more nets and basket traps were used, which do not preserve well in soil. It seems likely special fisherman huts were founded directly on the shore of the water and therefore fewer fishing items get into the human habitation places.

There is no direct archaeological evidence among everyday ware that eating fish required any special dish. However, for transport, some special "packages" can be identified. For example, in 2015, a fourteenth-century shipwreck was found in Tallinn, Kadriorg,[7] where a barrel of herring was found. Barrels were actively used for food storage and transport in medieval times and later, but no evidence is known from the earlier periods in the eastern Baltic. A small

---

5   See e.g. Lõugas, Lembi: "Stone Age Fishing Strategies in Estonia – What Did They Depend On?" *Archaeofauna* 5, 1996, pp. 101–109.
6   See e.g. Lõugas, Lembi: "Development of Fishery During The 1st and 2nd Millennia AD in The Baltic Region". *Estonian Journal of Archaeology* 5(2), 2001, pp. 128–147.
7   Roio, Maili et al.: "Medieval Ship Finds from Kadriorg, Tallinn". *Archaeological Fieldwork in Estonia 2015*, 2016, pp. 229–248.

box made of birch bark containing herring bones was found from the same shipwreck, which makes it possible to conclude that the box was also used to store fish. Such boxes are known in ethnographic collections as well.

Formation of Fish Fauna in the Eastern Baltic

For understanding the process of fish fauna formation in the eastern Baltic water system, we should have a look behind the millennia. Post-Glacial history of the Baltic Sea is quite short in the sense of the geological time frame and reaches as far back as 11–12 thousand years. During its history, the Baltic Sea had some freshwater as well as saline water stages, which both influenced the formation of aquatic fauna by making or breaking barriers on the way of fauna distribution within the sea. In the Late Glacial, when a gradual warming of the climate occurred, a huge freshwater body formed in the Baltic basin. It was melting ice, which filled up the depression area. According to Rosentau et al. (2009)[8] the first freshwater stage – the Baltic Ice Lake – existed between 13,300 and 11,600 years BP (Before Present). It was a huge body of water, which reached up to 140 metres at that time since glaciers still blocked the outflow to the Atlantic Ocean.

During the Ice Age, especially during the stage of maximum ice cover about 20,000 years ago, freshwater fish got isolated in different refuges in Europe. It was probably somewhat easier for marine fish to survive the cold period in oceans, than for freshwater fish that had to find a way to southern waters. The refuges where fish fauna could survive in Europe are hard to establish exactly, but can be detected according to the Late and Post-Glacial formation of bodies of water. Archaeological fish finds are always helpful in the detection of spatial-temporal history of fish, but their remains are rare among the Late Glacial Baltic finds. Characteristic material comes from the cave sediments of southern Poland. There, among other cold-adapted fish species like perch (*Perca fluviatilis*) and burbot (*Lota lota*), bones of whitefish (*Coregonus* sp.) and bullhead (*Cottus* sp.) were found.[9] The *Coregonus* group of fish is not common in that area nowadays, but was most probably spread to the north when glaciers retreated. Different forms of whitefish are found in the lakes of northern Poland (formed during the Late Glacial period) as well as in many other northern European lakes and the Baltic Sea nowadays. It seems that these mentioned cold-adapted freshwater

---

8   Rosentau, Alar et al.: "Development of the Baltic Ice Lake in the Eastern Baltic". *Quaternary International* 206, 2009, pp. 16–23. DOI: 10.1016/j.quaint.2008.10.005.
9   Lõugas, Lembi et al.: "Palaeolithic Fish From Southern Poland: A Palaeozoogeographical Approach". *Archaeofauna* 22, 2013, pp. 123–131.

fish were the first immigrants into the ice lake system of northern Europe. In addition, it has been supposed that smelt (*Osmerus eperlanus*) and four-horned sculpin (*Triglopsis quadricornis*) belong to this group of early immigrants, too.[10]

An event about 11,600 years ago, which marks the end of the Baltic Ice Lake, is called the "Billingen catastrophe". It means that at the Billingen plateau in central Sweden, water of the Baltic Ice Lake was drained into the Atlantic Ocean. The Baltic water level dropped about 25–30 metres in a short time and equalised with the ocean level.[11] Events in the Billingen and Post-Glacial land formations and hydrological conditions have changed the Baltic Sea many times. There was still a huge amount of water connected to the Scandinavian glacier, but it melted gradually with the rising temperature. Released water found its way over the huge area in central Sweden: over Närke Sound. A marine connection with the ocean was opened for the first time during a short period: from ca 11,600 to 10,800 years BP.[12] At that time, the outflow of freshwater was stronger than the inflow of saline water into the Baltic. Therefore, the salinity level was only few per mill in the Yoldia Sea – the newly formed sea in the Baltic basin, which got its name after the sea mollusc *Yoldia arctica*. As the marine influence was quite weak at that time, we cannot really tell how many fish species of marine origin entered the Baltic during this sea stage. One of them was most probably herring (*Clupea harengus*), which had to survive and adapt in freshwater conditions.[13] There are no archaeological herring bones known within the Baltic Sea, which could be dated to the Yoldia Sea stage. However, the bones of one species of marine mammal – the ringed seal (*Pusa hispida*) – are found from the Yoldia sediments, which is also confirmed by radiocarbon dating.[14] While the ringed

---

10 Paaver, Tiit / Lõugas, Lembi: "Origin and History of The Fish Fauna in Estonia". In: Ojaveer, Evald / Pihu, Ervin / Saat, Toomas (eds.): *Fishes of Estonia*. Teaduste Akadeemia Kirjastus: Tallinn 2003, pp. 28–46.
11 See e.g. Rosentau et al.
12 See e.g. Andrén, Elinor / Andrén, Thomas / Sohlenius, Gustav: "The Holocene History of The Southwestern Baltic Sea as Reflected in A Sediment Core from the Bornholm Basin". *Boreas* 29, 2000, pp. 233–250. DOI: 10.1111/j.1502-3885.2000.tb00981.x.
13 Munthe, H: "On the Development of The Baltic Herring in The Light of The Late Quaternary History of the Baltic". *Arkiv för Zoologi* 9, 1956, pp. 333–341; Lõugas, Lembi: *Post-glacial Development of Vertebrate Fauna in Estonian Water Bodies: A Palaeozoological Study*. (Dissertationes Biologicae Universitatis Tartuensis 32). University of Tartu Press: Tartu 1997; Ojaveer / Pihu / Saat 2003.
14 Ukkonen, Pirkko et al.: "An Arctic Seal in Temperate Waters: History of The Ringed Seal (Pusa hispida) in The Baltic Sea and its Adaptation to The Changing Environment". *The Holocene* 24(12), 2014, pp. 1694–1706. DOI: 10.1177/0959683614551226.

seal was more dependent on the whitefish populations as a food source, its occurrence in the Baltic was not dependent on herring.

After the short Yoldia period, land uplift in central Sweden closed a connection between the Baltic and Atlantic. The slightly brackish water in the Baltic was replaced by freshwater, which still flowed from the north, from the melting Scandinavian glacier, and from the south, from the large rivers, which descended into the Baltic. The formed freshwater body in the Baltic basin got the name Ancylus Lake. The name comes from the freshwater mollusc *Ancylus fluviatilis*, which was found from the sediments dated to 10,800–8800 years BP.[15] Different from the previous stages, conditions in the Ancylus Lake were more temperate, especially in coastal shallow zones. It was a good opportunity for freshwater fish to spread within the Baltic basin, and thanks to the relatively high water level, fish had access to many bays, which got isolated from the Baltic after the lowering of the water level. In the Ancylus Lake, many cold-adapted or moderately cold-tolerant fish species were probably widely distributed like the grayling (*Thymallus thymallus*), trout (*Salmo trutta*), pike (*Esox lucius*), roach (*Rutilus rutilus*), perch, burbot, ruffe (*Gymnocephalus cernuus*), etc. The other group of freshwater fish, which inhabited the Ancylus Lake, are those of Ponto-Caspian origin.[16] These are more thermophilic fish species, including asp (*Aspius aspius*), vimba bream (*Vimba vimba*), blue bream (*Abramis ballerus*), tench (*Tinca tinca*), wels catfish (*Silurus glanis*), and pikeperch (*Sander lucioperca*). Most of these fish species are represented in archaeological material within the Baltic Sea: e.g. the bone finds of pikeperch found from the Early Mesolithic human settlement site Pulli in Estonia prove that there was already a developed population of that fish in Pärnu Bay at the end of the Yoldia / beginning of Ancylus stage.[17] This is one of the oldest evident populations of pikeperch within the Baltic basin that still exists today.

The Ancylus Lake persisted until ca 8800 BP when a connection with the Atlantic Ocean was opened again in the region of the contemporary Danish

---

15 Jensen, Jörn B. et al.: "Early Holocene History of The Southwestern Baltic Sea: The Ancylus Lake Stage". *Boreas* 28, 1999, pp. 437–453. DOI: 10.1111/j.1502-3885.1999.tb00233.x; Andrén et al. 2000.
16 See e.g. Paaver / Lõugas 2003.
17 Lõugas 1997; Lõugas, Lembi: "Postglacial Development of Fish and Seal Faunas in The Eastern Baltic Water Systems". In: Benecke, Norbert (ed.): *The Holocene History of the European Vertebrate Fauna: Modern Aspects of Research*. (Archäologie in Eurasien 6). Marie Leidorf: Rahden 1999, pp. 185–200; Lõugas, Lembi: "Mõnedest mesoliitilistest faunakompleksidest Läänemere idakaldalt". *Muinasaja teadus* 17, 2008, pp. 253–260.

straits. The lake was transformed into the relatively salty Litorina Sea, which got its name from the marine mollusc *Litorina litorea*, found from concurrent sediments. This saline water sea lasted from ca 8800 until 4500 years BP. Thanks to the connection with the ocean, which developed starting with the sinking of land in the area of the Danish straits, the Baltic reached its maximum salinity ca 6500 years BP.[18] The salinity was 15–18 per mill even in the eastern Baltic at that time, exceeding the current level by 4–6 per mill. In addition, the mean temperature of water was a few degrees higher in the Litorina Sea than in the Baltic today.

A remarkable part of the true marine component of Baltic fish fauna originates from the Litorina stage. Marine fauna can be divided into permanent inhabitants and occasional visitors. The visitors are more frequent in the western part of the sea, while in the eastern part, their number diminish according to the distance from the Danish straits. The human settlement sites on the seacoast have yielded a significant number of fish and marine mammal bones from that period, proving that the diversity of sea organisms attracted people to exploit them (see Figure 1). At that time, marine fish like the cod (*Gadus morhua*), herring (*Clupea harengus*), flounder (*Pleuronectes flesus*) and turbot (*Scophthalmus maximus*) were actively fished. Anadromous eel (*Anguilla anguilla*) should be also added to that list.[19] In addition to humans, other predators competed for fish in the Litorina Sea. Namely, the grey seal (*Halichoerus grypus*), harp seal (*Phoca groenlandica*) and porpoise (*Phocaena phocaena*) appeared in the archaeological material of human settlements around the Baltic at that time.[20]

---

18 Gustafsson, Bo G. / Westman, Per: "On the Causes For Salinity Variations in The Baltic Sea During The Last 8500 Years". *Paleocenography* 17(3), 2002, pp. 12-1–12-14. DOI: 10.1029/2000PA000572.
19 Lõugas 1999; Paaver / Lõugas 2003.
20 Lõugas 1999; Storå, Jan: *Reading Bones. Stone Age Hunters and Seals in the Baltic.* (Stockholm Studies in Archaeology 21). Stockholms universitet: Stockholm 2001; Sommer, Robert / Benecke, Norbert: "Post-Glacial History of The European Seal Fauna on The Basis of Sub- Fossil Records". *Beiträge Zu Archäozoologie und Prähistorischer Anthropologie* 6, 2003, pp. 16–28; Schmölcke, Ulrich: "Holocene Environmental Changes and The Seal (Phocidae) Fauna of The Baltic Sea: Coming, Going and Staying". *Mammal Review*, 2008, pp. 231–246. DOI: 10.1111/j.1365-2907.2008.00131.x.

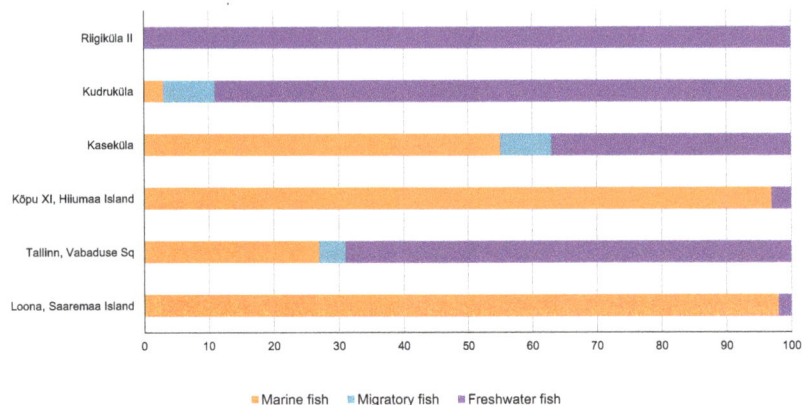

Figure 1. Relative amount of marine vs freshwater fish in some of the Mesolithic and Neolithic bone assemblages in Estonia.

Approximately 4500 years ago the inflow of saline water through the straits decreased and the Litorina Sea turned into the more brackish and colder Limnea Sea, which we know today as the Baltic Sea. Changes in the composition of the fish fauna of the Baltic basin did not end with the immigration of marine species. The later changes have been mostly quantitative, some species becoming rare and others increasing in numbers or broadening distribution area (see also Figure 2). In addition, some species had to adapt with the new conditions in the brackish Baltic Sea, and developed into different forms. Today, we talk about the Baltic herring (*Clupea harengus membras*), which in turn has at least two subforms (autumn and spring spawning herring), and the Baltic cod (*Gadus morhua callarias*).[21] A general characteristic that describes marine fish in the Baltic is that their growth rate is slower than their counterparts in the ocean. It means that fish of the same age can be remarkably different in size: bigger in the ocean and smaller in the Baltic. Different growth rates may be useful in the interpretations of archaeological bone finds since usually the size of a fish bone is not a good characteristic of how to separate subspecies. However, if we compare bone sizes of fish of the same age, we can differentiate between the Atlantic or Baltic origin of them quite well.

---

21 Ojaveer / Pihu / Saat 2003.

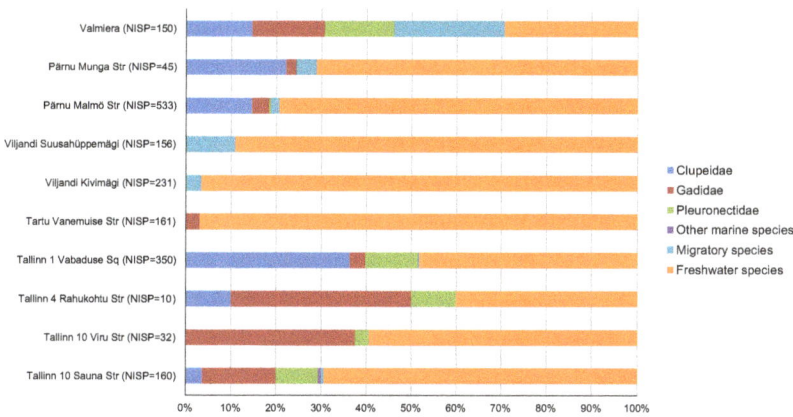

Figure 2. Relative amount of different fish groups in some of the medieval bone assemblages in Estonia.[22]

Case Studies

Studies on how to detect and distinguish medieval cod of local Baltic origin and those traded from the Atlantic have been in focus in Estonia for many years. The first time it was discussed was in 2001, in a paper about the development of fishery during the first and second millennia in the Baltic region.[23] There the evidence of traded fish was searched and it was stated that the main proof of the fish trade was unexpected finds of fish remains of species that could not have lived in the bodies of water surrounding the site. At the time of publishing, a find of cod bones from medieval material from Tartu, central Estonia, was the main proof of the cod trade from the coast to the inland. However, the origin from the ocean was not evident at that time. The trade is also evidenced by the finds of oyster shells in urban contexts of the eastern Baltic.[24] Another observation in the

---

22 Lõugas, Lembi: "Fishing and Fish Trade during the Viking Age and Middle Ages in the Eastern and Western Baltic Sea Regions". In: Barrett, James H. / Orton, David (eds.): *Cod and Herring: The Archaeology and History of Medieval Sea Fishing*. Oxbow Books: Oxford 2016, pp. 111–116.
23 Lõugas 2001.
24 See e.g. Lõugas, Lembi / Jürjo, Inna / Russow, Erki: "European Flat Oyster (*Ostrea edulis* L.) in the Eastern Baltic as Evidence of Long-Distance Trade in Medieval and Early Modern Times". *Heritage* 5, 2022, pp. 813–828. DOI: 10.3390/heritage5020044.

mentioned study concerns the large vertebrae of cod found from the medieval layers of Tallinn. As the cod in today's catches of the eastern Baltic is about 40–60 cm in body length, the vertebrae of some cod of medieval Tallinn are from 90–110 cm or even longer individuals. Therefore, these very large fish were presumed to have originated from the ocean and must be regarded as a result of trade. These large cod, represented only by vertebrae, are interpreted as remains of dried fish, i.e. specially processed and dried "stockfish", which is still common in Scandinavia today, especially in Norway (Figure 3). Finds of vertebrae of probable stockfish appeared first in the layers of the thirteenth century Danish residence in Toompea Castle, Tallinn. These finds are in good accordance with the medieval Danish records, which show substantial trade from North Sea suppliers of cod to the royal household, which primarily was located in the eastern Baltic parts of the realm.[25] Later, stockfish and other marine products became common trading objects of the Hanseatic League.

Figure 3. Drying the cod in northern Norway. Photo: Lembi Lõugas.

Estimating the body length of fish is not an absolute method for differentiation between the Atlantic and Baltic specimens. The growth of fish is dependent on

---

25 Holm 2016.

so many environmental and individual factors that in some circumstances the Baltic cod may grow as big as the Atlantic cod. Here, other methods appear to be more helpful in provenance analyses. Nowadays, biomolecular (DNA) and stable isotope researchers give very specific information about the origins of organisms. Only recently a few DNA analyses have been carried out from the fish remains of the Estonian archaeological sites, but detecting stable isotopes from bone collagen has been conducted on several bones.[26] The method is based on the knowledge about the circulation of carbon and nitrogen isotopes 12/13 (C) and 14/15 (N) within the environment. As the isotopes ^{12}C and ^{14}N are most common in nature, the ^{13}C and ^{15}N ratios in the organic compounds of organisms may vary according to the environment where they consume or in which trophic level they are in a food chain. In general, the ^{13}C value is lower in terrestrial than in marine environments, meanwhile the ^{15}N value is smaller in the lower trophic level (e.g. in plants) and larger in the higher trophic level (e.g. in carnivores). Carbon isotope ratios are commonly used to distinguish between marine and terrestrial resources, and nitrogen isotope ratios are useful in determining the trophic level of the organism but can often be used to distinguish between aquatic and terrestrial diets as well.[27] Since the cod has well-understood migration patterns in which many populations remain close to their spawning grounds, isotopic signatures can in principle be used to assign individual specimens to likely catch regions. Thus many cod bones found from Tallinn, and also Tartu, originate from fish of North Sea origin or even from Arctic Norway.[28] Fish bones and isotope analyses in combination with other archaeological finds shed light on the trade more efficiently than taxonomy or morphometrics alone.

In 2015, a new discovery was made at a construction site in Kadriorg (Pikksilma Street 2), Tallinn, Estonia. There, a fourteenth century shipwreck was found. Archaeological excavations of the shipwreck with a length of 15 m and width of 6 m were carried out, which have resulted in a large amount of

---

26 E.g. Orton, David et al.: "Stable Isotope Evidence For Late Medieval (14th–15th C) Origins of The Eastern Baltic Cod (Gadus morhua) Fishery". *PLoS ONE*, 2011. DOI:10.1371/journal.pone.0027568.

27 Schoeninger, Margaret J. / DeNiro, Michael J.: "Nitrogen and Carbon Isotopic Composition of Bone Collagen From Marine and Terrestrial Animals". *Geochimica et Cosmochimica Acta* 48, 1984, pp. 625–639; Bocherens, Hervé / Drucker, Dorothée: "Trophic Level Isotopic Enrichment of Carbon and Nitrogen in Bone Collagen: Case Studies From Recent and Ancient Terrestrial Ecosystems". *International Journal of Osteoarchaeology* 13 (1–2), 2003, pp. 46–53. DOI: 10.1002/oa.662.

28 Orton et al. 2011.

spectacular finds. According to the dendrochronology of oak wood of the board and typology of finds the wreck dates to the first half of the fourteenth century AD.[29] It is a medieval cog-like merchant ship (*Hansekogge*) that was discovered. Many everyday items made of metal, wood, birch bark, leather, textile, clay and stone, have been found both inside and around the ship. There were some signs of burning visible on the wreck, which gave reason to assume that there was a fire on board that caused the shipwreck. Probably the lower part of the ship, which was not damaged so much, sank and became covered by sandy sediments quite fast. This caused good preservation conditions for the organic compounds of the materials, including different animal products. In the frame of the current study the focus is on the fish remains found *in situ* from the bottom of this wreck. Fish remains had some clear distinguishable accumulation places within the ship: kitchen waste in the ship's stern, two separately located bundles of fish vertebral columns, the abovementioned wood barrel, and a box made of birchbark.

There, about 2675 fish bones were gathered and analysed.[30] The largest fish-bone assemblage comes from the stern of the ship, and could be associated with kitchen waste. The kitchen waste includes at least 47 individual specimens of pike. Among 1445 bones of pike there were only 135 vertebrae, which all came from the end of the cranial or caudal part of the vertebral column. The rest of the bones were from fish heads. This accumulation seems as typical waste, which is a result of fish processing – decapitation – and from where the fish trunks have been taken away, e.g. sold, somewhere else. In addition to the pike remains, 89 bones of cod, mainly vertebrae, and only one dental bone of turbot were found from this waste accumulation.

Two separately located accumulations of fish vertebral columns included only cod and were also found from the ship's stern, but more towards the middle part of the ship. One bundle of vertebral columns was found on the starboard side and the other the port. There were vertebrae of cods, 16 individuals in one bundle (Figure 4) and ten in another. Most probably their tails were tied together for transport. From the pectoral girdle, the most commonly represented bones are cleithrums. In most cases they were cut in half during decapitation, i.e. the posterior part of the cleithrum bone (sometimes the scapula is also connected) stays with the trunk.[31]

---

29 Roio et al. 2016.
30 Lõugas, Lembi / Bläuer, Auli: "Detecting Medieval Foodways in the Northeastern Baltic: Fish Consumption and Trade in Towns and Monasteries of Finland and Estonia". *Environmental Archaeology* 26(6), 2001, pp. 543–554. DOI: 10.1080/14614103.2020.1758993.
31 Roio et al. 2016.

Fish and Fish Trade by the Archaeological Sources 213

Figure 4. A bundle of cod vertebral columns in the shipwreck found in Kadriorg, Tallinn, in 2015. Photo: Lembi Lõugas.

The analyses of sediments from a wood barrel and a box made of birchbark gave quite a small amount of herring bones, as well as many scales, especially from a barrel. There were also two bones of 15 cm long perch in this barrel, which came from two different individuals. The body size of herrings refers rather to the Baltic herring, staying ca 18–25 cm in total length.

Even the cod finds from the shipwreck in Kadriorg, Tallinn, indicate long-range trade, but the pike and herring from this ship seem to be more of local origin. Accumulation of pike heads in the kitchen waste gives the impression that the processing of fresh fish took place on board and seemingly those pikes were fished not very far from Tallinn. Probably such kitchen waste formed during quite a short time just before the shipwreck: keeping stinky fish remains on the boat for a longer time seems doubtful. Here, the question arises of who got the pike's trunks. Most probably those fish, a minimum of 47 individuals, were not cleaned for food for the ship's crew, but rather for selling them to townsmen or drying before trading abroad. In the case of herrings, they also seem to be of local (Baltic) origin according to their sizes, but the studied material does not indicate whether they are imports or exports in Tallinn or rather food for the ship's crew. Another example of the accumulation of pike head bones comes from Kastre Castle, central-eastern Estonia, dated to the end of the fifteenth, beginning of the sixteenth centuries.[32] Additionally, the castle of Cēsis in Latvia yielded a lot of bones of pike heads.[33] These accumulations indicate the export of pike from the eastern Baltic probably to the West in medieval times.

Conclusion and Research Perspectives

There is a need for comparison of zooarchaeological data of fish and written records in future studies. In order to meet this aim, a selection of archaeological sites and other deposits reflecting relevant human activities will be chosen for zooarchaeological studies of fish. Potential collections are stored in Tallinn University and the University of Tartu, as well as in various museums. New archaeological excavations in urban contexts usually yield a great amount of bones. They should be included in the studies as soon as possible. Some samples for fish analyses are already collected all over the study region in order to establish wide spatiotemporal coverage of the material. Analysis of hand-collected and sieved vertebrate bones will focus on relative species abundance and size preferences, the level of fragmentation and related taphonomy, and skeletal abnormalities and modifications, particularly butchery and processing. This allows us to reconstruct changing dietary profiles for each site, species preferences, levels of fishing and import, and the introduction of new species (e.g. carp or crucian carp – typical species of fish breeding). The analyses of fish bones comprise a fundamental base for other studies, such as stable isotope and DNA as well.

References

Andrén, Elinor / Andrén, Thomas / Sohlenius, Gustav: "The Holocene History of The Southwestern Baltic Sea as Reflected in A Sediment Core From The Bornholm Basin". *Boreas* 29, 2000, pp. 233–250. DOI:10.1111/j.1502-3885.2000.tb00981.x.

Bocherens, Hervé / Drucker, Dorothée: "Trophic Level Isotopic Enrichment of Carbon and Nitrogen in Bone Collagen: Case Studies From Recent and Ancient Terrestrial Ecosystems". *International Journal of Osteoarchaeology* 13(1–2), 2003, pp. 46–53. DOI:10.1002/oa.662.

---

32 Lõugas, Lembi et al.: "Duty on Fish: Zooarchaeological Evidence From Kastre Castle and Customs Station Site Between Russia and Estonia". *International Journal of Osteoarchaeology* 29(3), 2019, pp. 432–442. DOI: 10.1002/oa.2764.

33 Lõugas, personal observations in 2018.

Gustafsson, Bo G. / Westman, Per: "On the Causes for Salinity Variations in The Baltic Sea During The Last 8500 Years". *Paleocenography* 17(3), 2002, pp. 12-1-12-14. DOI:10.1029/2000PA000572.

Holm, Poul: "Commercial Sea Fisheries in the Baltic Region c. AD 1000-1600". In: Barrett, James H. / Orton, David (eds.): *Cod and Herring: The Archaeology and History of Medieval Sea Fishing*. Oxbow Books: Oxford 2016, pp. 13-21.

Jensen, Jörn B. et al.: "Early Holocene History of The Southwestern Baltic Sea: The Ancylus Lake Stage". *Boreas* 28, 1999, pp. 437-453. DOI:10.1111/j.1502-3885.1999.tb00233.

Kahk, Juhan / Tarvel, Enn (eds.): *Eesti talurahva ajalugu* 1. [History of the Estonian Peasantry.] Olion: Tallinn 1992.

Lõugas, Lembi / Bläuer, Auli: "Detecting Medieval Foodways in the Northeastern Baltic: Fish Consumption and Trade in Towns and Monasteries of Finland and Estonia". *Environmental Archaeology* 26(6), 2001, pp. 543-554. DOI:10.1080/14614103.2020.1758993.

Lõugas, Lembi / Jürjo, Inna / Russow, Erki: "European Flat Oyster (*Ostrea edulis* L.) in the Eastern Baltic as Evidence of Long-Distance Trade in Medieval and Early Modern Times". *Heritage* 5, 2022, pp. 813-828. DOI:10.3390/heritage5020044.

Lõugas, Lembi et al.: "Duty on Fish: Zooarchaeological Evidence From Kastre Castle and Customs Station Site Between Russia and Estonia". *International Journal of Osteoarchaeology* 29(3), 2019, pp. 432-442. DOI:10.1002/oa.2764.

Lõugas, Lembi et al.: "Palaeolithic Fish From Southern Poland: A Palaeozoogeographical Approach". *Archaeofauna* 22, 2013, pp. 123-131.

Lõugas, Lembi: "Development of Fishery During the 1st and 2nd Millennia AD in the Baltic Region". *Estonian Journal of Archaeology* 5(2), 2001, pp. 128-147.

Lõugas, Lembi: "Fishing and Fish Trade during the Viking Age and Middle Ages in the Eastern and Western Baltic Sea Regions". In: Barrett, James H. / Orton, David (eds.): *Cod and Herring: The Archaeology and History of Medieval Sea Fishing*. Oxbow Books: Oxford 2016, pp. 111-116.

Lõugas, Lembi: "Mõnedest mesoliitilistest faunakompleksidest Läänemere idakaldalt". [Some Mesolithic Faunal Complexes From The Eastern Shore of The Baltic Sea.] *Muinasaja teadus* 17, 2008, pp. 253-260.

Lõugas, Lembi: "Postglacial Development of Fish and Seal Faunas in The Eastern Baltic Water Systems". In: Benecke, Norbert (ed.): *The Holocene History of the European Vertebrate Fauna: Modern Aspects of Research*. (Archäologie in Eurasien 6). Marie Leidorf: Rahden 1999, pp. 185-200.

Lõugas, Lembi: *Post-glacial Development of Vertebrate Fauna in Estonian Water Bodies: A Palaeozoological Study*. (Dissertationes Biologicae Universitatis Tartuensis 32). University of Tartu Press: Tartu 1997.

Lõugas, Lembi: "Stone Age Fishing Strategies in Estonia – What Did They Depend On?" *Archaeofauna* 5, 1996, pp. 101–109.

Munthe, H: "On The Development of The Baltic Herring in The Light of The Late Quaternary History of The Baltic". *Arkiv för Zoologi* 9, 1956, pp. 333–341.

Orton, David et al.: "Stable Isotope Evidence For Late Medieval (14th–15th C) Origins of The Eastern Baltic Cod (Gadus morhua) Fishery". *PLoS ONE*, 2011 DOI:10.1371/journal.pone.0027568.

Paaver, Tiit / Lõugas, Lembi: "Origin and History of The Fish Fauna in Estonia". In: Ojaveer, Evald / Pihu, Ervin / Saat, Toomas (eds.): *Fishes of Estonia*. Teaduste Akadeemia Kirjastus: Tallinn 2003, pp. 28–46.

Põltsam-Jürjo, Inna: "Kala tähtsusest kaubanduses, majanduses ning toidumenüüs 13.–16. sajandi Eestis". *Acta Historica Tallinnensia* 24, 2018, pp. 3–23. DOI:10.3176/hist.2018.1.01.

Roio, Maili et al.: "Medieval Ship Finds From Kadriorg, Tallinn". *Archaeological Fieldwork in Estonia 2015*, 2016, pp. 229–248.

Rosentau, Alar et al.: "Development of the Baltic Ice Lake in the Eastern Baltic". *Quaternary International* 206, 2009, pp. 16–23. DOI:10.1016/j.quaint.2008.10.005.

Schmölcke, Ulrich: "Holocene Environmental Changes and The Seal (Phocidae) Fauna of The Baltic Sea: Coming, Going and Staying". *Mammal Review*, 2008, pp. 231–246. DOI:10.1111/j.1365-2907.2008.00131.x.

Schoeninger, Margaret J. / DeNiro, Michael J.: "Nitrogen and Carbon Isotopic Composition of Bone Collagen From Marine and Terrestrial Animals". *Geochimica et Cosmochimica Acta* 48, 1984, pp. 625–639.

Sommer, Robert / Benecke, Norbert: "Post-Glacial History of The European Seal Fauna on The Basis of Sub-Fossil Records". *Beiträge Zu Archäozoologie und Prähistorischer Anthropologie* 6, 2003, pp. 231–246.

Storå, Jan: *Reading Bones. Stone Age Hunters and Seals in the Baltic.* (Stockholm Studies in Archaeology 21). Stockholms universitet: Stockholm 2001.

Ukkonen, Pirkko et al.: "An Arctic Seal in Temperate Waters: History of The Ringed Seal (*Pusa hispida*) in The Baltic Sea and its Adaptation to The Changing Environment". *The Holocene* 24(12), 2014, pp. 1694–1706. DOI:10.1177/0959683614551226.

Verliin, Aare / Ojaveer, Henn / Kaju, Katre et al.: "Quantification of the Early Small-Scale Fishery in the North-Eastern Baltic Sea in the Late 17th Century". *PLoS ONE* 8(7), 2013, pp. 1–11. DOI: 10.1371/journal.pone.0068513.

Kadri Tüür

Trawling for Atlantic Herring in Estonian Literature

Abstract: The article traces the consumption history of Atlantic herring in Estonia on the basis of literary sources. Herring as a food item and its cultural connotations are contextualised in the current debate of global commons and ocean resource use. Historical data about herring catches are juxtaposed with instances of herring representations in Estonian literature in order to map the stable periods, as well as the highs and the lows, of the catches. The appearance of herring in runo songs indicates that Atlantic herring as consumer goods can be associated with the trade networks of the Hanseatic League. Literary sources of the nineteenth and twentieth centuries help to trace the changing status of the herring as a food item – from a festive dish to sustenance for the poor. The twentieth century also saw several attempts of producing 'Estonian own' herring. When herring stocks in Northern Atlantic collapsed, herring disappeared from Estonian literature as well.

Keywords: Atlantic herring, blue humanities, runosongs, literary representations

Introduction

The present article is an attempt to contribute to the emerging research field designated as blue humanities.[1] Taking a sea-bound perspective in the study of cultural history might help to shift the "methodological nationalism" that almost goes without saying in land-based narratives toward a more entangled and inclusive approach. As the promoter of Atlantic history David Armitage wrote, shifting the focus from land to sea "stressed mobility and circulation and focused on exchange and hybridity, integration and communication".[2] The primary interaction of humans with the oceans is through fisheries. From those Atlantic cod and herring fisheries are probably the most well known and influential, especially in the context of European history. Herring as a subject

---

1 See Gillis John R.: "The Blue Humanities". In: *Humanities*, 34(3), 2013. Retrieved 10.07.2023 from https://www.neh.gov/humanities/2013/mayjune/feature/the-blue-humanities. This article has been supported by the PRG908 Estonian Environmentalism in the 20th century: ideology, discourses, practices.
2 Armitage, David / Bashford, Alison / Sivasundaram, Sujit: *Oceanic Histories*. Cambridge University Press: Cambridge, 2018, p. 87.

matter provides an opportunity to look into human–fish relations, as well as into the trade interactions of humans connected with fish, as well as the human–ocean interaction that occurs in the chase for herring.

Around 1500, the North Atlantic Fish Revolution, i.e. a dramatic increase in North Atlantic fish extraction that was likely to triple the availability of fish protein in the European market took place, as Poul Holm et al. suggest in their article.[3] The authors point out the necessity for multi-disciplinary research in order to understand the complex phenomenon of historical fishing, the dynamics of fisheries, and how it has shaped the course of human history on firm ground. The methods and data of environmental history and historical ecology are combined in their research. In the following, historical data about herring catches is juxtaposed with instances of herring representations in Estonian literature in order to map the stable periods, and highs and lows of the catches. The primary sources for the present survey are traditional Estonian folk songs and written literary texts. A transdisciplinary approach encourages bringing together a variety of sources to cast light on one central question from different disciplinary perspectives. This contribution looks at historical fisheries from a literary perspective. In order to bring together such seemingly far-apart fields, it has been necessary to combine methods from folkloristics, digital humanities, literary history, and semiotics. The combination is by no means clear cut, but it has led to at least some preliminary results to be shared in the form of the present article. My main question is how well the literary representations of herring coincide with the historical fisheries data about herring catches, and with the historical data about herring trade in the Baltics.

The article about the Fish Revolution presents a table of variables that need to be considered in the research of historical fisheries.[4] They can well be taken as the guideline for studying other historical periods and events as well. The proposed variables can be grouped as environmental, economic, social, political and cultural factors, and the authors differentiate between the application of these factors on the supply and on demand of fish. From the perspective of the present study, the cultural factors related to demand are central indeed. With the help of literary sources, we can look into fish-eaters' preferences and their esteem for herring, as well as the self-esteem, life, and working conditions of the herring fishermen and herring consumers in some cases.

---

3   Holm, Poul et al: "The North Atlantic Fish Revolution (ca. AD 1500)". *Quaternary Research* 2019, p. 3. https://doi.org/10.1017/qua.2018.153.
4   Holm et al. 2019, p. 3.

	Abiotic	Biotic	Economic	Social	Cultural	Political
Supply	Temperature, wind, currents, salinity, nutrients	Primary production, target fish abundance and distribution	Capital, technology, labour	Distribution of settlements	Esteem, migrant cultures	Subsidies, tariffs, war
Demand	Climate, shocks, hazards	Diseases	Integration, information flows	Human population	Preferences, religious prescriptions	Strategies (economy, settlement, navy), war

Figure 1. Proposed factors to be considered in historical fisheries research.
Source: Holm et al. 2013, p. 3.

The central themes of the current collection are trade, representations and relations, which can also be observed in the case of herring. For ecological reasons, Atlantic herring only reaches the shores of the Baltic Sea as a trade item, not as a live fish, and therefore it has for a long time been a part of the local economic history. Herring has been imported, sold and bought, it has even been used as a means of payment. The perception of a fish that is not seen alive often relies on representations. In oral poetry and literature, we must take into account the fact that herring appears there not at its "face value", but it is represented through certain poetic filters. Often such representations need to be carefully contextualised in order to be able to interpret them in a reasonable manner. Hands-on human–herring relations is a topic that appears in Estonian literature only in the 1930s, when the fishing fleets were organised to bring "our own herring" from the Atlantic ocean.[5] This, in turn, is related to fish as a trade object – or even more, in the Soviet context, herring is a means for fulfilling the

---

5   See Tüür, Kadri / Stern, Karl: "Atlantic Herring in Estonia: In the Transverse Waves of International Economy and National Ideology". In: *Journal of Baltic Studies*, 46(3), 2015, p. 396. DOI: 10.1080/01629778.2015.1073928. The article discusses Estonian endeavour of herring fishing in the 1930s and Atlantic herring as an item of international and domestic trade. We combined official and personal records of the same sequence of events in the 1930s, and demonstrated how both herring and its Estonian hunters were caught in the transversal waves of international trade and national aspirations. The approach of that article was definitely nation-centred, not sea-bound. Also published in Plath, Ulrike / Mincytė, Diana (eds.): *Food Culture and Politics in the Baltic States*. Routledge: London / New York 2017.

Five-Year Plans. In the framework of the Soviet plan-based economy, it is not the market that shapes the catches, but the Communist Party and its plans for achieving communism. Herring becomes a political instrument.[6] Through one species, such as herring, it is possible to put the individual histories into a larger context for further insights.

Herring as a Model Organism

A researcher of herring fisheries and fishing regulations, Mark Dickey-Collas,[7] writes that North Sea herring is a model species in fisheries research – it has been treated as a resource and overexploited by industrial means. It has been part of the menu of European people for at least the past 400 years; as a result of industrial catches that intensified at the turn of the nineteenth to twentieth centuries, the fish stock collapsed and was closed for exploitation for five years in 1977.[8] It is now recovering, but the fishing activities related to herring are closely monitored and controlled. Atlantic herring is also a fine model organism for literary research, as the amount of relevant texts is not ungraspable, and the existing ones form certain thematic groups – folk songs, belletrist texts, and fishing stories.

Estonia is located at a corner of the brackish Baltic Sea where a sub-species of herring, the Baltic herring, lives. Atlantic herring reaches those shores only in barrels, caught, salted and merchandised. Today, wooden barrels have been replaced by plastic trays, and refrigeration is implemented, but the main characteristic of herring has remained the same – it is bought and consumed here, not caught or admired alive. Thus the question arises, why should we expect to find herring in Estonian literature, and what would it tell us anyway? A possible answer might be that herring as a traditional consumer good and food item was an important part of local inhabitants' everyday life. Although

---

6   See Lajus, Julia: " 'Red Herring': The Unpredictable Soviet Fish and Soviet Power in the 1930s". In: Wormbs, Nina (ed.): *Competing Arctic Futures. Historical and Contemporary Perspectives*. Palgrave Macmillan: Cham 2018, pp. 73–94. DOI:10.1007/978-3-319-91617-0_4.

7   Dickey-Collas, Mark: "North Sea Herring. Longer Term Perspective on Management Science Behind The Boom, Collapse and Recovery of The North Sea Herring Fishery". In: *Management Science in Fisheries*. Routledge: London 2016. DOI:10.4324/978131 5751443.

8   Andrews, Jim / Nichols, John: *Assessment Data Sheet. PFA & SPSG North Sea Herring Fishery*. Final Report. Acoura Marine Ltd: Edinburgh, 2017.

widely known to all local classes, due to its somewhat exotic character it has left its traces in literary representations.

From a biological perspective, Atlantic herring (*Clupea harengus harengus*), as its name indicates, inhabits waters of Atlantic origin. It grows to be 37 cm, in the coastal waters of Iceland even 42 cm long. Its weight can reach 0.5 kilograms per individual. Herring may live up to 18 years.[9] It is also a source of vitamin D, having thus been an important addition to the predominantly grain-based menu of the peasants living in the Baltic region.

For the Baltic area, a local race of herring, *Clupea harengus membras*, is an important counterpart to the Atlantic herring. It lives in the brackish waters east of the Danish Sound. The growth of Baltic herring varies greatly between populations that inhabit different sea areas. On average Baltic herring are caught when they are 5–8 years old and 20–25 cm in length.[10] It has been one of the main food items for the local peasantry: salted Baltic herring was eaten as a daily "fast food" and it was affordable to all social classes.[11] Literary representations and cultural significance of Baltic herring will remain beyond the scope of the present article as the relevant material is considerably more ample than that concerning Atlantic herring.

From the perspective of the local languages, the difference between the Atlantic and Baltic herring lies not only in biology, but also in how they are called. The table presented in the ICES FishMap fact sheet[12] (see Figure 2) reveals that there are two different roots used to designate Atlantic herring in European languages. Quite generally the names can be grouped as Northern European *sill* area and Western European *herring* territories. Perhaps surprisingly, Estonia belongs to the latter group; the name for the fish has evidently been taken over via German, which used to be the local lingua franca from the thirteenth to twentieth centuries. It is also remarkable that the word designating Baltic herring is of a completely different root than that for the Atlantic one in Swedish, Estonian and Latvian. This, in turn, indicates the important position of Baltic herring in these cultures. It could also be that those two subspecies were not perceived as related

---

9   Dickey-Collas, Mark: *The current state of knowledge on the ecology and interactions of North Sea Herring within the North Sea ecosystem*. Report. WOT kennisbasis 3251229218, 2004.
10 Ojaveer, Evald / Pihu, Ervin / Saat, Toomas (eds.): *Fishes of Estonia*. Teaduste Akadeemia Kirjastus: Tallinn 2003.
11 Moora, Aliise: *Eesti talurahva vanem toit. Ilmamaa*: Tartu 2007, p. 365.
12 *ICES FishMap. North Sea Fish Species Fact Sheets. Herring*. Available at ICES website http://www.ices.dk/marine-data/maps/Pages/ICES-FishMap.aspx, retrieved 10.07.2023.

in vernacular knowledge, as they did not appear side by side in catches of the Baltic Sea. One was local, the other foreign. Linguist Mari Kendla writes in her study of Estonian fish names that Atlantic herring, cod and Baltic herring have all been metaphorically called *merehulgus*, "the one that roams the distant seas".[13]

common names			
Danish	Sild	*Icelandic*	Sild
Dutch	Haring	*Latvian*	Siļķe (Atl.) / Reņģe (Baltic)
English	Herring	*Norwegian*	Sild
Estonian	Heeringas (Atl.) / Räim (Balt.)	*Polish*	Sledz
Faeroese	Sild	*Portuguese*	Arenque
Finnish	Silli (Atl.) / Silakka (Balt.)	*Russian*	Сельдь (Atl.) / Салака (Baltic)
French	Hareng	*Spanish*	Arenque
German	Hering	*Swedish*	Sill (Atl.) / Strömming (Baltic)

Figure 2. Vernacular names of Atlantic and Baltic herring in European languages. Source: ICES FishMap fact sheet, p.1.

Periodisation

The article on the Fish Revolution that occurred ca 1500 in connection with the discovery of the Grand Banks fisheries off the coast of Newfoundland[14] points out that until then, fish had been a limited and high-priced resource in Europe. The main species that were consumed, either dried or salted, were cod and herring. Fresh marine fish could not be transported beyond the distances of 30–50 kilometres from the shore.

The authors offer a table depicting Dutch herring catches in the North Sea from 1580 until the beginning of the twentieth century. The red and green lines in the figure indicate the total allowable catches established for 1997–2007, i.e. the boundaries of ecologically sustainable fishing. It can be seen from the table that for the most part of the period, the catches have been relatively modest, dropping especially low in the beginning of the nineteenth century (due to Napoleon's wars and a shortage of salt), to rise then from the 1840s and rocketing beyond the estimated limits of sustainability in 1900 (see Figure 3).

---

13 Kendla Mari: "Eestikeelsed kalanimetused: teaduslikud versus rahvapärased". *Mäetagused* 57, 2014, pp. 53–68.
14 Holm et al. 2019, p. 1.

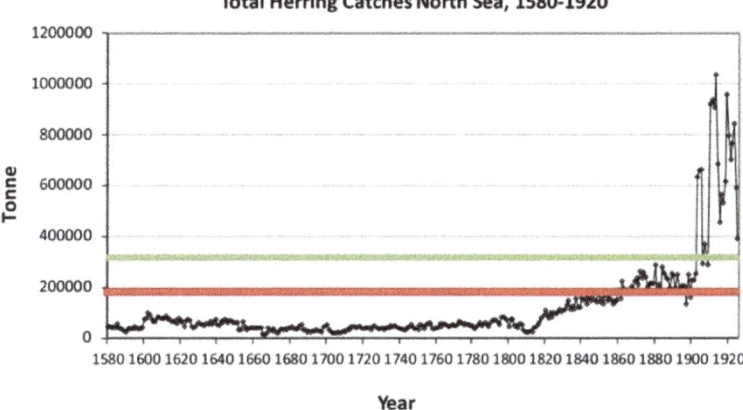

Figure 3. The dynamics of North Sea herring catches, 1580–1920. Source: Holm et al. 2013, p. 5.

It should also be noted that the present table reflects only Dutch catches, whereas other coastal communities have engaged in small-scale fishing for herring in the North Sea and elsewhere, too. It means that the total herring catches have probably been somewhat greater, but the graph still reflects the overall dynamics of herring fishing – until the nineteenth century catches were relatively stable and small. This had to do with the traditional, labour-intensive fishing techniques employed; those started to be modernised in the 1840s when steamers were taken into use for first transporting and then trawling for herring.[15]

Richard W. Gear[16] points out that small-scale fishing has not been sufficiently researched and taken into account in the history-writing of oceanic resource exploitation. On the example of Shetland's herring industry he distinguishes three larger periods that match the outline of the graph presented above: 1400–1700, Hanseatic trade period; 1700–1800, domestic merchandise; and 1800–1870, growth of the herring industry. He also writes that most of the herring exported from Shetland during the whole period were taken to various Baltic ports – and some of it probably eventually reached Estonia.

---

15 Coull, James R.: "The Development of Herring Fishing in the Outer Hebrides". *International Journal of Maritime History*, XV(2), 2003, p. 31.
16 Gear, Robert William: "Re-assessing Shetland's Herring Industry before the 1870s". *Journal of the North Atlantic*, 4, 2013, pp. 61–68, p. 63.

The period of relative scarcity and stability of herring production is reflected in Estonian runosongs that are discussed below. Stories and plays in Estonian from the nineteenth century also allow themselves to be mapped onto this graph: the catches grow, the availability of the fish on the market makes it more accessible and less prestigious food, and the alteration of the status of herring can also be observed in literary representations.

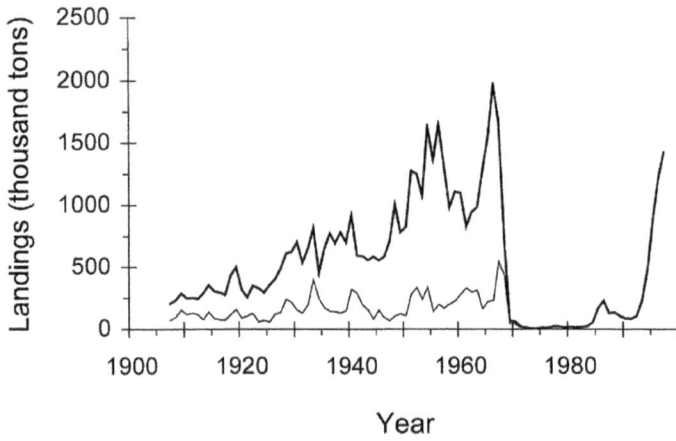

Figure 4. Landings of Norwegian spring-spawning herring in the twentieth century. Thick line – total landings; thin line – young herring. Source: Toresen, Østvedt 2000, p. 236.

Figure 4 features data about Norwegian herring catches during the twentieth century. The thick line follows the dynamics anticipated in the previous graph: the catches grow in the beginning of the century, to drop for a while during WWI, and then continue growing, making a leap in the 1930s, when the Estonian fleet, among others, started the endeavour of herring fishing in the North Atlantic. After WWII, the catches continue to grow, especially in the 1950s – that episode is documented in a play, a children's book and several poems by Juhan Smuul who participated in a Soviet Estonian herring fishing trip to the Norwegian Sea in 1955. When the herring stock collapsed and fishing was stopped by international agreements,[17] herring disappeared from Estonian literature, too.

---

17 Andrews, Jim / Nichols, John: *Assessment Data Sheet. PFA & SPSG North Sea Herring Fishery*. Final Report. Acoura Marine Ltd: Edinburgh, 2017. p. 25.

Herring Trade

Salted herring was already consumed in Central Europe in medieval times. Unlike many other seafood items, salted herring can be transported and stored over long distances and periods of time without dramatic loss of its nutritional and gustative qualities. The salting process itself is laborious and needs specific skill. Herring is a fatty fish. Before refrigerating became an option, the only way to preserve it was to seal freshly caught, gutted and salted herring into airtight wooden barrels so that it would not rot. Processing herring is more demanding than preserving cod, which can be salted and dried.[18]

A popular handbook of Estonian food states that herring arrived at the tables of Estonians at latest in the thirteenth century when Hanseatic merchants started to trade in the recently Christianised areas.[19] The Hanseatic League, a merchants' union specialising in maritime trade, was prominently present throughout the Baltic coast and inland from the thirteenth to seventeenth centuries. Coastal towns, such as Riga, Ventspils, Tallinn, Pärnu, and inland towns Viljandi, Tartu, Valmiera, etc. belonged to the League as full members. Viborg had a Hanseatic office; Narva was a trade centre that did not belong to the League.[20] The most important trade items imported to the Baltic Hanseatic towns included salt and herring.[21] Michael North points out that supra-national networks such as those of the Viking, Hanseatic, and Soviet powers are all instances of communication and cultural exchange that cannot be fully grasped from a land-based national perspective.[22] Moving the point of view to the maritime environment may help to reveal new details in the cultural history of the coastal communities.

From that perspective it is even possible to reconstruct the Baltic Sea as a "Dutch lake": after the gradual decline of the Hanseatic League, Dutch ships started to take over maritime transport, whereas local trading was left to Hanseatic merchants.[23] The Dutch had skills in shipbuilding and navigation, as well as in herring catching and processing. Historically, Dutch fishermen

---

18 Coull 2003, p. 25.
19 Bardone, Ester / Kannike, Anu / Põltsam-Jürjo, Inna / Plath, Ulrike: *101 Eesti toitu ja toiduainet*. Varrak: Tallinn 2016, p. 20.
20 Kiaupa, Zigmantas / Mäesalu, Ain / Pajur, Ago / Straube, Gvido: *The History of the Baltic States*. Avita: Tallinn 2002, p. 61–63.
21 North, Michael: "The Baltic Sea". In: Armitage, David / Bashford, Alison / Sivasundaram, Sujit: *Oceanic Histories*. Cambridge University Press: Cambridge, 2018, pp. 209–233.
22 North 2018, p. 209.
23 Ibid., p. 220.

invented the method of salting fresh herring in wooden barrels in the fifteenth century. That enabled fishers to perform this procedure immediately onboard the special herring ships (called loggers or busses). These boats fished on the open sea and brought back market-ready salted herring in barrels. This activity laid the basis for Dutch economical welfare during the fifteenth to sixteenth centuries.[24]

Salting fresh fish on the board was inevitable in the case of the open sea fishing far from the destination of the consumption of the product. It was crucial to ensure the quality of this work for long-term preservation. The barrels for herring were also made by skilled Dutch coopers, and the timber for the barrel staves often came from the Baltics.[25] In addition to Dutch herring, Scandinavian, Scottish, and Norwegian herring was traded here.

Historian Inna Põltsam-Jürjo writes that in Pärnu, a major Estonian coastal city and a member of the Hanseatic League, a trade policy from the end of the fifteenth century stated that when selling herring on the market, the place where it had been caught and salted had to be clearly indicated.[26] In the middle of the sixteenth century it caused complaints among the merchants that "salt, herring, iron, and all sorts of peasant goods" were sold to the local population in manors, evading official markets.[27] Hanseatic merchants certainly took into account the needs of local peasantry in addition to those of the upper-class city-dwellers of different ethnic origins and of the nobility, the so-called Baltic Germans. Herring was demanded by all social groups. Archival evidence demonstrates that herring has even been used as a part of a travelling craftsman's salary.[28]

Along with Baltic herring, Atlantic herring has been long known to Estonian peasants as a food item. "An old-established main course on the party table of Estonian peasants has been Atlantic herring," Aliise Moora writes.[29] For holiday seasons, salted herring was bought, even by the coastal people who caught and sold other fish species from the Baltic Sea. According to Moora, about 120 years ago Atlantic herring gradually started to replace Baltic herring in

---

24 See Pihu, Ervin (ed.): *Loomade elu IV. Kalad*. Valgus: Tallinn 1979, p. 100 and Sicking, Louis / Abreu-Ferreira, Darlene (eds.): *Beyond the Catch: Fisheries of the North Atlantic, the North Sea and the Baltic, 900–1850*. Brill: Leiden, 2009.
25 North 2018, p. 220.
26 Põltsam, Inna: *Liivimaa väikelinn varase uusaja lävel. Uurimus Uus-Pärnu ajaloost 16. sajandi esimesel poolel*. Tallinna Ülikool: Tallinn 2008, pp. 101–102.
27 Põltsam 2008, p. 104.
28 Ibid., p. 148.
29 Moora 2007, p. 367.

peasants' everyday diet. The shift in the fish preference initially spread among the wealthier peasants from the inland districts (Tartu, Mulgimaa). If the family was wealthier, herring was on the table on the weekdays, too; whereas poorer families still considered it a festive food item. Herring substituted for meat in the daily diet.[30] Salted herring was used up to the last bits: ground heads of both Baltic herring and Atlantic herring have been used for making covered pies.[31] Eating fried herring heads as a lent snack has been recorded from the Brotherhood of the Black Heads in Tallinn.[32] According to some sources, ground Baltic herring (including its head) was a remedy against scurvy.[33] On the other hand, Estonian folk belief forbids eating the eyes of any fish. Folklorist Mall Hiiemäe proposes that this has to do with mythological metonymic thinking – a part of a deceased living being can be used to grow a new being. Eating fish eyes could be regarded as an act of ultimate greed and an omen for looming famine.[34]

Representation of Herring in Runosongs

Poulsen, Holm and MacKenzie[35] write about the shifting baseline syndrome in fisheries research: estimations about the size and conditions of fish stocks depend on the observed time period and on the density of data available from particular periods. In order to make better judgements about historical fisheries, it is necessary to lengthen the timeline under study, including information about the catches of pre-industrial times. In order to do so, often the only option is to rely on so-called proxy data parallel to quantitative data on stock abundance, such as landing or commerce statistics, as well as traditional historical written sources, trade documents, anecdotal evidence, and literary sources.

One such unorthodox source for historical fisheries research is runosong (*regilaul*)[36]. It is estimated on the basis of linguistic research that runosongs

---

30 Moora 2007, p. 368.
31 Moora 2007, p. 399.
32 Põltsam-Jürjo, Inna: "Kala tähtsusest kaubanduses, majanduses ning toidumenüüs 13.–16. sajandi Eestis". *Acta Historica Tallinnensia* 24, 2018, p. 18. DOI:10.3176/hist.2018.1.01.
33 Weberman, Ernst Constantin: *Kala kui rikkalik vitamiinide allik*. Üleriiklne Kalanädala Peakomitee: Tallinn 1927, p. 8.
34 Hiiemäe, Mall: "Kakskümmend kaks kala eesti rahvausundis". *Mäetagused* 11, 1999, pp. 7–33.
35 Poulsen et al 2019, p. 1.
36 https://www.folklore.ee/regilaul/andmebaas. An overview of the rationale behind the creation of the database and the explanation of its content and functioning can be

emerged among the Finnish ethnic groups living in the Finnish Gulf area ca 2000 years ago.[37] They are characterised by a coherent poetical system that includes alliteration and parallelism. Mari Sarv and Janika Oras write: "In runosongs, mythological motifs, plots and figures are intertwined with the realities of peasant life (family life, work, social relations, the environment), and fantasies".[38] As Aliise Moora's above-referred remarks on the food history of Estonian peasants assure, herring could be considered among the realities of peasant life rather than a mythological creature. Therefore it should be justifiable to look into the appearance of herring in Estonian runosong as a reflection of the pre-industrial catch and trade situation.

Major public runosong collection campaigns in Estonia took place at the end of the nineteenth century, continuing throughout the twentieth century in the form of professional folkloristic activity. The number of recorded song texts is estimated to be ca 150,000, making it one of the biggest folklore collections in the world.[39] Not all of the collected runosongs are necessarily 2000 years old, but based on their type, distribution, and linguistic features such as syllabic and prosodic qualities and song melodies they can be grouped, and layers from different eras can be detected. It is possible to associate some of them even with archaeological data.[40]

We used[41] the digital online runosong database in order to find some Estonian proxy data from the more distant literary past. The database is searchable in a number of aspects: by location, performer, collector, genre, song type, or just by a word in the content. This is the option that we used, limiting the search to "runosong" only. However, it is not sufficient to search just for "heeringas", the contemporary correct form of the fish name. As Estonian is a language with a number of archaic dialects and the words shift their shapes in declination

---

found at Sarv, Mari / Oras, Janika: "From Tradition to Data. The Case of Estonian Runosong". In: *Arv. Nordic Yearbook of Folklore*, 76, 2020, pp. 105–117.

37 Sarv, Oras 2020, p. 105.
38 Sarv, Oras 2020, p. 106.
39 Sarv, Mari: "Towards a Typology of Parallelism in Estonian Poetic Folklore". In: *Folklore: Electronic Journal of Folklore*, 67, 2017, p. 87. DOI:10.7592/FEJF2017.67.sarv.
40 See http://hdl.handle.net/10062/56305, doctoral dissertation by Pikne Kama, about the possibilities of using data from runosongs in archaeology.
41 I am very grateful to Kanni Labi, an expert of runosong semantics and an editor at the Estonian Literary Museum, for her kind assistance and creative co-thinking in the database search and interpretation of the results.

(suffixes are added to the end of the word to indicate the case and the syntactic role of the word in a sentence), the search words must be multiple and the ending of the search word must be left open (hence the asterisk). We used the forms *heering**, *eering**, *ering**, *iering**, *hiering**. All together 51 texts were found mentioning herring that qualify as traditional runosongs and that can be divided into folkloristically established song types, such as wedding songs, wooing songs, songs about fishermen, songs about the power of singing, etc. In comparison – a rough search with the words "silk" and "räim" that denote Baltic herring (respectively processed and raw) produced eight times more results.

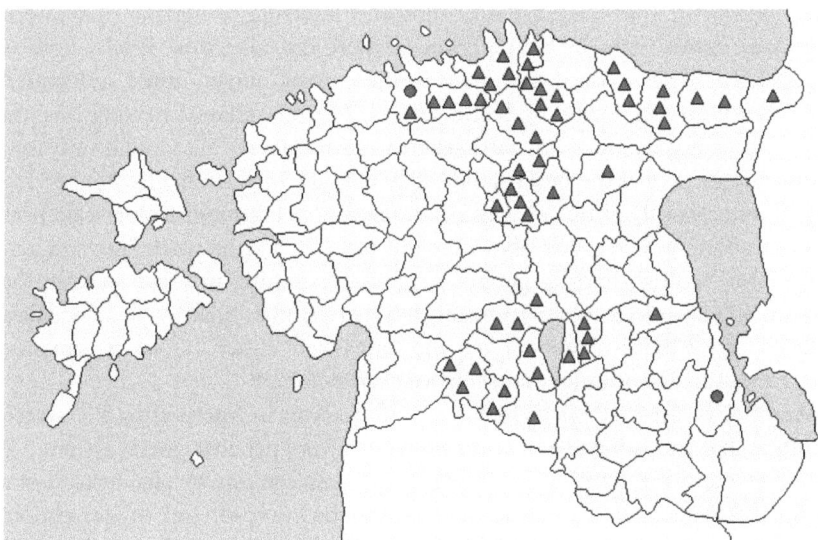

Figure 5. Distribution of runosongs containing references to herring as collected from Estonian parishes. Triangles mark runosongs; circles denote songs of unclear origin that mention herring. Map drawn by Kanni Labi with Kaardimasin app developed by Arvo Krikmann, www.folklore.ee/moistatused/kaardimasin.

Out of 106 Estonian historical parishes,[42] runosongs referring to herring have been collected in 26. Geographically, these parishes are located along the

---

42 The parish was the main administrative unit in Estonian territory during the thirteenth to twentieth centuries. Due to the feudal system, the local peasants were sedentary within the limits of one parish. As a result, distinct folk costumes, dialects, and runosong repertoires developed in each parish.

Northern coast of Estonia, around bigger towns, and in historical Mulgimaa (see the map in Figure 5) – i.e. in the areas where trading has played a more important role than elsewhere. One thing that needs to be taken into account is folklore collecting density – there are certain parishes where more songs have been written down than in others, be it because of their vicinity to Tartu (the centre of the collecting activities), because of some keen collector, or some other reason.[43] Of the parishes featuring herring in their runosongs, Kuusalu is certainly one where folklore has been collected relatively densely. Geographical distribution of our search results reveals that if herring is a topic in runosongs, there are usually three to four songs featuring them collected from one parish. It is noteworthy that there are no runosongs referring to herring in Western Estonian coastal areas, including islands, where the collection density itself is relatively high. The coastal fishermen in these areas caught, salted and traded their own local fish, including Baltic herring. Perhaps Atlantic herring was not part of the everyday life of coastal dwellers to such a degree as to find its way into their runosongs. On the other hand, daily contact with fresh Atlantic herring does not necessarily produce songs about the fish, as ethnomusicologist Frances Wilkins shows in her article about the musical contacts between Shetland and Greenland. Scottish herring gutters sang while working in order to keep the rhythm of the work (Estonian runosong has had a similar function in association with collective physical labour), but the songs were either hymns or romantic songs about treacherous love and shipwrecked sailors.[44]

Another factor that has to be taken into account in interpreting the search results is that runosong as a phenomenon has a very peculiar poetic system.[45] It is stichic instead of stanzaic, i.e. its lines are organised one by one, instead of a grouped organisation (stanzas) as in the rest of the European folk song tradition. The latter was introduced in Estonia only in the seventeenth to eighteenth centuries, parallel to the earlier system. The linear stichic organisation is based on euphony in the form of alliteration instead of the end rhyme. Linear organisation means that in order to express a meaning, lines with a certain structure can be added to each other ad libitum.[46] There is a repertoire of recurrent poetic

---

43 Sarv, Oras 2020, p. 113.
44 Wilkins, Frances: "'Da Merry Boys o Greenland': Explorations into the Musical Dialogue of Shetland's Nautical Past". *Folk Music Journal* 11(2), 2017, pp. 17–37, p. 32.
45 Sarv, Mari: "Regilaul kui poeetiline süsteem." [Runosong as a Poetic System.] In: Oras, Janika / Västrik, Ergo-Hart (eds.) *Paar sammukest XVII. Eesti Kirjandusmuuseumi aastaraamat.* Eesti Kirjandusmuuseumi Teaduskirjastus: Tartu 2000, pp. 7–122.
46 Sarv, Mari 2009: "Stichic and Stanzaic Poetic Form in Estonian Tradition and in Europe". *Traditiones*, 38(1), 2009, pp. 161–171.

formulae that are used as "building blocks" and combined with each other in order to express a more general (mythological) meaning. This brings along an important feature, namely parallelism: "The list of parallel elements altogether forms a concept of a more abstract level", as Sarv briefly puts it.[47] The meaning emerges in context, not in a single line. In case of herring it means that it usually appears among lines mentioning either other fish or more generally, valuable items, contributing toward more abstract categories of either "fish" or "valuables".

Poetic synonyms of Estonian substantives in Estonian runosong have been catalogued, systematised and studied by Juhan Peegel.[48] In his substantial work, completed far before the emergence of digital humanities, "herring" does not appear as a separate entry, but only as a synonym for "maiden". Peegel brings the following verses originating from Karksi parish in Mulgimaa as an example of the usage of herring in the poetic system of runosong: *Esä õiges eeringes, / ema kutse kulluses, / muu pere munasorase, / talu kutse tateresse, / külä kutse köömelesse.*[49] It is a song from the type where a maiden's golden life in her family home is contrasted to the subsequent harsh life in her husband's home. At her native home, she is called not by her given name, but by a number of nicknames referring to valuables, such as herring and small bell, as well as egg, buckwheat, and cumin, which, in turn, have mythological or folk medicine-related connotations. Songs featuring herring that belong to the same type have been collected from the neighbouring parishes that all belong to the historical Mulgimaa area: Halliste, Paistu, Tarvastu. This confirms Aliise Moora's remark that herring was widely known, bought and consumed in Mulgimaa, an area that is located far from the coast and that has historically been relatively wealthier than other Estonian rural areas.

The same prevalent idea, associating young women with herring as both are cherished and valuable, continues in the song types where a young man persuades a maiden to marry him and where she is promised a good life in her new home. In general, wedding and wooing songs are among the most archaic ones among runosongs. There herring appears as an indicator of wealth and a prospective comfortable life: *Meil et kuke kulda joova /Kuke kulda kana karda / Ani allasta õbedat /Meil et tsia silku söövä /Ärjä aina eeringida /Mullika muud kalada*[50] – poultry pecks on gold and silver, pigs eat Baltic herring, oxen eat

---

47 Sarv 2017, p. 80.
48 Peegel, Juhan: *Nimisõna poeetilised sünonüümid eesti regivärssides*. Eesti Keele Sihtasutus: Tallinn 2004.
49 H I 2, 144 (7) < Karksi khk., Polli v. – J. Tiidt (1888).
50 H III 16, 709/10 (4) – A. J. (1889).

Atlantic herring, and cattle munch on other fish. If domestic animals are already granted such luxury items, the lives of the human members of the farmstead must be even better.

The grim reality reflected in the song type "Beautiful / young wife needs fish", which forms ca 1/6 of all the herring-related runosongs, is, however, that not all men are able to provide their beloved women with fish. A young woman who has married an old and feeble man, regrets it to tears: *Noor nüid hakkas nutamaie, / ilus hakkas igatsemaie / kallike kahetsemaie. / Noor tahab süia nopet-näpet / kallis kalalihada / ilus süia heeringida, / peenikene püilileiba, / sile tahab sealihada, / kaunike kanamuneda.*[51] She cries and regrets it because she would like to eat fish, especially herring, but also bread made of fine flour, as well as pork and eggs, which her husband is not able to provide. The food items listed here are all among the main ingredients of traditional Estonian festivity dishes,[52] and herring is an established component of the festive meals along locally produced food items. A more general message of that song type could be that not all days are holidays anyway.

Herring also appears in the songs that describe the everyday chores of successful fishermen: *Minu vennake merela / Minu kaasake kalala / Esmaspäev toob heringida / Teisipäev toob tintisida / Kesknädal kena kalada.*[53] Here, both the singer's brother and husband (i.e. in a generalised sense, all the male family members) are fishing on the sea and bringing back different fish: Atlantic herring on Monday, smelts on Tuesday, other nice fish on Wednesday. Here, herring stands for "all sorts of fish" in the row of parallel verses. It has the same function, to be just one among a list of all the known fish, in similar songs that describe fishing in rivers, lakes, and even bogs, where actual catches of herring are ecologically impossible. However, Atlantic herring appears to be familiar enough for the ancient runo singers to be included in the nomenclature of fish along more common local species, such as Baltic herring, perch, eel, pike, ruff, ide, roach, and others that appear in the verses of those songs in the vicinity to the ones containing herring.

Herring, fishing, and distant seas are combined in some songs from Kuusalu parish that speak about the need to go further off the native coast in order to catch Atlantic herring. There are even some indications toward the origin of herring in

---

51 H III 3, 321/2 (19).
52 Moora 2007, p. 395.
53 EÜS IX 961 (27) < Peetri khk., Esna v., Embra k. – K. Viljak & W. Rosenstrauch < Leenu Steinberg, 72 a. (1912).

runosongs: *Minu oma Ollandissa, / Minu kaasa kaugeella, / Minu peigu Peipisissa, / Üvamees üle mereda, / Püüab aga kallile kalada.*[54] – My dear man is in Holland / far away / on Lake Peipus / across the seas, catching fish for his beloved. Holland, along with Lake Peipus and the city of Riga, are relatively frequently recurrent place names in runosongs. Associating (herring) fishing with the Holland area is historically sound. Historian Michael North writes, "During the late fifteenth and sixteenth centuries, the decline in Baltic and North Sea herring amplified the importance of Dutch herring fisheries".[55] There is no actual evidence, however, of Estonian men working on Dutch herring ships, but this possibility cannot be absolutely excluded. The weakening of the Hanseatic League during the sixteenth century made Dutch ships take over marine transport in the Baltic Sea.[56] The main goods transported by the Dutch at that time were herring and salt.

Access to foreign goods is praised in a wedding song from Kuusalu parish: *Sie tuond ruad Ruotsimaalta, / Piperad tuoned Piedarista, / Hieringed Vene rajalta, / Salatimed Saksamaalta, / Suolad tuonud Suomemaalta, / Tangud tuonud Taanimaalta.* The dishes on the wedding table have been made of ingredients, including salt and herring, brought from Sweden, St. Petersburg, Russia, Germany, Finland, and Denmark, i.e. from distant places overseas. It makes them especially valuable and the wedding party especially memorable. It also indicates that the family of the newlyweds must be well-off people with many functional ties across the Baltic Sea. This song could be sung from the late eighteenth century. When St. Petersburg was established as the new capital for Russia, it brought along more active maritime trade, as well as structural changes in the goods imported to this region: textiles and colonial products (such as pepper / spices mentioned in the runosong) replaced the earlier main articles of trade, herring and salt.[57]

There is a group of songs that could be brought together under the keyword "miracles". In the song type that describes the power of singing, runosong singing can transform entire landscapes: *Ma laulas mered maasse / Mere põhjad põrmandusse / Mere liiva lillakussa / Mere veered viidingisse / Mere ääred heeringisse.*[58] Seas are sung into lands, the seabed becomes a floor, and the coastal rim will be covered with bleak and herring. Here we encounter what Mari Sarv

---

54 H II 1, 279/80 (412) < Jõhvi khk., Puru k. < Paide ligidalt – M. Ostrow & O. Kallas < Leena Randmer, 35 a. (1888).
55 North 2018, pp. 217–218.
56 North 2018, pp. 218–224.
57 North 2018, pp. 225.
58 E 18801 (1) < Tartu-Maarja khk., Vesneri – August Feldmann (1895).

has designated "obscure parallelism" in her typology of runosong parallelism:[59] semantic connections between the lines are not clearly obvious, but they still contribute toward a common meaning, here – that something unreal (such as coasts covered with Atlantic herring) will happen as a result of singing.

Another group of runosongs tells about miracles that happen in the city of Viborg upon the release of a singer from the army: *Siis ma sain sõjast tulema. / Sain ma Viiburi linna alla, / Seal ma tegin teise tembu / Panin linna liiva jooksma, / Uulitsad udu sadama. /---/ Siis hakkas meri mürama, / Merihärga möurgelema, / Heering hakkas hilberdama, / Silgud tündris siputama.*[60] The song opens with the line "Then I was released from the war." In Viborg, the singer transforms the town into sand, streets into fog. Then the sea starts to roll, sculpins scream and herrings hover, smaller Baltic herrings kick about in barrels. Evidently a major mess is about, probably because the singer is overwhelmingly joyful to have been able to get out of the war alive. Herring and Baltic herring in barrels are mentioned in connection with the town of Viborg, which was a partner city of the Hanseatic League and a vibrant trade hub thereafter. Again, the combination of fantastic elements and references to realities is somewhat bizarre and definitely allows a number of interpretations of different degrees of conservatism, but the fact that herring is somehow involved in the young soldier's world is still there.

At last, it should be mentioned that in some cases, herring can acquire a directly obscene meaning. There is a small verse recorded as *carmina obscoenae* from Jõhvi on the north-eastern coast of Estonia in 1931,[61] containing a proposal for a young maiden: *Noorikuke, toorikuke, / laena mulle karvapatta: / ma tahan keeta kahte muna, / ühte eada eeringista* (translates roughly as "Oh young lady, would you lend me your kettle where I could boil two eggs and a sturdy herring"). A parallel can be drawn with a similar reference recorded in Shetland where among other herring-related records in The Court Book of Shetland from the early seventeenth century, "One of the more unusual references sees James Mouat of Ollaberry stealing a mare and 'leading his herring upon her'".[62] This record is left uninterpreted in the context of early fisheries research, but a parallel can certainly be drawn with the cases treated by Ken Ird in the present volume.

In conclusion it could be said that herring is present in Estonian archaic runosongs as a festive food, as a metaphor for something precise and valuable

---

59 Sarv 2017, p. 85.
60 H II 15, 525/7 (6) < Harju-Jaani khk., Kambi v. – M. Neumann (1889).
61 ERA II 37, 294 (27) < Jõhvi khk., Voka v., Pühajõe k. – Rudolf Põldmäe < Ann Sillenbach, 78 a. (1931).
62 Gear 2013, p. 62.

(such as young maidens), and as one among the many fish known to local people. Herring is associated with wealth and international trade. It can miraculously appear from the sea as a result of powerful singing, but its appearance in the texts can also be motivated by structural features of runosong, such as parallelism and alliteration. There is a diverse array of reasons and it is difficult to point out which of these is the most important one. What this brief inventory of runosongs reveals, however, is that Atlantic herring was known to the Baltic peasants even before industrial fishing began, and that it had a certain place in their daily lives. Herring does not appear in runosong as a class attribute differentiating peasants from landlords, or lower classes from upper ones, unlike the dynamics that are present in the case of herring in belletristic literature.

Representation of Herring in Pre-WWII Belletrist Sources

The background for this subset of samples is the herring boom that started in the 1840s and continued in a rising direction until WWI (see Figure 1). James R. Coull describes this process on the example of herring fishing in the Outer Hebrides.[63] Whereas the Dutch had been the leaders of the European herring market for centuries, British dominance gradually grew as well. Even in 1713 the peace treaty of Utrecht granted British dominance to the Grand Banks fisheries that had been discovered near Newfoundland in 1497, bringing about the Fish Revolution.[64] Dutch fishermen had invented a system where fresh herring was gutted and salted straight on the sea, onboard special herring vessels.[65] In the beginning of the nineteenth century, fishermen on the east coast of Scotland started to cure herring on shore, not in boats, and this paved the way to their subsequent success. By the middle of the century, more efficient fishing gear and steamers were taken into use. At the end of the nineteenth century it was possible to take advantage of the fast developing railway network, to get the freshly caught fish to consumers in bigger UK cities, as well as continental Europe.[66] Prior to WWI, Scottish herring was sold mainly to German and Russian markets; it was consumed in Estonia, too. The catches went up thanks to motorboats and steamers so that it was possible to send more herring overseas.

---

63 Coull 2003, pp. 21–42.
64 Holm et al. 2019, p. 2, 12.
65 Sicking, Abreu-Ferreira 2009.
66 Coull 2003, pp. 30–31.

The evidence of herring as a trade and food item can also be found in Estonian novels.[67] In the period of critical realism in Estonian literature, we find a recurrent motif of local general stores smelling of herring, petrol, and tar in the works of Estonian realist prose writers Eduard Vilde, August Kitzberg, Oskar Luts, Reed Morn, Peet Vallak, and many others.

Attitudes toward Atlantic herring consumption seem to have somewhat differed in towns and among peasants, as the following examples show. Since the thirteenth century, local peasants were ruled by the German-speaking upper class, designated as Baltic Germans. These two social groups remained distinct and never completely merged.[68] In towns, the situation was somewhat different. Burghers were German-speaking, but their ethnic origin was diverse and social advancement was, at least in theory, an option. Therefore, we can observe herring as an attribute of the urban upstart struggle in some major Estonian novels.

First, a depiction from the countryside. Eduard Vilde (1865–1933), one of the first professional Estonian writers and journalists, was a representative of a literary trend named critical realism. In his novels, he attempted to create characters typical to his age in their everyday milieu, and to generalise the real-life social factors that often led the lives of his fictional characters in undesirable directions.[69] In his novel "To the Frozen North" (*Külmale maale*), published in 1896, he depicts the gloomy life of landless Estonian peasants partly even in a rather naturalist way. At the end of the nineteenth century, cottage-dwellers, i.e. peasants who did not have a farm to live on, but just a hut to live in and landlords to work for, had to rely on their own bodily strength, and if this failed, there were no social institutions to provide any kind of support. This is what happens to the protagonist of the novel, Jaan Vapper. He falls ill, and is not able to continue hard physical work to support his elderly mother and three younger siblings. This eventually leads him to join a group of horse thieves, which, in turn, later leads to severe punishment and forced exile in Siberia. During one of his earliest contacts with the horse thieves, the criminals lend him some money in order to persuade Jaan to join them. We then see the main character squandering the criminal money on buying *silk*, herring, and black tea from the local parish store.[70] Tea is indicated to be a medicine for his mother to treat her respiratory problems. *Silk* (salted Baltic herring) is poor people's everyday food, and in this

---

67 I am grateful to Ulrike Plath, Elle-Mari Talivee, and the late Jaan Isotamm for their help in assembling the following references.
68 Kiaupa et al., 2002, p. 19.
69 Annus et al., 2001, p. 138.
70 Vilde, Eduard: *Külmale maale*. Eesti Riiklik Kirjastus: Tallinn 1960, p. 126.

context, Atlantic herring stands for pure affluence. Jaan buys two sorts of cured fish instead of just modestly spending money on some cheaper local food items to feed his family. His shopping decisions are driven by hunger and emotion, not rational calculation.

Vilde's character was by no means a man of exceptional fate. As another renowned Estonian writer and literary critic Friedebert Tuglas explains in his treatment of the novel, there was an agricultural crisis in the Baltic provinces in the 1890s. Landless people suffered the most; criminal records grew "to epidemic levels".[71] Vilde attempted to demonstrate that criminal behaviour in rural areas is caused by social injustice, not by the low morals of the villagers.

While herring is a sign of wealth among the countryside people of the end of the nineteenth century, things are a bit more complicated in city settings. A major literary figure of the Estonian National Awakening, Eduard Bornhöhe (1862–1923), published a long story "Bugbears. A Story of Life in Tallinn" (*Kollid. Jutustus Tallinna elust*) in 1903. It depicts the "city milieu of the 1880s with its class and nation-based barriers and anachronistic habits to be soon overthrown by the approaching stream of modernity," as the Estonian Writers Lexicon puts it.[72] The text opens with a clash between a Baltic German baron's family and a young man who is a citizen of the United States, but who is originally an Estonian who had emigrated to the States in search of personal freedom and income. The fates of the young protagonist and his Baltic German counterpart intertwine in the home of a Germanised tailor named Pohlig who is originally Estonian, but who has climbed higher on the social ladder in the town milieu of Tallinn by marrying a German girl. The tailor has three beautiful daughters, and the young men compete for the hand of the prettiest one, Helene. When the young baron comes to visit the Pohlig family, the tailor gives a feast that starts with herring and culminates with roasted duck, with all sorts of other fine dishes in between. Although not a wealthy man anymore, the old tailor is in the habit of receiving guests and showing off: "Seldom would an evening pass when there were no guests in the Pohlig house, and he always had something to offer them, be it sometimes merely a piece of bread and herring, and a drink."[73] The old man himself is a feinschmecker and devotes a lot of attention to eating, tasting, and enjoying the food. On one hand, herring is included in the menu of his

---

71 Tuglas, Friedebert: *Eesti kirjanduse ajalugu IV. Kriitiline realism*. Teaduslik Kirjastus: Tartu, 1947, p. 109.
72 Nirk, Endel: "Bornhöhe, Eduard". In: Kruus, Oskar (ed.) *Eesti kirjarahva leksikon*. Eesti Raamat: Tallinn 1995, pp. 65–66.
73 Bornhöhe 1903, p. 52.

decent feast; on the other hand, after the meal he explains: "See, I ate this salty wooden piece of herring only to prepare my palate for sardines and lobsters... Can you guess what else I have inside this fine box? Half a pound of fresh caviar straight from Astrakhan!"[74] Here the same kind of redundancy is depicted as in Vilde's novel: herring is accompanied with other seafood, which in this case is considerably fancier than Jaan's *silk*. So, Bornhöhe's novel features herring as a food frequently eaten by the upstart city-dwellers, although it is indicated that they do not consider it to be among the most decent ones in the possible array of snacks. The emergence of this attitude could be explained by the growing availability of herring at the turn of the twentieth century, as the catches grew and trade networks developed. Herring did not have an aura of exclusiveness about it anymore.

Herring as a marker of the barrier between distinct social groups is clearly present in the novel "I Loved a German" (*Ma armastasin sakslast*, 1935) by Anton Hansen Tammsaare (1878–1940). Tammsaare is among the most important writers responsible for the construction of modern Estonian identity with his epic prose. "I Loved a German" depicts "a tragic relationship between a young Estonian man and a Baltic German girl who prove unable to break the social barriers of the past in the name of their present love."[75] Ulrike Plath has argued that this novel can be read as a hunger novel – hunger for food, love, and culture are the central powers triggering the events.[76]

At the outset of the novel, set during the early years of the independent Republic of Estonia, we see the protagonist buying "four hundred grams of bread and a herring"[77] from a store. The main character, an Estonian parvenu student, is described as feeling inferior for buying such simple and cheap food, because he imagines it would be much more decent to walk the streets with a package, the shape of which would indicate pastries and be associated with the carrier as a gallant cavalier, rather than hurrying home with a single herring that may drip salt water and ruin one's clothes. He, however, cannot deny his poor financial

---

74 Bornhöhe 1903, p. 57.
75 Undusk, Rein: "Tammsaare, Anton Hansen". In: Egläja-Kristsone, Eva / Gasiliūnas, Virginijus / Mihkelev, Anneli: *300 Baltic Writers*. Institute of Lithuanian Literature and Folklore: Vilnius 2009, p. 319.
76 Plath, Ulrike: "Näljast ja näljapsühholoogiast A. H. Tammsaare romaanis "Ma armastasin sakslast"". In: Hinrikus, Mirjam / Undusk, Jaan (eds.) *Armastus ja sotsioloogia. A. H. Tammsaare romaan "Ma armastasin sakslast"*. Underi ja Tuglase Kirjanduskeskus: Tallinn 2013, pp. 50–71, p. 50.
77 Tammsaare, Anton Hansen: *Ma armastasin sakslast*. Eesti Raamat: Tallinn 1984, p. 14.

situation combined with purely physical nutrition necessities, and thus, herring is an optimal choice indeed.

This excerpt shows herring as a regular food item of the less well-off Estonians inhabiting towns. At that time, British herring was imported to Estonia in relatively large quantities. In 1927, 97 % of the Atlantic herring imported to Estonia came from the UK.[78] The 1920s were a golden era for herring imports: in 1925 and 1926, herring imports constituted 3–4 % of the total value of Estonian imports.[79] It means that herring was an important foodstuff, and also that it was widely available on the domestic market. Its prestige and symbolic value certainly diminished as a result. Therefore, we can see that Tammsaare's protagonist, a first-generation, urban, white-collar worker, is ashamed to consume herring because it is nothing exclusive any more. Herring does not help him to climb over the mental and habitual barriers between the two social classes – Estonians of peasant origin and Baltic German nobility. At the end of the nineteenth century, herring consumption had facilitated cross-class communication, as Bornhöhe's novel indicates. Tammsaare's novel demonstrates an opposite instance.

Representation of Herring in Post-WWII Belletrist Sources

After the war, much had changed, from the geopolitical status quo to the available fishing gear. Coull indicates that after WWII, the fishing pressure on herring, as well as on other fish species, increased sharply.[80] The new gear that was taken into use included specially constructed trawlers that were equipped with sonars, drift nets, purse seines, and/or trawling nets. In less than 20 years, all major herring stocks in the North East Atlantic Ocean collapsed.[81] The main states fishing for herring in that area during this period were Norway, Iceland, and the Soviet Union. Total yearly landings were close to 2 million tonnes in the mid-1960s. The collapse occurred in 1970, but commercial fishing was not banned until 1977.[82] A Soviet source on technology of herring fishing assures in 1962 that "There are

---

78 Pihlamägi, Maie: "Eesti kaubandussuhted Suurbritanniaga aastail 1918–1940". *Acta Historica Tallinnensia* 3, 1999, pp. 88–108, p. 92.
79 Tüür, Kadri / Stern, Karl: "Atlantic Herring in Estonia: In the Transverse Waves of International Economy and National Ideology". *Journal of Baltic Studies*, 46(3), 2015, p. 396. DOI:10.1080/01629778.2015.1073928.
80 James R. Coull. "The North Sea Herring Fishery in the Twentieth Century". In: *Ocean Yearbook Online*, vol. 7(1) 1988, pp. 115–131.
81 Thorir Sigurdsson. "The Collapse of the Atlando-Scandian Herring Fishery: Effects on the Icelandic Economy". In: *IIFET 2006 Portsmouth Proceedings* 2006, pp. 1–9.
82 Sigurdsson 2006, p. 2.

abundant resources of high-quality herring available in the North Atlantic Ocean" and provides data about Soviet catches of 1950s as follows:

1952–110,000 tonnes
1954–160,000 tonnes
1958–over 300,000 tonnes.

The State Plan required that in 1965, Soviet trawlers had the obligation to catch at least 220,000 tonnes of herring.[83] This was a utopian goal indeed, as the total biomass of adult herring already in the North Sea had dropped to less than one million tonnes in 1967 due to overfishing, as we retrospectively know.[84] The State Five-Year Plans were drawn by the Soviet central government and it was presumed by the authorities that these plans should be fulfilled and exceeded, or sanctions would be implemented upon the workers.[85] In practical life, including herring fishing, this brought along constant cheating about and exaggerating of the catch quantities, which can even be observed in literature.

The main Soviet harbours for North Atlantic fishing were Kaliningrad and Murmansk. Estonian herring fishing trawlers whose home harbour was Tallinn first joined the Soviet fishing fleets in 1955. Estonian writer Juhan Smuul (1922–1971) was taken along to the very first Soviet Estonian fishing expedition to the North Atlantic. He was born on Muhu Island and he had longed to become a sailor in his youth, but due to the family's financial situation and the war, he could never actually fulfil his dream. After WWII he gained recognition as a poet and quickly made a career in the Soviet Estonian Writers Union, being its long-term head from 1954 until his death in 1971.[86] Writers' trips to different industrial enterprises were a common measure at that time, which was hoped to bring the intelligentsia closer to "real life" and to increase literary production that would conform to the ideological guidelines of Socialist Realism. Evidently, Smuul was considered a suitable middleman by the officials between the readers and the Soviet herring flotilla. It must be added, though, that the captain of the trawler that Smuul was assigned to, Vaino Noor, was from the same island village as the writer himself. Strong personal ties between these men, and the outgoing, humorous personality of the writer resulted in prolific literary production: on

---

83 Ruljov, Nikolai: *Atlandi heeringa esmane töötlemine*. ENSV Rahvamajanduse Nõukogu Tehnilise Informatsiooni Büroo: Tallinn 1962, p. 3.
84 Andrews, Jim / Nichols, John: *Assessment Data Sheet. PFA & SPSG North Sea Herring Fishery*. Final Report. Acoura Marine Ltd: Edinburgh, 2017, p. 24.
85 Kiaupa et al. 2002, p. 182.
86 Tonts, Ülo: *Juhan Smuul. Lühimonograafia*. Eesti Raamat: Tallinn 1979.

the basis of the two months on the ocean, Smuul wrote a play, a children's book, a long feuilleton, and a handful of poems, that all brought a breath of fresh air into the frigidly stagnant literature of the Stalinist era.

The children's book "Murka the Sailor" (*Meremees Murka*, 1958) narrates daily events on the board of a herring trawler from the perspective of a ship dog, Murka. The central animal character holds together the feature-like chapters describing the work on the sea. The fishermen's daily routine is presented as follows: "Endless seaspace, 10 degrees of warmth in the middle of summer, nights without darkness, incessant swaying, swishing of wind in the shrouds. Heavy fogs near Iceland. This is what daily life on the sea feels like. Work days do not end at a certain time, but only when the work is done. Often wet clothes that do not dry overnight. The sternest saving of fresh water. Not to dream of sauna."[87] The harsh living conditions on the ships were mentioned in feature stories by Tammlaan in 1932.[88] In a news item from 1936, herring crew recruitment difficulties are mentioned with an addition that all the candidates have to be medically examined to see whether they are able to withstand the physical challenges in the chilly and damp climate of the North Atlantic.[89] A Soviet herring fishing handbook provides the average temperatures throughout the year in the main fishing grounds for the Soviet herring fleet, around the Faroe Islands and Jan Mayen Island.[90] Smuul was in the latter region where the average temperatures are oscillating between 0 and +5°C during the whole year. The North Atlantic experience inspired Smuul to create some of his most powerful literary figures, The Great Grey and the sense of enclosure on the ship, but these refer to general feelings of entrapment on the sea and do not have a direct relation to herring fishing.[91]

---

87 Smuul, Juhan: *Meremees Murka*. Eesti Raamat: Tallinn 1986, p. 55.
88 *Päevaleht*, 15.08.1932, p. 5. The first one among travelogues produced by Estonian writers onboard the actual herring ships. "Herring letters" by Evald Tammlaan (1904–1945) was published under the pseudonym Jänkimees in Päevaleht, from June to November, 1932. The herring letters provide ample and detailed information on the daily experiences of the first Estonian herring fishers of 1932. The rhetoric used in pre-trip newspaper reports about the establishment of a herring flotilla constructed by the entrepreneurs as "our vikings", who shall venture into the unknown in the quest for the "ocean vagabonds" (i.e. herrings, cf. fn 13, Kendla).
89 *Päevaleht*, 09.06.1936, p. 7.
90 Ruljov 1962, p. 6.
91 More about Smuul and seas in: Tüür, Kadri: "Water's three states in an Antarctic traveller's imagination". *Concentric: Literary and cultural studies*, 34(1), 2008, pp. 113–133.

As to the actual work on board, the play "Atlantic Ocean" (*Atlandi ookean*, 1957) by Smuul is one of the most illuminating pieces about the everyday life of the Soviet herring fishers. It has been criticised from a theatrical perspective for having too many characters and no clear storyline, but it can be regarded as truthfully reflecting the nuances of herring fishing, from the search for the shoals to the interpersonal relations overshadowed by strong ideological control.

The play opens with the preparations for departure of the herring ship from its home harbour, Tallinn. A radio journalist attempts to interview one of the crew members:

Reporter: "You are familiar with herring fishing already. Work on the sea is hard, I reckon?"

Mate: "Regular. If there is fish, we work 24 hours a day. If there is no fish, then just twelve." [...] (The mate realises that his replies are being recorded): "We, well, the collective, will fulfil the Communist Plan. We feel like front-rankers. We know the importance of every single herring..."[92]

The change from casual to official rhetoric is clearly visible. It creates a comic effect, but at the time of writing, the fulfilment of the Five-Year Plan indeed was a very important aim in all Soviet economies, including fishing. The sustainable use of resources was not a question to be discussed when the nutritional issues of one-sixth of the planet needed to be solved. The quality of the product was of secondary importance and it is referred to in the play only in the context of the prospective salary of the crew – lower quality herring was paid less for, influencing the final income of the crew members. If more herring was caught than the plan foresaw, bonuses were added to salaries. Thus, the Soviet ships certainly contributed to the fast over-exploitation of the North Atlantic herring stocks.

The emphasis on herring fishing can also be observed in the film chronicles of Soviet Estonia. In the years 1956–1962, altogether 9 chronicle pieces feature herring ships or the arrival of the catch to Tallinn harbour. After this period, there is one chronicle piece from 1968, a feature film "Night of Herring" (*Heeringaöö*) and an amateur film about life aboard a herring trawler dating to the same year.[93] Further analysis of these audiovisual historical sources will remain beyond the scope of the present chapter.

---

92 Smuul, Juhan: *Atlandi ookean*. Eesti Riiklik Kirjastus: Tallinn, 1957, p. 19.
93 https://www.eha.ee/fa/public/index.php, search "*heering**", limit to: films, accessed 10.07.2023.

After Smuul's works, herring disappeared from Estonian literature. It is mentioned neither as an object indicating status, nor as a foodstuff. Over the course of time and as a result of the wide availability of low-quality herring during the Soviet period, it seems to have lost its cultural significance. After the 1977 herring fishing ban in the North Sea, the Soviet herring flotilla was re-profiled to catch other fish near the African coast, especially in the territorial waters of the People's Republic of Angola, the government of which was supported by Soviet authorities.

This development is reflected in a graphic novel offering a retrospective story about Estonian ocean fishing, "Absolutely ordinary voyage" (*Täiesti tavaline merereis*, 2020)[94] by Meelis Kupits (1968). The story centres around the trips of one fishing vessel, SRT-R "Soela", built in 1961 in Stralsund, then the German Democratic Republic. The stories are based on the author's archival research as well as on his father's accounts of working on the ship as an electric engineer.[95] The fishing gear, herring search and catching process, as well as curing and handing the catch over to the base ships, are described in pictures and in accompanying text. An illegal secret trip to Norwegian coastal fjords in the pursuit of herring in order to fulfil the plan is included as an adventurous story (see Figure 6).

The majority of the book, however, deals with the ocean fishing in Mid-Eastern Atlantic waters where sardinillas, Atlantic and jack mackerels were caught. Whereas in the North the crew had no possibility to go ashore during a trip that lasted four to six months, as a rule, the southern waters were far enough from the Soviet homeland to necessitate trips to foreign harbours for fixing the ship or rotating the crew. Thus, the ocean fishermen themselves were not especially sad to have been forced to leave the herring and their spawning grounds. Herring stocks started to recover after the catching ban of five years. The ban had devastating effect on Icelandic economy,[96] but it resulted in restructuring of fishing activities in other countries, too. As we can see, it also had a slight effect on Estonian literature.

---

94 Kupits, Meelis: *Täiesti tavaline merereis*. Tänapäev: Tallinn, 2020.
95 The father of the author of the present article worked on the same ship as a radist (radio operator).
96 Sigurdsson 2006.

Figure 6. A page from the comic novel "Absolutely ordinary voyage" (*Täiesti tavaline merereis*), explaining the rationale behind the Soviet quest for herring. Some fishing ships were built in the USSR, but they were also ordered from Sweden, Finland, Poland, Germany. Ocean fishing was by nature an international endeavour even behind the Iron Curtain. Source: Kupits 2020, s.p. Reproduced with the permission of the author.

Conclusion

In my attempt to combine natural and cultural history, statistical data about historical herring fisheries was compared to the appearance of herrings in Estonian literature, both oral and written: runosongs, travel accounts, novels, as well as poems, plays, and a comic book. Quite generally it can be said that the dynamics of herring's presence in literature reflects the situation of herring in the ocean: during the period when it is a high-priced commodity, it stands for wealth in runosongs. As industrial fishing makes the fish more affordable, it can be seen to gradually lose its prestige as a festive food item. Estonian writers have also participated in the race and over-exploitation of the North Atlantic herring stocks, living, working and writing about life aboard a herring ship. After the herring population collapsed, herring disappeared from Estonian literature, only to resurface as a memoir in a recent graphic novel.

The answer to the central research question of whether the statistical data about herring fisheries is compatible with the appearance of herring in literary sources is affirmative: some correlations are certainly there. The novelty of the study lies in adding the dimension of cultural significance of the catch to historical fisheries research. Methodological difficulties must be admitted, though. The nature of the quantitative and qualitative data that need to be brought together requires careful assessment and application of a wide array of research and interpretation methods. The methodological approach needs to be refined in future studies, but the preliminary results provide a good platform to continue development in this direction. Some topics that could be covered on the basis of the same material, but that could not be addressed in the present chapter, include nuances of herring processing, its place in food history, herring in art and in audiovisual sources, political dimensions of international herring trade, and so on. A further insight that stems from the present study is that even a seemingly insignificant piece of information can become a meaningful part of the puzzle when it is taken out of its narrow disciplinary context and mapped to a larger background.

References

Andrews, Jim / Nichols, John: *Assessment Data Sheet. PFA & SPSG North Sea Herring Fishery*. Final Report. Acoura Marine Ltd: Edinburgh, 2017.

Annus, Epp et al.: *Eesti kirjanduslugu*. Koolibri: Tallinn 2001.

Armitage, David / Bashford, Alison / Sivasundaram, Sujit: *Oceanic Histories*. Cambridge University Press: Cambridge, 2018.

Bardone, Ester / Kannike, Anu / Põltsam-Jürjo, Inna / Plath, Ulrike: *101 Eesti toitu ja toiduainet*. Varrak: Tallinn 2016.

Bornhöhe, Eduard. *Kollid. Jutustus Tallinna elust*. G. Pihlaka raamatukauplus: Tallinn 1903.

Coull, James R.: "The North Sea Herring Fishery in the Twentieth Century." In: *Ocean Yearbook Online* 7(1), 1988, pp. 115–131. DOI: https://doi.org/10.1163/221160088X00084.

Coull, James R.: "The Development of Herring Fishing in the Outer Hebrides". *International Journal of Maritime History* XV(2), 2003, pp. 21–42.

Dickey-Collas, Mark: *The Current State of Knowledge on The Ecology and Interactions of North Sea Herring Within the North Sea Ecosystem*. Report. WOT kennisbasis 3251229218, 2004.

Dickey-Collas, Mark: "North Sea Herring. Longer Term Perspective on Management Science Behind The Boom, Collapse and Recovery of The North Sea Herring Fishery". In: *Management Science in Fisheries*. Routledge: London 2016. DOI:10.4324/9781315751443.

Gear, Robert William: "Re-assessing Shetland's Herring Industry before the 1870s". *Journal of the North Atlantic*, 4, 2013, pp. 61–68.

Gillis John R.: "The Blue Humanities". In: *Humanities* 34(3), 2013. Retrieved 10.07.2023 from https://www.neh.gov/humanities/2013/mayjune/feature/the-blue-humanities.

Hiiemäe, Mall: "Kakskümmend kaks kala eesti rahvausundis". *Mäetagused* 11, 1999, pp. 7–33.

Holm, Poul et al: "The North Atlantic Fish Revolution (ca. AD 1500)". *Quaternary Research* 2019, pp. 1–15. DOI:10.1017/qua.2018.153.

ICES FishMap: North Sea Fish Species Fact Sheets. Herring, retrieved 10.07.2023 from https://www.ices.dk/data/maps/Pages/ICES-FishMap.aspx.

Kama, Pikne: *Arheoloogiliste ja folkloorsete allikate kooskasutusvõimalused: inimjäänused märgaladel* [Combining Archaeological and Folkloristic Sources: Human Remains in Wetlands]. Doctoral thesis. University of Tartu Press: Tartu 2017. http://hdl.handle.net/10062/56305.

Kendla Mari: "Eestikeelsed kalanimetused: teaduslikud versus rahvapärased." *Mäetagused* 57, 2014, pp. 53–68.

Kiaupa, Zigmantas / Mäesalu, Ain / Pajur, Ago / Straube, Gvido: *The History of the Baltic States*. Avita: Tallinn 2002.

Kupits, Meelis: *Täiesti tavaline mereris*. Tänapäev: Tallinn, 2020.

Lajus, Julia: "'Red Herring': The Unpredictable Soviet Fish and Soviet Power in the 1930s". In: Wormbs, Nina (ed.) *Competing Arctic Futures. Historical and Contemporary Perspectives*. Palgrave Macmillan: Cham 2018, pp. 73–94. DOI:10.1007/978-3-319-91617-0_4.

Moora, Aliise: *Eesti talurahva vanem toit*. Ilmamaa: Tartu 2007.

Nirk, Endel: "Bornhöhe, Eduard". In: Kruus, Oskar (ed.) *Eesti kirjarahva leksikon*. Eesti Raamat: Tallinn 1995, pp. 65–66.

North, Michael: "The Baltic Sea". In: Armitage, David / Bashford, Alison / Sivasundaram, Sujit: *Oceanic Histories*. Cambridge University Press: Cambridge, 2018, pp. 209–233.

Ojaveer, Evald / Pihu, Ervin / Saat, Toomas (eds.): *Fishes of Estonia*. Teaduste Akadeemia Kirjastus: Tallinn 2003.

Peegel, Juhan: *Nimisõna poeetilised sünonüümid eesti regivärssides*. Eesti Keele Sihtasutus: Tallinn 2004.

Pihlamägi, Maie: "Eesti kaubandussuhted Suurbritanniaga aastail 1918–1940." *Acta Historica Tallinnensia* 3, 1999, pp. 88–108.

Pihu, Ervin (ed.): *Loomade elu IV. Kalad*. Valgus: Tallinn 1979.

Plath, Ulrike: "Näljast ja näljapsühholoogiast A. H. Tammsaare romaanis "Ma armastasin sakslast"". In: Hinrikus, Mirjam / Undusk, Jaan (eds.) *Armastus ja sotsioloogia. A. H. Tammsaare romaan "Ma armastasin sakslast"*. Underi ja Tuglase Kirjanduskeskus: Tallinn 2013, pp. 50–71.

Poulsen, Bo / Holm, Poul / MacKenzie, Brian R.: "A Long-term (1667–1860) Perspective on Impacts of Fishing and Environmental Variability on Fisheries For Herring, Eel, and Whitefish in The Limfjord, Denmark". *Fisheries Research* 87, 2007, pp. 181–195, Doi:10.1016/j.fishres.2007.07.014.

Põltsam, Inna: *Liivimaa väikelinn varase uusaja lävel. Uurimus Uus-Pärnu ajaloost 16. sajandi esimesel poolel*. TLÜ: Tallinn 2008.

Põltsam-Jürjo, Inna: "Kala tähtsusest kauborduses, majanduses ning toidumenüüs 13.–16. sajandi Eestis". *Acta Historica Tallinnensia* 24, 2018, pp. 3–23. DOI: 10.3176/hist.2018.1.01.

Ruljov, Nikolai: *Atlandi heeringa esmane töötlemine: teatmik püügilaevade kalameistritele*. Eesti NSV Rahvamajanduse Nõukogu Tehnilise Informatsiooni Büroo: Tallinn 1962.

Sarv, Mari: *Regilaul kui poeetiline süsteem*. Eesti Kirjandusmuuseum: Tartu 2000.

Sarv, Mari: "Stichic and Stanzaic Poetic Form in Estonian Tradition and in Europe". *Traditiones* 38(1), 2009, pp. 161–171.

Sarv, Mari: "Towards a Typology of Parallelism in Estonian Poetic Folklore". In: *Folklore: Electronic Journal of Folklore* 67, 2017, pp. 65–92. DOI:10.7592/FEJF2017.67.sarv.

Sarv, Mari / Oras, Janika: "From Tradition to Data: The Case of Estonian Runosong". *Arv. Nordic Yearbook of Folklore* 76, 2020, pp. 105–117.

Sicking, Louis / Abreu-Ferreira, Darlene (eds.): *Beyond the Catch: Fisheries of the North Atlantic, the North Sea and the Baltic, 900–1850*. Brill: Leiden 2009.

Sigurdsson, Thorir: "The collapse of the Atlanto-Scandian herring fishery: Effects on the Icelandic economy". In: *IIFET 2006 Portsmouth Proceedings*. Cemare: Portsmouth 2006, pp. 1–9.

Smuul, Juhan: *Atlandi ookean*. Eesti Riiklik Kirjastus: Tallinn 1957.

Smuul, Juhan: *Meremees Murka*. Eesti Raamat: Tallinn 1986.

Tammsaare, Anton Hansen: *Ma armastasin sakslast*. Eesti Raamat: Tallinn 1984.

Tonts, Ülo: *Juhan Smuul. Lühimonograafia*. Eesti Raamat: Tallinn 1979.

Toresen, Reidar, Østvedt, Ole Johan: Variation in Abundance of Norwegian Spring- Spawning Herring (Clupea harengus, Clupeidae) Throughout the 20th Century and The Influence of Climatic Fluctuations. *Fish and Fisheries*, 1, 2000, pp. 231–256. DOI: https://doi.org/10.1111/j.1467-2979.2000.00022.x.

Tuglas, Friedebert: *Eesti kirjanduse ajalugu IV. Kriitiline realism*. Teaduslik Kirjastus: Tartu, 1947.

Tüür, Kadri: "Water's Three States in An Antarctic Traveller's Imagination". *Concentric: Literary and Cultural Studies*, 34(1), 2008, pp. 113–133.

Tüür, Kadri / Stern, Karl: "Atlantic Herring in Estonia: In the Transverse Waves of International Economy and National Ideology". *Journal of Baltic Studies* 46(3), 2015, pp. 393–408. DOI:10.1080/01629778.2015.1073928.

Undusk, Rein: "Tammsaare, Anton Hansen". In: Eglāja-Kristsone, Eva / Gasiliūnas, Virginijus / Mihkelev, Anneli: *300 Baltic writers*. Institute of Lithuanian Literature and Folklore: Vilnius 2009.

Vilde, Eduard: *Külmale maale*. Eesti Riiklik Kirjastus: Tallinn 1960.

Weberman, Ernst Constantin: *Kala kui rikkalik vitamiinide allik*. Üleriikline Kalanädala Peakomitee: Tallinn 1927.

Wilkins, Frances: "'Da Merry Boys o Greenland': Explorations into the Musical Dialogue of Shetland's Nautical Past". *Folk Music Journal* 11(2), 2017, pp. 17–37.

Periodicals

Daily *Eesti Päevaleht*

15.08.1932

09.06.1936

Collections of the Estonian Folklore Archives

E 18801
ERA II 37
EÜS IX 961
H II 1
H II 15
H III 3
H III 16

Databases

Database of Estonian runosongs, https://www.folklore.ee/regilaul/andmebaas
Database of Estonian Film Archives, https://www.eha.ee/fa/public/index.php

III. REPRESENTATIONS

Tõnno Jonuks

Griffins in the Eastern Baltic Late Iron Age[1]

Abstract: This article is based on two objects from the 11th–12th-century Estonia, both depicting a creature with a massive beak, thus usually interpreted as an eagle. This interpretation is hereby challenged as on the basis of the presence of ears and other details in both figurines it can be suggested that the objects may rather depict griffins. Similar objects found from a large territory mostly on the Eastern side of the Baltic Sea demonstrate that knowledge of such a creature was widespread. The paper suggests that the spread of Christian culture did not mean only spreading of Christian religion. Together with that, various phenomena from Western European culture were imported to pagan countries, and interpreted further. The image of the griffin was probably associated with noble families across the Eastern Baltic, who shared a similar worldview and identity narratives.

Keywords: eagle, griffin, Christianisation, Late Iron Age

Introduction – Problems of Identifying Animal Species in Archaeology

Animals appear in archaeology as indicated by two major types of source material – either as osseous remains or as images of animals. When interpreting archaeozoological material, archaeologists with modern education and highly competent in biology tend to ascribe identifications emanating from the rational and well-proven biological data. The zooarchaeological methodologies are used not only to identify osseous collections but also to interpret human–animal relationships, animal husbandry, and human society in a broader perspective.[2]

Such reasonable and rational interpretations are also used extensively in interpreting (pre)historic animal images – pendants, figurines, animal-shaped details. Deriving from such a background, it is assumed that past people approached and organised nature similarly to our rational biology. This brings

---

1   The author is grateful to PhD Andreas Rau (ZBSA, Schleswig) for inspiring discussions. This study was supported by the European Union through the European Regional Development Fund (Centre of Excellence in Cultural Theory and Centre of Excellence of Estonian Studies, TK-145), by PRG908 Estonian Environmentalism in the 20th century: ideology, discourses, practices, and research grant of the Estonian Literary Museum EKM 8-2/20/3.
2   See, e.g Russell, Nerissa: *Social Zooarchaeology. Humans and Animals in Prehistory.* Cambridge University Press: Cambridge 2012.

us to the key issue of human–animal relationships in the past: how did people organise nature and animal species for themselves, what kind of animals were included into these taxons, and how do the stylistic features of some particular species influence the appearance of animals in art?

Organising and understanding "prehistoric taxonomy" should be the first step when identifying art objects.[3] Tim Insoll asked, "whether 'our' understanding of 'nature' is necessarily that of past populations?"[4] Archaeologists have often tried to bring modern taxonomy into discussions about animal images. Snakes form a characteristic example here, as depictions of specific snake species (grass snake, viper) are often looked for from the distant past. Borrowing interpretations from contemporary folk customs, qualitative attributes have been tried to be ascribed to prehistoric figurines, for example the grass snake is associated with female and positive chthonic powers, while the viper represents the opposite qualities.[5] The folksy interpretation about snakes is much simpler. Different biological species – the grass snake, viper, and a legless lizard slowworm – are all understood as one kind. Bo Jensen has also suggested the same approach to snakes in the Viking Age.[6] On the basis of linguistic and literary sources he suggests that "snake" was an indeterminate denominator, signifying all crawling and possibly dangerous animals like worms, snakes, dragons, etc.

As the previous example demonstrated, unlike the common scientific understanding, animals and birds were organised differently in folk taxonomy and their classification was rather based on some common characteristic or function. For instance, the ouzel is used in folk biology as a common denominator for most birds eating berries in gardens,[7] and the goshawk for most birds of prey.[8] Such a classification indicates the most important – both groups of birds were seen as dangerous to human properties, and more detailed identification

---

3   See also Ingold, Tim: *The Perception of Environment. Essays onLlivelihood, Dwelling and Skill*. Routledge: London / New York 2002.
4   Insoll Timothy: *Archaeology. The Conceptual Challenge*. (Duckworth debates in archaeology). Duckworth: London 2007, p. 99.
5   Loze, Ilze Biruta: "The Theme of The Grass Snake in Neolithic Material From Latvia". In: Loze, Ilze Biruta (ed.): *Art, Applied Art and Symbols in Latvian Archaeology*. Humanities and Social Sciences, Latvia. University of Latvia 2(39). University of Latvia: Riga 2003, pp. 48–59.
6   Jensen, Bo: *Viking Age Amulets in Scandinavia and Western Europe*. Archaeopress: Oxford 2010, p. 172.
7   Mäger, Mart: *Linnud rahva keeles ja meeles*. Eesti Raamat: Tallinn 1969, p. 43.
8   Ibid., p. 141.

was not necessary. From global anthropology even more bizarre examples can be found, like classifying all jumping animals (birds, frogs, etc.) as one kind.[9] Moreover, different animals did not have to represent distinct separate species in an animistic worldview. As an example, the bear is in most of the North Eurasian mythologies understood as man's wild brother, who is able to have a family and communicate with man.[10]

Considering the previous we should not expect prehistoric taxonomy to be organised in the same way as is our biological one, and in addition it could also include species that do not exist for us. This means that when approaching prehistoric systematics to nature and animals according to our scientific approach we most probably create artificial divisions as we do not follow past logic. When using the example of a snake, it is obvious for us to make a difference between a worm, a poisonous snake, a harmless snake, and a non-existing snake (dragon), while they all were probably the same crawling creature with differences only in nuances for people in the past.

Thus, it is debatable how useful modern zoology is as a starting point for interpreting prehistoric taxonomy at all. The way man in the Late Iron Age (or any other period in the past) systematised nature and the environment could have been significantly different. Rational zooarchaeology is justified for studying osseous material, but interpreting art objects and animal images requires an understanding of past people's worldview, mythology, and moral values. Ignoring these can result in rational interpretations, which seem to be in accordance with our modern thinking but may in fact be erroneous from the perspective of past culture.

Does the Absence of Narrative Cause the Absence of Mythological Beings in Interpretations?

Archaeologists always need some alternative source in addition to finds to create interpretations. The usage of these analogies is often problematic as they tend to come from different economic, social or time situations. In some cases, like Viking-Age Scandinavia, narrative sources, particularly the Eddic mythology and epics, have made it possible to see mythological beings, like dragons, elves, etc., in archaeological material. The subject of dragons and supernatural snakes, the

---

9   Lévi-Strauss, Claude: *Metsik mõtlemine*. Vagabund: Tallinn 2001, p. 235.
10 E.g. Helskog, Knut: "Bears and Meanings Among Hunter- Fisher- Gatherers in Northern Fennoscandia 9000–2500 BC". *Cambridge Archaeological Journal* 22(2), 2012, pp. 209–236.

world snake Jörmunganðr and the dragon Fáfnir from the Nibelungen saga, as well as the dragon in Beowulf, is a much-debated issue in Scandinavian religion and mythology.[11] Moreover, the image of the dragon is also widely used and discussed in medieval art across Northern Europe, particularly on weapons,[12] decorative panels,[13] and personal objects,[14] and also appears in obvious Christian contexts.[15]

The dragon also occurs occasionally in the narrative sources of the Eastern Baltic. The earliest example comes from the chronicle of Adam from Bremen from the 1070s–1080s, stating that people on an island called Aestland are like their neighbours, "Utterly ignorant of the God of the Christians", and "they adore dragons and other winged creatures (*dracones adorant cum volucribus*) and sacrifice to them living men whom they buy from the merchants."[16] Another example of a dragon from the period comes from the Brennu-Njáls saga, where an Icelander Þorkell hákr came across a flying dragon (*flugdreki*) in *Aðalsýsla* – probably Western Estonia – and killed it. The most possible interpretation of

---

11 Hedeager, Lotte: *Iron Age Myth and Materiality. An Archaeology of Scandinavia AD 400–1000*. Routledge: London / New York 2011; Brunning, Sue: "(Swinger of) the Serpent of Wounds'. Swords and Snakes in The Viking Mind." In: Bintley, Michael D. J. / Williams, Thomas J. T. (eds.): *Representing Beasts in Early Medieval England and Scandinavia*. (Anglo-Saxon Studies 29). The Boydell Press: Suffolk 2015, pp. 53–72; Symons, Victoria: "Wreonþenhilt ond Wyrmfah. Confronting Serpents in Beowulf and Beyond". In: Bintley, Michael D. J. / Williams, Thomas J. T. (eds.): *Representing Beasts in Early Medieval England and Scandinavia*. (Anglo-Saxon Studies 29). The Boydell Press: Suffolk 2015, pp. 73–93; and references therein.
12 E.g. Creutz, Kristina: *Tension and Tradition. A Study of Late Iron Age Spearheads Around the Baltic Sea*. (Theses and Papers in Archaeology 8). Stockholm. 2003.
13 E.g. Rybina, Elena: "Recent Finds from Excavations in Novgorod". In: Brisbane, Mark A. (ed.): *The Archaeology of Novgorod, Russia. Recent Results from the Town and Its Hinterland*. (The Society for Medieval Archaeology Monographs 13). Society for Medieval Archaeology: Lincoln 1992, p. 167.
14 Leimus, Ivar / Roio, Maili / Sarv, Krista: "Watertight Sources: Unique Find from the Bottom of Tallinn Bay". In: Roio, Maili (ed.): *Shipwreck Heritage: Digitalizing and Opening Access to Maritime History Sources*. (Muinasaja Teadus 23). Muinsuskaitseamet: Tallinn 2013, pp. 133–164.
15 Gräslund, Anne-Sofie: "Wolves, Serpents and Birds. Their Symbolic Meaning in Old-Norse Belief." In: Andrén, Anders / Jennbert, Kristina / Raudvere, Catharina (eds.): *Old Norse Religion in Long-term Perspective. Origins, Changes & Interactions*. (Vägar till Midgård 8). Nordic Academic Press: Lund 2006, p. 125.
16 Adam of Bremen: *History of the Archbishops of Hamburg-Bremen*. Columbia University Press: New York 2002, Book IV, chapter 17.

both stories is that dragons are one of the many labels to mark culturally strange and potentially hostile countries. Thus, neither of the examples describes the country nor its inhabitants' worldview but illustrates how Christian Western Europe denoted the pagan, strange, and hostile East, also referring to Christian world views and values.[17]

As the original mythology of the Finnish and Eastern Baltic Late Iron Age has not been preserved, it is difficult to speculate if and how the supernatural animals were (re)present(ed). A witch named Louhi turns herself into a winged dragon in Kalevala mythology but this is an individual example. The concept of giants is better known, with some attempts to date it to the Iron Age in Estonia.[18] However, these examples only suggest that the concept of supernatural beings could have existed in the Eastern Baltic but say nothing particular. While using local folklore as an analogy to the Iron Age archaeology, it has to be considered that in addition to centuries separating archaeological finds and records of folklore there is also a social difference. Folklore has been recorded since the nineteenth century from rural peasants but archaeological finds that depict real or supernatural beings are rather associated with the Iron Age nobility. Moreover, anthropomorphic supernatural beings, the most common in oral sources, are rarely depicted in art or on objects, or at least they have rarely been interpreted like that. Neil Price has demonstrated that dwarfs and possibly giants can occur more frequently in archaeological material but we do not identify them like that but rather as gods.[19]

Identification is the key term in interpreting figurative art and apparently we need narrative sources for that. If there is a lack of stories, figurative art is interpreted solely on the basis of biology or on records from other contexts, which

---

17 See also Jonuks, Tõnno. "A Few Additions to The Depiction of Estonia and The Eastern Shore of The Baltic Sea in Scandinavian Sagas". In: Zilmer, Kristel (ed.): *Dialogues With Tradition: Studying The Nordic Saga Heritage*. Tartu University Press: Tartu 2005, pp 45–63; Kaljundi, Linda: "Waiting for the Barbarians: Reconstruction of Otherness in the Saxon Missionary and Crusading Chronicles, 11th–13th Centuries". In: Kooper, Erik (ed.): *The Medieval Chronicle*, vol. 5. Rodopi: Amsterdam / New York 2008, pp. 113–127.

18 Loorits, Oskar: *Grundzüge des Estnischen Volksglaubens I*. Carl Bloms Boktryckeri A.-B: Lund 1949, p. 475.

19 Price, Neil: "What's in A Name? An Archaeological Identity Crisis for The Norse Gods (and Some of Their Friends)". In: Andrén, Anders / Jennbert, Kristina / Raudvere, Catharina (eds.): *Old Norse Religion in Long-term Perspective. Origins, Changes & Interactions*. (Vägar till Midgård 8). Nordic Academic Press: Lund 2006, pp. 179–183.

are sometimes uncritically exploited resulting in speculative interpretations. The first approach is apparently dominant in the archaeology of the Eastern Baltic. It is partly associated with the general valuing of rationality, which prefers rational and sensible interpretations while considering mythological, magical or religious ones speculative.[20] This approach has led to only biological interpretations and in some cases even clear examples of "unnatural" beings are ignored. On the other hand, uncritical studies have in some cases overemphasised mythology, and certain objects are associated with mythology without any indicative feature. It often goes for anthropomorphic figurines, which are interpreted as depictions of gods.[21]

By emphasising the importance of narrative in interpreting archaeological finds I wish to emphasise the critical approach to available narrative sources, and using these in chronologically, spatially and socially proper contexts. The main impact here is that narrative sources also make mythological beings visible in archaeological interpretations while the lack of one-time myths leads to rational interpretations. On the other hand, we should acknowledge the existence of narratives or moral values in the Iron Ages that are not preserved for us, but were nevertheless the background for producing these objects. The dragon is a good example here – in most of the narrative sources, available to us, the dragon represents uncontrollable nature and hostile powers to mankind, while representations of the dragon on archaeological finds refer to power and status in the Viking Age and in the medieval worldview.[22] Such a seemingly controversial representation refers on one hand to the complicated background of narrative sources but also simultaneously to lost narratives. Lost mythologies are always complicated subjects, even if perceived. There are few examples where disappeared mythologies are addressed on the basis of archaeology[23] but it is

---

20 See also Johanson, Kristiina / Jonuks, Tõnno: "Are We Afraid of Magic? Magical Artifacts in Estonian Museums". *Material Religion* 14(2), 2018. DOI: 10.1080/17432200.2018.1443894.
21 E.g. Kiudsoo, Mauri: *Viikingiaja aarded Eestist. Idateest, rauast ja hõbedast.* Imeline Ajalugu / Imeline Teadus / Eesti Meremuuseum: Tallinn 2016, p. 119.
22 For more details see Pluskowski, Alex: "The Dragon's Skull: How Can Zooarchaeologists Contribute to Our Understanding of Otherness in The Middle Ages?" In: Walker-Vadillo, Monica Ann / Picaza, María Victoria Chico / García, Francisco de Asís García (eds.): *Animals and Otherness in the Middle Ages. Perspectives Across Disciplines.* (BAR International Series 2500). Archaeopress: Oxford 2013, pp. 109–124; Brunning 2015.
23 E.g. Kaul, Flemming: *Bronzealderens religion. Studier af den nordiske bronzealders ikonografi.* Det Kongelige Nordiske Oldskriftselskab: København 2004.

also clear that a comprehensive narrative is not possible to reconstruct on the ground of archaeology only. Figurative finds do not tell the full story – these were produced to symbolise the story and refer to the most characteristic features only. Nevertheless, composing interpretations on the basis of finds and details on them, and not trying to find known mythologies, opens up a new direction to understand societies' prior preserved narratives.

Eagles or Griffins?

To illustrate the problem of rational archaeozoological interpretations, I chose three objects, which are usually described as bird or eagle-headed figurines. The first one is part of an antler handle, found under the rampart of the Lõhavere hillfort (see Figure 1). It is richly decorated with lines and "eyes" made of a ring and dot marks; one side of a nape is covered with a pleach ornament while the other side is heavily eroded. On the basis of the elaborated ornament Heidi Luik suggests that the object was probably not produced in Estonia but in some major craftsman centre, most likely in Novgorod.[24] However, the pleach ornament is not that common in the Novgorod area but occurs on multiple objects from present-day Estonia. This does not rule out the possibility that the object was originally crafted in Estonia or may be even in the Lõhavere hillfort, where high-quality handicrafts were produced.[25] The eyes of the creature are large and hollow and according to a recent study were originally filled with tin. However, it was also suggested that tin was only a binder and it was probably holding some shiny object, possibly a piece of gold or a semi-precious stone, to mark the eyes.[26] A similar ornament that covers the neck of the handle is used to decorate the head as well. It seems to be intentional that the ornament also covers the hook-like beak and forehead of the figure, thus leaving the impression of a diadem and giving a certain royal look to the whole figure. A feature that differs from a common bird interpretation and is mentioned only passingly in previous interpretations is that beside the eyes clear ears are depicted that question the interpretation of the object as an eagle or another bird of prey. The object has a cavity of small diameter under its bottom with a nail hole, indicating that it was

---

24 Luik, Heidi: "Linnupeakujuline sarvest käepide Lõhavere linnamäelt". In: Tamla, Ülle (ed.) *Ilusad asjad. Tähelepanuväärseid leide Eesti arheoloogiakogudest.* (Muinasja teadus 21). Ajaloo Instituut: Tallinn 2010, pp. 127–138.
25 Laul, Silvia / Tamla, Ülle: "Peitleid Lõhavere linnamäelt. Käsitöö- ja ehtevakk 13. sajandi algusest". Õpetatud Eesti Seltsi Kirjad 10: Tartu / Tallinn.
26 Luik 2010, p. 133.

originally attached to a larger implement, usually interpreted as a handle of some light object, like a whip. Part of the head and ornament is very well preserved, while another part is badly eroded. This is interpreted as a result of use/wear, when a whip was hanging and constantly rubbed against some harder material. However, in case of use/wear one could expect some sign of erosion of the entire object, but the rest of the ornament is perfectly preserved. Thus, instead of use/wear, this kind of erosion refers to post-depositional erosion while the object was already buried and only partially exposed to sun, rain, frost and sand.[27] The dating of the object is complicated as according to the original publication it was found from the "charcoal layer under the rampart and thus belongs to the earliest layer of the hillfort" (i. e. 11th century), but later the authors of the same article date the sculpture to the twelfth–thirteenth centuries.[28]

Figure 1. An antler object depicting a griffin from the Lõhavere hillfort (AI 3578: 626). Photo: Tõnno Jonuks.

---

27 Cf. Jonuks, Tõnno: "An Antler Object From The Pärnu River – An Axe, A God, or A Decoy?" In: Johanson, Kristiina / Tõrv, Mari (eds.): *Man, his Time, Artefacts, and Places.* (Muinasaja Teadus 19). Institute of History and Archaeology: Tartu 2013, pp. 225–246.
28 Moora, Harri / Saadre, Osvald: "Lõhavere linnamägi". In: Moora, Harri (ed.): *Muistse Eesti linnused. 1936.–1938. a. uurimiste tulemused.* Õpetatud Eesti Selts: Tartu 1939, p. 177.

Another bird-like figurine comes from the stone grave of Kolu, Western Estonia, dated to the twelfth century (see Figure 2).[29] Differently from the previous object this one is made of a copper alloy but the overall design is similar. It is also hollow, suggesting that it was once attached to something. The neck of the figure is thoroughly covered with spiral motifs, divided in three rows. The head is disproportionately small but a massive and partly opened beak is dominant. What makes the beak spectacular is a zig-zag motif, which leaves an impression that the beak also has teeth. Another zig-zag on the head resembles the comb of a rooster. This figure has ears too, although in this case the ears are marked with two circlets, where massive rings are attached.

Very similar finds of copper alloy heads with a massive beak and neck decorated with circles or spirals, are found from various sites on the Eastern side of the Baltic Sea – [30], Novgorod, Russia[31], Mezotne hillfort, Latvia[32], Räisälä cemetery, Russian Karelia[33], and the most similar analogy comes from the region of the Oyat River, NW Russia.[34] All these figures are commonly understood as depicting a predatory bird, despite the clear markings of ears (see Figure 5).

---

29 Mandel, Mati: *Läänemaa 5.–13. sajandi kalmed*. (Töid ajaloo alalt 5). Eesti Ajaloomuuseum: Tallinn 2003, p. 64.
30 Kivikoski, Ella: *Die Eisenzeit Finnlands. Bildwerk und Text*. Finnische Altertumsgesellschaft: Helsinki 1973, fig. 1224.
31 Koltschin, Boris Alexandrowitsch: *Drevnij Novgorod. Prikladnoe iskusstvo i arheologija*. Iskusstvo: Moskva 1985, p. 77.
32 Balodis, Francis: *Latviešu starptautiskie sakari ap 1000 gadu pec Kristus*. Latvijas Vēstures Institūta Žurnāls: Riga 1939, figure 2.
33 Kivikoski, Ella: *Die Eisenzeit Finnlands. Bildwerk und Text*. Finnische Altertumsgesellschaft: Helsinki 1973, fig.1224.
34 Aspelin, Johannes R.: *Muinaisjäännöksiä Suomen suvun asumus-aloilta.Rauta-aika V. Vepsäläisjä muinaiskaluja*. Suomalaisen Kirjallisuuden Seura, Helsinki: 1877, p. 231.

Figure 2. A bronze figurine (HM 8045: 68) from the Kolu stone grave, Western Estonia. Note the circlets symbolising ears, the zig-zag motif on the beak, and another on the head. Photo: Tõnno Jonuks.

There is also a third example of a bird with ears – this time attached to a bronze war horn, found in Otepää castle (Figure 3). It is suggested that the horn was made at the end of the twelfth–early thirteenth centuries somewhere in Central Europe (Germany?) and brought to Estonia by the first crusaders in the early thirteenth century.[35] The front edge of the horn is decorated with cross motifs, enthroned with a stylised figure of a large and half-opened beak. Small lines on the beak refer to the attempt to depict teeth. Similarly to previous examples this figure also has ears but contrary to the examples from Lõhavere and Kolu the ears of the figure of the Otepää horn are neither functional nor easy to carve. Producing of these ears was a special effort for the craftsman and should thus be considered meaningful and most likely also the key to identify the beaked figure.

---

35 Mäesalu, Ain: "Kas Otepää linnuselt leiti muinaseestlaste sõjasarv?" *Tutulus*, 2014, p. 29.

Figure 3. A detail of a bronze horn (AI 4036 II: 191) from the Otepää hillfort. Note the round ears and serrated beak. Photo: Tõnno Jonuks.

The interpretation of all the previous objects has been that of a bird, or more precisely an eagle or some other bird of prey. This understanding obviously comes from the most dominant feature, the beak, which really resembles the beak of an eagle or any other predator bird. However, another common feature, the ears, is ignored or mentioned only passingly as it does not fit with the bird interpretation. Certainly, some species of owls have ears, but the overall impression of the faces of figures discussed here is that the face does not resemble an owl. These two main characteristics, a massive beak and ears, indicate a specific creature, a griffin, that in medieval manuscripts is always depicted as a quadruped animal, with an eagle head and distinctive ears (Figure 4). This identification is further supported by other details, like tooth symbols on the beak or comb on the head of the Kolu example. In bestiaries the griffin is either portrayed or depicted as dominant over others to demonstrate its supremacy over other animals and humans.

Figure 4. A drawing of a griffin from 1607.³⁶ Note the prominent ears of the beaked monster. Its dominating position over the swine symbolises the supremacy over other animals, including humans.

The history of griffins reaches back to antiquity, and griffins have been important figures both in Western and Eastern European Christian art as well as in Islamic art.³⁷ It has also suggested that the figures of Lõhavere and Kolu originate from the east, either from Novgorod³⁸ or even further from the Volga region.³⁹ Multiple copper-alloy figures found around the Baltic Sea and in NW Russia rather suggest that the design was created and figures crafted in this region. Beaked heads without any marking of ears are also found from

---

36 Figure according to White, Terence, H. "The Bestiary: the Book of Beasts". Putnam: New York 1960, p. 23.
37 Adams, Noël: "Between Myth and Reality.Hunter and Prey in Early Anglo- Saxon Art". In: Bintley, Michael D. J. / Williams, Thomas J. T. (eds.): *Representing Beasts in Early Medieval England and Scandinavia*. (Anglo-Saxon Studies 29). The Boydell Press: Suffolk 2015, p. 22.
38 Luik 2010.
39 Mandel 2003, p. 64.

Novgorod.⁴⁰ In these cases the beak is often half-opened, carrying a ball,⁴¹ thus referring to the concept of an eagle stone (*aetites*). An eagle stone is a round or oval stone that derives from distant countries (most often India, the Caucasus), and is brought by an eagle to its nest. In folk religion *aetites* is often associated with healing magic and as the stone is believed to have another stone in it, it has been associated with pregnancy.⁴² In addition to the common association with female and pregnancy that derives from Aristoteles and Dionysius, the eagle stone is also associated with foreign wisdom and rulers, carrying thus broader symbolism. Indeed, the concept of a miraculous stone, associated with nobility, wisdom and magic, could be associated not only with the eagle but also with its mythical counterpart – the griffin.

On the other side it is debatable how sharply the division between griffins and eagles should be drawn. Drawing again the analogy with snakes, which encompass different slithering creatures,⁴³ we could suggest that the concept of a predator bird could include eagles, griffins and other powerful and winged creatures. This means that the exact difference between an eagle and a griffin could not have been so sharp, especially in the eyes of common people in the Eastern Baltic who did not know much about the one-time science of Western Europe and represented in bestiary books. However, certain details, like ears, can indicate that the master did not only have the eagle in mind, but its mythical counterpart. This also raises the question of how a character from Christian mythology reached the pagan Eastern Baltic? However, individual elements do not have to be dispersed with the whole packet of beliefs. In a similar manner, Christian cross pendants spread to Estonia since the eleventh century although officially Estonia was converted in the early thirteenth century.⁴⁴ Instead, mythological beings, not only griffins, but also dragons and others, could have been adopted as princely signs, and thus when spreading to other religious contexts could have

---

40 E.g. Koltschin 1985, pp. 88–91, 118.
41 Koltschin 1985, pp. 89, 118.
42 Hoffmann-Krayer, Eduard: "Adlerstein". In: Bächtold-Stäubli, Hanns / Hoffmann-Krayer, Eduard (eds.): *Handwörterbuch des deutschen Aberglaubens. Band 1*. Walter de Gruyter: Berlin / New York 2000, p. 190; Duffin, Christopher J.: "A Survey of Birds and Fabulous Stones". *Folklore* 123:2, 2012, pp. 189–190.
43 Jensen 2010, p. 172.
44 Jonuks, Tõnno / Kurisoo, Tuuli: "To Be or Not to Be… a Christian: Some New Perspectives on Understanding the Christianisation of Estonia". *Folklore* 55, 2013, pp. 69–98.

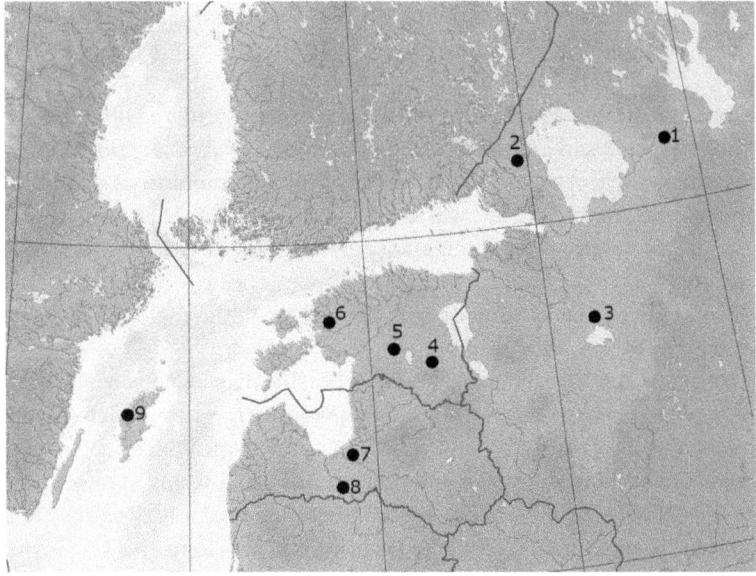

Figure 5. Sites mentioned in the text. 1. Oyat River, 2. Räisälä cemetery, 3. Novgorod town, 4. Otepää hillfort, 5. Lõhavere hillfort, 6. Kolu stonegrave, 7. Ikšķile hillfort, 8. Mezotne hillfort. Map by Tõnno Jonuks.

been perceived differently.[45] It is more likely that instead of the full medieval mythology, as represented in bestiaries, only the social meaning was ascribed to, and thus the foreign essence of the griffin was even more emphasised.

Discussion – What Do Griffins Want?

Griffins are not the only mythological species represented in Eastern Baltic archaeological material. Roberts Spirǧis has associated certain animal pendants with another supernatural being.[46] The wide mouth was used as an identifying

---

45 In more detail Jonuks, Tõnno: "Domesticating Europe – Novel Cultural Influences in The Late Iron Age Eastern Baltic". In: Selart, Anti (ed.) *Changing Aliens, Changing Natives. Baltic Crusades and Societal Innovation in Medieval Livonia*. Brill: Leiden / Boston 2022, pp. 29–54. DOI:/10.1163/9789004512092_003.

46 Spirǧis, Roberts: "Nahodki zoomorfnyh podvesok "smolenskogo" tipa na territorii Latvii i ih novaja interpretacija". *Stratum plus* 5, 2012, pp. 195–220.

feature and associated with leucrota or crocotta, a fantastic creature, joining a dog and a wolf, from the medieval bestiaries. The type of pendant is mainly distributed and possibly produced in Eastern Baltic areas in the eleventh to thirteenth centuries.[47] Objects representing dragons are less discussed, although they appear in central places around the Baltic Sea.[48] A detailed figurine of a dragon, with a crocodile-like snout, round nostrils and overwhelming canines, originates from the eleventh–twelfth-century Ikšķile hillfort at the lower reaches of the Daugava River, Latvia.[49] Additionally a corpus of replicas of tooth pendants, made of copper alloy, has been associated with mythological creatures. The form of these pendants does not resemble any organic teeth and thus the impressively large crown, combined with golden shine, may also refer to some supernatural snake or dragon.[50] Snakes and/or dragons are also widely represented on brooches and S-shaped pendants,[51] and in a more stylised manner also on bracelets.

The provenance of mythological creatures is little discussed. According to a common understanding the two Estonian figures of griffins originate from the east. However, as demonstrated above, the copper alloy examples of griffins have been collected as individual objects around the Baltic Sea, without referring to any particular centre. Additionally, the considerable similarities in ornament suggest the Eastern Baltic to be the origin of these objects. This suggests that also the antler sculpture of griffin from the Lõhavere hillfort is also of local origin, although represented by another material. Other kinds of mythological creatures may have had their own habitats. The origin of brooches and bracelets with "zoomorphic" additions, which are interpreted as dragons, is usually associated

---

47 Kurisoo, Tuuli: *"Adornment, Self-Definition, Religion: Pendants of the North-Eastern Baltic Sea Region 9th–13th Centuries"*. Studien zur Siedlungsgeschichte und Archäologie der Ostseegebiete, Band 19. Wachholtz, 2021, p. 177.
48 Kalmring, Sven: "The Dragonhead and the Wooden Jetty. Back in Birka's Black Earth Harbour". *Jahresbericht / Zentrum für Baltische und Skandinavische Archäologie*, 2015, pp. 58–61; Gräslund, Anne-Sofie: "Drakar I Uppåkra". In: Hårdh, Birgitta (ed.): *Fler fynd I centrum. Materialstudier i och kring Uppåkra*. (Acta Archaeologica Lundensia. Series in 8°; Vol. 45). Almqvist & Wiksell International: Stockholm 2003, pp. 179–188; Jonuks, Tõnno: "Bronze Tooth Pendants From The Late Iron Age: Between Real and Fictional Zooarchaeology". *Norwegian Archaeological Review* 50 (2), 2017. DOI: 10.1080/00293652.2017.1367838; see also references therein.
49 Jonuks 2017.
50 See in more detail Jonuks 2017.
51 Luik, Heidi: "S-kujulised ripatsid ja rihmakeeled". *Eesti Arheoloogia Ajakiri* 3/2, 1999, pp. 115–130; Kurisoo 2021, pp. 95–98.

with the Baltic countries,[52] and copper alloy tooth pendants derive from southwestern Finland.[53]

Thus, the knowledge and valuing of supernatural creatures was spread across the Baltic Sea region and should rather be approached semantically to understand the symbolism and agency and not be limited by the provenance of individual objects. The figures of Kolu or Lõhavere have been commonly understood as decorated parts of the handles of some light object, like a whip[54] or a ritual object.[55] The cavities of both objects are less than a centimetre in diameter and thus not suitable to be attached to extensively used tools (like a whip). The same goes for the dragon from Ikšķile with its cavity 1.5 cm in diameter. The same measurements also seem to apply to the griffin figures from other locations. Could it be that the unnatural animal heads were not part of handles, but rather some sort of ceremonial staff or position marker, which did not even have any practical function? As an analogy, dragon-headed staffs were also used in the mid-eleventh–thirteenth centuries to symbolise power in the greater Khurasan region in Persia.[56] The recently found figure of a dragon from Birka has been interpreted as a part of a decoration pin,[57] and dragon-headed decoration needles are broadly known across Western Russia.[58] To conclude, it seems likely that these unearthly creatures did not necessarily have to be part of a tool but rather of a symbolic object marking the rank and social position of its owner. Particularly, the Lõhavere figure with its possibly shiny eyes made of a piece of gold or precious stone[59] seems more suitable as a status symbol than a whip handle.

---

52 Iršenas, Marius: "Anthropomorphic and Zoomorphic Motifs on Balt jewellery". In Butrimas, Adomas (ed.): *Art of the Balts. The Catalogue of Exhibition*. Vilniaus dailės akademijos leidykla, Vilnius 2009, pp. 99–186.
53 Asplund, Henrik: "The Bear and The Female.Bear- Tooth Pendants in Late Iron Age Finland". In: Mäntylä, Sari (ed.): *Rituals and Relations. Studies on the Society and Material Culture of the Baltic Finns*. (Suomalaisen Tiedeakatemian Toimituksia, sarja Humaniora 336). Academia Scientiarum Fennica: Helsinki 2005, pp. 13–30; Kivisalo, Nora: "The Later Iron Age Bear-Tooth Pendants in Finland: Symbolic Mediators Between Women, Bears and Wilderness?" *Temenos* 44(2), 2008, pp. 263–291.
54 Luik 2010, p. 134.
55 Mandel 2003, p. 63.
56 Kuehn, Sara: *The Dragon in Medieval East Christian and Islamic Art*. Brill: Leiden 2011, pp. 45, 116.
57 Kalmring 2015, p. 61.
58 E.g. Koltschin 1985.
59 Luik 2010.

Supernatural creatures like dragons and griffins have often been interpreted as evil forces, partly influenced by Norse mythology, partly by Christian theology. Archaeological material, however, presents a different view. Oversized snakes and dragons are often associated with power and status in the Viking Age and in the medieval worldview.[60] Additionally, archaeological finds like the griffin-headed handles seem to be associated predominantly with the nobility and power centres, where these have been proudly on display, thus telling a different story than later mythology or Christian theology. The association with power and a high position in society can also be extended to other dangerous and supernatural creatures, such as predatory animals, but most likely also to the sophisticated ornament, which often hides a snake or a dragon in it. Handles in the shape of birds of prey with hooked beaks and ears from different sites all around the Baltic Sea[61] can also be added to these. It seems as if no particular snake, animal or bird species was important in itself, but rather that the idea of fearsome but noble birds of prey, dangerous predators, and mysterious animals formed the same mental concept to symbolise wealth, position, power, and authority. This interpretation could be taken somewhat further: the purpose of the figures was not to represent just some unearthly being, but these creatures represented certain moral and ideological values, shared by a common group. When examining the mythology where griffins and dragons exist, they are exclusively associated with wealth, prosperity, and richness, as well as aggressiveness and military power, which is well associated with the symbolism referred to by animal teeth pendants from the Late Iron Age.[62]

I have previously interpreted the bronze alloy tooth pendants as signs of relationships between ruling families in south-western Finland and the lower reaches of the Daugava River.[63] Accordingly the lower reaches of the Daugava represented one of the major trading centres in the Eastern Baltic, and bronze tooth pendants symbolise the links between the two communities, the nobility of which had close contacts.[64] Possibly we can also interpret these figures of

---

60 For more detail, see Pluskowski; Brunning.
61 Balodis 1939, p. 14; Luik 2010; Kivikoski 1973, taf 139: 1224.
62 Jonuks, Tõnno / Rannamäe, Eve: "Animals and Worldviews: A Diachronic Approach to Tooth and Bone Pendants From The Mesolithic to Medieval Period in Estonia". In: Madgwick, Richard / Livarda, Alexandra / Riera Mora, Santiago (eds.): *Bioarchaeology of Ritual and Religion*. Oxbow: Oxford 2018.
63 Jonuks 2017.
64 Pihlman, Sirkku: "Beziehungen mit den Liven? Gedanken über die Kontaktfelder der Raisio-bewohner am ende der Prähistorischen Zeit". In: Mäntylä, Sari (ed.): *Rituals and Relations. Studies on the Society and Material Culture of the Baltic Finns*. (Suomalaisen

griffin as signs of links of influential and ruling families who shared a common worldview, possibly kin relations and probably trading interests, despite distant centres. This would explain why the very similar kind of copper alloy griffin figures occur in distant locations as very similar, although isolated finds. The interpretation of mythological creatures as symbols of contacts between the ruling families opens up new ways of interpretation, not limited to broad trends of trading and migration, but allowing us to see smaller units, like families in a micro-archaeological sense.

References

Adam of Bremen: *History of the Archbishops of Hamburg-Bremen*. Columbia University Press: New York 2002.

Adams, Noël: "Between Myth and Reality. Hunter and Prey in Early Anglo- Saxon Art". In: Bintley, Michael D. J. / Williams, Thomas J. T. (eds.): *Representing Beasts in Early Medieval England and Scandinavia*. (Anglo-Saxon Studies 29). The Boydell Press: Suffolk 2015, pp. 13–51.

Asplund, Henrik: "The Bear and The Female.Bear- Tooth Pendants in Late Iron Age Finland". In: Mäntylä, Sari (ed.): *Rituals and Relations. Studies on the Society and Material Culture of the Baltic Finns*. (Suomalaisen Tiedeakatemian Toimituksia, sarja Humaniora 336). Academia Scientiarum Fennica: Helsinki 2005, pp. 13–30.

Balodis, Francis: *Latviešu starptautiskie sakari ap 1000 gadu pec Kristus*. Latvijas Vēstures Institūta Žurnāls: Riga 1939.

Brunning, Sue: "(Swinger of) the Serpent of Wounds'. Swords and Snakes in The Viking Mind." In: Bintley, Michael D. J. / Williams, Thomas J. T. (eds.): *Representing Beasts in Early Medieval England and Scandinavia*. (Anglo-Saxon Studies 29). The Boydell Press: Suffolk 2015, pp. 53–72.

Creutz, Kristina: *Tension and Tradition. A Study of Late Iron Age Spearheads Around the Baltic Sea*. (Theses and Papers in Archaeology 8). Stockholm 2003.

Duffin, Christopher J.: "A Survey of Birds and Fabulous Stones". *Folklore* 123(2), 2012, pp. 179–197.

Gräslund, Anne-Sofie: "Wolves, Serpents and Birds.Their Symbolic Meaning in Old-Norse Belief." In: Andrén, Anders / Jennbert, Kristina / Raudvere, Catharina (eds.): *Old Norse Religion in Long-term Perspective. Origins,*

Tiedeakatemian Toimituksia, sarja Humaniora 336). Academia Scientiarum Fennica: Helsinki 2005, pp. 207–223.

Changes & Interactions. (Vägar till Midgård 8). Nordic Academic Press: Lund 2006, pp. 124–129.

Gräslund, Anne-Sofie: "Drakar I Uppåkra". In: Hårdh, Birgitta (ed.): *Fler fynd I centrum. Materialstudier i och kring Uppåkra.* (Acta Archaeologica Lundensia. Series in 8°; Vol. 45). Almqvist & Wiksell International: Stockholm 2003, pp. 179–188.

Hedeager, Lotte: *Iron Age Myth and Materiality. An Archaeology of Scandinavia AD 400–1000.* Routledge: London / New York 2011.

Helskog, Knut: "Bears and Meanings Among Hunter-Fisher-Gatherers in Northern Fennoscandia 9000–2500 BC". *Cambridge Archaeological Journal* 22(2), 2012, pp. 209–236.

Hoffmann-Krayer, Eduard: "Adlerstein". In: Bächtold-Stäubli, Hanns / Hoffmann-Krayer, Eduard (eds.): *Handwörterbuch des deutschen Aberglaubens. Band 1.* Walter de Gruyter: Berlin / New York 2000, pp. 189–194.

Ingold, Tim: *The Perception of Environment. Essays on Livelihood, Dwelling and Skill.* Routledge: London / New York 2002.

Insoll, Timothy: *Archaeology. The Conceptual Challenge.* (Duckworth debates in archaeology). Duckworth: London 2007.

Iršenas, Marius: "Anthropomorphic and Zoomorphic Motifs on Balt Jewellery". In Butrimas, Adomas (ed.): *Art of the Balts. The Catalogue of Exhibition.* Vilniaus dailės akademijos leidykla, Vilnius 2009, pp. 99–186.

Jensen, Bo: *Viking Age Amulets in Scandinavia and Western Europe.* Archaeopress: Oxford 2010.

Johanson, Kristiina / Jonuks, Tõnno: "Are We Afraid of Magic? Magical Artifacts in Estonian Museums." *Material Religion* 14(2), 2018. DOI:10.1080/17432200.2018.1443894.

Jonuks, Tõnno: "Domesticating Europe – Novel Cultural Influences in the Late Iron Age Eastern Baltic". In: Selart, Anti (ed.) *Changing Aliens, Changing Natives. Baltic Crusades and Societal Innovation in Medieval Livonia.* Brill: Leiden / Boston 2022, pp. 29–54. DOI:/10.1163/9789004512092_003.

Jonuks, Tõnno / Rannamäe, Eve: "Animals and Worldviews: A Diachronic Approach to Tooth and Bone Pendants From The Mesolithic to Medieval Period in Estonia". In: Madgwick, Richard / Livarda, Alexandra / Riera Mora, Santiago (eds.): *Bioarchaeology of Ritual and Religion.* Oxbow: Oxford 2018, pp. 162–178.

Jonuks, Tõnno: "Bronze Tooth Pendants From The Late Iron Age: Between Real and Fictional Zooarchaeology". *Norwegian Archaeological Review* 50(2), 2017. DOI:10.1080/00293652.2017.1367838.

Jonuks, Tõnno / Kurisoo, Tuuli: "To Be or Not to Be... a Christian: Some New Perspectives on Understanding the Christianisation of Estonia". *Folklore* 55, 2013, pp. 69–98. DOI:10.7592/FEJF2013.55.jonuks_kurisoo.

Jonuks, Tõnno: "An Antler Object From the Pärnu River – An Axe, A God, or A Decoy?" In: Johanson, Kristiina / Tõrv, Mari (eds.): *Man, his Time, Artefacts, and Places.* (Muinasaja Teadus 19). Institute of History and Archaeology: Tartu 2013, pp. 225–246.

Jonuks, Tõnno. "A Few Additions to the Depiction of Estonia and the Eastern Shore of the Baltic Sea in Scandinavian Sagas". In: Zilmer, Kristel (ed.): *Dialogues With Tradition: Studying the Nordic Saga Heritage.* Tartu University Press: Tartu 2005, pp. 45–63.

Kaljundi, Linda: "Waiting for the Barbarians: Reconstruction of Otherness in the Saxon Missionary and Crusading Chronicles, 11th–13th Centuries". In: Kooper, Erik (ed.): *The Medieval Chronicle,* vol. 5. Rodopi: Amsterdam / New York 2008, pp. 113–127.

Kalmring, Sven: "The Dragonhead and the Wooden Jetty. Back in Birka's Black Earth Harbour". *Jahresbericht. Zentrum für Baltische und Skandinavische Archäologie*, 2015, pp. 58–61.

Kaul, Flemming: *Bronzealderens religion. Studier af den nordiske bronzealders ikonografi.* Det Kongelige Nordiske Oldskriftselskab: København 2004.

Kiudsoo, Mauri: *Viikingiaja aarded Eestist. Idateest, rauast ja hõbedast.* Äripäev: Tallinn 2016.

Kivikoski, Ella: *Die Eisenzeit Finnlands. Bildwerk und Text.* Finnische Altertumsgesellschaft: Helsinki 1973.

Kivisalo, Nora: "The Later Iron Age Bear-Tooth Pendants in Finland: Symbolic Mediators Between Women, Bears and Wilderness?". *Temenos* 44(2), 2008, pp. 263–291.

Koltschin, Boris Alexandrowitsch (Колчин, Борис Александрович): *Drevnij Novgorod. Prikladnoe iskusstvo i arheologija.* Iskusstvo: Moskva 1985.

Kuehn, Sara: *The Dragon in Medieval East Christian and Islamic Art.* Brill: Leiden 2011.

Kurisoo, Tuuli: *Adornment, Self-Definition, Religion: Pendants of the North-Eastern Baltic Sea Region, 9th-13th Century.* (Studien zur Siedlungsgeschichte und Archäologie der Ostseegebiete. Band 19). Wachholz: Kiel, Hamburg 2021.

Leimus, Ivar / Roio, Maili / Sarv, Krista: "Watertight Sources: Unique Find from the Bottom of Tallinn Bay". In: Roio, Maili (ed.): *Shipwreck Heritage: Digitalizing and Opening Access to Maritime History Sources.* (Muinasaja Teadus 23). Muinsuskaitseamet: Tallinn 2013, pp. 133–164.

Lévi-Strauss, Claude: *Metsik mõtlemine.* Vagabund: Tallinn 2001.

Loorits, Oskar: *Grundzüge des Estnischen Volksglaubens I.* Carl Bloms Boktryckeri A.-B: Lund 1949.

Loze, Ilze Biruta: "The Theme of The Grass Snake in Neolithic Material From Latvia". In: Loze, Ilze Biruta (ed.): *Art, Applied Art and Symbols in Latvian Archaeology.* (Humanities and Social Sciences, Latvia. University of Latvia 2 (39)). University of Latvia: Riga 2003, pp. 48–59.

Luik, Heidi: "Linnupeakujuline sarvest käepide Lõhavere linnamäelt". In: Tamla, Ülle (ed.) *Ilusad asjad. Tähelepanuväärseid leide Eesti arheoloogiakogudest.* (Muinasja teadus 21). Ajaloo Instituut: Tallinn 2010, pp. 127–138.

Luik, Heidi: "S-kujulised ripatsid ja rihmakeeled". *Eesti Arheoloogia Ajakiri* 3/2, 1999, pp. 115–130.

Mandel, Mati: *Läänemaa 5.–13. sajandi kalmed.*(Töid Ajaloo Alalt 5). Eesti Ajaloomuuseum: Tallinn 2003.

Moora, Harri / Saadre, Osvald: "Lõhavere linnamägi". In: Moora, Harri (ed.): *Muistse Eesti linnused. 1936.–1938. a. uurimiste tulemused.* Õpetatud Eesti Selts: Tartu 1939.

Mäesalu, Ain: "Kas Otepää linnuselt leiti muinaseestlaste sõjasarv?" *Tutulus*, 2014.pp.26–29.

Mäger, Mart: *Linnud rahva keeles ja meeles.* Eesti Raamat: Tallinn 1969.

Pihlman, Sirkku: "Beziehungen mit den Liven? Gedanken über die Kontaktfelder der Raisio-bewohner am ende der Prähistorischen Zeit". In: Mäntylä, Sari (ed.): *Rituals and Relations. Studies on the Society and Material Culture of the Baltic Finns.* (Suomalaisen Tiedeakatemian Toimituksia, sarja Humaniora 336). Academia Scientiarum Fennica: Helsinki 2005, pp. 207–223.

Pluskowski, Aleks: "The Dragon's Skull: How Can Zooarchaeologists Contribute to Our Understanding of Otherness in The Middle Ages?" In: Walker-Vadillo, Monica Ann / Picaza, María Victoria Chico / García, Francisco de Asís García (eds.): *Animals and Otherness in the Middle Ages. Perspectives Across Disciplines.* (BAR International Series 2500). Archaeopress: Oxford 2013, pp. 109–124.

Price, Neil: "What's in A Name? An Archaeological Identity Crisis for The Norse Gods (and Some of Their Friends)". In: Andrén, Anders / Jennbert, Kristina / Raudvere, Catharina (eds.): *Old Norse Religion in Long-term Perspective. Origins, Changes & Interactions.* (Vägar till Midgård 8). Nordic Academic Press: Lund 2006, pp. 179–183.

Russell, Nerissa: *Social Zooarchaeology. Humans and Animals in Prehistory.* Cambridge University Press: Cambridge 2012.

Rybina, Elena: "Recent Finds from Excavations in Novgorod". In: Brisbane, Mark A. (ed.): *The Archaeology of Novgorod, Russia. Recent Results From the Town and its Hinterland.* (The Society for Medieval Archaeology Monographs 13). Society for Medieval Archaeology: Lincoln 1992, pp. 160–192.

Spirģis, Roberts: "Nahodki zoomorfnyh podvesok "smolenskogo" tipa na territorii Latvii i ih novaja interpretacija". *Stratum plus* 5, 2012, pp. 195–220.

Symons, Victoria: "Wreonþenhilt ond Wyrmfah. Confronting Serpents in Beowulf and Beyond". In: Bintley, Michael D. J. / Williams, Thomas J. T. (eds.): *Representing Beasts in Early Medieval England and Scandinavia*. (Anglo-Saxon Studies 29). The Boydell Press: Suffolk 2015, pp. 73–93.

Anu Mänd

Visual Representation of Animals in Livonian Urban Space, ca. 1400–1550

Abstract: This paper discusses animal representations in medieval Livonian Hanse towns with a particular focus on limestone reliefs on the dwellings and guild houses. The paper explores the functions the animal figures had, the messages they conveyed, and the kind of emotions they might have evoked. In order to better understand the multiple symbolic layers of animal representations, the following subtopics are discussed: animals as attributes of saints, heraldic emblems, apotropaic signs, and characters in didactic-moralistic narrative scenes. It is argued that animal figures should not be studied in isolation but in the context of other visual and textual motifs used in the particular period of time.

Keywords: animals in medieval art, stone reliefs, Hanse towns, lions, dragons, burgher heraldry

Introduction

Animal representations played an important role in medieval visual culture: they were used as signs loaded with spiritual and moral meanings, as vehicles for communicating religious and allegorical messages, and as mnemonic tools that helped to memorise these messages.[1] While some animals developed into widely recognisable conventional signs (e.g. *Agnus Dei*, the Lamb of God), others could be "read" differently depending on the time and cultural context, the type of visual genre, and the intended audience.

This chapter will explore animal representations in late medieval Livonia – a historical region that roughly corresponded to present-day Estonia and Latvia.[2] Scholarly interest in animals in Livonian visual culture has emerged since the mid-1990s. The majority of studies on animal representations have focused on

---

1   Resl, Brigitte: "Beyond the Ark: Animals in Medieval Art". In: Resl, Brigitte (ed.): *A Cultural History of Animals in the Medieval Age*. (A Cultural History of Animals, vol. 2). Berg: Oxford / New York 2007, pp. 179–180.
2   Research for this chapter was supported by the Estonian Research Council grant PRG1276 and Tallinn University grant TF1620. The results have partially been published in Estonian: Mänd, Anu: "Loomad kunstis". In: Kreem, Juhan / Leimus, Ivar / Mänd, Anu / Põltsam-Jürjo, Inna: *Loomad keskaegse Liivimaa ühiskonnas ja kunstis*. Tallinn University Press: Tallinn 2022, pp. 115–172.

sacral art and discussed the meaning of animals on altarpieces, corbels, capitals and bases of church pillars, and so on.³ My aim is to turn from the sacred to the profane sphere and to study the visual representation of animals in urban public space. The focus is on the façades and portals of the guild houses and dwellings, with a glimpse into the interior of a guild house as a semi-public space. However, since church portals could serve as models or sources of inspiration for those of profane buildings, these will be surveyed as well. It has rightfully been pointed out that in the medieval urban sphere, the sacred and the profane, the "high" and the "low" intermingled and that it is not fruitful to draw a clear demarcation line between them.⁴

The artworks to be discussed are limestone reliefs on the imposts of portals or elsewhere on the exterior. Due to the scarcity of decorative architectural details surviving on or from late medieval dwellings and guild houses, the examples I am going to discuss originate predominantly from Tallinn where the amount of preserved "stone animals" is higher than in other towns of today's Estonia and Latvia.

The chapter will explore what kind of animals and birds were represented in the visual environment of a late medieval town and why. What functions did the animal figures have, what messages did they convey, and what kinds of emotions did they evoke? Was the choice and meaning of animals depicted on the dwellings and guild houses different from those in the sacred space? Can we talk about "animal fashion" in certain periods of time? Sometimes the stone reliefs were accompanied with an inscription in Latin or Middle Low German. Was

---

3   Kivimaa, Katrin: "Dualistlik maailmavaade eesti keskaegses loomasümboolikas". In: Alttoa, Kaur (ed.): *Ars Estoniae Medii Aevi Grates Villem Raam*. Eesti Muinsuskaitse Selts: Tallinn 1995, pp. 157–170; Mänd, Anu: "Bad Boys, Men, and Dogs in Bernt Notke's Tallinn Altarpiece". In: Markus, Kersti (ed.): *Bilder i marginalen. Nordiska studier i medeltidens konst = Images in the Margins. Nordic Studies in Medieval Art*. Argo: Tallinn 2006, pp. 305–320; Bome, Helen: "Kloostrielu kiusatused Padise piltkonsoolidel". *Kunstiteaduslikke Uurimusi = Studies on Art and Architecture* 21(1–2), 2012, pp. 7–36; Mänd, Anu: "Kass voodi all. Ühest motiivist Hermen Rode ja Bernt Notke Tallinna retaablitel". *Kunstiteaduslikke Uurimusi = Studies on Art and Architecture* 21(1–2), 2012, pp. 231–246; Mänd, Anu: "A Mouser in the Bedroom". In: Berggren, Lars / Landen, Annette (eds.): *Ecce leones! Om djur och odjur i bildkonsten*. Artifex: Lund 2018a, pp. 191–205.
4   See, e.g. Camille, Michael: "At the Sign of the 'Spinning Sow': the 'Other' Chartres and Images of Everyday Life of the Medieval Street". In: Bolvig, Axel / Lindley, Phillip (eds.): *History and Images: Towards a New Iconology*. Brepols: Turnhout 2003, p. 251, 276.

there a link between the images and the text or did they function independently from one another? How did the visual artistic sources reflect and shape human–animal relations and contemporary attitudes towards animals?

In order to better understand the importance of animal representations in urban space, it is not only relevant to study the images themselves but also to try and ascertain the identity of the commissioners, that is, in the case of the dwellings, the owners of the house. Due to the scarcity and nature of written medieval sources, this is not always possible, but even a few cases can shed light on the question on which social groups preferred certain themes and motifs, and why. Thus, this chapter takes an interdisciplinary approach and combines the sources and methods of visual culture with those of social history.

The time frame of my study runs from about 1400 to about 1550: the earliest images on Livonian profane buildings survive from the first decades of the fifteenth century and, as will be argued below, the amount and variety of animal depictions notably diminished after the Reformation.

Façade as the Principal Place of Visual Communication

Before proceeding to the animal representations in urban space, a few introductory words are needed about the decoration of the late medieval profane buildings. According to architectural treatises, a house had to express the owner's identity, his religious, ethical and moral beliefs – therefore, the façade can be regarded as the owner's public calling card.[5] The façade was the principal place of visual communication and self-representation. Most of the decorations were placed at the main entrance, which formed a transition zone between the exterior and the interior, between the public and the private space, both physically as well as symbolically. Thus, the images on and around the portal were of special significance.[6]

In Livonian towns, as in many other towns around the Baltic Sea, it was not only the portal that was decorated but, above all, the *Beischlagsteine* – a pair of

---

5   Kodres, Krista: *Esitledes iseend: tallinlane ja tema elamu varauusajal.* Tallinna Ülikooli Kirjastus: Tallinn 2014, pp. 61–62; Kodres, Krista: "Self-representation and Social Aesthetics: Wealthy Tallinn Burgher Homes in the Early Modern Period". In: Mänd, Anu / Tamm, Marek (eds.): *Making Livonia: Actors and Networks in the Medieval and Early Modern Baltic Sea Region.* Routledge: London / New York 2020, pp. 302–304.
6   Mänd, Anu: "Images and Inscriptions on Dwelling Houses in Livonian Towns, c. 1450–c. 1550". In: Leimus, Ivar (ed.): *Everyday Life in a Hanseatic Town.* (Varia historica 8). Eesti Ajaloomuuseum: Tallinn 2021, pp. 91–141.

vertical stone slabs framing the elevated terrace and the stairs from the street to the entrance. The shape of the slabs was similar to contemporary bench ends, and indeed, the narrow platform on either side of the stair was in warmer months used as a bench where the family could spend time, show themselves, and receive guests. Since the use of *Beischlagsteine* or *Wangensteine* was a "German" or "Hanseatic" phenomenon, there is no equivalent in English. In the following, I will use the term "doorside stones". Doorside stones were a typical element in front of late medieval town halls, guild houses, and dwellings in northern Germany, Prussia, Denmark, Sweden and Livonia.[7] Many of them have survived in Lübeck, Hamburg, Stralsund, Rostock, Gdańsk, and elsewhere in the Baltic Sea region, including the three largest Livonian towns Tallinn, Riga, and Tartu.[8]

In Livonia, the doorside stones were first recorded in front of the Tallinn town chancery in 1404 or 1405.[9] Most of the extant slabs or their fragments can be dated to the late fifteenth and the sixteenth centuries, with the latest ones originating from the seventeenth century. The last doorside stones and terraces were removed from the streets between 1825 and 1833, due to the need for space for traffic.[10] Some of them were re-used as tombstones.

The doorside stones, portals, reliefs on the gables and other decorated stones in Livonia were carved of local types of limestone, which are particularly abundant in northern and western Estonia. The extant slabs are monochrome, i.e. grey, but according to medieval documentary evidence, some of them were coloured, which increased their beauty and communicative power.

The most frequent motifs on the doorside stones were: a coat of arms or a house mark (usually that of the husband on the left slab, i.e. heraldically on the dexter, and that of the wife on the right slab); a Christian symbol, especially the

---

7   Saal, Walter: "Beischlagsteine und ihre Beziehungen zu Grabkreuzen und Sühnekreuzen". *Steinkreuzforschung* 4, 1982, pp. 30–35; Möller, Gunnar: "*Decora alta domus*. Mittelalterliche und frühneuzeitliche Wangensteine in Stralsund". In: Birli, Sonja et al. (eds.): *ene vruntlike tohopesate. Beiträge zur Geschichte Pommerns, des Ostseeraums und der Hanse. Festschrift für Horst Wernicke zum 65. Geburtstag*. Verlag Dr. Kovač: Hamburg 2016, pp. 355–372.
8   Üprus, Helmi: *Tallinna etikukivid*. Kunst: Tallinn 1971; Mänd 2021, pp. 94–116.
9   Kangropool, Rasmus: "Tallinna hilisgooti etikukividest". *Vana Tallinn* IV (VIII). Tallinn 1994, p. 5.
10 Üprus 1971, pp. 5–7. A nostalgic recollection of these outdoor "benches" and their removal can be found in: Sprengfeld, G.: *Meine Vaterstadt Reval vor 50 Jahren*. Verlag von Schnakenburg: Dorpat 1877, pp. 20–21. G. Sprengfeld was the pseudonym of Gotthard von Hansen (1821–1900), a Baltic German historian and Tallinn city archivist.

rose; a human figure (saint, angel); and an inscription (name, date, prayer, Bible quote, etc.). The main function of these images and inscriptions was to inform and to communicate: to identify the owner, express his social and marital status, and give testimony to his wealth and religious beliefs.[11]

In order to better understand the multiple symbolic layers of animal representations, I have divided them into the following groups: attributes of saints, heraldic emblems, protective (apotropaic) signs, and characters in didactic-moralistic narrative scenes. This classification is by no means strict because an animal could at the same time be a Christian symbol and a heraldic emblem, or a protagonist in a fable. This is above all true for high-status beasts, such as the lion – the king of all animals, which had multiple symbolic meanings, was popular in all forms of visual arts, featured prominently in heraldry, and was a frequent character in fables.[12]

Animals as Attributes

Animals played an important role in many saints' legends and became attributes of saints and Biblical figures in art. Indeed, many saints can be visually identified thanks to an animal figure standing next to or being held by him or her, for example, St John the Baptist and St Agnes with the lamb, St Jerome with the lion, and St George and St Margaret of Antioch with the dragon.[13] In the case of the Evangelists, the eagle, the winged lion and the winged ox were not only the attributes of St John, St Mark and St Luke, respectively, but also their symbols, which were often represented instead of the human figures. These attributes and symbols were well-known to medieval viewers and helped them to better learn and memorise the holy stories.

Saints and Biblical figures were depicted not only in the sacred but also in the secular space, including dwellings and guild houses. In Tallinn, a fragment of a doorside stone with St George slaying the dragon has been immured on the wall of 2 Mündi Street (Figure 1).[14]

---

11 Mänd 2021, pp. 139–140.
12 For the lion in Christian iconography, see Kirschbaum, Engelbert et al. (eds.): *Lexikon der christlichen Ikonographie* [hereafter LCI]. Herder: Rome et al. 2004, vol. 3, col. 112–119.
13 In the medieval context, one should not differentiate between "real" and "fantastic" beasts.
14 Üprus 1971, p. 25.

Figure 1. Fragment of a doorside stone with St George, ca. 1510–15. 2 Mündi St, Tallinn. Photo: Stanislav Stepashko.

It is not known if the slab had stood in front of the same building or not. Based on the details of the saint's armour, the relief can be dated to about 1510–1515.

Another St George with the dragon was depicted on the right doorside stone standing at Pikk Street. Its pair was decorated with the image of the Virgin Mary as the Madonna of the Apocalypse. The slabs have not been preserved, but they are known thanks to a drawing by Danish antiquarian and illustrator Søren Abildgaard, who in 1754 travelled in Estonia with his colleagues.[15] Based on the date 1493 above Abildgaard's drawing, and his comment about the location of

---

15 Grinder-Hansen, Poul: *Søren Abildgaard (1718–1791): Fortiden på tegnebrættet.* Nationalmuseet: Copenhagen 2010, p. 652, no. 847; Nationalmuseets Samlinger Online: *Bislagssten, Tallinn,* retrieved 10.07.2023, from https://samlinger.natmus.dk/DMR/asset/53694.

the house, it was possible to ascertain that the slabs had stood in front of 35 Pikk St, which belonged to merchant Hans Pawels, who acquired the house in about 1493.[16] Pikk (i.e. Long) Street was the main street of Tallinn and a very prestigious one, where the wealthy and the powerful lived, where the houses of the three major guilds were located and which was a route for processions and parades.[17] More will be said about Pawels and his home in the last subsection.

St George as an ideal Christian knight and a universal saint venerated by all social layers was a popular figure on doorside stones not only in Livonia, but also elsewhere in the Baltic Sea region: fifteenth and sixteenth-century slabs with St George have survived, for example, in Gdańsk and Hamburg.[18] In Livonian towns, he was particularly important for the two merchants' associations – the Great Guild and the Brotherhood of the Black Heads.[19]

The dragon in the legend of St George was an evil creature that symbolised Satan. However, as will be discussed below, the dragon was an ambivalent beast that, depending on the context, could also have a positive connotation.

Images on the façade and doorside stones of a guild house usually depicted the patron saint and the coat of arms of the organisation. In Tallinn, the figures of the patron saints of the artisans' guilds are both accompanied by an animal or animals. The mid-sixteenth-century doorside stones of St Canute's Guild (20 Pikk St) represented the arms of the guild on the left and that of the city on the right slab. On the arms of the guild one can see a shield with three lions and the bust of St Canute the King with the sceptre above the shield.[20] The three blue lions passant on a yellow background were the coat of arms of the Danish royal house and therefore also the attribute of King Canute.

---

16 Kangropool 1994, p. 7.
17 Mänd, Anu: *Urban Carnival: Festive Culture in the Hanseatic Cities of the Eastern Baltic, 1350–1550*. Brepols: Turnhout 2005, pp. 257–262.
18 Mänd 2021, p. 101, fig. 4a (Gdańsk). The Hamburg slab from the first half of the fifteenth century is displayed in the Hamburg Museum, inv. no. AB 1026.
19 Bruiningk, Hermann von: *Messe und kanonisches Stundengebet nach dem Brauche der Rigaschen Kirche im späteren Mittelalter*. Kymmel: Riga 1904, pp. 417–421; Mänd, Anu: "Saints' Cults in Medieval Livonia". In: Murray, Alan V. (ed.): *The Clash of Cultures on the Medieval Baltic Frontier*. Ashgate: Farnham 2009, pp. 204–205.
20 Mänd, Anu: "The Cult and Visual Representation of Scandinavian Saints in Medieval Livonia". In: Jensen, Carsten Selch et al. (eds.): *Saints and Sainthood around the Baltic Sea: Identity, Literacy, and Communication in the Middle Ages*. Western Michigan University: Kalamazoo 2018b, pp. 123–125, fig. 6.11–6.12. Today the upper parts of the doorside stones can be seen on the back wall of the guild house.

The gable of the ante hall of the former house of St Olaf's Guild (24 Pikk St) is decorated with a rectangular relief representing a standing figure of St Olaf (Figure 2).[21] The holy king is bearing a sceptre and his specific attribute – a battle axe. Under his feet is a hybrid creature – a winged dragon with a human head, bearing a crown. This widespread attribute of St Olaf has been interpreted as a symbol of the Antichrist.[22]

Figure 2. Relief with St Olaf on the gable of the ante hall of St Olaf's Guild in Tallinn, second quarter of sixteenth century. Photo: Tallinn City Archives, coll. R-242.

---

21 Mänd 2018b, pp. 109–110.
22 For details, see Lidén, Anne: *Olav den helige i medeltida bildkonsten. Legendmotiv och attribut*. KVHAA: Stockholm 1999, pp. 220–240.

As I have argued earlier, the representation of the patron saints on the façade of the guild houses had multiple functions: to identify the corporate owner, to manifest the saint's protection over the building and the guild members, and to evoke prayers to the saint.[23] Indeed, it even raises the question of whether the guild houses should be understood as profane buildings or rather spaces where the religious and the secular are closely interconnected.

The animals that were used as attributes of saints in Christian art are not passive elements of a composition but active visual agents. In a way, it is the dragon who defines St George: without this beast, he could be confused with any other military saint. The figure of the dragon leads the viewer's mind to the heroic battle of St George, and reminds us of the necessity to stand up to evil. Thus, although the primary function of animal attributes was to identify the saint, they were also memory-triggers that helped to recall and contemplate the saints' legends and other holy stories, or acted as religious signs and symbols that aroused emotions.

Animals in (Burgher) Heraldry

The subject of animals in Livonian heraldry, which was already mentioned above in connection with the arms of St Canute's Guild, is too large to be extensively discussed here; therefore, I will merely draw attention to two cases, illustrating the visual language of emblems by different social layers.

A pair of fragmentary doorside stones with a jumping dog has been preserved in Tallinn (Figure 3a and 3b). Today, they are found in secondary places but initially they stood in front of 19 Lai Street and belonged to Henninck Passow.[24]

---

23 Mänd 2018b, p. 109.
24 Kangropool 1994, pp. 7–8, Figures 2–4; Mänd 2021, pp. 109–113.

Figures 3a and 3b. Doorside stones of the knight Henninck Passow, ca. 1510, Tallinn. Photos: Stanislav Stepashko.

Passow (Passou) was a knight who, prior to settling in Tallinn in about 1510, had been in the service of Svante Nilsson (Sture) and the Swedish crown.[25] Unlike many other doorside stones, which depicted the arms of a husband and wife, those of Passow are both decorated with his arms. This was most probably due to the fact that Passow's wife was a daughter of a Tallinn merchant and town councillor, who, albeit belonging to the urban elite, was not equal to a nobleman.[26]

The dog on Passow's arms is unquestionably noble as well. First, it is a greyhound – a dignified hunting dog,[27] and second, it is wearing a distinctive collar – a sign of feudal loyalty.[28] Probably for the sake of symmetry, the image of the dog on the right mirrors that on the left.

---

25 Arbusow, Leonid (ed.): *Liv-, Est- und Kurländisches Urkundenbuch*, 2nd series, vol. 3. Deubner: Riga / Moskau 1914, no. 73.
26 For Passow, see Seeberg-Elverfeldt, Roland: *Testamente Revaler Bürger und Einwohner aus den Jahren 1369 bis 1851. Revaler Regesten* III. Vandenhoeck & Ruprecht: Göttingen 1975, no. 113; Mänd 2021, pp. 110–112.
27 For the greyhounds as markers of nobility, see Belozerskaya, Marina: "Good Dog: Model Canines in Renaissance Manuscripts". In: Hamburger, Jeffrey H. / Korteweg, Anne S. (eds.): *Tributes in Honor of James H. Marrow. Studies in Painting and Manuscript Illumination of the Late Middle Ages and the Renaissance*. Harvey Miller Publishers: London / Turnhout 2006, pp. 66–69.
28 For the meaning of a dog collar in art, see Friedman, John Block: "Dogs in the Identity Formation and Moral Teaching Offered in Some Fifteenth-Century Flemish

Only the upper part (the "head") of the left slab and a little more of the right slab have survived. However, in 1754 a drawing was made of the left slab by Søren Abildgaard,[29] and there one can see that the lower part of the slabs was provided with a two-line inscription in Gothic minuscule: *her hen[n]i[n]ck passou / ritter* (Lord Henninck Passou, knight). Thus, the knightly status of Passow was expressed with the image as well as with the inscription, being recognisable for the literate as well as the illiterate viewers.

In 1636, town councillor Matthias Poorten, who at that time lived at 19 Lai Street, had his arms and those of his wife Elisabeth Goldberg hewn below the arms of Passow, which is a fine example of how medieval doorside stones were re-used and treated with respect by later owners: they did not remove the original imagery but simply made additions to it.[30]

Animals and birds did not only occur as emblems on the coats of arms of the nobility (or as supporters of the shield) but also in burgher heraldry. Wealthy merchants, in particular town councillors, imitated noble lifestyles and designed coats of arms of their own. A post-Reformation doorside stone from Tartu is decorated with the head of a cock, the number 43 (i.e. the latter numerals of the year 1543) and an inscription, referring to the grace of God (Figure 4).[31]

Manuscript Miniatures." In: Gelfand, Laura D. (ed.): *Our Dogs, Our Selves: Dogs in Medieval and Early Modern Art, Literature, and Society*. Brill: Leiden / Boston 2016, pp. 353–362.

29 Kangropool 1994, fig. 2; Grinder-Hansen 2010, p. 651, no. 846; Nationalmuseets Samlinger Online: *Bislagssten*, retrieved 1.6.2020, from https://samlinger.natmus.dk/DMR/asset/54813.

30 Interestingly enough, Abildgaard, who was in search of Danish and Swedish heritage, did not include the seventeenth-century arms and other additions in his drawing.

31 Tartu City Museum, inv. no. TM 118 Aj 61. The slab was attached to the gateway in the yard of 11 Rüütli St.

Figure 4. Doorside stone with Honerjeger's arms, 1543, Tartu. Tartu City Museum. Photo: Stanislav Stepashko.

The emblem belongs among the *armes parlantes*, where a visual pun refers to the bearer's name. The arms with the head of a cock belonged to the Honerjeger/ Hünerjäger (literally, the hunter of hens) family, whose members lived in Tartu as well as in Tallinn.[32] The last numerals of the year 1543 indicate that this slab

---

32 Üprus 1971, p. 33; Nottbeck, Eugen von: *Siegel aus dem Revaler Rathsarchiv nebst Sammlung von Wappen der Revaler Rathsfamilien*. Rathgens: Lübeck 1880, plate 4, no. 67.

has stood on the right side, which means that it was most probably the wife who originated from the Honerjeger family.[33]

Animals and birds as heraldic emblems on doorside stones helped to identify the house owner and expressed his social and marital status. After the Reformation, increasing attention was paid to the Bible quotes (the Word of God), which demonstrated that the owner was a firm supporter of the Evangelical faith.

Animals as Apotropaic Signs

As mentioned earlier, the entrance was the most important element on the façade, a transition zone that required special protection, physically as well as symbolically. Animals and birds appeared first on the church portals, carrying complex symbolic meanings: referring to Christ, vices and virtues, or the everlasting battle between good and evil.[34]

Some animals, such as lions and dragons, were of an ambivalent nature: depending on the context, they could have positive or negative connotations. In Christian culture, the lion was, above all, a symbol of Christ and the Resurrection,[35] but as an animal, it is known to be strong, wild and dangerous. Dragons generally represented the Devil, evil forces, and were considered symbols of sin.[36] However, they were also guardians of treasure and fierce creatures that were believed to possess protective powers. For these reasons, lions and dragons were often used as apotropaic signs in the marginal places and transition zones, including portals where their main function was to frighten and avert demons.[37]

---

33 The most likely candidate is Jesche, a daughter of Tartu burgomaster Godeke Honerjeger, who was married to Heinrich von Wangersen, who acted as burgomaster in 1544–1555. Lemm, Robert Arthur von: *Dorpater Ratslinie 1319–1889 und das Dorpater Stadtamt 1878–1918*. Herder-Institut: Marburg / Lahn 1960, p. 8.
34 For Estonian examples, see Kivimaa 1995, pp. 158–161.
35 Ferguson, George: *Signs & Symbols in Christian Art*. Oxford University Press: Oxford et al. 1977, pp. 21–22; LCI, vol. 3, col. 116–117.
36 LCI, vol. 1, col. 516–523.
37 Mellinkoff, Ruth: *Averting Demons: The Protective Power of Medieval Visual Motifs and Themes*, vol 1. Ruth Mellinkoff Publications: Los Angeles 2004, pp. 60–61, 74–75.

Figure 5. Right impost of the main portal of the Dominicans' church in Tallinn, second half of the fourteenth century. Photo: Stanislav Stepashko.

On the right impost of the main portal of the Tallinn Dominicans' church, a lion and three winged dragons can be seen (Figure 5). The latter are carrying a snake, a well-known symbol of the Devil, in their mouth. The fact that the dragons are depicted in multiples, increases their effect as protectors of the sacred building. A tiny winged creature is depicted above the lion but outside the horizontal frames of the impost. Due to damages, the identity of this creature cannot be ascertained[38] but one possibility is to interpret it as a tiny demon that is trying to sneak into the church, but in vain – the lion and the dragons will not let it pass.

The portal of the parish church of St John in the town of Cēsis is decorated with two fantastic beasts in bas-relief (Figure 6a and 6b).[39] Although they look different, they can both be interpreted as dragons. The creature on the left has sharp ears whereas the head of the right creature resembles a dog, but the shape of its body, and long and frightful claws, leave no doubt that we are dealing here with a dangerous and powerful animal. The dragon on the left is unmistakably a positive one: its tail has a tripartite "floral" ending, which alludes to the Holy Trinity. Instead of flames, a branch with oak leaves comes out of its mouth. Oak was a symbol of Christ and the Virgin Mary, of faith and virtue.[40] The dragon

---

38 Earlier it has been interpreted as a sleeping dove. Kivimaa 1995, p. 159.
39 Tuulse, Armin: *Die Spätmittelalterliche Steinskulptur in Estland und Lettland*. Suomen Muinaismuistoyhdistys: Helsinki 1948, p. 39.
40 Ferguson 1977, p. 35.

Figures 6a and 6b. Imposts of the portal of St John's church in Cēsis, early fifteenth century. Photos: Stanislav Stepashko.

on the right (heraldically on the sinister) is not provided with such positive attributes; therefore, it is not clear if it is a positive or a negative character. It is possible that the creatures on the imposts represent a good and a bad dragon, which are fighting with each other, symbolising the battle between the forces of good and evil, but at the same time carefully guarding the entrance to the church.

Animals with protective powers were naturally not limited to lions and dragons. For instance, a partly surviving boar on the right impost of the portal of St Nicholas' church in Tallinn[41] falls into the same category.[42]

---

41 Lumiste, Mai / Kangropool, Rasmus: *Niguliste kirik*. Kunst: Tallinn 1990, p. 80, fig. 25.
42 For the boar as an apotropaic sign, see Mellinkoff 2004, pp. 75–76.

Until the last quarter of the fifteenth century, portals of the profane buildings in Livonian towns were not decorated – their main beauty was in their architectural form. A change in fashion was probably caused, among others, by the reconstruction of St Matthew's chapel of the parish church of St Nicholas in Tallinn. There, a magnificent portal was hewn in 1488 with the imposts depicting symbolic animals and plants.[43] The dominant creatures are lions who are represented in pairs: the first pair is holding a heart, the second is chasing a dog (Figure 7). The lions there can be regarded as guardians of the Christian church and faith and as protectors of the building.

Figure 7. Right impost of the portal of the chapel of St Matthew, 1488. St Nicholas' church, Tallinn. Photo: Stanislav Stepashko.

In the following decades, one can speak of the "animal fashion" in Tallinn: several portals with animal figures were hewn for the elite families. Some of them can still be found in the Old Town. A portal from 1498, which presently stands at 1 Suur-Karja Street, initially belonged to the same house but was placed on its façade facing the Old Market.[44] Its imposts bear an inscription, with two lions placed between the words. The lions are facing each other, with

---

43 Lumiste / Kangropool 1990, pp. 39, 78–79, fig. 22–23.
44 The portal was moved to its present location during the reconstruction works in 1923. For the original location, see Löwis of Menar, Carl von: *Die Städtische Profanarchitektur der Gothik, der Renaissance und des Barocco in Riga, Reval und Narva*. Nöhring: Lübeck 1892, plate XVII.

one paw raised and the tongue out (Figure 8). It is likely that they have a double function there: to act as protectors of the house and to symbolise the everlasting battle between good and evil.

Figure 8. Right impost of the portal, 1498, 1 Suur-Karja St, Tallinn. Photo: Stanislav Stepashko.

The inscription on the left has become partly illegible, but can be reconstructed with the help of the transcription by Löwis of Menar and the photo in his book: *an[n]o / m cccc / xcviii* (In the year 1498).[45] The inscription on the right is in a comparatively good condition: *help / ihesus / xpe* (Help, Jesus Christ). In the late Middle Ages, there developed a cult of the holy names of Jesus and the Virgin Mary. Inscriptions containing the holy names or monograms were believed to protect an object or a building from the Devil and his demons.[46] Thus, on the portal of 1 Suur-Karja Street, the lions and the invocation of Jesus both served as apotropaic signs and they were most likely meant to complement and strengthen one another.

The house on 17 Vene Street, which today houses the Tallinn City Museum, was enlarged and re-built by the influential town councillor and burgomaster

---

45 Löwis of Menar 1892, p. 21; Cf. Tuulse 1948, p. 45.
46 Blake, Hugo *et al.*: "From Popular Devotion to Resistance and Revival in England: The Cult of the Holy Name of Jesus and the Reformation". In: Gaimster, David / Gilchrist, Roberta (eds.): *The Archaeology of Reformation 1480–1580*. Routledge: London / New York 2003, pp. 175–177.

Hans Viant, who bought it from his father-in-law in 1501.[47] The portal has three pairs of lions on both imposts (Figure 9).

Figure 9. Right impost of the portal, early sixteenth century, 17 Vene St, Tallinn. Photo: Stanislav Stepashko.

The animals are not as masterfully carved as those on St Matthew's chapel, but their quantity (twelve) emphasises the elite context of the house. In addition to lions, there is a tiny face or a mask on the edge of the imposts. These, too, belong among the apotropaic signs[48] and, as such, work with the lions as averters of evil.

Animals with protective powers were depicted in the representative rooms of the interior as well. One of the most prestigious semi-public spaces in Tallinn was the grand hall of the Great Guild, which was used not only for the annual festivals of the guild but also for weddings and assemblies. The vaults of the grand hall are supported by three pillars, the octagonal capitals of which are decorated with symbolic plants, animals and birds (Figure 10a and 10b).[49]

---

47 Mänd, Anu: "Merchants as Political, Social and Cultural Actors: Tallinn Burgomaster Hans Viant (d.1524)". In: Mänd, Anu / Tamm, Marek (eds.): *Making Livonia: Actors and Networks in the Medieval and Early Modern Baltic Sea Region*. Routledge: London / New York 2020, pp. 258–264.
48 Mellinkoff 2004, pp. 103–107. See also Kodres, Krista: "Vene tn 17 – ühe maja ehitus-, arhitektuuri- ja kunstiajalugu". In: Ehasalu, Pia (ed.): *Keskaegsest kaupmehemajast muuseumiks. Tallinna Linnamuuseum 80 = From Medieval Merchant's House into a Museum. Tallinn City Museum 80*. Tallinna Linnamuuseum: Tallinn 2017, pp. 22–23.
49 Leimus et al. 2011, pp. 288–291; Üprus, Helmi: *Raidkivikunst Eestis XIII–XVII sajandini*. Kunst: Tallinn 1987, fig. 88–90.

Figures 10a and 10b. Two capitals of the grand hall of the Great Guild in Tallinn, 1410. Photos: Stanislav Stepashko.

The first capital is adorned with stylised plant motifs. The figures on the middle pillar include four pairs, facing each other: two dragons, a lion and a dragon, two eagles and two plant motifs. The third capital features a pair of dragons, three plant motifs, and an inscription, *anno d[omi]ni m cccc x* (In the year of the Lord 1410), which perpetuated the date of the construction of the new guild house.

These animals and birds are loaded with various meanings. On the one hand, the eagle and the lion are symbols of Christ and the evangelists St John and

St Mark. As such, they fit together with the stylised lilies and other tripartite plant motifs that refer to the Virgin Mary and the Holy Trinity. However, in the company of the dragons, the eagle and the lion also function as apotropaic signs. There are two different types of dragons on the capitals: those on the third pillar have a tail that ends with a lily-like motif. This pair of dragons is clearly benevolent. On the middle pillar, one of the pair has a similar tail, the other does not; therefore, it is likely that these two symbolise the battle between good and evil. The same goes for the pair that consists of a lion-like creature whose tongue is reaching out and of a winged dragon with an "ordinary" tail.[50]

The Great Guild, which united the mercantile and political elite of the town, was a wealthy and educated patron who must have paid a great attention to the symbolism of the decorations of their most important representative room. The apotropaic animals were chosen to guard the building and the guild members, their souls, and earthly possessions. The quantity of animals and the artistic quality of the reliefs was a marker of status, fitting to the elite context.

Animals in Didactic-Moralistic Narrative Scenes

Narrative scenes with animals occur comparatively rarely in medieval Livonian stone sculpture. The earliest example is a capital of a pillar found from the Teutonic Order's castle in Viljandi: it depicts a famous story "The Fox and the Stork" from Aesop's Fables.[51] Although the capital is partly damaged, one can recognise both episodes of the story: the stork visiting the fox and *vice versa*, with both tricksters left hungry (Figure 11).

---

50   Leimus et al. 2011, p. 291.
51   No. 426 in the Perry Index, Wikipedia: *Perry Index*, retrieved 10.07.2023, from https://en.wikipedia.org/wiki/Perry_Index. For the capital in Viljandi, see Tuulse, Armin: "Viljandi ordulossi kapiteelid". In: *Litterarum Societas Esthonica 1838–1938. Liber Saecularis* II. Õpetatud Eesti Seltsi Toimetised XXX (2). Õpetatud Eesti Selts: Tartu 1938, p. 758, fig. 2; Tuulse 1948, p. 13, fig. 3.

Visual Representation of Animals in Livonian Urban Space 295

Figures 11a and 11b. Capital with the fable "The Fox and the Stork" from the Teutonic Order's castle in Viljandi, ca. 1250–60. Photos: Herki Helves, Viljandi Museum.

The capitals of the Viljandi castle have been dated to the mid-thirteenth century, at the latest to the 1260s.[52] It cannot be ascertained if this particular capital originated from the chapel or the chapter hall.

---

52 Alttoa, Kaur: "Die Kapitelle der Ordensburg Fellin (Viljandi) – Dinge aus zweiter Hand aus Alt-Pernau (Vana-Pärnu)?". *Baltic Journal of Art History* 13, 2017, pp. 23–24. DOI 10.12697/BJAH.2017.13.02.

Representations of Aesop's fables were very popular in medieval visual culture: they were entertaining as well as didactic.[53] As demonstrated by a pair of fragmentary imposts in Tallinn, they also occurred on the portals of urban dwellings (Figures 12a and 12b).[54]

Figure 12a and 12b. Imposts of the portal with Aesop's fables, 1493, from 35 Pikk St. Carved Stone Museum, Tallinn. Photos: Stanislav Stepashko.

---

53 See, e.g. Lämke, Dora: *Mittelalterliche Tierfabeln und ihre Beziehungen zur bildenden Kunst in Deutschland*. Universitätsverlag L. Bamberg: Greifswald 1937, pp. 70–73.

54 The imposts are displayed in the Carved Stone Museum of the Tallinn City Museum, inv. no. TLM 7676 (right) and TLM 7830 (left).

Two fables are depicted on the imposts. On the left impost, there is again the story of "The Fox and the Stork". On the surviving fragment, only the second part of the story, that is, the stork's revenge, can be seen: the stork has invited the fox to a meal and the latter is left hungry because food is served from a high narrow-necked vessel. It is likely that the first part of the story was depicted on the missing half. On the right impost, one can see a sheep, a dog, a wolf holding a lamb and a lion seizing it from him. The group most likely represents the story "The Wolf and the Lion" where the wolf steals a lamb and the lion takes it from him.[55] The first two animals, a sheep and a dog, symbolically represent the flock.

There is yet another animal on the outer edge of the left impost – a ferocious-looking dragon with its teeth exposed. As discussed above, the dragon belongs among the animals who featured as apotropaic signs on the portals. In this particular case, the dragon shares this position with the lion depicted on the right impost. The only difference is that the lion is at the same time a character in the fable whereas the dragon has no direct link to the story of the fox and the stork. However, as Ruth Mellinkoff has argued, the portrayal of the fox and the stork could also serve as an apotropaic sign[56]; therefore, we cannot rule out that the animals on the Tallinn imposts had multiple functions – they were not only entertaining and instructive but also served as protectors of the house.

Above the animal figures, there is a partially surviving inscription. On the left impost, one can read: *[…] rex iudeorum*, on the right: *[…] cum pace 1493*. This is a version of a prayer (*ihesus nazarenus, rex iudeorum, veni cum pace*), which was frequently added to medieval church bells and which can also be found on a bronze door knocker of the Great Guild.[57] The inscription has no connection to the fables but it protects the house with the figures of the lion and the dragon.

The portal with Aesop's fables and the date 1493 most likely originates from the house at 35 Pikk Street, which belonged to merchant Hans Pawels, who also commissioned a pair of doorside stones with the figures of the Virgin Mary and St George, discussed earlier in this chapter. The dwelling was listed in the name of Pawels in the real estate book of Tallinn in 1494. In that book, records about the change of a house owner were often made some time after the actual transaction;

---

55 No. 347 in the Perry index.
56 Mellinkoff 2004, p. 87.
57 Carroll, Michael P.: *Catholic Cults & Devotions: A Psychological Inquiry*. McGill-Queen's University Press: Kingston / London 1989, p. 40. For the bell (1433) of the Holy Spirit Church in Tallinn, see Mänd, Anu: "Kirikukellad keskaegses Tallinnas". *Tuna: Ajalookultuuri ajakiri* 3, 2016, p. 54. For the door knockers (1430) of the Great Guild, see Leimus et al. 2011, pp. 279–281.

therefore, it is likely that Pawels acquired the dwelling in 1493. Hans Pawels was a wealthy merchant from Germany, possibly from Frankfurt am Main. He became a member of the Great Guild of Tallinn during the Christmas revels of 1493/94 and got married probably the same year. Thus, 1493 was significant in his life in many respects. Although Pawels was never elected to the town council, he was nevertheless an influential man who is mainly famous for his position as a warden of the parish church of St Olaf from about 1513 until his death in 1519: he supervised the rebuilding of St Mary's chapel and commissioned a magnificent dolomite cenotaph in his memory on the outer wall of this chapel.[58] The decorations of the chapel and Pawels's dwelling give testimony of his taste for art, his religious and aesthetic values, his desire to perpetuate his memory, and, last but not least, his economic possibilities.

Conclusion

The study of animal representations in urban public space enabled us to broaden the knowledge of various functions of animal imagery in the late medieval period, which, in turn, sheds light on contemporary attitudes towards animals in general and certain types of animals in particular.

Animal figures on the façades of dwellings and guild houses had multiple functions. The first was informative and communicative: animals (and other images) helped to identify the owner, be it an individual or a collective body. The patron saints, coats of arms and inscriptions expressed the owner's identity, his/their social status and wealth. In the medieval and early modern periods when the houses were not numbered but known after their owner's name, images and inscriptions on the doorside stones, the portal or gable also served as street signs, as mnemonic tools for "addresses". One can easily imagine that a visiting merchant from another town was led to the house of burgomaster Hans Viant with the guidance "Look for a portal with the twelve lions".

Animal figures were cultural codes with religious and moral meanings. They were Christian symbols and allegories, expressing the house owners' hope for salvation and the victory of the good forces over the evil ones. Closely linked to the previous one was the belief that certain animals possessed powers to protect

---

58 Mänd, Anu: "Church Art, Commemoration of the Dead and the Saints' Cult: Constructing Individual and Corporate *memoria* in Late Medieval Tallinn". *Acta Historica Tallinnensia* 16, 2011, pp. 11– 20. DOI 10.3176/hist.2011.1.01.

a building. In Livonia, it was primarily the lions and the dragons who served as apotropaic signs that guarded the immortal souls and earthly property of the owners and scared off the demons. Although the majestic lions and mighty dragons in this particular context were positive creatures, it was broadly known that in other contexts the same animals could represent the evil forces. Reliefs with popular fables, which in Livonian urban space did not occur very frequently, were amusing as well as instructive, conveying didactic-moralistic messages and reflecting human nature. Thus, the animal representations combined various religious, moral and social functions. This, in turn, can tell us a lot about the human–animal relations in the late medieval Christian culture.

The variety of animals on church portals was higher than on the profane buildings, and the iconography of the images in the sacred sphere was generally more complex. However, since the images in the urban public space were meant for "everybody", they had to be more easily decipherable than those in some other forms of visual media, which were targeted primarily to the learned elite. The decoration of the church portals undoubtedly influenced that of the dwellings and in all likelihood created a desire among the wealthy burghers to have a similar portal for their home.

Although the commissioners of the stone sculptures cannot always be identified, the examples discussed in this chapter indicate that usually buildings with reliefs belonged to affluent merchants, including town councillors and burgomasters, and to the nobility who resided in the town. Their houses were located on the most prestigious streets and had a significant impact on the visual environment of the town. The images, including the animal figures, reflect upon the needs and means of self-representation of the urban elites in general and merchants in particular.

After the Reformation, which in Livonian towns culminated with the iconoclastic outbursts in 1524–25, the representation of animals notably declined. The saints, with or without the animal attributes, gradually disappeared from the scene, with the exception of the patron saints of the guilds whose representations expressed the identity and long history of these associations. The belief in the magical protective power of animal figures was shaken by the growing knowledge on nature spread by printed media. From about the mid-sixteenth century, the animal depictions in Livonian urban space were mainly confined to heraldic emblems, Bible stories, and the symbols of the Evangelists.

In sum, animal figures were relevant in the urban public space in many ways. However, they should not be studied in isolation but in the context of other visual and textual motifs used in a particular region and period of time.

References

Alttoa, Kaur: "Die Kapitelle der Ordensburg Fellin (Viljandi) – Dinge aus zweiter Hand aus Alt-Pernau (Vana-Pärnu)?". *Baltic Journal of Art History* 13, 2017, pp. 11–37. DOI 10.12697/BJAH.2017.13.02.

Arbusow, Leonid (ed.): *Liv-, Est- und Kurländisches Urkundenbuch*, 2nd series, vol. 3. Deubner: Riga / Moskau 1914.

Belozerskaya, Marina: "Good Dog: Model Canines in Renaissance Manuscripts". In: Hamburger, Jeffrey H. / Korteweg, Anne S. (eds.): *Tributes in Honor of James H. Marrow. Studies in Painting and Manuscript Illumination of the Late Middle Ages and the Renaissance*. Harvey Miller Publishers: London / Turnhout 2006, pp. 65–74.

Blake, Hugo / Egan, Geoff / Hurst, John / New, Elizabeth: "From Popular Devotion to Resistance and Revival in England: The Cult of the Holy Name of Jesus and the Reformation". In: Gaimster, David / Gilchrist, Roberta (eds.): *The Archaeology of Reformation 1480–1580*. Routledge: London / New York 2003, pp. 175–203.

Bome, Helen: "Kloostrielu kiusatused Padise piltkonsoolidel". *Kunstiteaduslikke Uurimusi = Studies on Art and Architecture* 21(1–2), 2012, pp. 7–36.

Bruiningk, Hermann von: *Messe und kanonisches Stundengebet nach dem Brauche der Rigaschen Kirche im späteren Mittelalter*. Kymmel: Riga 1904.

Camille, Michael: "At the Sign of the 'Spinning Sow': the 'Other' Chartres and Images of Everyday Life of the Medieval Street". In: Bolvig, Axel / Lindley, Phillip (eds.): *History and Images: Towards a New Iconology*. Brepols: Turnhout 2003, pp. 249–276.

Carroll, Michael P.: *Catholic Cults & Devotions: A Psychological Inquiry*. McGill-Queen's University Press: Kingston / London 1989.

Ferguson, George: *Signs & Symbols in Christian Art*. Oxford University Press: Oxford et al. 1977.

Friedman, John Block: "Dogs in the Identity Formation and Moral Teaching Offered in Some Fifteenth-Century Flemish Manuscript Miniatures". In: Gelfand, Laura D. (ed.): *Our Dogs, Our Selves: Dogs in Medieval and Early Modern Art, Literature, and Society*. Brill: Leiden / Boston 2016, pp. 325–362.

Grinder-Hansen, Poul: *Søren Abildgaard (1718–1791): Fortiden på tegnebrættet*. Nationalmuseet: Copenhagen 2010.

Kangropool, Rasmus: "Tallinna hilisgooti etikukividest". *Vana Tallinn* IV (VIII). Tallinn 1994, pp. 5–13.

Kirschbaum, Engelbert et al. (eds.): *Lexikon der christlichen Ikonographie* [LCI]. Herder: Rome et al. 2004.

Kivimaa, Katrin: "Dualistlik maailmavaade eesti keskaegses loomasümboolikas". In: Alttoa, Kaur (ed.): *Ars Estoniae Medii Aevi Grates Villem Raam*. Eesti Muinsuskaitse Selts: Tallinn 1995, pp. 157–170.

Kodres, Krista: "Self-representation and Social Aesthetics: Wealthy Tallinn Burgher Homes in the Early Modern Period". In: Mänd, Anu / Tamm, Marek (eds.): *Making Livonia: Actors and Networks in the Medieval and Early Modern Baltic Sea Region*. Routledge: London / New York 2020, pp. 300–319.

Kodres, Krista: "Vene tn 17 – ühe maja ehitus-, arhitektuuri- ja kunstiajalugu". In: Ehasalu, Pia (ed.): *Keskaegsest kaupmehemajast muuseumiks. Tallinna Linnamuuseum 80 = From Medieval Merchant's House into a Museum. Tallinn City Museum 80*. Tallinna Linnamuuseum: Tallinn 2017, pp. 15–33.

Kodres, Krista: *Esitledes iseend: Tallinlane ja tema elamu varauusajal*. Tallinna Ülikooli Kirjastus: Tallinn 2014.

Lämke, Dora: *Mittelalterliche Tierfabeln und ihre Beziehungen zur bildenden Kunst in Deutschland*. Universitätsverlag L. Bamberg: Greifswald 1937.

Leimus, Ivar / Loodus, Rein / Mänd, Anu / Männisalu, Marta / Raisma, Mariann: *Tallinna Suurgild ja gildimaja*. Eesti Ajaloomuuseum: Tallinn 2011.

Lemm, Robert Arthur von: *Dorpater Ratslinie 1319–1889 und das Dorpater Stadtamt 1878–1918*. Herder-Institut: Marburg / Lahn 1960.

Lidén, Anne: *Olav den helige i medeltida bildkonsten. Legendmotiv och attribut*. KVHAA: Stockholm 1999.

Löwis of Menar, Carl von: *Die Städtische Profanarchitektur der Gothik, der Renaissance und des Barocco in Riga, Reval und Narva*. Nöhring: Lübeck 1892.

Lumiste, Mai / Kangropool, Rasmus: *Niguliste kirik*. Kunst: Tallinn 1990.

Mänd, Anu: "Loomad kunstis". In: Kreem, Juhan / Leimus, Ivar / Mänd, Anu / Põltsam-Jürjo, Inna: *Loomad keskaegse Liivimaa ühiskonnas ja kunstis*. Tallinn University Press: Tallinn 2022, pp. 115–172.

Mänd, Anu: "Images and Inscriptions on Dwelling Houses in Livonian Towns, c. 1450–c. 1550". In: Leimus, Ivar (ed.): *Everyday Life in a Hanseatic Town*. (Varia historica 8). Eesti Ajaloomuuseum: Tallinn 2021, pp. 91–141.

Mänd, Anu: "Merchants as Political, Social, and Cultural Actors: Tallinn Burgomaster Hans Viant (d.1524)". In: Mänd, Anu / Tamm, Marek (eds.): *Making Livonia: Actors and Networks in the Medieval and Early Modern Baltic Sea Region*. Routledge: London / New York 2020, pp. 251–278.

Mänd, Anu: "A Mouser in the Bedroom". In: Berggren, Lars / Landen, Annette (eds.): *Ecce leones! Om djur och odjur i bildkonsten*. Artifex: Lund 2018a, pp. 191–205.

Mänd, Anu: "The Cult and Visual Representation of Scandinavian Saints in Medieval Livonia". In: Jensen, Carsten Selch et al. (eds.): *Saints and Sainthood around the Baltic Sea: Identity, Literacy, and Communication in the Middle Ages.* Western Michigan University: Kalamazoo 2018, pp. 101–143.

Mänd, Anu: "Kirikukellad keskaegses Tallinnas". *Tuna: ajalookultuuri ajakiri* 3, 2016, pp. 47–60.

Mänd, Anu: "Kass voodi all. Ühest motiivist Hermen Rode ja Bernt Notke Tallinna retaablitel". *Kunstiteaduslikke Uurimusi = Studies on Art and Architecture* 21(1–2), 2012, pp. 231–246.

Mänd, Anu: "Church Art, Commemoration of the Dead and the Saints' Cult: Constructing Individual and Corporate *memoria* in Late Medieval Tallinn". *Acta Historica Tallinnensia* 16, 2011, pp. 3–30. DOI 10.3176/hist.2011.1.01.

Mänd, Anu: "Saints' Cults in Medieval Livonia". In: Murray, Alan V. (ed.): *The Clash of Cultures on the Medieval Baltic Frontier.* Ashgate: Farnham 2009, pp. 191–223.

Mänd, Anu: "Bad Boys, Men, and Dogs in Bernt Notke's Tallinn Altarpiece". In: Markus, Kersti (ed.): *Bilder i marginalen. Nordiska studier i medeltidens konst = Images in the Margins. Nordic Studies in Medieval Art.* Argo: Tallinn 2006, pp. 305–320.

Mänd, Anu: *Urban Carnival: Festive Culture in the Hanseatic Cities of the Eastern Baltic, 1350–1550.* Brepols: Turnhout 2005.

Mellinkoff, Ruth: *Averting Demons: The Protective Power of Medieval Visual Motifs and Themes*, vol. 1. Ruth Mellinkoff Publications: Los Angeles 2004.

Möller, Gunnar: "*Decora alta domus.* Mittelalterliche und frühneuzeitliche Wangensteine in Stralsund". In: Birli, Sonja et al. (eds.): *ene vruntlike tohopesate. Beiträge zur Geschichte Pommerns, des Ostseeraums und der Hanse. Festschrift für Horst Wernicke zum 65. Geburtstag.* Verlag Dr. Kovač: Hamburg 2016, pp. 355–372.

Nationalmuseets Samlinger Online, retrieved 10.07.2023, from https://samlinger.natmus.dk/.

Nottbeck, Eugen von: *Siegel aus dem Revaler Rathsarchiv nebst Sammlung von Wappen der Revaler Rathsfamilien.* Rathgens: Lübeck 1880.

Perry Index, retrieved 10.7.2023, from https://en.wikipedia.org/wiki/Perry_Index.

Resl, Brigitte: "Beyond the Ark: Animals in Medieval Art". In: Resl, Brigitte (ed.): *A Cultural History of Animals in the Medieval Age.* (A Cultural History of Animals, vol. 2). Berg: Oxford / New York 2007, pp. 179–201.

Saal, Walter: "Beischlagsteine und ihre Beziehungen zu Grabkreuzen und Sühnekreuzen". *Steinkreuzforschung* 4, 1982, pp. 30–35.

Seeberg-Elverfeldt, Roland: *Testamente Revaler Bürger und Einwohner aus den Jahren 1369 bis 1851. Revaler Regesten* III. Vandenhoeck & Ruprecht: Göttingen 1975.

Sprengfeld, G.: *Meine Vaterstadt Reval vor 50 Jahren*. Verlag von Schnakenburg: Dorpat 1877.

Tallinn City Museum, inv. no. TLM 7676; TLM 7830.

Tartu City Museum, inv. no. TM 118 Aj 61.

Tuulse, Armin: *Die Spätmittelalterliche Steinskulptur in Estland und Lettland*. Suomen Muinaismuistoyhdistys: Helsinki 1948.

Tuulse, Armin: "Viljandi ordulossi kapiteelid". In: *Litterarum Societas Esthonica 1838–1938. Liber Saecularis* II. *Õpetatud Eesti Seltsi Toimetised* XXX (2). Õpetatud Eesti Selts: Tartu 1938, pp. 755–769.

Üprus, Helmi: *Raidkivikunst Eestis XIII–XVII sajandini*. Kunst: Tallinn 1987.

Üprus, Helmi: *Tallinna etikukivid*. Kunst: Tallinn 1971.

Stefan Donecker (+)

Dogs of War: Wolves and Warfare in Early Modern Livonia (ca. 1555–1605)

Abstract: This paper argues that Livonia is particularly well suited for examining Early Modern representation of wolves and warfare as a country known for its high population of wolves, as well as one ravaged by continuous military conflict. The paper investigates three texts from sixteenth-century Livonia that represent very different genres and authorial positions: a semi-official chronicle by a local pastor (Balthasar Russow), a report by a foreign missionary (Antonio Possevino), and a private travelogue (Samuel Kiechel). All three share the basic assumption that wolves thrive in times of war, and, furthermore, bestow symbolic significance upon the animals' activities that extended far beyond their observable behaviour. In all three accounts, wolves embody unrest and disorder. This reflects the precarious, threatened socio-political order of wartime Livonia, which during the Livonian War (1558–1583) had fell victim to the expanding neighbouring states: Sweden, Poland, and the Muscovite. On a broader level, these texts also illustrate the discursive interdependency between narratives of war and narratives of wolf predation. Wolves are animals of crisis; they become visible in crisis situations, and it is during crisis situations that they are written about. On the one hand, the association between wolves and warfare was the direct result of the animals' behavioural patterns: during times of war, the otherwise reclusive animals became far more conspicuous. But the political symbolism of wolves extended far beyond their factual appearances. Accounts of lupine predation allowed early modern authors to express their anxiety over personal danger and the social disorder that resulted from warfare.

Keywords: Baltic history, early modern historiography, animal history, wolves, werewolves

The wolf follows the beat of the drum.[1]

Introduction

For centuries, an association between wolves and warfare has been firmly entrenched in the cultural perception of Europe's apex predator. Folklore

---

1 Attested as an early modern German proverb: *Der Wolf folgt der Trommel*. See Schattauer, Willi: "'Der Wolf folgt der Trommel': Wölfe und Wolfsjagd in der Nordpfalz". *Donnersberg-Jahrbuch* 24, 2001, pp. 134–136.

researchers have collected a vast amount of tales and traditions that depict the wolf as a grim creature of the battlefield and devourer of the fallen.[2] Early modern sources attested that wolf populations flourished during wartime: According to Johann Friedrich von Flemming's (1670–1733) *Der Vollkommene Teutsche Jäger*, a highly influential tract on the art of hunting, wolves migrated to countries ravaged by war, attracted by the opportunity to scavenge bodies of men, horses and livestock.[3] Such assessments were common in seventeenth and eighteenth-century treatises, which stressed that wolves were more numerous and also more aggressive during times of military strife.[4] Present-day zoology concurs, identifying war as one of the specific sets of socio-political circumstances – next to pandemics and famines – that increases the risk of predatory attacks by wolves on humans.[5]

In 1555, Swedish humanist Olaus Magnus (1490–1557) informed the readers of his highly influential *Historia de gentibus septentrionalibus* about the depredation caused by wolves in the lands along the Eastern Baltic littoral: "[T]he inhabitants of Prussia, Livonia, and Lithuania suffer grave setbacks from the ravages of wolves almost all the year round, inasmuch as large numbers of their livestock are torn to bits and devoured everywhere in the forests if they so much

---

2   See numerous examples in Peuckert, Will-Erich: "Wolf". *Handwörterbuch des deutschen Aberglaubens* 9, 1941, cols. 716–794, at cols. 743–744; as well as Pluskowski, Aleksander: *Wolves and the Wilderness in the Middle Ages*. Boydell: Woodbridge 2006, pp. 135–142.

3   Flemming, Hanns Friedrich von: *Der Vollkommene Teutsche Jäger* […]. J. C. Martini: Leipzig 1749, p. 106.

4   Cf. Schöller, Rainer G.: *Eine Kulturgeschichte des Wolfs. Tierisches Beuteverhalten und menschliche Strategien sowie Methoden der Abwehr*. (Rombach Wissenschaften. Ökologie 10). Rombach: Freiburg i.Br. et al. 2017, p. 48; Rheinheimer, Martin: "The Belief in Werewolves and the Extermination of Real Wolves in Schleswig-Holstein". *Scandinavian Journal of History* 20, 1995, pp. 281–294, here p. 283.

5   Linnell, John D. C. / Alleau, Julien: "Predators That Kill Humans: Myth, Reality, Context and the Politics of Wolf Attacks on People". In: Angelici, Francesco M. (ed.): *Problematic Wildlife. A Cross-Disciplinary Approach*. Springer: Cham 2016, pp. 357–371, here pp. 364–365. Predatory attacks on humans, i.e. instances where wolves regard humans as prey and assault them intentionally, as opposed to attacks by rabid animals or "defensive attacks" by wolves that had been cornered, are a rare phenomenon, but not unheard of. In such cases, the attacks generally target children or infirm individuals. See Linnell, John D. C., et al.: *The Fear of Wolves. A Review of Wolf Attacks on Humans*. NINA: Trondheim 2002, esp. pp. 4–5, 16, for a thorough discussion of such cases in a global perspective.

as stray a short distance from the herd [...]."[6] Wolves were not only abundant in the Livonian countryside but also known to venture into the city of Riga on numerous occasions, if anecdotal accounts from the following centuries are to be believed.[7] A substantial population of the predators persisted well into the nineteenth century and exerted a considerable strain on animal husbandry in the Baltic Provinces of the Russian Empire.[8] According to estimates published in 1803, in the areas most affected by lupine predation up to 16 % of sheep, pig and goat stocks and roughly 8 to 10 % of horses and cattle were lost to wolves each year.[9] At roughly the same time, wolf attacks on humans spiked with a staggering number of 45 fatalities in six Estonian parishes recorded in 1809 alone.[10] Accordingly, measures to reduce the wolf population or to eradicate the predators entirely were intensively discussed in learned journals of the time.[11]

6   Olaus Magnus: *Historia de gentibus septentrionalibus* [...]. De Viottis: Romae 1555, p. 642. This and the following translations are based on Foote, Peter (ed.) / Fisher, Peter (trans.) / Higgins, Humphrey (trans.): *Olaus Magnus: A Description of the Northern Peoples 1555*, 3 vols. (The Hakluyt Society: Works II, 182, 187 and 188). Hakluyt Society: London 1996-1998.
7   "Ein Wolf, der sich in der Stadt zu Tisch bittet". *Rigaische Stadt-Blätter* 1810, p. 142; "Noch ein Wolf; aber ein schlimmerer". *Rigaische Stadt-Blätter* 1810, pp. 166-167; "Lese-Früchte aus Stadts-Chroniken". *Rigaische Stadt-Blätter* 1815, pp. 5-6. The events described supposedly took place between 1618 and 1718.
8   Rootsi, Ilmar: *Hunt ja inimene: suhted Eestis XVIII sajandi keskpaigast XIX sajandi lõpuni* [Wolf and Man: Their Relationship in Estonia from the Mid-18th Century to the End of the 19th Century]. (Dissertationes Historiae Universitatis Tartuensis 24). Tartu Ülikooli Kirjastus: Tartu 2011; Hasselblatt, Arnold: "Die Wölfe in Livland. Eine culturhistorische Studie". *Baltische Monatsschrift* 29, 1882, pp. 659-678.
9   Friebe, Wilhelm Christian: "Wie ist die Viehzucht in Liefland zu verbessern?". *Abhandlungen der liefländischen, gemeinnützigen und ökonomischen Societät* 2(2), 1803, pp. 1-142, here pp. 136-137. Cf. Strods, Heinrihs: "Die Einschränkung der Wolfsplage und die Viehzucht Lettlands". *Ethnologia Europaea* 4, 1970, pp. 126-131.
10 Rootsi, p. 270. Cf. Linnell et al., p. 57; Kruuk, Hans: *Hunter and Hunted: Relationships Between Carnivores and People*. Cambridge University Press: Cambridge 2002, p. 72. Wolf attacks on humans occurred, albeit infrequently, throughout the twentieth century. The last fatality in the Baltic countries was, most likely, an elderly woman who was killed by a rabid wolf in the Estonian SSR in 1980. See Linnell et al., pp. 17, 21, 61.
11 Hupel, August Wilhelm: "Ueber die Ausrottung der Wölfe". *Nordische Miscellaneen* 1, 1781, pp. 229-231; "Probe einer ökonomischen Naturgeschichte Kurlands. Der Säugthiere. Der Wolf". *Geoponika eine ökonomische Monatsschrift für Kur- und Lievlands Bewohner* 1, 1798/99, pp. 593-620, here pp. 613-619; Friebe, pp. 137-142. Attempts to exterminate the wolf population were already undertaken by the Swedish

In 2006, Estonian historian Margus Laidre characterised the period between 1558 and 1661 as a "Hundred Years' War" in the eastern Baltic.[12] It is an appropriate label for a period of incessant political strife, when Sweden, Poland-Lithuania, and Russia fought a relentless struggle for control of the territory and hegemony in the region. Thus, Livonia seems to be well-suited to an examination of the early modern perception of wolves and warfare: as a country known for its high population of wolves, and ravaged by continuous military conflict. In the following, I intend to follow lupine tracks on the battlefields of late sixteenth-century Livonia, and to investigate three case studies that shed light on the image of these animals as "Dogs of War" and the embodiment of strife.

Portentous Wolves: Toolse, 1574

For any inquiry into the events of the Livonian War (1558–1583), and especially into the mindset of this period, it is well-advised to consult Balthasar Russow (1536–1600) and his *Chronica der Provinz Lyfflandt* (1578).[13] Russow, a Lutheran pastor in Tallinn, was the most formidable representative of an extraordinary generation of chroniclers[14] that sought to come to terms with the collapse of their society and the horrors of the war they had witnessed.

In Russow's chronicle, wolves take centre stage during the Swedish spring campaign of 1574. Operating from Tallinn, a Swedish army accompanied by large contingents of German-speaking and Scottish troops advanced eastwards against Russian-held territory around Rakvere.

> On March 15, the Swedes thought to try their luck on the castle of Tolsburg [Toolse], situated on the coast fifteen miles from Wesenberg [Rakvere]. But since they attempted to storm the castle without artillery support, they burned their fingers in the attempt.

---

    authorities a century before. See Liiv, Otto: *Die wirtschaftliche Lage des estnischen Gebietes am Ausgang des XVII. Jahrhunderts.* (Verhandlungen der Gelehrten Estnischen Gesellschaft 27). Õpetatud Eesti Selts: Tartu 1935, pp. 60–61.

12 Laidre, Margus: "Der Hundertjährige Krieg (1558–1660/61) in Estland". *Forschungen zur baltischen Geschichte* 1, 2006, pp. 68–81.

13 Among the extensive research literature on Balthasar Russow, Paul Johansen's posthumously published *Balthasar Rüssow als Humanist und Geschichtsschreiber.* (Quellen und Studien zur baltischen Geschichte 14). Böhlau: Köln et al. 1996, remains the most exhaustive treatment of the author and his work.

14 Cf. Raik, Katri: *Eesti- ja liivimaa kroonikakirjutuse kõrgaeg 16. sajandi teisel poolel ja 17. sajandi alul* [The Heyday of Estonian and Livonian Chronicle Writing in the Second Half of the 16th Century and the Beginning of the 17th Century]. (Dissertationes Historiae Universitatis Tartuensis 8). Tartu Ülikooli Kirjastus: Tartu 2004.

They suffered a shameful defeat and lost a number of men. During this siege, fourteen days before the withdrawal, there was an amazing and extraordinary portent of wolves. For several nights in a row masses of them gathered near the camp, in spite of the fact that there was a huge assembly of men there, making a great deal of noise. But the wolves began to howl and to bay in dreadful fashion and many a man's hair stood on end.[15]

The incident translated here as "an amazing and extraordinary portent of wolves" is described in the original Low German as *"ein seltzam vnde wünderlick gespenst mit den Wuluen"*. Middle Low German *gespens* has roughly the same meaning as the corresponding modern High German *Gespenst*: a spectre, phantasm or illusion.[16] Nevertheless, editors and translators of the chronicle have agreed on a somewhat different reading: in their opinion, Russow did not have a ghost-like apparition of lupine shapes in mind, but mundane, physical wolves acting in an unusual way. Only the allegedly portentous nature of their behaviour made these wolves stand out, and it was important enough to be included in the chronicle.

This interpretation of the wolf encounter at Toolse is, in my opinion, strongly supported by a near-contemporary passage from the Muscovite Chronicle of Conrad Bussow (ca. 1552–1617). Bussow, son of a Lutheran pastor from the vicinity of Hannover, served as a mercenary in Russia during the Time of Troubles. Upon his return to Germany, he compiled an account of the events he had witnessed. A planned publication of the text did not come to fruition, but different manuscript redactions under various titles circulated among the scholarly community.[17]

Before the rise of the pretender known as the first False Dmitry, a number of noteworthy portents were reported. According to Bussow, they heralded the turmoil of the years to come: sundogs and other celestial phenomena were observed in the skies, monstrosities were born, and meals lost their taste.

> Dogs devoured each other, and so did wolves. From whence the sword came, wolves set up so great a howling as no man could have thought possible. Wolves roamed around in such large packs that travellers could not move in small groups. [...] These and similar

---

15 Russow, Balthasar: "Chronica der Prouintz Lyfflandt [...]". *Scriptores rerum Livonicarum* 2, 1848, pp. 1–157, at p. 101. This and the following translations are taken from Smith, Jerry C. (trans.): *The Chronicle of Balthasar Russow & A Forthright Rebuttal by Elert Kruse & Errors and Mistakes of Balthasar Russow by Heinrich Tisenhausen*. (Wisconsin Baltic Studies 2). Baltic Studies Center: Madison 1988.
16 Möhn, Dieter (ed.): *Mittelniederdeutsches Handwörterbuch, Zweiter Band*. Wachholtz: Neumünster 2004, col. 89.
17 On Bussow's biography and work, see G. Edward Orchard's introduction to the English translation.

signs appeared, but the Muscovites paid no attention to what they portended, like unto the Jews at Jerusalem, who took evil omens for good auguries.[18]

The parallel between Bussow's account and that of his near-namesake Balthasar Russow is evident, although the portentous nature of the wolves' activity is emphasised by the former and only implied by the latter. Despite his profession as a soldier, Bussow had enjoyed some scholarly education and was well-read, and he could have been directly influenced by Russow's chronicle. Before joining the Muscovite cause, he had spent several years in Livonia in Swedish service, and it is perfectly plausible that he became familiar with the writings of the Tallinn pastor there. One of the existing manuscripts, Wolfenbüttel II, adds a marginal note to the above-quoted passage on wolves that draws a direct parallel to Livonia (although the reference is not to the events of the Livonian War but to occurrences some twenty-five years later during Bussow's time of service): "How the war in Livonia in the year 1600 between the crowns of Poland and Sweden raged, how the wolves set up so great a howling, also appearing in broad daylight in villages and public places doing great mischief."[19]

If the wolves' appearance was understood as portentous – explicitly by Bussow, implicitly by Russow – then what exactly did they portent? Again, the Muscovite Chronicle is more straightforward:

> Also since dogs were eating dogs and wolves were eating wolves, this contradicted the old proverb, "Wolves do not eat each other." One Tatar explained it thus: the Muscovites would betray each other, tear each other apart, and destroy each other like dogs. At the same time that these portents occurred, divisions and dissensions appeared in all orders of society, so that one man could expect nothing good from another.[20]

For his part, Balthasar Russow did not explicitly identify the events that had been presaged by the ominous howling of the wolves. But it is worth noting that the wolf episode at the siege of Tolsburg was followed by discord and betrayal only little more than two weeks later, as the Scottish troops in Swedish service turn against their allies:

> On March 17, 1574, discord arose between the Scots and the Germans in the camp before Wesenberg. This led to a dreadful melee in which over fifteen hundred Scots, but no more than thirty Germans, were slain and killed. [...] The remaining Scottish foot

---

18 Bussov, Konrad: *Moskovskaja chronika. 1584–1613*. Akad. nauk SSSR: Moskva 1961, p. 225. Translation by Orchard, G. Edward (trans.): *Conrad Bussow. The Disturbed State of the Russian Realm*. McGill-Queen's University Press: Montreal 1994.
19 Bussov 1961, p. 225.
20 Ibid.

soldiers, however, when they saw that their side had lost, fled with their company to the Russians in the castle of Wesenberg, where they were received with great joy. And thus did the Germans and Scots, subjects of the same lord, do battle with each other, but it was the Muscovite who gained a triumph and a victory from it.[21]

Balthasar Russow's prosaic, fact-based account is far more grounded than Bussow's grand spectacle of omens and portents. In Livonia, only the Swedish army is afflicted, while in Muscovy, dissent engulfs the entire realm. But in both cases, a collapse of order and an outbreak of internal strife is prefigured by the unusually bold and reckless behaviour of wolves.

Infernal Wolves: The Bishopric of Tartu, Early 1580s

On the opposing side of the denominational divide, the papal nuncio Antonio Possevino (1534–1611) left a somewhat similar, although far more sensational description of menacing wolves several years later. Possevino, a Jesuit father and spearhead of the counter-reformation efforts during the Livonian War[22], wrote an account of the situation in the eastern Baltic, ostensibly in the form of a letter to Eleonore, Archduchess of Austria and Duchess of Mantua, but clearly addressed at a wider audience. The Italian text, written in Tartu in August 1585, was published in Vilnius in the autumn of the same year, reprinted repeatedly in Italy in several editions and translated to French afterwards. Furthermore, major passages were incorporated into the chronicle of Michael van Isselt, which was published both in a Latin and a German version in 1586 and 1587.[23] Thus, the Jesuits' effective dissemination network ensured that Possevino's report on the state of the Catholic mission in Livonia was available to readers throughout Europe.

Throughout the letter, Possevino links the dire situation in the war-torn country to the lack of Catholic priests and sacraments. Many adults remain unbaptised, and even those lucky enough to have received baptism did so at the hands of Lutherans or Muscovites, whose imperfect rites lack the "exorcisms and adjurations against the Devil" employed by the Catholic Church. To stress the point that the land and its inhabitants have been left at the mercy of infernal powers, Possevino relates tales of unbelievably ferocious wolves. Along the road

---

21 Russow 1988, pp. 101–102.
22 Helk, Vello: *Die Jesuiten in Dorpat 1583–1625. Ein Vorposten der Gegenreformation in Nordosteuropa.* (Odense University Studies in History and Social Sciences 44). Odense University Press: Odense 1977, pp. 10–23, 51–77.
23 Ibid., p. 72.

leading to Pskov, in the vicinity of the castle of Vastseliina, four wolves, one of them with white fur, have supposedly killed and eaten 150 victims in a single year. They eschew sheep and other animals and target only human beings, especially pregnant women. At the fortress of "Fabino" (probably Vaabina), two hundred people have fallen victim to wolves. The commoners blame the Muscovites, whose incantations cause the animals to attack with such savagery.[24]

But now that the Catholic faith has returned to the land, Possevino asserts, the tide turns against the wolf menace. The predators are unable to hurt those who have been properly baptised and blessed in the Catholic tradition. Likewise, anyone wearing an Agnus Dei sacramental and believing in its sanctity is safe from harm. When a Calvinist officer complains to a Jesuit that his traps and wolf pits are ineffective against the animals, the priest sprinkles the pits with holy water. Subsequently, the traps begin to take their toll and many wolves are killed. But even after being a first-hand witness to the power of Catholic ritual, the Calvinist clings stubbornly to his heretic belief. His family and his entire household, however, acknowledge the miracle and convert to Catholicism.[25]

It quickly becomes apparent that Possevino's alleged on-site reports were influenced by the preeminent reference on Northern Europe available to the early modern scholarly community, the abovementioned *Historia de gentibus septentrionalibus* of Olaus Magnus. The arguably most dramatic motif – the wolves' gruesome preference for the flesh of pregnant women – was taken directly from there.[26] But Possevino was not just paraphrasing the Swedish humanist's description of northern wolves. He employed the accounts of quasi-demonic wolves to stress the power of Catholic sacraments vis-à-vis the impotence of Lutheran and Orthodox rituals and to reprimand the heretics for their obduracy in the face of apparent miracles. Both aspects were of eminent importance to the Jesuit agenda, but entirely unrelated to Olaus Magnus' treatise.[27]

---

24 Possevino, Antonio: *Kiri Mantova hertsoginnale / Lettera alla Duchessa di Mantova.* (Tascabile Maarjamaa 2). Maarjamaa: Roma 1973, pp. 32–34.
25 Ibid., pp. 34–36.
26 Olaus Magnus, p. 611: "Examples of their wicked ferocity are easy to verify, especially during the mating season and when temperatures are very low. At these times travellers must proceed armed to keep themselves and their pack animals from being molested, particularly when the party includes women in advanced pregnancy, for the wolves catch their scent and attack them with greater eagerness. For this reason no female is allowed to journey alone, but must have an armed escort at her side, as the above drawing illustrates."
27 Cf. Helk; Kurtz, Eduard: *Die Jahresberichte der Gesellschaft Jesu über ihre Wirksamkeit in Riga und Dorpat 1583–1614.* Gulbis: Riga 1925.

Shapeshifting Wolves: Rakvere, 1586

The third and final case was reported by one of the most well-travelled men of the late sixteenth century, Samuel Kiechel (1563–1619), scion of a prosperous merchant family from Ulm.[28] Between 1585 and 1589, he journeyed through most of Europe, visited Constantinople and the Holy Land, and even climbed the Pyramids in Egypt.

In August 1586, the young but experienced traveller found himself in an unpleasant situation on the road from Narva to Tallinn. His travelling companion had deserted him and taken the only available horse, leaving Kiechel alone in the war-torn village of Rakvere (in the very same area where Russow's portentous wolf-pack supposedly roamed more than a decade earlier): "The land is all devastated, desolate and deserted; all settlements, hamlets and castles have been pillaged by the Muscovite."[29] Fortunately, a Baltic German nobleman, Junker Jörg von Bergen, noted Kiechel's plight and invited him to his nearby estate. At this point, Kiechel's travelogue starts to read almost like a Gothic novel: Rain, darkness, a long coach ride along muddy roads, and finally a decrepit wooden mansion in the middle of nowhere.

> During this night, it occurred that wolves carried off six sheep from the yard. Seven more had already disappeared a few days previously, and the nobleman wondered whether they had wandered off, or whether they had been stolen. Possibly they had been taken by witches or sorcerers, who are very common in this land. There are people known to roam the land in lupine shape; they are known as werewolves, and they cause great harm.[30]

At dawn, the junker sent a young farmhand into the woods to look for the missing animals. Following werewolves into the forest, all on his own, was apparently a routine task for this brave young Estonian – Kiechel mentions neither fear nor

---

28 On Kiechel, see Taetz, Sascha: *Richtung Mitternacht. Wahrnehmung und Darstellung Skandinaviens in Reiseberichten städtischer Bürger des 16. und 17. Jahrhunderts*. (Kieler Werkstücke. Beiträge zur Sozial- und Wirtschaftsgeschichte 3). Peter Lang: Frankfurt a.M. et al. 2004, pp. 46–48; Koenig-Warthausen, Gabriele von: "Samuel Kiechel. 'Weltreisender' und Handelsmann aus Ulm 1563–1619". *Lebensbilder aus Schwaben und Franken* 11, 1969, pp. 23–47.
29 Haßler, K. D. (ed.): *Die Reisen des Samuel Kiechel* (Bibliothek des Litterarischen Vereins in Stuttgart 86). Litterarischer Verein: Stuttgart 1866, p. 126. The Livonian War had ended three years previously, with the Treaty of Plussa in August 1583. The uneasy truce between Sweden and Russia lasted for six and a half years before hostilities resumed in January 1590.
30 Haßler 1866, p. 126.

hesitation – and he managed to recover the head of one of the sheep. Meanwhile, a search of the estate had shown that there was no hole in the enclosure through which ordinary wolves might have slipped into the yard. With their suspicions of preternatural occurrences seemingly confirmed, the men at the estate "surmised that [the sheep] had been taken by the werewolves."[31]

Tales of lupine shape-changers were very common in Livonia during the late sixteenth and seventeenth centuries.[32] The second edition of Sebastian Münster's *Cosmographia*, printed in 1550, first introduced the motif: "In this land there are many sorcerers and witch women, who adhere to the erroneous belief [...] that they become wolves, roam about, and cause harm to all they encounter. Afterwards they transform back into human shape. Such people are called werewolves."[33] Confirmed and embellished, independently of each other, by Olaus Magnus and by Philipp Melanchthon (1497–1560) and his pupils, the abundance of werewolves in Livonia and the extent of their ravages were commonplace knowledge all over Europe.[34] Friedrich Menius (ca. 1593–1659), Professor of History in Tartu during the early 1630s, quipped that "the outrageous lycanthropy of the Livonians is so common that even those who can't tell the difference between their right and their left hands are usually familiar with it."[35] In local witch trials, the aspect of animal transformation was usually emphasised and tended to eclipse other components of the witch stereotype.[36] In short, the werewolf had become an iconic image readily associated with Livonia.

---

31 Ibid., p. 127.
32 Donecker, Stefan: "The Werewolves of Livonia: Lycanthropy and Shape-Changing in Scholarly Texts, 1550–1720". *Preternature: Critical and Historical Studies on the Preternatural* 1, 2012, pp. 289–322.
33 Munsterus, Sebastianus: *Cosmographei oder beschreibung aller länder* [...]. Henrichus Petri: Basel 1550, p. 930.
34 Donecker 2012, pp. 296–393.
35 Menius, Fridericus: "Syntagma de origine Livonorum". *Scriptores rerum Livonicarum* 2, 1848, pp. 511–542, here p. 524.
36 Blécourt, Willem de: "A Journey to Hell: Reconsidering the Livonian 'Werewolf'". *Magic, Ritual, and Witchcraft* 2, 2008, pp. 49–67, here pp. 61–62; Cf. also Vähi, Tiina: "Hexenprozesse und der Werwolfglaube in Estland". In: Dietrich, Manfred L. G. / Kulmar, Tarmo (eds.): *Die Bedeutung der Religion für Gesellschaften in Vergangenheit und Gegenwart*. (Forschungen zur Anthropologie und Religionsgeschichte 36). Ugarit: Münster 2003, pp. 215–238; Metsvahi, Merili: "Werwolfprozesse in Estland und Livland im 17. Jahrhundert. Zusammenstöße zwischen der Realität von Richtern und von Bauern". In: Beyer, Jürgen / Hiiemäe, Reet (eds.): *Folklore als Tatsachenbericht*. Estnisches Literaturmuseum: Tartu 2001, pp. 175–184; Madar, Maia: "Estonia

For Samuel Kiechel, the connection between war and werewolvery seemed evident: the ravaged countryside around Wesenberg provided the werewolves with the perfect setting for their predations: a land whose inhabitants lived like animals. "No one is willing to build and cultivate anything, apart from that which they need for daily nourishment and sustenance of the body, until a stable peace can be negotiated."[37] Kiechel implied that the breakdown of social order caused by the war was a precondition for the alleged shape-changers' activities.

But Samuel Kiechel was not the only one to reach this conclusion: One of the most influential early werewolf accounts, the report of Hubert Languet (1518–1581), a pupil of Melanchthon, was directly linked to the Muscovite invasion of 1558, which Languet had witnessed. Melanchthon and Jean Bodin (ca. 1529–1596), who disseminated the report, both retained the military context.[38] In many accounts, the werewolves themselves resembled military formations. They were said to be organised in large warbands and to follow a strict chain of command, with overseers and taskmasters who made sure that all werewolves followed the devil's orders.[39] The ever-imaginative Olaus Magnus envisioned a gathering that, to modern readers, resembled a werewolf boot camp: "On the borders of Lithuania, Samogitia, and Kurland there is a wall, still standing, of a ruined castle, where at a certain time of the year several thousand of these creatures congregate to test their agility in leaping. Any who fail to spring over this wall, as happens sometimes with the stouter ones, are lashed with whips by the leaders of the pack."[40]

Conclusion

Recent research on late medieval France and early modern Germany[41] has demonstrated the discursive interdependency between narratives of war and

---

I: Werewolves and Poisoners". In: Ankarloo, Bengt / Henningsen, Gustav (eds.): *Early Modern Witchcraft. Centres and Peripheries.* Clarendon: Oxford 1990, pp. 257–272.
37 Haßler 1866, p. 126.
38 Donecker 2012, pp. 298–299, 317.
39 Ibid., pp. 299–300.
40 Olaus Magnus, p. 642.
41 Kling, Alexander: "War-Time, Wolf-Time. Material and Semiotic Knots in the Chronicles of the Thirty Years' War". In: Masius, Patrick / Sprenger, Jana (eds.): *A Fairytale in Question. Historical Interactions between Humans and Wolves.* White Horse Press: Cambridge 2015, pp. 19–38, in particular pp. 27–32; Siemer, Stefan: "Wölfe in der Stadt. Wahrnehmungsmuster einer Tierkatastrophe am Beispiel des *Journal d'un Bourgeois de Paris*". In: Groh, Dieter / Kempe, Michael / Mauelshagen, Franz (eds.): *Naturkatastrophen. Beiträge zu ihrer Deutung, Wahrnehmung und Darstellung*

narratives of wolf predation. "Observed historically, wolves are animals of crisis; they become visible in crisis situations, and it is during crisis situations that they are spoken of and written about."[42] To a certain degree, the association between wolves and warfare was the direct result of the animals' behavioural patterns. During times of war, the otherwise reclusive animals became far more conspicuous. Since regular wolf hunts could not be sustained, the population increased. Wolves were commonly observed scavenging for bodies on battlefields, and a scarcity of prey after extensive foraging could incite the animals into unusually aggressive behaviour.[43] The wolf-related events narrated by Russow, Possevino and Kiechel are, to all intents and purposes, plausible under pre-modern wartime conditions.[44] Possevino is likely to have exaggerated the numbers of victims and heightened the drama by drawing his readers' attention to pregnant victims, but the core of his account seems realistic when compared to nineteenth-century statistics.[45] Russow's report lacks Possevino's disturbing details, but heightens the sense of drama even further with its evocative language. However, any siege is likely to attract scavenging wolves and it is easy to imagine that their audible presence aggravated the soldiers' anxiety. Kiechel's account of lupine predation on livestock, at last, remains absolutely mundane – as long as the locals' werewolf beliefs are disregarded.

But the political symbolism of wolves extended far beyond their factual appearances on battlefields. Accounts of lupine predation allowed early modern authors to express their anxiety over personal danger and the social disorder that resulted from warfare. In other words, the wolf was a symbol for war itself.[46]

---

 in *Text und Bild von der Antike bis ins 20. Jahrhundert*. (Literatur und Anthropologie 13). Gunter Narr: Tübingen 2003, pp. 347–365.
42 Kling 2015, p. 19.
43 Cf. Schöller 2017, pp. 40, 48–49; Pluskowski 2006, pp. 96, 103; Rheinheimer 1995, p. 283.
44 Cf. Linnell / Alleau 2016, esp. pp. 364–365; Kruuk 2002, pp. 69–73.
45 Possevino (1973, pp. 32–34) claims several hundred wolf fatalities in particularly affected areas in Livonia. According to Rootsi, p. 270, 56 children in six different parishes were killed between 1806 and 1810. Considering that Possevino refers to a wartime setting (whereas the Baltic Provinces were at peace between 1806 and 1810), his numbers are probably too high, but not entirely out of proportion.
46 In many wartime chronicles, the distinction between man and animal became blurred. Accusations of werewolvery could stress it even further, but pre-modern authors did not need to draw upon the supernatural to make their point. Even normal (that is, four-legged and non-shapeshifting) wolves were often likened to organised groups of combatants. Authors attested them an ability for tactical conduct and drew explicit

This observation holds true for the Livonian cases as well: the three texts represent very different genres and equally different authorial positions: a semi-official chronicle by a local pastor, a report by a foreign missionary and a private travelogue. Yet all three share the basic assumption that wolves thrive in times of war. Furthermore, the three authors bestowed symbolic significance upon the animals' activities that extended far beyond their observable behaviour. In all three accounts, wolves embody unrest and disorder, although the authors chose to emphasise different aspects. In Balthasar Russow's chronicle, wolves heralded discord among allies and ultimately betrayal. For Antonio Possevino, they were a symptom of a spiritual crisis that, in his opinion, could be only remedied by a return to Catholic faith. And for Samuel Kiechel – or rather, for the Estonian informers whom he met – wolves, resp. werewolves, represented the dismal state of a war-ravaged country where men had been reduced to a beast-like existence. Thus, the precarious, threatened socio-political order of wartime Livonia provided the ideal setting for wolf predation, in a factual as well as a metaphorical sense.

References

Blécourt, Willem de: "A Journey to Hell: Reconsidering the Livonian 'Werewolf'". *Magic, Ritual, and Witchcraft* 2, 2008, pp. 49–67.

Bussov, Konrad: *Moskovskaja chronika. 1584–1613*. Akad. nauk SSSR: Moskva 1961.

Donecker, Stefan: "The Werewolves of Livonia: Lycanthropy and Shape-Changing in Scholarly Texts, 1550–1720". *Preternature: Critical and Historical Studies on the Preternatural* 1, 2012, pp. 289–322.

Flemming, Hanns Friedrich von: *Der Vollkommene Teutsche Jäger* [...]. J. C. Martini: Leipzig 1749.

Foote, Peter (ed.) / Fisher, Peter (trans.) / Higgins, Humphrey (trans.): *Olaus Magnus: A Description of the Northern Peoples 1555*. 3 vols. (The Hakluyt Society: Works II, 182, 187 and 188). Hakluyt Society: London 1996–1998.

Friebe, Wilhelm Christian: "Wie ist die Viehzucht in Liefland zu verbessern?". *Abhandlungen der liefländischen, gemeinnützigen und ökonomischen Societät* 2(2), 1803, pp. 1–142.

---

parallels between wolf packs and bands of marauding mercenaries. See Siemer 2003, pp. 359–361.

Hasselblatt, Arnold: "Die Wölfe in Livland. Eine culturhistorische Studie". *Baltische Monatsschrift* 29, 1882, pp. 659–678.

Haßler, K. D. (ed.): *Die Reisen des Samuel Kiechel*. (Bibliothek des Litterarischen Vereins in Stuttgart 86). Litterarischer Verein: Stuttgart 1866.

Helk, Vello: *Die Jesuiten in Dorpat 1583–1625. Ein Vorposten der Gegenreformation in Nordosteuropa*. (Odense University Studies in History and Social Sciences 44). Odense University Press: Odense 1977.

Hupel, August Wilhelm: "Ueber die Ausrottung der Wölfe". *Nordische Miscellaneen* 1, 1781, pp. 229–231.

Johansen, Paul: *Balthasar Rüssow als Humanist und Geschichtsschreiber*. (Quellen und Studien zur baltischen Geschichte 14). Böhlau: Köln et al. 1996.

Kling, Alexander: "War-Time, Wolf-Time. Material and Semiotic Knots in the Chronicles of the Thirty Years' War". In: Masius, Patrick / Sprenger, Jana (eds.): *A Fairytale in Question. Historical Interactions between Humans and Wolves*. White Horse Press: Cambridge 2015, pp. 19–38.

Koenig-Warthausen, Gabriele von: "Samuel Kiechel. 'Weltreisender' und Handelsmann aus Ulm 1563–1619". *Lebensbilder aus Schwaben und Franken* 11, 1969, pp. 23–47.

Kruuk, Hans: *Hunter and Hunted: Relationships Between Carnivores and People*. Cambridge University Press: Cambridge 2002.

Kurtz, Eduard: *Die Jahresberichte der Gesellschaft Jesu über ihre Wirksamkeit in Riga und Dorpat 1583–1614*. Gulbis: Riga 1925.

Laidre, Margus: "Der Hundertjährige Krieg (1558–1660/61) in Estland". *Forschungen zur baltischen Geschichte* 1, 2006, pp. 68–81.

Liiv, Otto: *Die wirtschaftliche Lage des estnischen Gebietes am Ausgang des XVII. Jahrhunderts*. (Verhandlungen der Gelehrten Estnischen Gesellschaft 27). Õpetatud Eesti Selts: Tartu 1935.

Linnell, John D. C. / Alleau, Julien: "Predators That Kill Humans: Myth, Reality, Context and the Politics of Wolf Attacks on People". In: Angelici, Francesco M. (ed.): *Problematic Wildlife. A Cross-Disciplinary Approach*. Springer: Cham 2016, pp. 357–371.

Linnell, John D. C., et al.: *The Fear of Wolves. A Review of Wolf Attacks on Humans*. NINA: Trondheim 2002.

Madar, Maia: "Estonia I: Werewolves and Poisoners". In: Ankarloo, Bengt / Henningsen, Gustav (eds.): *Early Modern Witchcraft. Centres and Peripheries*. Clarendon: Oxford 1990, pp. 257–272.

Menius, Fridericus: "Syntagma de origine Livonorum". *Scriptores rerum Livonicarum* 2, 1848, pp. 511–542.

Metsvahi, Merili: "Werwolfprozesse in Estland und Livland im 17 Jahrhundert. Zusammenstöße zwischen der Realität von Richtern und von Bauern". In: Beyer, Jürgen / Hiiemäe, Reet (eds.): *Folklore als Tatsachenbericht*. Estnisches Literaturmuseum: Tartu 2001, pp. 175–184.

Möhn, Dieter (ed.): *Mittelniederdeutsches Handwörterbuch, Zweiter Band*. Wachholtz: Neumünster 2004.

Munsterus, Sebastianus: *Cosmographei oder beschreibung aller länder* [...] Henrichus Petri: Basel 1550.

Olaus Magnus: *Historia de gentibus septentrionalibus* [...]. De Viottis: Romae 1555.

Orchard, G. Edward (trans.): *Conrad Bussow. The Disturbed State of the Russian Realm*. McGill-Queen's University Press: Montreal 1994.

Peuckert, Will-Erich: "Wolf". *Handwörterbuch des deutschen Aberglaubens* 9, 1941, cols. 716–794, at col. 743–744.

Pluskowski, Aleksander: *Wolves and the Wilderness in the Middle Ages*. Boydell: Woodbridge 2006.

Possevino, Antonio: *Kiri Mantova hertsoginnale / Lettera alla Duchessa di Mantova*. (Tascabile Maarjamaa 2). Maarjamaa: Roma 1973.

Raik, Katri: *Eesti- ja Liivimaa kroonikakirjutuse kõrgaeg 16. sajandi teisel poolel ja 17. sajandi alul* [The Heyday of Estonian and Livonian Chronicle Writing in the Second Half of the 16th Century and the Beginning of the 17th Century]. (Dissertationes Historiae Universitatis Tartuensis 8). Tartu Ülikooli Kirjastus: Tartu 2004.

Rheinheimer, Martin: "The Belief in Werewolves and the Extermination of Real Wolves in Schleswig-Holstein". *Scandinavian Journal of History* 20, 1995, pp. 281–294.

Rootsi, Ilmar: *Hunt ja inimene: suhted Eestis XVIII sajandi keskpaigast XIX sajandi lõpuni* [Wolf and Man: Their Relationship in Estonia from the Mid-18th Century to the End of the 19th Century]. (Dissertationes Historiae Universitatis Tartuensis 24). Tartu Ülikooli Kirjastus: Tartu 2011.

Russow, Balthasar: "Chronica der Prouintz Lyfflandt [...]". *Scriptores rerum Livonicarum* 2, 1848, pp. 1–157.

Schattauer, Willi: "'Der Wolf folgt der Trommel': Wölfe und Wolfsjagd in der Nordpfalz". *Donnersberg-Jahrbuch* 24, 2001, pp. 134–136.

Schöller, Rainer G.: *Eine Kulturgeschichte des Wolfs. Tierisches Beuteverhalten und menschliche Strategien sowie Methoden der Abwehr*. (Rombach Wissenschaften. Ökologie 10). Rombach: Freiburg i.Br. et al. 2017.

Siemer, Stefan: "Wölfe in der Stadt. Wahrnehmungsmuster einer Tierkatastrophe am Beispiel des *Journal d'un Bourgeois de Paris*". In: Groh, Dieter / Kempe,

Michael / Mauelshagen, Franz (eds.): *Naturkatastrophen. Beiträge zu ihrer Deutung, Wahrnehmung und Darstellung in Text und Bild von der Antike bis ins 20. Jahrhundert.* (Literatur und Anthropologie 13). Gunter Narr: Tübingen 2003, pp. 347–365.

Smith, Jerry C. (trans.): *The Chronicle of Balthasar Russow & A Forthright Rebuttal by Elert Kruse & Errors and Mistakes of Balthasar Russow by Heinrich Tisenhausen.* (Wisconsin Baltic Studies 2). Baltic Studies Center: Madison 1988.

Strods, Heinrihs: "Die Einschränkung der Wolfsplage und die Viehzucht Lettlands". *Ethnologia Europaea* 4, 1970, pp. 126–131.

Taetz, Sascha: *Richtung Mitternacht. Wahrnehmung und Darstellung Skandinaviens in Reiseberichten städtischer Bürger des 16. und 17. Jahrhunderts.* (Kieler Werkstücke. Beiträge zur Sozial- und Wirtschaftsgeschichte 3). Peter Lang: Frankfurt a.M. et al. 2004.

Vähi, Tiina: "Hexenprozesse und der Werwolfglaube in Estland". In: Dietrich, Manfred L. G. / Kulmar, Tarmo (eds.): *Die Bedeutung der Religion für Gesellschaften in Vergangenheit und Gegenwart.* (Forschungen zur Anthropologie und Religionsgeschichte 36). Ugarit: Münster 2003, pp. 215–238.

Periodicals

"Ein Wolf, der sich in der Stadt zu Tisch bittet". *Rigaische Stadt-Blätter* 1810, p. 142.

"Lese-Früchte aus Stadts-Chroniken". *Rigaische Stadt-Blätter* 1815, pp. 5–6.

"Noch ein Wolf; aber ein schlimmerer". *Rigaische Stadt-Blätter* 1810, pp. 166–167.

"Probe einer ökonomischen Naturgeschichte Kurlands. Der Säugthiere. Der Wolf". *Geoponika eine ökonomische Monatschrift für Kur- und Lievlands Bewohner* 1, 1798/99, pp. 593–620.

Meelis Friedenthal

Cats, Allergy and Occult Powers in Early Modern Disputations

Abstract: In Early Modern academic texts about natural magic "the fear of cats" is often used as an example of occult (i.e. hidden) qualities. Descriptions of such hidden qualities were accepted in many physics textbooks by the middle of the seventeenth century, on the one hand preparing students to admit the existence of natural magic, and on the other hand attempting to find reasonable explanations for such phenomena. In Northern European university disputations, such interest in the occult is readily visible especially in the second half of the seventeenth century, particularly at the philosophical faculties where students would often discuss modern philosphical theories referring to the peculiar powers of animals. The aim of the article is to give an overview of the discussions regarding "the fear of cats" in philosophical and medical textbooks, disputations and compendia.

Keywords: cats, natural magic, occult powers, disputations, Early Modern, university

Introduction

When researching philosophical changes in Swedish universities during the early modern period I noticed cats mentioned in a 1692 disputation from the University of Tartu on *Sense errors*. The disputation was presided over by Professor of Theoretical Philosophy Gabriel Sjöberg (?–1704) and defended by Andreas Westerman (1672– ca. 1739). In the text the author is laying out Cartesian arguments against the Aristotelian world view and in the 23rd thesis of the disputation he argues that:

> Some are able to withstand the *effluvia* and smells coming from a cat very well, when others suffer and flee them and are even unable to remain in the same house with a cat. They have picked up this kind of aversion unknowingly when in the womb, or in their childhood, when their mother during pregnancy or they themselves in the cradle were injured by a cat, and the idea has been imprinted in their brain ever since.[1]

---

1   "Quidam effluvia et odores de fele exeuntes optime tolerant, quos tamen alii ut pessimum quodvis fugiunt, adeo ut in eadem cum fele domo manere nequeant. Hanc autem aversionem inscius vel in utero matris, vel in ipsa infantia contraxit, cum aut mater tempore gestationis aut ipse in cunis a fele laesus fuerit, quae idea cerebro tunc impressa contiunuo manet." Sjöberg, Gabriel / Westerman, Andreas: *Dissertatio philosophica de erroribus sensuum.* [Philosophical dissertation about sense errors.] Academia Dorpatensis: Dorpat 1692, thes. 23.

I would have left that curious remark without attention, but some time later I noticed again cats mentioned in a disputation from 1648 about *Natural magic*. The disputation was presided over by *magister* Johan Svensson Lagermarck (Wassenius) (1622–ca. 1691) and defended by Axelius Jonae Orraeus (d. 1676) in the University of Turku. In the third thesis of the disputation the author discusses antipathy, which is "the natural hatred of one body toward the other"[2] and enumerates a list of examples of natural hate, e.g. how newborn chickens are frightened by a bird of prey, but are not frightened by horses or oxen, or how sheep are terrified of wolves, and hares flee dogs, and how some people are unable to bear the presence of cats. Lagermarck argues that the causes of such antipathies are occult or hidden (*occultae*), i.e. caused by properties that we are not capable of sensing.

After some research, it turned out that cats appear quite often in disputations in the context of occult powers, unexplained antipathies, and sympathies. Here, of course, ancient authors are often quoted to characterise the inexplicable, and a passage from Martial is regularly cited where such feelings are expressed:[3]

> I do not love you, Volusius, and I can't say why.
> This only I can say, I do not love you.

While discussing sympathies and antipathies, many different examples are listed together to illustrate the peculiar natural relations between people, animals, and material objects. At the same time it begs the question of whether all such antipathies are alike. What kind of similarity could there be between the antipathy of butter against hot water with the antipathy of some people toward cats or Martial not liking Volusius? According to our current understanding a number of these cases could be explained by physical properties of materials, some with natural instincts, some with simple dislikes or phobias, and some of which seem to be cases of allergic reactions. In the early modern period the similarity between all such cases was that they were all unexplainable. It is uncertain whether there are some peculiar characteristics of the objects that

---

2 "[...] odium naturale corporis unius adversus aliud." Lagermarck, Johan Svensson / Orraeus, Axelius Jonae: *De magia naturali*. Academia Aboensis: Turku 1648.

3 "Non amo te, Volusi, nec possum dicere quare. Hoc tantum possum dicere non amo te." Riddermarck, Andreas / Eurelius, Torstanus: *Dissertatio inauguralis de magia*. Londini Scanorum: Lund 1685, § 10. The critical editions of Martial usually nowadays give the name as Sabidius. See: Ker, Walter Charles Alan (trans.): *Martial, Epigrams. I.* W. Heinemann / Harvard University Press: London / Cambridge, Massachusetts 1961, p. 48.

are at fault here or whether these are idiosyncratic temperaments, the effect of substantial form or perhaps there is some other underlying influence that causes such strange antipathies and sympathies.

In this article, I would like to look at the issue of occult powers through the lens of allergic reactions. What we today would describe as allergic reactions offer a good case study here, as in the early modern period many different mechanisms were proposed to explain these, and this permits us to get a good overview of the thinking associated with hidden forces and unexplained effects. Among such cases that could be allergic reactions, it seems that cats are most frequently mentioned, but reactions to foodstuffs, "rose fever", and bee and wasp stings are also listed. E.g. Professor of Mathematics in Uppsala University Johan Bilberg describes how some people have a "natural antipathy" toward cheese[4] and Professor of Physics and Botany in Turku, Georg Alanus, describes how some people are unable to eat fish.[5] What sets such cases apart from other or regular "antipathies" is that "loss of strength" (*deliquium animi*) or "profound sweating" (*largiter sudare*) is often mentioned in connection with them.[6] Additionally, the context shows that these are not some isolated instances of a dislike of certain foodstuffs or a phobia toward animals, since the examples are brought to illustrate some peculiar or inherited reaction. The condition of *deliquium animi,* according to the medical understanding of the period, denotes near-fainting or lipothymia,[7] and seems in the current context often indicative of a strong allergic reaction.

To my knowledge there has been no substantial discussion in recent literature about the connection between occult powers and allergic reactions toward foodstuffs or domestic animals.[8] Authors writing on the history of allergies

---

4   Bilberg, Johan / Prosperius, Gustaf: *De occultis qualitatibus.* [About occult qualities.] Academia Upsaliensis: Upsala 1687, p. 9.
5   Alanus, Georgius Christophori / Lagercrona, Johan Mattsson: *De magia naturali.* Academia Aboensis: Turku 1645, thes. 34.
6   Ibid.
7   Junius, Hadrianus: *Nomenclator omnium rerum propria nomina.* Egnolphus Emmelius: Francofurti 1620, p. 377, *defectus animi.*
8   Several discussions concern general occult properties of animals and cultural attitudes toward cats. See e.g.: Copenhaver, Brian P.: "A Tale of Two Fishes: Magical Objects in Natural History from Antiquity Through the Scientific Revolution". *Journal of the History of Ideas* 52(3), 1991, pp. 373–98; Paster, Gail Kern: "Melancholy Cats, Lugged Bears, and Early Modern Cosmology: Reading Shakespeare's Psychological Materialism Across the Species Barrier." In: Paster, Gail Kern / Rowe, Katherine / Floyd-Wilson, Mary (eds.): *Reading the Early Modern Passions: Essays in the Cultural History of Emotion.* University of Pennsylvania Press: Philadelphia 2004, pp. 113–129;

mention possible cases of cat allergy in the early modern period but the connection with occult powers is not explored further and mainly idiosyncrasies and temperament theories as explanations are mentioned.[9] The aim of the current article is to look at some of the descriptions of the "fear of cats"[10] and follow the discussions about the mechanisms of such an ailment. It seems that based on the reports of well-known medical and school authors the "fear of cats" becomes a certain *locus communis* and an oft cited example of an unexplained antipathy in early modern university texts. In Swedish universities we see such cases of "fear of cats" used to discuss occult qualities and hidden powers, natural magic, and new physical theories.

"Fear of cats" as Superstition or a Matter of Temperament

Occurrences of natural aversion and attraction have been known for centuries. Not all such cases can be attributed to allergies, and it is often difficult to make an unambiguous decision about the reasons for "antipathy", as authors often do not differentiate between the causes for such reactions. E.g. early modern authors regularly cite the passage by Plutarch where he (in the Latin translation) mentions that certain people hate (*oderunt*) cats, beetles, toads and serpents.[11] The animals mentioned here seem to be generally unpleasant, and cats do not seem to fit this list very well. However, it is not impossible that it is precisely because of their allergenic effects that they are included in this list.

It is noteworthy that the same animals were also commonly associated with demons and witchcraft.[12] The association with religious beliefs and such

---

Raber, Karen: "How to Do Things with Animals: Thoughts on/with the Early Modern Cat". In: Hallock, Thomas / Raber, Karen / Kamps, Ivo (eds.): *Early Modern Ecostudies*. Palgrave Macmillan: New York 2008, pp. 93–113.

9 Ring, Johannes: "History of Allergy in Middle Ages and Renaissance". In: Bergmann, K.-C. / Ring, Johannes (eds.): *History of Allergy*. Karger Medical and Scientific Publishers: Basel 2014, pp. 15–20; Schadewaldt, Hans: *Idiosynkrasie, Anaphylaxie, Allergie, Atopie: Ein Beitrag Zur Geschichte Der Überempfindlichkeitskrankheiten.* (Vorträge G 251). Westdeutscher Verlag: Opladen 1981; Cohen, Sheldon G. / Samter, Max: *Excerpts from Classics in Allergy, 3rd ed.* Symposia Foundation: Carlsbad, CA 2012.

10 In the disputations there are no other domestic animals (e.g. dogs or horses) mentioned in this context. Allergy as a specific condition was not known before the twentieth century; therefore, throughout this article "fear of cats" is often used instead.

11 Feles enim, cantharidesque nonnulli oderut & rubetas & serpentes. Plutarchus: *Opuscula Plutarchi Cheronei*. Ascensius: Paris 1521, d iii v, de invidia et odio.

12 Raber 2008, p. 102.

antipathies is corroborated by the Danish physician Thomas Bartholin, who records in his casebook on diseases that superstitious people believed that only those individuals over whose cradle a cat had jumped before they were baptised developed a natural antipathy toward cats.[13] By "superstition" Bartholin presumably has in mind the belief that unbaptised children were considered to be typically more susceptible to demonic influence. But the identification of superstition is not enough here, it was the task of philosophers to find scientific explanations for such diseases. Most early modern philosophers believed that even if the causes of such antipathy might be hidden, they were natural, not supernatural or demonic influences.

In the sixteenth century the Swiss physician and theologian Thomas Erastus (1524–1583) tells in his book *About the occult properties of medicines* a story about a certain Heroina who was an otherwise courageous and brave woman but had a terrible fear of cats.[14] The story of Erastus is repeated later verbatim by his pupil and well-known humanist Henrich Smet (ca. 1537–1614),[15] and subsequently finds its place in many texts that discuss occult properties and natural magic in medicine.[16]

Erastus writes that the "fear" was caused because when the woman's mother was pregnant, she was frightened by a cat, and after that she could not stand cats anymore. This fear was transmitted to her daughter. The description leaves an impression that foremost visual recognition or some other means of becoming aware of the cat was important for causing the reaction. The words that Erastus uses to describe the condition are "fear" (*terrere*), "abhorrence" (*abominare*) and "hate" (*odire*), and thus connect the antipathy to passions.[17] He considers this more an ailment of the soul than of the body – it is because of the imagination of the mother that the "fear" is imprinted into the soul of the foetus and this also changes the temperament of the child.[18] Erastus refers here also to other instances where effects and diseases that mothers were suffering were

---

13 Bartholin, Thomas: *Historiarum anatomicarum rariorum, centuria III et IV.* Vlacq: Hagae-Comitis 1657, pp. 42–43, cent. 3, hist. 28.
14 Erastus, Thomas: *De occultis pharmacorum potestatibus.* Perna: Basel 1574, pp. 61–62.
15 Smet, Hendrik de: *Miscellanea Henrici Smetii.* Jonae Rhodii: Francofurti 1611, p. 101.
16 Several authors credit especially Erastus for making the case of "fear of cats" well known. See e.g. May, Johannes Henricus: *Brevis et accurata animalium in sacro cum primis codice memoratorum historia.* Olffen: Francofurti 1686.
17 Paster 2004.
18 Erastus 1574, p. 62.

transmitted to the children, and his explanation is well in line with the general medical understanding of the time.

The mentioned changes in temperament refer to the four temperaments theory of Galen, which was one of the main bases of understanding diseases and health in the medieval and also the early modern periods – choleric, sanguine, phlegmatic and melancholic temperaments arose due to the mixture of bodily fluids (humours).[19] The temperament was considered to be idiosyncratic (peculiar) to each person due to the balance of the fluids, and therefore each person could have different reactions to medicines as well as food and animals.[20]

However, not all authors agree that temperament is the cause of the antipathy toward cats. Jesuit philosopher Laurenz Forer (1580–1659) firstly observes that the fear does not arise because of the imagination, as sweat and bodily change (*alteratio corporis*) are already felt when the cat is still not perceived by any of the senses.[21] He then examines the explanation that since cats are melancholic, the aversion could be caused by having an opposite temperament. He draws attention to the fact that people with sanguine temperament are not always affected negatively by cats. Furthermore, he argues, it seems that cats are not affected at all by persons who suffer from this ailment (which should otherwise be the case). He concludes that the temperament of the affected person can't be the culprit here and suggests some sort of other power that we don't know of.[22] Forer also explains that there are authors who deny the existence of any occult powers and try to explain even the property of magnetism to draw iron with the temperament of the magnet, but there are many arguments against it. He asks if antipathy and sympathy are explained by temperaments, then why does a magnet draw iron and not vice versa, as both the magnet and iron are equally similar to each other and thus should have equal attraction. Forer raises a related question of why people have antipathy toward cats but not for example toward stones or trees, with which we also have very dissimilar temperaments.[23]

---

19 Lindemann, Mary: *Medicine and Society in Early Modern Europe*. Cambridge University Press: Cambridge 2010, pp. 88–90; Gentilcore, David: *Food and Health in Early Modern Europe: Diet, Medicine and Society, 1450–1800*. Bloomsbury Academic: London 2016, pp. 11–15.
20 Schadewaldt 1981.
21 Forer, Laurenz: *Disputatio philosophica de sympathia et antipathia*. Universitas Ingolstadiensis: Ingolstadt 1618, thes. 5.
22 Forer 1618, thes. 5.
23 Forer 1618, thes. 5.

Additionally, the well-known Wittenberg philosopher Johann Sperling (1603–1658), whose works are often cited in Swedish universities, discusses the problem of the fear of cats in several places in his textbook. He agrees that this "fear" could not be caused by contrary temperaments as some authors state.[24] Sperling stresses instead that powers that are hidden and not directly perceptible are to be deduced here:

> Not because of taste, or smell, or touch or through any other sense could the particulars [of this influence] be exposed and explored, but only through observation and experience.[25]

Early Modern Physics and Occult Powers

Sperling refers here to a newer understanding of physics and his method of deduction is quite different from medieval attitudes. In the Aristotelian context it was axiomatically stated that there is nothing in our intellect that wasn't first in the senses, meaning also that we cannot know what we cannot perceive by our senses.[26] Hidden and occult qualities are not directly perceivable, hence they were also in principle unknowable. As such in medieval universities the existence of occult qualities was admitted, but generally not thematised, and only after the sixteenth century do we see more academic interest in occult.[27] The beginnings of such an interest are usually located in the writings of Marcilio Ficino, Pico della Mirandola, and other Neo-Platonists of the fifteenth century.[28]

---

24 Sperling, Johann: *Institutiones physicae*, 2nd ed. Johannis Bergeri: Witteberga 1649, pp. 932–933.
25 "Nec sapore, nec odore, nec tactu, nec ullo denique sensu explorari et indicari proprie possunt, sed una observatione et experientia." Sperling, Johannes / Willichius, Jodocus: *Disputatio physica de qualitatibus occultis*. Academia Wittebergensis: Witteberga 1639.
26 "Nihil est in intellectu quin prius fuerit in sensu." This quotation has been commonly attributed to Aristotle since medieval times, but originates from Thomas Aquinas' Quaestiones disputatae de ueritate (q 2, a 3, a 18). See Wood, Neal: "Tabula Rasa, Social Environmentalism, and the 'English Paradigm'". *Journal of the History of Ideas* 53(4) 1992, pp. 647–668, p. 651.
27 Hutchison, Keith: "What Happened to Occult Qualities in the Scientific Revolution?" *Isis* 73(2), 1982, pp. 233–253, p. 237.
28 See e.g. Walker, Daniel Pickering: *Spiritual and Demonic Magic: From Ficino to Campanella*. The Pennsylvania State University Press: University Park, PA 2000; Copenhaver, Brian P.: "Natural Magic, Hermetism, and Occultism in Early Modern Science." In: Lindberg, David C. / Westman, Robert S. (eds.): *Reappraisals of the Scientific Revolution*. Cambridge University Press: Cambridge 1990, pp. 261–302; Allen,

Their ideas about occult powers and natural magic did not immediately find adoption in universities, instead we see e.g. in Wittenberg protestant reformers like Luther and Melanchthon voicing strong criticism against such interest.[29] Also the Faust legend was typically used in universities as a negative example of what can happen when one starts to engage with magic.[30] In order to be accepted in universities, the ideas about occult qualities and natural magic had to be processed in textbook form, which happened during the period of increasing dissatisfaction of Aristotelian metaphysics during the latter part of the sixteenth century. This did not mean the immediate abandonment of the four-elements theory and Aristotelian physics, but instead in addition to the described and known qualities, more attention was increasingly paid to unknown qualities or imperceptible seeds that bring about these observed changes. There were many authors like Paracelsus and his followers who criticised Aristotle and talked about "powers", "mumia" or "spiritus mundi" in herbs and other materials.[31] The acceptance of the reality of hidden powers in nature however opened up possibilities that there might be many more hidden forces at work (benign, malevolent or neutral) than we could fully give an account of relying on traditional physics.[32] This in itself is not surprising since the "Yates thesis" has already established that magic, hermeticism and occultism played an important part in the development of philosophical sciences in the seventeenth century.[33] At

---

Michael J.B.: "At Variance: Marsilio Ficino, Platonism and Heresy." In: Hedley, Douglas / Hutton, Sarah (eds.): *Platonism at the Origins of Modernity: Studies on Platonism and Early Modern Philosophy*. (Archives Internationales d'histoire Des Idées 196). Springer: Dordrecht 2008, pp. 31–44.

29 Wels, Volkhard: *Manifestationen des Geistes: Frömmigkeit, Spiritualismus und Dichtung in der Frühen Neuzeit*. V&R unipress GmbH: Göttingen 2014, p. 76.

30 Van der Laan, James M. / Weeks, Andrew (eds.): *The Faustian Century: German Literature and Culture in the Age of Luther and Faustus*. (Studies in German Literature, Linguistics, and Culture). Camden House: Rochester, NY 2013, p. 50.

31 Copleston, Frederick Charles: *A History of Philosophy, Volume 3: Late Medieval and Renaissance Philosophy: Ockham, Francis Bacon, and the Beginning of the Modern World*. Image Books: New York 1993, pp. 265–274; Paracelsus: *Paracelsus (Theophrastus Bombastus von Hohenheim, 1493–1541): Essential Theoretical Writings*. Brill: Leiden 2008, pp. 6–25.

32 Levack, Brian P.: *The Witch-Hunt in Early Modern Europe*, 3rd ed. Routledge: London 2006, p. 63.

33 Yates, Frances Amelia: *Giordano Bruno and the Hermetic Tradition*. Routledge: London 1964; for a discussion of the influence and afterlife of the "Yates thesis" see: Hanegraaff, Wouter J.: "Beyond the Yates Paradigm: The Study of Western Esotericism Between Counterculture and New Complexity". *Aries* 1(1), 2001, pp. 5–37, pp. 15–17.

the same time this search for hidden powers is well in line with the observation of Brian P. Levack, that the mediaeval attitude toward magic was more sceptical and the early modern attitude more credulous.[34]

During the seventeenth century magic and occult powers were taken up as topics especially in philosophical faculties and often by professors of physics.[35] It is clear that at least in Swedish universities natural magic was foremost a philosophical and medical problem, rather than theological. The statutes of Uppsala University of 1626 specified that physics had to be taught according to the textbook of Johannes Magirus from Marburg (ca. 1560–1596), where the author devotes a chapter to occult qualities.[36] These Uppsala University statutes were also used (*mutatis mutandis*) in other Swedish universities: Academia Dorpatensis (established in 1632, present-day Tartu, Estonia) and Academia Aboensis (established in 1640, present-day Turku, Finland). The physics textbook of Magirus was not only used in the Swedish Empire but found widespread use in the universities of Europe and was also famously studied by Isaac Newton in Cambridge.[37] In the book Magirus cites the work on natural magic by the well-known Italian philosopher Giambattista della Porta, and also refers to the oft-cited passage of Julius Caesar Scaliger about the "saving refuge of occult

---

34 Levack, pp. 61–63; Clark, Stuart: *Thinking with Demons: The Idea of Witchcraft in Early Modern Europe*. Oxford University Press: Oxford 2006, p 191.
35 From the universities of Uppsala, Tartu, Turku and Lund there are no theological disputations from the period 1600–1700 that concern specifically magic or the occult. The sources for the list of disputations are available: Lidén, Johan Henrik (ed.): *Catalogus disputationum, in academiis et gymnasiis Sueciae atque etiam a Suecis extra patriam habitarum*. Edman: Uppsala 1778; Jaanson, Ene-Lille (ed.): *Tartu Ülikooli trükikoda 1632–1710: Ajalugu ja trükiste bibliograafia = Druckerei der Universität Dorpat 1632–1710: Geschichte und Bibliographie der Druckschriften*. Tartu Ülikooli Raamatukogu: Tartu 2000; Vallinkoski, Jorma: *Turun akatemian väitöskirjat 1642– 1828 = Die Dissertationen der alten Universität Turku (Academia Aboënsis) 1642–1828* I. Valtioneuvoston kirjapaino: Helsinki 1962.
36 On Magirus and his influence, see: Kusukawa, Sachiko: "Nature's Regularity in Some Protestant Natural Philosophy Textbooks 1530–1630." In: Stolleis, Michael / Daston, Lorraine (eds.): *Natural Law and Laws of Nature in Early Modern Europe: Jurisprudence, Theology, Moral and Natural Philosophy*. Ashgate Publishing: Abingdon 2013, pp. 105–122.
37 Ducheyne, Steffen: "Newton's Training in the Aristotelian Textbook Tradition: From Effects to Causes and Back". *History of Science* 43, 2005, pp. 217–37; Buchwald, Jed Z. / Feingold, Mordechai: *Newton and the Origin of Civilization*. Princeton University Press: Princeton 2013, pp. 13–14.

properties". Magirus argues that occult qualities are separate from primal and secondary qualities and are not reducible to those.[38] During the period, the primary qualities of hotness, coldness, wetness, and dryness were understood to stem from the four elements and to correspond to them.[39] Magirus also names fourteen secondary qualities (density, rarity, gravity, levity, hardness, softness, coarseness, fineness, aridness, slipperiness, viscosity, crumbliness, roughness, smoothness) that are the "offspring and children" of the primary qualities and the result of the combination and mixture of these.[40] He calls them tangible qualities, whereas the objects of other senses (like colours, tastes and smells) are dealt with under sense perception. Occult qualities, however, are not directly perceivable by the senses and do not arise from either primary or secondary qualities, but are instead dependent on the specific form of the object.[41] When discussing hidden qualities, Magirus references many of the usual examples of the occurrences of occult phenomena in ancient literature, like the antipathy of lions toward the "song of the rooster" and the hatred that elephants have of mice.[42] Additionally, he argues that the tides of the oceans are caused by some occult power (*qualitate occulta*) of the moon, which operates similarly to a magnet.[43] Magirus does not discuss "the fear of cats" in his textbook,[44] but Giambattista

---

38 Magirus, Johannes: *Johannis Magiri Physiologiae peripateticae libri sex, cum commentariis*. Johannis Berneri: Francofurti 1624, p. 251; for the discussion about the Scaliger's expression see: Copenhaver, Brian P.: *Magic in Western Culture: From Antiquity to the Enlightenment*. Cambridge University Press: Cambridge 2015, pp. 146–147.

39 For a discussion about primary and secondary qualities see: Pasnau, Robert: "Scholastic Qualities, Primary and Secondary." In: Nolan, Lawrence (ed.): *Primary and Secondary Qualities: The Historical and Ongoing Debate*. Oxford University Press: Oxford 2011, pp. 41–61.

40 "Densitas, Raritas, Gravitas, Levitas, Durities, Mollities, Grassities, Subtilitas, Ariditas, Lubricitas, Lentor, Friabilitas, Asperitas, Laevitas. Secundae [...] ex primarum temperatura et mixtione profluunt, atque ita quasi illarum propagines et soboles sunt." Magirus, p. 247.

41 About form and species see: Spruit, Leen: *Species Intelligibilis – from Perception to Knowledge. Renaissance Controversies, Later Scholasticism, and the Elimination of the Intelligible Species in Modern Philosophy*, vol. 2. (Brill's Studies in Intellectual History 49). Brill: Leiden 1995.

42 Magirus 1624, p. 252.

43 Ibid., pp. 374, 378; cf. Deparis, Vincent / Legros, Hilaire / Souchay, Jean: "Investigations of Tides from the Antiquity to Laplace". In: Souchay, Jean / Mathis, Stéphane / Tokieda, Tadashi (eds.): *Tides in Astronomy and Astrophysics*. Springer: Berlin 2013, pp. 37–39.

44 Though he takes interest in the keen vision of cats. E.g. Magirus, p. 669.

della Porta, whom he uses and cites, takes some interest in the question.[45] Della Porta's works were well known in Swedish universities and several disputations dealing with occult and magic mention him by name. Della Porta, in the section of his book *Phytognomonica*, devoted to antipathy and sympathy, argues that "cats injure many people with their fixed gaze and fascinate [i.e. cast an evil eye on] some people even only with their presence".[46] The existence of the evil eye or fascination was generally accepted among early modern scholars and similarly many theories were proposed to explain the phenomenon.[47] In essence, however, della Porta's argumentation admits that the exact mechanism is the occult, or unknown.

Antipathy Toward Cats as an Occult Problem

To admit the existence of occult antipathy means that there has to also be an explanation for how this antipathy works. The "fear of cats" became a convenient case to test different theories about the nature of such antipathies. The French surgeon Ambroise Paré mentions that:

> [...] any hair [of a cat] devoured unawares, may be enough to choak one, by stopping the instruments of respiration; yet the hairs of a Cat by a certain occult property, are judged most dangerous in this case. [...] yet I would not judge that to happen by the malicious virulency of the Cat, but also by the peculiar nature of the party, and a quality generated with him, and sent from heaven.[48]

The works of Paré were first written in French but were soon translated into Latin, German and English, and also became influential in Northern Europe at the beginning of the seventeenth century. Here the author stresses that besides the poisonous hair of the cat, the peculiar nature of some people is also the culprit. Paré remarks that such nature is "sent from heaven", which indicates an astrological predisposition. Throughout the whole period medics paid attention

---

45 E.g. Chesnecopherus, Johan / Andreae, Andreas: *Disputatio physica de occultis qvalitatibvs*. Academia Upsaliensis: Upsala 1625; Alanus.
46 "Feles nonnullos fixo oculorum obtutu laedunt, alios sola praesentia fascinant". Porta, Giambattista della: *Phytognomonica octo libris contenta etc.* apud Horatium Salvianum: Neapoli 1588, p. 299 (VII, 29).
47 See e.g. Friedenthal, Meelis: "Kuri silm: toimemehhanismid lähtuvalt antiiksetest ja keskaegsetest tajuteooriatest [The Evil Eye: Descriptions of Operation According to Ancient and Medieval Theories of Perception]". *Mäetagused* 51, 2012, pp. 7–20.
48 Paré, Ambroise: *The Workes of That Famous Chirurgion Ambrose Parey*. R. Coates and W. Dugard: London 1649, p. 482.

to the position of stars and planets and the importance of medical astrology started to wane only in the eighteenth century.[49]

We see that Paré identifies no single explanation of the "fear of cats", but several and even competing causes (direct physical contact, contrary temperament, astrological influence). This reflects the style of early modern medical books, where many possible explanations are listed, in the hope that some of them might be of some help to the reader. But the list of possible causes does not end there. Some, like the German physician and chemist Andreas Libavius (ca. 1550–1616), consider the culprit to be the "vile vapour" (*foetor et halitus*) of the cat.[50] In addition, the seventeenth century Wittenberg university physician Daniel Sennert (1572–1637), who was also influential in Swedish universities, proposes the theory of spiritual forms as a possible explanation about the "fear of cats". He discusses the occult powers in his widely used physics textbook *Hypomnemata physica* and remarks:[51]

> Next there are hidden qualities, which are with right named spiritual forms, what like the rays [of light] are continuously emitted from the body and are diffused in the environment and have their own sphere of activity and are sometimes even capable of penetrating other bodies. Thus a magnet is able to move iron even when there is a wooden tablet in between; a cat closed in a box is still capable of affecting a person who has an antipathy toward cats.

Here Sennert has presumably conducted an experiment and observed that despite the cat being enclosed in a box, a person with a natural aversion toward cats is still affected, and thus there has to be some hidden quality that is able to pass through the box.

---

49 Harrison, Mark: "From Medical Astrology to Medical Astronomy: Sol-Lunar and Planetary Theories of Disease in British Medicine, c. 1700–1850". *The British Journal for the History of Science* 33(1), 2000, pp. 25–48.
50 Libavius, Andreas: *D.O.M.A. Singularium Andreae Libauii ... pars secunda*. Impressa typis Ioannis Saurii: Francofurti 1599, p. 122, thes. 32.
51 "Deinde sunt qualitates occultae, quae species spiritales merito appellantur, quod quasi radius continuo promotus a corpore suo effluunt, in orbem diffunduntur, & suam certam activitatis sphaeram habent, & quaedam etiam per corpora alia penetrant. Ita magnes etiam tabula imposita ferrum trahit; felis etiam cistae inclusa alium, cui cum felibus antipathia est, afficere potest." Sennert, Daniel: *Hypomnemata physica I. De rerum naturalium principiis, II. de occultis qualitatibus, III. de atomis & mistione, IV. de generatione, viventium, V. de spontaneo viventium ortu*. Schleich: Francofurti 1636, p. 79.

The fear of cats had become toward the middle of the seventeenth century one of many common examples when illustrating occult antipathies and sympathies, and it also provided a way to test the explanatory power of new theories. This could also be the reason why René Descartes, who aimed to establish a new philosophy against the scholastics, took up the already famous example of the "fear of cats" in his *Passions of the Soul*. In discussing how the passions that are peculiar to certain persons produce their effects, Descartes remarks:

> [...] the strange aversions of certain people that make them unable to bear the smell of roses, the presence of a cat, or the like, can readily be recognised as resulting simply from their having been greatly upset by some such object in the early years of their life. Or it may even result from their having been affected by the feelings their mother had when she was upset by such an object while pregnant; for there certainly is a connection between all the movements of a mother and those of a child in her womb, so that anything adverse to the one is harmful to the other. And the smell of roses may have caused severe headache in a child when he was still in the cradle, or a cat may have terrified him without anyone noticing and without any memory of it remaining afterwards; and yet the idea of the aversion he then felt for the roses or for the cat will remain imprinted on his brain till the end of his life.[52]

This description is remarkably similar to the account given by Erastus, but here Descartes is carefully avoiding mentioning anything occult and proposes only bodily movements as the cause of the "fear". It is noteworthy that here Descartes fails to adequately take into account that certain people are "frightened" by cats even when they do not see the cat, and he does not even consider that cats themselves could also be contributing something (smell, occult effluvia or similar) to the effect.

Concluding Remarks

The phenomenon of the "fear of cats" succinctly illustrates the ambiguity and mixture of superstitious, astrological, philosophical and medical explanations that early modern philosophers were confronted with when dealing with the effects that had unknown or hidden causes. During the seventeenth century the domain of physics in academias expanded, sometimes natural magic was also included here and it also covered some theological topics (e.g. there are in universities philosophical works about angels and demons, spirits and souls).[53]

---

52 Descartes, René: *The philosophical writings of Descartes,* vol. 1. Cambridge University Press: Cambridge 1985, pp. 375–376, *The Passions of the Soul* 136, (AT XI 428–429).
53 Snellius, Rudolph / Valerius, Cornelius / Goclenius, Rudolph: *Rudolphi Snellii In Physicam Cornelii Valerii annotationes: Cum lectissimis aliorum observationibus*

Increasingly abandoning Aristotelian hylomorphic explanations, different kinds of new theories were proposed and employed to explain hidden influences. In a way, new seventeenth-century philosophies started to turn all previously known, observable and tangible qualities into occult qualities, postulating reasons for these phenomena (e.g. for explaining coldness, hotness) that are not directly apparent from sense perception.[54] When considering the disputations of Swedish universities as an example, it is evident that during the second half of the seventeenth century, the interest in these unexplained phenomena and occult qualities is considerably higher than in the first half of the century, and this also coincides with the influx of the new physical theories into northern universities. Presumably cats are discussed in the disputations based on the examples of the texts that were used in schools, as the descriptions closely echo the stories from the textbooks. In the disputations about antipathy it is also well observable how Aristotelian hylomorphic theories give room to explanations derived mostly from Cartesian authors at the end of the seventeenth century. However, the mechanism of the "fear" also remained unexplained in the next centuries, and only during the twentieth century was the allergen responsible for the phenomenon discovered.

References

Sources

Alanus, Georgius Christophori / Lagercrona, Johan Mattsson: *De magia naturali*. Academia Aboensis: Turku 1645.

Bartholin, Thomas: *Historiarum anatomicarum rariorum, centuria III et IV*. Vlacq: Hagae-Comitis 1657.

Bilberg, Johan / Prosperius, Gustaf: *De occultis qualitatibus* [About occult qualities]. Academia Upsaliensis: Upsala 1687.

Chesnecopherus, Johan / Andreae, Andreas: *Disputatio physica de occultis qvalitatibvs*. Academia Upsaliensis: Upsala 1625.

Erastus, Thomas: *De occultis pharmacorum potestatibus*. Perna: Basel 1574.

---

gnaviter collatae. Ad calcem adiectae sunt Notae Rudolphi Goclenii ad ipsum Physicos contextum pertinentes; Item Pneumatologia Snellii. Fischer: Francofurti 1596.

54 LoLordo, Antonia: "Gassendi and the Seventeenth-Century Atomists on Primary and Secondary Qualities." In: Nolan, Lawrence (ed.): *Primary and Secondary Qualities: The Historical and Ongoing Debate*. Oxford University Press: Oxford 2011, pp. 69–70.

Forer, Laurenz: *Disputatio philosophica de sympathia et antipathia.* Universitas Ingolstadiensis: Ingolstadt 1618.

Junius, Hadrianus: *Nomenclator omnium rerum propria nomina.* Egnolphus Emmelius: Francofurti 1620.

Lagermarck, Johan Svensson / Orraeus, Axelius Jonae: *De magia naturali.* Academia Aboensis: Turku 1648.

Libavius, Andreas: *D.O.M.A. Singularium Andreae Libauii ... pars secunda.* Impressa typis Ioannis Saurii: Francofurti 1599.

Lidén, Johan Henrik (ed.): *Catalogus disputationum, in academiis et gymnasiis Sueciae atque etiam a Suecis extra patriam habitarum.* Edman: Uppsala 1778.

Magirus, Johannes: *Johannis Magiri Physiologiae peripateticae libri sex, cum commentariis.* Johannis Berneri: Francofurti 1624.

May, Johannes Henricus: *Brevis et accurata animalium in sacro cum primis codice memoratorum historia.* Olffen: Francofurti 1686.

Paracelsus: *Paracelsus (Theophrastus Bombastus von Hohenheim, 1493-1541): Essential Theoretical Writings.* Brill: Leiden 2008.

Paré, Ambroise: *The Workes of That Famous Chirurgion Ambrose Parey.* R. Coates and W. Dugard: London 1649.

Plutarchus: *Opuscula Plutarchi Cheronei.* Ascensius: Paris 1521.

Porta, Giambattista della: *Phytognomonica octo libris contenta etc.* apud Horatium Salvianum: Neapoli 1588.

Riddermarck, Andreas / Eurelius, Torstanus: *Dissertatio inauguralis de magia.* Londini Scanorum: Lund 1685.

Sennert, Daniel: *Hypomnemata physica I. De rerum naturalium principiis, II. de occultis qualitatibus, III. de atomis & mistione, IV. de generatione, viventium, V. de spontaneo viventium ortu.* Schleich: Francofurti 1636.

Sjöberg, Gabriel / Westerman, Andreas: *Dissertatio philosophica de erroribus sensuum* [Philosophical dissertation about sense errors]. Academia Dorpatensis: Dorpat 1692.

Smet, Hendrik de: *Miscellanea Henrici Smetii.* Jonae Rhodii: Francofurti 1611.

Snellius, Rudolph / Valerius, Cornelius / Goclenius, Rudolph: *Rudolphi Snellii In Physicam Cornelii Valerii annotationes: Cum lectissimis aliorum observationibus gnaviter collatae. Ad calcem adiectae sunt Notae Rudolphi Goclenii ad ipsum Physicos contextum pertinentes; Item Pneumatologia Snellii.* Fischer: Francofurti 1596.

Sperling, Johann: *Institutiones physicae,* 2nd ed. Johannis Bergeri: Witteberga 1649.

Sperling, Johannes / Willichius, Jodocus: *Disputatio physica de qualitatibus occultis.* Academia Wittebergensis: Witteberga 1639.

Literature

Allen, Michael J.B.: "At Variance: Marsilio Ficino, Platonism and Heresy." In: Hedley, Douglas / Hutton, Sarah (eds.): *Platonism at the Origins of Modernity: Studies on Platonism and Early Modern Philosophy*. (Archives Internationales d'histoire Des Idées 196). Springer: Dordrecht 2008, pp. 31–44. DOI:10.1007/978-1-4020-6407-4_3.

Buchwald, Jed Z. / Feingold, Mordechai: *Newton and the Origin of Civilization*. Princeton University Press: Princeton 2013. DOI:10.1515/9781400845187.

Cohen, Sheldon G. / Samter, Max: *Excerpts from Classics in Allergy*, 3rd ed. Symposia Foundation: Carlsbad, CA 2012. DOI:10.1002/art.1780360529.

Copenhaver, Brian P.: *Magic in Western Culture: From Antiquity to the Enlightenment*. Cambridge University Press: Cambridge 2015. DOI:10.1017/CBO9781107707450.

Copenhaver, Brian P.: "A Tale of Two Fishes: Magical Objects in Natural History from Antiquity Through the Scientific Revolution". *Journal of the History of Ideas* 52(3), pp. 373–398. DOI:10.2307/2710043.

Copenhaver, Brian P.: "Natural Magic, Hermetism, and Occultism in Early Modern Science." In: Lindberg, David C. / Westman, Robert S. (eds.): *Reappraisals of the Scientific Revolution*. Cambridge University Press: Cambridge 1990, pp. 261–302.

Copleston, Frederick Charles: *A History of Philosophy, Volume 3: Late Medieval and Renaissance Philosophy: Ockham, Francis Bacon, and the Beginning of the Modern World*. Image Books: New York 1993.

Deparis, Vincent / Legros, Hilaire / Souchay, Jean: "Investigations of Tides from the Antiquity to Laplace". In: Souchay, Jean / Mathis, Stéphane / Tokieda, Tadashi (eds.): *Tides in Astronomy and Astrophysics*. Springer: Berlin 2013, pp. 37–39. DOI:10.1007/978-3-642-32961-6_2.

Descartes, René: *The Philosophical Writings of Descartes*, vol.1. Cambridge University Press: Cambridge 1985.

Ducheyne, Steffen: "Newton's Training in the Aristotelian Textbook Tradition: From Effects to Causes and Back". *History of Science* 43, 2005, pp. 217–237. DOI:10.1177/007327530504300301.

Friedenthal, Meelis: "Kuri silm: toimemehhanismid lähtuvalt antiiksetest ja keskaegsetest tajuteooriatest". [The Evil Eye: Descriptions of Operation According to Ancient and Medieval Theories of Perception.] *Mäetagused* 51, 2012, pp. 7–20. DOI:10.7592/MT2012.51.friedenthal.

Gentilcore, David: *Food and Health in Early Modern Europe: Diet, Medicine and Society, 1450–1800*. Bloomsbury Academic: London 2016.

Hanegraaff, Wouter J.: "Beyond the Yates Paradigm: The Study of Western Esotericism Between Counterculture and New Complexity". *Aries* 1(1), 2001, pp. 5–37, DOI:10.1163/157005901X00020.

Harrison, Mark: "From Medical Astrology to Medical Astronomy: Sol-Lunar and Planetary Theories of Disease in British Medicine, c. 1700–1850". *The British Journal for the History of Science* 33(1), 2000, pp. 25–48. DOI:10.1017/S0007087499003854.

Hutchison, Keith: "What Happened to Occult Qualities in the Scientific Revolution?" *Isis* 73(2), 1982, pp. 233–253. DOI:10.1086/352971.

Jaanson, Ene-Lille (ed.): *Tartu Ülikooli trükikoda 1632–1710: Ajalugu ja trükiste bibliograafia = Druckerei der Universität Dorpat 1632–1710: Geschichte und Bibliographie der Druckschriften*. Tartu Ülikooli Raamatukogu: Tartu 2000.

Ker, Walter Charles Alan (trans.): *Martial, Epigrams. I.* W. Heinemann / Harvard University Press: London et al. 1961.

Kusukawa, Sachiko: "Nature's Regularity in Some Protestant Natural Philosophy Textbooks 1530–1630." In: Stolleis, Michael / Daston, Lorraine (eds.): *Natural Law and Laws of Nature in Early Modern Europe: Jurisprudence, Theology, Moral and Natural Philosophy*. Ashgate Publishing: Abingdon 2008, pp. 105–122. DOI:10.4324/9781315597522.

Levack, Brian P.: *The Witch-Hunt in Early Modern Europe*, 3rd ed. Routledge: London 2006. DOI:10.4324/9781315685526.

Clark, Stuart: *Thinking with Demons: The Idea of Witchcraft in Early Modern Europe*. Oxford University Press: Oxford 2006. DOI:10.1093/acprof:oso/9780198208082.001.0001.

Lindemann, Mary: *Medicine and Society in Early Modern Europe*. Cambridge University Press: Cambridge 2010.

LoLordo, Antonia: "Gassendi and the Seventeenth-Century Atomists on Primary and Secondary Qualities." In: Nolan, Lawrence (ed.): *Primary and Secondary Qualities: The Historical and Ongoing Debate*. Oxford University Press: Oxford 2011, pp. 69–70. DOI:10.1093/acprof:oso/9780199556151.003.0004.

Pasnau, Robert: "Scholastic Qualities, Primary and Secondary." In: Nolan, Lawrence (ed.): *Primary and Secondary Qualities: The Historical and Ongoing Debate*. Oxford University Press: Oxford 2011, pp. 41–61. DOI:10.1093/acprof:oso/9780199556151.001.0001.

Paster, Gail Kern: "Melancholy Cats, Lugged Bears, and Early Modern Cosmology: Reading Shakespeare's Psychological Materialism Across the Species Barrier." In: Paster, Gail Kern / Rowe, Katherine / Floyd-Wilson, Mary (eds.): *Reading the Early Modern Passions: Essays in the Cultural History of*

Emotion. University of Pennsylvania Press: Philadelphia 2004, pp. 113–129. DOI:10.7208/chicago/9780226648484.003.0004.

Raber, Karen: "How to Do Things with Animals: Thoughts on/with the Early Modern Cat". In: Hallock, Thomas / Raber, Karen / Kamps, Ivo (eds.): *Early Modern Ecostudies*. Palgrave Macmillan: New York 2008, pp. 93–113. DOI:10.1057/9780230617940_6.

Ring, Johannes: "History of Allergy in Middle Ages and Renaissance". In: Bergmann, K.-C. / Ring, Johannes (eds.): *History of Allergy*. Karger Medical and Scientific Publishers: Basel 2014, pp. 15–20. DOI:10.1159/000358469.

Schadewaldt, Hans: *Idiosynkrasie, Anaphylaxie, Allergie, Atopie: Ein Beitrag Zur Geschichte Der Überempfindlichkeitskrankheiten*. (Vorträge G 251). Westdeutscher Verlag: Opladen 1981.

Spruit, Leen: *Species Intelligibilis – from Perception to Knowledge. Renaissance Controversies, Later Scholasticism, and the Elimination of the Intelligible Species in Modern Philosophy*, vol. 2. (Brill's Studies in Intellectual History 49). Brill: Leiden 1995. DOI:10.1163/9789004247000.

Vallinkoski, Jorma: *Turun akatemian väitöskirjat 1642–1828 = Die Dissertationen der alten Universität Turku (Academia Aboënsis) 1642–1828 I*. Valtioneuvoston kirjapaino: Helsinki 1962.

Van der Laan, James M. / Weeks, Andrew (eds.): *The Faustian Century: German Literature and Culture in the Age of Luther and Faustus*. (Studies in German Literature, Linguistics, and Culture). Camden House: Rochester, NY 2013.

Walker, Daniel Pickering: *Spiritual and Demonic Magic: From Ficino to Campanella*. The Pennsylvania State University Press: University Park, PA 2000.

Wels, Volkhard: *Manifestationen des Geistes: Frömmigkeit, Spiritualismus und Dichtung in der Frühen Neuzeit*. V & R unipress GmbH: Göttingen 2014.

Wood, Neal: "Tabula Rasa, Social Environmentalism, and the 'English Paradigm'". *Journal of the History of Ideas* 53(4) 1992, pp. 647–668. DOI:10.2307/2709942.

Yates, Frances Amelia: *Giordano Bruno and the Hermetic Tradition*. Routledge: London 1964.

Jaanika Anderson / Hilkka Hiiop

Fashion or Conceptual Choice: The Motifs of Animals, Birds, and Semi-Animals in Pompeian-Style Interiors in Estonia

Abstract: Since the eighteenth century, excavations have been carried out in the ancient cities of Pompeii, Herculaneum and their surrounding area and gradually a forgotten world has opened up to people. Pompeian-style wall paintings became fashionable and enjoyed peak popularity from the second half of the eighteenth century to the first half of the nineteenth century, and continued to be in vogue all over the world even in the early twentieth century. This fashion did not leave the region of Estonian area untouched either. The chapter is inspired by the fascinating discoveries of Pompeian-style murals during the investigation and conservation works carried out in Estonian manor houses during the last few decades, as well as the interior of the University of Tartu Art Museum in the university's main building, that was never hidden from people's gaze. The article focuses on the surprising findings of figurative paintings of animals, birds, insects and semi-animals in the Pompeian style. These examples are important in the context of the reception and appropriation of ancient art in the visual culture of Estonia.

Keywords: art of Classical Antiquity, Baltic German, mural painting, reception of art, manor houses

Introduction

The centrality of animals to human cultures is nothing new. As long ago as 30,000 BC, Paleolithic humans were expressing their awe and admiration for animals by painting their images on the walls of caves.[1] The lives of animals and humans were intertwined almost in every era in the ancient world: animals were loved as pets, watched in public shows, and used for all kinds of work and transportation, sacrificed to gods, and hunted as food. Visual material – artwork can illustrate the different relationships of humans and animals, and the attitudes of humans toward animals.[2] The history of ancient knowledge about terrestrial and marine animals as well as birds and insects has reached us via written

---

1 Kalof, Linda: "Ancient Animals". In: Kalof, Linda (ed.): *A Cultural History of Animals in Antiquity*. Berg: Oxford, New York 2011, pp. 1–16, here p. 1.
2 See more about the animals in antiquity in Fögen, Thorsten / Thomas, Edmund (eds.): *Interaction Between Animals and Humans in Graeco-Roman Antiquity*. De Gruyter: Berlin / Boston 2017, pp. 3–4; Lewis, Sian / Llewellyn-Jones, Lloyd

sources such as Pliny the Elder's *Natural History* (first century AD), which offers a thorough description of wildlife.[3] Natural disasters have also preserved for today several visual and material sources about animals and natural history as well as mythological scenes with gods, heroes and animals, semi-animals, and other creatures – Mount Vesuvius erupted in AD 79 and buried the area by the bay of Naples for centuries, conserving amounts of art, culture, and everyday life for further generations.

From the eighteenth century ancient cities such as Pompeii, Herculaneum, and their surrounding areas have been excavated, and a forgotten world has opened to the people step by step.[4] Today, the artefacts from the buried cities are preserved in the sites and museums (primarily Naples National Archaeological museum), and visualisations of large numbers of excavated objects were published in the books or as sheets of prints immediately after discovery.[5] The rediscovered Roman interior design, which had been buried for centuries under the volcanic settlements, fascinated people due to its colourful murals. Pompeian-style wall paintings came into fashion and enjoyed peak popularity from the second half of the eighteenth century to the first half of the nineteenth century, and continued to be en vogue even in the early twentieth century.[6]

The Romans were often derided as imitators of their predecessors even in antiquity. Scholars have now revised the picture and appreciate the Romans for their originality and their role as innovators in art and architecture.[7] The article is inspired by the fascinating discoveries of the Pompeian-style murals during the

---

(eds.): *The Culture of Animals in Antiquity: A Sourcebook with Commentaries*. Routledge: New York 2018.

3 See books 8–11 in Pliny the Elder: *Natural History*. Retrieved 15.1.2019, from http://www.perseus.tufts.edu/hopper/text?doc=Plin.+Nat.+7&fromdoc=Perseus%3Atext%3A1999.02.0138.

4 Amery, Colin / Curran, Brian: *The Lost World of Pompeii*. Frances Lincoln: Los Angeles 2011, pp. 30–47.

5 For example, the earliest publication series was *Le Antichità di Ercolano Esposte*, Parts 1–8. Regia Stamperia: Napoli 1757–1792.

6 See more for the reception and afterlife of Pompeian murals and architecture, Stackelberg, Katherine T. / Macaulay-Lewis, Elizabeth (eds.): *Housing the New Romans. Architectural Reception and Classical Style in the Modern World*. Oxford University Press: Oxford 2017.

7 Stackelberg, Katherine T. / Macaulay-Lewis, Elizabeth: "Introduction. Architectural Reception and the Neo-Antique". In: von Stackelberg, Katherine T. / Macaulay-Lewis, Elizabeth (eds.): *Housing the New Romans. Architectural Reception and Classical Style in the Modern World*. Oxford: Oxford University Press 2017, pp. 1–3.

investigation and conservation works[8] done in Estonian manor houses during the last few decades, as well as the never "hidden" interior of the University of Tartu Art Museum in the university's main building[9]. The article focuses on the surprising findings of the figurative paintings of animals, birds, insects, and semi-animals in Pompeian-style interiors in the manor houses of Vana-Võidu and Suure-Kõpu and the public interior of the University of Tartu Art Museum in Estonia.

These three cases are appropriations of the murals from Pompeian and Herculaneum houses. Those figurative motifs differ from murals inspired by antiquity found in other Estonian manors in their exactness in copying the originals.[10] Usually the murals are rather inspired by ancient models – neither accurate nor customised copies of original paintings – and are mainly represented in ornamental or architectonic style. In some rare cases ancient themes were used as a topic in the manors, but these are never exact copies of some Pompeian-style mural.[11]

The present paper is also a part of studies about the reception and appropriation of ancient visual culture in Estonia.[12] Reproductions of motifs of ancient animal representations are not observed specifically in Estonia and there is no corresponding literature on this particular topic. However, it is a peculiar and interesting phenomenon from the nineteenth century that deserves involvement in research of Estonian and even broader European visual cultures.

The objective of the following article is to find the direct models and meanings of the copies of Pompeian murals on the walls of Estonian interiors, to provide

---

8   Hiiop, Hilkka et al.: *Vana-Võidu mõisa peahoone söögisaali ning ümbritsevate ruumide (suur saal, vestibüül, ruumid 208/210) siseviimistluse uuringud ja osaline restaureerimine. Aruanne.* Eesti Kunstiakadeemia / H&M Restuudio: Tallinn 2016. Manuscript at the Archive of National Heritage Board, A-13143.
9   The museum was founded in 1803, the interior painting was made in 1868. See more in Kukk, Inge: "Vervielwältigtes Pompeji – von Graphikblättern bis zur Wanddekoration". *Baltic Journal of Art History* 3, 2011, pp. 181–213.
10 Hiiop, Hilkka: "In the Footsteps of Classical Antiquity. Influences of the Antique in Estonian Manor Murals". *Baltic Journal of Art History* 3, 2011, pp. 225–252.
11 See more in Hiiop, Hilkka "Antiigi peegeldus Eesti mõisamaalingutel = A reflection of Antiquity in the Paintings in Estonian Manors". In: *Eesti kunsti ajalugu. 3, 1770–1840. = History of Estonian Art. 3, 1770–1840.* Eesti Kunstiakadeemia: Tallinn 2017, pp. 288–298.
12 There is only one study about the reception of ancient visual heritage in Estonia, and it rather concerns collecting and museum issues. Anderson, Jaanika: *Reception of Ancient Art: The Cast Collections of the University of Tartu Art Museum in the Historical, Ideological and Academic Context of Europe (1803–1918).* University of Tartu Press: Tartu 2015.

their analysis in the aspect of original murals, their appropriation, allegorical meanings, and reception. In addition, the reasons for making copies of the antique interiors far from their original location are discussed. The article does not set ambition to analyse those murals in a broader context, but rather opens up the topic and provides information for further research.

At First There Were Pompeii and Herculaneum

Pompeii and Herculaneum, situated by the Bay of Naples, were buried under the volcanic ash during an eruption of Vesuvius in 79 AD. The extensive and systematic excavations began in the middle of the eighteenth century and the whole world was fascinated with the rediscovered domestic architecture with its paintings, mosaics, and interior decoration. In the second half of the eighteenth century there was huge public interest in Pompeii, and visiting the excavated site was an essential element of the Grand Tour, and insight into ancient life was a factor in the development of neoclassicism.[13]

From the second half of the eighteenth century new knowledge about ancient Rome influenced the visual culture throughout Europe and beyond[14], causing a real boom in the interpretation, adaptation, imitation and replication of antique architecture, interiors, applied art, and literature. Antiquity became an indirect source of inspiration for European visual culture, and was also directly quoted and copied.[15] The newly excavated houses of prominent Romans were often decorated with paintings and sculptures, like museums, and served as inspiration for modern people.[16] Art books and engraving techniques became the main mediators of the new discoveries.

Antiquity as a style may be an idealisation of the past. It may even be an attempt to dress up modern ideas in antique clothing. This reception of antiquity

---

13 Jones, Owen: *The Grammar of Ornament: A Visual Reference of Form and Colour in Architecture and the Decorative Arts*. Princeton University Press: Princeton and Oxford 2016, p. 120.
14 See more about Pompeian style interiors in America in Nichols, Marden Fitzpatrick: "Domestic Interiors, National Concerns. The Pompeian Style in the United States". In: von Stackelberg, Katherine T. / Macaulay-Lewis, Elizabeth (eds.): *Housing the New Romans. Architectural Reception and Classical Style in the Modern World*. Oxford: Oxford University Press 2017, pp. 126–152.
15 Werner, Peter: *Pompeji und die Wanddekoration der Goethezeit*. Wilhelm Fink Verlag: München 1970, p. 14.
16 Pappalardo, Umberto: *The Splendor of Roman Wall Painting*. The Paul Getty Museum: Los Angeles 2008, p. 12.

often includes a utopian aspect, whether looking to the past or the future. In many instances, it is possible to point to ancient models, but their influence may be more adaptive or eclectic.[17] Furthermore, it is important to remember that the reception of antiquity does not always follow a straight line from antiquity to the creator of a product inspired by antiquity but it can be a copy or appropriation of the copy.[18] When we speak about copying ancient art, including Pompeian interior design, we should keep in mind that even a copy is not normally completely identical with the original work of art. Differences can lie in the material, technique, integrity, size, style as well as in the details of the copy or combination of several aspects. At the same time the meaning of the original artwork can change or disappear in the process.[19]

Even Estonia, divided between the Estonian and Livonian Provinces of the Russian Empire in the nineteenth century, far away from the heart of Europe, was influenced by the interior décor discovered in the ancient cities of Italy. There are several interiors that have got inspiration from antiquity: the colours, decorative schemes, motifs or combinations thereof. The University of Tartu Art Museum, as well as the interiors of several private town and manor houses owned by the Baltic German nobility, were decorated with Pompeian-style wall paintings. Although the Baltic Germans travelled in Europe and visited the ancient cities, they were also often exposed to antique murals when they visited Germany, or the imperial capital St. Petersburg, which was the closest city brimming with ideas. A number of Baltic Germans served in the military and state administration at the Russian imperial court. Therefore, it is possible to follow the movement of the fashion from Russia to the Baltic provinces.[20] In addition, different printed sources were available to provide inspiration for decorating interiors in the Pompeian style.

---

17 Mathiesen, Hans Erik: "The Reception of Classical Antiquity: Some General Remarks". In: Nielsen, Marjatta (ed.): *The Classical Heritage in Nordic Art and Architecture*. Acts of the seminar held at the University Copenhagen 1st–3rd November 1988. (Acta Hyperborea 2). Museum Tusculanum Press: Copenhagen 1990, pp. 19–24, p. 20.
18 Mathiesen 1988, p. 23.
19 In this point of view the art copying is comparable to the translation of literary text. See in Anderson, Jaanika / Lotman, Maria-Kristiina: "Intrasemiotic Translation in the Emulations of Ancient Art: On the Example of the Collections of the University of Tartu Art Museum". *Semiotica* 222, 2018, pp. 1–24. DOI:10.1515/sem-2016-0118.
20 See more in Hellermann, Dorothee von: "Candidatus Pictura. Gottlieb Welté als Künstler des Rokoko und Klassizismus". In: Polli, Kadi / Raudsepp, Renita (eds.): *Maarjamaa rokokoo = Rokoko in Estland. Gottlieb Welté (1745/49–1792)*. Eesti Kunstimuuseum: Tallinn 2007, pp. 70–103, here p. 78.

The Triple Pompejanum of the von Stryk Family and the Fauna on the Walls

The three manor houses of the family von Stryk located in today's southern Estonia are united by the fact that during the last decades, murals in the Pompeian style have been discovered in all of them and that these have been partially or fully revealed and restored. However, those murals are similar by the style of painting, but differ by the conception. On two of them animals, birds and semi-animals are depicted that enrich the Estonian visual culture with the foreign animal world.[21] Representation of the animals or other creatures is rather unusual in the interior design of nineteenth-century Estonia.

Baltic German manor owner Bernhard Heinrich von Stryk served in the Army of the Russian Empire and retired as a lieutenant in 1768. Thereupon he bought several manors in Livonia.[22] He was blessed with 11 children and three of his sons became landlords of estates. Bernhard Heinrich bequeathed the Suure-Kõpu (Ger. Groß-Köppo) manor in Viljandi County to his eldest son Alexander Georg Gottlieb von Stryk (1787 or 1788–1845). After 1836 Alexander Georg began constructing a new two-storied late neoclassical manor house – one of the most prominent of its kind in southern Estonia.[23] Suure-Kõpu manor, designed by Emil Julius Strauss, the city architect of Jelgava (Ger. Mitau),[24] was completed in 1847, after Alexander Georg's death.[25] The dining room was probably decorated with huge centaurs and in the Pompeian style at the very end of the 1850s or 1860s.

---

21   Anderson, Jaanika / Hiiop, Hilkka: "The triple Pompejanum possessed by the von Stryk family: the manor houses of Vana-Võidu, Suure-Kõpu and Voltveti". *Baltic Journal of Art History* 13, 2017, pp. 165–192.

22   See for the history of the family Stryk in Stryk, Wolf-Dietmar von: *Riidaja von Strykide perekonna põliskodu Liivi- ja Eestimaal 1562–1919–2003: Ühe mõisa ajalugu sajandite tagant tänase päevani*. Põdrala vallavalitsus: Riidaja 2008, p. 35.

23   See about restoration in Volmer, Svea / Hiiop, Hilkka: "Salapärane ja ihaldatud Pompei". In: Matteus, Kais (ed.) *Muinsuskaitse aastaraamat 2007*. Muinsuskaitseamet, Tallinna Kultuuriväärtuste Amet, EKA muinsuskaitse ja restaureerimise osakond: Tallinn 2008, pp. 56–60.

24   Emil Julius Strauss was also the architect of the manor house of Alexander von Oettingen (1798–1846) in Courland (*Jensel*). The late neoclassical manor house was built in 1836–1844 and it is remarkably similar in appearance to the manor house of Suure-Kõpu. Ants Hein suggests that Strauss was also associated with the building of the Vana-Võidu manor house. Hein, Ants: *Viljandimaa mõisad*. Hattorpe: Viljandi 1999, p. 65.

25   Hein, Ants: *Eesti mõisaarhitektuur. Historitsismist juugendini*. Hattorpe: Tallinn 2003, p. 73.

Georg Constantin von Stryk (1797–1886), the youngest son of Bernhard Heinrich, purchased the Vana-Võidu manor (Ger. Alt-Woidoma), also situated in Viljandi County, in 1834. He wanted to make his new property more presentable,[26] as did his brother Alexander Georg in Suure-Kõpu and his closest friend and the brother of his wife Alexander von Oettingen in Courland. The manor house at Vana-Võidu was built in the 1840s and the architect was probably the aforementioned Emil Julius Strauss. Most likely the Pompeian-style murals in the small hall, obviously the dining room, were painted before 1864.

Ancient models have also been used for the interior paintings and décor of a third manor, which is the Voltveti or Tihemetsa (Ger. Tignitz) estate in Pärnu County. But the reception of antiquity has gone further and the interior décor actually bears a greater resemblance to the Adam style.[27] The style is named after three Scottish brothers, who introduced (1760s) the neoclassical interior design in which classical Roman decorative motifs were dominant, instead of presenting an exact copy of original ornaments.[28]

Centaurs of the Villa Cicero in a Southern Estonian Dining Room

The most interesting part of the Suure-Kõpu manor building in the context of Estonian animal history is the interior décor, especially the series of figurative Pompeian-style paintings in the former dining hall, which is unique in the manorial architecture of Estonia. The wall paintings were rediscovered in 2003 by the conservators and were fully revealed in 2006.[29] Under layers of paint, an enchanting Pompeian-style black-and-red interior was gradually revealed.

Three figurative scenes with the centaurs have survived in their entirety, one on the southern and two on the western wall. The door in the western hall was enlarged and thus the fourth panel lost a considerable part of its painted surface. The demolition of the northern wall probably destroyed one of the panels, which corresponded symmetrically to the painting on the southern wall. During the conservation project, the destructed wall, cavetto vault and architectonic division of the décor were reconstructed, although the missing figures in the painted panels were not repainted, even though the model was known (Figure 1).

---

26   Hein 1999, pp. 73–74.
27   The possible models for murals of the manor Voltveti may come from Zahn's album Zahn, Wilhelm: *Ornamente aller klassischen Kunst-Epochen nach den Originalen in ihren eigenthümlichen Farben*. Reimer: Berlin 1849, p. 145.
28   Anderson / Hiiop, pp. 173–175.
29   Hiiop 2011, pp. 243–244.

Figure 1. (a) Murals with centaurs in the dining room of Suure-Kõpu manor. Photo: Toomas Vendelin / Royal Norwegian Embassy, 2011. (b) Zahn, Wilhelm: *Die Schönsten Ornamente und merkwürdigsten Gemälde aus Pompeji, Herculaneum und Stabiae*. Band 3, Heft 8. Reimer: Berlin. 1852–1859, T. 74.[30]

---

30 Collection of the University of Tartu Art Museum, KMM GR 7663.

The mythological scenes depicting centaurs are copies of paintings from the dining room (Lat. *triclinium*) of the Villa Cicero in Pompeii. The originals were excavated and removed from the villa in 1749. Since that time they have been stored in the National Archaeological Museum in Naples.[31] The murals of the Villa Cicero inspired many eighteenth and nineteenth-century artists, who produced numerous black-and-white, and coloured, graphic copies. The models for the Suure-Kõpu artist supposedly come from Wilhelm Zahn's colourful albums.[32]

In the Villa Cicero, as well in the Suure-Kõpu Manor, the scenes with centaurs have a black background. But the figures of the centaurs in Suure-Kõpu are significantly larger than the originals in Villa Cicero. It is possible that von Stryk or the painter had not actually seen the original paintings or he preferred consciously to present the centaurs more dominantly in the manor's representative room to increase the spectacularity. However, the model engravings are very detailed and allowed the portrayal of the centaurs very precisely. In addition to the three centaurs, fragments of the central painting with a figure's head and wings on the western wall have also survived. The scene depicts Victoria, the Roman goddess of victory, with a spear and shield. The mural was excavated in *Casa dei Principi Russi* in Strada Stabiae in 1852.[33]

The overall composition, with its architectural articulations, caryatids between the windows, marbled Corinthian orders dividing the panels, painted dentil moulding that creates an optical spatial effect, and the imitation graining on the ceiling, come from the artist's own imagination.[34] The artistic quality of work suggests that it could not have been executed by a local master. This is also suggested by the hypothetical self-portrait of the artist, which is concealed in the marbling of a pillar framing a field of the painting, and seems to portray a southern type.[35] At any rate, it is remarkable that the Villa Cicero's small figurative scenes ended up as the large-scale elements of wall décor in an Estonian manor house interior.

---

31 Monaco, Domenico: *Les monuments du Musée National de Naples gravés par les meilleurs artistes Italiens*. Museo Nazionale di Napoli: Napoli 1877, p. 8, T. 28, T. 29, T. 30, T. 31; Nava, Maria Luisa / Paris, Rita /Friggeri, Rosanna (eds.): *Rosso Pompeiano. La decorazione pittorica nelle collezioni del Museo di Napoli e a Pompei. Catalogo della mostra (Roma, 20 dicembre 2007–31 marzo 2008)*. Mondadori Electa: Milano 2008, pp. 161–162.

32 Zahn, Wilhelm: *Die schönsten Ornamente und merkwürdisgsten Gemälde aus Pompeji, Herkulaneum und Stabiae: nebst einigen Grundrissen und Ansichten nach den an Ort und Stelle gemachten Originalzeichnungen*. Reimer: Berlin 1852–1859.

33 Zahn 1852–1859, T. 94.

34 Hiiop 2011, p. 246.

35 Hiiop 2017, p. 296.

Domesticated Creatures on the Dining Room Walls

In 2016, research was carried out in the Vana-Võidu manor to identify the original decorations under the posterior paint layers in the dining hall. The discovered murals copy the typical ancient wall diagram: a lower part of the wall in the bottom zone (here, marbled); large dark coloured panels in the central zone; and colourful painted figural compositions in the upper zone atop the windows and the door. Eight figural scenes with animals are placed into triangular frontons: motifs with birds and insects, a pigeon holding a letter, a parrot with a basket of treasure, a cat catching a bird; and four fields depicting a chariot driven by different creatures, such as a butterfly and beetle. By contrast, compared with the dado and panels, the figurative paintings in the zone of the frieze are unexpectedly well-preserved.[36] The fauna is painted with such great sensitivity and accuracy that identification of individual species is possible in some cases – the phenomenon was already known in Ancient Rome, where some mosaics were so precisely made that it is even today possible to determine the depicted species.[37]

One of the most complete and best-preserved motifs, depicting a parrot,[38] a chariot and a grasshopper, is atop the eastern door (Figure 2). On the mural of Vana-Võidu the parrot is pulling the chariot while the grasshopper is the charioteer. An example of this motif – a chariot-driving grasshopper with a parrot – was also found on 10 October 1745 in the souterrain of Resina, Herculaneum. It was detached, removed, and exhibited at the Naples National Archaeological Museum, but it no longer exists.[39] The motif of the painting was published as an engraving with text for the first time twelve years after the excavation[40]. The engraving is of high quality and the accompanying text, but

---

36 Hiiop et al. 2016, p. 14.
37 Harden, Alastair: "Animal in Classical Art". In: Campell, Gordon Lindsay (ed.): The Oxford Handbook of Animals in Classical Thought and Life. Oxford University Press: Oxford 2014, pp. 24–60, here p. 55.
38 The depiction of the birds has been thoroughly studied in Tammisto, Antero: Birds in mosaics: a study on the representation of birds in Hellenistic and Romano-Campanian tessellated mosaics to the early Augustan Age. Institutum Romanum Finlandiae, Rome 1997.
39 Jashemski, Wilhelmina Feemster / Meyer, Frederick G. (eds.): The Natural History of Pompeii. Cambridge University Press: Cambridge 2002, p. 232.
40 Le Antichità di Ercolano Esposte, Vol 1. Naples 1757, pp. 245–247. In 1782, the same motif was published once again in France – Saint-Non, Jean Claude Richard de: Voyage Pittoresque Ou Description Des Royaumes De Naples Et De Sicile (Band 2): Contenant Une Description des Antiquités d'Herculanum, des Plans & des Détails de son Théâtre, avec une Notice abrégé des différents Spectacles des Anciens. Les Antiquités de Pompeï [...]. Clousier: Paris 1782, 14a.

the book lacks the information on context that would be expected in today's archaeological work. The publication was designed more to impress its readers with the quality of the objects in the King of Naples' collection than to be used for research. The book provided impetus to the neoclassical movement in Europe by giving artists and decorators access to a huge store of ancient motifs. In this case, it is possible that a copy in Vana-Võidu is the only existing colour reproduction of the motif. Only black-and-white images have been preserved by the engravings in early publications. Therefore, we do not know the original colours of this figurative painting.

Figure 2. (a) The motif with the parrot and grasshopper above the dining-room door of the Vana-Võidu manor. Photo: Peeter Säre, 2017. (b) *Le Antichità di Ercolano Esposte*. Vol 1. Naples. 1757, p. 247.

The wall painting, a parrot drawing a go-cart driven by a grasshopper, is caricature-like. Rainbow colours, a squawk, a powerful hooked bill and the ability to mimic the human voice are hallmark traits of the parrot species that have fascinated people since antiquity. The ancient Greeks brought parrots from India – possibly both the Rose-Ringed Parakeet, which is depicted on the Vana-Võidu wall-painting, and the larger Alexandrine Parakeet. The ancient Romans kept both Rose-Ringed Parakeets and African Greys. The parrot was an expensive, much-loved household pet in Rome, because they had human tongues, and could be trained to speak by professional trainers. As well as everyday sayings, the parrots' vocabularies included greetings and obscenities. They are highly trainable, dexterous and acrobatic, using their powerful bill and zygodactyl toes to grasp and climb in search of fruit, seeds and buds in the wild.[41]

The Romans enjoyed teaching their pets tricks.[42] However an image preserved in lava from ancient Herculaneum, depicting a miniature chariot drawn by a parrot and driven by a grasshopper, was originally meant as satire, i.e. the stronger being driven by the weaker. It has been said that this referred to the influence that Seneca had over Nero.[43] It can be interpreted as a satiric allegory of Emperor Nero, who said that one of his principal talents was singing in public theatres and who placed greater importance on being a good charioteer than on running an empire, and of the philosopher Seneca, who is characterised as being strong in rhetoric but weak in action.[44] Others, with more reason, have

---

41 Warren Chadd, Rachel / Taylor, Marianne: *Birds: Myth, Lore and Legend*. Bloomsbury Publishing: London et al. 2016, p. 150.
42 For training animals and birds in ancient Rome explore further in Hammer, Jacob: "Trained animals in antiquity: a portion of a paper". *The Classical Outlook* 20(6), 1943, pp. 59–61.
43 Nero (37–68) became Roman Emperor at the age of 17. In the first year of his reign he was greatly influenced by his mother Agrippina and tutor Lucius Annaeus Seneca. Nero preferred Seneca's advice to his mother's.
44 Streeter, Colin: "Two Carved Reliefs by Aubert Parent". *The J. Paul Getty Museum Journal* 13, 1985, p. 57. The motif is also known from a vase carved by Aubert Parent (84.SD.76) acquired by the J. Paul Getty Museum (1789). A grasshopper in a triumphal chariot drawn by a parrot (*psittacus*) is depicted.

associated the grasshopper with the famous witch Locusta,[45] who provided Nero with the poison to murder his stepbrother Britannicus.[46] The chariot depictions may also be parodying the triumphant entries of rulers into Rome.[47]

Another motif with a chariot is fragmentary – the winged insect seems to be a charioteer whose chariot is being driven by a creature that also has wings. The original painting is preserved in the Naples National Archaeological Museum and the motif is published along with the parrot and the grasshopper – with the body, tail and legs of a lion and wings of an eagle griffin harnessed to the chariot.[48] On the model sheet is a winged object, which is without a doubt a butterfly, but on the wall of the Vana-Võidu manor house, it has taken more of the shape of a European mantis in flying position (Figure 3).

---

45 Compare the Latin *locusta* (f) with the English 'locust'. *Locusta* is also the Latin name for a certain species of short-horned grasshopper.
46 Monaco, Domenico: *A Complete Handbook to the National Museum in Naples*. William Clowes and Sons: London 1883, p. 3.
47 Jashemski / Meyer, p. 323.
48 Niccolini, Fausto / Niccolini, Felice: *Le case ed i monumenti di Pompei disegnati e descritti*, Band 3. Napoli 1890, T LVb. Also see the plate with a motif of a griffin alongside a motif of a parrot and grasshopper *Raccolta delle più interessanti dipinture e de più belli musaici rinvenuti negli scavi: di Ercolano, di Pompei, e di Stabia / che ammiransi nel Museo reale borbonico*. Museo nazionale di Napoli: Napoli 1843, p. 200.

Figure 3. (a) Chariot with griffin (partially destroyed) in the Vana-Võidu manor. Photo: Peeter Säre, 2017. (b) *Raccolta delle piu interessanti dipinture e de'piu belli musaici rinvenuti negli scavi di Ercolano, di Pompei, e di Stabia che ammiransi nel Museo reale borbonico*. Museo nazionale di Napoli. 1843, p. 200.

In addition to the scenes with chariots, there are three motifs with birds in the room. Another bird that could also be a pigeon has found its way to Vana-Võidu: a grey one standing on the edge of a jewellery basket holding a blue medallion in its beak. A model for the motif exists in Pompeii – in a floor mosaic in *casa del Fauno*.[49] Recently a figurative motif depicting a brownish cat catching a bird has been found. Unfortunately the figure of the bird is not preserved due to the renovation works in the manor house. The model for the motif of a cat catching a bird is found in Pompeii from *casa del Fauno* in the form of the floor mosaic[50] (Figure 4).

---

49 Niccolini, Fausto / Niccolini, Felice: *Le case ed i monumenti di Pompei disegnati e descritti*, Band 1. Napoli 1854, T II.
50 Niccolini / Niccolini 1854, T II; Monaco 1877.

Figure 4. (a) The motif of the bird with a jewellery basket and the cat catching a bird (partially destroyed) in the dining room of the Vana-Võidu manor. Photos: Peeter Säre, 2017, and the Estonian Academy of Arts, 2018, respectively. (b) Niccolini, Fausto; Niccolini, Felice. *Le case ed i monumenti di Pompei disegnati e descritti* Band 1, Neapel, 1854, T II. (c) Original mosaic from *Casa del Fauno* in Pompeii. Photo: Wikimedia Commons.

A couple of motifs have their models yet unidentified, such as a motif with a chariot that depicts a butterfly as the charioteer who is being pulled by an abstractly depicted stag beetle, and a motif with a pigeon and a sealed letter. Although pigeons were often shown on ancient mosaics and can be found on coins and gems, the images of pigeons mediating messages were rather rare but not unknown in antiquity. In 43 BC, when Decimus Brutus was besieged in Mutina by Antony, the pigeons (according to Frontinus Strat., III, 13, 7–8) took messages to Brutus from the consul Hirtius, who was preparing to relieve the town[51] (Figure 5).

Figure 5. (a) The fragment of the painting with a beetle in the Vana-Võidu manor (partially destroyed); (b) Pigeon with a letter above the dining room window of the Vana-Võidu manor. Photos: Peeter Säre, 2017.

---

51 It seems that the only known case of pigeons being used by Romans for communication outside of Egypt is Jennisson, George: *Animals for Show and Pleasure in Ancient Rome*. Manchester University Press: Manchester 1937, p. 103.

The visual sources for murals, fine painting techniques, and skilful style are very similar on the wall paintings of Suure-Kõpu and Vana-Võidu.[52] The Stryk brothers also built their manor houses at the same time span, so it is possible they also used the same master for the paintings. The wall paintings have a similar and elaborately retouched style, inherent more to the easel painting than wall decorations. Other similar elements include the marbled dado and dark wall panels. Suure-Kõpu motifs were painted after the murals of Pompeian heritage, and Vana-Võidu had murals and mosaic models at least partly from Herculaneum.[53] The choice of models is much clearer in the case of Suure-Kõpu – the elegant motifs of centaurs come from a single source, Villa Cicero. The choice of the figurative motifs in Vana-Võidu raises several questions, because there is no single source for the paintings. The models are mainly ancient, but in some cases seem to be quite modern ones (the pigeon with a letter, the beetle). There is no source that gives a clear answer to the questions of what the purpose of the scenes was, and if there was any special meaning of the motifs.

The diversity and abundance of creatures in the Vana-Võidu and huge centaurs in the Suure-Kõpu dining rooms raises several questions. What are the ideas behind the scenes? Why are many motifs combined in Vana-Võidu from different places and times? In the current research situation there is no definite answer to these and other questions. In addition, the transformation made in the process of reproducing a structure as a copy differs from the original model or context, and has an effect on the established reception of the original.[54]

Miniature Creatures – Academic Interior of the University of Tartu Art Museum

The Pompeian-style wall paintings of the University of Tartu Art museum were completed in 1868 under the direction of museum director Ludwig Schwabe. They were made by the local painter of Dorpat –Thomas Friedrich Redlin. The murals were created as a background for the exhibition of white plaster cast sculptures made after ancient originals, and it is also the first interior specially designed for an Estonian museum. In the case of the Art Museum, it is also remarkable that there is no creative freedom used as it has been done in the manor houses of Suure-Kõpu

---

52 For more about the self-portrait of the Suure-Kõpu artist, see Hiiop 2011, p. 246.
53 Hiiop 2017, p. 298.
54 For more about transformation and Roman "truth", see Rieche, Anita: "Reconstruction as Transformation: Reproductions of Ancient Roman Architecture". In: Bracker, Jacobus / Hubrich, Ann-Kathrin (eds.): *Art of Reception*. Cambridge Scholarly Publishing: Newcastle 2020, 191 ff.

and Vana-Võidu, and the model sheets have been copied almost fully with only some smaller changes. In the academic environment, the exact original has been followed for didactic purposes.

In the Art Museum seven halls are painted in the Pompeian style, and each of the halls is named after the main colour of the wall. Every hall has a model as a lithographic sheet: the model for the green hall is from the house of *Strada de' Mercadanti* in Pompei,[55] the red hall is painted after *casa del Poeta Tragico*, and the model of the dark red hall[56] and the small black room[57] were excavated in 1833 from the *casa de' Bronzi* in Pompeii. The blue hall is made after the *casa delle Vestali* in Pompeii[58], the model for the yellow hall was taken from Pompeian house number 57 in *Strada Stabiana*,[59] and the small red room is made after *casa del labirinto*, excavated in 1834.[60]

We can find some birds and semi-animals on the murals of the museum, but compared with the animals and creatures in Suure-Kõpu and Vana-Võidu their purpose is not to catch the eye – these are just details from the whole. In the four rooms small creatures are depicted, but in the dark red hall birds with apples and pears are depicted as directly taken from nature (Figure 6). In the blue hall geese and hippopotami are integrated into the mural as architectural elements. In the same way the geese and hippopotami are placed as well in the yellow hall, but in addition to those griffins are painted in a sculptural manner looking at each other. In the small red room several mythical winged horses walk on the walls but they figure as details rather than play a central role in the painting. The black hall has got a line of pattern made of decorative seashells.

Here we can see totally different types of creatures in the university rooms. The creatures don't carry a meaning or an idea, but they act as a part of the

---

55   Zahn, Wilhelm: *Die schönsten Ornamente und merkwürdigsten Gemälde aus Pompeji, Herculanum und Stabiae: nach einigen Grundrissen und Ansichten nach den an Ort und Stelle gemachten Originalzeichnungen*, Band 2, Heft 5. Reimer: Berlin 1842–1844, T. 44.
56   Zahn 1842–1844, Band 2, Heft 4, T. 34.
57   Zahn 1842–1844, Band 2, Heft 6, T. 55.
58   Zahn, Wilhelm: *Die schönsten Ornamente und merkwürdigsten Gemälde aus Pompeji, Herculanum und Stabiae: nach einigen Grundrissen und Ansichten nach den an Ort und Stelle gemachten Originalzeichnungen*, Band 1, Heft 10. Reimer: Berlin 1828–1829, T. 99.
59   Zahn, Wilhelm: *Die schönsten Ornamente und merkwürdigsten Gemälde aus Pompeji, Herculanum und Stabiae: nach einigen Grundrissen und Ansichten nach den an Ort und Stelle gemachten Originalzeichnungen*, Band 3, Heft 8. Reimer: Berlin 1852–1859, T. 79.
60   Zahn, Wilhelm: *Ornamente aller klassischen Kunst-Epochen nach den Originalien in ihren eigenthümlichen Farben*. Reimer: Berlin 1870, T. 34.

whole interior design system. At the same time, these motifs are not insignificant because they form a whole with other elements of the mural. Different disciplines and purposes deal with reconstructions in different ways – archaeology considers the accuracy of reconstruction. In the university's art museums the use of reconstruction is a part of the comprehensive educational program.

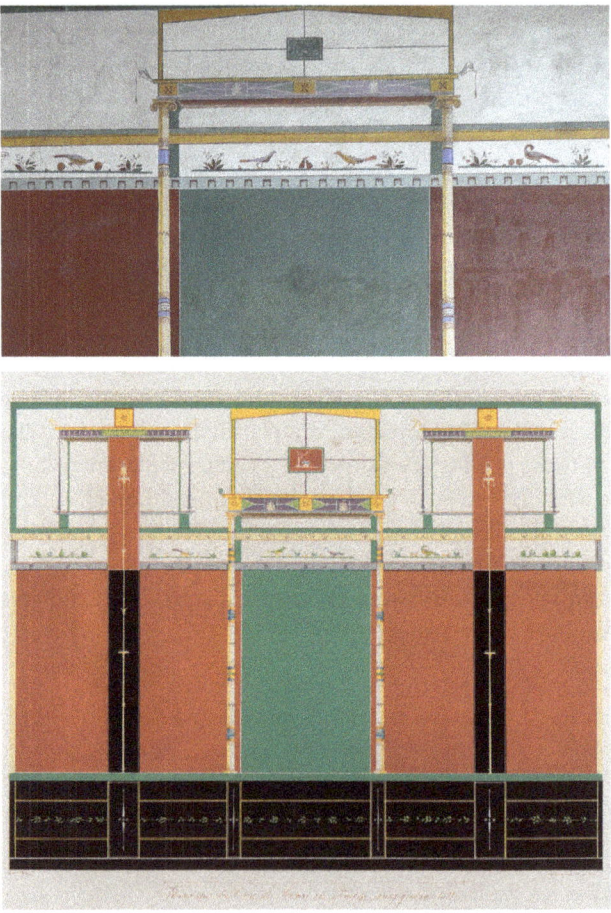

Figure 6. (a) Mural with birds and fruits in the University of Tartu Art Museum. Photo: Andres Tennus, 2018. (b) Zahn, Wilhelm: *Die Schönsten Ornamente und merkwürdigsten Gemälde aus Pompeji, Herculaneum und Stabiae.* Band 2, Heft 4, Reimer: Berlin. 1842–1844, T. 34.[61]

---

61 Collection of the University of Tartu Art Museum, KMM GR 7522.

Conclusion

In the 1850–60s the decorations of several Estonian manors and the University of Tartu Art Museum have got inspiration from the rediscovered ancient interiors in Pompeii and Herculaneum. Animal motifs of the Pompeian-style murals have enriched Estonian visual culture with strange creatures, exotic birds, insects, and animals. Here we can conclude that the story of the formation of the design of private and academic spaces is quite different.

In private manors figurative motifs are taken from the ancient models, and these are united and combined probably according to the vision of the master or customer. In the Suure-Kõpu manor figural motifs of centaurs are used, enlarged multiple times compared with the originals of Villa Cicero, making a colourful interior for the manor's dining room. It is largely based on the fashion flow of the era and the purpose of decorating the dining room attractively.

Murals in the Vana-Võidu manor are much more enigmatic – there are numerous figurative paintings of birds and other creatures, joined delicately in the dining room. In Vana-Võidu motifs from Pompeian and Herculaneum houses are combined, and in some cases the scenes have originally been part of wall paintings and floor mosaics, and in some cases the origins of the motifs are unknown. In the current state of the research there remains a mystery of how and why the choice of motifs for Vana-Võidu was made. It can impede viewers who are trying to find an allegory behind the pictures instead of enjoying the truthfulness of the animals and birds.

In the academic interior of the University of Tartu Art Museum, the lithographic model sheets that reflect the whole wall of the ancient house are strictly followed. The motifs of the semi-animals and the birds form an integral part of the wall pattern. The purpose of the murals of the academic interior are rather didactic than creative appropriation of ancient material.

To conclude, these examples are important in the context of the reception and appropriation of ancient art in Estonian visual culture. Undoubtedly, these are interesting and unique examples of interior design, which tried to imitate the ancient world, but also express the desire to follow the trends of nineteenth-century Europe. The reconstruction and reproduction of an ancient structure does not provide knowledge of historical truth but contributes to the construction of an image of antiquity.[62]

---

62 About the reconstruction and recreation of Roman architecture see Rieche, pp. 188–209.

Unfortunately, due to the current state of research and restoration works in Estonian manor houses, walls do not give any more material about animal culture in nineteenth-century Estonian interior design. Despite the fact that there are only a few examples of that type, these murals significantly enrich Estonian visual culture. On the one hand it testifies about the awareness of classical cultures in Estonia, which lies thousands of kilometres away from Pompeii and Herculaneum, but on the other side the opening of the topic could provide additional research interest to representatives of different disciplines.

References

Amery, Colin / Curran, Brian: *The Lost World of Pompeii*. Frances Lincoln: Los Angeles 2011.

Anderson, Jaanika / Hiiop, Hilkka: "The Triple Pompejanum Possessed by the Von Stryk Family: the Manor Houses of Vana-Võidu, Suure-Kõpu and Voltveti". *Baltic Journal of Art History* 13, 2017, pp. 165–192. DOI:10.12697/BJAH.2017.13.08

Anderson, Jaanika / Lotman, Maria-Kristiina: "Intrasemiotic Translation in the Emulations of Ancient Art: On the Example of the Collections of the University of Tartu Art Museum". *Semiotica* 222, 2018, pp. 1–24. DOI:/10.1515/sem-2016-0118.

Anderson, Jaanika: *Reception of Ancient Art: The Cast Collections of the University of Tartu Art Museum in the Historical, Ideological and Academic Context of Europe (1803–1918)*. University of Tartu Press: Tartu 2015.

Fögen, Thorsten / Thomas, Edmund (eds.): *Interaction Between Animals and Humans in Graeco-Roman Antiquity*. De Gruyter: Berlin / Boston 2017. DOI: 10.1515/9783110545623.

Hammer, Jacob: "Trained animals in antiquity: a portion of a paper". *The Classical Outlook* 20(6), 1943, pp. 59–61.

Harden, Alastair: "Animal in Classical Art". In: Campell, Gordon Lindsay (ed.): *The Oxford Handbook of Animals in Classical Thought and Life*. Oxford University Press: Oxford 2014, pp. 24–60. DOI:10.1093/oxfordhb/9780199589425.001.0001.

Hein, Ants: *Eesti mõisaarhitektuur. Historitsismist juugendini*. Hattorpe: Tallinn 2003.

Hein, Ants: *Viljandimaa mõisad*. Hattorpe: Viljandi 1999.

Hellermann, Dorothee von: "Candidatus Pictura. Gottlieb Welté als Künstler des Rokoko und Klassizismus". In: Polli, Kadi / Raudsepp, Renita (eds.): *Maarjamaa rokokoo = Rokoko in Estland. Gottlieb Welté (1745/49–1792)*. Eesti Kunstimuuseum: Tallinn 2007, pp. 70–103.

Hilkka Hiiop, "Antiigi peegeldus Eesti mõisamaalingutel = A reflection of Antiquity in the Paintings in Estonian Manors". In: *Eesti kunsti ajalugu 3, 1770–1840. = History of Estonian Art 3, 1770–1840*. Eesti Kunstiakadeemia: Tallinn 2017, pp. 288–298.

Hiiop, Hilkka et al.: *Vana-Võidu mõisa peahoone söögisaali ning ümbritsevate ruumide (suur saal, vestibüül, ruumid 208/210) siseviimistluse uuringud ja osaline restaureerimine. Aruanne*. Eesti Kunstiakadeemia / H&M Restuudio: Tallinn 2016. Manuscript in Archive of National Heritage Board, A-13143.

Hiiop, Hilkka: "In the Footsteps of Classical Antiquity. Influences of the Antique in Estonian Manor Murals". *Baltic Journal of Art History* 3, 2011, pp. 225–252.

Jashemski, Wilhelmina Feemster / Meyer, Frederick G. (eds.): *The Natural History of Pompeii*. Cambridge University Press: Cambridge 2002.

Jennisson, George: *Animals for Show and Pleasure in Ancient Rome*. Manchester University Press: Manchester 1937.

Jones, Owen: *The Grammar of Ornament: A Visual Reference of Form and Colour in Architecture and the Decorative Arts*. Princeton University Press: Princeton and Oxford 2016. DOI:10.2307/j.ctvc77mhw.

Kalof, Linda: "Ancient Animals". In: Kalof, Linda (ed.): *A Cultural History of Animals in Antiquity*. Berg: Oxford, New York 2011, pp. 1–16.

Kukk, Inge: "Vervielwältigtes Pompeji – von Graphikblättern bis zur Wanddekoration". *Baltic Journal of Art History* 3, 2011, pp. 181–213.

Le Antichità di Ercolano Esposte, Parts 1–8. Regia Stamperia: Napoli 1757–1792.

Le Antichità di Ercolano Esposte, Vol 1. Regia Stamperia: Napoli 1757.

Lewis, Sian / Llewellyn-Jones, Lloyd (eds.): *The Culture of Animals in Antiquity: A Sourcebook with Commentaries*. Routledge: New York 2018. DOI:10.4324/9781315201603.

Mathiesen, Hans Erik: "The Reception of Classical Antiquity: Some General Remarks". In: Nielsen, Marjatta (ed.): *The Classical Heritage in Nordic Art and Architecture. Acts of the seminar held at the University Copenhagen 1st–3rd November 1988*. (Acta Hyperborea 2). Museum Tusculanum Press: Copenhagen 1990, pp. 19–24.

Monaco, Domenico: *A Complete Handbook to the National Museum in Naples*. William Clowes and Sons: London 1883.

Monaco, Domenico: *Les monuments du Musée National de Naples gravés par les meilleurs artistes Italiens*. Museo Nazionale di Napoli: Napoli 1877.

Nava, Maria Luisa / Paris, Rita / Friggeri, Rosanna (eds.): *Rosso Pompeiano. La decorazione pittorica nelle collezioni del Museo di Napoli e a Pompei. Catalogo della mostra (Roma, 20 dicembre 2007–31 marzo 2008)*. Mondadori Electa: Milano 2008.

Niccolini, Fausto / Niccolini, Felice: *Le case ed i monumenti di Pompei disegnati e descritti*, Band 3. Napoli 1890.

Niccolini, Fausto / Niccolini, Felice: *Le case ed i monumenti di Pompei disegnati e descritti*, Band 1. Napoli 1854.

Nichols, Marden Fitzpatrick: "Domestic Interiors, National Concerns. The Pompeian Style in the United States". In: von Stackelberg, Katherine T. / Macaulay-Lewis, Elizabeth (eds.): *Housing the New Romans. Architectural Reception and Classical Style in the Modern World*. Oxford: Oxford University Press 2017, pp. 126–152. DOI:10.1093/acprof:oso/9780190272333.001.0001.

Pappalardo, Umberto: *The Splendor of Roman Wall Painting*. The Paul Getty Museum: Los Angeles 2008.

Raccolta delle più interessanti dipinture e de più belli musaici rinvenuti negli scavi: di Ercolano, di Pompei, e di Stabia che ammiransi nel Museo reale borbonico. Museo nazionale di Napoli: Napoli 1843.

Rieche, Anita: "Reconstruction as Transformation: Reproductions of Ancient Roman Architecture". In: Bracker, Jacobus / Hubrich, Ann-Kathrin (eds.): *Art of Reception*. Cambridge Scholarly Publishing: Newcastle 2020, pp. 188–209.

Saint-Non, Jean Claude Richard de: *Voyage Pittoresque Ou Description Des Royaumes De Naples Et De Sicile (Band 2): Contenant Une Description des Antiquités d'Herculanum, des Plans & des Détails de son Théâtre, avec une Notice abrégé des différents Spectacles des Anciens. Les Antiquités de Pompeïi* […]. Clousier: Paris 1782.

Stackelberg, Katherine T. / Macaulay-Lewis, Elizabeth (eds.): *Housing the New Romans. Architectural Reception and Classical Style in the Modern World*. Oxford University Press: Oxford 2017. DOI:10.1093/acprof:oso/9780190272333.001.0001.

Streeter, Colin: "Two Carved Reliefs by Aubert Parent". *The J. Paul Getty Museum Journal* 13, 1985.

Stryk, Wolf-Dietmar von: *Riidaja von Strykide perekonna põliskodu Liivi- ja Eestimaal 1562–1919–2003: ühe mõisa ajalugu sajandite tagant tänase päevani*. Põdrala vallavalitsus: Riidaja 2008.

Tammisto, Antero: *Birds in Mosaics: A Study on the Representation of Birds in Hellenistic and Romano-Campanian Tessellated Mosaics to the Early Augustan Age*. Institutum Romanum Finlandiae, Rome 1997.

Volmer, Svea / Hiiop, Hilkka: "Salapärane ja ihaldatud Pompei". In: Matteus, Kais (ed.) *Muinsuskaitse Aastaraamat 2007*. Muinsuskaitseamet, Tallinna Kultuuriväärtuste Amet, EKA muinsuskaitse ja restaureerimise osakond: Tallinn 2008, pp. 56–60.

Warren Chadd, Rachel / Taylor, Marianne: *Birds: Myth, Lore and Legend*. Bloomsbury Publishing: London et al 2016. DOI:10.1111/jofo.12192.

Werner, Peter: *Pompeji und die Wanddekoration der Goethezeit*. Wilhelm Fink Verlag: München 1970.

Zahn, Wilhelm: *Ornamente aller klassischen Kunst-Epochen nach den Originalien in ihren eigenthümlichen Farben*. Reimer: Berlin 1870.

Zahn, Wilhelm. *Die schönsten Ornamente und merkwürdigsten Gemälde aus Pompeji, Herculanum und Stabiae: nach einigen Grundrissen und Ansichten nach den an Ort und Stelle gemachten Originalzeichnungen*, Band 1–3. Reimer: Berlin 1828–1859.

Zahn, Wilhelm: *Ornamente aller klassischen Kunst-Epochen nach den Originalen in ihren eigenthümlichen Farben*. Reimer: Berlin 1849.

University of Tartu Art Museum, KMM GR 7522; KMM GR 7663.

Index

The Index presents an entangled selection of phenomena, places, persons, and other beings featured in this collection.

Agnus dei 275, 312
Arvidi, Andreas 80, 83
Atlantic herring 38, 217, 219–222, 226, 227, 230, 232, 234–237, 239, 242, 245
Auroxen 125, 127, 132

Baltic herring 205, 208, 213, 220–222, 226, 227, 229–232, 234, 236
Bartholin, Thomas 88, 325
Bear 20, 51, 255, 268, 290, 322, 323, 333, 345
Beaver 26, 30, 88, 167, 186
Beetle 324, 348, 354–356
Below, Jacob Friedrich 84–89, 155
Białowieża 121, 123, 124, 131, 135
Bison 26, 35, 37, 121–138
Black Heads (Brotherhood) 171, 227, 281
Boar 26, 48, 49, 54, 81, 131, 137, 289
Bohn, Johannes 86–88
Bornhöhe, Eduard 237–239
Bracke (brake) 49, 54, 356
Bussow, Conrad 309–311
Butterfly 348, 351, 354

Cat 39, 44, 77, 321, 324–326, 331–333, 348, 353, 354
Centaur 39, 344–347, 356, 359
Cēsis (Wenden) 169, 171, 214, 288, 289
Chicken 25, 47, 51, 67, 88, 322
Cock 114, 285, 286
Cod 34, 38, 201, 202, 207–213, 217, 222, 225

Courland (Kurland) 45, 46, 48, 60, 123, 176, 344, 345
Cow 33, 47, 50, 64, 67–69, 133
Crow 114, 115

Daugavpils 103, 108
Deer 81, 124, 131
Devil 44, 133, 287, 288, 291, 311, 315
Dobele 174
Dog 28, 31–34, 36, 44, 47–51, 53–55, 77, 78, 87, 97, 105, 108, 109, 135, 241, 267, 276, 283–285, 288, 290, 297, 305, 308–310, 322, 324
Dragon 33, 254–258, 265, 267–269, 275, 279–283, 287–289, 293, 294, 297, 299

Eel 207, 232
Eagle 38, 82, 114, 253, 259, 263, 265, 279, 293, 294, 351
Elk 122, 137
Eisen, Johann Georg 17, 32, 65
Erastus, Thomas 325, 333

Fáfnir 256
Forer, Laurenz 326
Fox 48, 49, 114, 173, 294, 295, 297
Fowl 66, 108
Friebe, Wilhelm Christian 146, 157, 307
Frog 87, 255

Goose 47, 48, 51, 78
Goat 67, 80, 156, 157, 307

God 13, 52, 61, 78, 127, 174, 256–258, 260, 275, 285, 287, 339, 340, 347, 350
Grasshopper 348–351
Guilds (Great Guild, Kanut, St Olaf's) 281, 289
Griffin 35, 38, 253, 259, 260, 263–270, 351, 352, 357

Haapsalu 46, 47, 51, 52, 176
Hamburg 17, 46, 190, 256, 278, 281, 284
Hanseatic League 145, 187, 191, 210, 217, 225, 226, 233, 234
Hein, Friedrich 80–83
Hound 24, 48, 49, 54, 284
Horse 17, 19, 24, 30, 31, 34–38, 44, 47–53, 55, 59, 67, 68, 77, 79, 80, 82, 87, 97, 105, 108, 125, 127, 147, 165–177, 183–195, 198, 236, 306, 307, 313, 315, 322, 324, 357
Hupel, August Wilhelm 64, 66, 95, 112, 189, 307

Jelgava (Mitau) 102, 104, 105, 344
Jörmunganðr 256

Kaliningrad (Königsberg) 240
Karksi 144, 169–171, 231
Kiechel, Samuel 305, 313, 315–317
Kihnu (Kyne, Kyna) 9, 10, 37, 141, 143, 148–158
Kupits, Meelis 243, 244
Kuldīga (Goldigen) 168, 169, 172, 176
Kuusalu 230, 232, 233

Lagermarck, Johan Svensson 322
Lamb 83, 153, 275, 279, 297
Lion 24, 33, 36, 275, 279, 281, 287–293, 294, 297–299, 330, 351
Lübeck 14, 46, 47, 50–52, 54, 167, 172, 185, 187, 278, 286, 290
Lüdinghausen-Wolff, Edmund von 110–112

Maasilinn (Soneburg) 169, 175–177, 188
Magirus, Johannes 76, 329, 330
Magnus, Olaus 28, 45, 132, 190, 306, 307, 312, 314, 315
Mammal 8, 32, 78, 114, 122, 124, 205, 207
Mare 67–69, 169, 176, 189, 234
Matthiae, Salomon 84–89
Menius, Friedrich 314
Mice 97, 330
Mouflon 131, 141–143
Muhu 189, 240
Müller, Ferdinand 64, 77, 113

Narva 46, 103, 169, 186, 187, 190, 191, 193, 194, 225, 290, 313
Novgorod 46, 186, 187, 190, 191, 256, 259, 261, 264–266

Orraeus, Axelius Jonae 322
Owl 114, 263
Ox 48, 82, 279

Pape 121–123, 125–130, 133–137, 140
Parrot 23, 348–351
Pawels, Hans 281, 297, 298
Peipisi (Peipus) 233
Perch 204, 206, 213, 232
Pig 23, 30, 47, 48, 51, 67, 79, 147, 231, 307
Pigeon 348, 353–356
Pike 201, 202, 206, 212–214, 232
Polyp 95, 112
Possevino, Antonio 305, 311, 312, 316, 317

Poultry 24, 25, 231
Predator 135, 207, 261, 263, 265, 269, 305–307, 312
Pskov 190, 191, 312
Pärnu (Pernau) 62–64, 68, 70, 148, 153, 186, 187, 206, 225, 226, 260, 295, 345
Põltsamaa (Oberpahlen) 171

Rakvere (Wesenberg) 39, 46, 171, 308, 313
Riga (Rīga) 19, 23, 29, 34, 37, 45–48, 51, 53, 64, 95, 97, 99–105, 107–115, 122, 124, 141, 146, 148, 171–173, 175–177, 184–186, 188–190, 198, 225, 233, 254, 261, 278, 281, 284, 290, 307, 312
Russow, Balthasar 209, 305, 308–311, 313, 316, 317

Saaremaa (Ösel) 30, 32, 38, 45, 146, 151, 152, 165, 176, 177, 188–190, 192
Sculpin 205, 234
Seal 31, 148, 205–207, 225, 234
Sennert, Daniel 332
Sheep 9, 10, 34, 37, 67, 80, 141–158, 297, 307, 312–314, 322
Sjöberg, Gabriel 321
Smelt 205, 232
Smuul, Juhan 224, 240–242
Snake 32, 81, 254–256, 265, 267, 269, 288
Sperling, Johann 327
St. Petersburg 103, 109, 110, 114, 233, 343
Squirrel 26, 87, 114
Stork 294, 295, 297
Stralsund 243, 278
Stryk, von family 344, 345, 347, 356

Tallinn (Reval) 8–10, 19, 23, 30–34, 40, 44, 46, 47, 50–52, 54, 65, 66, 68, 77, 78, 84, 97, 102, 103, 113, 147, 148, 153, 165–167, 171–173, 177, 183–188, 190, 191, 193, 194, 202, 203, 205, 210, 211, 213, 214, 221, 225–227, 231, 236–243, 254–256, 258, 259, 261, 275–284, 286, 288–293, 296–298, 308, 310, 313, 341, 343, 344
Tammsaare, Anton Hansen 238, 239
Tartu (Dorpat) 7–9, 17, 19, 20, 22, 26, 28, 29, 32, 33, 37, 39, 45, 46, 53, 68, 75–77, 80, 81, 83, 84, 86, 87, 89, 90, 102, 103, 107, 141, 144, 146, 147, 151, 153, 170, 186, 187, 189–193, 205, 209, 211, 214, 221, 225, 227, 230, 233, 237, 257, 259, 260, 278, 285–287, 294, 307, 308, 311, 314, 321, 329, 339, 341, 343, 346, 356, 358, 359
Tarvastu (Tarvast) 170, 171, 231
Teutonic Order 8, 18, 37, 38, 45, 165–167, 169, 171–173, 175, 177, 178, 186–190, 294, 295
Trakai 185, 188
Tuglas, Friedebert 34, 40, 237, 238
Turbot 207, 212

Valmiera (Wolmar) 192, 225
Valmiermuiža (Wolmarshof) 124
Vana-Võidu (Alt-Woidoma) 341, 344, 345, 348–357, 359
Ventspils (Windau) 114, 225
Viljandi (Fellin) 103, 144, 167, 170, 171, 177, 225, 294, 295, 344, 345
Viper 254
Viborg (Viiburi) 234
Vilde, Eduard 236–238

Wallander, Sequardus 80, 82, 83
Weasel 81, 114

Werewolves 21, 35, 39, 305, 306, 313–315, 317
Westerman, Andreas 321
Whale 77, 89
Whitefish 204, 206

Witch 257, 313, 314, 328, 351
Wolf 32, 39, 48, 49, 53, 56, 82, 267, 297, 305–307, 309, 310, 312–317, 344
Worm 78, 88, 114, 254, 255

Studies in Literature, Culture, and the Environment
Studien zu Literatur, Kultur und Umwelt

Edited by / Herausgegeben von
Hannes Bergthaller, Gabriele Dürbeck, Robert Emmett, Serenella Iovino, Ulrike Plath

Vol. / Bd.	1	Daniel A. Finch-Race / Stephanie Posthumus (eds): French Ecocriticism. From the Early Modern Period to the Twenty-First Century. 2017.
Vol. / Bd.	2	Alessandro Macilenti: Characterising the Anthropocene. Ecological Degradation in Italian Twenty-First Century Literary Writing. 2017.
Vol. / Bd.	3	Gabriele Dürbeck / Christine Kanz / Ralf Zschachlitz (Hrsg.): Ökologischer Wandel in der deutschsprachigen Literatur des 20. und 21. Jahrhunderts – neue Ansätze und Perspektiven. 2018.
Vol. / Bd.	4	Yi-Peng Lai: Eco*Ulysses*. Nature, Nation, Consumption. 2018.
Vol. / Bd.	5	Gabriele Dürbeck / Jonas Nesselhauf (Hrsg.): Repräsentationsweisen des Anthropozän in Literatur und Medien / Representations of the Anthropocene in Literature and Media. 2019.
Vol. / Bd.	6	José Manuel Marrero Henríquez (ed.): Hispanic Ecocriticism. 2019.
Vol. / Bd.	7	Tina-Karen Pusse / Heike Schwarz / Rebecca Downes (eds.): Madness in the Woods: Representations of the Ecological Uncanny. 2020.
Vol. / Bd.	8	Sébastian Thiltges / Christiane Solte-Gresser (Hrsg. / éds): Kulturökologie und ökologische Kulturen in der Großregion / Écologie culturelle et cultures écologiques dans la Grande Région. 2020.
Vol. / Bd.	9	Aurélie Choné / Philippe Hamman (Hrsg. / éds.): Die Pflanzenwelt im Fokus der Environmental Humanities. Deutsch-französische Perspektiven / Le végétal au défi des Humanités environnementales. Perspectives franco-allemandes. 2021.
Vol. / Bd.	10	Urte Stobbe / Anke Kramer / Berbeli Wanning (Hrsg.): Literaturen und Kulturen des Vegetabilen. Plant Studies – Kulturwissenschaftliche Pflanzenforschung. 2022.
Vol. / Bd.	11	Peggy Karpouzou / Nikoleta Zampaki (Hrsg.): Symbiotic Posthumanist Ecologies in Western Literature, Philosophy and Art. Towards Theory and Practice. 2023.
Vol. / Bd.	12	Linda Kaljundi / Anu Mänd / Ulrike Plath / Kadri Tüür (eds.): Baltic Human-Animal Histories. Relations, Trading, and Representations. 2024.

www.peterlang.com

Gehirn und Zauberspruch

Wolfgang Ernst

Gehirn und Zauberspruch

Archaische und mittelalterliche psychoperformative
Heilspruchtexte und ihre natürlichen Wirkkomponenten
Eine interdisziplinäre Studie
Mit einem Geleitwort von Prof. Dr. Volker Faust

Bibliografische Information der Deutschen Nationalbibliothek
Die Deutsche Nationalbibliothek verzeichnet diese Publikation
in der Deutschen Nationalbibliografie; detaillierte bibliografische
Daten sind im Internet über http://dnb.d-nb.de abrufbar.

Umschlagabbildung: Zeichnung Fritz Klier, Vornbach am Inn, 2010

ISBN 978-3-631-64591-8 (Print)
E-ISBN 978-3-653-03833-0 (E-Book)
DOI 10.3726/978-3-653-03833-0

© Peter Lang GmbH
Internationaler Verlag der Wissenschaften
Frankfurt am Main 2013
Alle Rechte vorbehalten.
PL Academic Research ist ein Imprint der Peter Lang GmbH.

Peter Lang – Frankfurt am Main · Bern · Bruxelles · New York ·
Oxford · Warszawa · Wien

Das Werk einschließlich aller seiner Teile ist urheberrechtlich
geschützt. Jede Verwertung außerhalb der engen Grenzen des
Urheberrechtsgesetzes ist ohne Zustimmung des Verlages
unzulässig und strafbar. Das gilt insbesondere für
Vervielfältigungen, Übersetzungen, Mikroverfilmungen und die
Einspeicherung und Verarbeitung in elektronischen Systemen.

www.peterlang.com

Geleitwort

Mit dem überraschenden, zunächst durchaus zwiespältig wirkenden Titel *Gehirn und Zauberspruch* schlägt Wolfgang Ernst ein Kapitel interdisziplinären Ansatzes auf, in dem er historische Heilspruchtexte in die neuzeitliche Neurobiologie transferiert, ohne dass die jeweiligen kulturellen und soziologischen Voraussetzungen vernachlässigt werden. So finden eine greifbare mythische rituelle ‚Verbaltherapie' Mesopotamiens und die ‚mythische Anatomie' beim Geburtsgesang von Schamanen der Mayas ebenso Berücksichtigung wie wichtige Aspekte christlicher Soteriologie des Mittelalters und damaliger medizinischer Versorgung.

Dr. Wolfgang Ernst, ein neurophysiologisch orientierter Nervenarzt, ist schon durch frühere medizinhistorische und ethnologische Arbeiten bekannt geworden. In Fachkreisen in guter Erinnerung sind beispielsweise *Heilzauber und Aberglaube in der Oberpfalz* (1991), *Zauber, Riten und Rezepte* (2007), *Oberpfälzischer Heilzauber* (2008), *Beschwörungen und Segen. Angewandte Psychotherapie im Mittelalter* (2011) u. a. In seinem neuen Buch hat er sich vor allem folgende Aufgaben gestellt:

Zum einen rein konzeptionell der Abschied von nebelhaften Vorstellungen über eine Art Magie im Gewerbe der alten Heilkundigen. Zum anderen die Erkenntnis, dass die bisherigen Deutungen der alten Texte allein mit der Wortakt-Theorie der Linguistik an ihre Grenzen gekommen sind. Und so gelingt es ihm mit kurzen Einblicken in sozialpsychologische Details, insbesondere in eine ‚Triade von Hilferuf, Helferzuspruch und Umweltsage' sowie in strikter Voraussetzung einer Notfallsituation mit Erwartungsdruck die notwendige wissenschaftliche Basis zu sichern. Und dies nicht zuletzt auf der Grundlage intermittierender interpersonaler Synchronisierung der Gehirntätigkeit, wie sie nach cerebralen Simultanableitungen in der Hyperscanningforschung nunmehr seit wenigen Jahren vorliegt.

Der Text braucht Eigenleistung, das geht schon aus dem interdisziplinären Fundament hervor, das diese eindrucksvolle Arbeit charakterisiert. Dafür wird aber auch rasch klar, dass beispielsweise der Kampf *der* Dämonen und der Kampf *gegen* die Dämonen im Gehirn spielt – und sich wechselseitig mit Kultur bedingt und tradiert. Dies wird besonders eindrucksvoll deutlich in dem Syndrom der ‚bösen Augen', eigentlich seit Jahrtausenden bekannt, jetzt aber vor allem bei Borderlinekranken von alltagsrelevanter Bedeutung was Diagnose und Therapie anbelangt.

In den Hauptkapiteln werden fünf Felder als Neuronenverbindungen erarbeitet, denen man allerdings wiederum Kommunikation und Überschneidungen unterstellen muss. Dafür wird durch diese schematische Unterteilung deutlich, welche Funktionen die junge Hirnforschung bieten kann, um den Verarbeitungsweisen näher zu kommen.

Die große Bedeutung der Metaphorik als übeltilgende Macht der Hirnleistung zieht sich durch viele Ausführungen in diesem Buch, besonders aber beim Beispiel des Alptraums. Denn es stellt sich seit Längerem die Frage, ob die Gehirne unserer Vorfahren nicht auch wussten, dass es triviale Ungeister nur in Träumen gibt. Hier spielt die Zusammenarbeit von limbischen und präfrontalen Gebieten, das ‚Labeling emotions', eine entscheidende Rolle.

In einem eigenen Kapitel behandelt der Autor die Ergebnisse der ereigniskorrelierten EEG-Signale, wie sie bei widersinnigen Sprachfiguren auftreten und als rhetorische Adynata zum beweiskräftigen Reaktionsmuster des Gehirns gehören. Dabei öffnet sich ein durch viele Kulturen bekanntes Quellenmaterial, sofern man es nutzt, was wiederum die Verdienste des Autors unterstreicht.

Ein weiteres Kapitel beleuchtet die Strategien imaginativer Induktionen bei Entbindungen: Für den Mediziner ist der erste Merseburger Spruch kein fiktiver nutzloser Fernzauber zur Befreiung von Kriegsgefangenen. Dabei werden auch weitere außereuropäische geburtsbegleitende Texte mit Regressionsstrategie aus Mesopotamien und Altamerika herangezogen.

Eine andere imaginative Strategie dient der Katharsis. Sie unterteilt der Autor nach den vorliegenden Textquellen und nach betroffenen Hirngebieten folgerichtig in introversive und extroversive Bereinigungsversuche. Konkret in eine leiderfüllte, modern gesprochen gewissensbetonte und eine eher aktive, expulsive, die Dämonen direkt packende Methode der Beschwörung. Dabei ist der Leser anhand von 5 Tabellen in der Lage, einen raschen Überblick zu gewinnen.

Vor allem die vielen Hinweise auf die mindestens punktuellen Parallelitäten antiker und moderner verbaltherapeutischer Methoden erlauben es, erfahrungsgeprägte kulturabhängige Urpsychotherapien für den Notfall anzuerkennen. So können sich für den Psychiater, Psychologen und Psychotherapeuten der Neuzeit die heutigen symbolbezogenen imaginativen Therapiemethoden wie etwa Hypnotherapie, katathymes Bilderleben und Psychodrama in einem neuen Licht historischer und hirnorganischer Synchronopsis darstellen. Für den Ethnologen und Historiker bieten sich Einblicke in die Verwicklungen von gesellschaftlicher Kultur mit individueller Natur: einer Beziehung, die in diesem Teilbereich nicht durch prothetische theoretische Brücken, sondern durch cerebrale sprachlich und rituell zwischenmenschlich fließende Abhängigkeiten geprägt wird.

Der Gewinn der Lektüre liegt neben den Erkenntnissen der einzelnen Teilgebiete vor allem im interdisziplinären Ansatz des Gebotenen. Das kann zwar auf den ersten Blick aufwendiger werden, macht sich aber als Erkenntnis-Synopsis bezahlt. Außerdem wird es in unserer Zeit der (auch erzwungenen) Spezialisierung immer seltener, dass sich ein einzelner Experte ein so breites Wissens-Spektrum erarbeitet. Und noch seltener ist es, dass er sich die Mühe macht, das erarbeitete Wissen auch zur allgemeinen Nutzung zur Verfügung zu stellen. Denn seit eine menschliche Kultur existiert, waren ihre Heilkundigen bemüht, Kranken auch mit geeigneten Worten zu helfen. Wenn auch archaische und mittelalterliche Heilspruchtexte bisher als magische oder per Wortakt performierende Instrumente gedeutet wurden, so ist es jetzt erstmals realisierbar, auch neurobiologisch denkbare Funktionsabläufe in die Diskussion einzubringen. Denn dadurch werden jetzt fließende reziproke Vermittlungen von Kultur zu Natur erkennbar.

Eine überfällige, wenn auch komplexe Aufgabe – durch das Buch von Dr. Wolfgang Ernst auf den rechten Weg von Forschung, Lehre und angewandter Heilkunde gebracht.

Volker Faust   Ravensburg, im August 2013

Inhaltsverzeichnis

Vorbemerkungen 15

A Eine kurze Forschungsgeschichte zu Entstehung, Anwendung
  und Wirkung des Heilspruchs und der Heilriten: Die Konstrukte
  des Unbegreiflichen 23
A1 Der Magiebegriff in Theorie und Praxis 23
   A1.1 Historische und historisierende Rezeptionen und Deutungen
       von Magie und ihr Fortbestand in die Gegenwart 25
   A1.2 Moderne Magie-Theorien 28
       A1.2.1 Rationalistische Erklärungen der Magie 28
       A1.2.2 Emotionalistische Erklärungen der Magie 29
       A1.2.3 Die strukturalistische Erklärung der Magie durch
             Claude Lévi-Strauss 31
             A1.2.3.a Ethnologie: Wildes und rationales Denken 31
             A1.2.3.b Das Gravitationsfeld 32
             A1.2.3.c Die Einbeziehung psychoanalytischer Theorien . 33
             A1.2.3.d Der Rekurs auf Therapien des Bilderlebens/
                   Imagination 34
             A1.2.3.e Neuropsychosomatische Zusammenhänge 35
       A1.2.4 Zwei Beispiele neuerer Magieforschung nach
             Lévi-Strauss 36
A2 Der linguistische Zugang zu den Heilspruchtexten:
   Sprechakttheorie und Performativität 38

B Die neuropsychosoziale Struktur einer Notfallbehandlung als Triade 45
B1 Der Kranke in Not 46
   B1.1 Hilferuf und Hilfserwartung 46
   B1.2 Aufmerksamkeit, Erwartung und Perzeption 52
B2 Heilkundige und Helfer 55
B3 Das Bündnis: Heiler und Patient in passagèrer Symbiose, Verkopplung
   und Entkopplung von Hirnsystemen zweier Individuen:
   Die neurale ‚Therapeutische Allianz' 62
B4 Soziales Umfeld, Kulturerwerb und ‚sozioneurale Hardware' 73

B4.1 Das europäische Mittelalter 75
  B4.1.1 Einige Aspekte der Gesellschaftsstruktur 76
  B4.1.2 Frömmigkeitsgeschichte – Vermittlungswege:
       Sehen und Hören und alle Sinne 76
  B4.1.3 Christliche ‚Medialität' 80
  B4.1.4 Christliche Krankenfürsorge und Krankheitsverständnis ... 81
  B4.1.5 Individuum und Ich-Verständnis im Mittelalter 82
  B4.1.6 Hirnbiologische Prämissen von Religiosität und
       Spiritualität 83
B4.2 Archaische Gesellschaften: Die mesopotamische Kultur 85
  B4.2.1 Einige Aspekte der Gesellschaftsstruktur 86
  B4.2.2 Aspekte geistiger Prägung 86
  B4.2.3 Krankheitsverständnis 87
  B4.2.4 Individuum und Ich-Verständnis in
       der mesopotamischen Kultur 90

C Psychoperformative Hirnleistung und Zauberspruch 93
C1 Das Bannen der Dämonen durch ihre Nennung und durch
die Schilderung ihres Wirkens – Strategie der internen und
externen Emotionsregulierung mittels Etikett und Metapher
am Beispiel des Alptraums und verwandter Affektionen 93
  C1.1 Zur Neurophysiologie der Emotionsverarbeitung 95
  C1.2 Psychologisch-psychiatrische Praxis von Emotionsverarbeitung ... 96
  C1.3 Versuch historisierender Einblicke in die Emotionsverarbeitung
      im Mittelalter 98
    C1.3.1 Einstimmung auf die durchlittene Nacht 101
    C1.3.2 Die Insinuation des Charakters von Träumen 102
    C1.3.3 Verbale Perduzierung einer verkörperlichten Simulation .. 104
    C1.3.4 Stimulierung sensomotorischer Schaltkreise 105
    C1.3.5 Aversive Stimuli und ihre neurochemische Wirkung 106
  C1.4 Versuch historisierender Einblicke in die Emotionsverarbeitung
      in den mesopotamischen Kulturen 107
    C1.4.1 Zur Frage der Emotionsverarbeitung am Beispiel
         von Traumgeschehen und Nachtgeistern in
         den mesopotamischen Kulturen 108
    C1.4.2 Die elementare natürliche Selbstverständlichkeit
         im Umgang mit Überirdischen 113
    C1.4.3 Begünstigung von Reappraisalprozessen durch
         Traumdeutung? 114
    C1.4.4 Rituelle Perduzierung einer verkörperlichten Simulation .. 114

C2 Die intrinsische neurale Reizüberflutung am Beispiel
des ‚Böse-Blick'-Komplexes 116
C2.1 Aspekte eines ‚subversiv' visuell getriggerten Kultsektors 116
C2.2 Die neurobiologische Grundlage des ‚Böse-Blick'-Syndroms 119
C2.3 Therapeutisches 122
C3 Die kognitive cerebrale Perturbation durch
sprachliche Inkongruenzkonstruktionen –
eine nicht-emotionale Konflikt-Provokation 123
C3.1 Zu den historischen Quellen von sprachlicher Inkongruenz 124
       C3.1.1 Funktionsschwankungen alter Inkongruenztexte:
       ‚Vogel federlos' und ‚Baum blattlos': Rätsel oder
       Heilspruch? 124
       C3.1.2. Am Grenzübergang zur Erzähltherapie: Adynata –
       weitere, Inkongruenz erzeugende Sprachfiguren 126
C3.2 Die neurophysiologische Darstellung der Signale von Kontrast
und Wider-Sinn bis z. ‚außergewöhnlichen Sensation' bei N400 .. 128
C4 Die Inszenierung von therapeutischen ‚Bildern' mittels Wortfiguren
und Erzählung 131
C4.1 Die regressionsfördernde verbale Bildgebung 141
       C4.1.1 Eine Regression analogisierende Imagination mit
       Verortung und Objektvorgabe (Ephesus, Mutterschoß
       und legendäre Siebenschläferhöhle) im europäischen
       Mittelalter 141
       C4.1.2 Die regressionsfördernde Imaginationen in
       der Geburtshilfe: Bindende Zeugung –
       Erhaltende Behütung – Lösende Entbindung 146
            C4.1.2.a Mu-Igala, der geburtskundliche Heilgesang
            eines Schamanen, erhalten durch Vermittlung
            des Cuña-Indianers Guillermo Haya aus Panama 148
            C4.1.2.b Die neosumerische Incantation vom
            ‚breeding bull' bis zu Gula's Schicksalsgriff 149
            C4.1.2.c Imagination der drei Idisigruppen
            des althochdeutschen ersten Merseburger
            Zauberspruchs als universale weise Heilfrauen .. 149
C4.2 Die verbale Imagination mit Wiederholungsfiguren
als rhetorisches Stilmittel, und ihre Hirnorganik:
Repetitionen, Aufzählungen, Wort- und Sinnketten,
Wort- und Sinnverknüpfungen 150
       C4.2.1 Die Bedeutung der Wiederholung von Worten und Sätzen
       in den Heilspruchtexten 150

      C4.2.2 Wiederholungsfiguren und ihre Beispiele 151
      C4.2.3 Ansätze zu einer Neurophysiologie
             der sprachlichen Wiederholungen 154
   C4.3 Strategien introversiver Katharsis 157
      C4.3.1 Leid- und Mitleid-Induktion mit Erinnerung
             an persönliche Schuld und an Schmerz unter
             Einbeziehung von überirdischen Helfern 157
             C4.3.1.1 Die kulturgeschichtliche Dimension
                    der introversiven kathartischen Imagination 158
                    C4.3.1.1.a Die therapeutische Induktion von
                         Leid und ‚Sünde' in babylonisch-
                         assyrischen Texten 158
                    C4.3.1.1.b Die zwei Typen kathartischer
                         Imagination (introversiv und
                         extroversiv) in zwei vedischen
                         Heilsprüchen: Sündenbekenntnis
                         vs. theriomorphe Vertreibung 160
                    C4.3.1.1.c Introversiv kathartische Elemente in
                         Texten des christlichen Mittelalters .. 161
             C4.3.1.2 Die neurobiologischen Effekte von introversiver
                    Katharsis: Redundante Aktivierung mit
                    Exposition von Schmerz und Leid 163
C5 Strategien der extroversiven Katharsis: Direkte Bannung des Übels
   durch Befehl und mittels überirdischer Helfer: Die Expulsion 169
   C5.1 Die Anwendung der expulsiven Methode im Kulturvergleich 170
   C5.2 Der Befehlsstand der Expulsion: Die Bedeutung des Helfers 172
   C5.3 Neurobiologische Effizienz durch extroversive Katharsis:
        Die Sensomotorik aktivierender Imperative und
        das homöostatische Gleichgewicht 174

D Nachbemerkungen 179

E Die Beispieltexte 185
E1 Europäisches Mittelalter bis Neuzeit; und Marcellus (4./5. Jh.) 185
E2 Mesopotamische Beispieltexte (Ausschnitte) 204
E3 Beispieltexte aus den altindischen Atharvaveden 212
E4 Texte aus der Tradition der Maya-Kulturen 215
E5 Beispieltexte aus Papyri graecae magicae 218

F	Register	221
F1	Literaturverzeichnis	221
F2	Verzeichnis der Beispieltexte	233
F3	Kurze Erläuterung zu einigen cerebralen Funktionsgebieten	237
F4	Nachweis der Abbildungen	239
F5	Sach- und Personenverzeichnis	242

Vorbemerkungen

Es mag verwegen erscheinen, zwei wissenschaftliche Fachgebiete von so unterschiedlichem Format angenähert zu sehen: das eine, noch in Kinderschuhen strauchelnd, das andere von altehrwürdiger, sich gleichwohl stets verjüngender Gestalt: frühe Kindheit neurobiologischer Forschung, fruchtbare Reifezeit philologischer Forschung. Jedoch sind die seit über 15 Jahren erzielten Befunde und Ergebnisse der bildgebenden Verfahren am menschlichen Gehirn ebenso wie die neuen EEG-Techniken der Messung ereigniskorrelierter Potentiale weit fortgeschritten. So weit, daß eine m. W. erste Bestandsaufnahme zur biologischen Anfrage an die Wirkungsprofile von Zaubersprüchen möglich ist.

Diese Studie richtet sich an alle diejenigen, die aus dem modischen Wust der gegenwärtigen Überdehnung und Vereinnahmung der Neurowissenschaften einen Ariadnefaden suchen und die gleichzeitig an Beispielen eines universalen menschlichen Phänomens magnetischer Kräfte von Sprache zu Bild und von Innenleben zum Sozialen interessiert sind. Über diese magnetische Beziehung hat Goethe geschrieben: „Wort und Bild sind Correlate, die sich immerfort suchen ...."[1]

Ich lade damit dazu ein, die rein fachspezifisch geprägten Schwellen zu überschreiten, an die die alte Magie- und die Sprechaktforschung mittlerweile gelangt sind. Das Gehirn ist das Zielorgan, für das Zaubersprüche geschaffen wurden, an dem Zaubersprüche sich vollenden können und das damit Kultur und Natur verbindet. Es geht also darum, Wege und Brücken zu konstruieren, die sich als vermittelnd-verweisend oder gar als komplementär verstehen können. Auf einem Nachbargebiet ist inzwischen eine beachtliche Pionierarbeit geleistet worden,[2] von der sich die hiermit vorgelegte Studie durch ihren am historischen Beispiel und am situativen Ereignis orientierten Bezug allerdings klar unterscheidet. Unser Thema verlangt nicht mehr nur interdisziplinäre, sondern notwendig multidisziplinäre Zugangsweise. Alle für den akut kranken und notleidenden Menschen jemals konzipierten und mannigfaltig aufgezeichneten tradierten ‚magischen', ‚performativen', ‚perturbierenden' ‚imaginativen' ‚zauberkundlichen' Formeln, Texte, Sprüche, Spells und Inkantationen übertreffen sich selbst. Sie weisen über sich weit hinaus, zumindest dann, wenn ihre einstige Funktion als

---
1   Goethe, Maximen und Reflexionen, Nr. 188.
2   Schrott, Raoul und Arthur Jacobs: Gehirn und Gedicht, München 2011.

Werkzeug des Medizinmannes, Schamanen und Arztes angemessen – und eben nicht allein als Kuriosum oder literarisches Produkt, als Dichtung – verstanden wird. Insofern wird Multidisziplinarität in unserem Rahmen neben Aspekten der Neurobiologie und Sprachwissenschaft auch einige kulturhistorische Hintergründe in der gebotenen Kürze einbeziehen müssen, weil die Therapie mit Worten nicht losgelöst von der jeweiligen sozialen Gemeinschaft denkbar ist. Ferner sind Seitenblicke auf einige Psychotherapiemethoden der Gegenwart wichtig, um das universal und zeitlos gültige Allgemeine aller verbalen Heilmethoden anzudeuten. Jedoch muß das ‚Wort' in seinen vielen Facetten Mittelpunkt und Ausgangspunkt im Dreiklang von Hilfe-Ruf, Helfer-Zuspruch und Umwelt-‚Sage' bleiben, weil es die größte Chance bietet, als Phänomen lebendiger Kommunikation aufzuscheinen. Dies entspricht der Erfahrung des Autors, der als Arzt in seiner Praxis stets dann am besten zu helfen vermochte, wenn er aus der Sprache des Kranken vor jeder Diagnose dessen ‚Weltbild' zu verstehen versuchte.

Dem Initium Johannis zur Menschwerdung des Wortes im Neuen Testament steht im Alten Testament verbindlich das Werk des Vaters gegenüber, der dem Menschen Odem einhaucht: „… und also war der Mensch eine lebendige Seele", was im 2. Jahrhundert in der Übersetzung des Targum Onkelos ins Aramäische, der Sprache des Sohnes, als kommunizierend-sprechender Geist übersetzt wurde, als ‚Adams discoursing spirit', worauf Miriam Faust,[3] die Herausgeberin des jüngsten Handbuchs zur Neuropsychologie der Sprache, hingewiesen hat. Die Worte der Handelnden wurden Motor oder Meteor wie viele andere Worte auch stets nur kommunikativ geformt und wirksam. Und Worte können bewegen, erleuchten und einschlagen, haben Völker erhoben oder erniedrigt. Mit Worten haben sich Liebende beglückt, haben sich Feindselige vernichtet und überzeugen oder verführen Redner und Geschäftige. – Es sei denn, die Worte fielen in eine Wüste. Es sei denn, Worten wurde kein Gewicht verliehen oder sie fielen zur falschen Zeit.

Die engen neuronalen Verbindungen der unsere Sprache repräsentierenden Gebiete mit den auditiven, visuellen und kinästhetischen Strukturen des Gehirns sind seit Beginn menschlicher Kultur auch Voraussetzung für die Einflußnahme durch Heilsprüche und all das, was gemäß den persischen Anekdoten des 12. Jahrhunderts über den Arzt Rhazes (ca. 854/860–925/935 nach Christus) als ‚ilaj-i-nafsani', als ‚Psychotherapeusis' bzw. Psychotherapie bezeichnet werden

---

3   Faust, Miriam, Einleitung zum Handbook of the Neuropsychology of language, 2012, Vol 1, S. XXVI.

konnte[4], was aber in Europa schon nachweislich als Verbaltherapie im 4. Jahrhundert vor Christus gefordert wurde[5].

Das dem Menschen eigene Sprachsystem, das nicht wurzellos ohne seine vorsprachlichen onto-phylogenetischen Voraussetzungen betrachtet werden kann, schafft die Vorlagen für die Erstellung der uns möglichen Weltmodelle. Anders als den Vogel das Fliegen, die Fledermaus das Radarsystem, die Bienen ihr Sonnenkompass, dem Kolibrigehirn sein Blütengedächtnis zeichnet uns Menschen Sprache und Gestik als wesentliches Mittel ihrer unentbehrlichen Kommunikation aus.

Sprache begleitet auch unsere verschiedenen Lebensbedingungen und schuf sich unter spezifischen Situationen – zum Beispiel im akuten Krankheitsfall – Worte mit Gewicht. Immer geht es darum, eingreifende, aufwühlende, möglichst strukturell nachhaltige Veränderungen im Gehirn zu erzielen, genau an der richtigen Stelle. Schon lange waren künstlerische Utopien darüber im Schwange, wie und wo die Entstehung und Umsetzung des gesprochenen oder gelesenen Wortes in Denken und Fühlen zu erkennen, zu kontrollieren und zu beeinflussen seien, nicht zuletzt als literarische Phantasie durch E. T. A. Hoffmanns Meister Floh mit seinem gedankenlesenden Minimikroskop, dem die neuen bildgebenden Verfahren weit unterlegen sind, und in Esoterik und Kunst mittels okkultistischer Sitzungen mit Gedankenfotografien sowie bei Surrealisten.

Die Frage, wie das tatsächlich nachweisbar funktioniert, hat Forscher mehrerer Disziplinen beschäftigt; in der Gegenwart eröffnet sich eine noch nicht übersehbare Baustelle für die Hirnforschung mit ihren neuen Methoden PET, fMRI mit der Möglichkeit des Neuroimaging, der Abbildbarkeit regionaler und interregionaler Aktivitäten sowie mit der Neurochemie der Transmitter und mit der Neuroimmunologie.

Ziel dieses Buches ist eine nach biologischen Aspekten geführte Analyse der Funktion von mittelalterlichen und archaischen Heilsprüchen, die nach den Quellentexten überwiegend in Akutsituationen (Kindbett, Krankheit, Schmerz, Verwundung, Angst, Überleben) eingesetzt wurden. Die Einbeziehung dieser Texte in eine neurobiologische Betrachtung ist bisher aufgrund von kategorialen Vorbehalten, Unkenntnis des tatsächlichen Textgebrauchs und archivalischen Quellenmängeln nicht begonnen worden.[6]

---

4   Browne, Edward G.: Arabian Medicine, Cambridge 1921, S. 83; Herrn Prof. Dr. Navid Kermani (Orientalist) danke ich für die bestätigende Mitteilung dieser Übersetzung (Brief vom 7.5.2013).
5   Platon, Charmides 157b.
6   Eine erste Annäherung habe ich versucht in: Beschwörungen und Segen – Angewandte Psychotherapie im Mittelalter, Köln Weimar Wien 2011.

Als Modelle verfügen wir einerseits über mittelalterliche Texte, die der Psychotherapie und psychosomatischen Therapie durch Heilberufe und verwandte Tätigkeitsfelder dienten. Es sind Beschwörungen von personalisierten Krankheitsdämonen und es sind Heilsegen. Sie wurden zum Teil nachweislich unter Mönchsärzten und Ärzten angewandt und erweisen sich damit als damals gezielt neuropsychosomatische Begleitinstrumente praktischer Behandlung. Es versteht sich von selbst, daß diese Texte ebenso wie ihre Anwender aus den Vorstellungen einer verallgemeinernden Zuordnung zu einer als System verstandenen Magie befreit werden können und daß somit auch archaische Texte mit heilkundlichem Inhalt, auch wenn ihre situativen Bedingungen nicht oder kaum bekannt sind, herangezogen werden können.

Eine biologische, also der Natur der Entfaltung, Übertragung und Wirksamkeit von Heilspruchtexten angemessene Sichtweise muß am Ort des Geschehens ansetzen, am menschlichen Gehirn. Die Theorien der geisteswissenschaftlichen Disziplinen sind in der Bearbeitung der funktionalen Elemente dieser Texte an ihr Ende gelangt. Das wird auch von den an diesen sogenannten magischen Texten arbeitenden Vertretern der Philologen und Linguisten inzwischen erkannt. Die Ironisierung der Wirkungsfrage als ‚Affenfrage' hielt unter ihnen nicht lange an. Vielfach sind die Grenzen und Reduktionsmechanismen rein philologischer, religionspsychologischer und soziologischer Forschung, zum Teil in Selbstzeugnissen belegbar.

Es „muss [...] rätselhaft bleiben, wie der konkrete Vollzug des magischen Aktes am Übergang von Text und materieller Welt zu denken sei."[7] – „Von der auf modernen Sprachzeugnissen fußenden Sprechakttheorie können Funktionsweise und Aktionsradius ‚magischen' Sprechens nicht adäquat abgebildet werden:"[8] – „Magie wird immer eine Restkategorie bleiben, vom wissenschaftlichen Beobachter geschaffen, um Handeln, das ihm unverständlich (irrational) erscheint, zusammenzufassen."[9] – „Wie kommt es, daß Menschen daran glauben konnten, ohne Arbeit durch das beschwörende Raunen eines poetischen Spruches direkt auf die Wirklichkeit wirken zu können? Zu Gutem und Bösem. Wir müssen Grenzen der Philologie und Linguistik überschreiten, um erklären zu können."[10] – „Sprechakttheoretiker vermuten die Kommunikationsmacht in der Einhaltung von Konventionen oder gar in einer (fast magischen) illokutionären Kraft, die aus der Sprache selbst hervor geht. Das scheint mir wenig plausibel."[11]

---

7 Haeseli, Christa M.: Magische Performativität, Würzburg 2011, S. 204.
8 Kropp, Amina: Sprachliche Betrachtungen zu den lateinischen Defixionum tabellae, in: Acta antiqua Academiae Scientiarum Hungaricae 49 (2009), S. 77–93, hier: S. 92.
9 Handbuch Religionswissenschaftl. Grundbegriffe 1998, S. 95, zit. nach Bamert, Martin, Magia, S. 13f.
10 Geier, Manfred: Die magische Kraft der Poesie. Zur Geschichte, Struktur und Funktion des Zauberspruchs, in: Dt. Vierteljschr. f. Literaturwsch. und Geistesgesch. Halle/ sp. Stuttgart 56 (1982), S. 359–385, hier: S. 378.
11 Reichertz, Jo: Kommunikationsmacht, VS Verlag für Sozialwissenschaften, 2009, S. 15.

Neben den großen technischen Fortschritten der letzten 20 Jahre haben auch diese und ähnliche Grenzerfahrungen der pragmabezogenen Geisteswissenschaftler die Naturwissenschaften und besonders die Hirnforschung angeregt. Zu einer generellen Hybris, den Stein der Weisen gefunden zu haben, ist ihre Zunft allerdings auch nicht gelangt. „Semantische Inhalte lassen sich im Gehirn nicht erfassen."[12] Und es bleibt die soziale Interaktion eine der Voraussetzungen für die Konstitution eines sich frei wähnenden Selbst.[13] Daß psychische Phänomene nichts anderes als feuernde Nervenzellen seien, ist auch von Gerhard Roth als Reduktionismus beschrieben; ebenso daß ein von seiner Umwelt isoliertes Gehirn keinen Geist entstehen läßt und daß die Forschungsergebnisse am Gehirn nicht global als objektive Wahrheit, sondern nur als plausibel zu verstehen sind.[14] Ähnlich beschreibt der philosophisch orientierte Psychiater Thomas Fuchs das Gehirn als Organ eines Lebewesens, einer lebendigen Person, das kein Eigenleben führen kann.[15] Der soziale Informationsfluß verbietet im Blick auf neurale Verarbeitung letztlich simple Schematisierung schon deshalb, weil er multidirektional und rekursiv ist.[16] Selbst ein Konstruktivist wie Ernst von Glasersfeld bezeichnet die Quintessenz des italienischen Philosophen Giambattista Vico im 18. Jahrhundert, „Gott ist der Schöpfer der Welt, der Mensch der Gott der Artefakte" als einen ‚wunderbaren Satz'.[17] Und erinnern wir an die Sichtweise Niels Bohrs zur Atomphysik: Verständigung ist nur durch Gleichungen oder Gleichnisse möglich; der Beobachter findet sich in den Atomen wieder; Naturwissenschaft versteht sich nicht logisch, sondern dadurch, daß innere und äußere Bilder zur Deckung gebracht werden.

Der vorliegende Versuch einer neurobiologischen Annäherung an historische und ‚archaische' medizinische Behandlungstexte trägt diesen Grenzen der Naturwissenschaften insofern Rechnung, als er einerseits *die situative Bedingung* von Notfall- und Krisenintervention, also strikt nur akute Maßnahmen als Auslöser eines verbaltherapeutischen Bündnisses unterstellend heranzieht. Diese Begrenzung entspricht den noch kontroversen neurophysiologischen Forschungsergebnissen über den Zusammenhang zwischen menschlicher Sprache

---

12 Singer, Wolf, FAZ 25.11.2004.
13 ders., Unser Menschenbild im Spannungsfeld zwischen Selbsterfahrung und neurobiologischer Fremdbeschreibung, in: Frühwald, Wolfgang u. a.: Das Design des Menschen, Köln 2004, S. 182–215, hier: S. 202.
14 Roth, Gerhard: Das Gehirn und seine Wirklichkeit, Frankfurt/ M., 1998², S. 285, 363.
15 Fuchs, Thomas: Das Gehirn – ein Beziehungsorgan, Stuttgart 2008/2013, S. 185–223.
16 Adolphs, Ralph, Cognitive neuroscience of human social behavior, in: Nature Reviews Neuroscience 4 (2003), 165–178, dort: Abb.1, S. 167.
17 Zit. nach Watzlawick, Paul und Giorgio Nardone (Hg.:) Kurzzeittherapie und Wirklichkeit, München 2012, S. 43–58, hier: S. 49.

und Erkenntnis. Anerkannt ist die Rolle der Sprache für ein Initialstadium der Information, möglicherweise emotional und durch nonverbale Signale gefördert.[18]

Andererseits wird von mir für diese Situation die *Struktur einer Triade* postuliert: Heilkundiger, Empfänger, kulturelles Umfeld. Denn es besteht kaum Zweifel, daß die Gehirne unserer Vorfahren und die unserer Zeitgenossen in anderen fernen, nicht modern wissenschaftlich geprägten Gebieten der Erde nach gleichen funktionalen Prinzipien arbeiteten und arbeiten wie die unseren, jedoch eine andere kulturelle Prägung erhalten haben. Aufgrund einer vergleichbaren, seit rund 40000 Jahren nur unwesentlich veränderten genetischen Ausstattung würde sich heute ein Steinzeitbaby genauso wie wir entwickelt haben, sofern es in unserer Gemeinschaft aufgewachsen ist.[19] Die ‚magischen' Dokumente der Spruchtexte sind Zeugnisse nicht nur einer Affinität zwischen Heiler und Heilungsuchendem, Helfer und Hilfsbedürftigem, sondern Relikte einer wichtigen Sonderform sozialer Interaktion. Signalvermittlung und Signalaufnahme, komplementäres und schwankendes Gegenüber von Phänomenen vermeintlicher Macht-Ohnmachtsbeziehung, Reflektierung der Signale und ihre Optimierung im genuin sozialen Spiegelsystem und eben darüberhinaus die Einbettung in die jeweilige gesellschaftliche Umwelt bilden dieses triadische Gravitationsfeld (**Kap. B**). Erst unter diesen Prämissen ist es möglich, an den verschiedenen Texten die jeweiligen neuronalen Verschaltungssysteme modellhaft abzubilden (**Kap. C**). Einige wissenschaftsgeschichtliche Szenen umreißen die bisherigen Versuche, all das, was man unter dem Megawort Magie verstand, zu klären (**Kap. A**).

Mit der Beschränkung auf einen solchen ebenso situationsbezogenen wie gesellschaftlich-sozialpsychologisch vorzustellenden teilweise systemisch orientierten Ansatz treten die theoretischen Fragen nach dem Weltbild der Hirnforschung und ihrer ‚schlechten Metaphysik' (Habermas), nach der Kluft zu den Geisteswissenschaften, aber auch die Fragen nach einem vermeintlichen und frei schaffenden Dirigenten des menschlichen Gehirns, dem Homunculus-Modell, in diesem Buch in den Hintergrund. Weder hirnbiologischer noch medizinmännischer Solipsismus prägen die hier unternommene Darstellung, sondern allein das Phänomen menschlicher Notlage mit ihrer Ermöglichung des Zusammenfließens von Wissenschaft und Gebrauchsdichtung in neuralen Kreisläufen. In menschlichem Leid und Not, in Unordnung und Desorientiertheit gelten andere Gesetze, die eben notgedrungen für statische Systeme und für ‚Disziplin' und ‚Diszipli-

---

18 Monti, Martin M. and D.N. Osherson, Logic, language and the brain, in: Brain research 1428 (2012), 33–42.
19 vgl. Singer, Wolf, Was kann ein Mensch wann lernen?, in: Universitas 56 (2001), S. 882.

nen' kategorial unmöglich werden, weil unerlaubte Vermischungen stattfinden müssen, die wir Kombinatorik nennen. Nur die Kombination historischer, neurobiologischer und philologischer Elemente ist in der Lage, ein Licht auf die Anwendung von Heilspruchtexten in den Kulturen zu werfen und Distanz zu einem ausgelaugten Begriff wie Magie zu nehmen.

Vorbehalte gegen ein zu Recht fragwürdiges „Alles-Neuro" sind uns von vielen Seiten entgegengeschallt. Reaktionen der Dichter reichen bis zur Technophobie und Entseeltheitsklage. **Botho Strauß:**[20] „... Echtzeit der Neurochemie, Geschwindigkeiten, die das Selbst nicht berühren, nichts angehen – es wüßte denn um sie". – Es wird die biologisch zu erschließende Gehirnwelt als etwas Unfassbares, als Zirkelschluß des Denkens bedichtet und die Aussicht auf Entgrenzung beschworen: **Hans Magnus Enzensberger:**[21] „Denk dir einen Baobab-Baum, riesenhaft reich verzweigt ... mit abertausend winzigen Affen ... Dann wieder springen sie, rasen behende, wimmeln elektrisch, taumeln und stürzen ab; [...] Lach, erschrick, wundere dich, doch hör auf, bevor du verrückt wirst, über das Nachdenken nachzudenken". – Der Soziologe **Günter Dux**[22] beschreibt wie viele seiner und anderer geisteswissenschaftlichen Zünfte das reduktionistische Vorgehen der Hirnforschung, ruft in höchster Bedrohtheit ein „Die Biologie ist ante portas" in seine Zeilen hinein und schlußfolgert in seinem theoretischen Werk: „Wenn man im Verständnis der menschlichen Daseinsform als Anschlußform an eine Evolution der Arten eine ‚Naturalisierung des Geistes' sehen will, so doch nur in dem Sinne, daß die geistigen Kompetenzen, die kognitiven, linguistischen, normativen und ästhetischen, auf naturalen Kapazitäten aufbauen, ohne in ihnen selbst schon enthalten zu sein." – Der Philosoph **Robert Spaemann**[23] spricht von einer Kategorienverwechslung. Die Hirnforschung möge sich auf Empirie beschränken. Er wirft den Neurologen vor, Befunde mit Folgerungen zu vermischen und ihrer eigenen Methode auf den Leim zu gehen: „Sie finden Hirnzustände und urteilen über Freiheit!" – Zuletzt wurde das MRT-Scanning als ‚magisches Objekt' gebrandmarkt, weil der Anspruch der Hirnforschung dahin gehe, sich gegenüber der Kultur „als ontologisch maßgebende erste Natur" zu betrachten.[24]

---

20 Strauß, Botho: Vom Aufenthalt, München 2009, S. 67.
21 Enzensberger, Hans Magnus: Neuronales Netz, in: Die Elixiere der Wissenschaft, Frankfurt 2002.
22 Dux, Günter: Historisch-genetische Theorie der Kultur, Weilerswist, 2000, S. 53, 67f.
23 Spaemann, Robert: „Ich und Gott ein Hirngespinst" Kooperationsveranstaltung 13. und 14. Februar 2009, Online-Mitteilung Bistum Augsburg.
24 Slaby, Jan: Die Objektivitätsmaschine, in: Mertens, K. und I. Günzler (Hg.): Wahrnehmen, Fühlen, Handeln, Paderborn 2012 (Onlineankündigung).

Die Seite der Hirnforschung kann dagegen im Wissen um ein weitgehend selbstbestimmtes System menschlicher Hirnleistung ihr Eindringen in die Domänen der Geisteswissenschaft als „fascinierende Konsequenz einer erneuten Annäherung von Kultur- und Naturwissenschaften" sehen[25] und verlangt ein „Mitspracherecht bei der Diskussion um Geist, Seele und Bewußtsein,"[26] weil sich „Physiologie und Neurochemie so nahe an die Entstehungsorte und Wirkmechanismen für bestimmte Gefühle und Wahrnehmungen herangetastet" haben. – „Ein eingehenderer Blick in die Werkstatt der Natur vermag das Bild des Menschen deutlich zu verändern und es vor allem zu präzisieren, finden wir doch manches wieder, das in Philosophie und Literatur in seinen Außenaspekten längst formuliert, aber unverstanden war."[27] – Sigmund Freud (1920, S. 65) hatte zu seiner Zeit den „Sprung" vom Seelischen zum Biologischen vorausgesehen: „Die Mängel unserer Beschreibung [des psychischen Geschehens] würden wahrscheinlich verschwinden, wenn wir anstatt der psychologischen Termini schon physiologische oder chemische einsetzen könnten."

Jenseits unserer Betrachtungen liegen auch berechtigte, aber immer nur unvollkommen lösbare Zweifel an der Treffsicherheit, mit der wir uns in diesem Buche ‚dem Geist der Zeiten' annähern. So bleibt die Parallelisierung heutiger Psychotherapiemethoden mit historischen und ethnologischen Textquellen natürlich unvollkommen. Hingegen würde die Illusion einer absolut angemessenen Versetzungsmöglichkeit in einen historischen Rahmen, wie sie zum Beispiel Michel Foucault mit ‚Wahnsinn und Gesellschaft' lieferte – ähnlich der Ge- und Befangenheit eines autistischen Savants in seine *eine* Bibliothek – unverkennbare Züge eines Romans tragen.

---

25 Singer, Wolf: Der Beobachter im Gehirn, Frankfurt 2002, S. 33.
26 Bauer, Roman, Marburg, Universitas-online, o. J.
27 Dichgans, Johannes, Mimik, Gesten und Sprachmelodie. Medien sozialer Kommunikation und ihre neuronalen Grundlagen, in: Frühwald, Wolfgang et al. (Hg.): Das Design des Menschen, Köln 2004, S. 217–231, hier: 229.

A Eine kurze Forschungsgeschichte zu Entstehung, Anwendung und Wirkung des Heilspruchs und der Heilriten: Die Konstrukte des Unbegreiflichen

A1 Der Magiebegriff in Theorie und Praxis

Die bisherigen Versuche, Entstehung, Praxis und Folgen der Anwendung psychoperformierender Sprüche zu erklären, stehen für Laien und einen Teil der Wissenschaftler auf dem ungewissen Boden eines Allerweltsbegriffes: Noch immer gilt ‚Magie' als schlagkräftige Antwort auf alle betreffenden Fragen.

Aber was war und was ist Magie? Der Begriff gehörte und gehört zu den am schwierigsten definierbaren Begriffen. Jedem, der sich auf den Versuch einließ, ihn zu klären und zu fassen, drohte ein Gespenst. Es drohte mit Verschlingen, Blitzschlag oder Wasserflut, einen ewigen Zauberlehrling zu fesseln, niederzustrecken oder zu ersäufen.

Denn ‚Magie' ist vielfältig und wandelbar in ihren Phänomenen wie Proteus und kaum schlagbar regenerationspotent wie die Hydra. So sammelte ein Doktorand der Neuzeit allein zum Phänomen „böser Blick" bis zu seinem Tode 18 Millionen Belege und vererbte sein Vermögen dem Doktorvater; allein die Sammlung blieb verschwunden.[28] Andere Sammler waren unter Beherrschung ihrer anankastischen Persönlichkeitsanteile oft bis zum Versuch gekommen, Ordnungsprinzipien in all dem zu schaffen, was jeweils als Magie oder magieverdächtig verstanden wurde.

Mit Beginn des Buchdrucks erschienen die Werke von Agrippa von Nettesheim und Johannes Trithemius. In den folgenden Jahrhunderten begegnen Sammlungen und Lexika, die ihrerseits wiederum Sammlung und Deutungen veranlaßten, wie sie Will-Erich Peuckert 1936 und 1967 mit Pansophie und Gabalia anhäufte. Zu Ende des 18. Jahrhunderts hatte Johann Samuel Halle die ‚Fortgesetzte Magie oder die Zauberkräfte der Natur' vorgelegt und am Beginn des 19. Jahrhunderts Georg Conrad Horst seine ‚Zauberbibliothek'. Immer wieder neue Auflagen teilweise obskurer Verleger für ein spezielles Marktsegment erzielten unter anderen die Mosis-, Albertus Magnus-, Glorez- und Staricius-Bücher. Das ‚Handbuch des deutschen Aberglaubens' (1927–1942) war eine Teamleistung der germanistisch-volkskundlichen Zunft.

---

28 Hauschild, Thomas: Der böse Blick, Berlin 1982², S. 3.

‚Magie' geht mehr oder weniger obskure Bündnisse, ja Legierungen und Konfundierungen mit Forschungsgegenständen vieler Disziplinen der hohen Wissenschaften ein, mit Sprache, Bild und Verhalten, mit Gebet und Therapie und Machtanspruch, mit Traum, Phantasie und Weltanschauung, woraus ganz verschiedene Aspekte von Magie resultieren, ganz zu schweigen von allen Gelegenheiten des Alltagslebens, bei denen der Begriff poetisch bis polemisch eingespannt wird. „Die Bemühung um eine allgemeingültige Definition (scheint) ins Unerreichbare gerückt zu sein."[29]

> Beispiel für diese Versammlung unvergleichbarer Zuordnungen zu Magie findet sich bei Daniel Lawrence O'Keefe 1982 (Stolen Lightning. A social theory of Magic, New York 1982); er subsummiert medizinische, schwarze, weiße, zeremoniale, religiöse Magie, Okkultismus, Paranormales, magische Kulte und Sektenwesen.[30] Beispiele für eine instrumentelle Ausweitung des Begriffes oder – schwer zu trennen – von Legierungen liefern auch Soziologen: ‚Magie' wird als Vorläuferin von Technik in die Nähe affektiver Besetztheit, Faszination am Automatismus und damit in die Nähe von Zweckmäßigkeitsdefinition alter Prägung gerückt und als Zwangshandlung menschlicher Mängelwesen, so Arnold Gehlen (1904–1976). Sie betrachten ‚Magie' als Konsumobjekt unserer Zeit im Strom der irrationalen Fluten aus Medien und entdecken ihre dauerhafte Janusköpfigkeit gebunden an einen Macht-Ohnmacht-Komplex. „Das Magische ist eine Form der Beziehung des Individuums zur Außenwelt, bei der Angst und (oder) der Wunsch, die äußere Welt unter Kontrolle zu bringen, das viel feinere Verlangen nach Erklärung und Rationalisierung überwiegt."[31] Selbst Erklärungen blieben noch an Zwecke gebunden. – Aber ähnlich jenem ewigen Doktoranden droht dem Janusergreifer eben dieser mit seiner List, Vor- und Rückblick in einem zu wagen. Und so erliegt der Soziologe seiner rück-und-vor-sichtsvollen Ehrlichkeit, indem er einräumt, daß seine eigene Arbeit [gegen einen sich ausweitenden Gesellschaftsbegriff] polemische Bedeutung gewinnen könne.[32] Eine Parallele zur modernen Welt findet auch Claude Lévi-Strauss, wenn er eine Ähnlichkeit von mythischem Denken und politischer Ideologie feststellt.[33] – Heute seien magische Praktiken nicht Beeinflussungsversuche höherer Mächte, sondern „individuelle Abwehrmechanismen."[34]

Mit diesen ihren Proteus-, Hydra- und Hetären-Eigenschaften haben die schillernden Phänomene der Magie über die ihrer Bändigung dienenden Sammlungen und Kategorisierungen hinaus viele Theoretiker gefunden, nicht zuletzt weil ihre subversiv bedrohlichen Kräfte nicht nur immer wieder empfunden, sondern auch weil sie genutzt werden konnten. Und dies ebenso für zentrale Fragen nach

---

29 Angst, Beatrice E.: Magische Praktiken des Menschen unserer Zeit, Bern u. a. 1972, S. 16.
30 Zit. nach Petzoldt, Leander, Magie und Religion, in: Dinzelbacher/ Bauer (Hg.:) Volksreligion, S. 483.
31 Mongardini, Carlo, Über die soziologische Bedeutung des magischen Denkens, in: Zingerle/ Mongardini, Magie und Moderne, S. 14.
32 Ebd., S. 55.
33 Lévi-Strauss, Claude: Strukturale Anthropologie, I, Paris 1958, (dt.) Frankfurt/M. 1981², S. 230.
34 Angst, Beatrice E.: Magische Praktiken, S. 131,135.

menschlichen Erkenntnismöglichkeiten, -abwegen und -chancen. Und weil sie heute in eine Epoche fallen, die die Annahme einer absoluten, objektiven Wahrheit aufgegeben hat und die Realität und Irrealität einer rein subjektiven Instanz unterstellt.

A1.1 Historische und historisierende Rezeptionen und Deutungen von Magie und ihr Fortbestand in die Gegenwart

Im 6. Jahrhundert vor Christus kamen die Begriffe ‚magus' und ‚magein' zusammen mit dem Wissen um den Kult der persischen Priesterkaste der Magi oder Magoi in die griechische Sprache. ‚Magh' als indoeuropäisches Suffix bedeutet Fähig-Sein und Kraft-Haben und soll etymologisch mit ‚Macht' zusammenhängen. Für antike Autoren verband sich mit den Magoi die Vorstellung von übersinnlichen Fertigkeiten und verborgenen Kräften, von Wissen um Astrologie (Matthaeus 2,1) und um Heilriten. Wie unterschiedlich und schwankend allerdings heutige rückblickende Sichtweisen sind, zeigt eine Übersicht über einige religionshistorische Standartwerke.[35] Mit dem frühen Christentum behielten ‚Magier' und ‚Magie' ihren schon zuvor suspekten Klang und wurden nicht nur mit Heidentum, sondern auch mit Dämonischem gleichgesetzt. Die Lehren des Augustinus (354–430) und des Isidor von Sevilla (ca. 560–636) blieben bindend. Im karolingischen Zeitalter gewann die Abgrenzung der Religion von Magie zunehmend auch politische Bedeutung im Rahmen der Missionierung und Kultivierung der Völker Europas.

Im 12. Jahrhundert zeigen sich im westlichen Europa wieder erste Anzeichen einer Differenzierung im Verständnis des Begriffes, nachdem zuvor die ‚Künste' Geomantie, Hydromantie, Aeromantie, Pyromantie bis hin zum Sprechen von Beschwörungen über Kranke offiziell oftmals als Teufelswerk galten. Scholastische Wissenschaften der Universitäten unterscheiden im Bildungssystem neben den sieben freien Künsten und den unfreien mechanischen Künsten eine bedeutende, im Verborgenen existierende Gruppe der verbotenen Künste, die Artes illicitae, nämlich Magie und Mantik.[36] Albertus Magnus (ca. 1200–1249) eröffnet in seinem naturkundlichen Werk den Blick auf eine nichtdämonische natürliche Magie (magia naturalis, schon der Begriff gab einen ganz neuen Denkanstoß), wie sie von den Geheimnissen um Naturgesetze ausgeht. Sie, die Magia naturalis, wird der Magia diabolica entgegengesetzt und erfährt in den folgenden Jahrhunderten hohe Beachtung und Wertschätzung, besonders bei Giovanni

---

35 Bamert, Martin: Magia – Maga – Makha, Zwischenprüfungsarbeit Leipzig WS 06/07.
36 Weddige, Hilkert: Einführung in die Mediävistik, München 1992², S. 53.

Baptista della Porta (1535–1615). Eine erste Übersicht über verschiedene Formen vermittelte Heinrich Cornelius Agrippa von Nettesheim (1486–1535), der sich wie die meisten weitgehend des sozial separierenden abhebenden Charakters seiner Wissenschaft bedient. Für ihn besteht im Menschen eine schöpferische imaginative Kraft und ein Lebensgeist, der Wunderbares schafft. Und es verrät sich der Nettesheimer, der diese eine Kraft der Magie als apollinische Begeisterung beschreibt: „Ich habe diese Wissenschaft so vorgetragen, daß den Klugen und Verständigen nichts davon verborgen bleiben soll. Den Schlechten und Ungläubigen dagegen soll der Zugang zu diesen Geheimnissen verschlossen bleiben."[37]

Aber der Haupttrend, Magie in ihren vielfältigen Facetten stets als Gegensatz und Gegenspieler von Religion oder Hochreligion möglichst sauber zu trennen, ja zu definieren, beherrscht mit Zähigkeit viele literarische Niederschläge bis in die Neuzeit. So wurde etwa die Dämonologie des Konvertiten und Päpstlichen Geheimkämmerers Egon von Petersdorff (1892–1963) noch 1995, also nachkonziliar in dritter Auflage verlegt, während der Wiener Mediävist Helmut Birkhan (geb. 1938) in einer Nachgeburt alter Magiesammlungen frei heraus bekennt: „Aberglaube und Magie kann man in unserem Kulturkontext einfach dadurch bestimmen, daß sie im Widerspruch zur aktuellen christlichen Lehre, aber auch zum Erkenntnisstand der (Natur-) Wissenschaft und zur menschlichen Vernunft stehen." [38]

Da waren andere aufmerksamere Zeitgenossen schon weiter. Die alte Aufteilung Religion/ Magie war als ‚künstliches Problem'[39] erkannt worden. Die drei Unterscheidungsschablonen Bitte/Zwang, Demut/Zweckdenken, Personalisierung Gott/ Satan waren überwunden. Einige haben in Magie die Reaktion auf mangelhaftes empirisches Wissen, auf menschliche Begrenztheit sehen wollen, also eine Ersatzfunktion. Zuletzt war eine sozialpsychologische Sichtweise diskutiert worden. So hält der Germanist und Volkskundler/Ethnologe Leander Petzoldt (geb. 1934) mit dem Historiker und Religionspsychologen Karl Beth[40] (1872–1959), die Wunschvorstellung, die ‚einfache und unverhüllte Objektivierung des Wünschens' als ‚anthropologische Grundkomponente' für wesentlich und interdisziplinär konsensfähig.[41] Kein geringerer als Elias Canetti hat ziemlich freimütig und treffsicher in diese Kerbe geschlagen: „Zuwenig Zaubersprüche gelesen. Gestern nacht nahm mich ein altes Zauberbuch der Inder, der

---

37 Agrippa von Nettesheim: Der Geheimen Philosophie drittes Buch, S. 379.
38 Birkhan, Helmut: Magie im Mittelalter, München 2010, S. 11.
39 Pettersson, Olof (1957), dt. Übersetzung in: Petzoldt, L.: Magie und Religion, S. 386.
40 Herausgeber einer Zeitschrift für Religionspsychologie von 1927–1938, u. a. mit dem Thema Psychologie des Unglaubens.
41 Petzoldt, L.: Magie und Religion, in: Dinzelbacher/Bauer (Hg.:) wie oben, S. 473.

Atharvaveda, gefangen. Es stehen unheimliche Dinge darin. Nirgends sind die Wünsche des Menschen unverhohlener ausgedrückt. Es ist eine ganz elementare Welt, und wer etwas Wirkliches über die Menschen erfahren will, muß zu den Mythen die Zaubersprüche, die nackt sind, dazunehmen."[42] Wir werden bei der Diskussion neurobiologischer Ergebnisse der Erwartungs- und Aufmerksamkeitsforschung (Kap. **B1**) sehen, wie sich diese anthropologische Theorie bestätigt hat.

> Zu vermuten ist, daß die Vernachlässigung der seit 100 Jahren mit den britischen und französischen Ethnologen vorgedrungenen wissenschaftlichen Magieforschungen bei manchen Mediävisten mit der Gewichtung aufsehenerregender Dämonologie seit dem ‚Hexenhammer' zusammenhing. Die Konkretisierung der ursprünglichen theoretischen Dämonologie der Gelehrten und ihre juristische Instrumentalisierung beim Versuch der Bearbeitung der Krisenbewältigung im europäischen Spätmittelalter wurde weithin zum Hauptthema. Mindestens seit der Zeit der Aufklärung waren auch die mittelalterlichen Heilspruchtexte nur noch bei Volksheilern und in Volksbüchern in meist korrumpierten Formen verbreitet, während die von der Kirche allein noch bewahrten Exorzismen als Relikte der Dämonenvertreibung Aufmerksamkeit erregten.

Die Ärzte des europäischen Mittelalters standen der Magie kaum anders gegenüber als die Gesellschaft; nach vielen schriftlichen Zeugnissen konnte gegenüber den Theologen meist wenig eigenständige Haltung zu Magie und Dämonen gefunden werden. Magie war auch ein Begriff der Polemik. Am Dämonenglauben wurde bis in die Neuzeit festgehalten, wenn auch viele zu einer gemäßigten Magia naturalis neigten wie Paracelsus, der den Teufelsglauben bezweifelte und die ‚Natur' als Fundament seines alchemistischen Systems darstellte. Paracelsus hat allerdings überschwenglich die hohe Bedeutung der Imaginationskraft als ‚*Sonne im Menschen*', als kosmische Kraft beschrieben: der „ganz himel ist nichts als imaginatio",[43] und damit neue Impulse gegeben. Und wenn die dämonenzwingenden Kräfte auch als abzulehnen erkannt wurden, so waren damit meist die Kräfte der konkurrierenden Exorzisten und Zauberer gemeint. Selten sind aus dem Hochmittelalter Warnungen vor denjenigen naturwissenschaftlich orientierten Ärzten notiert, die dämonische Besessenheit leugneten.[44]

---

42 Canetti, Elias: Nachträge aus Hampstead. Aufzeichnungen 1957–1959, München 1994, S. 17.
43 Godet, Alain: „Nun was ist die Imagination anderst als ein Sonn im Menschen", Zürich 1982, S. 35f.
44 Vgl. Beispiel bei Franz, Adolph: Benediktionen, II, S. 520.

A1.2 Moderne Magie-Theorien[45]

A1.2.1 Rationalistische Erklärungen der Magie

Promotoren einer erstmals rationalistischen Erklärung von Magie waren die Briten Edward B. **Tylor** (1832–1917) und James G. **Frazer** (1854–1941). Ihr Zeitalter stand unter dem Eindruck der englischen Aufklärung, besonders aber der biologischen Evolutionslehre Charles Darwins (1809–1882) und der Begegnung mit fremden Völkern während der Victorianischen Epoche des Kolonialismus. Zunächst entwarfen Tylor und Frazer ein evolutionistisches fortschreitendes Stufenkonzept zur menschlichen Denkweise aus Magie über Religion zur Wissenschaft und stießen damit teilweise auf Widerstand. Nicht so sehr die weltzeitliche Abfolge, doch aber diese Dreiteilung, schon von Hippocrates angelegt (Die heilige Krankheit), blieb manchem Wissenschaftler bis heute maßgebend, obwohl stets die Gefahr von Werturteilen inhärent ist. Besonders ihre Analyse des magischen Denkens erwies sich als richtungweisend: Tylor ersetzte Kausalität und Effizienz der magischen Aktion durch ‚similarity' und ‚contagion'; Magie wird als vorwissenschaftlich gedeutet, aber von Okkultismus unterschieden. Frazer entwickelte das Gesetz der ‚sympathischen Magie': Der Glaube an die Aktionskraft von Dingen in magischen Riten ist danach das unbeabsichtigte Resultat von Ideenassoziationen, also einer menschlichen mentalen Struktur, womit „magische Prozeduren als spezifische kognitive Prinzipien"[46] vorgestellt werden können.

Der französische Soziologe Emile **Durkheim** (1858–1917) hat in Reaktion auf Tylor und Frazer mehr die kollektiven Elemente in den Ethnien hervorgehoben. Man begann nun, die Erfahrungen des engen Gemeinschaftslebens und seiner Ereignisse als wesentliches Movens für kulturelle Strukturen mit ‚magischen' Erscheinungen zu erfassen. Während also ein Schamane mit den Zeichnungen auf seiner Bekleidung die Buschsorten repräsentiert, die eine Raupe in ihren Stadien durchmacht und dann das Fliegen des Schmetterlings darstellt, sitzen alle Mitglieder im Halbkreis und haben Blickkontakt mit ihm, der diese Wandlungen vollzieht. Dies ist für Durkheim ein kommunikatives Geschehen, das nicht durch die sympathetische Imitation, sondern nur durch die Gemeinschaftserfahrung effektiv werden kann. „In Durkheim's work individual errors of thinking were replaced with social modes of association"[47], womit für uns Spätere zugleich ein Seitenblick auf Morenos Psychodrama gewagt werden könnte.

---

45 Einteilung in Anlehnung an Sørensen, Jesper: A cognitive Theory of Magic, Aarhus 2000/01.
46 Ebd., S. 11.
47 Glucklich, Ariel: The end of magic, N. Y. u. a. 1997, S. 40f.

Aufbauend auf diesen Forschungen steht für Bronislaw **Malinowski** (1884–1942) die Frage nach der Bedeutung der Instrumente der magischen Aktion und ihrer Unmittelbarkeit im Vordergrund, womit situativer Kontext und kultureller Rahmen bedeutsam werden. Er war Sozialanthropologe und verbrachte viele Jahre bei Eingeborenen auf den Trobriandinseln. Als erster hat er eine allgemein übersichtliche auf psychische, ökologische, sozialreligiöse und ökonomische Faktoren setzende Forschung vorgelegt. Insbesondere „betont er die pragmatischen Effekte von Sprache und antizipiert damit die spätere performative Wende der Sprechakttheorie"[48] von John L. Austin (How to do things with Words, postum Oxford 1962) und John R. Searle. Er behauptet eine der Magie eigene ‚sacred language' gegen die gewöhnliche Sprache und den Einfluß dieser Sondersprache des Spruchtextes auf die Motivation der Teilnehmer und den Effekt der Prozedur. „Damit ist nun Magie weder falsche Wissenschaft noch primitive Mentalität."[49]

In Reaktion auf Tylor und Frazer hatte es auch Stimmen gegeben, die zwischen Religion und Magie lediglich instrumentelle Differenzen sahen und die betont die Ähnlichkeit der intellektuellen Prozesse bei Wissenschaft, Religion und Magie heraustellten; deren Suche nach erklärender Theorie wird als Versuch beschrieben, einem chaotischen Weltall Ordnung aufzudrängen (Robin **Horton** 1970).

A1.2.2 Emotionalistische Erklärungen der Magie

Gefühle als Ursprung magischer Aktion und Funktionsbedingung heranzuziehen legt den Schwerpunkt auf individuelle psychische Phänomene. Für Robert R. **Marett** (1866–1943), der im Schatten von Frazer stand, hat Magie ihren Ursprung in emotionalen ‚outbursts' wie Liebe, Angst und Hass, die sich nicht auf ein Objekt, sondern auf Mana oder Gott als Projektionen beziehen. Die Funktion der Magie wird als elementare Katharsis zur Spannungsabfuhr beschrieben, wie dies auch von anderen Magieforschern (s. u. Lévi-Strauss) geäußert wurde. Dabei soll aus einem ‚naiv belief' ein ‚make belief' werden. Marett nähert sich der Theorie der abreagierenden Redekur Breuers und der Psychoanalyse Freuds, der ihn seinerseits in Zusammenhang mit seinen Animismusthesen heranzieht.

Diesem individualpsychologischen Ansatz stand mit dem ‚Gesetz der Teilhabe' von Lucien **Lévy-Bruhl** (1857–1939) eine mehr soziale Theorie gegenüber, wie sie sich schon bei Emile Durkheim ankündigte. Fremden Kulturen wird ein

---

48 Sørensen, ebd., S. 19.
49 Ebd., S. 20.

prälogisches eigenes Denken zuerkannt, eine Art Netzwerk kollektiver Konnektion. Am Beispiel sozialer Unternehmungen wie der Jagd der Eingeborenen stellt Lévy-Bruhl fest: „Jagd ist eine essentielle magische Operation und von ihr hängt alles ab, nicht vom Zielen und von der Kraft des Jägers, sondern von der mystischen Kraft, die das Tier in ihre Gewalt nimmt."

> Die weltweiten Befundaufnahmen werden jeweils als kollektive Repräsentationen der sozialen Gruppe beschrieben. Die kollektive Aktion unterstehe festen mystischen Beziehungen. Vor der Jagd finden gemeinsame Maskentänze, Gesänge mit Beschwörungen der Jagdgeister statt. Fastenbräuche, Lanzenwurfsprüche, Haarabschneiden der Frauen während der Jagd, Opfer zur Zähmung der Jagdgeister nach der Jagd, alles in ‚operativer' magischer Strategie.[50]

Nach diesem Gesetz der Teilhabe würde magisches Denken in die Nähe des Traumes rücken und würden Wesenheiten, Phänomene, Gegenstände, alles gleichzeitig selbst und anders sein. Man versteht dann Seelenwanderung, Multipräsenz und Diffusion von Raum und Zeit, Belebtem und Unbelebtem. Dieses Gesetz, auch als ‚participation mystique' benannt „fand in der wissenschaftlichen Diskussion weite Beachtung, führte es doch bereits in die Richtung des von C.G. Jung untersuchten ‚kollektiven Unbewußten'."[51] Später hat Lévy-Bruhl seine Thesen neu formuliert und auch den modernen Gesellschaften mystisch-prälogische Mentalität zuerkannt. Diese Ideen sind in letzter Zeit aufgegriffen worden, insbesondere durch den Soziologen Günter Dux in seiner historisch-genetischen Kulturtheorie und durch Jean Piaget. Zuletzt hat der Psychologe Steve Stewart-Williams[52] eine partiell ‚in certain circumstances' [!] gültige evolutionär geprägte Hypothese entworfen, die angeborene Ideen als ‚content and principles' zu naturalistischen Quellen metaphysischen Wissens macht. Diese Einschränkung auf eine *situationsgebundene* Betrachtung der Zauberspruch-Theorie wird uns bei der neurobiologischen Einschätzung begleiten.

Ebenfalls mit diesbezüglich emotionalen Kausalattributionen hatte Sigmund **Freud** (1856–1939) die Magie auf einen narzisstischen Glauben an die Omnipotenz des Denkens zurückgeführt und mit dem Denken und Verhalten zwangsneurotischer und wahnkranker Patienten verglichen. Magische Praktiken funktionieren als Angst-Stress- und Schuld-Linderung.[53] Beeinflußt von Taylor und Frazer im evolutionistischen Ansatz schreibt Freud: „Im animistischen primitiven Stadium schreibt der Mensch sich selbst die Allmacht zu; im religiösen hat er sie den Göttern abgetreten ..." – „Die Geister und Dämonen sind ... nichts als die

---

50 Lévy-Bruhl, Lucien: Les fonctions mentales dans les sociétés inférieures, Paris 1928, S. 262–275.
51 Petzoldt, Leander (Hg.:) Magie und Religion, Darmstadt 1978, Einleitung S. XV.
52 Stewart-Williams, S., Innate ideas as a naturalistic source, Biology and Philosophy 20 (2005), S. 791–814.
53 Freud, Sigmund: Totem und Tabu, Frankfurt/M. und Hamburg 1964, S. 98ff.

Projektionen seiner Gefühlsregungen; er macht seine Affektbesetzungen zu Personen, bevölkert mit ihnen die Welt, und findet nun seine inneren seelischen Vorgänge außer seiner wieder, ganz ähnlich wie der geistreiche Paranoiker Schreber ...."[54] In der Gegenposition zu Freud beharrt Malinowski auf einer ‚account of emotions', auf menschlicher pragmatischer Insuffizienz, die sich eine Ersatzaktivität schafft, eben die Magie.[55] Mit Sørensen muß auf die zu unterscheidenden Ursprünge von Magie und Zwangsneurose, die Freud klärend heranzog, hingewiesen werden. Im hier vorgelegten neurobiologischen Ansatz wird sich zeigen, daß bei Benennung von Geistern immer die Bedeutung metaphorischer, meist anthropomorpher und theriomorpher Verarbeitungen bei Kranken, Helfern und Gesellschaft zu bedenken ist.

A1.2.3 Die strukturalistische Erklärung der Magie durch Claude Lévi-Strauss

Einen wesentlichen Beitrag zum inter- und multidisziplinären Verständnis der universalen Zusammenhänge nach den von ihm und anderen vor Ort erhobenen völkerkundlichen Befunden lieferte der Ethnologe und Anthropologe Claude **Lévi-Strauss** (1908–2009). Seine Sichtweise bildet nach Wolfgang Schmidbauer[56] die vierte der großen narzistischen Kränkungen, nach denjenigen durch Kopernikus, Darwin und Freud: Sie beraube uns des Dünkels, daß Schriftkulturen die besseren seien. Man kann in Bezug auf die uns hier interessierenden Fragen etwa folgende Schwerpunkte und Ergebnisse der Forschung von Lévi-Strauss überblicken:

A1.2.3.a Ethnologie: Wildes und rationales Denken

Die sorgfältige Analyse der Verwandschaftsverhältnisse und der Systematiken und Klassifikationen im Alltagsleben der Eingeborenen führten ihn zur Beschreibung einer unserer Wissenschaft formal vergleichbaren, nicht unterlegenen Strukturierung auch im Leben und in den Gesellschaften fremder Völker.

> „Die Eigenart des mythischen Denkens besteht, wie die der Bastelei (bricolage) auf praktischem Gebiet darin, strukturierte Gesamtheiten zu erarbeiten ... durch Verwendung der Überreste von Ereignissen: „odds and ends" würde das Englische sagen, Abfälle und Bruchstücke, fossile Zeugen der Geschichte des Individuums oder der Gesellschaft. In gewissem Sinne ist also das Verhältnis zwischen Diachronie und Synchronie umgekehrt: das mythische Denken

---

54 Ebd., S. 104.
55 zit nach Sørensen, Cognitive theory, S. 28.
56 Schmidbauer, Wolfgang: Vom Umgang mit der Seele, München 1998, S. 238.

dieser Bastler erarbeitet Strukturen, indem es Ereignisse oder vielmehr Überreste von Ereignissen ordnet (*aus zweiter Hand), während die Wissenschaft ... sich in Form von Ereignissen ihre Mittel und Ergebnisse schafft, dank den Strukturen, die sie unermüdlich herstellt und die ihre Hypothesen und Theorien bilden. Aber täuschen wir uns nicht, es handelt sich nicht um zwei Stadien oder um zwei Phasen der Entwicklung des Wissens, denn beide Wege sind gleichermaßen gültig."[57]

In allen Kulturen gibt es ‚wildes Denken' und rationales Denken. Lévi-Strauss verharrt nicht in der alten Kontrastkoppelung von Ratio vs. Magie und bietet der weiteren Forschung damit eine Überbrückung der Kluft zur Biologie. Das Denken in diesem Gegensatzpaar wird als allgemein menschliche komplementäre Dichotomie erkannt. Der grundlegende Gegensatz sei die Opposition zwischen ‚Natur' und ‚Kultur'.[58]

A1.2.3.b Das Gravitationsfeld

Von besonderem Interesse für die hier zur Diskussion stehende gehirnbezogene Betrachtung heilkundlicher Sprechaktionen ist die von Lévi-Strauss als ‚eine Art Gravitationsfeld' beschriebene Beziehung zwischen Zauberer und der /dem von ihm Bezauberten und der öffentlichen Meinung. Es ist die offene schauspielartige Darstellung der ‚Adreaktion' und ‚Abreaktion' durch den Magier unter Einschluß des Publikums, das diese wechselseitige Beziehung wegen der Effizienz der Verfahren wirksam belebt, ja überhaupt ermöglicht. Vergleichsweise handele es sich nach Lévi-Strauss gemäß Psychoanalyse um einen individuellen Mythos, den der Kranke mit Hilfe von Elementen seiner Vergangenheit ohne Beisein der Öffentlichkeit errichtet, im Schamanismus ist es ein gesellschaftlicher Mythos, den der Kranke von außen empfängt und der keinem früheren persönlichen Zustand entspricht.

Wir können folgern, daß ein solches Gravitationsfeld im europäischen Mittelalter die religiös-kulturelle Situation der theistischen Gesellschaft einschließt und ebenfalls keinen individuellen Mythos. In allen Fällen bildet die Heilungsaktion eine Triade. Am Beispiel der fremden Kulturen deutet Lévi-Strauss darauf hin, daß die unbezweifelbare Wirkung magischer Praktiken erstens dem Magier Glauben an seine Technik impliziert, zweitens dem Kranken Glaube an die Macht des Zauberers und drittens der öffentlichen Meinung Vertrauen und Anspruch. Damit – so seine Schlussfolgerung – ist „die magische Situation ein Phänomen des Konsenses."[59]

---

57 Lévi-Strauss, Claude: Das wilde Denken. Frankfurt/M. 1986 [Paris 1962], S. 35.
58 Ders.: Strukt. Anthropologie, S. 199.
59 Ebd., S. 183–185.

A1.2.3.c Die Einbeziehung psychoanalytischer Theorien

Zur Beschreibung des Verhältnisses vom Zauberer zum Kranken hat Lévi-Strauss Übereinstimmungen und Unterschiede im Verhältnis zwischen Psychoanalytiker und Arzt herausgearbeitet. Er greift den Freudschen Begriff der Abreaktion als Katharsis affektiver Besetzung auf und stellt die Heilverfahren wiederholt und aus wechselnden Perspektiven gegenüber. (Man beachte, daß Lévi-Strauss seine Feldforschung bei Eingeborenen der Lehranalyse gleichstellt.) Der Schamane vollzieht nach schauspielartiger Reproduktion mit Nacherleben von Originalität und Gewalt des Konfliktes die Abreaktion und kehrt schließlich zum Normalzustand zurück. Er ist „professioneller Abreagierer."[60]

Der Schamane spricht also für den Kranken, der passiv schweigt, wohingegen in der Psychoanalyse der Kranke spricht und gegen den Arzt, der ihm zuhört, abreagiert. „Aber obgleich das Abreagieren des Arztes nicht mit dem des Kranken zusammenwirkt, ist es dennoch erforderlich, denn der Arzt muß eine Analyse durchgemacht haben, um Analytiker zu werden. Die von den beiden Techniken [Schamanentum – Psychoanalyse] der Gruppe vorbehaltene Rolle ist weit heikler zu definieren, denn die Magie paßt die Gruppe mit Hilfe des Kranken wieder an vordefinierte Probleme an, während die Psychoanalyse den Kranken mittels eingeführter Lösungen wieder an die Gruppe anpaßt."[61] Kritisch zur Psychoanalyse äußert sich Lévi-Strauss: „Die PsA. wandelt [jetzt] ihre Behandlungen in Bekehrungen um ...." Die Gefahr sei groß, daß die Behandlung sich nicht auf die Krankheit, sondern auf die Neuordnung des Weltbildes des Kranken reduziert.

Am Beispiel eines alten Beschwörungsrituals der Cuña-Indianer in Panama hat Lévi-Strauss die psychologischen Übertragungsmechanismen präzisiert: (siehe auch **Kap. C4.2**).

„Der Schamane hat die selbe Doppelrolle wie der Psychoanalytiker: In der ersten Rolle – der Psychoanalytiker als Zuhörer, der Schamane als Redner – wird eine unmittelbare Verbindung mit dem Bewußtsein (und eine mittelbare mit dem Unbewußten) des Kranken hergestellt. Das ist die Rolle des eigentlichen Beschwörungsgesanges. Aber der Schamane spricht nicht nur die Beschwörung: er ist auch der Held dieses Gesanges, da er an der Spitze des überirdischen Bataillons der Geister in die bedrohten Organe eindringt und die gefangene Seele befreit. In diesem Sinne wird er wie der Analytiker zum Objekt der Übertragung, um dann – durch die dem Kranken eingegebenen Vorstellungen – zum tatsächlichen Protagonisten des Konflikts zu werden, den dieser auf der Schwelle zwischen der organischen und der psychischen Welt ausficht. Der Neurotiker überwindet einen individuellen Mythos, indem er sich mit einem wirklich vorhandenen Psychoanalytiker konfrontiert; die gebärende Eingeborene überwindet eine

---

60 Ebd., S. 198f.
61 Ebd., S. 201.

organische Störung, indem sie sich mit einem zum Mythos gewordenen Schamanen identifiziert."[62]

An anderer Stelle wird der Vergleich zwischen Magier und Psychopathen und zwischen Magier als individueller Psyche und Gesellschaft als sozialer Struktur ganz im Sinne von Komplementarität, also von gegenseitig abhängiger Ergänzung gesehen, ebenso wie ‚normales' und ‚spezielles' Verhalten.[63]

A1.2.3.d Der Rekurs auf Therapien des Bilderlebens/Imagination

Mit der Beachtung der Thematik des Erlebens und Gestaltens innerer Bilder tritt eine entscheidende universale und elementare Kraft ins Blickfeld des Anthropologen. An Hand der Psychoanalyse Freuds hatte er die Rollen und Übertragungsverhältnisse von Schamanen und Analytikern gegenüberstellen können. Als wesentlich interessanter, weil näher liegend für den Vergleich mit Schamanismus, entdeckt Lévi-Strauss eine andere psychotherapeutische Methode, bei der der Therapeut auf einer symbolischen Metaebene dem Patienten begegnet. Der Therapeut „spricht also mit dem Kranken in Symbolen, die aber noch sprachliche Metaphern sind"[64]: Die Therapie mit dem Wachtraum, rêve éveillé dirigé, beschrieben 1961 von Robert Desoille (1890–1966). Diese Therapie ist weniger bekannt als die Psychoanalyse, hat aber seit der zweiten Hälfte des 20. Jahrhundert differenzierte Ausarbeitung gefunden. Der Kranke wird – z. B. im katathymen Bilderleben nach Leuner – eingeladen, Assoziationen nach vorgegebenen zu verinnerlichenden Bildern zu entwickeln. Es ist eine Einladung zum Phantasieren und Dichten, die die Macht des Träumens und der Ein-Bildung zugrundelegt. Ein erster Ansatz findet sich bei dem Arzt Carl Happich 1932, der auf einer Ruheübung nach Art des Autogenen Trainings und der Annahme eines ‚Bildbewußtseins' aufbaut, in welchem der Patient sorgfältig geführt werden muß.[65] Diese Therapie sucht den Abstand vom kritischen Denken und ist philosophisch in ihrer lebensnahen Stofflichkeit ausgearbeitet von Gaston Bachelard, der in ‚Poetik des Raumes' (München 1957) die ‚transsubjektive Geltung des Bildes' beschreibt und zuletzt von Ernesto Grassi; sein Buchtitel ist Programm: ‚Macht des Bildes: Ohnmacht der rationalen Sprache. Zur Rettung des Rhetorischen' (Köln 1970). Das Mittelalter hatte diese ‚vis imaginativa' für die Entstehung innerer seelischer und körperlicher Ein-Bildungkräfte verantwortlich gemacht,

---

62 Ebd., S. 218.
63 Lévi-Strauss, Claude: Marcel Mauss – Sociologie et Anthropologie, Introduction, S. XVIII – XXIII.
64 Levi-Strauss, Str. Anthrop., S. 219.
65 Happich, Carl, Das Bildbewußtsein als Ansatzstelle psychischer Behandlung, in: Zentralblatt für Psychotherapie und ihre Grenzgebiete 5 (1932), S. 663–677.

hat ihr aber darüber hinaus auch Wirkungen auf die Umgebung und auf die Welt zuerkannt. Damit diente sie zur Erklärung der Magie.[66]

Dabei gibt es in der praktischen Anwendung heute eine methodische Skala zurückhaltend abwartender bis stärker führender Haltung des Therapeuten. Und es gibt Bildvorgaben vom vieldeutigen Assoziationsfeld ‚Wiese' bis zum dramatisch herausfordernden ‚Tiefseegrund', zum ‚Höhleneingang' oder zur ‚Hexensuche'. Die teilweise Vergleichbarkeit dieser Therapie der ‚réalisation symbolique' mit Schamanismus ist von psychotherapeutischer Seite anerkannt worden.[67] Insbesondere haben sich viele Therapiezweige, seit 1975 auch wieder die Hypnotherapie entwickelt, die der Anwendung von Metaphorik hohen Stellenwert einräumen. Dazu gehört auch das Neurolinguistische Programmieren (NLP). Diese Therapiemethoden werden wir aufgrund ihrer imaginativen Strategie als Vergleichselemente bei der Einschätzung der hier betrachteten Texte heranziehen. Mit ihnen wird klar, worin sich die therapeutischen Zugangswege zum Menschen von einst und von heute, von archaischen und religiösen Gesellschaften und den Gesellschaften des Individualismus gleichen und worin sie sich unterscheiden.

A1.2.3.e *Neuropsychosomatische Zusammenhänge*

Neuropsychosomatische Zusammenhänge im magisch-heilkundlichen Handeln beschreibt Lévi-Strauss mit dem schlagkräftigen Befund der Beobachtung von Todesfällen, dem Tabu-Tod, infolge von Beschwörungen und Verhexungen in aller Welt und der Schlussfolgerung, daß diese wie Angst und Wut auf das sympathische Nervensystem wirken, z. B. auf Blutdruck, Blutvolumen und Gefäßpermeabilität.[68] Insgesamt bleibt festzuhalten, daß sich Lévi-Strauss angesichts der ihn stark interessierenden biologischen Befunde, der „unbezweifelbaren Wirkung" von Magie, (ähnlich wie Freud[69] weit vor ihm) in den 50er Jahren des letzten Jahrhunderts nicht scheute, die Konsequenz fachlicher Grenzüberschreitung zur Klärung der Zusammenhänge zwischen Psychotherapie und Magie zu prognostizieren. Während noch manche heutige Geisteswissenschaftler

---

66 Godet, Alain, „Nun was ist die Imagination ..", S. 1.
67 Vgl. z. B. Boesch, E.E.: Ritual und Psychotherapie, in: Z. für klinische Psychologie und Psychotherapie 30 (1982), S. 214–234.
68 Lévi-Strauss, Str. Anthropologie, S. 183.
69 Freuds Confiteor: „Es ist die Absicht, eine naturwissenschaftliche Psychologie zu liefern, das heißt psychische Vorgänge darzustellen als quantitativ bestimmte Zustände aufzeigbarer materieller Teile und sie damit anschaulich und widerspruchsfrei zu machen". In: Entwurf einer Psychologie (1895 – ed. posthum 1950), vgl. dazu Jung, Richard: Neurophysiologie und Psychiatrie, S. 603.

gegenüber Fragen von neuralen Wirkprinzipien geradezu hermetische Burganlagen errichten, hat Lévi-Strauss mit dem Hinweis auf erste biochemische Entdeckungen am Nervengewebe von Psychotikern auf mögliche bevorstehende Neukonzeptionen auch im Sprachgebrauch in Richtung auf physiologische oder biochemische Termini hingewiesen, nicht nur auf die Struktur von Neurosen und Psychosen, sondern auch auf die allgemeine Wirkungskraft der Symbole auf verschiedene Ebenen des Lebewesens, einschließlich organischer Vorgänge.

Verstehen wir den Lévi-Strauss'schen Strukturbegriff in seiner ganzen Schärfe. Im erweiterten Blick auf die Verbindungen zwischen Natur und Kultur, Bios und Geist, schreibt er, „daß sich die traumatisierende Kraft einer beliebigen Situation nicht aus den zugehörigen Merkmalen ergeben kann, sondern daraus, daß bestimmte Ereignisse, die in einem psychologischen, historischen und sozialen Zusammenhang auftreten, eine affektive Kristallisierung herbeiführen können, die in der Form einer vorher bestehenden Struktur erfolgt. […] Die Gesamtheit dieser Strukturen würde das sogenannte Unbewußte bilden."[70] Womit Geist an Struktur gebunden wird und Struktur an Geist. Dem entgegen kommt von der Seite des Biologen und Neurologen Eric Kandel die Erforschung einer simplen Schnecke, der Aplysia, die uns nach den überzeugenden Versuchen des Nobelpreisträgers lehrt, daß Verhaltensänderungen (ebenso wie die genannten Traumatisierungen) zu Strukturveränderungen der Nerven führen können. Der Strukturbegriff – ob praktisch oder theoretisch verstanden – schließt also das Gehirn nicht nur in eine kalkulierte epistemische Ordnung ein. Das Postulat universeller Strukturen des menschlichen Geistes in Kultur, Wissenschaft und Wildnis führt zwangsläufig zur Heimkehr an die Quellen dieses Geistes. Dieses Postulat ist ein Zirkelschluss und mag ein Kategoriensprung mit umfangreichen Schleifenbildungen sein, der den Menschen als Teil der Natur, also seine Gehirnstruktur als Organ von Sprache und Denken, Emotionalität und Intentionalität begreift. Aber es ist kein Kurzschluß. Damit wird die vom strukturalistischen Denken bevorzugte Gegensatzpaarung von Kultur und Natur als dynamische Kraft der Forschung im geistes- wie im naturwissenschaftlichen Bereich wirksam und vermag ungehindert tragfähige Brücken zwischen Wissenschaften zu bauen.

A1.2.4 Zwei Beispiele neuerer Magieforschung nach Lévi-Strauss

In neuerer Zeit wurden Malinowskis, Tambiahs, Lévy-Bruhls und Lévi-Strauss' Ideen mehrfach aufgenommen; ich verweise auf Ariel **Glucklich**,[71] der den Ver-

---

70 Levi-Strauss, Str. Anthropologie, S. 223.
71 Glucklich, A., The end of magic.

gleich Magie zu Religion und den Gebrauch des Begriffes Magie auf seine allein historisch bedingten und allein regionalen europäischen Wurzeln reduziert, während das Phänomen weltweit nur als ‚magic experience' existiere.[72] Gemäß seiner Analyse einer Reihe von Begegnungen im heutigen Indien kommt er zu dem Schluß, daß die magische Erfahrung als Erlebnis keine objektivierbare, sondern eine persistierende ‚psychologische Qualität wie Imagination' sei. „Keine Aktion und kein Objekt ist immer intrinsisch magisch; sie werden nur in ein System von Perzeptionen und Antworten inkorporiert, ein System, das zum Teil eines magischen Ereignisses wird." Damit kommt Glucklich zu einer Überbetonung der Dyade zwischen Heiler und Klient.[73] Seine Versuche, Magie von Kultur abzulösen zugunsten ihrer universalen Gegebenheit und die Begriffsverwendung ‚Magie' unter Wissenschaftlern für rückgängig zu erklären, verleiten ihn zu seinem auffälligen Buchtitel. Bemerkenswert ist die Einbeziehung der psychoneuroimmunologischen Aspekte in seine Analysen.[74]

In Deutschland hat Monika **Schulz** anhand der umfangreichen Sammlungen des Corpus der deutschen Segens- und Beschwörungsformeln eine ‚Theorie des magischen Wortes' und seiner ‚magischen Semiotik' komponiert mit der Frage, ob es eine magische Geheimsprache gebe.[75] Dabei sucht sie nach kausalen psychodynamischen Elementen der ‚attractio similitudinis', d. h. nach der Wirkung von Analogie und Kontiguität (wie Frazer mit similarity und contagion) und verweist auf den optionalen Aspekt,[76] wie ihn Karl Beth herausgestellt hatte, auf persuasive Kräfte[77] der Ähnlichkeitskonstruktionen („Das Wie ist .. persuasiv zu lesen") und auf die „grundlegenden Vorstellungen einer die Teile eines Ganzen vollständig durchziehenden vis vitalis,"[78] jener schwer definierbaren Kraft, wie sie verbreitet seit Hufeland, Hahnemann und später bei Wilhelm Reich vorkommt. Schulz bezieht sich auf Lévi-Strauss' Anliegen, in den Konzepten der Magie nach Erkenntnissen über Funktionsweisen des menschlichen Geistes zu suchen[79] und auf Lévi-Strauss' Zurückführung magischer Handlungen auf Wiederherstellung einer Einheit;[80] also ordnungsverpflichteten Gesetzlichkeiten archaischer Denkstrukturen. Als ‚doppelte Selektion' kombiniert Schulz den

---

72   Ebd., S. 211f.
73   Ebd., S. 235.
74   Ebd., S. 68f.
75   Schulz, Monika: Magie oder Die Wiederherstellung der Ordnung, Frankfurt u. a. 2000, S. 176, S. 194f.
76   Ebd., S. 225.
77   Ebd., S. 252f.
78   Ebd., S. 244.
79   Ebd., S. 17.
80   Ebd., S. 389.

Schwebezustand einer intrapersonalen Assoziationsdiffusion (im Sinne von Frazer) mit einem zielfixierenden letztlich performierendes Element (im Sinne von Malinowski): das „Hilfsinstrument ... des auswählenden, richtungsweisenden begrenzenden Wortes."[81] Immer wieder wird Sprache und Wort als zielführendes Element von therapeutisch angewandter Magie herausgehoben. Man könne nicht erklären, warum eine Hasenpfote unter dem Kissen einer Schwangeren die Geburt erleichtern soll. Zwar verbürge das Gesetz der mystischen Teilhabe mit seiner Garantiezusage eines Gehalts auch winzigster Partikel der Objekte die (erwünschten) Qualitäten Flinkheit und Fruchtbarkeit, aber die unerwünschten, wie könnten sie ausgeschlossen werden?

A2 Der linguistische Zugang zu den Heilspruchtexten: Sprechakttheorie und Performativität

Der Begriff Performativität aus der Sprechakttheorie der Philologen John L. **Austin** (1911–1960) und John **Searle** (geb. 1932) wurde von Tambiah (s. u.) in die Magieforschung eingeführt. Mit dem Begriff soll zum Ausdruck gebracht werden, daß es nicht nur konstative, also feststellende Aussagen, Zustandsbeschreibungen gibt, die Austin lokutionär nennt, sondern auch wirkmächtige, ausführende, handelnde und konkretisierende. Letztere gelten als performativ, als sozial eingreifend. Austins Vorlesungsreihe (1955) titelt Aufsehen erregend als Veröffentlichung posthum 1962 mit „How to do things with words" und rechnet zunächst konstituierende Sprechakte wie jene während Heirats- und Tauf- und Ernennungszeremonien, beim Schwören, in Befehlen und Warnungen zu sog. illokutionären Akten. Diese Handlungen werden also durch Äußerungen, und zwar durch ‚very utterances', vollzogen. Im weiteren hat dann Austin die Umstände einer Äußerung in seine Theorie einbezogen. Jede Äußerung kann danach drei verschieden Akte beinhalten. Ich nehme als Beispiel den Satz: „Ich gebe es dir zurück". Wer dies seinem Freund sagt, mit entsprechend sachlicher Betonung, Wortklang und Stimme kann ein geliehenes Buch meinen, dann ist die Aussage locutionär, eine simple Feststellung. Es kann sich aber auch im Falle gespannter sozialer Beziehung um eine Warnung handeln, um die Idee einer Revanche, dann ist eine Handlung in den Raum gestellt, die ausgeführt wird oder werden kann; diese Äußerung ist illokutionär. Wenn der Hörer einer solchen als feindlich zu verstehenden Äußerung, die mit außersprachlichen Mitteln kombiniert sein kann, auch noch Wirkung spürt und beeindruckt ist, wird von einem perlokutionären Akt gesprochen. Nun kann dieser Sprechakt primär performativ

---

81   Ebd., S. 246f.

sein: „Warte nur, ich gebe es dir zurück!" oder explizit performativ: „Nimm dich in Acht, ich gebe es dir zurück!". Der Sprechakt kann verunglücken, falsch oder ehrlich sein, er kann einen propositionalen, intentionalen, kommisiven Gehalt haben.

30 Jahre nach Malinowskis empirischen Ergebnissen (s. o.) greift Stanley J. **Tambiah** (geb. 1929) die Austinschen Arbeitsergebnisse auf und verweist auf die Unterschiede unter den rituellen Sprachformen, die oft nicht einmal von den Exorzisten verstanden wurden, sodaß Worte auch als konventionelle und kommunikative Akte zu verstehen seien. Erst Tambiah lenkte die Aufmerksamkeit der Ethnologen auf die unauflösliche Verbindung von Wort und Akt im magischen Ritual. Er verzichtete auf die Magie-Religion-Unterscheidung und führte den Begriff der Performativität in seine Disziplin ein. Er erkennt im Bezug auf die Arbeiten von Evans-Pritchard über die Zande in Zentralafrika und seine Unterscheidungen zwischen ‚empirischer Analogie' = Wissenschaft und ‚persuasiver Analogie' = Magie, daß es sich dabei um ein Beobachterproblem handelt. Andererseits verankert Tambiah den Gebrauch von Metaphern und Metonymen in die Theorie der Magie. Die Sprachfiguren erlauben „abstrahierendes Denken auf der Basis analoger Aktionen."[82]

Darüberhinaus erweitert Tambiah die Theorie Lévy-Bruhls (s. o.) von der prälogischen Mentalität der „mystischen Partizipation" und kombiniert sie mit Austins Theorie: Alle Menschen besitzen komplementäre Arten der Weltannäherung und damit sowohl distanziert kausales rationales Denken als auch Performativität und performative Äußerungen. Magie ist damit ‚konventionelles intersubjektives Verstehen' und ‚performative Wirkkraft des kommunikativen *Aktes*'.[83]

Auch der Soziologe und Ethnologe Eike **Hinz** (geb. 1945) beschreibt nach Kenntnis der Aztekenforschung und mit Rückgriff auf Malinowski eine Magie mit Relevanz der kognitiven Komponente des Wortzaubers.[84] Er erweitert die Funktionalität von Magie um psychologische Aspekte von Verhalten, Gedächtnis, Aufmerksamkeitssignalen und Sprechakttheorie. Bedeutungsvoll wurde vor allem seine Betonung des situations- und chancenabhängigen Charakters magischer Strategie, d. h. der situativen wunscherfüllten Mangelsituation, in der Verbalmagie einsetzbar und erfolgversprechend wird.

---

82 Tambiah, Stanley Jeyaraja (1968) The magical power of words, Man 3, 175–208 (189); Tambiah (1970): Form und Bedeutung magischer Akte, in: Kippenberg/ Lucchesi S. 259–300, hier: S. 284ff.
83 Ders., Magic, Science, Religion, and the Scope of Rationality, Cambridge 1990, S. 109, zit. nach Sörensen.
84 Hinz, Eike, Cognitive Structures and Processes in Verbal Magic, in: Sociologus 28 (1978), S. 122–148.

Im Laufe der letzten 20 Jahre hat dann die Erarbeitung der im Sprechen liegenden Kräfte selbst immense institutionsgenerierende Kräfte freigesetzt. Die Theorien des Performativen boten den geisteswissenschaftlichen Fakultäten nach den erst idealisierenden, essentialistischen, danach strukturalistischen Arbeitsweisen des 20. Jahrhunderts neuen Auftrieb. Sie sind aber an ihre Grenzen gelangt. Ich bezeichne folglich die hier besprochenen Texte als ‚psychoperformativ', wobei im Wortteil ‚psycho-, ihre neurale Heim- und Wirkungsstätte bedingungslos mitzudenken ist.

Es kam zur Bildung zahlreicher interdisziplinärer Forschungsprojekte. Beteiligte Fächer sind u. a. Theaterwissenschaft, Philosophie, Pädagogik, Soziologie, deutsche und romanische Philologie, Kultur-, Musik-, und Filmwissenschaft (Wikipedia – Performativität). Ein Sonderforschungsbereich 447 der DFG hatte sich bis 2010 der Untersuchung von ‚Kulturen des Performativen' gewidmet. Die Ich-Psychologie z. B. kann sich dieser Theorien für Selbstgespräche bedienen, um Verinnerlichung, Selbstdisziplinierung, Selbstkreation zu studieren. Eine Ikonologie des Performativen mit der Einbildungskraft des Imaginären[85] erregte großes Aufsehen. Man beobachtete soziale Phänomene des Performativen bis hin zu Alltag und Familie, Riten, somatischen Inszenierungen und zum „doing gender", meist mit der These vom strukturierenden Vollzug. Der Begriff und seine Verwendungen wuchsen quasi „rhizomatisch" und erleiden „derzeit [...] konzeptionelle Überdehnung, die die Distinktivität des Konzepts und damit dessen deskriptive Leistungsfähigkeit einschränken."[86]

Frühe Anwender der Sprechakttheorie auch für den Zauberspruch waren behutsam vorgegangen. So hat Roderich **Feldes**[87] in seinem Bestreben, Poesie, Zauberformel und Werbeformel auf einen Nenner zu bringen, den Terminus performativ streng auf ‚Formel' und auf ‚Verdichtung und Unverständlichkeit', also vor allem Fremdartigkeit der Worte eingeschränkt. Er nennt Beispiele aus Agrippas Texten, das ‚Thezay lemach ossanlamach azabath azach azarc ...', nennt Wortgerölle, Amulette und Siegel, Schwundformeln, Reihen und Visualisierungen wie im Bereich der Werbung, schließt aber epische Elemente wie etwa den gesamten zweiten Merseburger Zauberspruch aus.[88]

Dagegen wird die Idee des performativen Wortes in neueren Arbeiten sehr weit gespannt. Ein Beispiel für die linguistische Arbeit zum Zauberspruch in der Antike gibt die Germanistin Amina **Kropp** (2009). Es werden die performativen Anteile antiker lateinischer Verwünschungstafeln (Defixiones) untersucht. Eine elektronische Datenbank konnte historische weitgehend ureigen originelle Sprechsituationen darstellen, um an 391 Stücken die ihnen zugeschriebene ‚magische' Macht zu analysieren. Die Austin-Searlesche Sprechakttheorie gab Wortsorten vor. Eingeschlossen waren z. T. auch rituelle („ich durchbohre sie, damit sie zugrunde gehe") und situative Elemente (Konflikte, Prozesse, Wettkämpfe) im Sinne einer pragmalinguistischen Studie. Alle diese Verwünschungen waren als Fernzauber ohne Wissen des Opfers zu verstehen (z. B. erwünschte Zungenlähmung vor Gericht), also als aggressive Eingriffe in das Leben eines anderen verbal und rituell. Sie galten also Personen, die nicht

---

85 Wolf, Christoph, Digi 20 der Bayerischen Staatsbibliothek.
86 Hempfer, Klaus, in: Hempfer, K.W. und J. Volbers (Hg.:) Theorien des Performativen, Bielefeld 2011, S. 38.
87 Feldes, Roderich: Das Wort als Werkzeug, Göttingen 1976, S. 60–63.
88 Ebd., S. 163.

anwesend und nicht eingeweiht waren und die diese Worte weder hören noch lesen, vielleicht einmal ahnen konnten. Die Autorin kommt zu dem schon oben zitierten ernüchternden Ergebnis.[89]

Ein verallgemeinerndes Performativitätsverständnis fließt teilweise auch in die Dissertation zu einigen althochdeutschen Zaubersprüchen von Christa M. **Haeaseli**[90] ein. Eine gleichsam mikroskopische Analyse untersucht sorgfältig den ‚Vollzugscharakter' der kommunikativen Akte von Heilspruchtexten des Hochmittelalters. Dabei wird der offenbar dem Theologischen entnommene Begriff des Vollzugscharakters – Sakramente wirken ex opere operato allein durch ihren Vollzug ohne Mitwirkung des Segensempfängers – gleichermaßen für Magie wie für Performativität als zentrales Anliegen verwendet. Ebenso wird das Ziel einer ‚konkreten Veränderung der Welt' bzw. ‚realweltlicher Veränderungen' sowohl der Magie als auch den Mechanismen performativer Texte zugeschrieben. Es werden im Hinblick auf die spezifisch schriftliche Performativität des historischen Materials auch latente Wirkmächtigkeiten, dynamische Umschlagspunkte, liturgische Rahmungen der Texte und die Textumgebung der Codices in die Untersuchung einbezogen.[91]

Bei der Beschreibung der Doppelfunktion des Sprechers in Nähe und Distanz zur Gottheit[92] im Trierer Pferdesegen beleuchtet die Autorin eine der therapeutischen Haltung zum Patienten parallele Fertigkeit, der wir im **Kap. B3** über das Bündnis in neuraler therapeutischer Allianz weiter nachgehen. Der Heiler spricht als Gott, vergleichbar dem Priester im Zentrum der Messe, ohne eine direkte Identifikation zu behaupten, womit höhere Hilfe und aktueller Heilungsvorgang mittels expliziter Analogie zusammenfließen. Ferner sind assertive Sprechakte als Wirkungsgarantie erkannt wie z. B. im Pariser Epilepsiesegen und deutlicher im Chiasmus des Regensburger Hiobsegens (➔30), wo die Heilwirkung nicht nur garantiert, sondern auch schon vorwegnehmend festgestellt ist.[93]

Im Blick auf die intensive Anwendung der Spechakttheorie entsteht allerdings die Frage, was eigentlich noch locutionär, was nicht performativ sei, wenn die Gesamtheit aller Textinhalte, Textformalien und ihrer Umgebung, ja sogar ihrer Grenzauffaserungen, ihrer oft wohl unwillkürlichen semantischen Unvollkommenheiten und des oft lexikalisch-mnestischen Charakters involviert wird. Selbst die Lage der Texte im Codex, die priesterliche Funktion der Gottesnähe, die Gebete, alle Gebrauchsanweisungen der Texte werden an eine magische Performativität angelehnt und als solche interpretiert. Können die Grenzen zu den Sakramenten, zur Stellvertreterfunktion des Priesters und zur kirchlichen Weihe, zur medizinischen Praxis mit ihren psychosomatischen Techniken und zu Archivierungszufällen noch abgesteckt werden? In welchem Maße ist eigentlich pragmalinguistische Deutung noch möglich?

89	Kropp, Amina, Sprachliche Betrachtungen.
90	Haeseli, C. M., Mag. Performativität.
91	Ebd., S. 25.
92	Ebd., S. 112.
93	Ebd., S. 175, s. a. Haeseli, C.M.: Sprachmagische Texte des Clm 536, in: Herberichs, Cornelia und Christian Kiening (Hg.): Literarische Performativität, Zürich 2008, S. 63–81.

Die Grenzen der Sprechakttheorie werden deutlich, wenn an ihre Anfänge erinnert wird. Austin hatte an eine illokutionäre Kraft gedacht, die man „an ihren Wirkungen" erkenne, damit „der Hörer versteht, was der Sprecher getan, indem der Sprecher die Äußerung getätigt hat, daß der Hörer versteht, was der Sprecher vom Hörer will. Die wichtigste Komponente ist der Zweck." „Wird verstanden, was verstanden werden sollte?"[94] Mit der Frage nach Ursache und Wirkungsprinzip des Sprechens wuchs die Grenzerfahrung der Sprechakttheorie und die Erkenntnis, daß sie „eigentlich eine philosophische Disziplin" ist.[95] Mit ihren Grenzüberschreitungen könnte sie einen Beitrag zum Verständnis des Zen-Kundigen leisten, der im rechten Handeln ohne etwas zu tun ein *Großes* Handeln übt und sein Ich neu ordnet. Der im Bogenspannen lernt, daß es nicht ums Treffen geht, sondern um das Werden der Seele, wie beim Schießen, so auch beim Sprechen.

Die Sprechakttheorie leistet einen wichtigen Beitrag zum Verständnis der Details auf der Ebene der Wortbedeutung, auch für die Einschätzung von performativen Texten zu therapeutischen Zwecken. Sie muß aber schon selbst im Rahmen der Sprachwissenschaft durch andere sprachliche Forschungsfelder, insbesondere deiktische, d. h. die situative, personale und soziale Situation, das sog. Hier und Jetzt betreffende Ergänzungen, präzisiert werden. Für eine umfassendere Betrachtung des triadischen Gravitationsfeldes (s. u.) kann sie eine sektorale Bedeutung gewinnen.

„Im Anfang war das Wort; bis Ende währt das Wort", könnte man für die Sprechakttheorie auch titeln. Die Theorie hat das Wort beseelt, ohne nach dem Sprecher zu fragen, geschweige denn nach dem Empfänger. Und man könnte auf die „Ephesia grammata" zurückgreifen. Als ob sie der Sprechakttheorie Flügel gegeben hätten und den Worten selbst ihre Motorik. „Ephesische Zettel", die Zauberworte, der alte Buchstabenzauber mag selbst Pate gestanden haben. Denn, so wurde es Carl Wessely zugeschrieben, „das ist der Glaube, daß einigen Worten eine besondere Kraft inne wohnt, welche denen zugute komme, die so glücklich wären, diese Lautgebilde zu kennen."

Es ist zeitlich und gedanklich ein weiter Sprung zu den berühmten sechs Zauberworten der griechisch-römischen Antike, die – weil von diesem großen Kulturzentrum kommend – damals als „Ephesische" zum Musterbegriff für die ganze Vielfalt ihrer Art auf den Zauberpapyri wurden, weil sie einst einer Kultfigur der dortigen Artemis eingraviert gewesen sein sollten (Apg. 19.19: Christen schwören ihrem Kult ab): aski kataski lix tetrax damnameneus aision. Auch ihnen hat man heute Reim, Alliteration und Assonanz[96] zugesprochen, während sie mindestens vom 2. Jahrhundert vor bis zum 5. Jahrhundert nach Christus vielfach als heilig galten.

---

94   Staffelt, Sven: Einführung in die Sprechakttheorie, Tübingen 2008, Glossar u. a., S. 37, S. 9.
95   Ebd., S. 9.
96   Önnerfors, Alf: Antike Zaubersprüche, Stuttgart 1991, S. 7f.

Ihre faktisch antidämonische Macht scheint allerdings zeitweise weniger bezweifelt worden zu sein als jene Kraft, die den illocutionären und perlucotionären performativen Verben von Seiten der Sprechakttheorie beigegeben wurde. Schon John R. Searle hatte im Wirken des illocutionären Aktes ‚ein Element von Magie' entdeckt. Vertreter der Sprechakttheorie haben den konnotativen Sog von „Kraft" zur Physik (und zur Biologie) übersehen.[97]

Das Austinsche „How to do things with words" wirft für die therapeutischen Spruchtexte letztlich drängend die entscheidende Frage auf nach konkreter funktioneller Realisierung von Sprachfiguren und den mit ihnen zusammenhängenden rituellen Gebärden im soziopsychosomatischen Ist-Zustand des menschlichen Gehirns. Die Entwicklung der *Neuro*linguistik war unausweichliche Konsequenz.

---

[97] Vgl. Reichertz, Jo: Kommunikationsmacht, S. 209f.

B Die neuropsychosoziale Struktur einer Notfallbehandlung als Triade

In diesem Kapitel sollen die Grundlagen vorgestellt werden, die dem Prozess jeder Notfall- und Krisenintervention eignen und die uns dazu berechtigen, die mittelalterlichen, frühkulturellen und archaischen Heilspruchtexte auf ihre Wirkpalette zu untersuchen. Der „Not-" bzw. „Krisenfall" wird hier überwiegend nur als Sammelbegriff für jedes dringende akute medizinische Management unabhängig von einer Diagnose verstanden. „Not- und Krisenfall" sind ebenso wie „Krankheit" relative Begriffe, abhängig von kulturellem Kontext, Beobachtung und Erwartung.[98] Die Ausführungen müssen belegen, daß wir heute erste neurophysiologische Kriterien besitzen, allein anhand der alten Texte Folgerungen auf Wirkungsmöglichkeiten dieser Verbaltherapie zu ziehen, obwohl wir die situativen Bedingungen quellentechnisch nicht exakt zu reanimieren vermögen.

Aus vielen Schriftzusammenhängen wissen wir allerdings von einem gemeinsamen und oft aufeinander abgestimmten Gebrauch medikamentöser und chirurgischer Maßnahmen mit Spruchtexten und Riten. Nachweislich war das Interesse der Ärzte für rein pragmatische Medizinschriften gering; die Medizin des ‚Thesaurus pauperum' eines Petrus Hispanus (13. Jahrhundert, Papst Johannes XXI?) fand erst Verbreitung, als sie mit Heilsegen und Zaubersprüchen ausgestattet wurde.[99] Für die altägyptische Medizin wird vermutet, daß der Heilkundige „alle medizinischen Vorgänge mit Zauberpraktiken" „kombinierte."[100] Für die mesopotamischen Verhältnisse verweise ich auf die Zusammenstellung in **Kap. B4.2**.

Theorie und Praxis dieser frühen Ganzheitsmedizin für Leib und Seele konnten nicht stets neu erfunden werden. Medizinkunst wurde nach den Erfahrungen tradiert. Das schließt keinesfalls aus, daß der Arzt kulturell als von Gott oder Göttern berufen vorgestellt wurde und daß deshalb für das christliche Mittelalter von einer ‚religiösen Medizin'[101] gesprochen werden kann, deren Ärzte dem Vorbild des ‚Guten Samariters' und des ‚Christus medicus' verpflichtet waren. Die Arzt-Patient-Beziehung unter Einschluß ihrer verschiedenen Methoden zu-

---

98 Vgl. Maturana, Humberto R., in: Riegas, Volker und Christian Vetter: Zur Biologie der Kognition, Frankfurt/M, 1993³, S. 28.
99 Telle, Joachim: Petrus Hispanus in der altdeutschen Medizinliteratur, Heidelberg 1972, S. 169–171, 367.
100 Schott, Heinz und Rainer Tölle: Geschichte der Psychiatrie, München 2006, S. 20.
101 Vgl. Gerlitz, Peter, in: Theologische Realenzyklopädie (TRE) Bd. XIV (1985), S. 738.

mal in einem Notfall hat jedoch auch für historische Betrachtung eine eigene umschreibbare und definierbare soziale Bedeutung. Sie kann sowohl vom Traumdeuten, Segnen, Beten und Dämonenbeschwören als auch von rein technischer zweckfreier Nachrichtenübermittlung und Routine unterschieden werden. Sie ist immer auch oder vorwiegend als situativ bedingte menschliche vertrauliche Verhaltensabstimmung und damit als Spezialfall von Kommunikation zu erkennen.[102] Und ein Zweites ist sicher, ich wiederhole: Die Gehirne der diese Texte sprechenden und der die Texte empfangenden Menschen arbeiteten funktionell nach gleichen neuronalen Prinzipien wie die unseren im 21. Jahrhundert, wenn auch kulturelle Prägungen vermittels der Wertordnungen unterscheidbar sind (s. u.).

B1 Der Kranke in Not

B1.1 Hilferuf und Hilfserwartung

Die Erwähnung des Allgemeinplatzes, daß Menschen soziale Wesen seien, ist nur noch erträglich, wenn er beharrlich auf seinen Gehalt und seine Dimensionen befragt wird. Wir müssen dann feststellen, daß zwischen der existentiellen Abhängigkeit des Säuglings und den hybrid vorgetragenen Formen von Selbstbestimmtheit scheinbar ein Ozean liegt. Hat einer sein sturmfestes Schiff gezimmert, um unterwegs lernend von Ufer zu Ufer zu gelangen, kann er sich z. B. praktisch des Kompasses von ‚Handlungsfreiheit' bedienen: Die Annahme einer wahren Freiheit im kommunikativen Handeln setzt dreierlei voraus: Eine Handlung muß originell, also selbst erdacht und erfunden sein; sie muß unter gleicher situativer Bedingung auch anders ausgeführt werden können und sie muß vernünftig sein.[103] Während diese drei sehr harten Bedingungen zur Annahme der Möglichkeit von Entscheidungsfreiheit selten zusammentreffen können – selbst das Genie bedarf derer, die es ‚nähren' – ist beim akut schwer kranken Menschen tatsächlich kaum einmal ein Ansatz von Entscheidungsfreiheit gegeben. In der Notlage wäre er gemäß unserem Bilde ans Ufer totaler Abhängigkeit geworfen wie ein Säugling. In tatsächlicher oder gefühlter Lebensbedrohung erfolgt Unterordnung unter einen Helfer. Dann entfällt selbst die dritte der Bedingungen: Rationales Handeln unterliegt dann dem Diktat der Hilflosigkeit im Zustand seelisch-körperlicher Not. Das heißt auch: Die assoziativ bewußt arbeitende Hirnrinde gibt Kompetenzen ab, weil sie zu „langsam und fehleranfällig" ist; das

---

102 Vgl. Reichertz, Jo, S. 98f.
103 Vgl. Walter, Henrik: Neurophilosophie der Willensfreiheit, Paderborn u. a. 1997, S. 93.

Hirn neigt, hier im Gegensatz zu Meditation und Kontemplation, zu unbewußten und vegetativen Lösungen.[104] ‚Gefühle' übernehmen das Dirigat über einen körperlichen Resonanzboden.

Bereits in primitiven biologischen Zellsystemen spielen korrelierende Aktionen eine bedeutende Rolle zur Aufrechterhaltung der Funktion. Es können nur Zellen überleben, die evolutionär biochemisch in der Lage sind, Signale an andere Zellen auszusenden, beispielsweise bei Pilzen das Cyclo-AMP, um ‚Notrufe' gleichsam wie Hungerschreie zu vermitteln. Mit enormem Aufwand hat Evolution bei höheren Lebewesen Signalstoffe und entsprechende Signalempfänger-Rezeptoren aufgebaut bis hin zu inneren Botenstoffen (z. B. Hormone, Transmitter) und äußeren Reizen akustischer, olfaktorischer und visueller Art. Ähnlich wie in diesen primitiven Systemen behält auch das Gehirn von seinen in Überfluß angelegten Nervenzellen bei seiner Entwicklung nur diejenigen, die die richtigen und verstetigten Verbindungen gefunden haben.[105]

Für den allein unter allen Lebewesen sprachbegabten Menschen haben sich trotz dieser herausragenden Fähigkeit die motorischen nonverbalen und vorintentionalen Signale, welche evolutionär die Vorläufer der Sprache sind, bestens erhalten. Jede Geste des Kleinkindes, seine Mimik, sogar seine prosodischen Reize vermögen elterliche Empfänger zu stimulieren und sich zu Verhaltensmustern auszubilden, die rekurrente, also gegenseitig systematisierte Interaktionen durch Wachstum spezieller Synapsen zur Folge haben. Zahlreich sind die Untersuchungen zur Wirkung des ‚Kindchenschemas' und der ‚babyish features'[106] mit ihren prägnanten Rundungen an Augen und Gesicht. Bei ihrem Anblick fand sich unter bildgebenden Verfahren eine erhöhte Aktivität des Nucleus accumbens im vorderen Basalhirn mit seinem Belohnungssystem und von 130 ms Dauer in der orbitofrontalen Rinde. Entladen sich innerer Druck, Angst, Schmerz, Hunger oder Verlassenheit im Weinen von Säuglingen und Kleinkindern, dann ist ein unvergleichbar hoher Grad an Hilfsbedürftigkeit und sehnsuchtsvoller Erwartung ausgedrückt, der unter gewöhnlichen Umständen eine Resonanz erzwingt. Wimmern und Weinen werden ritualisiert und sind Wurzel des frühen imperativen gestischen und sprachlichen Aufforderns.[107] Technische Alarmsigna-

---

104 Roth, Gerhard: Fühlen, Denken, Handeln. Frankfurt/ M. 2001, S. 231; der von Roth gemeinte ‚Notfall', der umgekehrt diesen Cortex bemüht, bezieht sich auf kognitive Kompetenz bei sozialen Herausforderungen gesunder Menschen.
105 Vgl. Lohse, Martin J., Notrufe und andere Signale. Die Sprache der Zellen, in: Universitas 55 (2000), S. 718f.
106 Z. B. Kringelbach, L. Morton, Neural basis of mental representations of motivation, emotion and pleasure, in: Berntson/ Cacioppo, Handbook II (2009), 807–828 (814f und Fig. 41.3).
107 Tomasello, Michael: Die Ursprünge der menschlichen Kommunikation, (dt.:) Frankfurt/M. 2009, S. 150.

le gleichen in ihrer Frequenz um 3,5 Kilohertz genau der Frequenz von Babyschreien, es ist der Bereich höchster, unwiderstehlicher Äußerung. Bildgebende Verfahren an der Amygdala mütterlicher Gehirne haben überraschend eindeutig gezeigt, daß dieses Schreien gegenüber dem Schreien eines Erwachsenen und gegenüber der Reaktion eines Mannes um 900 % stärkere Reaktionen bewirkt.[108]

Öffentliche Empörung über das ‚Wegschauen' (look-away-society) und die Flüchtigkeit von Passanten bei brutal Verletzten in gewissen Zonen von Großstädten könnte ein verbliebenes Empfinden des ‚homo sociale' über alarmierende Signale des Hilflosen belegen. Wo immer zu allen Zeiten ein Schmerzgeplagter schon mit Körpersprache der Hinfälligkeit seine Umgebung ‚alarmiert', ein Verwundeter sich ‚mit letzter Kraft' zum nächst möglichen Helfer ‚schleppt' oder Paniker und Histrionische ihre Mitmenschen ‚erregen', beschränkt sich zumeist rationale gezielte Planung und konzentriert sich Intentionalität des Leidenden eingleisig auf das Funktionieren und die Erfüllung einer einzigen, die Befreiung suchenden Erwartung. Dabei kommen im Falle panischer Grundstimmung Gesten der Hilflosigkeit bis hin zu Totstellreflexen mit Erstarrung, mit Halskloß, Brustdruck, weichen Knien, Zittern, Durchfall, hängende Schultern auf Basis parasympathischer Überreaktion zum Ausdruck.

Dem Menschen des Mittelalters und dem Menschen anderer früherer Kulturen erging es nicht anders, gleichgültig, welcher Gewichtung er einer ihm von Historikern nach der Quellenlage oft zugesprochenen Dichotomie von biologischem und religiös-spirituellem Heil zustrebte. Auch dann, wenn er Krankheit und Not als ‚Unterwegssein zum Heil' verstand und wenn die Medizin im Hochmittelalter ‚eine Magd der Theologie'[109] war, ihn ergriff in körperlicher und seelischer Bedrängnis jene Erwartung, die sich in erster Linie an einen Mitmenschen und dessen empfangsbereites Signalsystem richtete.

Legen wir den Schwerpunkt zunächst ungeachtet der Rekurrenz der Interaktionen auf die Signalaussendung. Im Tierreich wurden äußere Merkmale wie lange Hörner, Farben sowie unwillentliche und unkontrollierbare emotionale Zustände als ‚kommunikative Displays' von den eigentlich kommunikativen Signalen abgegrenzt. Nach Michael Tomasello werden nur letztere als intentional gedeutet, als zielgerichtet, um andere Artgenossen zu beeinflussen. Im Tierreich sind Rufe und Gesten, selbst der Muttertiere bei ihren Jungen, nicht an Artgenossen adressiert, sondern erwachsen streng auf eigener Emotionalität und unflexibler Vokalisierung. Kommunikative intentionale Signale sind im Tierreich

---

[108] Sander, Kerstin et al.: fMRI Activations of Amygdala, Cingulate Cortex and Auditory Cortex by Infant Laughing and Crying, in: Human Brain Mapping 28 (2007), 1007–1022.
[109] Vgl. Vanja, Christa, Krankheit im Mittelalter, in: Dinzelbacher, Mentalitätsgeschichte, S. 195–200.

wahrscheinlich nur bei Menschenaffen[110] zu finden. Sie zielen auf Erkenntnis bei einem Empfänger. „Um als kooperative Kommunikation zu gelten, muß das Ziel irgendwie Hilfe oder Teilen sein ...."[111] Immer aber und absolut entscheidend für höhere Formen von Kommunikation ist die nur Menschen eigene, die Sprache einschließende „Fähigkeit, einen gemeinsamen begrifflichen Hintergrund zu schaffen – gemeinsame Aufmerksamkeit (joint attention), geteilte Erfahrung und gemeinsames kulturelles Wissen". Dabei bewegt sich ‚Begriff' und ‚begrifflich' stets auch zum sensomotorischen ‚Greifen' und zum abstrakten ‚Ergreifen' und ‚Begreifen'.

Mit anderen Worten: Es „wird die menschliche Kooperation durch ‚geteilte Intentionalität' oder ‚Wir-Intentionalität' strukturiert."[112] Damit sind auch institutionelle und kulturelle Bereiche involviert, und wir können ergänzen: In herausragender Weise werden medizinische, psychotherapeutische und sozialmedizinische Sofortmaßnahmen oder Interventionen, ob sie durch Signale Gesunder oder schon Kranker als Notrufe initiiert werden, fast als Paradigmen verständlich. Diese Sonderform von Kommunikation schuf dem Menschen unter allen Lebewesen einen gewichtigen Anpassungsvorteil.

Neurobiologische Ansätze mit bildgebenden Verfahren zur Objektivierung von Hilfsverlangen und seinen unmittelbaren Folgen widmeten sich Systemen, die als neurale Korrelate von Erwartung (‚response expectancy') und Aufmerksamkeit gelten können. Es hat verschiedene Forschungsansätze gegeben:

Es ist eine schon die primäre **Sehrinde** betreffende separate Aktivierungsleistung je nach Wünschbarkeit, Indifferenz oder Unerwünschtheit als Auslesefunktion für verschiedene Bildkategorien nachgewiesen worden. Das heißt, nicht das Auge entscheidet darüber, was wir sehen, sondern unser Hirn wählt und gestaltet unsere Bilder und ihre Bedeutung. Denn diese Aktivierungen verbanden sich mit ebenso unterscheidbaren Aktivierungen von superior-orbito-frontalen Gebieten und von mittleren und anterioren Cingulumgebieten.[113]

Auch die primäre **Hörrinde** erzeugt in der Erwartungssituation eine erste Auslese auf die Nützlichkeit einkommender Signale. Tierversuche zeigten, daß sich Wechsel, Frequenz und Zeitbezogenheit akustischer Erwartung, z. B. das Abwarten bestimmter Momente (vorgesehener Beginn eines Ereignisses, Ab-

---

110   Yamamoto, S. Humle, T. and Tanaka, M.: Chimpanzees help each other upon request (2009) PlosONE 4(10): e7416, doi:10.1371: Sie „führen im Gegensatz zum Mensch selten Akte von willkürlichem Altruismus aus."
111   Tomasello, Ursprünge, S. 25.
112   Ebd., S. 17 mit Verweis auf Searle 1995; Bratman 1992; Gilbert 1989; vgl. auch Zaboura, Nadia: Das empathische Gehirn, Vs-Verlag 2009, S. 86f.
113   Kawabata H, Zeki S (2008) The neural correlates of desire. PLoS One 3(8): e3027 doi: 10.1371/j.

stimmung musikalischer Einsätze, Startsignale) als kraftvoller und effizienter Funktionsvorteil für die Hörrinde erweisen kann.[114] Verstärkt konzentrierte Aufmerksamkeit und Erwartung steigern die Verarbeitung von Signalen in der Hörrinde, wobei dafür gesonderte, mehr laterale Gebiete fungieren, die nicht tonotop organisiert sind.[115] Es gibt auch erste Anzeichen dafür, daß akustische, verbale Signale bei der Perzeption von Bildervorstellungen den Darbietungen von visuellen Signalen überlegen sein könnten.[116]

Involviert in das cerebrale **Erwartungssystem** ist besonders und übereinstimmend unter Neurobiologen der bereits erwähnte Nucleus accumbens des ventralen Striatums im Bereich der Stammganglien als Schwerpunkt des Belohnungssystem. Mit experimentell ansteigenden Erwartungsschritten stieg dessen Aktivierung und stieg damit die Ausschüttung von Dopamin in Placeboversuchen. Vieles spricht dafür, daß bei sogenannten Placebowirkungen neben dem Nucleus accumbens ein umfassendes erwartungsabhängiges Hirnresiliencesystem in Aktion tritt.[117]

Der Grad der Schmerzlinderung, der nunmehr durch die bildgebenden Verfahren messbar wurde, so ergaben andere Untersuchungen, war von verbalen Instruktionen zum Aufbau der Erwartungshaltung[118] bzw. von Suggestionen abhängig, wenn bewußte Aspekte wie Schmerz oder Körperbewegung (bei Parkinsonkranken) angesprochen werden.[119] Neben diesen Effekten der Erwartung, die auch endogene Opiate mobilisieren, erwies sich der Effekt von Konditionierung als ebenso wichtig, d. h. zum Beispiel daß die vor Placebogabe verabreichten wirksamen Präparate einen günstigen Erwartungsdruck vorbereiten.[120] Für die Anregung der endogenen Opiatsysteme unter Placebobedingung spielt der

---

114 Jaramillo S, Zador A M, The auditory cortex mediates the perceptual effects of acoustic temporal expectation, Nature Neuroscience (2010), doi: 10.1038/nn.2688.
115 Woods D L et al. (2009) Functional maps of human auditory cortex: effects of acoustic features and attention, doi: 10.1371/journal pone 0005183.
116 Lupyan, Gary and Michael J. Spivey, (2010) Making the invisible visible: Verbal but not visual cues enhance visuel detection, PLoS One 5(7), doi: 10.1371/j.pone.0011452.
117 Scott D J, Zubieta J K et al., (2007) Individual differences in reward responding explain placebo-induced expectations and effects, in: Neuron 55,325–336 doi 10.1016/j.
118 Pollo A, et al. Response expectancies in placebo analgesia and their clin. relevance, in: Pain 2001 Jul; 93(1):77.
119 Benedetti F Pollo A et al. Conscious expectation and unconscious conditioning in analgesic, motor, and hormonal placebo/nocebo responses, in: The journal of neuroscience 2003, 23(10): 4315–4323; De la Fuente-Fernandez R et al., Expectation and dopamin release: mechanism of the placebo effect in Parkinson's disease, in: Science 2001, 293 (5532), 1164–1166.
120 Amanzio M, Benedetti F (1999), Neuropharmacological dissection of placebo analgesia: Expectation-activated opioid system versus conditioning-activated spec. subsystems, in: J. of neuroscience 19(1), 484–494.

anteriore cinguläre Cortex eine Schlüsselrolle.[121] Dieses Gebiet ist aufgrund seiner Direktverbindungen zu vielen sensorischen Eingängen, zu Gedächtnisinhalten, zu Motivation und Emotion eine für erwartungsabhängigen Placeboeffekt prädestinierte Schaltstelle.[122]

Diese neuronalen Pfade der Kooperation von **Erwartung und Konditionierung** bedürfen noch der Klärung; eindeutig hat die Forschung aber bereits einen kollektiv vernachlässigten Allgemeinplatz bestätigt, wonach Gesundheit durch Änderung unserer geistigen Einstellungen nachhaltig, u. U. strukturell gefördert[123] und durch verbale Maßnahmen geformt werden kann.[124] Denn diese sogenannten Placeboeffekte – ein Placebo per se wirkt ebensowenig gesundheitsfördernd, wie ein Schraubenzieher jemals eine Schraube bewegt hat – sind weitgehend unabhängig von der Art der Persönlichkeit des Hilfsbedürftigen, seiner Krankheit, seinem Alter und – überraschend auch unabhängig von seiner individuellen Suggestibilität. Große Statistiken lehren uns, daß vielmehr Angst und Sorge, verbunden mit der Erwartung der lindernden Behandlung und besonders die Attraktivität und Erfahrung des Therapeuten wirksam werden können.

Geprüft wurde ferner mit bildgebenden Verfahren die Wirkung bewußter zeitlicher Verzögerung bei Gewissheit einer bevorstehenden Erfüllung der erwarteten Antwort. Wir schließen dabei eine Hilfeleistung durch einen zuverlässigen Heilkundigen ein. Hier erwies sich die Zeitdauer geradezu als kumulativer Motor für das Erwartungspotential einer Wahrscheinlichkeit: Je länger die Vorperiode, umso größer wuchs die Aktivität präfrontaler, parietaler und weiterer Rindengebiete, woraus geschlossen wurde, daß damit auf spezifische auditive, visuelle oder olfaktorische Erfahrungen zurückgegriffen wird.[125]

Schließlich sind auch neurophysiologische Experimente zu erwähnen, die Ansätze zur Erkundung von Vertrauensabschätzung im Gehirn bieten.[126] Daß diese psychische Fähigkeit, eigene Vertrauensseligkeit und eigenes Mißtrauen auszubalancieren, einen hohen Stellenwert für jede kommunikative Aktion hat, wird

---

121 Petrovic P, Kalso, E, Petersohn K M und Ingvar M (2002), Placebo and opioid analgesia – Imaging a shared neuronal network, in: Science 295 (5560), 1737–1740.
122 Weiss, Thomas, Psychophysiologische Aspekte des Placeboeffekts bei Schmerz, in: Zeitschrift für Neuropsychologie 15(2), 2004, S. 99–110; dort Übersicht Forschungsgeschichte.
123 Wager T D, Nitschke J B, Placebo effects in the brain: Linking mental and physiological processes, in: Brain, Behavior and Immunity 19 (2005),281–282.
124 Stewart-Williams S, Podd J (2004) The placebo effect: dissolving the expectancy versus conditioning debate, in: Psychol Bull 130(2):341–3.
125 Coull J T (2009) Neural Substrates of mounting temporal expectation, PLoS Biol 7(8): e1000166. doi: 10.1371/j
126 Dzhelyova, M. et al., Temporal dynamics of trustworthiness perception, in: Brain Research 1435 (2012), S. 81–90; Ito, A. et al., The contribution of the dorsolateral prefrontal cortex to the preparation for deception and truth-telling, in: Brain Research 1464 (2012), S. 43–52.

oft vergessen; für die Notsituation dürfte sie meist nicht von übergeordneter Bedeutung sein.

Diese Ergebnisse könnten einen ersten Blick in die Verarbeitungsweise von Konditionierung jeglicher verbal und/oder rituell aufgeladenen Heilmethode geben.

Die ganze **Komplexität einer existentiellen Notlage** wird neurophysiologisch allerdings erst dann adäquat abbildbar, wenn nicht nur emotionale Teilfunktionen, sondern die Sichtweise eines aufs Überleben gerichteten neuronalen Kreislaufs bedacht wird. Es werden dann über Emotionen hinaus auch Motivationen, variable Verstärkerfunktionen, intrinsische Fahndung nach psychophysischen Ressourcen und konditionierte (erlernte) wie unkonditionierte Antworten auf Herausforderungen u. a. m. eingeschlossen. Sie betreffen dann neben Amygdala und Nucleus accumbens auch weitere Hirngebiete bis hin zum dorsalen Höhlengrau, zum Hypothalamus, also vegetative und hormonregulierende Zentren.[127]

Kommen wir jetzt zurück auf die **Unfreiheit des Notfalls** mit seiner *scheinbaren* geistigen Ohnmacht und seiner aufgrund der situativen Umstände nicht im Einzelfall experimentell neuronal messbaren, aber nach allen körperlichen und psychiatrischen Befunden feststellbaren intentional hochgespannten Erwartung. Sie ebnet den Weg für eine von uns Außenstehenden als Täuschung gegenüber der Realität vorstellbare Beeinflussbarkeit des Gehirns. Diese entspricht mit anderen Vorzeichen der Vorbereitung und Einleitungsphase von Suggestion, Trance und Hypnose, indem sie rational assoziative Verarbeitung bremst und Aufmerksamkeit auf Erwartung einengt und umleitet. – Nimmt der Patient alles an, was Hilfe verspricht? Was heißt in diesem Fall Annahme? Bestimmt seine Hirnstruktur die Bedeutung des Gebotenen?

B1.2 Aufmerksamkeit, Erwartung und Perzeption

Die Begriffe Aufmerksamkeit und Erwartung sind sich sehr nahe und in ihrem international wissenschaftlichen Gebrauch gelegentlich austauschbar. Beide Begriffe sind einander in ihrer zugleich intentionalen und beim Beobachter nicht als aktiv scheinenden Eigenschaft verwandt. Mit ihnen wird vielfältig psychologisch experimentiert. Sie dienen als Grundlage zahlreicher neurologischer Versuchsanordnungen.

In unserem Rahmen wird Erwartung eng als unmittelbare psychophysische Reaktion nach einer unerwarteten Notsituation konkret mit Hilferuf und Hilfser-

---

[127] LeDoux, Joseph, Rethinking the emotional brain, in: Neuron 73 (2012), 653–676.

wartung verbunden und bezeichnet. Im Gewahrwerden bzw. in erster Annäherung an eine mögliche Hilfsquelle erwächst aus der Erwartung zunächst eine sich konzentrierende, sich sammelnde Aufmerksamkeit (**Abb.2**), die neurobiologisch von verarbeitenden intracerebralen Perzeptionsmechanismen nicht mehr zu unterscheiden ist. Angelpunkt mag in diesem Moment das „Aha"-Erlebnis sein, wenn der Erfolg der Hilfesuche die Richtigkeit der Signalaussendung bestätigt und hormonell belohnt. Dabei lenkt die gezielte Erwartung auch das Ziel der Aufmerksamkeit, wobei nach den grundlegenden Untersuchungen von Michael Posner[128] die Entdeckung von verdächtigen, brauchbar scheinenden äußeren und inneren Stimuli auf die Sinnesorgane, mit der Be- oder Verhinderung der Aufnahme anderer Reize einhergeht. Wie ein Lichtkegel (spotlight) versammelt Aufmerksamkeit ausgewählte Signale, je präziser, umso enger.

Cerebrale Gebiete, die für die Steigerung und Ausrichtung der Aufmerksamkeit fungieren[129]:

– dorsolateraler präfrontaler Cortex als Warnsystem kritischer Entscheidungsfindung
– rechter vorderer Schläfenlappen für u. a. lexikosemantische Vorgänge
– vorderer cingulärer Cortex als spotlight, Prüfung von Signalen auf Neuigkeit und Nutzen
– Praecuneus des Scheitellappens, aktiviert bei erfolgreicher Begriffsfindung
– linkes unteres Stirnhirn als wichtig bei Bildung von Assoziationen
– untere Sehrinde und Kleinhirn mit Umstrukturierung visueller Reize und Aufmerksamkeitsaktivierung

Wir werden sehen, welch eminente Bedeutung diese Aufmerksamkeitsfesselung und deren selektive Verarbeitung für die Wirkung der Heilspruchtexte hat. In Zusammenarbeit von Hirnstamm- und Rindengebieten kann z. B. den narrativen Imaginationspartikeln (**Kap. C4**) gezielte erwünschte Relevanz zuteil werden, können unter Oszillieren von Aktivität zwischen Hirngebieten inkongruente Metaphern wie Oxymora und Adynata lexikalisch geprüft werden (**Kap. C3**) und können bedrohende Signale mittels Hypervigilanz bzw. Emotionsregulierung (**Kap C1**) entschärft werden. Das gilt auch für die begleitenden Riten, etwa die obsessiv anankastischen Ketten- und Stufen-Techniken. Die Experimente mit der Aufmerksamkeit boten (vor der neuen Hyperscanning-Technik) einen Schlüssel für das Verständnis von Gedächtnismanipulationen, Illusionistenerfolgen und allen Arten von ‚Performativität', besonders auch für Suggestion, Trance und hypnoide Verfahren.

Nach den weiteren Untersuchungen von Posner und seiner Arbeitsgruppe[130] sowie der die Magie, das Überlisten und Verwirren einschließenden Arbeit von

---

128 Posner, Michael I. et al., Attention and the detection of signals. In: Journal of experimental psychology, general. 109 (1980),160–174.
129 Modifiz. nach Zusammenstellung Schrott, R. und A. Jacobs: Gehirn und Gedicht, S. 138.

Amir Raz[131] wurde die Aufmerksamkeit in drei anatomisch und physiologisch disparate Kontrollsysteme unterschieden:

1. Ein <u>exekutives</u> Netzwerk verarbeitet Konflikte und Inkompatibilitäten in Denken, Fühlen und Antworten. Es überwacht warnend und entscheidet auflösend. Eingebunden sind der frontale Teil des Cingulum und die lateralen praefrontalen Rindengebiete. Biochemischer Modulator ist Dopamin. (Yikes! How do I resolve this CONFLICT?
2. Das <u>alarmbereite</u> Netzwerk unterhält eine hohe Sensitivität für exogene Signale. Es scheint ein entscheidender Zusammenhang mit Transmitter-Modulatoren der Neuralaktivität (Norepinephrine) zu bestehen. Dieses System arbeitet mit Thalamus und frontalen und parietalen Regionen. („ALERT?" But, I don't know where, just when!)
3. Das <u>orientierende</u> System trifft die Auswahl von Informationen aus dem sensorischen Input. Es arbeitet auf der Basis superiorer und inferiorer Parietalgebiete, des Thalamus, der Colliculi und sensorischer Rindengebiete (z. B. frontales Augenfeld) und modulatorisch mit Acetylcholin. (Hmm, I know where to ORIENT, but when will things happen?)

Zusammenfassung Kap. B1: Der Kranke in Not: Hilferuf und Hilfserwartung, Aufmerksamkeit und Perzeption

Menschliche Notsituationen mit dem Anschein von Ohnmacht und den Symptomen von Hinfälligkeit täuschen zwar eingeschränkte Freiheitsgrade vor. Sie können aber im Falle der Entdeckung von möglicher Hilfsbereitschaft oder der Begegnung mit einem Heilkundigen in dieser sozialen Sondersituation durch zwingende Signalaussendung von Erwartung hohe Beachtung einfordern. Der Suchprozess involviert gleichzeitig einen Selektionsprozess. In der Psychoanalyse wurde das allgemeine Prinzip struktureller Beziehung zwischen Beobachter und Beobachtetem etwa so beschrieben: „Wir sind nie logisch; unser Realitätssinn ist von unseren Bedürfnissen durchdrungen. Wir sehen bewußt im Grunde nie, wie es ist, sondern immer nur wie wir es wollen."[132] In den letzten 15 Jahren haben die bildgebenden Verfahren am Gehirn diese Thesen experimentell belegen können, sodaß der neurobiologisch arbeitende Verhaltensphysiologe ähnlich formuliert: „Es ist der Empfänger, der Bedeutung konstituiert."[133] oder wie der kanadische Philosoph Paul M. Churchland: „Das Gehirn übt eine entscheidende Kontrolle darüber aus, wie es etwas sieht oder hört."[134] Die oben genannten Kontrollsysteme für Aufmerksamkeit modulieren unser Erkennen, Denken und Handeln, während gleichzeitig Einflüsse der Stammganglien und des limbischen

---

130 Abdullaev, Y. and M. I. Posner. Attentional mechanisms, in: Berntson/ Cacioppo, Handbook, I, 370–382.
131 Raz, Amir, Varieties of attention, in: Berntson/ Cacioppo I, 361–369.
132 Fenichel, Otto: The collected papers, New York 1953, S. 11.
133 Roth, Gerhard: Das Gehirn und seine Wirklichkeit, Frankfurt 1998², S. 107.
134 Churchland, Paul M.: Die Seelenmaschine, Heidelberg u. a. 1997, S. 133.

Systems, insbesondere des Corpus striatum mit dem Nucleus accumbens und seinem Belohnungssystem und der Amygdala selektierend emotionale Einflüsse ausüben, die bremsend eine überhöhte oder überdehnte Aufmerksamkeit zu lenken oder zu schwächen vermögen. Die Verbindungen zwischen der Hirnrinde und dem Basalhirn unterliegen dynamischen Feedbackschleifen und operieren als Verbände (Assemblies). Akte der Wahrnehmung (Auswählen, Aufteilen, Nah- oder Ferneinstellung) und ihrer Interpretierung sind ebenso wie Emotion und Kognition nicht klar zu trennen. Das bedeutet, daß Wahrnehmungen gerade unter Umständen einer Notfallsituation stark von der äußeren Realität abweichen können. ‚Aufmerksamkeit' und ‚Erwartung' sind in der modernen Neurobiologie letztlich zu einem Problem von Aktivität und Zeitpunkt von Neuronensynchronisation, zu einem Bindungsproblem, geworden. Nach Antonio Damasio ist diese Synchronisation überlebenswichtig; ohne sie gäbe es keine Vorstellung, kein assoziatives Denken, keine Inspiration, keine dem Menschen mit seinem größer entwickelten Gehirn eigentümliche Planung und Erfahrung. Zugleich bildet unser Wissen über die Bedeutung der hochkomplexen hirnorganischen Synchronisation und Integration einen wichtigen Gegenpart zu den alten, zunächst die Forschung anregenden Ideen von Franz Josef Gall (1758–1828) und seiner Phrenologie bis zu manchen daraus ableitbaren pseudowissenschaftlichen Charakterverortungen in überspitzten cerebralen Fixpunkten. Auf die Begabung des Menschen zur Schaffung sowohl interpersonaler ‚geteilter Aufmerksamkeit' als auch ‚gemeinsamen kulturellen Wissens', als sich hier ankündigende Vorraussetzungen für das Gelingen hilfreicher Aktionen wird in folgenden Kapiteln (**B3**, **B4**) eingegangen.

B2 Heilkundige und Helfer

Zu allen Zeiten und in allen Kulturen hat es Einzelne gegeben, die sich komplementär zum Hilfsbedürftigen als Hilfskraft für die Nöte anderer entdeckten oder vielmehr entdeckt wurden, sei es vom ‚Geist der Ahnen', sei es vom geistlichen Führer oder in der mesopotamischen Kultur aus familiärer und priesterlicher Tempel-Elite. Denn der dem Menschen immanente Mangel an rationalen Begründungs- und Regulierungs-Funktionen für psychische und psychosomatische Störungen, ja für alle, selbst für zugefügte traumatisch bedingte Leiden und für die Rätsel um seine Träume und Phantasmen hält den Bedarf an Rettungskräften aufrecht und führt permanent zur Rettungssuche außerhalb eigener Instanz und Kompetenz. Zahlreich waren die Motive, die gesellschaftlichen Zwänge und

Aufgaben, die die Helferin oder den Helfer dazu veranlaßten, sich in den Dienst einer Gruppe oder eines Kranken zu stellen.

Die **Schamanen** fremder Kulturen waren für vieles zuständig: Mißernten und Wetter, Viehsterben und Verbrechen, Krankheiten und Geburtshilfe. Im Gegensatz zu den Mönchsärzten des Hochmittelalters versetzten sie sich vielfach unter Drogen, mit Tänzen, Musik, Feuerzauber und Fasten in Rausch, Trance und Ekstase. Damit übergaben sie sich mit ihrer Gruppe und gemeinsam mit dem Kranken der spiritualisierten Natur eines kollektiven Mythos, oft unter Einschluß der Ahnen. In den Initiationsriten für Pubertierende wurde eine rituelle Rückkehr in den Mutterschoß mit Neugeburt in eine neue Ebene zum Erwachsenen simuliert, eine Regression als Zündstoff großer Kräfte. Unter dem Einsatz guter und böser Mächte wurde das Leid des Kranken rituell und verbal abgebildet. Immer wieder belebten diese Heilkundigen gleichsam als Amtsträger die Ordnung stiftende ‚participation mystique' (**Kap. A1.2.2**). Vieles im Wirken der Schamanen war einerseits neurotischem oder schauspielartigem, andererseits psychostrategischem grenzüberschreitendem Verhalten vergleichbar; einige Begabte vermochten sich an gesellschaftliche und individuelle Desorientheit anzunähern und arbeiteten mit durch Performation gelenkten Korrekturen, wie von Lévi-Strauss dargestellt (**Kap. A1.2.3a–c**).

Die **Mönchsärzte**, Mönche und Ärzte des Hochmittelalters, wenn sie Heilsprüche anwendeten, waren meist spezialisiert und ausgebildet unter strenger Kontrolle. Sie lernten einen Beruf im Rahmen christozentrischer Gesellschaften. Ihre monotheistisch orientierte Konzeption griff nur teilweise, und zwar zur Behandlung von Psychosen, Wahn- und Hysterie-Syndromen auf extrem dualistische, archaische und antike Muster zurück, die als Exorzismen mit Gott und mit Teufel und Dämonen operierten (**➔ 38 und 61**). Daß diese Gegenüberstellung von Gegenspielern unter günstigen Umständen entsprechend der Psychopathologie und Ätiologie mancher Krankheitsformen als dynamisierendes sprachliches Element wirkte (vgl. Saussures Theorien) und zur Entwicklung von metonymischen Ketten und transformativen Aktivierungen im sensomotorischen System des Gehirns der Betroffenen beitragen konnten, liegt auf der Hand. Nach heutiger Auffassung bediente man sich in diesem Krankheitsbereich eines ‚Rapports mit dem Symptom' und näherte sich dem Kranken indirekt an. Im allgemeinen bezogen sich die Mönchsärzte in Imitatio Christi auf die Heilwunder Jesu und die Wundertaten der heiligen Gestalten der Kirche, womit oft auch eine All-Sympathie natürlicher Kräfte angesprochen war. Weil Erlösung dem Helfer als Belohnung für Krankenversorgung galt, lag die optimale Zuwendung zu ihm in ureigenem Interesse. Im Blick auf dieses eigene Seelenheil im Glauben an das ewige Leben und an das Richteramt Christi und auf die verpflichtenden Ordensregeln darf der empirische und praktisch-methodische Arbeitsaufwand an Diagnostik, Chirurgie, Geburtshilfe und Phytotherapie, den sie leisteten, bei und neben Beschwörungen und Heilsegen nicht vergessen werden.

Priesterärzte der mesopotamischen Kultur. Es ist die ältesten Kultur, die mit ihren Keilschrifttexten Zeugnis von ärztlichem Wirken seit 4 Jahrtausenden vor Christus hinterlassen hat. Die Medizin – wie wir sie aus den erhaltenen Tontafeln kennen – lag wohl einerseits in den Händen der privilegierten, speziell in Tempeln in Geheimwissenschaft gebildeten und in Familien tradierten Priesterschaft, die zum engsten höfischen Kreis gehörte. Daneben ist vermutet worden, daß es eine sozial niedriger gestellte Gruppe von eššebu-Ekstatikern und Schlangenbeschwörern gab, die vor allem Abwehrzauber betrieben, „ambivalente Gestalten, ... die zum Straßenbild der babylonischen und assyrischen Städte gehört haben."[135] Über sie gibt es keine eindeutigen Zeugnisse.

Unter den angesehenen Gruppen von Heilern gab es solche mit überwiegend prophetischer Funktion, die auch ‚Diagnosen' und ‚Prognosen' am Kranken erstellten und es gab Geisterbanner, die Beschwörungen vollzogen, aber auch Sühneriten und Opferritualien einleiteten. Eine weitere Gruppe unter den Priestern widmete sich schwerpunktmäßig der praktisch-physischen Krankenversorgung, auch mit Amuletten, Naturalien (Pflanzen, Tierteile) und mit Operationen; allerdings gab es Überschneidungen und Kooperationen dieser Gruppen; eine strikte Separierung in religiös, wissenschaftlich und magisch therapierende Spezialisten, wie das modernem Denken naheliegt, war jedenfalls konzeptionell nicht gegeben. Zwar hat „die ideale Unterteilung der Heilkunst in ašiputu, die Exorzistenkunde, und in asûtu, die ‚physische' Ärztekunde, die Organisation und Transmission der babylonischen magischen und medizinischen Texte tief beeinflußt" aber die beiden „Basisstrategien" wurden „als komplementär betrachtet."[136]

Gemeinsam hatten **Mönchsarzt** wie **Schamane** und **mesopotamischer Priesterarzt** im Ansehen der Hilfsbedürftigen und der Gesellschaft Autorität oder Charisma – Bedeutungen also, die von manchen Soziologen als Hauptquellen von Kommunikationsmacht betrachtet werden, selbst im Blick auf die Ausübung sprachlicher Überzeugungskraft.[137] Gemeinsam besaßen diese ‚Berufenen' eine Fülle kulturell unterschiedlicher, aber technisch gut vergleichbarer empirischer Methoden, die es uns erlauben, anhand der Texte und auf der Basis elementarer neurobiologischer Prinzipien eine gemeinsame Darstellung vorzunehmen. Vorstellbar ist es, daß die Empathie- und Transienz-Fähigkeiten der Helfer über eine

---

135 Schwemer, Daniel: Abwehrzauber und Behexung, Wiesbaden 2007, S. 246.
136 Ders.: Magic Rituals: Conceptualization and performance, in: Radner/ Robson, Handbook, S. 423.
137 Vgl. Reichertz, Jo: Kommunikationsmacht, S. 211–217 unter Bezug auf Pierre Bourdieu.

fixierte affektive Nähe zum Hilfsbedürftigen, über reinen Samariterdienst oder strikte professionelle Routine oft hinausgingen. Das bei den meisten der alten Heilkundigen zu vermutende Changieren zwischen Nähe und Distanz, Emotion und Ratio mit Rückbeziehung auf die jeweils kulturgegebenen höheren Wesen, basiert auf verschiedenen inzwischen funktional durch neurophysiologische und bildgebende Verfahren belegte Kompetenzen integrierter Neuronensysteme (**Kap. B3**).

Gemeinsame Techniken lassen sich besonders aus Sicht der Hypnotherapie erfassen. Wir kennen die historisch relevanten Riten mit Augen-Fixationsmethoden der frühen Zeit bereits aus dem 2. bis 4. Jahrtausend vor Christus im Papyrus Ebers. Der Tempelschlaf im antiken Griechenland umfaßte Tranceförderung[138] und gezielte Verbalsuggestion. Das feste Anfassen beim Exorzismus, das Handauflegen und andere körperliche Berührungen sowie die Fixierung der Augen auf das Kreuz haben eine lange Tradition. Die Schaukelrhythmen und die ekstatischen Tänze sind seit ältester Zeit weltweit bekannt und gebräuchlich.[139] Prosodische, also die sprachliche Lautgebung betreffende Techniken, etwa das monotone Murmeln unverständlicher Worte, sich wiederholende Vokalformen (**Beispieltext ➔60**) oder der Singsang waren wohl allgemein so oder so üblich. Die begleitenden rituellen Handlungen, etwa das Zweigspalten und Wiederverbinden (**➔31**) bei Blutgefäßschäden, das Umgürten der Organe (**➔23**) oder, im Mesopotamischen und unter Schamanen besonders gepflegt, der Einsatz geformter Figuren zur stellvertretenden Durchführung von Straf- oder Heilhandlungen (**➔43, Zeile 15; ➔44, Zeile 135; ➔50, Z. 132; ➔58, Z. 65 u. a.**) fügten sich als den Erfolg simulierendes Anschauungsmaterial ein in die suggestive Ausrüstung der Heilkundigen und wirkten teilweise nach Art einer Konditionierung (s. o.) oder aus einer anderen Sicht mit hoher Bedeutung und weitgehend üblich für alle ehemaligen Heilkundigen als Techniken, die heute als regressiv bezeichnet werden.

Während Psychologen zunächst unter dem Einfluß Sigmund Freuds bei Schamanen und im Seelenleben der ‚Wilden' mit ihren ‚archaischen Vorstellungs- und Denkformen im Verhalten' (Freud 1914, S. 523ff) die sog. **regressiven Tendenzen** als unerwünscht entdeckten, ein ständiges ‚Sich-gehen-lassen' im Erlauben von Phantasien, haben sich bald psychotherapeutische Methoden etabliert, die diese Regressionen für ihre Zwecke zu nutzen vermochten.[140] Für C.G. Jung kann Regression – verstanden als Inversion oder Introversion – psychodynamische Prozesse zur Ichfindung fördern. Sandor Ferenczi weist auf eine ‚archaische Regression' hin, die bei schweren Krankheiten und Traumen ein Notprogramm bieten kann, das den

---

138 Ein umfangreiches Beispiel in Papyri graecae mag. IV, 850–929, vgl. Betz, Hans Dieter, S. 55.
139 Vgl. Hole, Günter: Direkte Induktionen, in: Revenstorf/ Peter: Hypnose, S. 182.
140 Vgl. Schmidbauer, Wolfgang: Helfen als Beruf. Hamburg 1983, S. 68.

Schutzengel einbezieht. Manche Psychoanalytiker hatten bald erkannt, daß Regression, zum Beispiel per Hypnose, ins Kindesalter führte, die bewußte ICH-Instanzen schwächte, aber damit auch den Zugang zum Unbewußten verbesserte oder daß sogar oft ‚Regression um der Progression willen'[141] erfolgen kann. Moderne Therapie vieler Schulen hat längst erarbeitet, daß der Rückgriff auf frühe persönliche Erfahrungen, wenn er nicht in den Bereich des Krankhaften gehört, sowohl verbal, als auch in visuellen und haptischen Techniken genutzt werden kann. Historisch wird sich dies nicht nur in schamanischen, sondern auch in babylonischen und christlichen Beschwörungen zeigen, wenn etwa Figurinen, Amulette und Pflanzenteile anthropomorphisierend oder theriomorphisierend Verwendung finden (**Kap. C4.1.2**). An regressionsfördernden Texten, die von Jonas im Walfisch, von Hiob im Mist oder von der Auferstehung des Lazarus erzählen, wird die Möglichkeit sehr deutlich, umgehend wieder Progression durch Errettung und Neuerweckung zu Licht, Raum und Leben, zu vermitteln.

Auch die verbale Seite früherer Suggestionen ist aus hypnotherapeutischer Sicht in ihren Imaginationen, ihrer Metaphorik und dem Einsatz ihrer Anthropomorphismen, ihrer Geister und überirdischen Heilsgestalten ebenso wie der Synchronisierung von Vergangenheit Gegenwart und Zukunft gut nachvollziehbar. All die figuralen Elemente können zwanglos als Attraktionsziele der im Konzept der Psychoanalyse als Affekt-Übertragung an eine dritte Person bezeichneten Vorgänge verstanden werden. Diese dritte konstruierte oder biografisch und kulturell bedeutsame Gestalt oder oft die Gegenüberstellung handelnder Gegengestalten, auf die Therapeut und Patient sich gemeinsam beziehen, hat soziokulturell verschiedene Prägungen und versetzt den Heiler in eine Funktion der Vermittlung, die seine Kräfte weniger übersteigen kann als seine bloß direkte autoritäre Macht. Moderne Hypnotherapie hat dieses figurale therapeutische Konstrukt als intrapersonale Form des Unbewußten und als kluge wohlwollende Instanz insinuiert. Sie hat damit auf das Faktum der innerseelischen Ressourcen und deren neuronale Erinnerungsspuren verwiesen.[142] Es kann die damit verbundene therapiestrategische Erwartung auch auf konfliktreich bedrohliche und damit bearbeitbar werdende seelische Quellen treffen.

Zu den therapeutisch wirksamen Strategien der Heilkundigen gehören beim Gebrauch von Wort und Ritus auch die Techniken der Zeitverknappung und der Transienz. Jede erfolgreiche Begegnung dieser Art setzt eine Begrenzung der Zeitvergabe voraus, weil Autorität auch von begrenzter Verfügbarkeit lebt. Wir streben umsomehr nach Gütern und Gaben, je seltener sie erreichbar sind. Die rituellen Anwendungen tragen dabei zur respektheischenden Zeitstruktur ebenso bei wie seinerzeit Sanduhren an den Kanzeln der Kirchen oder wie der Blick auf die Uhr durch Manager und Politiker. Selten sind den Texten der Beschwörungen und Heilsegen Angaben über Zeitmaße beigegeben. Die Maqlû-Beschwö-

---

141 Balint, Michael: Therapeutische Aspekte der Regression, Stuttgart 1997², S. 161.
142 Vgl. Peter, Burkhard, Therapeutisches Tertium und hypnotische Rituale, in: Revenstorf/Peter, S. 70–77.

rungen erstrecken sich über eine Nacht. Für die Durchführung von Exorzismen nach den Regeln der Kirche wird ausnahmsweise hoher zeitlicher Aufwand zu berechnen sein. Kompakte Exorzismen standen wahrscheinlich seltener im Spektrum der Akuttherapie. Der Notfall als unerwartetes Ereignis war und ist Präzedenzfall für die Begegnung von Heilkundigem und Hilfsbedürftigem mit der äußerlich gegebenen Rollenverteilung in Führer und Geführten im Ausnahmezustand. Ein solcher Zustand kann und darf nicht von ungesteuerter Dauer sein.

Andere Effekte heilkundlicher Strategie wie die Macht der Augen beim Blickkontakt, dessen differente Botschaften nach heutigen Untersuchungen von großer Bedeutung sind (**Kap. C2**), ebenso wie die Stimmung und die offene oder die oben genannte prosodische Ausstrahlung von Motivation seitens der Heilkundigen, können von uns für die Beurteilung ihrer früheren Heiltätigkeit nach den Textquellen nur bedingt herangezogen werden.

Aber alle Macht- und Altruismusvorstellungen über Heilkundige, selbst die ihnen oft zugewiesene Leidenschaftlichkeit und Liebenswürdigkeit, ihr Einfühlungsvermögen oder der Idealismus bei der Arbeit für den leidenden Nächsten sind notwendig beschränkt und von den Beteiligten jeweils geteilt erlebbar durch die Komplementarität im sozialen und neurobiologischen Rollenspiel. Ein Germanist hat das Thema an einem Beispiel der Literatur (Effi Briest) konsequent aufgegriffen: „Um Parteinahme zu ermöglichen, muss/ kann die Affizierung des anderen nötig werden." „Wir wollen das Leiden des anderen, um Empathie mit ihm empfinden zu können. Diese, in struktureller Hinsicht konsequente Perversion ...."[143]

Der Sozialpsychologe P.R. Hofstätter hat ausgehend von Freuds Theorie der Projektion von Allmacht auf idealisierte Objektrepräsentanzen das Verhältnis von Hypnotiseur zu einem Medium als ‚dynamische Sequenz' bezeichnet und damit das Verhältnis von Führer zu Geführtem wie von Mutter zu Kind mit Wunscherfüllung und Idealisierung verknüpft. Die Dynamik entwickle sich an der Hilflosigkeit und ihren Phantasien von Größe und der Teilhabe an Macht. Auf dem Niveau des Führers aber spiele sich die gleiche Sequenz ab wie auf Seiten des Geführten: „Der Führer bedarf der Geführten kaum weniger dringend als dieser seiner ...."[144] Ein Akutwerden dieses Führerproblems sei „wohl stets" ein Ausnahmezustand, da es nur bei unerwarteten Zuständen eines originalen Führers bedarf; nur in einer solchen Situation bestehe auch das Gefühl der Hilflosigkeit, das die Führerrolle voraussetzt. Diese Rolle ist daher prinzipiell von passagèrer Art.[145]

---

143 Breithaupt, Fritz: Kulturen der Empathie, Frankfurt/ M. 2009, S. 116.
144 Hofstätter, Peter Robert: Einführung in die Sozialpsychologie, Stuttgart 1963, S. 355.
145 Ebd., S. 363.

Darüberhinaus müssen wir von der Vorstellung fixierter statischer Zustände neuraler Systeme und Organismen Abstand nehmen zumal bei Betrachtung von Begegnungen mit einem anderen Organismus und seinem neuralen System. Das legen sowohl die Theorien zur Übertragungsbearbeitung der Psychoanalyse als auch die Hypothese der Autopoiese des Gehirns, also des sich selbst schaffenden und entfaltenden Organs im Sinne von Maturana nahe. Operative Vorgänge vertauschen Beobachter und Beobachteten, beeinflussen gegenseitig Kreativität in nicht vorhersagbarer Weise und verändern zumindest potentiell ständig die Erfahrungs- und Wahrnehmungsebenen. Damit ermöglichen sie erst den konsensuellen Gebrauch der Heilspruchtexte und ihrer Metaphern und Geschichten mit der Tendenz, auch einer dritten Person, etwa dem göttlichen Wesen näher zu treten. Der Spruch fungiert also auch als eine Art Leitschiene gegen ein mögliches Zerfließen oder Abdriften der therapeutischen Intention ins Vage; er gewinnt in diesem Zusammenhang noch mehr an Bedeutung. Er ermöglicht eine konsensuelle identifikationsmindernde Zielperspektive.

Am Text des Trierer Pferdesegens zur Heilung der Pferderähe ist das Nebeneinander von Identifikation und Distanz dargestellt worden, d. h. der Heiler spricht als Gott wie ein Priester bei der Wandlung; und sein Gesprochenes hat Zitatcharakter:[146] „Christus und der heilige Stephan kamen in die Stadt saloniun. Da wurde das Pferd des heiligen Stephan verfangen. So wie Christus dem Pferd … das Verfangensein heilte, so heile ich dieses Pferd mit der Hilfe Christi … Wohlan, Christus, geruhe durch deine Gnade diesem Pferd das Verfangensein … zu heilen". Ähnlich ist etwa im Bilde der Flammen über dem Kopf der heiligen Hildegard von Bingen (Rupertsberger Codex Wiesbaden Hs. 1) jene grenzüberschreitende Annäherung durch die Visionseingabe zitiert, die die Heilige selbst als körperlich erlebt kommentiert hat: „Es durchströmte mein Gehirn."

Zusammenfassung Kap. B2: Heilkundige und Helfer

Die Tätigkeit der mittels Beschwörungen und anderen Heilsprüchen operierenden Schamanen, Priester- und Mönchsärzte entsprach komplementär der Hilflosigkeit von psychisch, psychosomatisch und auch körperlich Kranken, weil Ursachen von Leiden und Schicksalsschläge dem Menschen nur mangelhaft rational erkennbar sind. Während sich Schamanen selbst durch Drogen und Askeseformen rauschhaft einem kollektiven Mythos anzunähern suchten und damit auf Urerlebnisse der ‚mystischen Teilhabe' zurückgriffen, bezogen sich Mönchsärzte und Heilkundige des Mittelalters auf die Wundertaten Christi und deren erzählerische Ausgestaltung, auf Heiligenlegenden und auf meist adaptierte

---

146 Haeseli, Christa M.: Magische Performativität, S. 108–112.

außerkirchliche Formeln der Antike oder der germanischen Welt. Sie übten sich in Geboten der Nächstenliebe und der Ordensregeln. Die archaischen Priesterärzte des alten Zwischenstromlandes arbeiteten noch in einer durchgehend ‚magisch-sympathetischen' Gesellschaft, die uns einen optimalen Rückblick in die Ursprünge der von uns als triadisches System aufgefaßten Beziehungen vermitteln kann.

All diese Helfergruppen konnten – soziologisch und psychologisch gesehen – durch Charisma, paternalistische Haltung und Empathiefähigkeit therapeutische Wirkungen erzielen, wobei neben den begleitenden Riten besonders auch sprachliche Bildgebungen mit Metaphorik und Metonymien die oftmals nach therapeutischer Regression erreichbaren Ressourcen des Patienten ansprachen. Die Heilsprüche dienten außerdem den zur Ausübung Berufenen als Leitschienen gegen eine Diffusion der Zielperspektive, sprich: zur Vermeidung von Ermüdung und Abdriften der Konzentration bei ihrer Tätigkeit.

Soziopsychologische Darstellungen von Funktionen der Führer und Geführten, von Machtausübung und Machtteilhabe, von Nähe und Distanz sowie die Grenzüberschreitungen zur Berührung mit göttlichem Wesen und Numinosum werfen Fragen auf nach den physischen Bedingungen, besonders nach einer neurobiologischen Basis effektiver interpersonaler Kooperation im Bereich des Krisenmanagements. Können Ärzte wirklich Schmerzen ihrer Patienten fühlen und über eine Aktivierung von rostralem Cingulum und rechtem vorderen Praefrontallappen des Patienten lindern?[147] Diesen Fragen ist im Folgenden (**Kap.B3**) nachzugehen.

B3 Das Bündnis: Heiler und Patient in passagèrer Symbiose, Verkopplung und Entkopplung von Hirnsystemen zweier Individuen: Die neurale ‚Therapeutische Allianz'

Was geschieht im Miteinander des worttherapeutischen Prozesses?

Zum Verständnis der Interaktion von Heilkundigem und Hilfsbedürftigem ist ein weiterer Aspekt erforderlich. „Wir wollen das Leiden des anderen", „Der Führer bedarf des Geführten" – mit dieser oben genannten, einer landläufigen Vorstellung über therapeutische Arbeit und Wirkung schon zuwiderlaufenden Rollen- und Erwartungsumkehrung ist es noch nicht getan. Greifen wir auf ein Gespräch mit dem amerikanischen Hypnotherapeuten Milton H. Erickson zurück, das einer der beiden Verfasser eines Buches über neurolinguistische Pro-

---

147 Kaptchuk, Ted and Jensen, Karin, Physicians' brain scans indicate doctors can feel their patients' pain – and their relief, EurekAlert 29–Jan–2013.

gramme, dieser weit ausgearbeiteten Methode suggestiver Führung[148] einem Kapitel über das Ankern anfügt. (Wir nehmen zuvor an, der Referent hat paradoxartig ein extremes understatement vorgegeben, um den verehrten Meister herauszufordern; der Referent ist natürlich schon ein Therapeut.) Er schreibt: „Das letzte Mal, als ich bei M. E. war sagte er etwas zu mir, [...] wofür ich ziemlich lange brauchte, um dahinter zu kommen. Milton sagte zu mir: ‚Du selbst hältst dich nicht für einen Therapeuten, aber du bist ein Therapeut.' Und ich sagte: ‚Nun, eigentlich nicht.' Darauf sagte er: ‚Laß uns so tun, als ob ... du ein Therapeut bist, der mit Leuten arbeitet. Das Allerwichtigste, wenn du so tust, als ob, ... ist, zu wissen, ... daß du es eigentlich nicht bist ... Du tust nur so, als ob ... Und wenn du in dem So-tun-als-ob wirklich gut bist, dann werden die Leute, mit denen du arbeitest, so tun, als würden sie sich verändern. Und sie werden vergessen, daß sie nur so tun, als ob ... für den Rest ihres Lebens. Hauptsache, du läßt dich dadurch nicht täuschen.'"[149] Diese auf den ersten Blick verschlungene und verwirrende Beratung durch den ‚Meister' verharrt nicht im Entwurf kühler und reiner technischer Distanz und Darstellungskunst des Therapeuten, nicht einmal im oszillierenden Schweben seiner Aufmerksamkeit zwischen Selbst und Anderem oder zwischen Distanz und Nähe. Sie weist vielmehr letztlich auf ein Phänomen hin, das wir als die Reflektierung einer transienten Symbiose und ihrer Wirkung erkennen. Wir verstehen hier unter Symbiose eine dem Zweck nach primär lebensnotwendige oder unter vitaler Not erlebnismäßig unvermeidbare Zusammenkunft von Individuen in einer situativ komplementären Rollenverteilung. Wer seine Kinder gelegentlich und *dosiert* nachäfft, mit „Als-ob"-Spielen, schenkt ihnen die Möglichkeit, diese Gefühle den Eltern zuzuschreiben und erfolgreiche Affektregulation als interaktionelle Kompetenz zu erleben. Es ist ein Teil der frühen erzieherischen Emotionsregulierung.[150] Die Fundierung dieser psychologischen Mechanismen, deren erste neurobiologische Bausteine im weiteren beschrieben werden, scheint inzwischen im Ansatz nachweisbar. Nur im Ansatz, es sind Bruchstücke eines derzeit intensiven Forschungsfeldes.

Zunächst sei auf die Ebene der Dichtung hingewiesen. Es gibt ein einleuchtendes Sinnbild für die höchstmögliche Nähe menschlicher Übereinstimmung.

---

148 Das Neurolinguistische Programmieren NLP entwickelt sich zu einer perfektionistischen Methode der alle motorischen, mimischen, sprachlichen, vegetativen Äußerungen eines Patienten kommunikativ spiegelnden Feedbackschleifen; die Betreiber halten diese ihre Methode für wirkungsvoll; unbewußt werde Spiegeln von allen Menschen eingesetzt, um effektiv zu kommunizieren.
149 Bandler, Richard und J. Grinder: Neue Wege der Kurzzeit-Therapie, Paderborn 1994, S. 166.
150 Fonagy, Peter und Mary Target, Neubewertung der Entwicklung der Affektregulation vor dem Hintergrund von Winnicotts Konzept des ‚falschen Selbst', in: Psyche 56 (2002), 839–862.

Im West-östlichen Diwan hat uns Goethe im Bild vom Ginkgoblatt diesen geheimnisvollen Kerngedanken nähergebracht. Im Gefühl und im Wissen um den poetischen Gleichklang, die weitgehende Gemeinsamkeit mit seiner Geliebten, das Zusammenfließen ohne Ich-Verlust beschreibt er sein ‚Eins und doppelt'-Sein. Es ist ein außergewöhnliches menschliches Erleben, sich mit einem anderen auf der selben ‚Wellenlänge', der selben Frequenz im Fühlen und Wollen zu finden, aber zur gleichen Zeit auch voll bewußt Selbst-Ich zu bleiben und darüber hinaus diese nach logischen Gesetzen fast als unmöglich daherkommende Konstellation auch noch beschreibend zu reflektieren. An anderer Stelle hat Goethe das Phänomen als „Duodrama" bezeichnet. Gleiche Erlebnisse mögen Künstlern, die gemeinsam mit Instrumenten oder mit ihren Stimmen ein Musikwerk interpretieren nicht fremd sein. In vielen kommunikativen Situationen ließe sich nach diesem Phänomen fahnden. In der Langzeit-Psychotherapie gilt manchen eine solche bewußt gewollte Ich-Aufteilung mit der Einübung in Metaebenen der ‚Übertragungen' als Garant für ereignisschaffende Fortschritte für Therapeut und Patient. Schon für antike Rituale mit der Methode der Trauminkubation wurden reziproke psychische Mechanismen im Sinne von Gaetano Benedetti[151] zwischen Patient und Heilkundigem oder Exorzist angenommen, wenn etwa Priester Gottesbotschaften durch Träume für ihren König empfingen, sei es bei Fragen um Krieg oder Frieden.[152] Das Sinnbild des Ginkgoblattes kann auch zum Verständnis der mitmenschlichen Situation der Notfalltherapie und Notfallhilfe und einer Heilspruchatmosphäre beitragen. Betrachten wir das Alsob als Entschlossenheit für eine Flexibilität von Selbst- und Fremddistanz praktisch und als Kontrast zu und Vorbeugung vor einem sog. Helfersyndrom und einem sich Verlieren im Engagement. Dabei ist hier mit ‚Als-ob' nicht Maskierung und künstliche Theaterwelt gemeint, noch vorwiegend der rational erlernte ‚Willensakt der Distanzierung',[153] schulisch oder gar erst universitär erlernt, sondern gleichermaßen die wohl allein menschliche natürliche neuronale Anlage von Empathiefähigkeit einerseits und deren Blockademöglichkeit andererseits. Damit gerät die historisch wichtige soziopsychologische und psychoanalytische Akzentuierung von Macht-Ohnmacht-Verhältnissen, von Freuds ‚Entdeckung' der Identifikation mit dem Heiler und der Partizipation an seiner Macht durch Unterwerfung, in den Bereich der Ideologie. Zumindest erliegt diese Vorstellung dem Bereich einer puren Fremdbetrachtung durch Dritte, dies besonders im Hinblick auf die neuronalen kommunikativen Vorgänge am menschlichen Zentralorgan. Ebenso entfällt die von der Sprechakttheorie hervorgehobene Intentiona-

---

151 Benedetti, Gaetano: Die Botschaft der Träume, Göttingen 1998, S. 141.
152 Zgoll, Annette: Traum und Welterleben im antiken Mesopotamien, Münster 2006, S. 413ff.
153 Schrott, R. und A. Jacobs: Gehirn und Gedicht, S. 33.

lität des persuasiven Aktes. Denn die Wirkung des Notfall- und Krisen-Therapeuten besteht im **aufgerufenen Aufrufen** biochemischer Systeme und damit im Aktivieren und Mobilisieren ihrer Ressourcen.

Wie wir seit der Entdeckung der **Spiegelneurone** Ende der 90er Jahre durch Giacomo Rizzolatti und seiner Arbeitsgruppe und von vielen darauf folgenden Ergebnissen wissen, besitzt unser Gehirn Neuronensysteme, die Bewegungen anderer – nicht nur die eigenen – selbst intern zu repräsentieren, nachzuempfinden und zu ‚verstehen' vermögen, ohne eine rational intendierte Überlegung und ohne erhöhte Aufmerksamkeit. Das heißt: <u>Die Durchführung einer Aktion einerseits und die Beobachtung einer Aktion andererseits aktivieren die selben inneren Aktionsrepräsentanten.</u>[154] Und das heißt ferner: Es gibt Neurone, die bimodal arbeiten, also sowohl motorische Reize als auch Wahrnehmung der Sinnesorgane vermitteln. Es ist eine andere, eine erweiterte Form des Erkennens, die von der Aktivierung des eigenen motorischen Wissens über die Handlungen anderer angeregt wird. Die Spiegelneurone ermöglichen Als-ob-Erfahrung. Die Beobachtung des motorischen Ereignisses führt zu einer „Einbeziehung des Beobachters in erster Person, die es ihm gestattet, es unmittelbar zu erleben, als ob er selbst der Ausführende wäre."[155] Wir verwenden eine Vielfalt von metaphorischen Redensarten, um innere Simulation zu beschreiben, das ‚in die Schuhe eines anderen steigen', das ‚mit anderer Augen sehen', ‚sich in jemanden hineinversetzen'; sie alle drücken eine motorische Bewegung des Zueinander-Kommens aus und verweisen damit auf das ursprüngliche, vorsprachliche Element, auf Körpersprache und Sensorimotorik, auf die Ursache des Herzklopfens, das wir mit einem Seiltänzer im Zirkus erleben oder den Brechreiz beim Anblick eines sich Übergebenden. Wer noch glaubt, daß Sprechen und Sprachverstehen das Eine und Hand und Handlung das Andere seien, der bedenke die Fülle der Redensarten um unsere Hand, die in allen menschlichen Bereichen vom Alltag und Rechtsbrauch bis zu Handel und Psychologie unser Leben bestimmen.[156] Dabei ist das ‚Mit Händen und Füßen Reden' nicht nur vom Zweck der Unterstreichung eigener Worte, sondern auch von seinem neuronalen Ursprung her

---

154 Eine klinisch neurologische Forschung konzentriert sich deshalb zur Zeit auf die therapeutische Anwendbarkeit etwa in Form indirekter, visueller und sogar akustischer Übungsbehandlung per Bildschirm oder Live-Darbietung bei Lähmungen und schmerzhaften Hemmungen. ‚Planmäßiges wiederholtes bewußtes Sich-Vorstellen einer Bewegung ohne deren gleichzeitige praktische Ausführung' (Jeannerod 1995) ist schon in der Sportmedizin und im Training angewandt. – Über eine visuomotorische Imaginationstherapie bei visuell-räumlichen Neglecten berichtet z. B. Anouk Welfringer 2010 (online).
155 Rizzolatti, Giacomo und Corrado Sinigaglia: Empathie und Spiegelneurone. Die biologische Basis des Mitgefühls, Frankfurt/ M. 2008, S. 142f.
156 vgl. Röhrich, Lutz: Lexikon der sprichwörtlichen Redensarten, Freiburg, Bd.2, S. 638–659.

erklärbar. In der Tat konnten mit bildgebenden Verfahren Beziehungen zwischen visuell-räumlicher Metaphern-Verarbeitung und Empathievermögen im selben parietalen Hirngebiet vermutet werden, schon wenn man Versuchspersonen aufgab, die Körperposition eines anderen nachzuvollziehen.[157]

Während bisherige Sprachforschung eine Separierung von Perzeption und Aktion, d. h. von Aufnahme der Sinneseindrücke und deren gesonderter assoziativer Verarbeitung annahm, haben interdisziplinär der Neurologe **Vittorio Gallese** (Parma) und der Linguist **George Lakoff** (Berkeley) aufgrund der modernen neurobiologischen Ergebnisse im Jahre 2005 die Theorie einer inhärent multimodalen Sprache vorgestellt. Danach baut Begriffswissen auf den vielen eng verbundenen Sinnesqualitäten auf: Hören, Sehen, Tasten, Riechen, Schmecken und Bewegen nutzen nach dieser Theorie das sensomotorische System, das den semantischen Gehalt aufgrund unseres Weges mit unserem Körper in der Welt liefert. (Wir wagen den strukturell gemeinten Vergleich mit einer Konzentration des fiktiven Weltwissens innerhalb der UNO versus einem fiktiven Weltwissen als Globalisierungseffekt.) <u>Etwas imaginieren und etwas tun haben das selbe neurale Substrat.</u> Das sensomotorische System vereint alle Sinneseindrücke, ist multimodal und schafft Bedeutung, Grammatik und Konnotationen. Zelluläre Basis ist die angeborene Multimodalität der Spiegelneurone, sodaß etwa ein einziges feuerndes Neuron sowohl mit Zusehen beim Greifen als auch mit Ergreifen selbst korrelieren kann. Die bisher von Neurologen als Assoziationsgebiete vorgestellten praemotorischen Areale sind die Träger dieser Multimodalität, ebenso die parietalen, bisher für Areale der Verarbeitung sensorischer Informationen gehaltenen Netzwerke. Beide Gebiete konstruieren die Vielfalt motorischer Handlungen, deren Verortungen und die Codierung von Sprache, z. B. der lokativen Verben. „Die semantische Beziehung zwischen einer Handlung und ihrem Ort ist Teil der Begriffsstruktur". Und: Dieses integrierte Netzwerk konzipiert und produziert Cluster möglicher Handlungs- und Planungs-Subcluster auch komplizierter Abläufe in zeitlicher Abfolge. Wir können also in Clustern Entwürfe erstellen auch ohne particuläre Spezialisierung, womit ein Fundament für Begriffe gegeben ist, wiederum auf der Basis der multimodalen Funktionen vieler Neuronen. Anblick oder Klang eines Objektes triggern die Spiegelneurone und zwar jene 70%, die nicht handeln, sondern simulieren. Es findet verkörperlichte Simulation (**embodied simulation**) statt. Die Autoren weisen auf entsprechende stichhaltige Versuchsreihen bei Menschen und Menschenaffen hin. Diese verkörperlichte Simulation ist zugleich eine Art Schmelztigel unserer Bildersprache, unserer Metaphorik. Die Vorstellungsbreite nie gesehener Welten und nie vollzogener Taten wird unendlich. Indem also eine Trennung von Imagination einerseits und Perzeption und Aktion andererseits nicht besteht, gleicht etwa die bloße geistige Vorstellung von Muskelspannung, Herzschlag und Atemfrequenz im Effekt weitgehend diesen realen Aktionen. – Letztlich – und dies ist für das Verständnis von allen sogenannten performativen Akten wesentlich – schließen Gallese und Lakoff, daß der selbe Schaltkreis, der den Körper und die Struktur-Perzeptionen bewegt, auch die abstrakten Denkstrukturen bewegt. Damit wird Imagination zu einem gleichsam materiellen, jedenfalls physisch fassbaren Wirkprinzip und unsere Sprache wird teilweise zum Produkt der Evolution, bleibt weniger Produkt menschlicher Erfindungskraft.[158]

---

157 Thakkar, N.K. et al., (2009) Exploring empathic space, PLoS ONE 4(6): e5864.doi:10.1371/j.
158 Gallese, Vittorio and George Lakoff: The brain's concepts: The role of the sensory-motor system in conceptual knowledge, in: Cognitive Neuropsychology, 22 (2005), 455–479; zur kritischen Reaktion und Weiterentwicklung bzgl. Theorie von Gallese/ Lakoff siehe z. B.

Diese erweiterte, andere, nicht kognitiv-blasse, nicht kalte Form des ‚Erkennens' gilt ebenso für die Wahrnehmungen von Gefühlen anderer. Sie werden von den visuellen Arealen her nicht über die bewußte somatomotorische oder somatosensorische Rinde des Großhirns, sondern direkt zur Insula und zur cingulären Rinde geleitet. In der Insula wird ein ‚autonomer spezifischer Spiegelmechanismus' aktiviert, der auch viszerale Reaktionen, vages Unwohlsein, mitfühlende Übelkeit und Erbrechen hervorrufen kann. Jedoch genügt eine insuläre Repräsentation solcher Reaktionen, um die Emotionen anderer in eigener Person zu verstehen.[159] Das gilt für die Insula neben Cingulum und praefrontaler Rinde besonders dann, wenn der Schmerz anderer als affektive Äußerung signalisiert ist.[160]

> Die schon lange bekannte Potentialantwort kontralateraler Handmuskeln auf transkranielle Magnetreizungen (TMS) der motorischen Hirnrinde konnte 1995 in Experimenten dann als erhöht gemessen werden, wenn die Versuchspersonen gleichzeitig jemanden beobachteten, der Objekte mit seiner Hand ergriff. Und nur beim Menschen zeigte sich auch ein solcher Effekt auch bei Beobachtung intransitiver Akte, also bei Gesten und Gebärden. Besonders beeindruckend war aber, daß gegenüber Affen beim Menschen Einzelteile der Bewegungsabläufe fraktioniert registriert werden konnten. Man hat daraus auf die interne Beeinflußbarkeit der Bedeutungsvarianz rezeptiver Prozesse an jedem Punkt seiner Afferenz geschlossen. – Mit bildgebenden Verfahren konnte dann die Lokalisation dieser Spiegelneurone im parietopraezentralen Rindengebiet und im hinteren Teil des unteren Frontalhirns, also teilweise im Broca'schen Sprachareal festgestellt werden. Mit einer anderen Methode, der Ableitung von ERPs (Ereigniskorrelierte Potentiale als Kopfhautableitung) fanden sich ebenfalls verstärkte Aktivierungen (P50) bei Beobachtung von Bewegungen anderer, ob schmerzhaft oder nicht.[161]

Die zeitweilige funktionelle „An-Wesenheit" eines anderen nicht nur metaphorisch im „Sinn", sondern auch materiell im Hirn und in den Hirnen des/ der Kommunizierenden wird aber noch durch weitere Befunde belegt. Man kann zweifeln, ob Studien an Einzelpersonen vor Bildschirmen in der Lage sind, die dynamischen Aspekte von Kommunikation zu ergründen. Die Forschung der letzten Jahre ist deshalb zunehmend auf Simultanableitungen der elektrischen Tätigkeit der Gehirne übergegangen.

Die Überraschung war groß: Elektrisch in buchstäblich gleicher Wellenlänge verschaltet und ‚sympathetisch' cocreativ verbunden können Partner sein. Darauf

---

    Arbib, Michael A.: From grasp to language: Embodied concepts and the challenge of abstraction, Journal of physiology (Paris) 102(2008), 4–20, es werden von Arbib breitere Einflüsse besonders subtiler kommunikativer, phonologischer und ontologischer Elemente über embodied simulation hinaus auf die Entwicklung der Sprache diskutiert.
159   Rizzolatti/ Sinigaglia, Empathie, S. 188.
160   Decety, Jean and Claus Lamm: Empathy and Intersubjectivity, in: Berntson/ Cacioppo, Handbook, 940–957.
161   Martinez-Jauand, M. et al., Somatosensory activity modulation during observation of other's pain and touch, in: Brain Research 1467 (2012), 48–55.

weisen gleichzeitige EEG-Ableitungen bei sozial interagierenden Versuchspersonen in der **Hyperscanning**-Forschung hin (**Abb. 5**). Verschiedene Bedingungen von Interaktionen der Partner, z. B. Beobachtung sinnloser Handbewegungen, spontane und andererseits direkt gegenläufig induzierte Imitationsbewegungen wurden mit multiplen Vergleichsprozeduren untersucht. Es ergab sich, daß die Teilnehmer in hohem Grade (78% der Zeit) im α-μ-Frequenzband und zentro-parietal synchronisiert waren. In statistisch signifikanter Koppelung zeigten die EEG-Wellen eine Phasen-Synchronie in den Perioden des imitativen bzw. reziproken Austausches. Damit wurde bewiesen, daß während Verhaltensinteraktionen zweier Partner gleiche Hirngebiete in übereinstimmender teils paralleler und teils sich gegeneinander abwechselnder kooperativer Frequenz arbeiten. Die dabei abgeglichenen Hirngebiete gelten als entscheidende Teilnehmer im Gesamtablauf sozialer Hirntätigkeit und als diesbezügliche Repräsentanten des Spiegelneuronsystems.[162]

Ähnliche Ergebnisse wurden auch mittels bildgebender Verfahren (fMRI) bei Gesicht-zu-Gesicht-Kontakten[163] gefunden. Die neurale Aktivität eines bestimmten Netzwerkes des Empfängers konnte exakt am entsprechenden Netzwerk des Senders vorausgesagt werden und war vom kommunizierten Affekt abhängig. Und auch bei zwischenmenschlichen Gestiksignalen fanden sich derartige soziale Hirnkorrelationen.[164] Ebenfalls mit bildgebenden Verfahren (fMRI) wurde mittels eines Doppel-Video-Systems die interaktive Imitation von freien, vorgegebenen und selbstreflexiven Episoden untersucht und konnten sozial kooperierende Areale ermittelt werden.[165]

Besonders interessieren uns die Ergebnisse der gemeinsamen Sprachverarbeitungen im Verhältnis zwischen Sprecher und Hörer. Bei solchen Simultanableitungen mittels fMRI unter Erzählung von Geschichten fanden sich räumliche und zeitliche Koppelungen der Gehirnaktivitäten, wobei sich natürlich im allgemeinen eine Verzögerung beim Hörer widerspiegelte, aber in manchen Gebieten des Hörerhirns auch ‚voraussagende' Aktivierungen. Deutlich war die strikte Abhängigkeit von zwischenmenschlichem Kontakt im Sinne kommunikativem Verhaltens für die Entstehung von Emergenz bei der Koppelung von Sprecher-

---

162 Dumas, Guillaume et al.: Inter-brain synchronisation during social interaction, PLoSONE 5(8):(2010) e12166.doi:10.1371/j.pone.
163 Redcay, Elizabeth et al., Live face-to-face interaction during fMRI: A new tool for social cognitive neuroscience, Neuroimage 50 (2010), S. 1639–47 sowie Anders, Silke et al. Neuroimage 54 (2011),439–446.
164 Schippers, Marleen B. et al. Mapping the information flow from one brain to another during gestural communication, doi: 10.1073/pnas.1001791107 (2010).
165 Guionnet, Sophie et al.: Reciprocal imitation: Toward a neural basis of social interaction, in: Cereb. Cortex (2011) doi: 10.1093/bhr177.

und Hörer-Hirnen. Lokalisatorisch gekoppelt waren Hörrinde, oberer Temporalgyrus, Gyrus angularis, die Temporo-parietal-Verbindung (= Wernicke-Feld), inferiorer Temporalgyrus (= Broca-Feld) und die Insula, also ausgedehnte Gebiete, die zum Teil linguistischer Verarbeitung dienen. Auch weitere dem Spiegelneuronsystem zuzurechnende Gebiete sozialer Verarbeitung waren beteiligt.[166]

Nun ist eine solche als zugleich emotional und kognitiv vorzustellende Symbiose zwischen agierenden Menschen in der Regel eine recht kurzlebige Verbindung, wenn man krankhafte Veränderungen ausschließt. Monotopie und Monochronie in der Werkstatt der Zellverbände von Spiegelneuronen als permanente Simultanfunktionen von Beobachten und eigenem Handeln widerspricht realistischem und gesundem Verhalten. Und in der Tat ist einem ungezügelten Ineinanderstürzen von Individuen verläßlich vorgebeugt. Für die Synchronisation und Äquivalenz der Tätigkeit neuraler Zellsysteme zweier Individuen muß schon auf Grund klinischer Befunde ein dem Spiegelneuronsystem übergeordneter Mechanismus angenommen werden, der unter dem Diktat der Erhaltung von Ichgrenzen und von Subjektstabilität arbeitet.

Beweis dafür ist einerseits die pathologische enthemmende Entgleisung von Ich-Stabilisatoren bei schizophrenen und traumatischen Stirnhirndefekten in Form von Echolalie und Echopraxie. Man könnte von einer Spiegelsaal-Existenz sprechen. Schizophren Kranke halten eigene Gedanken für von Fremden, von außen eingegeben: die sogenannte ‚Ich-Störung'. Besonders eindrucksvoll bei einer Folie à deux. Bei dieser induzierten wahnhaften Störung (ICD10-GM-2012: F24) zweier im Alltag sehr emotional verbundener Personen ist fast stets nur der eine nachhaltig krankhaft gestört, der andere, bei Trennung aufhebbar, nur ‚angesteckt'. Zu dieser Folie à deux (shared psychotic disorder) gibt es eine große Anzahl von Fall-Publikationen, jedoch m. W. noch keine gezielte simultane fMRI-Studie, mit der wichtige Schlüsse zur Induktion von paranoiden Störungen und deren cerebrale Repräsentationen zu gewinnen wären, eine bedauerliche Lücke. Sprache, Gestik und Verhalten, bei letzteren auch Wahnideen, werden in solchen Fällen reflexartig, völlig ungebremst nachgeäfft, sodaß dabei eine andauernde auch motorisch wirksame Dominanz in der Art der Spiegelneurone zu vermuten ist.[167]

Dies muß nicht nur für den Übergang von reiner zellulärer Repräsentation in motorische Aktion, sondern auch im Bereich von ‚Therapie' für ‚innere' Umsetzungen gelten, insbesondere für das Gefühl, selbst die Ursache oder der Auslöser und der Ausführende einer Aktion zu sein bzw. gewesen zu sein. Dieses Gefühl

---

166 Stephens, Greg J. et al., Speaker-listener neural coupling underlies successful communication, PNAS 107 (2010), 14425–14430; doi:10.1073/pnas.1008662107.
167 vgl. Rizzolatti/ Sinigaglia, Empathie und Spiegelneurone, S. 153.

ist nach Meinung mancher Untersucher in der supplementär-motorischen Rinde, im Kleinhirn und in Teilen parietaler Rinde lokalisiert.[168] Involviert ist der Fall therapeutischer Bemühung, etwa selbst vermittels Heilspruch und Ritus den Erfolg einer Therapie zu bewirken. Hier haben die Vertreter der Spiegelneurontheorie hypothetisch kontrollierende Prozesse der Einschätzung eigener Möglichkeiten eingefügt. In Experimenten mit Verhaltensmodifikationen wurden zwei gesonderte Netzwerke, ein führendes und ein zögerndes, vermutet.[169] Die ventrolaterale praefrontale Rinde vermag eigenes Erleben zu blockieren und trägt damit zu einer Unterscheidung von Selbst- und Fremdempfindung bei.[170]

Andererseits wird nicht allein die Funktion der Spiegelneurone für die Fähigkeit, sich in andere hineinversetzen zu können, heranzuziehen sein. Eine strikte Vergleichbarkeit von menschlichen und Makakenspiegelneuronen[171] oder eine weitgehend durch Spiegelneurone erklärbare Verarbeitungsweise von Aktionsworten im motorischen Sprachzentrum (Broca) wurden ohnehin in Frage gestellt.[172] Neben der durch die Spiegelneurone gestützten sog. Simulationstheorie hat sich eine Forschungsrichtung der sogenannten Theorie-Theorie gebildet. Sie nimmt an, daß der Mensch in seinem Leben ein besonderes Wissen entwickelt, das es ihm ermöglicht, die Gefühle anderer abzuschätzen. Die Forscher dieser Richtung weisen auf die Differenz von Beobachten und Handeln hin, auf die das Gehirn – wie gesagt – zur Erhaltung der Ichgrenzen strikt achten muß. Dafür haben Untersuchungen unter Bedingungen der Ich-Perspektive bei Erzählungen im Vergleich zur Theorie-of-mind-Theorie ergeben, daß unter Ich-Perspektive nicht nur der praefrontale Cortex medial und der temporoparietale Cortex links, sondern auch zusätzlich die schon erwähnte das Ich-Selbst abgrenzende temporoparietale Region rechts und der mediale parietale Cortex aktiviert werden (**Abb. 1**). Diese Einbeziehung des rechten Parietotemporalbereichs (temporoparietal junction) aber bedeutet, daß sich der Betreffende in eine Ich-Perspektive begibt, die von der Simulation einer dritten Person unterschieden ist.

---

168 Yomogida, Yukihito et al.: The neural basis of agency: An fMRI study, in: Neuroimage 2009.12.054, doi:10.1016.
169 Nahab, Fatta B. et al., The neural processes underlying self-agency, in: Cerebr. Cortex (2011) 21(1): 48–55, doi: 10.1093/cercor/bhq059.
170 Breithaupt, Fritz: Kulturen der Empathie, Frankfurt/M., 2009, S. 75 mit Zitaten neurologischer Ergebnisse; s. a. Jeannerod, Marc and Thierry Anqueitl: Putting oneself in the perspective of the other: A framework for self-other differentiation, in: Neuroscience 2008, 3(3–4), 356–367, die diese Funktion der Area 19, der parieto-occipital-junction zuordnen.
171 Turella, M. et al., Mirror neurons in humans: consisting or confounding evidence?, in: Brain & Language 108, 1 (2009), S. 10–21.
172 De Zubicaray, G. et al., Mirror neurons, the representation of word meaning, and the foot of the third left frontal convolution, in: Brain & Language 112,1 (2010), S. 77–84.

Eine Arbeitsgruppe um Kai Vogeley in Köln hat dabei auf eine weitere klinisch bekannte Schädigung dieser Region hingewiesen, ein sog. Neglect-Syndrom, bei dem der Kranke die linke Hälfte seines Körpers und seines ihn umgebenden Raumes nicht mehr recht wahrnehmen kann, also keine eigene Mitte mehr hat. Vogeley hat auch deshalb auf die Annahme einer Möglichkeit, sich in Perspektiven anderer hineinzuversetzen, eine Kombination beider, der Simulation per Spiegelneurone und einer speziellen Ich-Perspektive gefolgert.[173] Diese Hypothese einer gesonderten Eigenperspektive erfordert allerdings weitere differenzierte Abklärung.[174] Bedenkenswert jedenfalls ist der hohe räumliche Grad zellulärer funktioneller Differenziertheit, der die Ergebnisse der bildgebenden Verfahren am Gehirn einschränken kann: Z. B. ist in unserem Zusammenhang bekannt, daß Menschen mit der Unfähigkeit, Gesichter einer bekannten Person zuzuweisen (Prosopagnosie nach Schädigung des hinteren temporalen Assoziationsfeldes) trotzdem meist in der Lage sind, Gesichtsausdrücke von Mitmenschen nach ihrem affektiven Eindruck zu interpretieren.

Es versteht sich ja, daß die Gehirne der Beteiligten, des Therapeuten wie des Hilfebedürftigen, die eigene ‚Leistung' via Kontrollmechanismus der Spiegelneurone und funktionsverwandter Gebiete der Abgrenzung von Fremd- und Selbstempfindung sowie vielleicht einer gesonderten Ich-Perspektive als Erfolg nicht ‚selbstlos' versanden lassen. Die Balancewirkung der kooperierenden Systeme zielt auf vitale Interessen und aktiviert erfolgsmeldende Botenstoffe (Endorphine) für eigenes Erleben: psychologisch für Ansehen und Selbstbehauptung, gemeisterte Angstreduktion und Schmerzlinderung; ökonomisch für Verdienst und Krafterwerb, psychohygienisch für Selbsterneuerung und biosoziale Schadensminimierung und Nichtansteckung. So entsteht auch vermittels Selbstbelohnung ein emergenter kreativer Kreislauf. Das Spiegelneuronsystem kann insofern nicht nur als Brücke zwischen einer individuell als intentional erlebten und einer nicht intentionalen Kooperation betrachtet werden, sondern auch als Basis sozial geteilter Intentionalität und Kommunikation.[175]

Zusammenfassung Kap. B3: Das Bündnis: Heiler und Patient in passagèrer Symbiose. Neurale ‚therapeutische Allianz'

Während die inneren Selbstverdrahtungen der Großhirnrinde beim Menschen alle anderen Strukturen und damit Funktionen weit übertreffen, sich also das Gehirn überwiegend mit sich selbst beschäftigt, und obwohl die Efferenzen des

---

173 Vogeley, Kai und Albert Newen, Ich denke was, was du nicht denkst, in: Könneker, Carsten (Hg.): Wer erklärt den Menschen? Frankfurt/M, 2006, S. 59–73.
174 Vgl. dazu auch David, Nicole et al., Differential involvement of the posterior temporal cortex in mentalizing but not perspective taking, in: Soc.Cogn.Affect. Neuroscience (2008) 3(3):279–289, doi:10.1093/scan/nsn023.
175 Vgl. Atmaca, Silke et al.: Action co-representation: The joint SNARC effect, in: Social Neuroscience 2008 3(3–4), 410–420.

Großhirns die neuronalen Zugänge etwa um das fünffache übertreffen,[176] zeigen neuere Forschungsergebnisse mit interindividuellen Simultanableitungen doch auch gewichtige Anzeichen für flüchtige synchron, äquivalent und äquitendent mit anderen Individuen ablaufende cerebrale Aktivierungen in bestimmten sozialen Situationen. Damit ist ein Erspüren oder gar Erkennen von fremdem Bewußtsein, fremden Gedanken, ‚Gedankenlesen', eine Doppelsubjektivität, eine natürliche und gesunde Folie à deux, ein Naturphonophor i. S. Ernst Jüngers oder der erwähnte ‚Meister Floh' E.T.A. Hoffmanns oder dergleichen Phantastisches weder vermutet noch grundsätzlich einkalkuliert. Jedoch geben die Prozesse der neuralen Koppelung wie der Entkoppelung und insbesondere die Theorie der ‚embodied simulation' und die ihr zugehörige Balancefunktion von Selbst- und Fremd-Empfindung und -Unterscheidung wichtige Hinweise auf die Biologie von extremer Nähe und Distanz. Die Annahme einer Anwendbarkeit der metaphorisierenden Emotionsetiketten der archaischen und mittelalterlichen Heilspruchtexte in der Notfall-Krisensituation entzieht sich damit dem Vorwurf der Spekulation und erweitert die philologischen Arbeitsansätze, die in ‚Magie' und ‚Performativität' an eine Grenze geraten sind, um eine Dimension.

Ohne auf den vielfältigen und spannenden Diskurs und auf die europäische Geschichte der Intersubjektivität einzugehen,[177] verbleiben uns im Hinblick auf die Bedeutung der Spiegelneuronentdeckung und die Gefahren ihrer Überbewertung für die Performativität von Heilspruchtherapie einige Fragen, die wir den Philosophen überlassen müssen und die nicht neu sind: Ist neurale therapeutische Allianz im hier gegebenen Sonderfall nicht eigentlich eine ‚natürliche Selbstverständlichkeit', eine eingepaßte Schlüssel-Schloss-Situation, die keiner Argumente für die Möglichkeit eines Nutzens bedarf? Wird der Patient auch einmal zum Therapeut? – Welche Ideologien können ein therapeutisches Bündnis, eine von uns als passagère Symbiose verstandene Kommunikation nach derzeitigem vulgärem Verständnis korrumpieren: Vermittelt die doppelte Zweigleisigkeit neuronaler Reziprozität ausreichend oder zu viel Vorbehalt zu allein menschlich kulturellen Werten: Liebevolle Hilfsbereitschaft vs. heischende Erpressung, professionelles Dienen vs. Ver-Dienen, Täuschung vs. Rechtschaffenheit, Vertrauen vs. Mißbrauch?

---

176 Status 1989: Roth, Gerhard: Das Gehirn und seine Wirklichkeit, S. 63.
177 Siehe in unserem Zusammenhang die gestraffte Übersicht bei Zaboura, Nadia: Das empathische Gehirn, Wiesbaden 2009.

B4 Soziales Umfeld, Kulturerwerb und ‚sozioneurale Hardware'

Unter Menschen ist weder eine solitäre Existenz noch eine rein dyadische Beziehung möglich. Die Vertreter des dialogischen Prinzips hatten dem Ich-Du-Verhältnis einen sehr bedeutenden Rang eingeräumt. Ob als Mutter-Kind-Beziehung oder als Arzt-Patient-Beziehung, die Dyade galt ihnen als entscheidende menschliche Dimension. Nicht nur kulturhistorische und soziologische Betrachtungen, auch die cerebralen Entwicklungskriterien verlangen jedoch nach Beachtung dessen, was den Boden für jede Beziehung bereitet.

Der Neurologe Wolf Singer weist in Zusammenhang mit der Theorie menschlicher Lernprozesse auf die Verzahnung von Hirnentwicklung und Umwelt hin: „Das Nervensystem tastet die Umwelt aktiv ab, sucht nach Mustern, die den Selektionsvorgang unterstützen können, und erlaubt diesen Aktivitäten nur dann, Verschaltungen zu verändern, wenn sie in einem weiteren Kontext als adäquat identifiziert werden." „Die Existenz interner Bewertungssysteme ist nun von herausragender Bedeutung für die Beurteilung umweltabhängiger Entwicklungsprozesse. Das Gehirn entscheidet, gesteuert von seinen eigenen Bewertungen, welche Aktivitätsmuster Veränderungen der Verschaltung induzieren dürfen. Das hierfür benötigte Vorwissen liegt in der funktionalen Architektur der Bewertungssysteme und ist festgelegt, also angeboren. Ein verwandter Mechanismus sorgt dafür, daß Sinnessignale nur dann strukturierend auf die Entwicklung einwirken können, wenn sie Folge aktiver Interaktion mit der Umwelt sind, bei denen der junge Organismus die Initiative hat." Dabei ist gegenüber den tierischen Primaten die Größe des menschlichen Gehirns und die lange Erziehungszeit als formender Einfluß mit Schaffung differenter neuraler auch generationenübergreifender Strukturen als kulturprägend zu bedenken. Dies gilt auch besonders für den Erwerb der Erstsprache, die durch ihre Prosodie (Akzent und Melodie) bei neuraler Verarbeitungsroutine lebenslang prägend wirkt.[178] Erkenntnisse über den allerersten Spracherwerb bei Kindern sind inzwischen auch näher an grammatikalische Strukturen herangekommen. Das Gehirn von 4 Monate alten Kleinkindern orientiert sich bereits an Regularitäten im Sprach-Input; es vermag nach einigen Versuchsreihen z. B. falsche Hilfsverben und Flexionsformen von Verben zu unterscheiden, wobei dies allein an EEG-Ableitungen (ERP, event-related bzw. EKP ereigniskorrelierte Potentiale), noch nicht an Verhaltensreaktionen erkennbar ist.[179] All dies deutet auf feste und frühe hirn-

---

178 Singer, Wolf, wie oben, Universitas 56 (2001), S. 888; Online: Vortrag 1. Werkstatt-Gespräch Initiative McKinsey der Deutschen Bibliothek Frankfurt/M am 12. 6. 2001.
179 Friederici, Angela, in: Spektrum der Wissenschaft, Januar 2010, S. 66–71.

organische Imprägnierungen und die Verflechtungen von Sprache und Umwelt, ohne welche eine adäquate Lebenstüchtigkeit nicht erreichbar ist.

Stets erwachsen Lebensfähigkeit und Kommunikation auf gegebenen gesellschaftlichen Bedingungen. „Sobald Menschen kommunizieren, sind nicht nur diejenigen anwesend, die anwesend sind, sondern zugleich die Gemeinschaft, die ihnen die Sprache und die Bedeutung gegeben hat und diese auch verbürgt." „Intersubjektive Bedeutung ist sozial erworben."[180] Gehirne sind Träger der Kultur, und Kultur ist ihrerseits Gehirnprodukt. Genetik und Kulturevolution, ‚nature and nurture', bilden in ihrer Verflochtenheit die elementare Matrix bei der Schaffung menschlichen Geistes.

> Eine Anzahl von fMRI-Untersuchungen hat sich dem Nachweis kultureller Einflüsse auf Funktion und Struktur des menschlichen Gehirns gewidmet, wobei Unterschiede zwischen asiatischen und westlichen Kulturprägungen auffielen.[181] Kulturelle Primärwerte einer Gesellschaft, nach heutigen Vorstellungen insbesondere persönliche Unabhängigkeit vs. gesellschaftliche Abhängigkeit, Individualismus vs. Kollektivismus, die sich durch aktives alltägliches Engagement fixieren, werden als neuronale und genetische Muster verkörperlicht.[182]

Eine umfassende Zugangsweise, die biologische und psychologische Gesichtspunkte auf einen Nenner bringt, zielte auf die Erkennung neuronaler Prägungen von sozial angepasstem Verhalten.[183] Dabei war die Erkenntnis wichtig, daß in erster Linie emotionsverarbeitende Hirnstrukturen auf verschiedenen Ebenen soziales Verhalten schaffen: Höhere perzeptive Rindengebiete sind ebenso wie Amygdala und Striatum in eine mit der sozialen Gruppe kontextuelle Verarbeitung involviert. Davon hängen Emotionen wie Furcht und Angst, Glück und Stolz, Schuld und Scham, Ekel und Verlegenheit, Neid und Eifersucht sowie Stimmungen der Traurigkeit oder der maniformen Abgehobenheit als moralische oder soziale Gefühle ab. Es besteht ein rekursiver und multidirektionaler Informationsfluß unter Einbeziehung der individuellen Erfahrungen. Über die basalen Verarbeitungen hinaus sind eher willentliche Detailverarbeitungen im Gyrus fusiformis (Gesichtserkennung) und Gyrus temporalis superior (sensorisches Sprachzentrum), im Operculum frontale (motorisches Sprachzentrum) und Hippokampus (Archivverwalter des Gedächtnisses) repräsentiert und können als

---

180 Reichertz, Jo, Kommunikationsmacht, S. 247, 167.
181 Jinkyung Na and Shinobu Kitayama, Spontaneous trait inference is culture-specific, in: Psychological Science 22,8 (2011), S. 1025–1032.
182 Chiao, Joan Y. and Genna M. Bebko, Cultural Neuroscience of Social Cognition, in: Han, Shihui/ Pöppel, S. 19–39 mit Abbildung zur Verteilung der Serotonin-Transporter-Gene auf einer Weltkarte. Kollektivistische Nationen zeigten erhöhte Prävalenz von S-Allel-Trägern mit weniger Angst und Depression.
183 Adolphs, Ralph, Cognitive Neuroscience of human social behavior, in: Nature Reviews 2003; ders. und Michael Spezio: Social cognition, in: Berntson/ Cacioppo, Handbook II, Kap.47.

soziales Nachdenken praefrontal aufgegriffen werden. Alle diese sozioneuralen Verhaltens-Funktionen sind inzwischen vielfach untersucht und bestätigt.

„Cultures are stored in people's brains. And people's cultural niche reciprocally affects their brain development" formuliert die Sozialpsychologin Susan Fiske. Und wer sich sozial ausschließt, muß mit einer Zunahme der Aktivität im vorderen Cingulum (ACC), dem Zentrum für Konfliktlösungen rechnen.[184] Ein voraussagender Irrtums-Monitor, der ebenfalls den ACC und auch den Nucleus accumbens umfaßt, wurde postuliert, um sozial konformes Verhalten zu erklären. Wir neigen dazu, unsere Meinung anzupassen, um störungsfrei lernen zu können.[185] Auch ist die Abgrenzbarkeit der individuellen und der sozialen Regulationsmechanismen im Gehirn dargestellt worden.[186]

> Neben den Neuroimage-Experimenten zu Unterschieden zwischen den Kulturen und zu den Zeichen sozialer Angepasstheit wurde eine Fülle von Untersuchungen bei verschiedenen psychischen Krankheiten vorgelegt, bei denen Veränderungen im sozialen Verhalten erwartungsgemäß selbstverständlch sind, die aber auch sehr variabel zu bewerten sind. Eine spezifisch sozial relevante Störung, die nicht mit Minderung oder qualitativer Änderung des sozialen Verhaltens, sondern mit Übersteigerung sozialer Zuwendung einhergeht, wurde ebenfalls untersucht, das Williams-Syndrom, eine Erbkrankheit mit auch Störung der Herztätigkeit und der Raumerkennung. Nicht nur für die Neurophysiologie, sondern für grundlegende ethische Fragen könnte sich mit diesem Syndrom ein einzigartiges Fenster in genetische Einflüsse auf soziales Verhalten öffnen. Diese Kranken sind extrem an anderen interessiert, was sich in Sprache, Gefühlen, gerichteter Aufmerksamkeit und Vertrauensseligkeit manifestiert. Man fand zunächst Veränderungen an Amygdala, Hippocampus und praefrontalem Cortex bzw. deren Verbindungen und hat geschlossen, daß die hier fehlende Fremdenangst mit der mangelnden Aktivierung der Amygdala auf furchterregende Gesichter zusammenhängt. Jedoch haben weitere Untersuchungen Zweifel an einer ursächlichen Amygdalavolumenveränderung ergeben, weil beim Williams-Syndrom letztlich neurodegenerative Entwicklungsstörungen die gesamte Mittellinienstruktur des Gehirns betreffen.[187] Die cerebralen Mittellinienstrukturen sind aber letztlich die früh entwickelte ‚Hardware' eines selbstbewußt empfindenden und zugleich emotional flexiblen lernfähigen Individuums.

B4.1 Das europäische Mittelalter

Unsere Kenntnis des europäischen Mittelalters stützt sich neben den Resten der Baukunst und Plastik im wesentlichen auf Handschrift- und Bildquellen. Es

---

184 Fiske, Susan T., Cultural Processes, in: Berntson/ Cacioppo, Handbook II, Kap. 51.
185 Klucharev, Vasily, in: EurekAlert 14–Jan–2009.
186 Krendl, Anne C. und Heatherton Todd F., Self versus Others/Self-Regulation, in: Berntson/ Cacioppo, Handbook II, Kap. 44.
187 Capitão, Liliane et al., MRI amygdala volume in Williams Syndrome, in: Research in Developmental Disabilities 32(2011), 2767–2772; doi:10.1016/j.ridd.2011.05.033.

standen sich mündliche Laienkultur und schriftliche lateinische Kultur der Klöster gegenüber. Die Scriptorien der Klöster haben manches aus der Volkskultur aufgezeichnet, darunter auch einen Teil der uns bekannt gewordenen Beschwörungs- und Segenstexte, die nicht dem orthodoxen kirchlichen Bereich entstammten und nicht immer genehm waren. Die Kirche und die Orden hatten ein Bildungsmonopol, noch weit über die Gründung der ersten Universitäten im 13. Jahrhundert hinaus, das sie aufgrund ihrer nördlich der Alpen handwerklichen und geistigen Überlegenheit gründlich zu nutzen vermochten.

B4.1.1 Einige Aspekte der Gesellschaftsstruktur

Betrachtet man die verschiedenen Ansätze der Mittelalterforschung, Medialität, Sozial-, Frömmigkeits- und Mentalitätsgeschichte, so findet man für das Früh- und Hochmittelalter weitgehende Übereinstimmung unter den Historikern, daß allmählich alle Lebensbereiche von klösterlichem, kirchlichem bzw. biblischem Gedankengut durchtränkt waren. Ob als asketische Lebensform oder als vita communis, die Klöster – ständig sich aus Retardierungen reformierend – mit ihrem Einfluß auf den Adel, ja teilweise und zeitweise als deren Exponent – und mit gesellschaftlichen Verflechtungen, entfalteten gewaltige Kräfte auf die kommunikativen Strukturen der Gesellschaft.[188] Im Hinblick auf Tendenzen mancher Forscher, eine klassenmäßige und bildungsabhängige Dichotomie der Gesellschaft zu beschreiben, hat vor allem Klaus Schreiner betont, daß stets ein „Ethos schichtenübergreifender Solidarität und Brüderlichkeit" unbeschadet der Verteilung von Gütern und Lebenschancen bestand.[189] Die Durchdringung der Gesellschaft mit christlichen Grundsätzen betraf alle beruflichen Sparten bis hin zum Handeltreibenden, dessen Gewissen wegen Gewinnstreben in Nöte kam, da die Kirche bis ins 14. Jahrhundert die Zinsnahme als Wucher ansah.[190]

B4.1.2 Frömmigkeitsgeschichte – Vermittlungswege: Sehen und Hören und alle Sinne

Nicht nur aus dem öffentlichen, religionspolitischen, sondern auch aus dem persönlichen Bereich der Einzelnen gibt es mannigfaltige Bearbeitungen von Zeugnissen, die uns zeigen, daß sich die Christen in seelischen und in körperlichen

---

188 Vgl. die Übersicht bei Michael Borgolte: Sozialgeschichte des Mittelalters, München 1996, S. 333–337 mit Verweis auf Klaus Schreiner, Friedrich Prinz, Otto Gerhard Oexle und andere.
189 Ebd., S. 356.
190 Ebd., S. 450, Bezug auf Erich Maschke.

Nöten verschiedener Praktiken der Frömmigkeit bedienten. Viele Erlebnisberichte sind als zugleich „Symbole kollektiver Identität" und „Sache religiös ergriffener Individuen"[191] eingeschätzt worden. In diese übungs- und verhaltensgeprägten Erfahrungen wurden neben den Augen bisweilen auch der ganze Körper durch Berühren, Küssen, christ-imitierende Gestik bis hin zu vegetativen Symptomen einbezogen. Beispielsweise im Kirchenraum und in der heiligen Messe: „Die ganze Komplexität der christlichen Heilsgeschichte und die Fülle ihrer Heiligen sind repräsentiert durch Zeugnisse, die als [...] Spuren der Vergangenheit den Abstand der Zeiten überbrücken und die Präsenz des Heilsgeschehens demonstrieren. Die Dichte dieser Zeugnisse ist überwältigend."[192] Alle Sinnesorgane wurden liturgisch eingebunden und in Erinnerungsarbeit als Zeichensystem überführt. Das galt fürs leseunfähige Volk gleichermaßen wie für den Adel, für den die Messe fester Bestand des höfischen Zeremoniells war.[193]

Es war ein festes Zeichensystem entstanden, dessen Funktion als Ressource und als kognitive Basis uns für die Analyse der Verarbeitung von Emotionen in den Köpfen der Menschen als elementare Richtschnur gelten muß. Fast jede sinnliche Erscheinung konnte letztlich als Zeichenhaftigkeit des Wirkens Gottes verstanden werden: Leben und Tod, Brot und Wasser als Werke und Geschenke, Blut und Wunden und jede Krankheit als Mahnung und Weisung, Kriege und Seuchen als Gerichte, alles sichtbar gewordene Sprache Gottes. Der Zisterziensermönch Alanus ab Insulis (ca.1120–1202) hat diesen überaus dynamischen Zusammenhang in seine berühmten Verse gefaßt: Omnis mundi creatura quasi liber et pictura, nobis est et speculum: nostrae vitae, nostrae mortis, nostri status, nostrae sortis, fidele signaculum ...

Den Bildern wurde Sprachfähigkeit zuerkannt; Beter fanden in der darstellenden Kunst ihre Dialogpartner. Der paradoxe und antithetische Aspekt, den der Afrikaner Tertullian im 2./3. Jahrhundert in die beginnende Geschichte christlicher Lehre einschrieb, zeigte seinen Niederschlag im Bilderleben: Er schreibt: „Gegenstand unserer Verehrung ist der eine Gott. Er ist unsichtbar, obwohl man ihn sieht; ungreifbar, obwohl er durch die Gnade gegenwärtig ist; unfaßbar, obwohl die menschlichen Sinne ihn fassen können – deshalb ist er wahr und groß."[194] Darüberhinaus besaßen nonverbale zeremoniale Kommunikationsfor-

---

191 Schreiner, Klaus: Soziale, visuelle und körperliche Dimensionen mittelalterlicher Frömmigkeit, in: Schreiner, (Hg.) Frömmigkeit, S. 13.
192 Wenzel, Horst: Hören und Sehen, Schrift und Bild. Kultur und Gedächtnis im Mittelalter, München 1995, 101.
193 Ders., S. 95.
194 Zit. nach Benedikt XVI, Generalaudienz 30. Mai 2007; zum Bilderleben s. Krüger, Klaus, Bilder als Medien der Kommunikation, in: Spieß, Karl Heinz (Hg.): Medien der Kommunikation, S. 155–204.

men, wie z. B. bei der Verehrung der Heiligengebeine und Reliquien einen hohen Stellenwert als Medien zwischen Mensch und Gott.[195] Nur selten scheinen Ärzte und Intellektuelle auf die krankhaften Elemente bei Visionären und Ekstatikern hingewiesen zu haben, ihre Meinung wurde unterdrückt;[196] ein Dämonologe warnt vor solchen Ärzten; der schlesische Jurist Witelo nennt das Hören und Sehen von Teufeln [nur von diesen!] geisteskrank.[197]

Die andere Seite der physisch-sinnlichen Inkulturation, die mit Stimme und Gehör, war mindestens auf philosophischen und theologischen Ebenen dem Mittelalter durch Aristoteles und Paulus ebenso bedeutungsvoll wie das Bild. Aristoteles hatte die Hörfähigkeit zur Vorbedingung gesellschaftlicher Teilnahme, und Paulus hatte den Glauben vom Hören abhängig gemacht. Und so haben spätere interpretierende Rückblicke gern auf die höhere Unmittelbarkeit der akustischen Kommunikation, die im Mittelalter ohne den Umweg über Schrift und Bild direkt von Mensch zu Mensch verläuft und auf die Verbindung von Hören und Gehorchen gewiesen.[198] Das gilt ganz unmittelbar für die Seelsorge, die Beichte, das individuelle Schuldbekenntnis, das Eingeständnis von Schwäche und Grenze, ein dem heutigen Denken besonders widerstrebiges Thema. Die Technik der medizinisch genutzten christlichen Kultur des Leidens und des Mitleidens in Buße und Reue, die wir entgegen anderen, besonders psychologischen und philosophischen Begriffsdefinitionen im Rahmen der Vermittlung von Heilsprüchen als introversive Katharsis bezeichnen, findet in **Kapitel C4.3** Niederschlag.

Es mag dahin gestellt bleiben, inwieweit es Divergenzen gegeben hat über das Primat vom Hören des Wortes oder mehr dem Sehen des Bildes bei der katechetischen Vermittlung. Jede Imagination, die über die rein sinnliche, physisch-optische Ebene hinaus zur mentalen Visualisierung führte, sollte den Kern des Menschen erreichen und in einer ars memorativa geformt werden. Es ging um die Tiefe und Nachhaltigkeit der Glaubensverwurzelung. Neurobiologische Ansätze zur Prüfung des Einflusses von emotional gehaltvollen Bildern auf das Gedächtnis haben erste differenzierende Zusammenhänge erarbeitet.[199] Ob in den

---

195 Röckelein, Hedwig, Nonverbale Kommunikationsformen- und -medien beim Transfer von Heiligen im Frühmittelalter, in: Spieß, Karl-Heinz (Hg.) Medien, S. 83–104.
196 Vgl. Franz, Adolph: Benediktionen, II, S. 516.
197 Vgl. Dinzelbacher, Peter, Religiöses Erleben von bildender Kunst, in: Schreiner, K., Frömmigkeit, S. 313.
198 Brenner, Peter J., Hörkulturen. Stimme und Schrift, Hören und Lesen in der abendländischen Kultur, in: Universitas 61 (2006), 224–235.
199 Baeken, Chris et al., The influence of emotional priming on the neural substrates of memory. A prospective fMRI study using portrait art stimuli, in: Neuroimage 61 (2012), 876–883, doi.org/10.1016/j.

Exerzitien des Ignatius von Loyola (1491–1556), die in ihren Stufenplan alle Sinne vom Auge bis zum Geruch und Tastsinn einbeziehen[200], ob in der Seelenanthropologie des Johannes Nider (ca.1385–1438), die Erkenntnislehre des Thomas von Aquin (ca. 1225–1274) behielt eine entscheidende Bedeutung. Sehe ich das nicht ganz verquer, so bietet der Aquinate auch heute noch plausible Teileelemente einer Einführung in perzeptive und verarbeitende Funktionsabläufe des menschlichen Gehirns, man denke allein an die Annahme einer vis aestimativa, einer Einschätzungskraft, die das imaginativ Vorgestellte der phantasia auf das Nützliche und Schädliche überprüft. Wir können das heute als eine Funktion von assemblies cortiko-limbischer Neurone, vor allem der bedeutungsmeldenden emotional orientierten Amygdala, benennen. Im Rahmen unserer speziellen Kapitel zur textumsetzenden Neurobiologie werden wir auf die Bedeutung der Emotionen, der ihnen zugeteilten Einschätzungskraft (appraisal) und ihrer therapeutischen Bedeutung einzugehen haben (**Kap. C1**).

Allgemein wird der innere Mensch als Projektionsfläche von Bildern vorgestellt, wobei Betrachtung wie Gebet und Predigt[201] zum inneren Visualisierungsprozess führen. Auch das „Gebet wird [...] als ein Vorgang der Memoria beschrieben. In der Memoria des Menschen entscheidet sich sein Heil und es liegt ganz bei ihm, welchen Bildern er bei der Imagination und Memoria im Gebet folgt. Zudem gilt: Das Gedächtnis des Heils ist primär Bildgedächtnis im Inneren des Menschen." [202] Damit ist die prozedurale, nicht die inhaltlich-thematische Nähe zu den heutigen Therapien des katathymen Bilderlebens exakt beschrieben. Erst mit dem Philosoph und Humanisten Gianfrancesco Pico della Mirandola (1469–1533) gerät Phantasia (Imaginatio) stärker ins Visier der Irrtumsverdächtiger und später der Reformatoren bis hin zu den Bilderstürmern. War das körperliche Erleben mit gemalten Bildern und geformten Figuren als meditative Technik und als Instrument zur inneren Bildverankerung, zur Kommunikation und zur Empathie ein „epochetypisches"[203] Verhalten für den Betrachter wie für Künstler und ihre Auftraggeber, so markieren die Klagen mancher Theologen

---

200 Man beachte die ausdrückliche Maßgabe für den Begleiter, sich der Ratschläge zu enthalten – wie in der modernen Psychotherapie.
201 Vgl. z. B. Steer, Georg, Bettelorden-Predigt als ‚Massenmedium', in: Heinzle, Joachim (Hg.): Literarische Interessenbildung im Mittelalter, S. 314–336: Predigt als Kanzel- und Schriftpredigt in erweiterter Sicht.
202 Lentes, Thomas: Inneres Auge, äußerer Blick und heilige Schau, in: Schreiner, K. (Hg.): Frömmigk., S. 185f; zur Einübung sequentieller Bildvorstellungen mit Inhaltsörterlehre vgl. auch Knape, Joachim, Rezension zu Heimann-Seelbach, Sabine, Ars und scientia. Genese, Überlieferung und Funktionen der mnemotechnischen Traktatliteratur im 15. Jh., in: Zeitschrift f. dt. Altertum und dt. Lit., 134 (2005), S. 123–128.
203 Dinzelbacher, Peter, Religiöses Erleben von bildender Kunst, in: Schreiner, Klaus, Frömmigk., S. 329f.

des 15. und 16. Jahrhunderts über aufkommendes *ästhetisches* Interesse an Kunst das Ende dieser Epoche, das Ende des Mittelalters.

Den Inhalten und Praktiken der Frömmigkeitszeugnisse steht eine Fülle von Schriftzeugnissen der Dämonenliteratur und der gegen Häresien gerichteten Anstrengungen von Klöstern und Kirchen gegenüber. Ein nicht messbarer Teil des Volkes entzog sich ihren Bemühungen und benannte, beachtete und verehrte ihre eigenen Geister. Dabei bleibt unter den Forschern umstritten, welchen Anteil die Quellen am gesellschaftlichen Leben hatten, germanische Mythologie, tradiert im Leben des Volkes oder antike Mythologie, transportiert in den Schriften der Gelehrten, von ‚unten' oder von ‚oben'. Texte der Kirchensynoden, Missionierungsanweisungen, Bußbücher seit dem 8. Jahrhundert dokumentieren die Sorge um Zaubereisünden. Besonders in den Dekreten des Bischofs Burchard von Worms im 11. Jahrhundert wird die ‚Dummheit des Volkes' beklagt. In diesen Schriften finden manche mythische Namen Erwähnung, die wir in den Merseburger Heilsprüchen mit den weisen Frauen (➔23) und mit Wotan (➔26) wiederfinden und später im Münchner Nachtsegen (➔1) und in den Alb- und Elbentexten (➔3 und 4). Das Gros der mittelalterlichen Beschwörungen und Heilsegen, die wir als Zaubersprüche bezeichnen, erwächst auf überwiegend christlicher Basis, auf die man Zusätze installierte, woran der Klerus Anstoß nehmen konnte, aber ihre Anwendung und Notierung generell nicht zu vermeiden vermochte.

Diese Zusätze an Legende, Mythos und Volksdichtung, an Phantasie und außergewöhnlichen Sprachfiguren sind es, die als ‚Aberglaube' abgetan unsere Zaubersprüche charakterisieren. Sie sind der in keiner Kultur fehlende Versuch, dem Kanon des allgemeingültigen Wortes zu entrinnen, um erhöhte Aufmerksamkeit zu erzeugen, in heutiger Sprache der Neurobiologen, Psychotherapeuten und Sprachwissenschaftler: Performativität und Perturbation.

B4.1.3 Christliche ‚Medialität'

Abgesehen von den bekämpften Resten tradierter mythologischer Residuen lebten die Menschen des Hochmittelalters in einer christozentrischen Gesellschaft. Man spricht heute vom ‚Verlorenem Paradies' (Hermann Hesse), von ihrer Literatur als ‚Wahrheit der Fiktion' (Walter Haug), von ‚Verzauberter Welt'[204] und vom ‚Zeitalter des Glaubens' (Peter Dinzelbacher).

---

[204] Allgemein; und siehe die Entzauberungsmetapher als prozessualer Gegensatz zu magischen Gesellschaften bei Max Weber – Winckelmann, J., Die Herkunft von Max Webers „Entzauberungs"-Konzept, in: Kölner Zeitschrift für Soziologie und Sozialpsychologie 32 (2005), S. 12–53.

Zwar gibt es eine Reihe diverser Zeugnisse aus allen Ständen, einschließlich und akzentuiert auch von Ärzten, vom Atheismus über Gleichgültigkeit bis hin zu Zweifelsäußerungen, besonders an den Dogmen der Kirche: Es „handelte sich nur um punktuelle Widersetzlichkeiten gegen die sehr weitgehend das gesamte Leben durchdringende Frömmigkeit der Epoche."[205] Obwohl in der Folge von Dinzelbacher in den letzten Jahren weitere Anzeichen für atheistische Tendenzen gesammelt wurden, mag diese Einschätzung bei Fehlen jeglicher Statistik und jeglicher personaler Analysen nicht an Gewicht verlieren.

Aus der Sicht einer mittelalterlichen Medialität im Sinne einer speziellen Ebene der heute sog. Kommunikationsmittel ist – nach dem Selbstverständnis personalisierten Mittlertums jener Epoche – an Jesus Christus als einzigem ‚Leitmedium'[206] nicht zu zweifeln. Christus verkörpert als Erlöser mittels Bildhaftigkeit und Sprachlichkeit den Weg zu Gott. Mit Augustinus wird durch Christus Jesus als Mensch und als Teil der Dreifaltigkeit eine Analogie zwischen imago dei und imago rei corporalis, also spiritueller und sinnlicher Wahrnehmung bewirkt. Nicht Repräsentation von Bild- und Sprachbegriff, sondern die „Ununterscheidbarkeit von Abbild und Urbild"[207] werden erstrebt, wie wir dies auch von der Spiritualität der Hildegard von Bingen kennen oder aus der Erkenntnistheorie des Philosophen Boëthius (4./5.Jahrhundert).

B4.1.4 Christliche Krankenfürsorge und Krankheitsverständnis

Seit dem 7. Jahrhundert hatten Klöster die Versorgung von Kranken übernommen. Sie trugen in Übereinstimmung mit den Herrschafts- und Kulturstrukturen Sorge für körperlich und seelisch Kranke. Im Mittelpunkt stand aufgrund der Zeichenhaftigkeit von Krankheit die christliche Ethik der Nächstenliebe und der Jenseitsvorsorge, sowohl irdische Heilung (cura corporis) als auch religiöses Heil (cura animae) versprechend. Ich verweise auf das in **Kapitel B2** Ausgeführte über die Heilkundigen. Im Blick auf die Einstellung zur Krankheitsursache wird sich nicht sicher und nicht zeit- und situationsgebunden feststellen lassen, inwieweit und wo die strenge asketisch orientierte Lehre Augustinus' von der Erbsünde und ihren Folgen, von Adams Schuld als Ursache jeden Übels die Gemüter beherrschte oder inwieweit vielmehr die gnadenvolle Entbindung von steter personaler Mitverantwortung an Krankheit durch Christus (nach Evangelium Johannes 9,3: „Weder er selbst noch seine Eltern haben gesündigt") den

---

205 Dinzelbacher, Peter: Unglaube im „Zeitalter des Glaubens", Badenweiler 2009, S. 151.
206 Kiening, Christian, Mediologie – Christologie, in: Das Mittelalter 15 (2010), S. 16–32, hier S. 17, 23f.
207 Kiening, ebd., S. 18.

Einzelnen stärker seelisch beeinflußte. Hält man sich an die hier in diesem Buche angestrebte Betonung der Not- und Krisensituation, so bekommen Todesangst und Hoffnungsverlust mit gleichzeitigem Erlösungshorizont (**Kap. C4.3.1.2**) und suchender Erwartung einen sehr dezidierten Stellenwert, zumal das Christentum gegenüber den anderen Hochreligionen als die einzige Religion der Therapie verstanden werden kann.

Dem Zeichensystem von religiösem Wort und Bild entsprachen in der Verbaltherapie bei der Sorge um den Kranken vorrangig diejenigen Organsysteme, die eine Brücke schlugen zwischen Evangelium und Krankem. Entsprechend am häufigsten und vielgestaltigsten finden wir Texte, die bei einer traumatischen Blutung gesprochen wurden (➜**21 und 33**) und solche, die gegen Schmerzen die Passion Christi imaginierten (➜**36**) und andererseits solche, die den Wunderheilungen nachgebildet werden konnten, für psychisch Kranke, für Augenleiden und für Epileptiker.

B4.1.5 Individuum und Ich-Verständnis im Mittelalter

Im allgemeinen können wir davon ausgehen, daß die zwischenmenschliche Bedeutung von Gefühlen wie Schuld und Scham, Angst und Ekel, Selbst- versus Andere-Regulierungen in hirnbildgebenden Verfahren wahrscheinlich ebenso nachweisbar gewesen wäre wie heute mit den vergleichenden Untersuchungen asiatischer und westlicher Denkart (s. o.). Vielleicht hätte es sogar Parallelen zu außereuropäischen sozialkonformen Akzentuierungen gegeben. In den folgenden speziellen Kapiteln zur heilspruchperformierenden Hirnleistung wird uns die Frage beschäftigen, inwieweit demutsfördernde, Leid und Angst perzipierende und selbst den Tod einkalkulierende, heute als reinigend, kathartisch, zu verstehende Textgestaltungen (**Kap. C4.3**) in ihrer christlichen Zeichenhaftigkeit einen ‚Königsweg' ehemaliger Therapie waren, entgegen einem nur ‚positiven' Denken nach heutigem Sprachgebrauch. Appraisal und Reappraisal jedenfalls, wie die inneren Deutungs- und Verarbeitungsvorgänge heute in der Kognitions- und in der Neurowissenschaft genannt werden (**Kap. C1**), hatten völlig andere inhaltlich-thematische Grundlagen und Begrifflichkeiten. Darauf ist immer wieder hingewiesen worden:[208] Die Seele ist von Gott geschenkt und im Menschen somit ein Teil Gottes; Begriffe wie Intention, Subjekt und Sinn sind nur aus diesem Zusammenhang zu verstehen. Niemals jedenfalls wäre totale Absage an Gesellschaft, Elternhaus und Gemeinde, Entankerung existentieller Abhängigkeit von dieser mittelalterlichen Gesellschaft und ihrer kollektiven Identität als selbstauf-

---

208 Reichertz, Jo, Kommunikationsmacht, S. 49, 63.

bauende hilfebietende Strategie verzeichnet worden. „Im Mittelalter war ein Einzelmensch verloren."[209] Und wissen wir doch auch, daß die Literatur des Hochmittelalters selten das ‚Ich' darbot, erst in seiner Spätphase. Das gleiche gilt für Segen und Gebete; erst ab 14. Jahrhundert tauchen in ihnen Ich-Erwähnungen in Form von Eigennamen auf.[210] Literaturgeschichtlich ist bekannt „... daß dieser individualistische Ich-Begriff auch mit dem Aufkommen von Metaphern korreliert,", wenige Metaphern bei Walther von der Vogelweide, umso mehr bei Shakespeare.[211] Danach wäre die Hypothese zu wagen, daß die im Hochmittelalter im allgemeinen zu beobachtende Rarität an außergewöhnlichen Sprachfiguren dem an Metaphern und Analogien reichen Heilspruch und der Legende eine einst attraktive Stellung verliehen hat.

B4.1.6 Hirnbiologische Prämissen von Religiosität und Spiritualität

Für die Menschen im europäischen Mittelalter brachte Religiosität als elementare individuelle Grundhaltung sowie als Zugehörigkeit zur Gemeinschaft der Getauften offensichtlich große Vorteile und war dem Einzelnen schon als soziale Organisationsbasis nützlich. Ich verweise auf den Aspekt der neurobiologischen Anpassung (**Kap.B1.1**) und warne vor der Zuordnung der genannten Kategorien Vorteil und Anpassung in diesem Zusammenhang an ethische Betrachtungen; die Biologie des Gehirns arbeitet nicht primär nach ethischen Prinzipien.

Außerdem muß auch der Neurobiologe grundsätzlich zunächst eine Unterscheidung treffen: Auf der einen Seite gibt es die laue, taube oder nur latente Religiosität mit ihrer Tendenz zu Indifferenz oder Gleichgültigkeit, die sich mehr oder weniger in kultureller Zugehörigkeit zur entsprechenden Gruppe aufgrund der Bestimmung durch Eltern und Umwelt versteht, äußert und erschöpft. Und es gibt die aktivierte innerseelische oder spirituelle Religiosität, die bis zu einem Überschießen kommen kann auf der anderen Seite und es gibt dazwischen die nicht fassbare und wechselhafte Menge. Eine Neurobiologie ‚der Religion' wird immer bodenlos bleiben, wenn sie die Komplexität der Begriffe ‚Religion' und ‚Religiosität' übersieht. Selbst die Differenzierung in 1. Organisatorische Bindung mit materieller oder/und psychologischer Vorteilsnahme, 2. Ideologische konservative Einstellung, 3. Gewohnheitsmäßige Bindung und Verpflichtung, 4. ‚Spirituelle' persönliche Bindung an ‚Gott' oder höheres Wesen sowie 5. in den heutigen Experimenten eine Einkalkulierung der artefiziellen Laborsitua-

---

209 Fossier, Robert: Das Leben im Mittelalter, München 2008, S. 283.
210 Ernst, W.: Beschwörungen, S. 294.
211 Schrott, R. und A. Jacobs: Gehirn und Gedicht, S. 202.

tion,[212] kann keine befriedigenden Ergebnisse liefern. Die hier auf das Mittelalter und auf frühe und archaische Kulturen bezogene Betrachtung ist im Blick auf gelebte Religiosität eine andere Art von Laborversuch, der – wie ich noch einmal betone – fragmentarisch bleibt.

Sicher ist lediglich, daß eine lebensgeschichtlich fixierte unwiderrufbare Religiosität, ob sie einer orthodoxen oder frei flottierenden Spiritualität nahe steht, wenn sie mit Berufungsideen, überwertigen irrealen, erfahren übernatürlichen Beziehungen, passagèren oder evtl. andauernden Halluzinationen, Visionen oder Akoasmen und entsprechendem sozialem Verhalten einhergeht, eine materielle Basis hat: Sie zeigt in der Regel Volumenveränderungen im temporalen Cortex.[213] Das betrifft den Punkt 4 der genannten Differenzierung. Ebenso hat man bei traumatischen Hirnschäden und Tumoren im Bereich der postparietalen Rinde mit Transcendenzerfahrungen, Weltallerlebnissen etc. zu rechnen.[214] Beide schon oben betrachteten Hirngebiete – postparietaler und temporaler Cortex – berühren sich im Gyrus angularis = TPJ, temporoparietal junction, die wichtige Gegend für Selb-Ander-Unterscheidung und Sozialverhalten,[215] bei deren Gestörtheit es zu Out-of-body-Erfahrungen und Doppelgängererlebnissen kommen kann. Der inferiore temporale Bereich ist für Erlebnishalluzinationen und polysensorische visuelle und auditorische Rückblenden mit Versetzung in andere Aufenthaltsorte schon lange bekannt, wie man von chirurgischen Eingriffen weiß.[216]

Derartige heute in westlichen Kulturen meist als krankhaft angesehene Persönlichkeitsvarianten können auch genetische Wurzeln haben, wie man an monozygoten, getrennt aufgewachsenen Zwillingen fand.[217] Die Zwillingsforschung hat deutlich eine genetische Komponente von – nicht christlich zu verstehender – ‚Spiritualität' entdeckt. Die verschieden angelegten Konzentrationen neuraler Botenstoffe, der Monoamine Serotonin, Adrenalin und Dopamin haben einen Einfluß auf irrationale Äußerungen und Emotionen und können zusammen mit krankhaften Veranlagungen (z. B. erbliche Epilepsie) zu Hyperreligiosität, Missionierungssucht und Geisterglauben führen, zu Phantasien und Fanatismen.

---

212 Bradshaw, Matt et al.: Do genetic factors influence religious life? in: Journal for the scientific study of religion 47 (2008), S. 529–544 doi org/10.1111/j.1468–5906.
213 Kapogiannis, Dimitrios et al., Neuroanatomical variability of religiosity (2009), PLoSOne 4(9): e7180. doi:10.1371/j.pone.
214 Urgesi, Cosimo et al: The spiritual brain: Selective cortical lesions modulate human self-transcendence, in: Neuron 65 (2010), S. 309–319.
215 Decety, J. und Lamm,C., The role of the right temporoparietal junction in social interaction, in: Neuroscientist. (2007) 13(6): 580–93.
216 Roth, Gerhard: Das Gehirn und seine Wirklichkeit, S. 181.
217 Bouchard, Jr, Thomas J et al., Intrinsic and extrinsic religiousness: genetic and environmental influences and personality correlates, in: Twin Research 2(1999), S. 88–98, doi.org/10.1375.

Ich erinnere an Dostojewskis Selbstschilderungen.[218] Psychische Devianzen, die häufig von Historikern an augenfälligen Beispielen (Visionäre, Ekstatiker) herangezogen und verallgemeinernd überdeutet werden, sollten nicht die Einschätzung mittelalterlicher Religiosität in ihren verschiedenen und schwerlich zu reanimierenden Facetten vorschnell bestimmen.

> Utopische Gehirnvorstellungen über ein „Gottes-Modul" (V.S. Ramachandran), eine „Ausrüstung für religiöse Erfahrung" (Andrew Newberg), ein „Jesus-Syndrom" (Wolf Singer, Interview 2008) oder über eine (undifferenzierte!) „Spiritualität" haben viel Popularität erreicht; sie gewähren uns aber – schon weil sie die kulturellen und biografischen Prägungen wegen Komplexität nicht berücksichtigen können – keinen Blick auf gelebte Religiosität, schon garnicht auf Fragen nach Wahrheit und Gott. – Untersuchungen von Trainingsprogrammen für unspezifische Aufmerksamkeit haben – ob unter religiöser Etikettierung oder als Vergleich mit nonreligiösen Versuchspersonen – ebenfalls keine für unsere Diskussion relevante Bedeutung. Letztlich gibt es ernüchternde verhaltenspsychologisch-neurologische Untersuchungen (mit fMRI), die uns für die Menschen unserer Zeit belegen, daß Nachdenken über Gottesglauben viele derselben Hirnregionen aktiviert, die auch dann aktiv werden, wenn über privaten „Ego-Glauben" nachgedacht wird.[219]

B4.2 Archaische Gesellschaften: Die mesopotamische Kultur

Nicht anders als für das christliche Mittelalter ist für die Heilverfahren in den verschiedenen archaischen Gesellschaften über eine duale Situation hinaus ein Blick in das kulturelle Milieu unentbehrlich. „Das Vertrauen und die Forderungen der öffentlichen Meinung, die ständig eine Art Gravitationsfeld bilden, in dem die Beziehungen zwischen dem Zauberer und denen, die er verzaubert, liegen," stellen einen dritten mitwirkenden Teil beim Erfolg magischer Praktiken dar.[220]

Unser Wissen über eine der ältesten Kulturen der Erde stützt sich auf archäologische Funde, insbesondere auf die Tontafeln aus verschiedenen Tempelbibliotheken. Besonderer Stolz verband sich mit dem Besitz einer Schrift, der Keilschrift auf diesen Tontafeln, die zugleich Herrschaftsinstrument und Kulturmotor war von ca. 3200 bis zum 1. Jahrhundert vor Christus.

---

218 Jaspers, Karl: Allgemeine Psychopathologie, Berlin u. a. 1948, S. 72, 97, 120.
219 Eply, Nicholas et al., Believer's estimates of God's beliefs are more egocentric than estimates of other people's beliefs, in: PNAS (2009), doi: 10.1073/pnas.0908374106.
220 Lévi-Strauss, Strukturale Anthropologie, S. 184.

Schreiben galt – ebenso wie dem Mönch des Scriptoriums – als Schlüssel zum Überleben nach dem Tode.[221] Anders als im nordeuropäischen Mittelalter hatte sich diese Schrift aus zunächst numerischen Zeichen im Lande selbst entwickelt.

B4.2.1 Einige Aspekte der Gesellschaftsstruktur

Die mesopotamischen Kulturen waren Gesellschaften, die eine Gleichheit vor dem Gesetz nicht kannten. Schon aus Registraturen der protocuneiformen Ära kann abgelesen werden, daß junge Nutztiere, etwa Schweine, mit denselben numerischen Zeichen aufgelistet wurden wie junge ‚Sklaven'.[222] Die Fruchtbarkeit des Landes zwischen den Strömen erlaubte eine wirtschaftliche Entwicklung und Zivilisation, die von der Urbanisierung ab mittlerem vierten Jahrtausend vor Christus ausgehend, zu früher Technik, Astronomie, Astrologie, Mathematik und Kalendererfindung führte. Was die Herrschaftsstruktur angeht, hat man ein Dreiecksverhältnis König – Götter – Priesterschaft angenommen.[223]

B4.2.2 Aspekte geistiger Prägung

Die Kulturen des Zwischenstromlandes vom älteren südlichen Teil Sumer bis zu den späteren Reichen der Akkader und Assyrer waren von dämonistischen Kulten und Polytheismus beherrscht. Dämonenvorstellungen gehörten zum Alltagswissen und Alltagsleben und waren legitim. Die Dämonen unterstanden in der Vorstellung der Menschen wiederum dem Diktat von Göttern, vor allem Anu, dem Himmelsgott, Enlil, dem Gott des Sturmes, Nintu und Enki für Erde und Wasser. Zeitweise fungierten Ea, Marduk und Assur. Diese Welt- und Götterbilder waren selbstverständlich und umumstößlich strukturell verankert. Und „während die Realität von Hexen, Geistern, Gottheiten und Dämonen nie in Frage stand, war sich das Volk der Außergewöhnlichkeit dieser Wesen bewußt." [224] Großen Einfluß auf das Selbstverständnis des Volkes hatten die Mythenerzählungen von Gilgameš und Atramhasis, letzterer gilt als der mesopotamische Noah. Die Götter bedürfen der Menschen, um sie für sich arbeiten zu lassen und bewirken damit eine gegenseitige Abhängigkeit. Mythen der Sintflut und der

---

221 Foster, Benjamin, R., The person in Mesopotamian thought, in: Radner/ Robson Handbook, S. 130.
222 Englund, Robert K., Accounting in Proto-cuneiform, in: Radner/Robson, Handbook S. 32–50, hier S. 46f.
223 Waerzeggers, Caroline, The Pious King: Patronage of Temples, in: Radner/Robson, S. 746.
224 Schwemer, Daniel, Magic Rituals: Conceptualization and performance, in: Radner/Robson S. 420.

Welterschaffung in sieben Tagen ebenso wie manche kulturellen Erwerbungen lebten als Bruchstücke in einigen anderen Kulturen weiter.

B4.2.3 Krankheitsverständnis

Gesellschaftlich gesehen war der einzelne, ob er einen Tempelpriester in Anspruch nehmen konnte oder sich aufgrund seiner sozialen Stellung an Straßenbeschwörer halten mußte, einer nachhaltigen Prognostik ausgeliefert. Ausgeliefert dem Schicksal, ob er als Kranker und Unreiner noch zur Gesellschaft gehörte oder ausgestoßen wurde. Das Schicksal vieler Kranker in dieser Kultur kann wohl insofern selbst schwerlich dem verglichen werden, was im europäischen Mittelalter dem Zwangszustand der Besessenheit [225] entsprach und exorzistisch angegangen wurde, weil die Attribution einer Besessenheit, das heißt einer Besetztheit von Dämonen offenbar im alten Zweistromland nach den vielen Zeugnissen nicht auf neuropsychiatrische Krankheiten beschränkt war.

Gemäß einem Teil der Zeugnisse galt Krankheit – eindeutiger als etwa zu Pestzeiten im Christentum – als Bestrafung für Sünde, nicht nur für eigene, sondern auch für Verfehlungen der Familie und Sippe, aber auch als Eingriff übernatürlicher Mächte. Damit war sie Krise, Ruptur und Absonderung ins Außirdische einer anderen Welt. Nur der Fachmann konnte den Kranken zurückholen. Gebetsartige Beschwörungen berichten vom Befall durch böse Geister und von Reue und Bußübungen. Ein anderer Teil der Schriften ist der Abwehr der durch Hexen und Zauberer verursachten Schäden gewidmet. Gerade diese letzteren Texte, denen man in der frühen Mesopotamien-Forschung eine marginale und esoterische Rolle zuwies, „bilden ein bedeutendes Prisma zum Studium der mesopotamischen Kultur. Sie berühren physische, psychologische und soziale Probleme."[226]

Während sich der Christus des Evangeliums dem Kranken zuwendet und ihn der Isolierung entreißt und während sich die mittelalterliche christliche Gesellschaft dem Kranken verpflichtet fühlt, gibt es Anzeichen dafür, daß zeitweise mancher Kranke in der mesopotamischen Gesellschaft isoliert wurde und sein Zustand als Schande und Fluch angesehen wurde.[227] Manche Forscher haben diesbezüglich eine Korrespondenz mit semitischen Vorstellungen vermutet und die Parallelen zum Alten Testament herangezogen. In der althebräischen Kultur war Krankheit mit Unreinheit verbunden. Im Vergleich zum mittelalterlichen

---

225 Vgl. Ernst, W. (2011) S. 211f.
226 Abusch, Tzvi, Einleitung zu Schwemer, Daniel: Abwehrzauber und Behexung, Wiesbaden 2007, S. XIII.
227 Sigerist, Henry E.: Der Arzt in der Mesopotamischen Gesellschaft, Zürich 1963, S. 49.

Mönchsarzt, der sich als spiritueller Mittler zum erlösenden Christus versteht und der sich von weltlicher Macht zu distanzieren sucht, erscheint der babylonische beschwörende Priesterarzt als elitärer Standesvertreter, der sich eines Machtanteils bewußt ist und ihn einsetzt. Er läßt insofern an die Tätigkeit des Exorzisten denken, der im Mittelalter allerdings eher eingeschränkt auf psychotische Krankheiten (Besessenheit) wirkte (**Kap. B2, B4.2.3, C1, C5**).

Zu welchen Anliegen und Themen die Beschwörungs- und Gebetsformeln insgesamt verwendet wurden, ergibt sich aus der bis heute noch weitgehend anerkannten Einteilung durch Falkenstein[228] 1931:

1.) Vorbereitende auf göttlichen Auftrag verweisende Legitimierung des Beschwörenden als Selbstschutz vor Dämonenangriffen. Dabei kann – nicht anders als im Münchner Nachtsegen – das Treiben der Geister, das sich im allgemeinen gehäuft nachts abspielt und (**→40 und 41**) alptraumgemäß den Träumer in Netzen fängt – geschildert werden; die Geister werden direkt angesprochen: „Magst du ein böser Alû[229] sein …", „Der böse Utukku …".[230] Es wird die Tätigkeit des Beschwörers minutiös geschildert (**→42**: „Wenn ich an den Kranken herantrete …"), es werden die Schädiger oder die personalisierten Krankheiten aufgezählt und es wird ihnen Einhalt befohlen, alles zunächst zum Schutz des Priesters.

2.) Ein zweiter ähnlicher Typ der Formeln ist oft prophylaktisch zu verstehen und soll dem Dämon, der oft als Totengeist benannt wird, die Annäherung nun speziell an den Kranken verbieten: „… Du sollst dein schreckliches Wort nicht zu ihm sprechen, deine wilde Stirn nicht auf ihn richten, dein wütendes Auge nicht auf ihn werfen …" „… Der Mensch möge ins Haus eintreten – [aber] du sollst nicht eintreten! …" Hier werden oft die Körperteile des zu behandelnden Menschen aufgezählt (**→46**).

3.) Der Marduk-Ea-Typ ist nicht mehr prophylaktisch und dient der Vertreibung der Übeltäter (**→45 u. a.**). Hier sind es oft ausführliche Klagen über die verhexten Glieder, den Spuk und bösen Zauber, der den Menschen aufgewühlt, gequält und verknotet hat. Um Macht zu erlangen, hat sich der Dämon einer

---

228 Falkenstein, Adam: Die Haupttypen der sumerischen Beschwörung, Diss. Leipzig 1931 sowie dazu Schramm, Wolfgang: Ein Compendium sumerisch-akkadischer Beschwörungen, Göttingen 2008, S. 16–21.

229 Alû ist einerseits ein Angehöriger der sieben utukke der Sumerer, ein Totengeist, andererseits der Himmelsstier, den Gott Anu auf Wunsch der Ištar erschafft, damit er gemäß dem Epos gegen Gilgameš kämpfe. Der Held aber erlegt ihn mit Hilfe seines Freundes Enkidu.

230 akkadisch utukku gilt sowohl als Sammelname für die sieben Geister wie auch als selbständiger Totengeist, als Gerippe. Er ist es, der dem Gilgameš als toter Enkidu erscheint, vgl. Ebeling, Erich, Artikel Dämonen im Reallexikon der Assyrologie, Berlin und Leipzig 1938 II,107–115.

figürlichen Tonnachbildung des zu Quälenden bedient, im Abwehrzauber revanchiert sich der Geschädigte ebenso. Ich verweise auf die speziellen Ausführungen zur mesopotamischen Traumausgestaltung und zu Nachtgeistertum im **Kap. C1.4.1.** Es wird Revanche geübt, es werden Vernichtungsverben verwendet, der Zauberer soll im Feuer verbrennen (= maqlû), soll in Rauch vergehen, Tamarisken und Datteln als verbreitete apotropäische Mittel (➔43) werden eingesetzt und schließlich auf die Hilfe von Marduk und Ea vertraut (➔**49 und 52**). Kraftvoll sowohl introversiv in Trauer- und Leidbekundung (**Kap. C4.3**) als auch extroversiv in Verwünschungs- und Vernichtungsworten entwerfen viele der Texte kathartische Erzählungen, die nach modernen Vorstellungen einer psychosomatischen Betrachtung zugänglich sind (**Kap. C5**).

Ein Teil der Beschwörungen und Rituale, die schädigende ‚schwarze' Magie, der Schadenzauber, galt als ‚Kapitalverbrechen'.[231] Dabei sind juristische Anklagen (im Gegensatz zum antiken Rom) selten, weil falsche Beschuldigung hart bestraft wird und weil man unklare Fälle mit Ordalien prüfte, selbst Eide unterlagen dieser ‚göttlich' geregelten Wahrheitsfindung. Und so soll es direkte Anweisungen zum Schadenstiften, also die ‚schwarze Magie' späterer Zeiten nicht gegeben haben,[232] immer nur den Verdacht und die Furcht vor ihnen. Hilfe boten die entsprechenden gegenzauberischen Riten und Beschwörungstexte. In diesem Bereich sind Hexer und Hexen, Zauberer und Zauberin, Widersacher, Prozessgegner und Feinde aller Art, also Mitmenschen verdächtigt, meist weibliche. „Die grundsätzlich als böser Mitmensch vorgestellte Kaššaptū nimmt in der Beschwörungsliteratur teilweise dämonische Züge an,"[233] eine m. E. ähnliche Entwicklung, wie wir sie aus dem späten Mittelalter und der frühen Neuzeit in Europa kennen.

Wir werden sehen (**Kap. C4**), daß nun auch erzählerisch dialogische Elemente in diesem Marduk-Ea-Typ zur Anwendung kommen:[234] Marduk [oder Asalluhi] geht zu seinem Vater Ea [Enki] und berichtet ...(➔**49, Zeile 17–19 und ab Zeile 35**) ebenso wie in einem Zahnwurm-Text mit Ea und dem Sonnengott Schamasch (➔**48**).

---

231 Schwemer, Daniel: Rituale und Beschwörungen gegen Schadenzauber, Wiesbaden 2007, S. 1.
232 Thomsen, Marie-Louise: Zauberdiagnose und schwarze Magie in Mesopotamien, Museum Tusculanum Press, 1987, S. 9f.
233 Schwemer, Daniel: (2007), S. 2.
234 Falkenstein, Haupttypen, S. 53.

B4.2.4 Individuum und Ich-Verständnis in der mesopotamischen Kultur

Zeugnisse aus den Keilschriften über dargestellte Individualität im heutigen Sinne, also biografische Schilderungen gab es nur in wenigen kurzen Ansätzen. „The person as story was developed into a fictional autobiographical genre of Accadian literature in which a ruler of past reflected on certain events in his life [...], but none of the thousand of letters from Mesopotamia contains extend autobiographical reflections or anecdotes."[235] Folgen wir weiter dieser Analyse von Foster, der vor Vergleichen mit der europäischen Betrachtungsweise warnt, dann gab es durchaus nach den Schriften selbstbewußte, planende und berechnende, auch mit den Göttern um Wohlleben, Genuß und Gesundheit handelnd abrechnende Lebenseinstellung. Wie gesagt, hatten die Götter ja den Menschen für sich zum Arbeiten geschaffen und waren von ihnen ebenso abhängig wie die Menschen von den Göttern. Jedoch gibt es wenig Hinweise für eine Leib-Geist-Seele-Trennung. Gegenüber den Göttern gab es ein ‚wie du mir, so ich dir', eine Reziprozität um Opferkult und persönliche Lebenskraft. Aber „der Körper ist die Person"[236] und nur der Tod schafft die genannten Geister. Niederschlag dieser Vorstellungen sind die Totenkulte mit ihren Opferungen und ihren ausgedehnten Klageprozeduren mit Tanz, Gesang und Instrumenten, um die Götter an ihre Verantwortung zu erinnern. Ursächlich nimmt man an, daß es nach einem dichterischen Motiv auf die Abgrenzung der zivilisierten Gesellschaft von dem seine Toten verlassenden Nomadentum ankam und auf die starke Identifizierung des einzelnen mit Familie und Verwandtschaft und ihrem Namen. So bestand nach Karen Radner und Eleanor Robson doch nur eine ‚geteilte Identität', die so weit ging, daß im Totenkult die Verstorbenen als Geister und Phantome wiederkehren und Lebende plagen konnten, wenn ihrer nicht gebührend gedacht ward. Solche Unterlassungen hielt man denn auch „verantwortlich für alle Sorten von Unglück, individuell und gemeinschaftlich.[237] Der Beschwichtigung und der Wiedergewinnung ihres Wohlwollen dienten denn auch zahlreiche exorzistische und ‚magische' Rituale. Und das Gericht der Götter im Moment des Todeseintritts, wenn zornige Verstorbene in die Unterwelt eintraten, war zugleich der Befreiungsschlag für die Lebenden, denen sie im Traum Schrecken eingejagt hatten.[238]

---

235 Foster, B. R. in: Radner/Robson, Handbook, S. 117–139, hier: S. 127.
236 Foster, ebd., S. 120.
237 Radner/Robson, in Oxford Handbook, S. 114.
238 Demare-Lafont, Sophie, Judicial decision-making: Jugdes and arbitrators, in: Radner/Robson, S. 335.

Zusammenfassung Kap. B4: Soziales Umfeld, Kulturerwerb und ‚sozioneurale Hardware'

Kultur formt menschliches Denken, Fühlen und Verhalten. Biologische und soziologische Theorien zur kulturellen Evolution und ethnologische historische Vergleiche haben Bedingungen differenter genetischer und sozialer Verankerungen darzustellen versucht. An Vergleichen bei lebenden Kulturen konnten auch cerebrale Entwicklungsdifferenzen zwischen Asiaten und Amerikanern, kollektivistischen und individualistischen Prägungen nachgewiesen werden. Könnten wir diesbezüglich die Bewohner des alten Babylons oder einer mittelalterlichen Reichsstadt mit modernen Methoden auf Aktivierungen des praefrontalen Cortex mit seinen wichtigsten Verbindungen untersuchen und ihre geistigen Führer und Medizinmänner auf Aktivierung von autoritätsstabilisierenden Funktionen (rechte vordere Insel, linke untere Temporalgegend, rechten Hippocampus und vorderes Cingulum) dann hätten unsere Modellversuche zu den angewandten Spruchtexten eine etwas festere Basis. Aber ohne den Versuch wenigstens einer Beachtung der soziokulturellen Bedingungen – soweit eben möglich – und über die Zweierbeziehungen von Mutter und Kind sowie Medizinmann und Krankem hinaus, wären Schlussfolgerungen über die Angemessenheit der Heilspruchtexte obsolet. Schon aufgrund des Spracherwerbs ist Gemeinschaft frühzeitig im Leben neurophysiologisch imprägniert

Die mittelalterliche gesellschaftliche Struktur, zumindest von Früh- und Hochmittelalter, war durch die führende Kraft der Klöster und ihrer Verflechtung mit dem Adel geprägt. Die Zeugnisse von Bild, Schrift und Plastikwerken lassen erahnen, daß tiefe sozusagen psychosomatische Schichten, ‚bis in Fleisch und Blut' jedes einzelnen, einer kollektiven Identität verhaftet waren. Ein Single-Dasein als individueller Lebensentwurf dürfte für gesunde Menschen kaum vorgekommen sein. Dabei fungierte gesellschaftsbildend ein heute umfassend erforschtes und beschriebenes Symbol- und Zeichensystem, das aus christlichen Wurzeln alle Lebensbereiche bis in den Alltag hinein erfasste. Entsprechend galt der Kranke nach christozentrischem Weltbild als Mitbruder, der nach den gegebenen Möglichkeiten mit empirischen Mitteln und auch mit Heilsprüchen behandelt werden sollte. Beides hatte den doppelten Zeichencharakter von Heil und Heilung. Die vom biblischen Kanon abweichenden Sprachfiguren und semantischen und erzählerischen Neuerungen, Überraschungen und Zusätze an den Heilspruchtexten haben verschiedene Quellen, legendär, antik- und germanisch-mythologisch, und wurden nicht immer geduldet und oft unterdrückt.

Einige derzeitige hirnbiologische Hypothesen über Religiosität und Spiritualität, die zu einer Klischierung von Kulturen verleiten könnten, bietet das **Kapitel B4.1.6.**

Die mesopotamische Kultur war von polytheistischen und dämonistischen Vorstellungen geprägt, die nicht nur als selbstverständlich galten, sondern auch strukturell und in der Rechtssprechung verankert waren. Dabei wurden die Götter schon vom Schöpfungsmythos her weitgehend vermenschlicht, sodaß sie quasi im Handel um Schuld und Buße, Krankheit und Gesundheit als Partner angesehen werden konnten. Dementsprechend gab es ein ‚Wie du mir, so ich dir', wie wir das übrigens in altägyptischen Zaubersprüchen sogar in erpresserischer Manier finden (→62). Auf Grundlage der mesopotamischen Keilschriftquellen und der archäologischen Funde besteht ein starkes Übergewicht an Opferkulten mit ausgedehnten Riten. Zwar dienten auch im christlichen Mittelalter heilige Vorbilder und Gestalten und ihre Symbole und Riten als Vermittler zur Gottheit, aber es bestand hier eine therapieimmanente Konvergenz auf Christus als Wunderarzt und als verläßlicher Erlöser. Kathartische Elemente in der mesopotamischen Heilspruchpraxis und Dämonenabwehr streuen viel stärker die Affekte von Zorn und Wut, von Vergeltung und (ohnmächtigem?) Hass ein, zumal die Vielzahl der von den Göttern abhängigen Dämonen wahrscheinlich einem zwiespältigen Denken über Schicksalsfragen Vorschub leistete.

C Psychoperformative Hirnleistung und Zauberspruch

C1 Das Bannen der Dämonen durch ihre Nennung und durch die Schilderung ihres Wirkens – Strategie der internen und externen Emotionsregulierung mittels Etikett und Metapher am Beispiel des Alptraums und verwandter Affektionen

Die hier vorgenommene Kapiteleinteilung weicht von den bisherigen Übersichten der bearbeitenden Germanisten und Kulturforscher über die Spruchtexte bei weitem ab. Die neuropsychologischen Wirkfaktoren stimmen mit den kulturgegebenen inhaltlichen und den sprachformalen Textinhalten nur selten überein – das ‚Böse-Augen'-Syndrom ist eine Ausnahme. Kriterien, die die Zaubersprüche in Gebet, Segen und Beschwörung[239] einteilen oder in Befehl, Analogie und Erzählung[240], haben für die Bedingungen, unter denen das Gehirn arbeitet, nur eingeschränkte Bedeutung. Zum Beispiel unterliegt die Wirkung des Münchner Nachtsegens nicht allein den Befehlsformen, sondern auch der Dämonenbenennung und der ausgiebigen Psalmenlesung und ist zugleich Erzählung. Ebenso griffe eine Kategorisierung des Heilgesangs des Schamanen mit Mu-Igala für Gebärende in seiner archaischen Physiologie nach Kriterien von Erzählung, autoritärem operativen Befehlsstand und Klagelied zu kurz. Eine neurophysiologische Betrachtung ist zunächst nur im Blick auf Einzelelemente dieser Texte möglich. Außerdem ist in den vorausgehenden Ausführungen deutlich geworden, welche natürlichen Grundlagen die Kommunikation zwischen Heilkundigen, Kranken und Gesellschaft auch in der Vergangenheit geprägt haben. Behandelt wurden nicht Geister, sondern Kranke in Not.

Zu den psychohygienisch für alle Lebenslagen wichtigsten Voraussetzungen zur Erhaltung des inneren Gleichgewichts gehört die Registrierungspotenz für jegliches ankommende Übel im Sinne eines Sich-bewußt-Machens eigener Gefühlsreaktionen und – à la longue – das Durchdenken des jeweils eingetretenen

---

239 Hampp, Irmgard: Beschwörung Segen Gebet, Stuttgart 1961.
240 Holzmann, Verena: „Ich beschwör dich wurm vnd wyrmin ...". Bern u. a. 2001, S. 52–55.

Psychische Taktik und Strategien
moderne Begriffe:
- Etikettierung der Gefühle (Labeling emotions)
- Umwandlung vager Angst in gerichtete, verhandel- und erzählbare Phobie
- Appraisal = Primäre Evaluierung von Sinn und Bedeutung aller Reize auf Seele und Körper; Reappraisal = bewußte Emotionsregulierung mit Neuinterpretation der Stimuluswirkung zur Herstellung eines seelischen Gleichgewichts
- Stressbewältigung (Coping)
- Benennung des Ungewissen
- Repetitive Exposition des Angstobjekts und Konditionierung: siehe Kap. C4/2

historisch/ volkstümlich/ alltäglich:
- Erstellung eines Arsenals von (geheimen) Waffen gegen Schicksal oder Peiniger, offene Wortkanonaden, verbaler Befreiungsschlag, Abwehr- und Gegenzauber
- Befreiung aus „Horror vacui" (seit Aristoteles),
- Verfluchung, Verbannung und Beschwörung eines Übels, personalisierbar, personalisiert oder versinnbildlicht
- exorzistische Techniken und kirchlicher Exorzismus s. Kap. C4.3 und C5
- Klagelieder (Lamenting) an die Götter (Mesopotamien)

Anwendungsfelder (Indikationen) für den Einsatz von präventiven und therapeutischen emotionsregulierenden Instruktionen
Alle Formen der Angst, Erregungszustände, Paniksyndrom, Schlafstörungen, Parasomnien, Alptraumsyndrom (mesopot. Elinitu), akute dissoziative Symptomatik, akuter Schmerz, histrionischer Anfall, traumatische Blutung; (Burn out, Mobbing)

Neurophysiologie der Verarbeitung und der therapeutischen Verfahren
- Entlastung des limbischen Systems, Aktivierung medio-frontoventraler Gebiete und des Cingulums (obere Anteile: appraisal und expression, untere: Regulation) nach Übersicht Etkin, Amit (2011)
- Reziproke Beziehung zwischen rechts praefrontolateralem Gebiete u. Amygdala nach Lieberman, Matthew D. (2011)
- Embodied Simulation per adäquater Metaphorik via sensomotorischem System als Funktion der Spiegelneurone

Tabelle 1: Kognitive Verarbeitung negativer emotionaler Stimuli

Gefühls. Der Gesunde vermag Beleidigung und negative Kritik meist kognitiv, d. h. unter Einsatz der cingulären und frontalen Rindenregionen, zu verarbeiten. Er nimmt sprichwörtlich den Stier bei den Hörnern, indem er den *eigenen* Stier packt, er wendet Elefanten in Mücken. Es scheint zwar selbstverständlich, daß die Herkunft der Bedrohung und der Hort der Dämonen ins Panikergehirn zu verlegen sind; man kann also etwa als geübter Beter oder Meditierer jeden von außen kommenden Angriff als Echo oder Revanche auf eigene discordante Hirnaktivitäten verstehen. Aber aus dieser Sicht verlangt die Praxis oft allzu Heroisches.

C1.1 Zur Neurophysiologie der Emotionsverarbeitung

Unter Neurophysiologen hat sich nach circa einer Dekade der Forschung zur Analyse der Verarbeitung von Gefühlen die Theorie einer wechselseitigen (reziproken), also einer Wippenfunktionalität zwischen Amygdala und limbischem System einerseits und rechter praefrontaler Rinde andererseits entwickelt,[241] wobei die Rolle von Cingulum und medialer Praefrontalrinde im Sinne einer Vermittlung vermutet wird.[242] Es überwiegen Ergebnisse, die eine Betonung der Emotionsverarbeitung in der rechten Praefontalrinde darstellen, wobei auch klinische Befunde bei rechtsseitigen Apoplexien heranzuziehen sind, die häufiger mit Alexithymie, d. h. mit einem Verlust der Wahrnehmung der Gefühle anderer einhergehen. Man weiß heute, daß ein Teil seelischer Krankheiten auf ererbten oder erworbenen Leitungsstörungen im Bereich dieser limbisch-corticalen Verbindungen zurückzuführen ist und daß die individuellen Unterschiede in der Verarbeitungsmöglichkeit von Psychostress durch Differenzen in der Konstruktion dieser Amygdala-Praefrontal-Koppelung beruhen.[243]

Im Gegensatz zur aktiven, den Störungsfall überdenkenden Emotionsverarbeitung korreliert eine forcierte Unterdrückung von Gefühlen mit Aktivierung und Volumen der dorsomedialen praefrontalen Rinde.[244] Gegenüber einer solchen Gefühlsunterdrückung ergaben bei Untersuchung von Prüfungsstress die emotionsverarbeitenden Strategien bessere Möglichkeiten, negative Emotionen zu beeinflussen, weil das limbische System damit stärker deaktiviert ist.[245] Wer

---

241   Ochsner K N, Bunge S A, Gross J J, Gabrieli J D E, Rethinking feelings: An fMRI study of the regulation of emotion, Journal of cognitive neuroscience 14 (2002), 1215–1220, doi:10.1162/08989290 2760807212; Lieberman et al. (2007) Putting feelings into words: Affect labeling disrupts amygdala activity to affective stimuli, in: Psychological Science 18 (2007), 421–428; ders. et al., (2011) Subjective responses to emotional stimuli during labeling, reappraisal and distraction, in: Emotion 11, 468–480.
242   Etkin, Amit et al. (2011) Emotional processing in anterior cingulate and medial prefrontal cortex, in: Trends in cognitive sciences, doi:10.1016 /j.tics.2010.11.004.
243   Lee, H. et al., Amygdala-prefrontal coupling underlies individual differences in emotion regulation, in: Neuroimage 62 (2012), 1575–1581.
244   Kühn, Simone et al. (2011), „Keep calm and carry on", in: PLoS One 6(1): e16569.doi: 10.1371/j.pone.
245   Baron, Katja: Der Zusammenhang zwischen kognitiver Emotionsregulierung und der Fähigkeit, negative Emotionen zu beeinflussen, eine fMRI-Studie, Diss. Ulm 2011.

Emotionen unterdrückt, bekommt eine stärkere Amygdalaaktivierung, wenn er von traurigen Bildern oder Ereignissen loskommen möchte,[246] während eine aktive Verarbeitung von persönlichen Erlebnissen (Recalling) im Reappraisal, auch per Imaginationen, zu gezielten neuralen Aktivierungen in kognitiv verarbeitenden Regionen messbar war.[247]

Wir können insgesamt folgern, daß die angelegte neurale Hardware-Architektur dieser Verbindungen bei der Entwicklung und Reifung des Menschen qualitativ und quantitativ durch Reappraisal-Aufgaben und Herausforderungen kulturell angemessen formbar ist und geformt wird. Daß andererseits individuelle Verarbeitungsstile für Erzählungen vorkommen – jedenfalls für den Hörer unserer Zeit – hat kürzlich eine italienische Forschergruppe nachgewiesen.[248] Gegenüber einer Placebowirkung nimmt Reappraisal lediglich andere neurale Pfade; gemeinsam ist ihnen die Abschwächung der Amygdalaaktivität rechts.[249]

Schließlich scheint auch die Wirkung von cognitive reappraisal auf den Immunstoffwechsel nachweisbar zu werden.[250]

C1.2 Psychologisch-psychiatrische Praxis von Emotionsverarbeitung

Im Praxisalltag begegnen heute dem Arzt akut Patienten nach einer Mobbing-Attacke im Betrieb oder nach familiärer seelischer Verletzung (Scheidungsandrohung) in unterschiedlicher Weise: Einerseits schildert ein abrupt entlassener Mann zornig: „Mir fielen gleich alle erdenklichen Waffen ein, ganze Arsenale!" Andererseits läßt eine seelisch niedergeschlagene Frau und Mutter ihre Rat- und Sprachlosigkeit spüren; sie verarbeitet ‚passiv'. Diese beiden Gegenpole möglicher Reaktionsmechanismen lassen im ersten Fall das automatische Etikettieren des Zorngefühls mittels einer Metapher (alle Waffen) erkennen, hier wird praefrontal und unter Einsatz des sympathischen Nervensystems kämpferisch

---

246 Vanderhasselt, Marie-Anne et al., Interindividual differences in the habitual use of cognitive reappraisal, in: Biological Psychology (2012) doi:org/10.1016/j.
247 Holland, Alisha C. and Elizabeth A. Kensinger, The neural correlates of cognition reappraisal during emotional autobiographical memory recall, in: Journal of Cognitive Neuroscience 25 (2013), S. 87–108.
248 Benelli, E. et al., Emotional and cognitive processing of narratives and individual appraisal styles, Front. Hum. Neurosciences 2012; 6:239.
249 Wencai Zhang et al., Neural mechanism of placebo effects and reappraisal in emotion regulation, in: Progress in Neuro-Psychopharmacology and Biological Psychiatry 40 (2013), S. 364–379.
250 Diess.: (2011) The integrative effect of cognitive reappraisal on negative affect, PLoS One 7(2):e30761.

aktiviert mit u. U. Blutdruck- und Pulserhöhung, verstärkter Muskelspannung und erhöhter Konzentration, im Doppelsinn von ‚Putting feelings into words' (Lieberman, M.D.). Im zweiten Fall, einer protrahiert perzeptiven Reaktion, zeigt sich eher die persistierende Aktivierung des limbischen Systems, also eine leidbetonte abwartende Gefühlsreaktion unter Belastung des parasympathischen Systems, das zu Hilflosigkeit, Erstarrung und Halskloß führen kann.

Die Selbstregulierung von Gefühlen – deren Minderung, Erhaltung oder Förderung – umfaßt eine Flut potentieller Verarbeitungsvorgänge. Nach einer der Theorien zur Emotions-Regulierung[251] werden 5 Stadien des Prozesses unterschieden, die zur Verbesserung und Ausbalancierung als bedrohlich erlebter Ereignisse führen sollen: Situationsauswahl (Person, Ort usw.), Situationsmodifizierung, Aufmerksamkeitsentwicklung, kognitiver Wechsel, (d. h. Bedeutungszuweisung) und Antwortmodulation. Diese Ordnungsprinzipien können für eine Notfall/ Krisen-Situation nur bedingt gelten, da eine freie Wahl der Situation entfällt. Dafür tritt eine Außenlenkung ein.

Wo eine einmal eingetretene Krisen-Notfallsituation mit Erschöpfung eigener emotioneller Regulationsmechanismen einhergeht, da setzt die therapeutische Unterstützung eben in diesem Bereich an. Sie muß mit Hilfe verbaler und/oder ritueller Methoden eine entsprechende ersatzweise quasi stellvertretende Strategie einleiten. Der Therapeut bleibt – ob er mehr will oder nicht – Stellvertreter in der Funktion der Unheilstaxierung. Er betreibt in der situativ definierten passagèren cerebralen Symbiose die Einschätzung der Bedeutung und die Benennung des Übels für ein geschwächtes neurales Teilsystem, indem er die gefühlsprovozierenden Eigenschaften (Zorn, Angst, Ekel, Entmutigung), zu erkennen und konspirativ zu benennen hilft, ob abstrakt als Gefühlsbenennung oder metaphorisch mit kulturgängigen Ausdrücken. Die Verbalisierung, das Aufschreiben oder das Erlesen von Emotionen und deren abgrenzende Unterscheidung von Kognition und Sprache (emotional awareness) sind heute wichtige Strategien und Spezialisierungen in vielen Formen von Psychotherapie geworden. Von der Förderung retardierter Kinder über die Schmerztherapie, die Behandlung von Verstimmungen reaktiver Art (nicht endogene Depressionen), die Therapie gewaltbereiter Jugendlicher bis hin zur Briefschreibtaktik an den personalisierten Tinnitus und zu Führungsaufgaben in der Wirtschaft werden Methoden der Emotionsregulierung weithin seit langem angewendet.

<small>Theoretisch beziehen sich psychologische Arbeiten zur kognitiven Bewertung von Emotionen zumeist auf das transaktionale Stressmodell von Richard Lazarus (1974). Nicht mehr die objektive Reizbeschaffenheit, sondern die subjektive Bewertung des Reizes ist zum Angelpunkt der Stressforschung geworden. Ein personaler Wahrnehmungsfilter selektiert Reize, führt sie</small>

---

[251] Gross, James J. (Hg.) Handbook of Emotions³, NY London 2008.

primärer und sekundärer Bewertung zu und analysiert ggf. die verfügbaren Ressourcen. Verschiedene Stressbewältigungsstrategien (Coping) entscheiden über Umgangsformen mit jeder Herausforderung. Lazarus gilt als Pionier der Emotionsforschung. Er hatte zunächst seinen Versuchspersonen Filme über Genitalverstümmelungen bei Aborigines vorgeführt, was zu starken Stressreaktionen führte. Bei verharmlosenden Kommentaren oder bei vorheriger Erklärung blieben dagegen die Emotionen gedämpft. Diese Ergebnisse sind immer wieder bestätigt worden. Sowohl für Prävention als auch für Therapie. Z. B. wurde bei depressiven Syndromen experimentell gezeigt, daß die zu Depressionen neigenden Personen genauso erfolgreich mit Instruktionen gegen eine Verstimmung unter Filmclips getriggert werden konnten wie Gesunde. In den Versuchen wurde etwa eine Instruktion zu Reappraisal angeregt: „Bitte versuchen Sie, eine neutrale und unemotionale Haltung einzunehmen, wenn Sie den Film [Unfalltod und Trauerreaktionen] sehen. Objektivieren Sie das Gesehene. Stellen Sie sich vor, ein Direktor zu sein und prüfen Sie die technischen Aspekte, wie man Stimmungen produziert und welche Kameraeinstellungen verwendet werden müssen. Meiden Sie dabei Gefühle!"[252] Zuletzt wurde auch der Vergleich realer und fiktiver grausamer externer Vorabbildungen geprüft: Selbst fiktive Reappraisal-Strategie führte zu einer bilateralen Aktivierung des ventrolateralen Cortex bei gleichzeitiger Minderung der (emotionalen) Amygdala- und Insel-Aktivierung.[253]

C1.3 Versuch historisierender Einblicke in die Emotionsverarbeitung im Mittelalter

Nicht viel anders als in der Gegenwart haben die Gehirne der Menschen des Mittelalters Bedrohung und Katastrophe erlebt. Feuersbrände, Überschwemmungen, Krankheiten und Verwundungen waren nicht durch merkantile Versicherungen abgefedert. Familiäre Ereignisse, kriminelle Überfälle, Kriegseinflüsse, Isolierung und andere äußere Anlässe und Störungen im Lebenslauf konnten nicht immer bewältigt werden und waren oft existenzbedrohend, nicht nur in materieller Hinsicht. So konnte eine Krisenintervention[254] mit dem Mittel der emotionalen Regulierung lebenserhaltend sein. Selbst- oder Fremdaggression, Flucht ins Niemandsland aus der Gesellschaft oder psychosomatisch-vegetative Schockreaktion bei Panikspannung konnten eine Krisenintervention erfordern. Damals wie heute gelten für den Helfer Akzeptanz des Betroffenen, Empathie

---

252 Ehring, Thomas et al., Emotion and vulnerability to depression: Spontaneous versus instructed use of emotion suppression and reappraisal, in: Emotion 10(2010), 563–572.
253 Mocaiber, T.A. et al. Antecedent descriptions change brain reactivity, (2011).doi.org/10.1016/j.neuroscience.
254 Wir verwenden diesen Begriff ungeachtet seiner modernen Bedeutungsfixierung. Heute wird zwischen Notfall und Krise unterschieden. Psychosoziale Notfälle werden als akute gesundheitl. Gefahr eingeschätzt, „im Gegensatz zur Krise." (Die Notfallmedizin, Hg.: Burkhard Dirks, Heidelberg 2007, S. 314–319).

und Echtheitsinszenierung sowie der Versuch, die Situation zu strukturieren als Voraussetzung für einen wirkungsvollen Einsatz.

Beim Verarbeiten allerdings hat kulturelle Prägung inhaltlich – in wieweit auch im Gehirn strukturiert, wissen wir nicht – wesentlich andere Vorgaben bereitgestellt als die Gegenwart in der westlichen Welt. Der Mensch des Mittelalters konnte, wie oben beschrieben, vieles oder alles in seinem Dasein als Zeichen verstehen. Daraus entstand eine Form der Emotionsregulierung, die erlittenes Leid, Schmerz und Angst, eigenes Versagen und von anderen verursachte Versagung („Frust") in eine elementar demütige christliche Haltung einbetten konnte. Eine habituelle emotionale und zugleich soziale Konditionierung seit der frühesten Kindheit konnte bis hin zum inneren Dialog mit Märtyrern und mit Christus gehen, als eine religiöse Form der Versicherung. War also bei einem kranken Menschen im Mittelalter die Fähigkeit zu eigenständiger Verarbeitung (coping) gesenkt, sei es anlagemäßig, sei es infolge hirnorganischer Veränderung oder bei Extrembelastung, so lag eine religionsgeprägte ‚Nachhilfe' durch Heilspruchtexte genauso nah wie heute eine der vielen Individualtherapien. Allerdings waren Psalmen, Gebet und kirchliche Segen zunächst nur Basisebene für die kulturelle Habituation, darüber hinaus mußten weitere Stilmittel in diese gesprochenen Texte genommen werden, die zu Psychoperformation oder Perturbation führten.

Als erste Beispiele für viele Krankheiten wählen wir eine auch heute noch namentlich bekannte und hirnorganisch gut definierbare Erscheinung: das Alptraum-Syndrom,[255] Phänomen einer Implosion psychoinflammatorisch aufgestauter Emotionen mit Angst und mit somatischen Sensationen. Im **Kapitel C4.3** folgen Syndrome, die den Schmerz behandeln und damit der zentralen christlichen Idee der Leidensüberwindung noch näher kommen. **Kapitel C5** widmet sich der beschwörenden und bannenden Expulsion der Dämonen.

> Das Alptraum-Syndrom wird heute – wie oben für die allgemeine Emotionsverarbeitung ausgeführt – als eine Funktionsstörung in präfrontalen und frontolimbischen Verbindungen mit erblicher Komponente aufgefaßt. Viele Untersucher haben vermutet, daß Tagesreste im REM-Schlafstadium emotional nicht richtig verarbeitet werden können. Bei häufig von Alptraumstörungen betroffenen Personen ließen sich zuletzt außerdem durch Testuntersuchungen Veränderungen finden, die zusätzlich auf Verlangsamung exekutiver kognitiver Funktionen hinweisen, als ob auch eine Hemmung semantischer Verarbeitung vorliege.[256]

Neben den großen Solitärtext des Münchner Nachtsegens (**Beispieltext ➔1**), der mit der ergiebigen Zusammenstellung von über zwanzig einheimischen Dämo-

---

255 Vgl. Ernst, Wolfgang: Beschwörungen (2011), S. 250–263.
256 Simor, Péter et al., Impaired executive functions in subjects with frequent nightmares, in: Brain and Cognition 78 (2012), S. 274–283.

nennamen und deren Tätigkeiten, die sorgfältig in religiöse Zeilen zuvor und danach eingebunden sind, eine Sonderrolle spielt, stellen wir ein kurzes gewöhnliches Abendgebet (➔2) als einen der Nachtangst vorbeugenden Text aus einer Gebetssammlung, um auch die oben beschriebene allgemeine christozentrische Haltung anzudeuten. In zwei weiteren Nachtsegen (➔3 und 4) findet sich keine größere Geisteraufzählung. Beispiel 3 entstammt allerdings einem Sammelband mit einem gewaltigen Beschwörungshandbuch,[257] das 189 nicht einheimische böse Geisternamen, gefallene Engel und Zirkelmagie verzeichnet, Exorzismen, Alchemie, Astrologie und Nekromantie, also magisch-naturbezogene Texte, während der Münchner Nachtsegen (➔1) in einem Codex mit physikalischen, medizinischen und phytotherapeutischen Texten ruht. **Beispieltext ➔4** ist einem rein medizinischen Buch entnommen.

Die ältesten Belege für die Verbannung der Alpgeister finden sich in exorzistischen Texten, z. B. in einem Amulett gegen Besessenheit um 800: „Adiuro te satanae diabulus, aelfae, per deum uiuum ac uerum, et per trementem diem iudicii, ut refugiatur ab homine illo ...."[258]

In fast allen Methoden der modernen Psychotherapie spielen Träume und Alpträume theoretisch wie praktisch eine große Rolle. Die Gestalt-Therapy nach Friedrich S. Perls, die Funktionale Psychotherapie und die Transaktionsanalyse nach Eric Berne zählen Träume zu den ‚Abbildern' und ‚nächtlichen Übungen' der Emotionen. Sie werden heute auf verschiedene Weise bearbeitet, teilweise durch Ausspielen des Traumes im Psychodrama als Rollenspiel, teilweise durch verbale Vergegenwärtigung und in Konfrontation. Die Symptome des Alptraumsyndroms und seine Behandlungsversuche sind aus allen Kulturen seit Jahrtausenden bekannt. Es wird sich zeigen, daß einige Parallelen zwischen modernen und alten Therapiemethoden verblüffend sind.

Kasuistik (Modell, 13./14. Jahrhundert): Bei der Frau eines Schustermeisters, der seine Werkstatt unmittelbar an der Stadtmauer hat, sind in den letzten Jahren schon wiederholt nächtliche Ängste aufgetreten. Sie fühlte ein Einschnüren der Brust, Luftnot wie Erstickung und hatte das Gefühl, von einer groben bedrängenden Masse oder einem Wesen, das sich auf sie setzt und sie krabbelt, belästigt zu werden. Ihr Mann hatte sie beruhigt, so etwas komme auch bei der Nachbarin vor, es sei nur der Mahr, der Alp, der sie aufdringlich drücke. –

Eines Tages kommen fremde randalierende Horden, überfallen die Wachen am Stadttor und das Geschäft des ehrbaren Schusters, der sich nur verteidigen

---

257 Kieckenhefer, Richard: Forbidden Rites, A Necromancer's Manual, Pennsylvania State Univ. Press 1979: „The compiler was far from squeamish about invoking clearly and explicitly fallen angels" (S. 155).
258 Franz, Adolph, Benediktionen II, S. 578.

kann, indem er Werkzeuge und Stiefel auf die Eindringlinge wirft und der von den Verbrechern schwer verletzt wird. Während nun ihr Mann einige Tage im Spital versorgt werden muß und die Söhne zur Wache abgestellt sind, ist die Frau auf sich selbst gestellt allein im Haus. Die eingeschlagenen Fenster und Türen sind noch nicht repariert. Die Talglichter sind aufgebraucht. Im Schlaf wähnt sie diesmal zusammen mit der sie erstickenden Masse auch Scharen von tierisch fauchenden und stechenden Gestalten, die über sie steigen und wähnt träumend schließlich, daß ihr Mann, der ja Gespenster seiner Nachbarin kennt, sie jetzt mit dieser betrügt. Andere Nachbarn hören sie schreien, ein Symptom, das über den einfachen Alptraum hinausgehend zum Syndrom des „Schlafterrors"bzw. des Angsttraums gehört, und rufen einen Franziskaner, der sich ja auf Geisteraustreibung verstehe. Der Pater zündet im Morgengrauen drei Lichter an, setzt sich neben ihr Bett, fordert sie auf, nur das Kruzifix zu betrachten, spricht die üblichen Gebete und einige Psalmen. Er kennt eine spezielle geisterabwehrende Beschwörung und flicht sie zwischen die Psalmen ein.

Der Mönch hat eine mehrstufige ‚operative' Verbaltherapie gegen den Alptraum zu leisten, die er zu seiner Zeit wahrscheinlich als Benedictio, als Segen mit exorzistischem Beiwerk und nicht als ‚Traumbearbeitung' im heutigen Sinne benannt und verstanden hat. Er arbeitet nach heutiger Sicht auf mehreren Ebenen synchron: Einsatz von spiritueller Zeichenhaftigkeit des Bedrohlichen und des Erlösenden, Identifikation mit der/dem Betroffenen mittels Empathie sowie ‚Psychotherapie' im Sinne einer persönlichen Stellvertretung für die externe Rationalisierung von Emotion. Speziell wird vom helfenden Mönch mit seinem Text autoritär ein konfrontatives und zugleich imitatives ‚archaisches' Geisterdrama inszeniert. Dabei geschieht gemäß historischem Hintergrund Folgendes:

C1.3.1 Einstimmung auf die durchlittene Nacht

Die Nacht ist Symbol und Lebenswirklichkeit für all das, was Finsternis und Unheimlichkeit bedeuteten. Für den Mensch des Mittelalters war die Nacht Abgrund und Nichts, Unwissen und Chaos, eine geheimnisvolle und numinose Macht. Sie konnte ihn in seine Blindheit und Verworfenheit verstricken. Im Alten Testament und im Judentum ist Finsternis ein Reich des Satans so wie in der mesopotamischen Kultur eine Nacht der Dämonen. Das Christentum hat aus der Antithese zum Tag die Weihnachts- und die Osternachtsliturgie geschaffen. Der Ostermorgen mit seinem ersten Sonnenlicht ist auch heute die heiligste Stunde des Jahres für viele Christen. Eine Verbannung schriftlich tradierter, volkstümlicher und heidnischer Geister ist im Mittelalter der oft allzu bemühten Kirche in Jahrhunderten nicht gelungen. Zum Schutz vor den Gefahren der Nacht und ihren Geistern, seit Augustinus theologisch in ihrer Existenz unbestritten, hat das Mönchstum Psalmen aus dem Judentum übernommen und Gebete verwendet, die ein fester Bestand ihres Lebensrhythmus wurden. Besonders Psalmen prägten die Stundengebete zur Nacht – so auch Ein- und Ausgang des Münchner Nachtsegens –, weil mit ihnen gegen die Versuchungen des Bösen und der Phantasie um Hilfe gebetet wurde. Psalmen hatten eine große und weite Bedeutung und wurden auch von Ärzten (Arnald von Villanova, Nicolaus Myrepsus) als Behandlung empfohlen und angewandt. Ich zitiere einen der bekanntesten Psalmen, der im Münchner Nachtsegen (➔**1, Zeile 67**) zum Einsatz kommt und der die Erwartung des Morgens mit einem Nachtwächter einschließt: „Bi dem ‚De profundis'"

> Psalm 130, 1–8 nach der Einheitsübersetzung
>   Aus der Tiefe rufe ich, Herr, zu dir.
> Herr, höre meine Stimme! Wende dein Ohr mir zu
> Achte auf mein lautes Flehen!
>   Würdest du, Herr, unsere Sünden beachten,
>   Herr, wer könnte bestehen?
> Doch bei dir ist Vergebung, damit man in Ehrfurcht dir dient.
>   Ich hoffe auf den Herrn, es hofft meine Seele,
>   ich warte voll Vertrauen auf sein Wort
> Meine Seele wartet auf den Herrn mehr als die Wächter auf den Morgen
> Soll Israel harren auf den Herrn. Denn beim Herrn ist die Huld,
>   bei ihm ist Erlösung in Fülle.
> Ja, er wird Israel erlösen von all seinen Sünden.

C1.3.2 Die Insinuation des Charakters von Träumen

Dualistische Zeichen sowie Zitierung (Evokation) der möglichen Traumgeister in Form von Metaphern versorgten eine (in unserem Modell) reaktiv ausgelöste emotionale Krise mit phobischen und paranoiden Inhalten. Die Gefahren der Nacht lagen wider alle historische Gewichtung verschriftlichter mythologischer und kultureller Ausgestaltung ihrer cerebralen Produkte im Reich der Träume und Phantasien. Ob die Inhalte von Alp- und Angstträumen Ursache oder Folge von Geisterglaube sind, entspricht der Frage zu Henne oder Ei. Sicher ist, daß der Traum neben den mit Halluzinationen und anderen Hirnstörungen einhergehenden Krankheiten die wichtigste und einzig natürliche – nicht literarische – Geisterspedition über Jahrtausende war. Der Traum konnte als katechetisches Instrument immer auch zu den von höherer Macht gegebenen Zeichen eingesetzt werden. Seine biblischen Geschichten und seine Darstellungen in der Kunst waren ein wichtiges Mittel der Christianisierung ehemals heidnischer Vorstellungen. Man denke etwa nur an Jacobs und Josephs Träume, an den Traum der

Drei Könige, der Frau des Pilatus oder vieler Heiligen der Kirche. Selbst päpstliche Politik hat Träume in ihren Dienst gestellt, um ihr Gebäude zu stabilisieren.[259] Nichts lag zugleich dem persönlichen Erleben jedes Menschen näher als sein Auftauchen und nichts schien pragmatisch unerklärbarer und geheimnisvoller. Dabei standen die vom Traum vermittelten Zeichensetzungen immer an einem Schnittpunkt. Denn auch der Satan, der Dämon, das personalisierte Böse konnte sich einmischen und schuf den Alptraum mit seinen negativen Emotionen, die zugleich im Volk als Elbenglauben und als Dichtung eine oft ausgedehnte Erzähltradition begründeten.[260]

> Während in der Antike Traumdeutungen zur Diagnostik gehörten, gibt es in der mittelalterlichen Medizin kaum ein Beispiele für eine Verwendung unter Ärzten. Nur in literarisch-lexikalischer Tradition, die über die arabische Medizin mit Rhazes wieder nach dem Westen gelangte und aus griechisch-römischer Antike (Hippokrates und Galen) stammte, war eine Traumlehre überliefert. Sie war allerdings strikt an die Vier-Elemente-Lehre gekoppelt.[261]

Im Nachvollzug einer vom Helfer extern dargebotenen Geisterlitanei konnten Alptraumgeplagte durch Nacherleben ihrer schmerzhaften Horror-Phantasien, jetzt in einem anderen Umfeld, auf die ersten Schritte zu einer Klärung geführt werden. Dies entspricht der oben beschriebenen Etikettierung und der Umdeutung/Umwertung (reappraisal) des Emotionalen.

Auch in der modernen Hypnotherapie wird z. B. ‚Depression' zum ungebetenen ‚Hausgast' in verschiedenen Gestalten imaginativ personalisiert.[262] In unserem Nachtheilsegen wird eine ganze Palette von Gefühlen angeregt, kann vom Franziskaner vollständig oder ausgewählt gezielt dargeboten werden und wird mittels der Einkleidung in Geisternamen metaphorisiert: Zorn über die Dreistigkeit der Quälgeister, Scham über deren Annäherung, Angst vor deren vielen Kräften etc.. Als Metaphern können die Gefühle in den praefrontalen und sensori-motorischen Hirngebieten verarbeitet werden.

> Der Dichter Franz Werfel hat unter Rückgriff auf ‚archaisches' Vorgehen treffend die genannte Taktik (C1.3.2 und C1.3.3) als Umkreisung durch Benennung geschildert:
> Wart! Ich ermanne mich,   Die Silben, die dich sagen,
> Vogel und nenne dich,   Werden, Falken und Dohlen,

---

259 Deremble-Mannes, Colette, Die Traumwelt der Legenden in den Glasmalereien von Chartres, in: Bagliani/Stabile: Träume im Mittelalter, S. 41–54; Gardner, Julian, Päpstliche Träume, in: ebd., S. 113–124.
260 Siehe z. B. Grunewald, Eckard: „Der túfel in der helle ist úwer schlaf geselle", in Dinzelbacher/Bauer (Hg.): Volksreligion, S. 130–143.
261 Palmer, Nigel F., Von den natürlichen troymen, in: Janota, Joh. et al. (Hg.): FS Haug/Wachinger, S. 769–792.
262 Meiss, Ortwin, Hypnotherapeutische Verfahren in der Arbeit mit depressiven Patienten, in: Revenstorf, D. und B. Peter: Hypnose, S. 502.

Nennend verbrenn ich dich,	Dich fremden Krächzer niederholen.
Böser, und banne dich!	Oder aus meinem Garten jagen.
Mein Wort hat Macht die ich nicht weiß,	Worte, fliegt aus, die mein Nachtschlaf
Ich zieh um dich den starken Nennungskreis.	gewann:
	In den Bann mit dir, in den Bann,
	in den Bann!!

Die Versuche zum Verständnis der kaum hoch genug einzuschätzenden kulturhistorischen Bedeutung von Namensgebung, Namensnennung und zum Namensnativismus sind ein sehr altes Thema, das hier nur angedeutet wird: Für Michel Foucault nimmt der archaische Mensch ein Wort für ein Ding. Wort und Sache werden infolge einer von Gott gegebenen ersten ‚raison d'être' als radikal ähnlich betrachtet und Worte ähnlichen Klanges können folglich auch einen sachlichen Zusammenhang haben. Als die Namen von Gott gegeben wurden, „war die Sprache ein absolut sicheres und wahres Zeichen der Dinge, weil sie ihnen ähnelte. Die Namen waren auf dem von ihnen Bezeichneten deponiert, wie die Kraft in den Körper des Löwen."[263] – Nicht anders hatten es die frühen Philosophen von Aristoteles bis Augustinus beschrieben: Immer wieder wird die nativistische Fiktion der Namengebung als gemeinsame Existenzbasis für Schriftzeichen und gesprochenes Wort betrachtet, wobei Aristoteles ihr Konvergieren als Abbild von den Dingen und als Zeichen für die ‚Affektionen in der menschlichen Seele' (!) beschreibt. – Insofern konnte jeder Name als Seinsqualität aufgefaßt werden, und mit jeder neuen Benennung wuchsen die Attribute und wuchs die Macht alter Herrscher. Und wuchs die Kraft von Dämonen und von denjenigen, die deren Namen geheimhielten oder preisgaben. Benennen war zugleich Schöpfungs- und Ermächtigungsakt, im Johannesevangelium (In principio erat verbum) wie in den Ephesia grammata. Und für den Magier: Nach Malinowski sucht er Archaismus im Rekurs auf die heilige Sprache und bedient den ‚Koeffizienten des Unheimlichen'.

C1.3.3 Verbale Perduzierung einer verkörperlichten Simulation

Symptome des Brust- und Halsdrucks mit verschiedenen mobilen Aufdringlichkeiten und mit motorischer Körperlähmung bilden die somatische Seite des Alpdrucksyndroms. Der Münchner Nachtsegen und die meisten der exorzisierenden Texte stellen nicht nur die Namen der Quälgeister bloß – als Emotionsmetaphern, sondern auch deren Tätigkeiten. Das Gewimmel der Alben, Mahren und Wichtel erlaubt sich Drücken und Kneifen, beschreitet den Schläfer, reitet auf ihm, bläst und haucht ihn an, beräuchert und betastet ihn, verwirrt ihn und – besonders typisch für jedes Alptraumsyndrom – lähmt ihn durch Anlegen des Fußkrampfes (vuzspor = Fußsparr, Fußsperre). Der Segen beinhaltet damit auch das neuropathologische Phänomen der Trennung von Motorik und Psyche, dem motorischen Gefesseltsein im Schlaf bei psychischem Halbwachsein. Als Gegensatz ist das Schlafwandeln vorzustellen mit motorischer Entfesselung und psychischem Schlaf. Und auch hier finden sich wieder Parallelen in der modernen

---

263 Zit. aus Schulz, Monika: Magie oder, S. 176.

Hypnotherapie, wenn Patienten im Dialog mit Therapeuten ihre somatischen Beschwerden berichten und mit dem schon genannten ‚Hausgast' zu kommunizieren beginnen: „Was willst du, warum bist du da? Wieso tust du das.... ?" Der sich bedrückend als ein lähmender Fremdkörper breit machende Hausgast, der die Luft zum Atmen nimmt, wird wie im Exorzismus ausgefragt und auf diesem Wege entmachtet.

Der mittelalterliche wie der moderne Therapeut wiederholt verbal auch die durchlittenen somatischen Beschwerden und überführt in passagerer Symbiose mitfühlend die Sensorimotorik des Erlebten in einen neuen Kontext, in die Zone des Nachgemachten, der Simulation. Er wirkt als Katalysator, indem er eine Situation schafft, in der der Leidende das von ihm gemiedene Schmerzliche nacherfahrend verarbeiten kann. Wie oben (**Kap. B3**) ausgeführt, aktivieren Worte wegen der multimodalen Eigenschaft vieler Spiegelneurone die selben neuronalen Schaltkreise des sensomotorischen Systems wie eine Bewegung. Begriffe mit ihren Metaphern und Konnotationen sind im sensomotorischen System prämotorisch-parietal kartografiert und können in dieser therapeutischen Situation ihren Einfluß auf Körperfunktionen ausüben: Die vom Therapeuten perzipierte und gleichzeitig dem Leidenden reziprok induzierte szenische Vorstellung wird zur Aktion. Auf dieser Basis geschieht

C1.3.4 Stimulierung sensomotorischer Schaltkreise

Eine **verbale und rituelle Stimulierung sensomotorischer Schaltkreise**, die bisher als exorzistische Geisterbannung mittels Befehlen durch autoritäres Wort unter Berufung auf höhere göttliche Gewalt benannt wurde. Ich ordne sie einer extroversiven Katharsis zu und verweise auch auf **Kap. C5**. Im Crescendo der Befehle gegen die Alpgeister, vom anfänglichen einfachen Wunsch, „ir sult von hinnen gangin" über die Verbote, „ich verbite dir aneblasen ... kruchen unde anehuchen" bis schließlich zum „Ich beswere dich ungehure", steigert sich der Text (➔1) in einen Kraftakt.

Und während der Dichter Franz Werfel seine Nennungs-Arbeit als zwingend kraftvoll, als jagend, schildert, aber auch als unverstehbar, können wir modellhaft das etikettierende Abstempeln der Dämonen mit Namen als sinnvolle und erfolgversprechende psychoperformierende Taktik zur Erzielung einer emotional regulierenden Hirnleistung betrachten. Oder auch nach alter Vorstellung – ich verwende einmal bewußt diese Metapher – als einen ersten, inneren Zauberkreis. Gleichgültig, ob der Alptraumgeplagte diese Namen kennt und ob er sie als spirituelle Kräfte mit Aufforderung zu Buße, Einsicht und Bekenntnis versteht oder als Fehlalarm erträumter böser Außenseiter, er hat im Durchleben der nächtlichen Attacke eine massive Emotionsaufwallung erlitten, mobilisiert hormonell Cortisol und Oxytocin und sensibilisiert auch damit seine sensomotorischen Schaltkreise für die Perzeption und Verarbeitung der **aktionsbezogenen Sätze und Worte**. Der Therapeut verbindet die Dynamik des Ereignisses mit neuer Kraft und neuer Bewegungsrichtung zum einstmals zauberhaft verklärten zweiten Nennungskreis des Dichters. Dies ist der Wendepunkt, an dem moderne Psychotherapie über die Benennung der bösen Aufdringlinge im o. g. hypnotischen Prozess hinaus eine ‚kreative Aggression'[264] einsetzen könnte, jene autoritär geballte Kraft, die wir als extroversive Katharsis beschreiben und die in die vegetativen homöostatischen Regulierungen eingreifen kann (**Kap. C5**).

C1.3.5 Aversive Stimuli und ihre neurochemische Wirkung

Mit der aufzählenden Konfrontation von Nachtgespenstern und ihren Untaten vor der Kranken geschieht eine aversive, d. h. unangenehme oder schmerzhafte Exposition und Einflußnahme, die ganz im Gegensatz zu einer freudigen Nachricht steht. Es mögen auch schlafgestörte Kinder, denen die Eltern bei hartnäckigen Alpträumen[265] Monster imaginieren lassen, um sie dann zu bekämpfen, zunächst eher zu leiden haben. Derartige Stimuli wirken gemäß Tierversuchen aktivierend auf dopaminbildende Neuronenanteile des ventralen Tegmentums wie bei einer Akupunktur. Die Dopaminbildung wird im Nucleus accumbens und im dorsalen Striatum allerdings erst dann freigegeben, wenn der Schmerzreiz beendet wird, sodaß sich die Linderung aversiver Bedingungen letztlich als mit Belohnung vergleichbar erweist.[266] Daß parallel zu körperlichem Schmerz eben auch Schmerz durch Worte ausgelöst werden kann und daß beides in cerebralen Zentren überlappende Aktivierungen findet, wird in **Kap. C4.3.1** (introversive Katharsis) weitergehend beschrieben.

Ein Blick vergleichsweise in moderne Dichtung zeigt uns deren Nähe zu den ‚alten' Zaubertexten, wenn wir noch einmal die Benennung und die Personifizierung des Unheimlichen betrachten. weil uns nämlich „… nur ein Akt der Benennung über Dinge verfügen läßt. Wenn wir einen Namen für sie haben, werden sie denkbar und verhandelbar, weil erst die Lautgestalt eines Wortes die dazugehörigen, individuell gebildeten und abgespeicherten Konzepte auf- und herbeizurufen imstande ist. Bleibt etwas namenlos, so sind wir so stumm wie blind."[267]

---

264 Vgl. Bach, George R., in: Corsini, Handbuch, S. 571–586.
265 Faust, Volker und Günter Hole: Der gestörte Schlaf, Ulm 1991, S. 271.
266 Budygin, E.A. et al.: Aversive stimulus differentially triggers subsecond dopamin release in reward regions, in: Neuroscience 201 (2012), 331–337.
267 Schrott, R. und Jacobs, A. Gehirn und Gedicht, S. 223.

Diese Personifikation schafft Geschlechtsmerkmale, Farben und Eigenheiten, verleiht Präsenz und macht lebendig durch Vermenschlichung. Dem Grauen wird Gesicht, Knöchel und Emotion zugeschrieben, wie in Sylvia Plaths ‚Der Mond und die Eibe':

> Der mond ist keine tür. Er hat ein recht auf ein gesicht,
> weiß wie ein knöchel und so fürchterlich bestürzt.
> Es zieht das meer hinter sich her wie ein dunkles verbrechen; es ist still
> mit dem aufgerissenen O völliger verzweiflung.

All diese dichterischen, uns wie archaisch wirkenden Wortfiguren gelingen deshalb, weil wir „mit Humanem am meisten vertraut sind", weil wir „damit vereinnahmen können, was wir beschreiben" und letztlich weil wir „eine Projektion von sensomotorischen Schemata auf unpersönliche Situationen – einen metaphorischen Transfer" – leisten.[268]

C1.4 Versuch historisierender Einblicke in die Emotionsverarbeitung in den mesopotamischen Kulturen

„Es ist geradezu schauerlich zu lesen, wie diese Dämonen über Menschen und Tier herfallen, um ihnen ohne Gnade den Garaus zu machen." So schreibt 1938 Erich Ebeling, der Verfasser des Artikels ‚Dämonen' im Reallexikon der Assyrologie angesichts der von dem englischen Archäologen R. Campbell Thompson (1876–1941) 1903/04 herausgegeben Darstellung der umfangreichen britischen Tontafel-Sammlungen. Schränkt man Ebelings emotional getönten Rückblick auch nur auf kranke oder wesenhaft furchtsame Mesopotamier ein, dann lebten viele Bewohner des Landes zwischen Euphrat und Tigris über Jahrtausende inmitten einer großen Fülle von Geisterfiguren, denen man spezialisierte körperteilbezogene Funktionen und einen totalen Vernichtungswillen unterstellte. Ich verweise auf das oben im **Kapitel B4.2.3** Ausgeführte. Das Besondere an diesen Dämonen war neben ihrer (nach den Keilschrifttexten) Grausamkeit aber noch mehr ihre durchgehende Charakteristik als Totengeister, die sich meldeten, Wiedergänger, die jedem im Diesseits schaden konnten, die aber offenbar auch eine lebensbedrohliche Stimmung verbreiten konnten. Inwieweit die Rituale gegen diese Geister ein wesentlicher Bestandteil des alltäglichen Lebens waren und Gefühlsleben dauerhaft formten, muß dahin gestellt bleiben, weil wir aus den Keilschriften nur über Fiktionen individueller Biographien unterrichtet werden (**Kap. B4.2.4.**).

---

268 Ebd., S. 224, sowie 221–228.

C1.4.1 Zur Frage der Emotionsverarbeitung am Beispiel von Traumgeschehen und Nachtgeistern in den mesopotamischen Kulturen

Wenn die Schriftquellen auch nach einer heutigen intimitätsaffinen Betrachtungweise kein lebensnahes individuelles Gefühlsleben enthüllen, so kann die Fiktion der Gefühle zumindest als Resonanz oder Spiegelung herangezogen werden, zumal sie nach Meinung von Foster zwar nicht reflektiert, aber implizit erscheinen.[269] Die nach Foster „vielleicht" am besten belegbaren Gefühle im öffentlichen und rituellen Bereich waren Zorn und Hass der Götter mit ihren Dämonen gegen Öffentlichkeit oder gegen Einzelne für deren Missetaten sowie umgekehrt ebenfalls Zorn und Hassvermutung einzelner bei ihren Schädigern. Dass diese Dämonen wie auch die dämonisierten Schädigergestalten unter der Macht eines mythologischen Götterhimmels standen, mag nach heutiger Sicht zur Entstehung zwiespältiger Emotionen geführt haben. Dabei scheint allerdings die Spannung zwischen diesen performativen Mächten und dem „Handeln der Gottheit als freier Person [...] durch ein konsequent anthropomorphes Gottesbild gemildert" worden zu sein.[270] Was die hier speziell herauszustellenden Traumgestaltungen angeht, so hat sich die Forschung[271] teils ihrer wahrsagenden (divinatio, omina, Symbolik), teils ihrer natürlichen ‚lebensweltlichen' Ausprägung zugewandt.

> Unter strikter Betonung der Wahrsagekomponente hat Annette Zgoll nach linguistischen Analysen[272] auf das große Interesse der Mesopotamier am Traum hingewiesen. Sie hält nach ihrer Materialgrundlage eine zuvor[273] gegebene Einteilung in a) Offenbarungsträume der Gottheiten, b) Symptomatische Träume geistiger und körperlicher Zustände und c) mantische Träume (Wahrsageträume) für problematisch. Denn der Erlebnischarakter des altorientalischen Menschen beim Träumen sei generell nicht anders als bei seiner Welterfahrung im Wachen mit der von ihm erfaßten bzw. postulierten Struktur. Und es lasse sich im Vergleich nachweisen, daß mesopotamische Träume nicht in ihrer individuell geträumten, sondern in ihrer gedeuteten Fassung niedergeschrieben waren (S. 41, 43). Das aber – so ihr Ergebnis – bedeutet, daß ein Traum im alten Mesopotamien entweder Relevanz für die Zukunft besitzen mußte oder irrelevant gewesen sei, dies also ganz entgegengesetzt heutigen Traumtheorien (S. 234). Der Traum sei also nicht als Hirntätigkeit sondern als Botschaft höherer Mächte vorgestellt worden und auf die dahinter stehende Botschaft reduziert gewesen (S. 245). Der Traum wird damit als per-

---

269 Foster, B. R., The person in Mesopotamian thought, in: Radner/Robson: Handbook, S. 132.
270 Schwemer, Daniel: Rituale und Beschwörungen gegen Schadenzauber (2007) S. 2.
271 Herrn Prof. Dr. Daniel Schwemer, Würzburg, verdanke ich wichtige Hinweise auf die zugrunde liegende Literatur.
272 Zgoll, Annette: Traum und Welterleben im antiken Mesopotamien, Münster 2006.
273 Oppenheim, A., Leo: The interpretation of dreams in the ancient Near East, Philadelphia 1956.

sonale Begegnung mit höheren Wesen, mit Gottheiten aufgefaßt und als „nicht weniger real erlebt und empfunden als die Geschehnisse der Wachwelt" (S. 263). Folglich ist ‚Traumtherapie' im alten Mesopotamien für die Autorin nicht Therapie mit Träumen, sondern Therapie der Träume (S. 399). Diesem ihrem Ergebnis ordnet Zgoll auch alle Indizien für möglicherweise symptomatische neurophysiologische Indizien weitgehend unter, nicht nur visuelle und akustische, sondern auch vegetative und haptische Phänomene, geisterhaften Lufthauch, erlebte numinose Erschütterungen, Fesselungen, Starre, Angst- Schmerz- und Zornerlebnisse. Bis auf Ausnahmen sei deutlich, „daß der Träumer in mesopotamischen Sprachen nie ‚aktiver Täter' seines Traumes sein kann" (S. 76), zumal es zwar ein Wort für Traum, aber keines für ‚Träumen' gibt (S. 74). Sein Traum sei Außenereignis, weil der Mensch besucht werde.

Im mythischen Denken lag allerdings die moderne deutende Annahme eines Alptraums immer nahe, wenn etwa wie im Gilgamešepos Enkidu von grauenhaften Wesen gepackt in die Unterwelt verschleppt und mißhandelt wird.

Trotz der vermutlich hohen Bedeutung der zukunftsgerichteten Traumdeutekunst der Priester kann nicht außer acht bleiben, daß Träumen als natürliche aktive Tätigkeit des Gehirns bei allen Menschen und zu allen Zeiten auch unabhängig von kulturellen Prägungen seine eigene Dynamik entfaltet hat und persönlich-intimem Experiment und persönlicher Traumarbeit zur Lebensgeschichte, also zur persönlichen Vergangenheit unterlag. Denn kulturelle Prägungen waren in keiner Kultur eindeutig und monokausal, und prognostische Träume dürften auch in frühen Kulturen seltener vorgekommen sein als die naturgegebenen. Aus ethnologischen Untersuchungen an Völkern Polynesiens und Amerikas wurde zwar ein Phänomen deutlich, daß man psychoanalytisch als Deutungswiderstand im indigenen Denken benannt hat und das vor allem an der *Botschaft* der Träume auftrat, nämlich die Deutung als Prophetie und deren Chiffrierung nach einem Schlüssel. Und nur selten werden in Volksstämmen konventionelle individuelle den Alltag betreffende Zusammenhänge bei Traumdeutungen mitgeteilt. Es überwiegen kollektive Bedeutungssysteme, die dann einer normierten Symbolik entsprechen können.[274] Zu bedenken ist aber eben auch, daß selbst Götter zwiespältige Kundschaft übermitteln lassen, daß Dämonen lügnerische oder undurchsichtige Botschaften einflüstern wie im Mittelalter der schillernde Satan. Mindestens für unsere Zeit lag es für die Psychoanalyse nahe, die vielfältigen Doppelgesichtigkeiten und den Doppelsinn der Träume zu enthüllen.[275] Geträumt wurde im Gehirn schon immer, auch dann, wenn es keine Worte für diese Tätigkeit dieses in seinen Funktionen einst kaum entdeckten Organs gab.

Es interessiert hier der dem neurophysiologischen Geschehen näher liegende natürlich-symptomatische Traumaspekt, wenn seine Symptome auch nur indirekt erschlossen werden können. Im Gegensatz zu den als Traumdeutung zu bewertenden Texten scheinen vor allem jene Texte für die Ermittlung von Alptraum-

---

274 Vgl. Ahrens, Ullrich, Fremde Träume, Berlin 1996, S. 168–194.
275 Benedetti, Gaetano: Botschaft der Träume, Göttingen 1998.

phänomenen heranzuziehen sein, die als Abwehrrituale gegen Hexen und Zauberer und ihre Untaten dienten, also nicht von Verstorbenen, sondern von Lebenden aus dem Alltag, aus der Lebenswelt ausgelöst. Ich erinnere an die von Tzvi Abusch geäußerte Einschätzung ihrer sozialpsychologisch heuristischen Relevanz (**Kap. 4.2.3**). Dafür sprechen 1. der Gehalt an Klagen der sich von Übeltätern geschädigt fühlenden Kranken, d. h. psychosomatisch persönliche Klagen, mit denen man sich den Göttern oder dem persönlichen Gott hörbar machen konnte, nicht die kollektiv-normierten abstrakten Klagen über gesellschaftlich Verdrängtes, 2. die emotionale Komponente von Argwohn bis zu Furcht und Angst – oder anders: von Wut und Zorn sowie 3. die in dieser Art der Texte weitgehend fehlende Nennung konkreter Krankheitssymptome, was uns eine weniger voreingenommene Verschriftlichung vermuten läßt. 4. Diese Träume unterlagen zwar oft soziokulturellen Zwängen, wurden als unwichtig, schlecht und trivial oder als schambesetzt unterdrückt oder schwarzgefärbt[276], aber ließen sich insgesamt am Symbol der Hexerei abheften.

Entgegen den Untersuchungen von Zgoll hatte Sally Butler[277] argumentiert, daß nichtprognostische Träume, die in der Forschung als ‚symptomatische' (A.L. Oppenheim) oder allgemein als ‚irrationale' Träume benannt werden, vom Zorn der Götter, von den Zauberern, von Dämonen oder bei Krankheiten ausgelöst werden konnten. Sie seien aber in den Keilschriften im Details nur selten beschrieben worden (S. 6). Dabei habe es zwar viele Begriffe für ‚böse' Träume gegeben (Tabelle S. 28, es werden 11 Begriffe registriert, in Übersetzung: confused, unfavourable, obscure, very obscure, incorrect, bad, meaningless, frightening, false, bewildering, baleful), doch sei in der Forschung strittig, welche Kriterien über gewöhnliche und Angstträume hinaus einem Alptraum zuzuordnen seien. Verdächtig auf einen Alptraumhintergrund sind für Butler:

– ein Bericht über das Beworfensein mit Erde bei Hexerei, was den Reinigungsritus veranlaßt (S. 41, 114; 166); er mag auf einen Traum hindeuten, der an eine Selbstbeerdigung denken läßt.
– eine Inkantation gegen einen unbekannten Dämon (Lamaštu?) mit dem Eingang: „(Regarding the demon) who crossed the edge (?) of my bed; (who) frightened me (and) caused me to panic; (and who) showed frightening dreams to me, they deliver him (the demon) Biduh, the chief door-keeper of the Underworld ..." (S. 50; SpTU 3, Nr. 82 u. a.),
– wenn einer unter dämonischem Einfluß träumt, daß eine Stadt wiederholt auf ihn fällt und er seufzt, aber keiner hört ihn und „a very evil šedu is fastened to his body", sodaß der Exorzist den Dämon von seiner Beute vertreiben muß (VAT 7525,col. III, Z. 28–35).
– Neben dem Wirken von Lamaštu und Sedu, – beide in diesen Fällen nicht als heilbringende Geister – ist auch Alû als möglicher Auslöser von Nachtmahrträumen genannt (➜**40 und 41**).
– Ferner erwähnt Butler eine Art diagnostischen Text des Britischen Museums, den Campbell Thompson dem medizinischen Teil zugerechnet hat. Er belegt eindeutig, daß Alp-

---

276 Vgl. Ahrens, U. ebd., S. 192f.
277 Butler, S.A.L.: Mesopotamian conceptions of dreams and dream rituals, Münster 1998.

träume im modernen Sinne – nennen wir die Symptome modern Angst, Brustdruck, Fußkrampf und vegetative Reaktion – bekannt waren und als Zaubereifolge aufgefaßt wurden: „If a man's head constantly afflicts him; (in his sleep); his dreams are frightening, he spoke in his sleep; his knees are paralyzed; his chest has paralyses; (and) his skin (flesh) is constantly covered with moisture – this man is bewitched". (S. 54; AMT 86/1)
- Beim Zorn der Götter, insbesondere Marduks weist sie (S. 57f) auf das ‚Poem of the righteous sufferer' hin, eine Erzählung eines Kassiten, wo in Tafeln 1 und 3 ‚Träume' berichtet werden („I lay down, and my dream was frightening during the night"; „I wail day and night alike. [I am] equally wretched (in) dream (or) waking dream".)
- Spezifische Alpträume finden sich nach Meinung Butlers in spätbabylonischen Texten aus Uruk: „His dreams are bad; he repeatedly sees dead people in his dreams; he panics; his dream does not seize the eye; in his dreams he is like one who has sex with a woman and his sperm flows out." (S. 59 ; SpTU 2, Nr. 22, col.I Z. 19'–21). Und mit der häufigen Erscheinung von Totengeistern, die wie oben beschrieben, ihren Tribut fordern, zeigen sich quälende Träume auch in weiteren Tafeln:
- O Shamaŝ! The ghost, the one who frightens me, which has been fastened to my back for many days, and will not be loosened, constantly pursues (persecutes) me all through the day, (and) repeatedly frightens me throughout the night" (S. 59f; CT 16, col.V, Z. 34–53). Aber im allgemeinen erscheinen mesopotamische Geister eher in Träumen, als sie zu verursachen.
- Es erscheinen in Träumen durch Hexen oder persönliche Götter verursacht immer wieder auch einzelne direkt benannte persönliche Totengeister, ein Gott, ein König, ein Prinz u. a.:
„O Clod, the evil of the dream which I saw repeatedly (during) the first watch (of the night), the middle watch, (or) the third watch – (in) which I saw my dead father; (or in) which I saw my dead mother; (or in) <which I saw> a god; (or in) which I saw the king; (or in) which I saw an important person; (or in) which I saw a prince, (or in) which I saw a dead person; (or in) which I saw a living person" (S. 61; Clod-Incantation, Ashur dream ritual compendium ADRC, ed. here, col.II, Z. 2–4).
- Schließlich fügt Butler auch sexuelle Träume an, die sich nach Art von Incubus und Succubus die Mesopotamier mit Lilit- oder Lilitu-Figuren als Nachtmahr-Träume vorgestellt hätten (S. 62–65) (**Beispiel ➔44**).

Wie schon oben im **Kap. B4.2.3** gesagt, gab es eine Fülle derartiger Abwehrzauber (Oberbegriff: Kišpu). Das umfangreichste Ritual mit diesem Ziel ist Maqlû, übersetzt: das ‚Verbrennen'. Maqlûtexte waren weit verbreitet und wurden bisher in über 100 Exemplaren gefunden, davon zahlreiche Fragmente in der Bibliothek des Assurbanipal in Ninive. Sie entstammen überwiegend dem 1. Jahrtausend vor Christus. Das große Hauptritual dieses Typs mit seinen 1600 Zeilen auf 8 Keilschrifttafeln war zur Durchführung innerhalb einer Nacht vorgesehen, aber es gab auch kürzere Anteile und Varianten, sodaß nicht nur ein prophylaktischer Gebrauch zur Zeit des Neumonds zu vermuten ist. Die Zeremonie beginnt nach Sonnenuntergang. „Der Patient verbündet sich mit den Göttern der Unterwelt, die er bittet, die Hexen einzufangen und mit den Göttern des Himmels, die ihn reinigen wollen. Der Exorzist zieht einen magischen Zirkel zum Schutze eines

Schmelztiegels [...] und der gesamte Kosmos wird um Einhalt gebeten und um Hilfe für den Patienten, um seiner Klage abzuhelfen. Und es folgt eine lange Serie von Verbrennungsriten, in deren Verlauf die Figurine des verräterischen Lügners [orig.: warlock] im Tiegel verbrannt wird". [278]

Es lag nach all diesen Befunden aus den Keilschriften und den naturgegebenen modernen Kausalattributionen nahe, in manchen nicht eindeutig übersetzbaren oder definierbaren Personalnomina, die in Zusammenhang mit Schreckensträumen auftreten, den Alptraum zu erkennen. Und in seinem ersten Entwurf hat Gerhard Meier 1937 eine solche Übersetzung vorgelegt, die heute als veraltet gilt.[279] So war ‚e-li-ni-tu' zunächst zum Alp geworden und es gab Nachtmännchen, Nachtweibchen und Nachtmägde in der Maqlû-Beschwörung[280] statt lilu und lilitu. Diese wurden zumeist als verführerische Dirnen den mittelalterlichen Succubi verglichen (➔44). Die Alptraum-Benennung bot sich zwar nach den Torturen der niederen nächtlichen Hexen an. Die neuere Übersetzung nennt „Elinitu" nicht mehr Alp, sondern Lügnerin[281], womit – wie ich meine – eine nun zur ‚Schwindlerin' personalisierte als bedrohlich erlebte Eingebung im Traum, um die es sich in beiden Übersetzungen handelt, nicht mehr modern europäisiert ist; konnotativ zur Schwindlerin mögen Abstrakta wie falsche Eingebung, Täuschung und wahnhafte oder (be)trügerische Delusion darauf hindeuten, daß nicht nur emotional, sondern auch kognitiv eine kritische Distanz als Vorraussetzung für Abwehrmaßnahmen in Funktion trat. Die Bedrängung wurde als boshafte Farce enttarnt und verarbeitet. Ziehen wir daneben den im selben Text genannten Lamaštu heran, er wird als ‚Greifer' und ‚Packer' verstanden sowie Alû, den Totengeist der sieben utukkē der Sumerer, dann kann das Ausmaß der psychosomatischen Angriffe und Erfasstheiten durchaus der Symptomatik unserer ‚Alpträume' verglichen werden.

Als Beispiele für die Umwertung der emotionellen Belastung wählen wir mesopotamische Texte (➔43, 44, 45, 46), die dem Münchner Nachtsegen vergleichbar sein können – mit ihrem „Nennend verbrenne ich dich" im Sinne von Franz Werfel als Bannung arbeiten und auch als „Verbrennung" (= Maqlû) des Bedrohlichen bezeichnet und verstanden wurden.

---

278 Schwemer, Daniel, Magic rituals: conceptualisation, in: Radner/Robson: Handbook, S. 418–442.
279 Freundliche Mitteilung durch Herrn Prof. D. Schwemer, 1.X.2012.
280 Meier, Gerhard: Die assyrische Beschwörungssammlung Maqlû, Berlin 1937, in: Archiv für Orientforschung, Beiheft 2, Neudruck Osnabrück 1967, S. 7,12,23,42,44,45,61.
281 Abusch, Tzvi und Schwemer, Daniel: Das Abwehrzauber-Ritual Maqlû, in: TUAT NF4 (2008), S. 136; s. a. Schwemer, Daniel: Abwehrzauber und Behexung, Wiesbaden 2007, S. 142f mit Erläuterungen zu Elenitu.

Auch den Patienten der mesopotamischen Kultur boten sich Methoden der Emotionsregulierung an, deren Basis in den Keilschriften in enger Verbindung von Text und Ritual erkennbar wird. Sie wurden teils vom Kranken selbst, aber im Falle seiner Behinderung vom ašipu (Exorzisten) gesprochen und per Figurinen stellvertretend am Übeltäter vollzogen. Die strikten gesellschaftlichen Bedingungen einer Vermenschlichung der wirkenden Schicksalskräfte, sprich Polytheismus und Anthropomorphismus gaben dem Wirken der Methode nur Vorschub. Was wir ‚Zauber' nennen, war bekannt, legitim, ja gehörte zum Alltag; und Gegenzauber war ‚natürliche' Reaktion und entsprach einem höheren Handelspakt.

C1.4.2 Die elementare natürliche Selbstverständlichkeit im Umgang mit Überirdischen

Die elementare natürliche Selbstverständlichkeit im Umgang mit Überirdischen entspricht der Gestaltung der mesopotamischen performierenden Texte mit ihren parallelen Ritualen. Eine moderne psychopathologische Interpretation kann ihnen nicht adäquat gerecht werden. Auch wenn in ihnen eine Kommunikation mit höheren und niederen Wesen und mit einer oberen und unteren Welt mit allen Sinnesorganen aufscheint, so wäre eine dem Symptom des Paranoiden oder Sensitiven zugeordnete Betrachtungsweise nicht richtig, es sei denn man würde diese gesamte Kultur als paranoid sezernieren. Somit wird die Personalisierung der Emotionen als regulativer Prozess um vieles deutlicher und als freier liegend zu registrieren sein als in den mittelalterlichen nicht kanonischen Texten und Prozeduren, deren Gebrauch zu ihrer Zeit umstritten war. Das alte Mesopotamien bietet seine Verbaltherapie – zumindest schematisch – im Tagebau. Drängende Gefühle konnten mittels Zitierungen ausgewählter, teilweise persönlich beanspruchter Schutzgötter und Gestalten evoziert werden, nicht Schicksalsergebenheit und Demut gegenüber einem in allem waltenden über den Geistern stehenden Gott, sondern Zorn und Hass, sehr ‚menschliche' Gefühle werden notiert und scheinen fast ökonomisch instrumentalisiert worden zu sein. Methodisch nicht viel anders als in der schon erwähnten modernen Therapie der ‚kreativen Aggression' mit ihrer organisierten kathartischen Freisetzung aversiver Emotionen im Kampftraining waren sie hier allerdings eingebettet in der umfassenden kulturellen All-Sympathie. Zum besseren Verständnis könnte als Kontrast auf das christliche Ethos im Liebesgebot Jesu gemäß Markus 12,28–34 gewiesen werden: „Das vornehmste Gebot vor allen: Du sollst Gott, deinen Herrn lieben ... das ist mehr denn Brandopfer" – mit Rekurs auf AT Samuel 15,22: „Hat der Herr an Brandopfern ... Gefallen?"

C1.4.3 Begünstigung von Reappraisalprozessen durch Traumdeutung?

Wird den mesopotamischen Texten und ihren Anwendern gemäß modernen Interpretationen eine sehr starke Tendenz zu Traumdeutung und Wahrsagekunst unterstellt, die nach festen Regeln verlief, dann ist gleichwohl eine *starre* Beeinflussung cerebraler Aktivierungen bei Not- und Krisenfällen nicht anzunehmen. Denn einerseits steht auch hier die Deutungshoheit der limbischen und praefrontalen Systeme entgegen, die von Kindheit an von einem durchaus ambivalenten Verhältnis zu ihren vermenschlichten Göttern geprägt waren. Andererseits war die Unberechenbarkeit und Vieldeutigkeit der Omina, der Prognostik und Orakelkunst bekannt. Wir müssen vermuten, daß mit dieser psychosomatisch-psychologisch in Wirklichkeit ungewissen und unwägbaren (im heutigen Sinne) Imaginationstechnik, die dem kulturellen ‚mythisch-religiösen' System entsprach – oftmals jeweils sehr persönliche – nicht dem Traumbuch und seinem Schema anzupassende Gedanken über eigene Konflikte – mobilisiert wurden. Wie gesagt: Auch die Götter und Dämonen zeigten Gefühle. Textworte und deren Assoziationen haben deshalb unter Umständen individuell im Reappraisal-Prozess wohl hohe Wirkungspotentiale erzielt, weil sie als höhere, ‚göttliche' Inspiration mit individuell ausrichtbarer ‚göttlicher' Offenbarungsvorstellung legiert und verarbeitet werden konnten. Respektiert Überirdisches und gewohnt Alltägliches betrafen ein und denselben neuropsychosomatischen Haushalt: Es begegneten sich „archetypische versus biografische Botschaft in <u>einem</u> Traum."[282] Gleichzeitig wird damit das schon verinnerlichte kulturelle und gesellschaftliche Gravitationsfeld am therapeutischen Prozess beteiligt.

C1.4.4 Rituelle Perduzierung einer verkörperlichten Simulation

Der Einsatz von Ritualhandlungen kann als Vergegenständlichung von Emotionen verstanden werden, die auf dem Boden einer kulturell verankerten Leib-Seele-Geist-Einheit fungiert. Es kommt zu einer totalen, gleichsam neurometaphorischen Simulation: Der fremde Körper z. B. der Hexe *ist* die (produzierte) Figurine; die eigene Seele des Beschwörenden oder des Patienten *ist* die Emotion des Zornes; der Geist *ist* zur Kraft einer Intention mit Vernichtungswillen per Verbrennung oder Köpfung geworden und alle drei vereinigen sich nach alter Vorstellung ‚magisch-performierend', nach den neurophysiologischen Erkenntnissen im sensomotorischen System multimodaler Neuronen. Körperbewußtsein und Geist-Geister-Vorstellung fließen – besonders bei Heranziehung der Toten-

---

282 Benedetti, Gaetano: Träume, S. 67.

kulte mit ihrer Vorstellung substantieller Sukzession – zusammen. Mit der neurophysiologischen Interpretation kann letztlich belegt werden, was philologische Forschung für den ‚archaischen Menschen' formuliert hat: „(für ihn) sind die Sachen und die Worte nicht unterschieden, die Ähnlichkeit der Struktur der Sprache entspricht der Ähnlichleit der Struktur der Wirklichkeit" – „Zeichen und Gezeigtes fallen (für ihn) zusammen, ... Wort und Sache werden als kongruent gedacht." [283] Der nach unserer Vorstellung bestehende Unterschied zu den sogenannten physiko-chemisch wirksamen Heilmitteln scheint damit ebenso minimiert wie es auch die mangelhafte Unterscheidung der priesterärztlichen Berufsgruppen ist.

Moderne Begriffe:
- Dysregulation unbewußter optischer Stimuli durch fremde Augen
- Excessive Reading mind in the eyes; Superior mentalizing ability; Mental state discrimination; Inaccurate hypermentalizing

Historisches/ volkstümliches/ alltägliches Begriffsfeld:
- Schadenzauber; unterschwellige Blick-Botschaften; Augenzauber,
- igi hul (mesopotamisch); böser Blick, böse Augen, eye evil
- Augensprache; Malocchio
- Gedankenleser; Geisterseher, Spökenkieker

Ursachen:
- intrinsic: – Borderline-Syndrom, alle Formen der Angst, sensitive und paranoide Syndrome,
- extrinsic (hypothetisch, auslösend): kulturspezifische Triggerung (Feindbilder, Monumente, Abwehrmuster etc); Intoxikationen

Therapie:
- Emotionale Regulierung negativer intrinsischer Stimuli (s. a. Tabelle 1),
- themenzentrierte Erzähltherapie zur ‚Aufdeckung' der Übel,
- übertreibende Übelabtragung (exaggerating evil-transfer),
- Revanche-Mechanismen,
- Aufdeckung der Reizsituation, Kunsttherapie, dialektische Verhaltenstherapie, Selbstschutzmechanismen

Neuropathologie:
Überreagibilität rechter Temporalpol und mittlerer temporaler Gyrus mit stärkerer Verbindung zu Amygdala und praefrontalem Cortex u. a.

Tabelle 2: Das unbewußte Blicktriggersystem und seine Störungen

---

283 Schulz, Monika: Magie, wie oben, S. 178,176 nach Zitaten Oehler, Foucault u. a.

C2 Die intrinsische neurale Reizüberflutung am Beispiel des ‚Böse-Blick'-Komplexes (s. Tabelle 2)

Der uralte und immerwährende Schadenszauber mit bösem Blick und sein Abwehrzauber erfordern eine eigene Betrachtung, obwohl Phänomene, Symptome und Therapieversuche den unter **C1** genannten umfangreicheren Affektionsformen verwandt sind oder sich mit ihnen überschneiden. Man könnte diese Art der scheinbar blickgetriggerten Überflutung mit negativen Emotionen und Imaginationen als einen Sonderfall gestörter Emotionsregulierung verstehen. Jedoch unterscheidet sich das hier besprochene Syndrom durch 1. seine unbewußte und auch nicht traumbedingte Entstehung, 2. durch seine andere neuropathologische und hirnlokalisatorisch wahrscheinlich eng umschreibbare Ätiologie, also in seinen kausalen Wirkfaktoren. Besonders erzwingt 3. seine breite kunst- und kulturhistorische Bedeutung und Gestaltung eine gesonderte Heraushebung.

C2.1 Aspekte eines ‚subversiv' visuell getriggerten Kultsektors

Zumeist sind wir uns sicher, daß uns als soziale Wesen am ehesten das verbinde, was vom Tier abhebt, Denken und Verstand – und als deren Explicitum gelten Sprache und Wort, gesprochen vom einen zum anderen. Gewiss, das Wort kann Motor oder Meteor sein, Kraftakt des Herrschens oder Schrei der Verzweiflung, kann Botschaften des Mitfühlens oder des Erkennens vermitteln. Aber es gibt noch ganz andere Botschaften, die vom Gehirn nicht übersehen werden oder sogar überbewertet werden können. Erst sie begründen oder vervollständigen unsere Kommunikation. Schriftsteller und Philosophen haben auf das Ungenügen der Sprache hingewiesen. Nach Schätzungen werden 55 % der Informationen über die Gefühle eines anderen per Gestik und Mimik, 38 % über die Sprachmelodie und nur 7 Prozent durch die Semantik der Sprache vermittelt.[284]

Wir hatten in Kapitel **B1.1** schon die Funktion von Mimik, Gestik und Prosodie für Kleinkinder erwähnt. In diesem Kapitel wird ein Organ erfasst, das an Signalkraft und Breitenwirkung für ausgesendete emotionale und kognitive Botschaften kaum überschätzt werden kann: das Auge mit seinem Bewegungs- und Umgebungsapparat; aber gleichermaßen oder mehr noch interessiert hier jene Empfangsstation, die aus den Augen eines anderen zu ‚lesen' vermag.

Treffend hat der Neurologe Johannes Dichgans den Kleinen Prinzen von Antoine de Saint-Exupéry herangezogen. Der einen Freund suchende Fuchs sagt

---

284 Dichgans, J., in Frühwald, W. (2004), Das Design, S. 220.

zum kleinen Prinzen: „Man sieht nur mit dem Herzen gut ... du setzt dich zunächst ein wenig abseits von mir ins Gras. Ich werde dich so verstohlen, so aus dem Augenwinkel anschauen und du wirst nichts sagen ..., die Sprache ist die Quelle der Mißverständnisse."[285]

Vor jeder sprachlichen Äußerung und jeder Überlegung wecken uns die strahlenden Augen eines Kindes, hören wir von oder erleben wir ‚Liebe auf den ersten Blick', berühren uns die stechenden Auges eines Zornigen und der Blick in die Ewigkeit eines Sterbenden. „Das Auge ist des Leibes Licht", schreibt der Evangelist (Matth. 6,22). „In den Augen liegt das Herz", sagt der Dichter (v. Kobell), vom „Glanz in den Augen der Mutter" schreibt der Psychoanalytiker (Heinz Kohut). Und es gibt in weiteren Redewendungen das alles erkennende Auge des Gesetzes, den Röntgenblick und nicht zuletzt das Auge Gottes. Allerdings sind viele Zeugnisse zur Allegorie von Blick und Auge kulturell, künstlerisch oder gar katechetisch genutzt und nicht immer mit den hier zu beobachtenden individuell hirnorganisch verursachten Phänomenen zu vergleichen.

Eine besonders markante Spur in Symbolik und in Texten aller Völker seit Beginn von Kultur bis in die Neuzeit zeichnen die Phänomene böser Augen und Blicke, in den Texten oft verbunden mit bösen Zungen und Mäulern. Es sind bei weitem alles Überzeichnungen, deren wissenschaftliche Klärung seit der Antike erfolglos versucht wurde. Es gibt manigfaltige Gestaltungen an Tempeln und Burgen, an Toren und Stalltüren, Waffen und Amuletten, an Gemmen, Siegeln und Brakteaten, am Bug von Schiffen und an den Fratzen der Dome. Die indischen Atharvaveden (VI, 20) kennen den tausendäugigen Gott, der das Kraut ‚Allsehend' spendet. Auch heute wird mancher indische Rikscha wegen Einwirkung ‚böser Blicke' zum Heiler geschickt.[286] Es war ein Augenarzt,[287] der als erster eine weltweite Sammlung abwehrender, bedrohender und schützender Kultformen zusammentrug. Verhaltensforscher und Kulturhistoriker haben unter Hervorhebung von Angstverarbeitung und Vertrauensaufbau auf jene in uns liegenden Programme und Strategien hingewiesen, die als stammesgeschichtliches, kulturübergreifendes Erbe der Gefahrenmeidung dienen. Signalsender und Signalempfänger kommunizieren mit wechselseitigem Vorteil oder Nachteil über meist kaum wahrnehmbare Pupillenformen, Augenmotilität, Lid- und Augenringmuskelaktivität und Augenbrauen.[288]

---

285 Ders., S. 218.
286 Glucklich, A.: The end of magic, S. 92.
287 Seligmann, Siegfried: Der böse Blick, Berlin 1910; ders.: Die Zauberkaft des Auges und das Berufen, Hamburg 1922.
288 Vgl. z. B.: Eibl-Eibesfeldt, Irenäus und Christa Sütterlin: Im Banne der Angst, München 1992, S. 55ff; für literar. Quellen des Mittelalters: Meisen, Karl, Der böse Blick und anderer Schadenzauber, in: Rheinisches Jahrbuch für Volkskunde 1(1950), 3(1952).

Unter Menschen beginnen erste kindliche Signale an die Mutter mit Augenaufschlagen und Blickzuwendung auf eine Schallquelle, wobei selbst Blindgeborene sich so verhalten, als würden sie Blickkontakt mit der Mutter suchen. Oft wird daher Blindheit Neugeborener erst spät erkannt. Die kindlichen Signale werden von der Mutter positiv bewertet und schaffen gegenseitige Bindung und Urvertrauen. Freundliche Begegnung unter Erwachsenen beginnt meist und unter allen Kulturen mit einem Augengruß, dem kurzen Anheben der Brauen für 1/6 Sekunde und nur kurzer direkter Blickberührung.

Vor allem aber hat sich unsere Phantasie durch innere Bilder und sprachlich durch Metaphorik eine eigene kleine Welt geschaffen. Sie erzählt in den Sagen der Antike mit der Meduse Gorgo, bei deren Anblick man versteinert, mit dem einäugigen Zyklopen, mit Salomons Augen, mit einem ‚Augengott' und mit Dämonen und Wolfsaugen ergreifende und schreckliche Schicksale. Künstlerische Gestaltung schuf stellvertretend für uns selbst die wirksamsten Figuren.

Beschwörungen und Volkssegen in Zaubersprüchen ersannen Formen der Abwehr.

Eine der frühen Beschwörungen (**Beispieltext** ➔47) gegen den zur Hexe personifizierten bösen Blick entstammt den Keilschrifttontafeln der Bibliothek des neuassyrischen Königs Assurbanipal (668–627 vor Christus). Der böse Blick, igihul, spielt in sumerischen und akkadischen Abwehrformeln eine ‚prominente Rolle'.[289] Im europäischen Mittelalter finden sich erste Belege innerhalb von volkstümlichen Gebeten seit dem 12. Jahrhundert (➔5). Es tauchen viele Beschwörungen auf, die die guten Augen Gottes gegen die bösen Augen der Feinde einsetzen, wobei nicht nur die Kraft des göttlichen Blickes, sondern auch eine Überzahl seiner dreifaltigen Augen als Gegenwaffe fungiert (➔6). Sie zeigen im Verlauf der folgenden Jahrhunderte immer mehr die Tendenz zur Revanche des bösartig Angetanen (➔7), zu Personalisierung und Gesinnungsdeutung (➔8), und zu Gegenfluch und Gegenbann.[290]

Bei soviel Attraktivität außergewöhnlicher geheimnisvoller Zeugnisse, die Kulturen und Zeiten überdauern, die in den Abwehrtexten oft ganze Katastrophenkataloge kompilieren, wurden die Fragen immer drängender. Der Ethnologe Thomas Hauschild hat diese Zeugnisse wieder zusammengestellt und hat biologisch-hereditäre, sozialpsychologische, ethnologische und psychoanalytische Deutungen vorgelegt. Seine Eingangsfrage, „Werden wir je wissen, wie es ist, um den bösen Blick zu wissen?" blieb indessen unbeantwortet. „Die sozialpsy-

---

[289] Schwemer, Daniel (mit Quellenhinweisen): Abwehrzauber und Behexung, Wiesbaden 2007, S. 68.
[290] Sammlung für die letzten Jahrhunderte aus dem Corpus der deutschen Segens- und Beschwörungsformeln Dresden, s. bei Spamer, Adolf: Romanusbüchlein. Aus seinem Nachlaß bearbeitet von Johanna Nickel Berlin 1958, S. 109–157.

chologische Analyse kann unser Verständnis des Glaubens an den bösen Blick vertiefen, aber nicht die Bedingungen der An- und Abwesenheit dieses Glaubens in verschiedenen Kulturen restlos erklären .... eine Lücke im funktionalistischen Weltbild". [291]

C2.2 Die neurobiologische Grundlage des ‚Böse-Blick'-Syndroms

Seit den 70 Jahren des vorigen Jahrhunderts wurde eine Reihe von psychologischen Testuntersuchungen zur Erforschung der Labilität von Borderline-Kranken, der häufigsten Persönlichkeitsstörung, früher unter ‚Psychopathie' genannt, durchgeführt. Diese Kranken haben neben verschiedenen Verhaltens-, Denk- und Gefühlsstörungen[292] meist eine große Unsicherheit bei der Einschätzung der Gefühle und Absichten anderer, was zu sozialen Brüchen und zu Selbstbeschädigungen bis zu Selbstmordtendenzen führen kann. Als ein Merkmal für die Borderlinestörung – mit dieser Benennung seit den 70er Jahren erfolgte auch eine Revision der Deskription des alten Psychopathiebegriffes – wurde schon bald ein Paradox angenommen.[293] Man erkannte eine Diskrepanz zwischen der erhöhten emotionalen Sensitivität und scharfsinnigen Weitsicht für die Gefühle anderer in deren Gesichtsausdruck und den gestörten zwischenmenschlichen Beziehungen dieser Kranken. Die Untersuchungen am *gesamten* Gesichtsausdruck ergaben aber zunächst widersprüchliche Resultate.

Es zeigte sich auch ein Zusammenhang zwischen individuell erhöhter Ängstlichkeit und Sensitivität und der Wahrnehmung von hintergründigen Signalen, die nicht bewußt und nicht gezielt etwa aus dem peripheren Gesichtsfeld, aus dem ‚Augenwinkel', geboten werden.[294] Und auch tachistoskopische Reize von weniger als 33 Millisekunden Dauer alarmierten das limbische System in besonderer Weise.[295]

Eine isolierte Analyse der engen Augenregion durch einen „Reading the mind in the eyes"-Test (RMET) hatte schon seit den Experimenten von Baron-Cohen (2001) und seinen Mitarbeitern ganz allgemein (nicht Borderline) die entschei-

---

291 Hauschild, Thomas: Der böse Blick, S. 6, S. 180f.
292 Vgl. Faust, Volker: Von Amok bis Zwang, Heidelberg u. a. 2013, Band II, S. 37–49.
293 Krohn, A.J., Borderline ‚empathy' and differentiation of object representation: a contribution to the psychology of object relations, in: Intern. J. of Psychoanal. Psychoth. 3 (1974), 142–165.
294 Bayle, Dimitri J. et al. Unconsciously perceived fear in peripheral vision alerts the limbic system, (2009) PLos One 4(12) e8207. doi:10.1371/j.pone.0008207.
295 Hirsch, Joy et al., Fleeting images of fearful faces reveal neurocircuitry of unconscious anxiety, EurekAlert, 17–Dec–2004.

dende Rolle allein der Augen für soziale und emotionale Verarbeitungen eines Gegenüber gezeigt. Die Augen decodieren bei Begegnungen als erstes automatisch fremde Mentalzustände, weit vor jeder bewußten mehr kognitiven Verarbeitung. Die Augengegend übermittelt Informationen über eine Reihe komplexer Sinnesmodalitäten, über Motivationalität, Gesinnung und Interessenlage. Fokussierungen auf einen Mund allein und auf ein Gesamtgesicht erbrachten dagegen im Vergleich zum Auge keine eindeutigen Ergebnisse. Die Arbeitsgruppe um Eric Fertuck[296] hat diesen RMET-Test auf Borderline-Kranke angewandt.

> Unter Berücksichtigung auch sozialer Parameter und psychologischer, suchtbezogener, schizophrener, endogen depressiver und hirntraumatischer Nebenfaktoren wurden die Borderlinekranken mit dem RMET-Test untersucht. Sie erhielten Fotos mit unterschiedlichen emotionalen Blickvarianten vorgelegt mit je 4 Worten zur Auswahl, wovon nur bei **einem** Foto Übereinstimmung zwischen Augenausdruck und dessen Beschreibung bestand. Dabei zeigte sich statistisch relevant, daß Borderlinekranke gegenüber Gesunden eine bessere Diskriminierungsleistung für die Mentalitäts-Unterschiede performierten. Dies ergab sich bei neutralen ebenso wie bei positiven Blickvarianten. Lediglich bei negativen Augenausdrücken lagen sie mit Gesunden in einem Trend.

Eine Arbeitsgruppe von Forschern aus Heidelberger, Tübinger und Erlanger Instituten hat diese Ansätze von Baron-Cohen (1997, 2001), und Fertuck (2009) und ihren Mitarbeitern aktuell (2012) unter Einsatz des RMET fortgesetzt und durch bildgebende Verfahren am Gehirn ergänzt.[297] Dabei bestätigte sich die erhöhte, präzisere und beschleunigte Diskriminierungsfähigkeit der Borderlinepatienten für affektgetränkte Blicke anderer. Das gilt hier sowohl für die Erkennung böser als auch gutwilliger Blicke, nur nicht für den neutralen Bereich.

Was die Hirnlokalisation betrifft, so ergaben sich bei den fMRI-Untersuchungen der Borderlinepatienten während negativer und positiver Augenausdrücke stärkere Aktivierungen des rechten Temporalpols und des mittleren temporalen Gyrus. Der Temporalpol (**Kap. B4.1.6**) wird als visuelles Gebiet höherer Ordnung angesehen, der sehr eng mit der dabei ebenfalls stärker aktivierten Amygdala und dem praefrontalen Cortex verbunden ist. Er vermittelt komplexe Eingänge auch mit visceralen emotionalen Antworten und spielt eine Schlüsselrolle bei Gesichtsverarbeitungen. „Die zusätzliche Aktivierung des Temporalpols und des medialen Frontalgyrus bei Borderline verglichen mit Gesunden mag deshalb als ein Anzeichen tieferer emotionaler Verarbeitung visueller Stimuli interpretiert werden", so die Deutung der Forschergruppe.

---

296 Fertuck E.A. et al., Enhanced ‚Reading the Mind in the Eyes' in borderline personality disorder, in: Psychological Medicine 39(2009), doi: 10.1017/S003329170900600X.
297 Frick, Carina et al., Hypersensitivity in Borderline Personality Disorder during Mindreading, PLoS One 2012, doi 0.1371/journal.pone0041650.

Hinzu kommt, daß Borderlinekranke nach Meinung einiger Untersucher reduzierte Aktivitäten in Hirngebieten zeigen, die dem Spiegelneuronsystem angehören, d. h. ihre Gehirne arbeiten eher intuitiv und automatisch, sie spiegeln die Mentalzustände anderer einfach zurück, während Gesunde im Anblick des anderen auch Gebiete mobilisieren, die mit bewußten emotionalen Repräsentanten die Mentalzustände anderer empathisch assoziieren (Insula, obere temporale Rinde). „Borderlinekranke mögen eine überaktive und gesteigerte Resonanz mit fremden Mentalzuständen haben"; „sie sind hochvigilant auf soziale Stimuli" und „geraten mit anderen in Resonanz."[298]

Es ist klar, daß die Hirnorganik des Borderline-Syndroms nicht auf die Dysregulation des visuellen Systems beschränkt ist und daß dem Rechnung getragen wurde. Neben den oft eng auf RMET-Methode abzielenden Experimenten wurde auch auf andere, besonders auf die im **Kap C1** hervorgehobenen emotionsregulierenden Parameter eingegangen. Mittels aversiver Stimuli wurden zwei unterscheidbare Verarbeitungsstörungen gefunden: Die schon erwähnte erhöhte affektive Reagibilität und die Defizite bei der Emotionsregulation infolge orbitofrontaler Störung.[299]

Außerdem stand immer zur Frage, inwieweit noch andere neuropsychiatrische Syndrome mit der abnormen intrinsischen Blicktriggerung behaftet sind. So haben manche Untersucher bei schizoiden Prodromalstadien und bei Menschen mit erhöhtem Schizophrenie-Risiko ähnliche Befunde wie bei Borderline mitgeteilt, die auch teilweise Aktivitätssteigerung in temporoparietalen und praefrontalen Gebieten zeigen.[300] Manche Experimente haben auch die erhöhte Diskriminierungsfähigkeit vom Ausdruck fremder Augen bei paranoiden im Vergleich zu den anderen Schizophrenien festgestellt. Wegen der unterschiedlichen und vielfältigen symptomatischen, ätiologischen und hirnlokalisatorischen Parameter für die Gruppe der Schizophrenien und auch für Autismus ist die Forschung aber noch im Fluß.

Erinnert sei schließlich an jene in der Gesellschaft und den Medien gern bestaunte, wohl auch für Begabte trainierbare Fähigkeit, ähnlich wie als Zauberkünstler die Gedanken anderer zu lesen oder beruflich als lebendiger Lügendetektor im Dienst der Polizeifahndung zu arbeiten. Die Emotionspsychologie hat mit dem „Facial Action Coding System" eine Methode zur Prüfung von Gefühlen im Gesicht entwickelt. Die Fähigkeit erhöhter affektiver Mentalisierung dürfte in leichterer und nicht devianter, eng begrenzter Ausprägung nicht krankhaft sein.[301]

---

298 Frick, C. ebd.
299 Schulze, Lars et al. Neuronal correlates of cognitive reappraisal in Borderline patients, Biolog. Psychiatry (2010), doi:10.1016/j.biopsych.
300 Brüne, Martin et al., An fMRI study of ‚theory of mind' in at-risk states of psychosis, in: Neuroimage 55(2011, S. 329–337, doi.org/j. 2010.12.018.
301 Scott, L.N. et al., (2011) Mental state decoding abilities in young adults with borderline personality disorder traits. Personality Disorders: Theory, Res Treatment 2:98–112.

C2.3 Therapeutisches

Gemäß einer Reihe von Übersichtsarbeiten zu bisherigen Therapieversuchen haben sich keine *spezifischen,* nachhaltig heilsamen Methoden ergeben.[302] Dies stimmt mit der historisch weitreichenden Permanenz des ‚Böse-Blick'-Syndroms mit seinen kulturellen Ausgestaltungen bis in den neuen Kunstbetrieb überein. Insofern liegt es nah, die ‚diagnostischen' Leistungen dieser Folklore-Psychiatrie – also die Ausmalung, Aufstellung und Bedichtung des ‚bösen Auges' – mit einem therapeutischen Anliegen zu verknüpfen, einem Selbstschutzmechanismus. Die ‚Diagnose' ist zugleich eine Art notdürftige Therapie.

Die Kranken sehen sich vielfach durch die Augen von ‚Verfolgern' oder ‚Aggressoren'. Ein Betroffener soll sich demnach geäußert haben: „Als Borderline-Patient bringe ich mein Gegenüber dazu, sich als Role-Giver (Täter) zu verhalten, der mich zum Role-Receiver (Opfer) macht. Einerseits demonstriere ich damit symbolisch, was mir von früheren Personen meines sozialen Symptoms angetan wurde (Rekonstruktion symbolisierter Rollenkonserven), andererseits schaffe ich es so, meine ‚bösen' Rollen (Täter)-Introjekte an eine andere Person zu delegieren. Dort kann ich sie dann bekämpfen. […] Auf diese Weise kann ich mit Wut, Hass, maligner Aggression, Ohnmacht und Hilflosigkeit sowie Scham- und Schuldgefühlen umgehen." [303] Diese Äußerung gibt natürlich entgegen der Behauptung der sie Zitierenden keine spontane Erkenntnis oder Reflexion des Patienten wider, sondern weist allzu deutlich auf die Fachsprache der Psychotherapeuten. Sie ist als Abklatsch sprachlicher Vorgaben erkennbar und muß daher als Symptom gestörter Spiegelneurontätigkeit verstanden werden, als undurchdachte Rückspiegelung fremder Worte.

In der Psychodrama-Therapie des Syndroms werden u. a. auch Figuren und Puppen angewendet, wie in archaischen Ritualen.

Schließlich gehört der Blick jedes Anderen in seinem Verhältnis zum Begehren, zur Begierde und zur Scham zu den tiefsten und berührendsten Signalen der Selbstwahrnehmung und Selbstbewertung. Ein Weg zur Entkrampfung von Scham kann nach psychoanalytischer Sicht erst dann gelingen, wenn die Scham nicht als von außen begründbar, sondern als ein „Selbstverhältnis des Analysanden" verstanden wird,[304] das heißt, ‚böser Blick' ist nur als Phänomen eines von

---

302 Vgl. z. B. Stoffers et al., Psychological therapies for people with borderline personality disorder, Cochrane Database Syst Rev. 2012 Aug 15;8:CD005652.
303 Bender, Wolfram und Christian Stadler: Psychodrama-Therapie, Stuttgart 2012, S. 146/147.
304 Strassberg, Daniel, Scham als Problem der psychoanalytischen Therapie, in: Schweizer Archiv für Neurologie und Psychiatrie 155 (2004), S. 225–228.

diesem Blick Getroffenen selbst ausgelösten Kreislaufs personaler Interaktion zu deuten und zu behandeln.

C3 Die kognitive cerebrale Perturbation durch sprachliche Inkongruenzkonstruktionen – eine nicht-emotionale Konflikt-Provokation

Eine der spitzfindigsten Strategien vieler Heilspruchtexte – als Erzählpartikel oder als Wortdoppel eingebaut – ist die Konstruktion von Unmöglichkeiten als Zaubereffekt und die Konstruktion unmöglicher Aufgaben zur Irreleitung und Erledigung der Krankheitsdämonen. Bereits der kaiserlich römische Hofbeamte oder Hofarzt Marcellus von Bordeaux notiert im 5. Jahrhundert die Methode, die in einem Spruchtext gegen Magenbeschwerden einen Baum in Meeresmitte imaginieren läßt: Stabat arbor in medio mare, woran ein Gefäß mit menschlichen Därmen hängt (**Beispieltext ➔9**).

Psychische Taktik des Spruchtextes

moderne Begriffe:
- Inkongruenzeffekt, Unmöglichkeitskonstruktion, semantische/ syntaktische Anomalie bzw. Verletzung, Evidenzvorgabe, Aufmerksamkeitsfesselung
- Rhetorisch: Adynata, Oxymora; Aporie (Auswegslosigkeit, Ratlosigkeit)
- Kontrastkoppelungen; performative Inkongruenz; Ambiguität
- Minimal kontraintuitive Erzählpartikel

historische/ volkstümliche/ alltägliche Begriffe:
- Holzwegsätze, Überraschungszauber, Unsinnssignale, Extraordinäres;
- ‚Kamel durchs Nadelöhr', Verwirrung stiften, Unpassende Bemerkung, sprachliches Mißverhältnis, Widersinnigkeit.

Anwendungsfelder für den Einsatz:
- allgemein

Neurophysiologie:
- Ereigniskorrelierte EEG-Potential-Ableitung (EKP) zur Registrierung von semantischen und pragmatischen Kontextinformationen;
- Zentrencephal ableitbares N400-Potential (ohne bisher exakt festgestellte neuronale Quellen)

Tabelle 3: Die cerebrale Registrierung der Eingabe nicht-emotionaler Inkongruenzsignale

Die Reaktion eines Gehirns auf solche rhetorischen ‚Spiele' ist seit ca. 20 Jahren neurophysiologisch nachweisbar, nicht nur bei Oxymora, der Extremform der Antithese: ‚schwarze Kreide' ‚schwarze Milch', ‚felix culpa', sondern sie wäre es auch in seltsamen Sätzen wie dem alten römischen. Moderne Forschung mit dem EEG registriert im Labor sowohl Satzbaufehler wie „Die Kirche wurde im gelehrt" als auch inhaltliche Kuriositäten wie „Das Gewitter wurde gebügelt". Auf alle syntaktischen und semantischen Abstrusitäten signalisiert das gesunde Gehirn Fehlermeldungen und läßt damit erkennen, daß sprachliche Überraschungen ‚Wirkung zeigen', eine herausragende Wirkung, ob gehört oder gelesen.

C3.1 Zu den historischen Quellen von sprachlicher Inkongruenz[305]

C3.1.1 Funktionsschwankungen alter Inkongruenztexte: ‚Vogel federlos' und ‚Baum blattlos': Rätsel oder Heilspruch?

Betrachtet man die Forschungen der Germanisten zu einigen literarischen Gestaltungen wie etwa dem Spruch vom ‚Vogel federlos' (→11), so sieht es aus, als hätten sie vorweggenommen, was neurophysiologisch in der EEG-Zacke N400 zum Indikator von kognitiver Provokation wurde. Lange Jahre stand zur Frage, ob es sich ursprünglich um ein Rätsel handele, also zur Sorte der spielerisch unterhaltsamen, zeitweise sozial prüfend-taxierenden[306] (Halslöserätsel!) kognitiven Herausforderung gehörig, oder ob es ein Heilzauber mit medizinischer Anwendung sei.[307] Die Befunde waren selbst bei Beachtung des Kontextes nicht immer eindeutig. Einerseits standen die Texte des altrömischen Marcellus (→10) und des mittelalterlichen Ägypters Abu Bekr ibn Bedr nachweislich Ärzten und Tierärzten zur Verfügung.[308] Auch ist ein altkoptischer Heilspruch mit Überresten eines Isiskultes aus Altägypten bekannt: ‚Ein Mittel gegen Leibschmerzen eines Kindes: „Hor, der Sohn der Isis, ging auf einen Berg, um zu schlafen ... er stellte seine Netze auf und fing einen Sperber. Er zerschnitt ihn ohne Messer, kochte ihn ohne Feuer und aß ihn ohne Salz ..."'.[309] Und es gibt einige mehr in der Volksmedizin der Neuzeit, selten im Mittelalter verzeichnete Heilspruchtexte

---

305 Mit ‚Inkongruenz' ist hier nicht der von Psychologen als personale Diskrepanz zwischen Eigenkonzept und Eigenerfahrung bei Neurosen verstandene Begriff gemeint.
306 Bässler, Andreas, Die Funktion des Rätsels im Lalebuch, in: Daphnis 26 (1997), S. 53–84, hier: S. 58f.
307 Schupp, Volker (Hg.:) Deutsches Rätselbuch, Stuttgart 1972, S. 371.
308 Eis, Gerhard: Altdeutsche Zaubersprüche, Berlin 1964, S. 67–76.
309 Zit. aus: Leipoldt, Joh. und Siegfried Morenz: Heilige Schriften, Leipzig 1953, S. 187.

mit derartigen Sprachfiguren gegen verschiedene Übel (➔12,13). Andererseits stammt die früheste europäisch mittelalterliche Fassung aus einer Rätselsammlung der Insel Reichenau (➔11) des 10. Jahrhunderts; und über die Jahrhunderte bis in die Gegenwart gelten viele dieser Texte mit ihren Varianten als Rätsel.[310]

Es wurden und mußten bei der Tradierung funktionswechselnde Übertragungen angenommen werden. Man sieht, daß Verbaltherapie und Rätsel als geistige Hirnprovokateure gemeinsame Wurzeln haben. Beide kennen die Techniken des Versteckens wie Paradoxie, Metaphorik und Doppeldeutigkeit (Ambiguität).

Und nicht nur das: Die Anwendung von Rätselkurzformeln z. B. nach Art der genannten ‚Ohne-Sprüche' in der modernen Therapie hat durch die Ausarbeitung von Reframingstrategien zumindest theoretisches Interesse gefunden. So werden im Neurolinguistischen Programmieren Reizformen wie die ‚Kirsche ohne Stein', das ‚Huhn ohne Knochen', das ‚Baby, das nicht schreit' eingesetzt, um eine Erweiterung oder Erneuerung der Rahmengrößen des Denkens und der Wahrnehmung zu üben. Aus solchen kleinen engen und vagen, aber sehr reizvollen Bildern sollen im therapeutischen Prozess durch Umdefinieren neue Bildfolgen erarbeitet werden, die etwa in diesen drei Beispielen zu ‚Blüte', ‚Ei' und ‚schlafendem Baby' führen[311].

Auch Vertreter anderer Psychotherapiemethoden haben die Technik der Verwirrung durch eine Paradoxe Intervention angewandt, nicht nur innerhalb der Praxis wie in folgenden Begegnungen: Der Hypnotherapeut Milton H. Erickson mit seiner Instruktion über etwas Zwielichtiges am Selbstverständnis des Therapeuten an einen Kollegen (**Kap. B3**) oder mit seiner absurden Reaktion auf den körperlichen Zusammenstoß mit einem Passanten. „Bevor er sich von seinem Schreck erholt hatte [...], sah ich umständlich auf meine Uhr, als hätte er sich nach der Zeit erkundigt und sagte höflich: ‚Es ist genau zehn Minuten vor zwei', obwohl es fast vier Uhr war, und ging weiter. Mehrere Häuser weiter drehte ich mich um und sah, daß er mir immer noch nachblickte, offenbar noch immer verwirrt."[312]

---

310 Siehe Tomasek, Tomas: Das deutsche Rätsel im Mittelalter, Tübingen 1994, S. 119–122; Richter, Gisela, „Auf dem Birnbaum ohne Blätter ...", in: Innsbrucker Beiträge zur Kulturwissenschaft, Germanistische Reihe Bd. 57 (1997), S. 375–378.
311 Dilts, Robert: Sleight of mouth. The magic of conversational belief change (1999/2006), deutsch u. d. T. Die Magie der Sprache, Paderborn 2008², S. 39.
312 Zit. nach Watzlawik, Paul: Wie wirklich ist die Wirklichkeit?, München 1976, S. 39.

C3.1.2. Am Grenzübergang zur Erzähltherapie: Adynata – weitere, Inkongruenz erzeugende Sprachfiguren

Auch heute noch vermögen begabte Rhetoriker besonders mit Sprachfiguren der Unmöglichkeit, des Niemals und des Nimmermehrs erfolgreich zu überzeugen oder wenigstens zu verwirren. Der CDU-Politiker Heiner Geißler etwa, um eine Satzvollendung gebeten, ‚ich werde in die SPD eintreten, wenn …' bemerkt treffsicher: „wenn der SPD-Vorsitzende Kurt Beck zur deutschen Schönheitskönigin gewählt worden ist." (Fernsehinterview ntv Juni 2008). Ähnlich hatten die Schöpfer der therapeutischen Texte oft Unmöglichkeiten – von Sprachspezialisten Adynaton, Mehrzahl Adynata, genannt – als Signaturen wie Stoppschilder oder mindestens Sackgassen zur Eindämmung von Krankheit, Unheil und Dämonen in ihre Sprüche eingebaut.

> Die Fragen nach Unmöglichkeit, Dialektik und Kontrastkoppelungen in ihren vielen räumlichen, zeitlichen und seelischen Aspekten haben Philosophie und Theologie immer tiefgründig bewegt (siehe z. B. den unlängst erschienenen Sammelband: Dalferth, Ingolf U., Philipp Stoellger und Andreas Hunziker: Unmöglichkeiten. Zur Phänomenologie und Hermeneutik eines modalen Grenzbegriffs, Tübingen 2009) und haben Prediger und Dichter beflügelt.
>
> Von Vergils Bucolica-Bildern mit den im Äther weidenden Hirschen, den nackt am Strande lebenden Fischen und dem Germanen am Tigrisstrom – gemeint hat der Dichter damit die Unmöglichkeit, einen bestimmten Gott seiner Sympathie je zu vergessen – einerseits, über die christlichen Antithesen der ‚felix culpa' im Exsultet der Osternacht und der schon oben genannten hohen Dialektik des Tertullian – des unsichtbar-sichtbaren und ungreifbar und doch gegenwärtigen Gottes andererseits – ziehen sich breite Spuren durch die Geschichte. Minnesänger und Meistersinger waren geübt darin und betrieben es als eigenen kreativen Zweig ihrer Künste. Volkslieder, Märchen und Sagen haben bis in die Neuzeit Spannung, Humor und ‚Verkehrte Welt' durch Adynata vermittelt. Immer wieder standen sich Banales und Ernsthaftes krass gegenüber. Heute sind es auf hochwissenschaftlicher Ebene ganz offiziell und säkular das Weltall, das als endlich-unendlich und der Mensch, der als schuldig-unschuldig ‚entlarvt' sein soll.

Die Konstrukteure mancher der alten Heilspruchtexte haben sich einer reichhaltigen Palette der schlichten meist biblischen Bilder und der Legenden bedienen können, um – ohne es zu ahnen – neuronal N400 zu triggern und um damit, soweit es derzeitige Forschung ermittelt hat, Nervenzellkomplexe zum Blättern im eigenen Lexikon anzuregen. Hinter manchen der vielfältigen Ideen mag die Rede Jesu in Matth. 19,24 und Lukas 18,25 gestanden haben: „Denn es ist leichter, daß ein Kamel durch ein Nadelöhr gehe, als daß ein Reicher in das Reich Gottes komme", selbst wenn die Bibelübersetzung falsch sein sollte und es um ein Schiffstau statt ein Kamel oder/und um ein enges Stadttor statt eines engen Nadelöhrs gehe. Und so sind in der Volksmedizin Sprüche überliefert, die die absolute Vergeblichkeit der Aktion des Schädigers imaginieren, wenn ihm, im Bilde der mystischen Dichtung ‚Hymelstrass', die Himmelstür verschlossen bleibt (➔14), eine versperrende Aktion, die auch als Analogie mit verruchten Personen wie Pilatus, Judas, Pfaffenweiber, Hexen und korruptem Müller (➔15) konstruiert werden konnte.[313] Oder wenn von ihm vielleicht nach dem Vorbild der Legenden um die Kreuzauffindung der heiligen Helena verlangt wird, mit dem heiligen Kreuz in Händen wiederzukommen (➔3) oder Krippe und Windeln des Jesuskindes beizubringen (➔16).

Besonders mag die Vertreibung böser Geister nach den Evangelien der Wunderheilungen und nach Matth. 12,43 und Lucas 11,24: „Wenn der unsaubere Geist von dem Menschen ausgefahren ist, durchwandelt er dürre Stätten" Pate gestanden haben. Gerade in Zeiten der Gegenreformation und des Barock hatten diese Texte Inflation und bauten kleine Litaneien zum Beispiel in Abwehr von Alpträumen durch Truden aus, denen man befahl, durch alle Wasser zu waten, alle Bäume zu entlauben, alle Grashalme oder alle Fichtennadeln zu pflücken, die Sterne, Schneeflocken, Regentropfen und Sandkörner am Meer zu zählen oder gar alle Kinder, die seit Christi Geburt zur Welt kamen. Einem ‚kaltes Gesicht' wird befohlen, Kieselsteine zu essen (➔17), ein auch schon neurologisch untersuchtes Adynaton (s. u.). Oder, weil die Truden ja durch fehlerhafte Taufe entstanden sind, Johannes den Täufer mitzubringen.[314] Manche weise Frauen oder Heilkundige haben drastische Bilder entworfen, wie jene Alptraumabwehr gegen die Mora aus Dalmatien (➔18). Diese Texte mit Kurzkontrastreihen wurden seit dem 15. Jahrhundert von Tausenden von heilkundigen Helfern im Volke angewandt. Im 11. Jahrhundert finden sich erste Andeutungen der Taktik in der Roßarznei (➔19).

Ein ausgedehnter erzählerischer Spruchtyp geht von einem Sinnkontrast aus, der dem Wirken des Schädlings beschwörend Einhalt gebieten soll und der meist auf heilige Personen und deren Gefühle zurückgreift. Es handelt sich hier um gegensinnige Bilder. Analog gesetzt wird dem Wurmtreiben etwa der falsche Spruch des ungerechten Richters (➔20) und das Leid der Gottesmutter unterm Kreuz. Dem Wurm soll also sein Nagen und Fressen ebenso *zuwider* sein wie das Urteil des Pilatus, eine Konstruktion, die seit dem 14. Jahrhundert meist in Blutsegen vorkommt.[315]

In der Volkskunde und Volksmedizin wurden all diese überraschenden Konstruktionen als Analogien bearbeitet und gemäß der historischen Kontrastkoppe-

---

313 Ernst, Wolfgang, Das Kunstbuch des Johannes Zahn von Dürnberg bei Wunsiedel, in: Archiv für Geschichte von Oberfranken 76 (Bayreuth 1996), Seite 187f.
314 Ernst, Wolfgang: Oberpfälzischer Heilzauber, Pressath 2011², S. 99–103,239.
315 Ohrt, Ferdinand, Ungerechter Mann (Segen) in: Handbuch des dt. Aberglaubens VIII,1416; zahlreiche Beispiele bei Monika Schulz: Beschwörungen im Mittelalter, Heidelberg 2003, S. 50f.

lung Magie versus Religion in empirische, magische und religiöse Vergleiche[316] unterteilt, speziell die Adynata-Aufgaben in für Dämonen angenehme und unangenehme,[317] die dem Dämon ablenkend, zeitraubend oder sinnlos-unendlich sein können. Dagegen hat der Altphilologe Alf Önnerfors mit der Definition „Parallelisierung von etwas nicht (mehr) Vollziehbarem und dem erwünschten Nicht-Entstehen eines Unheils"[318] auf den zeitübergreifenden Aspekt der Unmöglichkeiten hingewiesen. Die Germanistin Monika Schulz[319] erinnert an die Homerischen Vergleiche und verweist auf die Bedeutung des magischen Wortes mit seiner Fusionierung „präteritalen und präsentischen Anliegens", wo das Gestern ins Heute eindringt: „Der heilsgeschichtliche Kontext hat mit dem präsentischen Anliegen der gegenwärtigen Beschwörung von der Sache her nichts gemein" und gehört als „heilige Geschichte" einem „analogischen Konstrukt der Historiola" (so die Überschrift) an. Allerdings gehört nicht jeder Vergleich in den Bereich der N400-Inkongruenz, sodaß die Unterscheidung von Irmgard Hampp[320] von ‚negativer Beglaubigung' als diskrepanter Vergleich zu beachten ist: daß ein Krankheitsdämon zu verschwinden hat, bis die Gottesmutter einen anderen Sohn gebärt, bis der Jüngste Tag des Gerichts anbricht oder bis ein neues Evangelium geschrieben ist. ‚Positive Vergleiche' wie wir sie z. B. aus den Fieberbeschwörungen mit den Siebenschläfern kennen (**Kap. C4.1.1**), gehören in den Bereich der gleichsinnigen Analogie-Strategie. Damit kommen wir zur beabsichtigten Grenzüberschreitung, die das *Konstrukt* der wie immer gearteten oder beabsichtigten Transformationen und Adaptationen der Spruchanalogien aus dem Blickwinkel neurobiologischer Wirksamkeit untersucht.

C3.2 Die neurophysiologische Darstellung der Signale von Kontrast und Wider-Sinn bis zur ‚außergewöhnliche Sensation' bei N400

Nach seiner Entdeckung der elektrischen Potentiale am menschlichen Gehirn hatte der Psychiater Hans Berger 1929 sogleich zur Frage gestellt, ob man damit den Einfluß intellektueller Arbeit nachweisen könne. Jahrzehntelange Forschung und technischer Fortschritt haben dann das EEG (Elektroenzephalogramm) zu einem unverzichtbaren diagnostischen Instrument gemacht, aber erst in neuer Zeit wurden gezielte Ableitungen für die Neurolinguistik interessant (s. a. oben

---

316 Hampp, Irmgard: Beschwörung Segen Gebet, Stuttgart 1961, S. 152–158f.
317 Ebd., S. 98f.
318 Önnerfors, Alf: Antike Zaubersprüche, S. 21.
319 Schulz, Monika: Beschwörungen, S. 30, 50.
320 Hampp, ebd., S. 173.

Kap. **B3** und **B4**). Technische Finessen wie das Averaging erlauben nun bei dieser nichtinvasiven Methode hohe zeitliche Auflösbarkeit, also die Zuordnung zum aktuell Gehörten oder Gesehenen. Und erlauben vermittels Ableitung von Latenz, Amplitude, Polarität und Topografie von Summenpotentialen von je etwa 10000 synchron feuernden Zellen des Neocortex auch qualitative Unterscheidungen kognitiver Prozesse. Insbesondere fand sich bei einer Latenz von 400 Millisekunden ein Indikator für semantische Anomalien: Ist ein Wort im Satz falsch, z. B. „Der Honig wurde ermordet" [321] oder: „Er aß eine Schüssel Steine" zeigte sich ein vergrößertes Potential (**Abb. 6**), wenn man das Ergebnis mit einem sinnvollen Satz, etwa „Er aß eine Schüssel Chips" verglich.[322] Diese N400-Potentialveränderungen wurden folgerichtig auch zur Diagnostik der Schizophrenie verwendet. Das Fehlen dieser den Empfang einer Widersinnigkeit anzeigenden Neuronenreaktion kann auf krankhaften Bedeutungswandel im Denken hinweisen, falls nicht bloß eine isolierte anerzogene oder anlagemäßige Inkongruenztoleranz vorliegt oder – damit verwandt – eine double-bind-geprägte Entwicklungsstörung ohne evident pathologische Auswirkungen.

Seit den diesbezüglichen Forschungen von Gregory Bateson haben verschiedene Fachrichtungen über die Zusammenhänge zwischen sprachlich doppeldeutigen (ambigen) paradoxen kommunikativen Einflüssen auf Persönlichkeitsveränderungen und magischem Realitätsverlust – individuell und gesellschaftlich – hingewiesen.[323] Es wurde die Bedeutung der N400 als „kognitiver Mehraufwand bei der Desambiguierung eines mehrdeutigen Wortes" erwogen.[324] Und auch die Einflüsse von seelischer Stimmungslage auf die Wortverarbeitung bzw. umgekehrt die Wirkung positiv und negativ konnotierter Worte konnten mittels Messung der ereigniskorrelierten Potentiale differenziert werden. So sollen Reizworte wie ‚Liebe' und ‚Tod' eine unterschiedliche Verarbeitung haben.[325]

Der N400-Komplex reagiert allerdings nicht nur auf unsinnige Worte und Satzgebilde, sondern ist auch abhängig vom Überraschungseffekt: Je unerwarteter ein solches auditives oder auch visuelles Signal, umso größer die Amplitude

---

321 Friederici, A. D., Menschliche Sprachverarbeitung und ihre neuronalen Grundlagen, in: Meier/ Ploog (Hg.): Der Mensch und sein Gehirn, München u. a. 1998², S. 137–156, hier S. 145.
322 Vgl. dazu und zum Folgenden die Übersicht bei Drenhaus, Heiner und Peter beim Graben, Ereigniskorrelierte Potenziale (EKPs), in: Zeitschrift für germanistische Linguistik 40 (2012), S. 68–96.
323 S. z. B. von germanistischer Seite: Hermann, Fritz, Double-bind und Linguistik. Zur quasimagischen Zerstörung von Persönlichkeit durch kommunikative Paradoxien, in: Lange-Seidl, Annemarie (Hg.:) Zeichen und Magie, S. 57–69.
324 Schrott, R. und Jacobs, Arthur, Gehirn und Gedicht S. 173 mit Abb.
325 Nikki L. Pratt and Spencer D. Kelly, Emotional states influence the neural processing of affectiv language, in: Social Neuroscience 3 (3–4), (2008), S. 434–442.

von N400. Es fanden sich zudem Hinweise dafür, daß dabei nicht der Wahrheitsgehalt entscheidet, sondern das Ausmaß an Arbeit der Neuronen für die Aktivierung der lexikalischen Einträge im Gehirn. So konnten beim Vergleich zwischen kategorial verwandten und nicht verwandten Reizpaaren sowohl semantischer als auch phonetischer und visueller Art signifikante Differenzen gefunden werden; stets war die Amplitude von N400 bei nicht verwandten Reizpaaren höher. Da jedoch eine Ähnlichkeit der Peaks in hirntopografischer Hinsicht bestand mit einer Herkunftslokalisation bilateral mittlerer und superiorer Temporalgyrus (weniger deutliche und mit leichten Differenzen auch frontal und insulär), wurde N400 generell als Zeichen von ‚mismatch' gedeutet, das kognitiv beim Aufruf verschiedener Zusammenhänge erwächst.[326]

Besonders interessieren uns auch Ergebnisse der neurolinguistischen Märchenforschung. Sie lassen erkennen, daß in den Grimm'schen Märchen vor allem die unerwarteten Wendungen wie „Der dürre alte Baum redet mit dem Mädchen" mehr als etwa „ein alter wieder ergrünender Baum", also eine Art Wunder, zur Auslösung der N400 führen. Märchen mit solchen erstgenannten totalen Unmöglichkeiten haben eine viel größere Verbreitung gefunden.[327]

Darüberhinaus reagiert N400 nicht nur auf der Ebene von Sätzen, sondern auch bei der Integration größerer pragmatischer diskursiver Textinformationen. Mit einer Geschichte, in der sich eine Erdnuss in eine Frau verliebt, wurde zunächst durch diese semantische Verletzung („Die Erdnuss war verliebt") eine vergrößerte N400 ausgelöst; die Geschichte wurde dann so gestaltet, daß das Verliebtsein der Erdnuss im Verlauf akzeptabler wurde („Eine Geschichte über eine verliebte Erdnuss"). Am Ende der Geschichte stand ein Satz, der entweder das Gesalzensein oder das Verliebtsein der Erdnuss ausdrückte. Die Messungen im Vergleich von gesalzen zu verliebt ergaben dann eine größere N400 für gesalzen, sodaß der Schluss nahe lag, daß Kontextinformationen antizipiert werden und in der Folge sogar überschrieben werden können.[328]

Notieren wir an dieser Stelle bei Betrachtung der ereigniskorrelierten Potentiale noch eine weitere Einflußnahme auf das Nervensystem, die inzwischen durch diese Potentialableitungen nachgewiesen ist, die **Reimverarbeitung**. Sie

---

[326] Khaleb et al.: On the origin of the N400 Effects: An ERP waveform and source localization analysis in three matching tasks, in: Brain topography 23 (2010), 311–320.
[327] Norenzayan, Ara et al., Memory and mystery: The cultural selection of minimally counterintuitive narratives, in: Cognitive Science 30 (2006), S. 531–553.
[328] Drenhaus, H. und b.Graben, P, S. 79 nach Nieuwland, M.S. und van Berkum, J.J.A., When peanuts fall in love: N400 evidence for the power of discourse, in: J. of Cognitive Neuroscience 18 (2006), 1098–1111.

wurde erstmals 1984 entdeckt.[329] Mag ein veritabler Niederschlag von Endreim, ja sogar wie später noch festgestellt, von Alliteration im menschlichen Gehirn zwar nicht überraschen, mag ein exaktes konsekutives Verarbeitungscluster auch noch nicht gesichert sein, – rein phonologisches Signal oder schon lexikalischer Zugang? – so deutet sich doch mit dem medial und rechts hochparietal betonten N450-Potential ein Merkmal und Zeichen dafür an, welcher Stellenwert jeder Reimdichtung im allgemeinen und speziell in den Heilsprüchen auch körperlich-biologisch zukommt.

Ergänzend zu unserem Ausgangspunkt einer hypothetisch rein kognitiven Inkongruenzreaktion und im Blick auf die erwähnten double-bind-Theorien muß bedacht werden, daß *emotionale* Einflüsse auf diese ereigniskorrelierten Potentiale (EKP) nicht von vorn herein ausschließbar sind und in letzter Zeit auch untersucht wurden. Emotional wirkende linguistische Stimuli – nicht nur Film-, Gesichts- und Bilder-Reize – zeigten für syntaktische Verletzungen einen deutlichen Einfluß auf eine linke anteriore Negativität (LAN), auf ein P600-Potential und auf weiteres komplexes Potentialverhalten.[330] Ferner werden auch der Hippocampus und das vordere Cingulum als stets für Neuigkeitssignale reagibler Komplex auf Diskrepanzen zwischen Erwartetem und tatsächlich Rezipiertem aktiviert.[331] Darauf wird im Rahmen der narrativen Performation noch zu kommen sein.

C4 Die Inszenierung von therapeutischen ‚Bildern' mittels Wortfiguren und Erzählung

Dieses Kapitel widmet sich einer breiten Palette von therapeutischen verbalen Angeboten an das Gehirn des Patienten in der Krisen- und Notsituation und den damit verbundenen mehr oder weniger einfühlsamen bis provozierenden Einflussnahmen auf das Gehirn. Es handelt sich um zugleich kognitiv und emotional wirksame Eingaben, die dem Ziel dienen, innere Vorstellungskräfte, Imaginationen aus verschiedenen inneren ‚Bibliotheken', anzuregen und neuropsychosomatische Umstimmungen zu bewirken. An der einen Seite der therapeutischen Möglichkeiten steht a) die bloße Vorgabe eines Bildrahmens z. B. im Sinne einer

---

329 Rugg, M.D., Event-related potentials in phonol. matching tasks, in: Brain & Language 23 (1984), 225–240.
330 Jimenez-Ortega, Laura et al., How the content of discourse affects language comprehension (2012) PLoS ONE 7(3): e33718. doi:101371/j. pone.0033718.
331 Kumaran, Dharshan und Eleanor A. Maguire, A unexpected sequence of events: mismatch detection in the human hippocampus (2006) doi: 10/1371/j.pbio. 0040424.

suggestiven situativen Verortung der inneren Vorstellungen des Patienten mit
– ob vom Therapeuten bewußt gewollten oder ungewollten – Anregung zum
eigenen Phantasieren, wie wir das von vielen historischen Spruchtexten bis zur
heutigen Methode des katathymen Bilderlebens kennen. Auf der anderen Seite
stehen b) ‚exorzistische' Verfahren, wo wir den Therapeuten im Intensivrapport
mit einem Symptom personalisierter bzw. metaphorischer Dämonisiertheit sehen
können. Dazwischen betrachten wir c) die Fülle der Darbietung von Kurzerzäh-
lungen, den Historiolae durch einen gleichsam allwissenden, auktorialen Erzäh-
ler, die sowohl an Erlerntes und Erfahrenes beim Notleidenden anzuknüpfen und
Ressourcen zu mobilisieren versucht, als auch andererseits mit überraschenden
Neuigkeiten Umstimmungen erreichen möchte. Dabei ist jedoch eine strikte
Separierung dieser Zugriffsmethoden kaum möglich, zumindest für den Bereich
historischer Kulturen.

Psychotherapeutische Strategie

moderne Begriffe:
- Erzähltherapie, Narrative Imagination, guided affectiv (mental) Imagery
- Katathyme Bildgebung (Leuner), suggestive Imagination, Rêve éveillé dirigé (Desoille)
- Reframing (NLP), Narrative Exposition, Verbildlichung von Sprache (Linguistik), Schamanisch-mythische Anatomie und Physiologie (Lévi-Strauss) – Ressourcenmobilisierung

historische/ alltägliche/ volkstümliche Begriffe:
- gezielte Ablenkung, Umstimmung durch Themenwechsel, Lenkung der Einbildungkraft – Beeinflussung durch Suggestion, Kettenmärlein und Litaneien – Katharsis (nach historischem Verständnis kultische Reinigung)

Anwendungsfelder/ Indikationen für den Einsatz verbalisierter Bildgebung
Schmerz, Angst, psychosomatische und histrionische Störungen, Geburtsbegleitung, Traumafolgen, Schlafstörungen

Neurophysiologie
Aktivierung gezielter Aufmerksamkeit in verschiedenen kognitiven und emotionalen Gebieten zur Wiedererinnerung und Modulation gespeicherter innerer Bilder und zur Konstruktion neuer Ideen und Bilder

---

Tabelle 4: Narration und narrative Elemente in der psychoperformativen Akuttherapie: Induktion von Regression und Katharsis und die Bedeutung von Sprachfiguren (Tropen)

Historisch frühzeitig finden sich Erzählungen und erzählerische Elemente, auf denen Heilspruchtexte aufbauen konnten in der assyrisch-babylonischen Schöpfungsgeschichte, etwa zur Entstehung des Wurmes (**Beispieltext ➔48**) und in Heilsprüchen des Marduk-Ea-Typs (➔49). Als Typen der Erzählungsgestaltung im europäischen Bereich wurden die Analogieerzählungen in ihrer Bedeutung erkannt und im Begegnungsspruch (➔17, 35) und Wanderschaftsspruch (➔24) gefunden.[332] Allgemein war für die Entstehung der narrativen Elemente „eine Art Götterwelt" zur „Fusion von Vergangenheit und Zukunft, die als heilbringend gesetzt wird" vorzustellen.[333] Dabei kann vor allem diese alte Marduk-Ea-Formel mit der Frage des Menschen-Boten und Göttersohnes, was für den Kranken zu tun sei (➔49: „what will soothe her?"), herangezogen werden. Auf diesem Boden einer heiligen Geschichte über die Taten der Götter für Krankheiten habe sich die Erzählung (Historiola, epischer Teil des Zauberspruchs) über die Funktion einer Analogie hinaus entwickelt. Und aus einer „sakralen und segenspendenden Atmosphäre der mythischen Zeit"[334] sei eine „Charta des magischen Tuns in der Gegenwart" geworden.

Die hervorragende **Bedeutung der Erzählung** erwuchs für den religiösen Menschen aus der Inszenierung einer fiktiven Teilnahme am Schöpfungsgeschehen, an einem originalen ersten Gestaltungsprozess aus diffusem gestaltlosen Raum wie aus Himmel und Erde (➔49, 60). Dabei konnte dann über Ereignisse mythischer, legendärer und phantastischer Provenienz ebenso wie über kanonische und real-säkulare Schilderungen ein Sachverhalt mitgeteilt werden, den das Gehirn auf seine mentalen kognitiven und emotionalen Anteile und seine lebenspraktische Bedeutung prüfend abtastete. Somit kommt es immer wieder zu Verbindungen zwischen Irdischem und Überirdischem: „Zauberin ... die du den Himmel beschmutzt" (➔45) „Die Krankheit fiel vom Himmel" (➔10).

Jeder narrative Komplex besteht aus Ketten von Signalen, deren einzelne ‚Perlen' für die Verarbeitungsprozesse des Gehirns unterschiedlich in Abhängigkeit von ihrer lebensweltlich-situativen Nützlichkeit perzipiert werden. Das Gehirn wird nach seinen lexikalischen Fertigkeiten und seinen Erinnerungsprädilektionen verortende, verwirrend-ambivalente, gegenstandsbezogene oder personenbezogene ‚Bilder' aus einem Text heranziehen, ohne daß der vorgestanzte Erzählkomplex sofort als ganzer ‚verstanden' und ‚erwünscht' ist. Insofern sind die in der Krisen-Notfallsituation in Heilspruchtexten angebotenen verschiedenen Erzählelemente klar von Unterhaltungsdarbietungen abzugrenzen, die einen Zuhörer oder Leser ‚beschäftigen', denn dabei sind immerhin Tendenzen zu vermuten, die den einzelnen Perlen einer Erzählkette ähnlichgewichtete Perzeption und Aufmerksamkeit widmen. Auch im Bereich des scheinbar bloßen Er-

---

332 Vgl. Hampp, Irmgard, S. 175ff, S. 216ff.
333 Schulz, Monika: Magie oder, S. 287ff.
334 So nach Lévi-Strauss: Das wilde Denken, S. 273.

zählens ist die situative Bindung einer triadischen Kombination also nicht zu vernachlässigen.

> Wir wissen aus den bildgebenden fMRI-Untersuchungen, daß das Verstehen von Erzählungen sich auf die Fähigkeit des Gehirns stützt, Absichten und Erkenntnisse verschiedener in der Erzählung interagierender Vermittler, einem Ziel oder Problem näherzuführen. Man hat diese Verarbeitung mit einer ‚theory of mind' verglichen, einem „reasoning about the state of mind of another person, real or fictional". Und man hat die Angelpunkte eines interagierenden Netzwerkes in der rechten temporoparietalen Junction (**Abb. 2**) und im dorsomedialen praefrontalen Cortex gesucht, wobei eben auch die Erfassung von aufeinanderfolgenden Signalketten hineinspielen muß. Diese Funktion ist bei Autismus gestört.[335] – Wichtig war auch die Unterscheidung des Verstehens fixer Formelphrasen („Formuleme") von all dem, was Neuigkeit bietet. Das erstere findet zusammen mit Gesten auch Verarbeitung in den Speichern der Basalganglien und ist bei Parkinsonkranken behindert,[336] das Neue einer Erzählung oder Formulierung wird über Hippocampus und ACC vermittelt. Neues und Formelhaftes stehen in einem „Wechselspiel".

Bei der Vermittlung und Perzeption von Heilzaubertext finden eben nicht lexikalisch inhaltlich ‚ausgewogene', ‚neutrale' Zugriffe, sondern scharfe Selektionen statt, die die Bildeingabe subjektiv in jenem Ausmaß ‚färben' und ‚filtern' können und ihre Qualität und Quantität dem eigenen Bedarf anpassen.

Diese scheinbare Einschränkung der Bedeutung des narrativen Elementes verbürgt die Möglichkeit, dem Gehirn performative Trigger nach jeweiligem situativem Kontext und kulturellem Gehalt sozusagen differentialtherapeutisch als kommunikative elektive Prozedur anzubieten. Es ist deshalb immer von den Einzelelementen des Narrativen auszugehen. Ziehen wir einen Blutungssegen heran, der einleitend die kleine Geschichte einer Wundverletzung des Christuskindes beim Spielen andeutet (➔21), dann könnte das Gehirn des Patienten vielen verschiedenen Afferenzen eine Rangfolge zuordnen: Erinnerung an eigene ‚Verletzung', physisch, psychisch, blutig, unblutig, als Kind beim Spielen, als Soldat im Kampf, aktuell oder früher. Erinnerung an ein Bild oder eine Statue Christi als Wundmalweiser mit dem Daumen an der Seitenwunde. Erinnerung an die Legende um Longinus, den Erzeuger der Seitenwunde, an seine Lanze als Verletzung oder an seine Lanze als Reichsinsignie des Kaisers oder doch daran, daß der ‚blinde Jude' seine Sehkraft durch Christi Blut wiedererlangte. Oder gar an eine selbst verschuldete Wunde bei einem anderen und so weiter. Man sieht, daß der kleinste Erzählteil unabschätzbare Richtungen nehmen kann und daß die

---

335 Mason, R.A. und J.A. Just, The Role of the Theory-of-Mind cortical network in the comprehension of narratives, in: Lang. Linguist. Compass (2009) Jan 1;3(1):157–174; Abstract PMID: 19809575.
336 Van Lancker Sidtis, Diana, Two-Track Mind: Formulaic and novel language support a dual-process model, in: Faust, Miriam (Hg.:) Handbook, S. 342–367, hier: S. 358f.

Summation von thematisch gleichgerichteten Erzählpartikeln und gleichgerichtetem Ritus zur Verstärkung von großer Bedeutung war.

Zur Auswahl von Metaphern als intersubjektive sprachliche Kraft

Werfen wir einen Blick auf Sprachbilder, auf das, was im Verständnis von Imagination immer schon mitschwingt. Von hoher nicht zu unterschätzender Bedeutung nicht nur für jede gelingende Therapie, sondern auch für alle menschliche Verständigung durch Sprache und Schrift gilt allgemein die Metapher. Sie begleitet uns überall. Ohne Metaphern, ohne Sinnbilder, ist menschliche Kommunikation nicht vorstellbar. Und wer ein Sprechen ohne sie versuchen würde, wäre rasch als Apparat erkennbar, er wäre unbelebt oder zum Schweigen verurteilt. Ohne die Verwendung von Sinnbildern vermögen wir es nicht, verschiedene Objektbereiche zu verbinden, zu erkennen, uns vorzustellen oder zu vergleichen. Mit Metaphern treffen wir Auswahl und Entscheidung, und schaffen damit subjektive Lichtblicke und Lichtkegel; man bedenke nur etwa die geometrischen und mathematischen Fachbegriffe und Maße, mit denen jetzt hier ein Lichtkegel zu beschreiben wäre.

Wir filtern aber mittels Metaphern auch ab und verbergen bestimmte Aspekte. Das bildet eine wichtige Strategie beim Unterlaufen psychologischen Widerstands gegen Anmaßung von Therapeuten, beispielsweise ihrer unausgesprochenen Beherrschung abstrakter Kategorien mit Verhaltenskodices, Intellektualismus und moralischen Rechtsprechungen. Das ‚Unterlaufen' kann zum ‚Überfliegen' werden. Andererseits: Im Verbergen und Betonen, im Verlagern und Hervorheben von Sprachbildern unterläuft der Therapeut auch die bewußten oft allzu stabil gespeicherten ‚Vormeinungen', vermag sich brach liegenden seelischen Gebieten zu nähern und Ängste zu entschärfen. Ressourcen können aktiviert werden, um die vergessenen positiven Seiten der Biografie wiederzubeleben. Die Suche nach geeigneten zündenden Sprachbildern, anknüpfenden Erinnerungen und ihren Kombinationen ist entscheidende Arbeit im therapeutisch imaginativen Prozess. Die Fundgruben für diese Zündungen mögen in alten Kulturen mehr den im Kollektiv erlebbaren Ereignissen entsprechen. Für die alten und vormoderne Kulturen war die Konvergenz der von außen therapeutisch gegebenen Bilder mit religiösen oder mythischen Sinnbildern der jeweiligen Zeit zum glatten Transport auf einem ‚Königsweg' ebenso wichtig wie die Beachtung der unendlich vielen Königswege der Individualität westlicher Kulturen im 21. Jahrhundert.

In den Eingangskapiteln war schon auf die große zeitlose Bedeutung der imaginativen Besprechungs-Techniken und deren Anwendung hingewiesen worden, auf die diesbezügliche Gemeinsamkeit von Priester- und Mönchsärzten und

Schamanen (**Kap B2**), auf die über Wort- oder Bilddarbietung und Ritus gleichermaßen anregbare mentale Visualisierung einer ars memorativa (**Kap. B4.1**), auf die neurale Verflechtung imaginativer, perceptiver und aktionstriggernder Materialisierung ins Psychosomatische (**Kap. B3**) und auf die Beziehungen zu heutigen Verfahren (**Kap. A1.2.3.c/d**). Die neurobiologischen Untersuchungen können in Ergänzung zu den bisherigen Forschungen auch dazu beitragen, den Verlauf akustischer Darbietung von narrativen Suggestionen in neurale Verarbeitung zu ‚inneren Bildern' und ihre Modulierungen zu verfolgen.

Zur neuralen ‚Installation' von Tropen (vgl. auch Kap. B2 und B3)

Grundsätzlich ist davon auszugehen, daß bei den meisten Menschen besonders in Notsituationen und bei Krankheit gute Ratschläge, plumpe Ermahnungen, sich zu beruhigen oder gut gemeinte Aufforderung, keine Angst zu haben, nicht zum Ziel führen und sich als sinnlos oder kontraproduktiv erweisen. Derartige dabei eingesetzte abstrakte Worte kommen schwer an. Das haben nicht nur ärztliche und nichtärztliche Erfahrungen, sondern auch neurophysiologische Untersuchungen gezeigt.[337] Vielmehr kann der Hilfsbedürftige vermittels *konkreter* Bilder und Sinnbilder am sichersten Schritt für Schritt geführt werden.

Besonders in Trance erbrachten die bildgebenden Untersuchungen (fMRI) mit Gegenüberstellung hoch bildhafter Wortpaare wie Affe oder Kerze gegen abstrakte Wortpaare wie Buße und Moral eindeutige Ergebnisse: Bei suggestiblen Personen ist die Reproduzierbarkeit der hoch bildhaften Worte erhöht, während Abstrakta zu verminderter Reaktion führten.[338] Außerdem scheint Hypnose „einen direkten Einfluß auf das implizite, nicht bewußt ansteuerbare Gedächtnis zu haben; es werden unbewußte Suchvorgänge evoziert." [339] Der sogenannte innere Widerstand, die Vermeidungshaltungen, werden unterlaufen, indem einfach vorstellbare konkrete Objekte genannt werden, etwa Pflanzen oder Körperteile (**Kap. C4.2**), zu denen das Arbeitsgedächtnis unschwer naheliegende Assoziationen über erinnerbares Aussehen, Geräusch, Anfühlen oder auch Geruch herstellt. Das gleiche gilt aber im Krisenfall auch schon für Imagination ohne direkte Trance oder Hypnose, weil die indirekten Anregungen zur inneren Bildvorstellung nicht aufdringlich sind.

---

337 Tolentini, L.C., Are pumpkins better than heaven? An ERP investigation of order effects in the concret-word advantage, in: Brain & Language 110,1 (2009), S. 12–22; West, W.C. Holcomb, P.J., Imaginal, semantic, and surface-level processing of concret and abstract words: an eletrophysiological investigation, in: Journal of Cognitiv Neuroscience 12, 6 (2000), S. 1024–37.
338 Vgl. Halsband, Ulrike, in: Revenstorf/ Peter, S. 813.
339 Ebd., S. 815.

Unterstellen wir also eine meistens imaginative Hypnoidtherapie, mit Vorgabe von Zeit und Ort verändernden Bildern einer einst bekannten Legende, Erzählung, oder einer abgewandelten biblischen oder mythischen Geschichte. Die Geschichte konnte im erwartungsvoll geöffneten und doch unerwartet durch die mit Tropen arbeitende Modulation im überraschten Gehirn einen komplexen elektrochemischen Erregungszustand mit folgenden weitgehend synchronen Vorgängen herstellen:

– Das Arbeitsgedächtnis ist zum ersten flüchtigen Empfang pragmatischer Signale aufgerufen und gibt eine erste Skizze der flexiblen Planung.
– Es kommt zum Abgreifen lexikalisch gespeicherten Materials über den Hippocampus, dem Archiv für Gedächtnis, unter Einschluß präfrontaler Gebiete, die über eine zielgerichtete Relevanz bezüglich Wiedererinnerung mitentscheiden.
– Das anteriore Cingulum (ACC) überprüft auf kognitiven Neuigkeitswert.
– Es wird festgestellt, daß die neuen Signale der zu installierenden Bilder (Ort-, Zeit- und Ereignisvorgaben) den gespeicherten inneren Bildern nicht total fremd sind, sodaß sie mittels elektrophysiologisch turbulenter Reizflutung verarbeitet und assoziierend angenähert werden können.
– Infolge gleichzeitiger Reizausbreitung in die limbischen Gebiete, besonders die Amygdala, werden die neuen Signale auch auf ihre emotionale Verwertbarkeit überprüft hin auf eine Entscheidung über Ablehnung oder Annahme. Je nachdem erfolgt selektive Ausschüttung von Botenstoffen.
– Diese Botenstoffe können die Aufnahmebereitschaft steigern und sowohl in höheren Zentren präfrontal (–> mögliche bewußte Weiterverarbeitung) als auch in vegetativen Schaltstellen (–> Durchblutung, Atmung, Blutdruck, Herzfrequenz) wirksam werden.

Erste Nachweise für einige der genannten Funktionsabläufe unter Voraussetzung einer durch Trance gebahnten Psychoperformation ist in den letzten Jahren durch die bildgebenden Magnetresonanzverfahren erbracht worden. Zum Beispiel zeigten Untersuchungen mit dem Indikator Farbwahrnehmung eine Aktivierung des für komplexe visuelle Stimulusverarbeitung zuständigen Gyrus fusiformis links, (unterer occipito-temporaler Cortex), der zum Sehzentrum gehört. Es konnte nachgewiesen werden, daß in Trance je nach verbaler Instruktion der hochsuggestiblen Versuchspersonen graue Muster als farbig und farbige Muster als grau interpretiert wurden. Die verbale Suggestion visueller Bilder bewirkt außerdem „ein komplexes zeitliches Zusammenspiel des Farbsehzentrums mit dem ACC und dem parietalen Cortex". „In der Hypnose setzen sich verbale Instruktionen leichter in innere Bilder um und rufen somit eine veränderte Realitätswahrnehmung hervor."[340] Das gilt ebenso für die Imagination anderer Sinnesqualitäten und beweist die Möglichkeit einer besseren „Nutzbarmachung multimodaler Verarbeitungsweisen unter Hypnose."[341]

So konnten neben der Farbmodulation bei Versuchspersonen auch Formveränderungen durch Imagination hervorgerufen und mit bildgebenden Techniken nachgewiesen werden: Fordert man unter Experimentbedingungen dazu auf, mit geschlossenen Augen den Umriß eines vorher gezeigten, also perzipierten und im Gedächtnis gespeicherten Gegenstandes zu vergrößern, dann läßt sich diese Größenveränderung auch in Area 17 der Sehrinde nachweisen. Das heißt, daß eine von außen erwünschte ‚innere' Bildgestaltung Aktivierungen der selben Areale bewirkt wie die direkte Perzeption mit den Sinnesorganen.[342] Stephan Kosslyn und sein Team erinnern an das Sehen mit dem inneren Auge [wir folgern: an den blinden Seher Teiresias und all seine Nachfolger] wie ebenso an das Hören mit dem inneren Ohr, wie wir es vom ertaubten Beethoven kennen. Daß auch Verbindungen durch innere oder äußerlich angeregte Freudens- oder Schreckensbilder via Visualisation emotionalen Materials ins limbische System aktiviert werden und in die Insula als Feedbackstation des autonomen Nervensystems, haben wir schon oben gezeigt und wird uns bei der Betrachtung narrativ kathartischer Imaginationen erneut begegnen (**C4.3.1.2**). Innere Bildvorstellung ebenso wie Metaphorik sind aus dem Vulgärgebrauch etwa des Begriffes ‚Ein-Bildung' als blöde Naivität oder der politischen Instrumentalisierung von ‚Vision' – als allein krankhaft – herauszulösen. Vielmehr vermögen sie generell fremd- oder selbstgesteuert in vielen Bereichen des Lebens für manches Hindernis und manchen Mangel einzuspringen.

Im Hinblick auf die Dominanz der Hirnhälften wird allgemein für die linke Seite beim Rechtshänder ein Schwerpunkt für logische Entscheidungen angenommen. Die Erforscher der Korrelate von Tropenempfang anhand der bildgebenden Verfahren waren sich jahrelang unsicher über eine mögliche Hemisphärendominanz dieser Funktion bei Metaphern und Metonymen. Zuletzt hat die Arbeitsgruppe um Alexander Rapp eine differenzierte Studie zur Metonymie vorgelegt.[343] Verglichen wurden unter anderem literale („Afrika ist trocken"), metonymische („Afrika ist hungrig") und sinnlose Kurzsätze („Afrika is wollig"). Danach geschieht metonymisches Verstehen per Lesen in einem links lateral betonten frontotemporalen Netzwerk mit Maxima links, aber auch rechts inferior temporal. Man hat schlußfolgernd einen Rückgriff auf ‚Weltwissen' über den links inferior frontalen Bereich des Broca'schen Sprachzentrums erwogen aber auch dazu alternativ für diese Aktivierung den erhöhten kognitiven

---

340 Ebd., S. 809f.
341 Ebd., S. 811.
342 Kosslyn, S M et al., Mental Imagery, in: Berntson/ Cacioppo, Handbook, S. 383–394, hier: 386.
343 Rapp, A. et al., Neural correlates of metonymy resolution, in: Brain & Language 119 (2011), S. 196–205.

Verarbeitungsbedarf zur Integration des nonliteralen zum allgemeinen Satzsinn. Metonyme werden rhetorisch als Tropen der Kontiguität verstanden, weil sie von der Gottheit zu einem Funktionsbereich, von einer Handlung und Person zu deren räumlichen und zeitlichen Hintergründen, vom Konkretum zum Abstrakten und umgekehrt oder vom Körperteil zur Eigenschaft führen; damit wird ihre neurale Verortung und Interaktion von Interesse.[344] Ich erinnere an Frazers und Tylors ‚contagion' (Kap. A1.2.1). Unsere Beispieltexte sind voll von ihnen.

Andererseits fanden Untersucher für Metaphern und Metonyme, daß beim Hören eigentlich keine gesonderten Regionen aktiviert werden, die kulturell Erworbenes speichern, sondern daß der Empfang über das ganze Gehirn verteilt sein kann abhängig vom jeweiligen Begriff im domänenspezifischen Rindenbereich. Das war wohl nicht anders zu erwarten. So werden etwa Metaphern, die mit Berührung zusammenhängen über entsprechende sensorische Gebiete verarbeitet, ebenso wie Metaphern der Motorik über sensorimotorische Netzwerke perzipiert werden. Eine berührungsrelevante Metapher wie „This is a hairy situation" aktivierte im parietalen Operculum [rechts], einem Gebiet, das mit Berührung betraut ist, während ein taktil-neutraler Satz („This is a precarious situation") dieses Gebiet nicht beeinflußte.[345] Dies entspricht der schon mehrfach geschilderten Vorstellung einer ‚embodied simulation' (Kap. B1, C1) und konnte bestätigt werden durch transkranielle Magnetstimulation der zentralmotorischen Rinde für Beinmuskulatur: Die Muskelpotentiale dort zeigten die höchste Modulierbarkeit dann, wenn in den Sätzen beinbetreffende Verben in literalen oder metaphorischen Konstruktionen eingesetzt wurden („La ragazza attraversa la strada trafficata"; „La signora corre con la fantasia spesso"), weniger bei semantisch abstrakten Satzbildungen.[346]

Im Hinblick auf die kontrovers diskutierte Bedeutung der Hemisphärendominanz scheint letztlich für die Verarbeitung von Sprachfiguren ganz allgemein die Ansicht am plausibelsten, daß die rechte Hirnhälfte Schwerpunkt für semantische Regelverletzungen und für Emergenz ist, während die linke Hirnhälfte eher eine Basis für bekannte semantische Regeln bietet.[347]

Die prosodische und gestische Komponente bei der ‚Installation' von Sprache

In vorangegangenen Kapiteln (Kap. B1.1 und B2, C2) zu Fragen der Hilferufe und der Applikation von Spruchtexten durch den Helfer in früheren Kulturen hatten wir schon auf die hohe Bedeutung der nichtsprachlichen Kommunikationsprozesse hingewiesen. Wie Worte in einer Notlage bei einem Kranken ‚ankommen', hängt – abgesehen von den performativen Sprachgestaltungen – nicht nur von ihrem Inhalt ab. Melodie, Intonation, Betonung und Artikulation von Sprache, verbunden mit Mimik und Gestik spielen eine gewaltige Rolle. Pros-

---

344 Plett, F. Heinrich: Einführung in die rhetorische Textanalyse, Hamburg 1991, S. 77f.
345 Lacey, Simon et al., Metaphorically feeling: Comprehending textural metaphors activates somatosensory cortex, in: Brain & Language 120,3 (2012), S. 416–421.
346 Cacciari, C. et al., Literal, fictive and metaphorical motion sentences preserve the motion component of the verb: A TMS study, in: Brain & Language, 119 (2011), S. 149–157.
347 Faust, Miriam, Thinking outside the left box: The role of the right hemisphere in novel metaphor comprehension, in: dies. (Hg.:) Handbook, S. 425–448, hier: S. 441.

odie wird von Kindheit an auch mit Umwelt verzahnt in der Hirnentwicklung angelegt (**Kap. B4**). Daß dies so ist, belegen jüngste fMRT-Untersuchungen, die uns zeigen, daß die Produktion von Prosodie und die Perzeption von Prosodie in den gleichen prämotorischen Regionen, besonders im linken inferioren Frontalgyrus sich überlappen, ebenso wie Aspekte der sozialen Kommunikation, Empathie und Verständigung.[348] Bei einer Untersuchung mit Wort-Gesten-Paaren erwies sich eine kongruente begleitende Geste für die Verarbeitungsaktivität eindeutig fördernd.[349] Man denke auch an die bekannte unter Umständen hohe Bedeutung des Lippenablesens für Sprachverständnis, ganz zu schweigen von der Gebärdensprache für Taubstumme. Diese Ergebnisse kommen angesichts der stammesgeschichtlichen und ontologischen Entwicklung von Sprachträgern nicht unerwartet;[350] wir hatten im Zusammenhang mit dem Verhalten von hilfebedürftigen Kleinstkindern (**Kap. B1**) schon darauf hingewiesen: Gestik kommt vor Sprache. Dementsprechend bauen viele Heilspruchtexte mit zugehörigen Handlungsanweisungen auf der psychosomatischen Ur-Konvergenz von Wort einerseits und Geste und deren kulturell fixierten Surrogaten, den Riten, andererseits auf. Beispielsweise ist es die Behandlung des Wurmes Talpa mit Umgehungen und Fußtreten (➔29) oder das Schienen mit der Haselnussrute bei Frakturen (➔31), was eine Wiederverbindung der vermeintlich zerrissenen Lebenselemente Psyche und Soma und eine Verkörperlichung von Handlung hervorruft. Das ist nicht in einem vermuteten ‚magisch' intransparenten Sinne, sondern als kommunikative intersubjektive Hirnkorrelation von Gestiksignalen über diese vorsprachliche menschliche Reizvermittlung des Spiegelneuronsystems zu sehen.[351] Das gleiche gilt für Texte, die Pflanzen- und Kräutermittel einsetzen, weil die Spiegelneurone an der Anthropomorphisierung und Theriomorphisierung der Welt durch Projektion des Menschseins in eine als *be-greifbar* und zugleich verstehbar vorgestellte Welt beteiligt sind.[352]

Wie schon erwähnt, können nach Schätzungen im alltäglichen Gespräch 38 % über die Sprachmelodie und 55 % als Mimik und Gestik ein anderes Gehirn erreichen. Wenn auch in dringenden Erwartungshaltungen die Aufnahmeaktivität der Hörrinde wächst, so kann etwa das Gehirn einer Kranken unserer Zeit im Falle eines ‚sachlichen' ärztlichen Herunterstammelns von Labordaten als Überfülle an kognitiven weit vor emotionalen Erregungsmustern eine nonverbale

---

348 Aziz-Zadeh, Lisa et al., Common premotor regions for the perception and production of prosody, (2009) PLoS One 5(1): e8759. doi: 10.1371/j.pone.0008759.
349 Josse, Goulven et al., The brain's dorsal route for speech represents word meaning: Evidence from gesture (2012), PLoS One 7(9): e46108. doi: 10.1371/j.pone.0046108.
350 Tomasello, Ursprünge, S. 32, 121.
351 Vgl. Schippers, M. B. et al., Mapping the information flow, (**Kap B3**, Hyperscanning).
352 Vgl. Schrott/ Jacobs, Gehirn und Gedicht, S. 30.

Information vermissen, was zu Unstimmigkeiten führt. Andererseits können Ärzte oft mit Erstaunen berichten, daß Patienten von aufklärend begründenden Erläuterungen ihrer Zustände nichts verstanden haben, daß sie jedoch immer wieder an ihren Lippen und Augen hingen oder am Telefon an ihren Stimmen, wenn diese auch nur Gelassenheit und Zuversicht aussandten. Worte müssen also inszeniert werden, müssen rituell präsentiert oder/und emotional gleitend und in Übereinstimmung mit den Inhalten übermittelt werden. „Nur wenn die drei Instrumente [Mimik, Gestik und Sprachmelodie] der nonverbalen Kommunikation das gleiche mitteilen, können sich Vertrauen und Zuneigung entwickeln"[353], um ein Empfängergehirn so zu erreichen, damit beispielsweise eine akute Alptraumstörung entschärft wird.

Ziehen wir etwa unseren Münchner Nachtsegen (**Kap. C1**) heran, so ist, wie so oft bei dieser Textart, in den mittelalterlichen Dokumenten ein Anwendungsmodus nicht beschrieben. Der situative Kontext, der Zeitaufwand, oder gar die jeweiligen Subjekte in ihren Eigenheiten und Gesinnungen sind nicht bekannt. Aber es darf vermutet werden, daß sein Sprecher, sei es ein Arzt oder ein Mönchsarzt gewesen – der Segen steht in seinem Codexband mit medizinischen Schriften und Heilkräuterlisten zusammen – den Text nicht seelenlos ablas. Die Einbindung in Psalmen spricht für einen Ritus und soweit unser Wissen über allgemeinen Gebrauch dieser performativer Texte reicht, wurden sie meist als Sprechgesang vorgetragen. Neben der Benennung der Geister ist also eine nonverbale Komponente für eventuellen Erfolg bei dieser Psychotherapie strikt vorauszusetzen. Ebenso wird auch ein Spruch zur Unterstützung von Entbindung, ja werden viele der genannten Texte nicht ohne angemessene gestische, mimische und prosodische Begleitung vollzogen worden sein.

C4.1 Die regressionsfördernde verbale Bildgebung

C4.1.1 Eine Regression analogisierende Imagination mit Verortung und Objektvorgabe (Ephesus, Mutterschoß und legendäre Siebenschläferhöhle) im europäischen Mittelalter (Beispieltext ➜22)

Die Legende

Unser erstes Beispiel (➜22) für die erzählerische ‚Installation' eines Bildes in das erwartungsvolle Gehirn eines akut Notleidenden – sei es ein fieberkranker

---

353 Dichgans, J. in: Frühwald, Das Design, S. 220.

Schlafloser oder ein panisch Erregter – hat seelische und körperliche Beruhigung zum Ziel. Dafür dient hier ein Legendenbruchstück. Ich setze eine Kenntnis der Siebenschläferlegende von Ephesus für den mittelalterlichen Menschen voraus, weil sie von Theologen verbreitet wurde. Die Siebenschläfer sind im dritten Jahrhundert nach Christus wegen ihrer Verweigerung einer Verehrung heidnischer Götter eingemauert worden. Als nach fast zwei Jahrhunderten die Höhle geöffnet wird, entsteigen sie ganz lebendig und kommen in ein inzwischen christliches Ephesus. Die Legende diente der Glaubenserziehung als Zeichen für die Uneinholbarkeit christlichen Märtyrertums gegen Polytheismus.

Ganz anders nun für den sensorischen Apparat eines einzelnen Patienten: Die Höhle am Berg Celion bei Ephesus wird einem Embryonalzustand analogisiert. Der Therapeut versetzt a) den Kranken an einen heiligen Ort, weg von der aktuellen Leidensstätte. Er leitet ihn b) zur zweifachen zeitlichen Regression – nicht auf, sondern expressis verbis eindeutig in den Schoß, den Uterus, einer leiblichen Mutter sowie den Schoß eines legendären dramatischen Ereignisses und erweitert danach c) den Horizont der Märtyrergestalten von ihren lateinischen Eigennamen in eine Welt, die andere Namen nennt, völkerübergreifend in Universalität, „quacunque nationis". Eingebettet in christliche Gebete und vielleicht auch Psalmen und rituelle Bräuche mit Fixierung eines Kruzifix konnte eine trancegeförderte Imagination wirksam werden. Schließt man einen effektiven Trancezustand aber theoretisch aus, so kann kompensierend die aktivierende Kraft der hohen Erwartungshaltung eintreten.

Was geschieht dabei im Gehirn?

Die Aktivierungen des Sehzentrums durch Trance und Spruchdarbietung haben nicht nur einen wissenschaftlich interessanten verwertbaren Effekt, der uns Einblick in natürliche körperliche Abläufe gibt, sondern diese Aktivierungen können auch praktisch nutzbar mit dem Grad der vom Patienten empfundenen Beruhigung und der objektivierten Hypnosetiefe korrelieren.[354] Mit dem Siebenschläfertext und seiner bildbahnenden Wirkung konnten also sowohl posteriorparietale raumverortungsbezogene (Ephesus), als auch eher objekt-bezogene (Gebärmutter/ Höhle) ventrale visuelle Bahnverläufe aktiviert werden, also beide, die untere ‚Was'- und die obere ‚Wo'-Bahn des Gehirns.[355] In der synchronisierten Verbindung von auf Topografie, Gestalt und Zeit bezogenen Versetzungen des Patienten leistet die Textgestaltung mit all ihren Sprachfiguren eine

---

354 Halsband, Neurobiologie Hypnose, in: Revenstorf/ Peter, S. 811.
355 Zum Ungerleider-Mishkin-Modell vgl. Roth, Gerhard: Das Gehirn und seine Wirklichkeit, S. 184–187.

auch oft traumimmanente, grenzüberschreitende Vermittlung ins Archaisch-Mythische. Immer wieder ist es diese hohe fast alles verbindende und reziprok auf Nutzen abgleichende Synchronisierung mit Erinnerungslegierung der neuralen Systeme, die in ihrer Dynamik lebenserhaltend zu wirken vermag.

Für die Versetzung eines ‚Hier und jetzt' in ein entferntes ‚Dort und jetzt' in Erzählungen mit ihren Szenarien wurde in fMRI-Versuchen ohne Trance die Beteiligung des posterioren Cingulums und der rechten temporalen Junction im Verbund mit einem größeren Netzwerk der Mentalisierung angenommen. Diese Untersuchungen zielten über die oben genannte Parzellierung der Erzählelemente und Sprachfiguren hinaus auf ein Verständnis des Gesamtsinnes der Erzählung und ihrer Perspektive, wobei auch der praefrontale Cortex einbezogen ist.[356]

Für die zusätzliche Bedeutung der Namensnennungen – hier die sieben heiligen wunderbar erlösten Gestalten – verweise ich auf **Kap. C1.3.2**. Die Geschichte bot Voraussetzung für einen wirkungsvollen Einsatz als Entspannungbeitrag, auch weil sie in ihrem Ausgang emotional positiv eingeschätzt und somit vom limbischen System akzeptiert werden konnte.

Die kulturelle Seite der Mutterschoß-Imagination

Einen besonderen Stellenwert müssen wir dabei der Visualisierung der Höhle als Gebärmutter einräumen. Diese zum Urgrund eines beginnenden Lebens zurückgreifende Idee berührt Fragen nach unserer je eigenen Existenz sowie nach dem Geheimnis des Lebens überhaupt. Wenige Ideen verbinden unser Selbst derart tief und intensiv mit der Schöpfung, mit einem biokosmischen, religiösen oder archaisch-mythischen Anfang, zumindest mit einer vorgestellten Geborgenheit und einer elementaren Bindung.

Für den Dichter des deutschen Idealismus erinnert der Mutterschoß an eine ‚heilige Insel' (Friedrich Schiller: Gedichte, Der spielende Knabe, 1795: „Spiele Kind, in der Mutter Schoß! Auf der heiligen Insel/ findet der trübe Gram, findet die Sorge dich nicht"). Wissenschaftler haben emphatische Begeisterung für eine „zeitlose und unendliche Wunderwelt des pränatalen Lebens" formuliert, wie Gustav H. Graber (1893–1982), der Pionier der Pränatalforschung. Oder wie der Neurobiologe Humberto Maturana, geb. 1928, der jede Erkenntnis als „Spiegelung der Ontogenese des Erkennenden" in struktureller Verbindung ansieht, eine nativistische Überhöhung, wie wir sie schon den platonischen Dialogen verdanken: „Und falls es so ist, daß wir unser Wissen vor der Geburt erhalten und es im Moment der Geburt verloren haben, und dann durch das Entwickeln unserer Sinne gegenüber den Sinnesobjekten dieses ehemalige Wissen wiederentdecken, so ist das, was wir Lernen nennen, nur die Wiederherstellung unseres eigenen Wissens" (Phaidon, 75e, Kap. 19–20 zur Präexistenz der Seele).

---

356 Mano Y et al., Perspective-taking as part of narrative comprehension, Neurophysiologia 2009 Feb ;47(3).

Im Rahmen des therapeutischen Textvortrages im Mittelalter kann davon ausgegangen werden, daß ‚Mutterschoß' als ehemals schützend umgebendes und ernährend-entwickelndes Daseinsfluidum einen hohen metaphorischen Sinngehalt hatte. Es gab und gibt wohl kaum Sinnbilder, die einen höheren Grad an Schutz und Sicherheit versprechen. Das Sinnbild gehört zum Bestand an gemeinsamem ‚kulturellem Wissen', das als Voraussetzung für geteilte Erfahrung in Kommunikation gilt (**Kap. B1**) und das im Kulturerwerb via Spiegelneurone in sozial geteilter Intentionalität und fruchtbarer Symbiose verbindet (**Kap. B3**), hier nicht einer passagèren flüchtigen Symbiose, sondern als tragfähiges Kontinuum. Und obwohl heute psychologische Fahndung nach kausalen Faktoren für pränatal bedingte Kernneurosen und Traumatisierungen voranschreitet, scheint der Gedanke an ein beschützendes, ideales Mutterwesen weitgehend erhalten.

Die Möglichkeiten und die Grenzen der Rückerinnerungen

Biologisch ist allerdings zu bedenken, daß direkt *visuelle* perzeptive intrauterine Reize und Speicherungen für den Fötus – abgesehen von Hell-Dunkel-Signalen – nicht gegeben sind, sodaß eine retrograde rein visuelle Imaginierung auf dem Wege über die frühe Sehrinde nicht möglich scheint. Obwohl in manchen ‚Therapie'-Methoden vorgegeben wird, daß ein ‚Rebirthing' oder ein ‚Urschrei' zu simulieren ist, und in ‚Aquaenergetic' dieses frühe Materialmedium mit Erlösungsphantasie befrachtet wurde, entzieht sich eine Objektivierung der frühen individuellen Erlebnisse jedem Nachweis. Inwieweit jedoch die sicher gegebenen akustischen, kinästhetischen, gustatorischen und olfaktorischen frühen Einflüsse und Speicherungen aus den letzten Fetalmonaten[357] mit Hilfe der Multimodalität mancher Neurone und über damalige hormonelle Einflüsse zu einer Ermöglichung von situativer Wiedererinnerung auch mit visueller Imagination führen können, steht außer Frage.

Sicher gibt es eine mindestens subcortical gespeicherte ‚Bindungserfahrung' aus prä- und postnataler Zeit als Ressource zur Bewältigung von Krisen: Neuronenverschaltungen werden in Abhängigkeit von mütterlichen biochemischen und neurophysiologischen Gegebenheiten entwickelt. Mütterlicher Stress und entsprechende Ausschüttung von Botenstoffen je nach emotionaler Lage sowie z. B. Alkoholabusus verändern die Neuronenverschaltungen des Fötus. Das implizite, ‚unbewußte', an limbische Strukturen gebundene Gedächtnis reift vor dem expliziten, über den Hippocampus nach frontal geleitete Gedächtnis und bewahrt früheste Erfahrungen; es nimmt seine Arbeit einige Wochen vor der Entbindung

---

357 Vgl. Spitzer, Manfred: Lernen. Heidelberg 2002, S. 201f: Nachweise von Tongedächtnis, Experimente mit gefärbtem Zucker, Tastverarbeitungen, Anisgeruch.

auf.[358] Intrauterine Erinnerungen sind also in der Speicherung von Zellstrukturen in Gehirn und Körper erhalten, kommen somatisch zur Erscheinung, können aber nicht als innere Bilder mitgeteilt werden, weil diese imaginative Fähigkeit sich erst später und nach dem Spracherwerb entwickelt.[359]

Sicher ist, daß publizierte Spekulationen über eine direkte gar ereignisreproduzierende Wiedererinnerungen mit unbewußten, gelegentlich auch bewußt *erwünschten, rückwirkenden* Färbungen und Verfälschungen kontaminieren. Das in der Hypnosegeschichte bekannte Phänomen der Altersregression, oft verbunden mit phantastischen Reinkarnationserlebnissen oder mit der Illusion reeller Erinnerungen wurde oft unkritisch interpretiert.[360] Meist kreieren die in Trance befindlichen Personen Pseudoerinnerungen, die allein den gegenwärtigen Wunschvorstellungen entsprechen. Den Berichten über prae- und perinatale Rückerinnerungen, oft handeln sie von Höhlen-, Erdbeben-, und Unterwassererlebnissen aus Träumen ist eine sekundäre Prägung aus späteren Erinnerungsebenen zu unterstellen. Die Möglichkeit unbeabsichtigter Einflussnahme auf Erinnerungen unter Hypnose wird immer wieder betont.[361] Semantische Inhalte lassen sich im Gehirn ohnehin wissenschaftlich nicht nachweisen.

> Die seriöse und wissenschaftlich anerkannte heutige Imaginationstherapie verwendet bei akuten psychischen Traumen mit Destabilisierung teilweise weniger konkrete Bildangebote, wie z. B.: „Sie haben gesagt, daß Sie sich nirgends mehr sicher fühlen. Stellen Sie sich einen Ort vor, an dem Sie sich ganz sicher und geborgen fühlen." [362] oder zur Altersregression die Formel: „Wenn ich von zehn bis eins gezählt habe, werden Sie wieder klein sein und sich fühlen, wie sie sich als Kind gefühlt haben".[363] Jedoch werden von erfahrenen Therapeuten im Laufe der katathymen Imagination auch ‚Höhleneingang', ‚Sumpf' und ähnliches als Ausgangspunkt angeregt (vgl. Kap. A1.2.3.d). In der Geburtshilfe wurden Versuche mit Metaphern aus der Natur durchgeführt (**Kap C4.1.2**).

In jedem Fall aber vermag die narrative Imagination des Mutterschoßes erhöhte Aufmerksamkeit per Aktivierung unbewußter Speichersysteme subcorticaler als auch praefrontaler Zielgebiete zu bewirken. Der präfrontale Cortex verwirft Irrelevantes und bewahrt und remobilisiert relevante Erinnerung, wann und auf

---

358 Vgl. Leuzinger-Bohleber, M., Roth, G. und A. Buchheim: Psychoanalyse, Neurobiologie, Trauma. Stuttgart u. a. 2008, S. 19–31; vgl. Roth, Gerhard und Nicole Strüber, Pränatale Entwicklung, in: Cierpka, Manfred (Hg.:) Frühe Kindheit, Heidelberg 2012, S. 4ff.
359 Hüther, Gerald, Pränatale Einflüsse auf die Hirnentwicklung, in: Krens, Inge und Hans (Hg.): Grundlagen einer vorgeburtlichen Psychologie, Göttingen 2005, S. 61.
360 Vgl. Revenstorf / Peter, Hypnose, Kapitel 19: Altersregression, S. 288f.
361 Kandyba, Kristina: The Modification and Creation of Memories in Regression to early Childhood and Uterus. A Thesis, Montreal 1996, S. IV.
362 Stein, Claudius: Spannungsfelder der Krisenintervention, Stuttgart 2009, S. 191.
363 Revenstorf, D. und U. Freund, in: Revenstorf/ Peter, Hypnose, S. 211.

welchem Wege auch immer erworben.[364] Psychoanalytisch gesehen können unbewußte Erinnerungen zusammen mit der entfesselten ‚phantastischen' Legierung von individuellen und mythischen Vergangenheiten, von Gegenwart und Zukunft im Sinne von Sigmund Freud „wie an der Schnur des durchlaufenden Wunsches aneinandergereiht" werden.[365] Hier fungiert eine in der modernen Hypnotherapie klassische, seit Milton Erickson geübte Imagination des erreichten Zieles, wenn das Überleben der fast 200 Jahre dauernden Höhlengeborgenheit, die Aussicht auf das Licht am Ende des Tunnels vom Therapeuten eingestellt wird.

Den neurobiologischen Beweis für die Erfolgschancen garantiert die Plausibilität der Theorien: Psychotraumen führen zu verschiedenen psychosomatisch auffälligen Bindungsstörungen bei Kindern infolge Destabilisierung von Cortisol-sensitiven Neuronen in Hippocampus, limbischem System und präfrontalem Cortex, mitunter sogar zur Degeneration von Neuronen.[366]

C4.1.2 Die regressionsfördernde Imaginationen in der Geburtshilfe: Bindende Zeugung – Erhaltende Behütung – Lösende Entbindung

Waren wir schon im vorigen bei einem imaginativen Szenario des Rückblicks auf frühe unbewußte Lebensereignisse im Mutterschoß verwiesen, so erzählen uns Texte (**→23, 49, 58**), die der Geburtshilfe bei schwierigen Entbindungen dienen wollen, von Ereignisabfolgen, die nun der Natur der Menschwerdung und ihrem biophysiologischen Ablauf auf verschiedenen symbolischen Ebenen Gestalt geben. Die Einladung an die Gebärende zu innerer Bildgebung in Richtung ‚Vor-Geschichte', medizinisch gesprochen Anamnesis und zur ‚Ur-Sache' des aktuell gegebenen Entbindungsdramas versucht eine gezielte Rückkehr zur Ur-Natur unter Abkehr von gesellschaftlichen Zwängen mit all ihren Erfahrungen von Furcht, Spannung und Erfolgserwartung. Sie entspricht scheinbar einer eingrenzenden, zeit- und ereignisbezogenen Regression, ist aber gleichwohl an perspektivisch weitreichende kulturelle Botschaften, eben an die Symbolik der Teilhabe am Kollektiv-Archaischen und an Traumhaftes im Sinne von Lévy-Bruhl geknüpft (**Kap. A1.2.2**). Im Fall der Cuña-Indianer (**→58**) und der neoassyrischen Formel (**→49**) geschieht eine Symbol-Verkörperlichung via embodied simulation (**Kap. B3, C1**), Im Falle des nur als Bruchstück erhaltenen

---

364 Vgl. Kuhl, Brice A. and Wagner, Anthony D., Forgetting and Retrieval, in: Berntson/ Cacioppo I, S. 586–605.
365 Freud, S.: Der Dichter und das Phantasieren (1908), S. 174, zit. nach Ahlers, Ullrich: Fremde Träume, Berlin 1996, S. 195.
366 Brisch, Karl Heinz, Online, Universitätskinderklinik München 2012.

europäischen Dreifrauentextes (➔23), von dem uns keine praktische Anwendung überliefert ist und nur knappe aktionsgebundene Beschwörungen, sind es allein bekannte Symbolgestalten, die mit ihren Befehlen den Ablauf zu regulieren suchen und via Spiegelneuronen, sensomotorischer Metaphorikverarbeitung und Wiedererinnerung zur Archaisierung beitragen können. Kurzum: Die heilkundigen Sprecher dieser Texte arbeiten mit der Zweigleisigkeit von physiologischer und mythischer Darstellung im Sinne von Lévi-Strauss (siehe **Kap. A1.2.3**) und bedienen somit eine Triade soziopsychosomatischer Heilstrategie.

Wie bereits oben dargelegt: Es handelt sich bei dieser Vorgehensweise um eine gelenkte, passagère und situationsbezogene Therapie, die unter Ausnutzung regressiver Verortung psychosomatische Reaktionen hervorruft. Sie ist imaginativ sowohl durch verbale als auch durch figurativ-rituelle Bildgebungen (Figuren, Pflanzen) aktiv, indem sie eine anthropomorphe oder theriomorphe Verwandlung der Welt unter Einschluß meist archaischer Elemente betreibt. Den Holz- und Tonarbeiten wird quasi Leben eingehaucht, um Schmerzen, Befunde, Wünsche und Prognosen parallel mit dem Verbalen darzustellen.

> Wer einen Bogen zu moderner Anwendung der Hypnostrategie für Geburtshilfe bildet, findet säkular angepasste Bildgebungen. Es wird in der Geburtsvorbereitung das winzige Samenkorn imaginiert, das aus der Erde gedeiht und zum Licht wächst. Es wird eine Blumenmetapher, die Lieblingsblume der Schwangeren muß es sein, genommen mit all ihren Entfaltungsvorgängen bis zur Öffnung der Blütenblätter als Sinnbild des Muttermundes. Für die Strapazen der werdenden Mutter dient die Ausmalung einer Bergwanderung mit den Etappen zwischen Ausruhen und Erschöpfung, Kraftsammeln und Gipfelsturm als Verbildlichung der Wehenarbeit des Uterus. Nicht mehr die Kraft der Götter oder Dämonen, sondern die Kräfte der Natur sind metaphorisch eingesetzt. Psychosomatisch vergleichbar ist manches, etwa die narrativen Elemente der Ruheorte, hier die Bergwiese, in der mythischen Kultur die heiligen Orte wie etwa das rituelle ‚Wohnen' im neuassyrischen Text, hier die Zielvorgabe der Lichtwerdung oder der vorausgesagten Berührung mit dem Föten, im Mythischen das schicksalbildende Nabelschnurlösen durch eine Göttin (➔49, letzte Zeile), bei beiden die Bilder von fließenden Naturgewässern als Nachbildung des Fruchtwassers und seiner Kanalisierung. Erfolge der hypnotherapeutischen Methode im geburtshilflichen Bereich zeigten sich nach dieser statistischen Untersuchung für gut suggestible Frauen mit internaler Kontrollüberzeugung in einer günstigen Vorbeugung gegen Wochenbettdepressionen und in der Schmerzwahrnehmung während der Austreibungsphase der Geburt.[367]

Erinnern wir uns an die Funktion der Spiegelneurone, etwas imaginieren und gleichzeitig dasselbe tun unter Verwendung eines gleichen neurales Substrats (**Kap. B3**), dann aktiviert suggestive Imagination nicht nur Gebiete des Sprachverstehens, sondern auch diejenigen Gebiete, die das Imaginierte motorisch auszuführen vermögen. Dabei sind hier deren biochemische, hormonelle Komponenten involviert, in der Geburtshilfe insbesondere die Bereitstellung des wehen-

---

367 Vgl. Ripper, Kathrin: Hypnose und Geburt, Diss. Tübingen 2003, S. 59.

fördernden Oxytocins, sodaß vermittels Wortkategorien mit motorischen, (nach der Sprechakttheorie performativen) Verben auch eine Koaktivierung der betreffenden motorischer Zentren – zumindest in Bereitstellung – erfolgen kann. Zusammen mit der erreichten Trance oder einem hypnotischen Zustand, für den die Schwangere im allgemeinen hochgradig suggestibel ist, wird also ein archaischer Zustand angestrebt, womit vermittels Angstminderung die Ausschüttung von Stresshormonen verringert wird.

C4.1.2.a Mu-Igala, der geburtskundliche Heilgesang eines Schamanen, erhalten durch Vermittlung des Cuña-Indianers Guillermo Haya aus Panama (Beispieltext ➔58)

Ein Sprechgesang schildert in 535 Versen zunächst minutiös die Verwirrung der Hebamme, ihre Schritte und Wege, nachdem sie erfahren hat, daß die Gebärende bereits zwei Tage lang kreißt und sich Fieber (,heißes Gewand') eingestellt hat. Ebenso detailliert wird ihr Gang zum Schamanen beschrieben, dessen Aufbruch und Ankunft in der Hütte der Gebärenden. Die Verbrennung von Kakaobohnen, womit die Kleider des Schamanen bei Begegnung mit Muu, der foetenbildenden Macht gestärkt werden und die Schnitzarbeit kleiner Nuchu als Schutzgeister auf der Suche nach diesem Muu. Denn Muu, an sich eine positive Kraft, hat hier nun seine Befugnisse weit überschritten und sich der Seele der werdenden Mutter bemächtigt. Lang und kompliziert ist der Weg gegen böse Mächte aller Art. – Die Bedeutung des Gesanges liegt in der Doppelgleisigkeit eines sowohl mythischen als auch physiologischen Weges, denn Weg und Ort von Muu sind buchstäblich Vagina und Uterus, die der Schamane erforscht und in denen er den Kampf besteht, und ist zugleich eine Welt „mythischer Anatomie". In der fast den gesamten Text beherrschenden Detailfreudigkeit sieht Lévi-Strauss die obsessive Erzähltechnik zur Herstellung von Aufmerksamkeit, in der manipulierten Beziehung zwischen Ungeheuern und Krankheit (hier: schwere Geburt) die Beziehung zwischen Symbol und symbolisiertem Gegenstand, zwischen Signifikant und Signifikat.[368] So legt der Schamane z. B. den Aufstieg nach oben in die Vagina mit Hilfe von Zauberkappen zurück, bei der Expression des Föten nach unten erfolgt die Lenkung der Aufmerksamkeit auf die Füße der Gebärenden. Im Sinne der archaisch-schamanischen Triade ist die Dorfbevölkerung einbezogen. Und Mythos und erzählte Handlung bilden ein Paar.

---

368 Levi-Strauss, Str. Anthropologie, S. 204–217.

C4.1.2.b Die neosumerische Incantation vom ‚breeding bull' bis zu Gula's Schicksalsgriff (Beispieltext ➔*49)*

Der ‚Geburtshelfer' beginnt seine imaginative mytho-psychosomatische Reisebeschreibung mit einem kraftvollen Theriomorphismus zum Liebesakt. Ein Zuchtbulle hat diese Frau bestiegen. Der gewöhnlich naturhaft und ‚professionell' unfehlbare Samengeber vertritt die Urkraft der Mannheit, verwickelt reales Geschehen mit mythischer Vorzeit und versetzt in sie. Es ist der Schöpfergott Enki, auch er einst vom Bullen gezeugt, auf den sich der Sprecher beruft. Und es ist dessen Sohn Asalluhi, der auf die leidende entbindende Frau – keine Normgeburt, die Gebärende ist ja ‚verwirrt' (Zeile 15) – aufmerksam wird und zu seinem Vater Enki geht. Dem will er berichten, über den Schrei der Frau, der sich an Himmel und Unterwelt (Zeile 10) richtet und damit an die Heilige Hochzeit erinnert, einen Kult der Hohepriester des Marduk und der Oberpriester der Ištar. In diesem Kult wird der Himmel erotisch stimuliert, um Regen zu erzeugen und Fruchtbarkeit. Es geht um universale Kräfte, die auf das Werden und die Entwicklung des Föten Einfluß nehmen bzw. genommen haben. Ob mit der Nennung der Schiffe an Figurinen gedacht ist, die wie der Fötus innerhalb der Amnionhülle in Wasser schwimmen oder ob damit die Gebärende selbst gemeint ist, die auf einem Schiff den tiefen Ozean passiert, ist nicht sicher.[369] Der Heilkundige bedient sich direkt der Einführung einer Narration, die zum Dialog zwischen Gott Vater und Sohn führt, zu der in vielen älteren Texten bekannten Traditionsformel des erwähnten ‚Marduk-Ea-Typs'. Er führt seinen Sprechgesang bis hin zum Einsatz der Heilgöttin Gula, die schließlich die Nabelschnur versorgt und durch endgültige Ent-Bindung das Schicksal der Frucht bestimmt. Der sumerische Asalluhi wird später unter den Babyloniern zu Marduk als Sohn von Enki/Ea.

C4.1.2.c Imagination der drei Idisigruppen des althochdeutschen ersten Merseburger Zauberspruchs als universale weise Heilfrauen (Beispieltext ➔*23)*

Für die Deutung des ersten Merseburger Zauberspruchs als einen die Entbindung begleitenden Sprechgesang vs. einen Zauber zur Befreiung von Kriegsgefangenen haben sich im Laufe der Bearbeitungen immer mehr Argumente gefunden.[370] Aus ärztlicher Sicht muß das Wirken der drei Idisigruppen symbolisch den drei Etappen einer Menschwerdung entsprechen: die Empfängnis als Bindung mit

---

369 Cunningham, Graham, ‚Deliver me from evil', Roma 1997, S. 72ff.
370 Ernst, W.: Beschwörungen und Segen, S. 122–132.

Knüpfung von Knoten, als *hapt heptidun,* das Austragen der Schwangerschaft als antiabortives Festhalten und als Zurückhalten der Verknotungen, als *heri lezidun* und schließlich das freigebende Entbinden, das Entwirren der verknüpften Fäden als *clubodun umbi cuonio uuidi* durch die dritte Idisigruppe. Diese letzteren sollten mit ihrem Befehl ‚Entspringe den Haftbanden, entfahre den (geburtsbehindernden fesselnden) Dämonen' besonders mobilisiert werden. Bei Placenta praevia, fixierten Querlagen und anderen Komplikationen war eine Geburt früher meist ein Todesurteil für Mutter, Kind oder beide.

C4.2 Die verbale Imagination mit Wiederholungsfiguren als rhetorisches Stilmittel, und ihre Hirnorganik: Repetitionen, Aufzählungen, Wort- und Sinnketten, Wort- und Sinnverknüpfungen

C4.2.1 Die Bedeutung der Wiederholung von Worten und Sätzen in den Heilspruchtexten

Dieses Kapitel widmet sich einem sprachmorphologischen (rhetorischen) Schwerpunkt in der Strategie der erzählenden Texte, einer Strategie, die mit mehr oder weniger ausgedehnten Wiederholungen, vom Auf-der-Stelle-Treten bis zu kalkulierten Schrittfolgen vorgeht. Sie ist ebenso simpel wie wirksam. Sie kann ihre Dynamik durch Reihenbildungen erzielen; und einige ihrer Vertreter – diejenigen, die Götter- und Heldentaten aufzählen – wurden von germanistischer Seite auch einst als ‚Aufreihlieder'[371] bezeichnet, um ihre weltweiten Verwandtschaften aufzuzeigen. Moderne wie alte Rhetorik mag ihre Wirkung getan haben, man denke an Churchills Kriegsrede: „We shall fight on the beaches, we shall fight on the landing grounds, we shall fight in streets and fields ..." und an die Redekunst des Weißen Hauses.[372] Aber wir müssen uns mit diesem Begriff nicht auf reine Ereignisreihung beschränken. Die ketten-, perlen- und stückweise vorstellbaren Erzählteile erinnern formal an Rosenkränze und Litaneien, ja selbst an anankastische Ritualbildungen aller Art. Sie sind inhaltlich und formal heterogen, haben nur erzählerisch verschiedene Ziele, und es eint sie am Heilspruch das Ziel performativer Therapie. Sie können sich an ein rettendes oder an ein Unheil stiftendes mythisches, fabulöses oder religiöses Wesen richten. Oft sind

---

371 Schröder, Franz Rolf, in: Germanisch-Romanische Monatsschrift 4 (1954), S. 179–185.
372 Lubrich, Oliver, Figuralität und Persuasion, in: Paragrana 20 (2011), (zu Obamas Rhetorik) S. 248–265.

es einfach Aufzählungen von Körperteilen, die zu schützen sind, wobei zwanghaft darauf geachtet ist, daß nichts am Körper vergessen werden darf. Diese Texte bilden gelegentlich technisch-mathematisch wirkende Perpetua, die mit Gleichheit und Konstanz der Formen Langeweile und Dämmern erzeugen, sog. Redundanz, also einen Überfluss der Wortglieder. Damit verhelfen sie durch ihre Langatmigkeit dem Ungleichen und dem Neuen zu erhöhter Beachtung und Aufmerksamkeit. Auch in den Heilsprüchen findet immer wieder einmal dieser Abbruch des Gleichförmigen statt, wie er von Hans Carossa in seinem Gedicht vom alten Brunnen dargestellt ist: „… Das helle Plätschern setzt auf einmal aus ….", ein Geschehen, das im Gehirn erfahrungsgemäß nicht ohne nachweisbare Wirkung bleibt. An Tierversuchen wurde gezeigt, daß für den Beginn eines Geräusches und das Ende eines Geräusches jeweils separate Bahnen fungieren, was für Wortverständnis und Lautabgrenzung von hoher Bedeutung ist.[373] Andererseits bieten sie auch oft Überleitungen vom Kleinen zum Großen, vom Konkretum zum Abstraktum, worauf bereits hingewiesen wurde.

> Dabei stehen viele dieser Figuren in diesen unseren medizinischen Gebrauchstexten einer poetischen, d. h. verdichtenden Gestaltung geradezu entgegen. Zu beachten ist allerdings, daß Verschriftlichungen und Übersetzungen besonders im Blick auf Wiederholungsfiguren, Alliterationen und auf phonologische Details in ihren Wiedergaben z. T. beschränkt sind. Wenn die Aufzeichnungen editorisch exakt wiedergegeben sind, wird man auch konsonantische und vokalisch betonte Textteile unterscheiden können, die hirntopografisch und zeitlich verschieden rezipiert werden. Eine Besonderheit in dieser Hinsicht bietet etwa **Beispieltext →60** mit seinen langvokalischen Zauberworten aus den griechischen Papyri.

C4.2.2 Wiederholungsfiguren und ihre Beispiele

Im Rahmen einiger Beispieltexte des Kapitels **C4.1.** waren derartige Anteile bereits zu bemerken. Die Verkettung der Schutz- und Vertreibungsanliegen im Münchner Nachtsegen (**→1, Zeilen 6,8,11–16; Zeilen 22,24,26,28 sowie Z. 67–73**) und die Aufzählung der Hinterlisten des bösen Alû-Geistes und der Hexe in mesopotamischen Texten (**→40 und 41**) sind lauter Übeltaten und Übelandrohungen, die durch ihre Aufdeckung und Ausmalung entschärft werden können. Es sind mehrfach repetitive Anaphora, Epiphora und häufig Isocola, letztere etwa im Beispieltext **→43**: „Die Tamariske, die im Wipfel üppig ist, möge mich reinhalten; die Dattelpalme, die allen Wind standhält, möge mich lösen…". Oft kommt es auch zu Vermischungen der Typen oder zu effektvollen Stellungsvarianten dieser synonymen Parallelismen wie im Beispieltext **→45**: „Hexe … die du des nachts jagst, …. die du die Männer tötest – jetzt sahen sie dich, packten

---

373 Wehr, Michael et al., in: EurekAlert 10–Feb–2010.

dich, veränderten dich ... Ea und Marduk" und im Beispieltext ➜46: „meine Augen, mit denen ich sehen konnte, hat sie gepackt; meine Füße, mit denen ich einhergehen konnte, hat sie gepackt"

So finden sich in mittelalterlichen Beschwörungssprüchen sehr häufig Aufzählungen von personalisierten Gichtern und Würmern als scheinbar an Organe gekoppelte Sorten (➜24), die angegangen werden. Ähnlich treffen wir auf Reihungen der zu schützenden Körperteile, die von der personalisierten streunenden Bärmutter – dem sonst auch inneren Kolikerzeuger – schon im altgriechischen Zauberpapyrus (➜60) beschworen werden und nicht okkupiert werden sollen (➜25), wo die sehr einfache Repetitio exzessiv betrieben wird, jedoch durch die natürliche Begrenzung Kopf bis Fuß doch auf ein Ende zustreben kann. In den Atharvaveden (➜54) sind ebenfalls gründlich aufgezählte Körperteile von Kopf bis Fuß installiert, aus denen der Medizinmann vermittels seiner ‚Kraft' eine personalisierte Krankheit vertreiben will.

Davon zu unterscheiden sind die kürzeren Ketten von Organnennungen, die imaginativ eine reparative Zusammenheilung von Frakturen und Gewebsschäden anzielen, indem sie mobilisierende Verben oder Adverben einfügen wie „Mark wandere zu Mark, Bein zu Bein, Haar an Haar ..." (➜53 und 26) oder die ebenfalls funktionssimulierend einen Wurm beschwören und ihn sukzessive „aus dem Mark in die Ader, von der Ader in das Fleisch ..." (➜27) zu befördern trachten, wie in einem der ältesten althochdeutschen Zaubersprüche des 10. Jahrhunderts. Dabei fallen viele Sprüche auf, die kleine Gegenstände, meist eine Pflanze imaginieren und auch praktisch im begleitenden Ritus als Apotropäon einsetzen, um danach jene Funktion der Wiederherstellung in gemeinsamer Zuwendung an einen Schöpfergott vollziehen zu können, an eben den Gott, der gemeinsame Kreatürlichkeit von Pflanze und Mensch vermittelt hat. Im altindischen Atharvaveda ist es die Heil-Schlingpflanze Rohani/Arundhati, die in manchen dortigen Texten als mit Gott verbündet, die als Kraft- und Duftspender, Heiler und Schützer gilt, die besonders gegen Frakturen angesprochen wird (➜53), erst nach ihr wird auch Gott Dhatar angerufen.

Deutlicher noch entfaltet sich bei ernsthaften Erkrankungen das Ritual der Schamanen bei den Lakandonen (➜59), Nachfolger der Maya in Mexiko, die mit Zweigen, mit anthropomorphen [?] Figuren und Weihrauch- und Blütenduft arbeiten. Das Ritual beginnt mit dem Pflücken von Palmenblättern als ‚Hocker der Götter' und einem Götterhaus mit Opferfiguren, um dann im Beschwörungstext anaphorisch „ich setze die Knochen ein, ich setze das Herz ein, ich setze die Lunge ein" zu imaginieren. Hier treffen wir auch auf sogenannte absolute Wiederholungsfiguren (Zweimal: „Das sind die Streben des Hauses..."), die sich durch gleiche Wortkörper und Wortinhalte auszeichnen, die in den Nachdrucken aus Platzgründen oft nicht wiedergegeben werden und die für die Performativität der Heilsprüche so wichtig sind, wie die fortwährenden gleichbleibenden Wiederholungen ohne etwas Neues während einer Hypnose oder einem Autogenen Training. Diese Form der Wiederholung kann mit ihrer die Aufmerksamkeit einschläfernden trancefördernden Dynamik an die Seite der Prosodie gestellt werden, obwohl wir darüber in ihren historischen Phänomenen nur vermuten können. Von Sprachwissenschaftlern werden diese absoluten Wiederholungsfiguren gelegentlich als überflüssig, als redundant, bezeichnet.

Im europäischen Mittelalter wird die göttliche Kokreation von Mensch und Pflanze an die heilige Dreifaltigkeit geknüpft, deren drei Personen im Anapher je einem Teilaspekt der Kräuterbeschaffung zugeteilt werden. (➔28)

Der genannte Mayavolksstamm nutzt nun auch neben diesen Redundanzen eine Strategie gegen Angstzustände, die sehr augenfällig die Parzellierung eines Ritus aufbaut, der nur an Kürbisflaschen, Hängematten, Tragenetzen und Taschen ansetzt, also an der materiellen Kultur des Volkes. Ein Schamane veranlaßt das Deponieren all dieser Gegenstände auf dem Boden, bespuckt den Kranken dreimal, um dessen „Körper für die Gestalten der unsichtbaren Welt zu öffnen."[374] Denn durch das Ablegen der zuvor hängenden Gegenstände wurden deren Seelen von ihren Körpern getrennt und mit dem Sprechen eines Textes „fährt (die Seele) des Wissenden in die unsichtbare Welt." Er imaginiert simultan einen Kettenritus der Stück für Stück wieder zusammensetzt, was in der Vorstellung der Menschen zerrissen war: „... Die Kürbisflasche, dort erhebt sie sich, die Kürbisflasche, ich habe sie zertrümmert; die Kürbisflasche, dort erhebt sie sich, die Kürbisschale, ich habe sie zertrümmert, ich habe sie umarmt ... die Hängematte, sie schaukelt und sie tanzt, dort erhebt sie sich ... die Opferschale, ich habe sie zerdrückt, zerrieben, sie schaukelt, dort erhebt sie sich, dort erhebt sie sich ...." [375]

Ebenfalls als Angstbehandlung sind die Schilderungen der Schritte und Bewegungen der Hebamme bei den schon oben genannten Texten der Cuña-Indianer (➔58, Zeile 7–14) einzuschätzen mit ihrer durchgehenden Wegesstückelung, dem Merkmal einer obsessiv suggestiven Strategie, die mit bewußter Überperfektion den Heilungsvorgang einleitet, wie im Zeitlupentempo. Es ist die Methode der vielen kleinen zum großen Ziel gelangenden Schritte, die Parzellierung des Gipfelanstiegs. Sie kann im ganzen gesehen „von der banalen Realität zum Mythos, vom physischen zum physiologischen Universum" führen.[376] Sie unterläuft hier durch Vortäuschung einer Verzögerung die Dramatik des Entbindungs-Notstandes, der schon allein mit der Herbeirufung eines Medizinmannes

---

374 Rätsch, Christian: Bilder aus der unsichtbaren Welt, München 1985, S. 265.
375 Ebd., S. 266.
376 Levi-Strauss, Str. Anthr. I, S. 212.

durch die Hebamme provoziert ist: „Wir müssen zum Medizinmann, du weißt" – das war und blieb eine unheilschwangere Mitteilung. Die massive Parzellierung erinnert auch an die psychotherapeutische Technik der paradoxen Intention, der Symptomverschreibung. Hier nun wird das möglicherweise Verhangen-Sein der Leidenden am Irdisch-Objekthaften, entsprechend ihrer Verzögerung des Geburtsverlaufes, ihrer ‚oknophil' (Balint) zwanghaften Verklammerung geradezu eingesetzt, um Spannung zu lösen. Fortschreitende dynamisierende Aktion vermitteln auch jene Texte, die in Schrauben- und Schnecken-Wendeln auf ein Ziel hinführen, etwa einige Wurmbeschwörungen der letzten Jahrhunderte in Europa.[377]

C4.2.3 Ansätze zu einer Neurophysiologie der sprachlichen Wiederholungen

Wie wirken diese repetitiven Imaginationen im Gehirn des Kranken?
Wir hatten bereits oben (**Kap C4 Vorspann und a. a. O.**) auf die neurobiologischen Ergebnisse zu Metaphorik und Metonymie-Verarbeitung hingewiesen, wobei sich zeigte (**Kap. C1**), daß bisher weitgehend emotionale Merkmale zur Ausgangsbasis für Experimente dienten, besonders im rationalen Durcharbeiten und Bezeichnen der Gefühle (reappraisal und labeling emotions).

Ausgangspunkt einer eher *primär kognitiven* Betrachtung war die Priming-Forschung, die die Einflüsse nicht nur affektiver, sondern auch semantischer Vorgaben, zumeist nach subaudio-visuellen Reizen untersuchte und mit ihr bei Versuchen mit mehrfachen Wiederholungsreizen die herausfordernde Entdeckung einer wiederholungsbedingten intrazerebralen Reizabschwächung (repetition suppression, R.S.) machte. Die Primingforschung hatte zunächst Beweise für diese Reizabschwächung vorgelegt und sie als Zeichen erfüllter Erwartungshaltung gedeutet. Zum Beispiel ging man von Laborbedingungen mit Wahrscheinlichkeitsbestimmung aus: Wenn eine weitere Wiederholung von Reizen für ein Gehirn unwahrscheinlich, also unerwartet war, reduzierte sich die Aktivität der entsprechenden sensorischen Gebiete.[378]

Demgegenüber zeigten viele Untersucher auch geradezu eine der Wortwiederholung folgende Aktivitätsverstärkung (repetition enhancement) und zwar besonders dann, wenn nicht gleiche, sondern semantisch nur verwandte Worte eingesetzt wurden, z. B. unter Anwendung von Lesetexten. Hier war repetitives Priming mit Abschwächungen (R.S.) im linken Frontalbereich und bilateral im

---

377 Ernst, W.: Beschwörungen und Segen, S. 321.
378 Summerfield, C., et al., Neural repetition suppression reflects fulfilled perceptual expectations, in: Nature Neuroscience 11 (2008), doi:10.1038/ nn.2163.

Parahippocampalbereich verbunden, während Aktivitätszunahme im rechten medialen Frontalgyrus und im rechten medialen Temporalgyrus beobachtet wurde. Diese Lokalisierung konnte einem Wiedererkennungsvorgang entsprechen. Die Untersucher nahmen an, daß Wiederholung beim semantischen Priming von verschiedenen cognitiven Verarbeitungen abhängt.[379] Auch die Untersuchung mit wiederholten und stellvertretenden (coreferenting) Namen erbrachte nicht durchgehend einen Primingvorteil.[380] Ebenso bei Höraufgaben mit Worten und Pseudoworten zur Frage, ob eine Wortwiederholung semantische Verarbeitung lexikalischer Begriffe involviert, fand sich kein klares Modell.[381]

Schließlich hat eine Studie zeigen können, daß neben den Eigenheiten der Stimuli auch vielmehr die Anzahl ihrer Wiederholung von Bedeutung ist. Man beobachtete, daß die selben visuellen Reize sowohl eine Verminderung als auch eine Verstärkung der neuralen Aktivierungen bewirken konnten, ganz in Abhängigkeit von der Zahl ihrer Wiederholung. Bei der Reizung mit Niederkontrastbildern erfolgte vom ersten bis zum fünften Reiz eine anwachsende Aktivierung der Parahippocampal-Area (Teil des limbischen Systems, Funktionen für erkennendes Erinnern) und des occipitalen Sulcus transversus (Teil der Sehrinde für Erkennen von Szenen), danach eine Verringerung der Antwortpeaks über die Zahl der weiteren Wiederholungen (‚inverted U-shape'-Kurve). Neue Reize erzielten grundsätzlich stärkere Aktivierungen als wiederholte Reize. Die Autoren schließen daraus auf zwei separate Verarbeitungsweisen, eine für die Verbesserung der Stimulus-Repräsentation selbst, die andere für die Unterscheidung neuer von bekannten szenischen Reizen.[382]

Übertragen auf unsere Heilspruchtexte kann davon ausgegangen werden, daß Untersuchungen auch der auditiven Funktionsverläufe mit differenten Wortwiederholungsfiguren eine vergleichbare cerebrale Mobilisierungs- und Modifikationsverarbeitung in den entsprechenden Gebieten erbringen könnten. Trotz der noch wenig aussagekräftigen Ergebnisse der bisherigen Gehirnstudien zu Reizwiederholungen wird deutlich, daß sich ebenso wie andererseits im emotionalen Bereich Möglichkeiten einer differenzierteren ‚Abbildung' dieser wichtigen rhetorischen Figuren ankündigen. Die ersten Spuren sprechen dafür, daß allein schon diese spezielle Taktik der Heilspruchtexte eine vielschichtige Dynamik mit erhöhter Aufmerksamkeit, mit Rückgriff auf Erinnerungen und mit

---

379 Raposo, A. et al., Repetition suppression and semantic enhancement: An investigation of the neural correlates of priming, in: Neuropsychologia 44 (2006), S. 2284–2295.
380 Ledoux, Kerry, (2008) Online studies of repeated names and pronominal coreference.
381 Bihler, S.M.E. et al., Auditory repetition of words and pseudowords: An fMRI study, in: Alter, K. et al. (Hg.:) Brain Talk: Discourse with and in brain, Lund 2009, S. 3–9.
382 Müller, Notger et al., Repetition Suppression versus Enhancement – It's Quantity that matters, in: Cerebral Cortex 23 (2013), S. 315–322).

Appraisal-Reaktion auf ‚überraschende' Neuigkeitssignale zu entfalten vermag. Darüberhinaus können prosodisch, mimisch und gestisch kongruente Einwirkungen zur Summierung der performierender Reize beitragen. Es versteht sich, daß die nonverbalen Signale gerade bei Wiederholungsformeln ihren Beitrag zur Psychoperformativität leisten.[383]

Eine in der aktuellen neurobiologischen Psychotherapieforschung untersuchte Methode ist ebenfalls im Funktionsbereich der Wiederholungsfiguren anzusiedeln: die Expositionstherapie mit Konditionierung angstauslösender Elemente. Es wird ein neutraler Stimulus mit einem aversiven Stimulus solange abstimmend dargeboten, bis auch der neutrale Stimulus allein schon ängstliche Antworten erregt. Weitere Konditionierung kann dann durch wiederholte Darbietungen des neutralen Signals die ängstliche Reaktion löschen. Man hat dabei z. B. bei Spinnenphobie mit der Annäherung an chilenische Tarantellen über bleibende Erfolge berichtet und diese nach fMRI-Befunden auf eine Dämpfung der Ansprechbarkeit des angstsensiblen Netzwerks von Amygdala, Insula und Cingulum und gleichzeitiger erhöhter Repräsentanz des präfrontalen Cortex zurückgeführt.[384] Die hirnorganischen Mechanismen behaupteter Neuorganisation des limbisch-frontalen Systems sind allerdings noch zu bestätigen.[385] Es wird berichtet, daß mit dieser Therapie behandelte bis dato schwerkranke Phobiker nach 2–3 Stunden eine Tarantel anfassen konnten und nach einem halben Jahr bei Nachuntersuchung immer noch stabil angstfrei waren. Eine zu bedenkende Übertragung dieser modernen Therapieerfolge einer differenzierten standartisierten Kurzzeitmethode auf archaische Textinduktionen ist nur punktuell möglich, muß als spekulativ eingeschätzt werden, ist aber als empirische Praxis vorwissenschaftlicher Konditionierung auch nicht auszuschließen.

---

383   Josse, Goulven et al., The brain's dorsal route for speech represents word meaning: Evidence from gesture (2012), PLoS One 7(9): e46108. doi: 10.1371/j.pone.0046108.
384   Hauner, K.K. et al., Exposure therapy triggers lasting reorganization of neural fear processing, in: Proc. Natl. Acad. Sci USA 2012 Jun 5; 109(23): 9203–8. Epub 2012 May 23.
385   Vgl. Wendt, J. et al. Mechanisms of change: effects of repetitive exposure to feared stimuli in the brain's fear network, in: Psychophysiology (2012) Oct; 49(10): 1319–29.

C4.3 Strategien introversiver Katharsis[386]

C4.3.1 Leid- und Mitleid-Induktion mit Erinnerung an persönliche Schuld und an Schmerz unter Einbeziehung von überirdischen Helfern

Neben die direkte emotionsperturbierende Behandlung durch Etikettierung von Gefühlen (**Kap. C1**), neben die die kognitiven Funktionen perturbierende konfliktprovozierende Darbietung von sprachlichen Inkongruenzfiguren (**Kap. C3**) und neben die gezielt regressionsfördernde Imaginierung (**Kap. C4,1**) setzen wir hier den Schwerpunkt auf Formen narrativer kathartischer Methode. Der Begriff der Katharsis geht auf die klassische griechische Literatur und die Darstellung von Rührung und Klage zur Reinigung in den Tragödien zurück. Allerdings hatte Aristoteles den Begriff aus der hippokratischen Medizin übernommen; es ging um die Ausräumung schädlicher Körpersäfte.

Entgegen der in der Geschichte der Magieforschung (**Kap. A1.2.2**) als Spannungsabfuhr verstandenen Katharsis muß bei differenzierender Betrachtung der psychoperformativen Texte ein breiteres Begriffsprofil angewendet werden. Unter introversiv kathartisch verstehen wir Textteile, wie sie nicht nur, aber vor allem für die christliche Heilspruchkultur bekannt sind und hier der Versenkung in Christi Leiden dienen; sie sind nun als Reue- und Bußübung zum Zwecke der inneren Reinigung zu benennen". Sie werden zur „erczney um der tewfel geselleschaft ledig (zu) werden.[387] Den strategischen Gegensatz dazu bilden in den Spruchtexten jene Verbannungs- und Verfluchungsstrategien zur *direkten* Austreibung des Übels. Sie setzen mittels Befehlen einen imaginativ extroversiven Angriff in Szene und entsprechen den emotionsetikettierenden Texten durch ihre oft theriomorphe Metaphorik und entsprechen der neurophysiologischen Leistung durch ihre besondere Sensomotorik (**Kap. C1.3.4 und C5**). Dabei ist für unsere Annäherung auch an diesen Themenbereich von einer Deutung, Bewertung und Ethik religiöser Inhalte abzusehen.

---

386 Der Begriff Katharsis wird hier weder nach kunstästhetischen noch nach psychologischen Maßstäben von Entsorgung und Aggressionsabfuhr in Psychoanalyse, Bioenergetik, Psychodrama u. a. modernen Methoden verstanden. Der Begriff erstreckt sich in somatische, geistige und geistliche Felder und eignet sich damit vorzüglich für die Anwendung bei Heilspruchelementen. Die Unterteilung in introversive und extroversive Methodik bezieht sich auch nicht auf Charaktereigentümlichkeit wie in der Psychologie von C.G. Jung, sondern beschreibt die strategischen Angriffsrichtungen von Heilspruchpartikeln.

387 Vgl. z. B. Werbow, Stanley Newman: Martin von Amberg. Der Gewissensspiegel, Berlin 1958, S. 69.

Bei den mit Leidenstönung arbeitenden kathartischen Textteilen können provozierte passagère Stimmungsabsenkungen auch cerebrale Funktionsabläufe erzielen, sodaß eine gesonderte Herausstellung im Rahmen der narrativen Imaginationen gerechtfertigt und erforderlich ist. Dies umso mehr, als wiederum auch die mitwirkenden emotionalen Teil-Elemente dieser Katharsis mit bildgebenden Verfahren nach ihren Verläufen differenzierbar scheinen. Wir verstehen therapeutische leidbetonte Katharsis im hier historischen an den Heilspruchtexten orientierten Sinne. Sie erzielt – da können wir sicher sein – ein initial nicht angenehmes, meist auch schmerzhaftes Sich-Zurücknehmen, ein Verinnerlichen im Durchleben und Durchleiden eines gestauten Affektes. Dieser kann mit der Überwindung zu Schuldbekenntnis und eigener Fehlbarkeit zusammenhängen. Und immer kann damit eine Zunahme von Angst, depressiver Stimmung und Schamgefühlen provoziert werden. So verstandene therapeutische Katharsis erinnert Menschen an Endlichkeit und Sündhaftigkeit, Unerlöstheit, Versäumnisse und Verluste und kann bis zu Vorstellungen von Todesnähe führen. Religiös verstandene und empirisch psychoperformativ verstandene kathartische Kraft sind dabei nur für den Beobachter voneinander zu trennen.

C4.3.1.1 Die kulturgeschichtliche Dimension der introversiven kathartischen Imagination

C4.3.1.1.a Die therapeutische Induktion von Leid und ‚Sünde' in babylonisch-assyrischen Texten

Die babylonische Beschwörungstafel Maqlû VII (→**43–47**) bietet wie der gesamte Maqlû-Text keine Formulierung eines klaren individuellen Schuldbekenntnisses. Zwar haben die Übersetzer den Begriff der Sünde verwendet: „Die Ersatzfigur trage meine Sünde, diene als mein Stellvertreter", … „Erhebe dich, Morgen, mach auf, Erde, empfange meine Sünden!", aber es ist nicht zu erkennen, worin eine tatsächliche oder innerlich vorgestellte Schuld besteht, worin der Patient gefehlt hat, oder auch, was ihm von seinem Beschwörer oder von den Göttern als Ursache für seine Krankheit vorgeworfen wird. Im Gegenteil: Entsprechend dem Grundanliegen von Maqlû-Beschwörung und -Ritual überwiegt Anklage gegen Übeltäter, Hexen, Zauberer, böse Machenschaften, ja Vorwurf gegen die bestrafenden Götter. Sollte es innerlichen Schmutz geben, im Kopf oder am Körper, so kommt er jedenfalls nicht nur vom Selbst. Aber wenn die ‚Freude zur Trauer geworden' ist und ‚die Göttin entfernt' erlebt werden muß (→**43**) oder wenn in kathartischer Klage über die Machenschaften der Hexe, die den Patienten erkannt hat und ‚hinter mir her kam' (→**47**) das Übel zupackt, dann werden Leid und Unheil durchlebt. Schicksalsschläge werden theriomorph am Beispiel von Bedrohungen durch Vögel, Fuchs und Löwen nachgeschildert (➔**40 und 41**).

Allgemein hatte mesopotamische Krankheitsursachenvorstellung die Schuld und damit die Sünde relativiert und auf mehrere Schultern verteilt: Neben dem Betroffenen selbst konnten Götter, Angehörige und Verstorbene herangezogen werden, ähnlich wie in vedischen Texten (➔**56**). Und so richtet sich das Ansinnen der **Beispieltexte** ➔**44, 45, 46, 50** auf Reinigung mittels Übertragung von ‚Sünde' auf die stellvertretenden Ersatzfiguren oder auf beliebig Begegnende auf der Straße; ‚Sünde' erscheint wie etwas Verhandelbares. Oft kommt es auch zu Revanchegelüsten (➔**43**). Gleichwohl betont der Reinigungsritus des Beschwörers mit dem exzessiven anankastischen Händewaschen (➔**50**) ausdrücklich die Unsauberkeit des Kranken, seine Befleckheit. Die assyrologische Forschung hat immer wieder über die Frage von Sünde, Schuld und Bekenntnis reflektiert und die Vorstellung von Krankheit als Sündenstrafe für die verschiedenen Epochen und Textarten nicht einseitig beantworten können. Im Vergleich zu den hethitischen Krankengebeten etwa, in der der Kranke mit dem Sonnengott rechtet, seine Unschuld beteuert und wissen will, was ihm vorgeworfen wird (**Beispieltext** ➔**51**), ähnlich einem Hiob, waren mesopotamische theologische Denksysteme komplizierter.[388]

Zieht man nämlich ein anderes Textgenre aus Mesopotamien heran, dann findet sich bei manchen unter den wenigen erhaltenen Klage- und Buß-Liedern der sumerisch-akkadischen Schriften das Bekenntnis zu Sünde und Schuld deutlich vorgetragen, obwohl es sich ja um ‚öffentliche Liturgien' in speziellen kultischen Psalmen handeln soll, nicht um persönliche Erlebnisäußerungen von einzelnen. In einem der Gebete werden sogar spezielle Vergehen genannt: „Ich, dein Knecht, | habe immer wieder in allem gesündigt, trat vor dich hin, suchte aber immer wieder nach Unrechtem. | Lügen sprach ich immer wieder, | tat leichthin ab meine Sünde, sagte immer wieder Unheilvolles: | alles dies sollst du wissen! | Was dem Gott, der mich schuf, ein Greuel ist,| nahm ich zu mir, betrat freventlich Unbetretbares, | tat immer wieder Böses. | Auf deinen weiten Besitz | richtete ich (begehrliche) Blicke, | nach deinem Silber | ging meine Gier. [...][389] „Die meisten [unter diesen Texten] verbinden die Klage mit dem Sündenbekenntnis und die Bitte um Heilung mit dem Gebet um Sündenlösung". Sie mußten öffentlich „‚gerufen', d. h. mit lauter Stimme gebetet werden."[390] Schlußfolgerungen über kollektivistische vs. individualistische gesellschaftliche Tenden-

---

388 Ünal, Ahmet, Hethitische Hymnen und Gebete, in: Kaiser, Otto (Hg.) TUAT, S. 791.
389 Falkenstein, A. /W. v. Soden (Hg.:) Sumerische und akkadische Hymnen und Gebete, Stuttgart 1953, S. 272.
390 Ebd., S. 44.

zen sind aus diesen Quellen nicht möglich. Mit unserem **Beispieltext** ➜52 konnte beispielsweise ein Fieberkranker durch den Priesterarzt nach vorliegendem Spruchschema behandelt werden. Der Arzt erleidet vielleicht mitfühlend mit dem Kranken Schrecken, Not und Elend als dessen Sündenstrafe; er beschreibt den finsteren Blick des Patienten und dessen Gebundensein, während der Kranke selbst seine Sünden eingesteht. Krankheit und Sünde stehen also in einem Teil der Keilschriften in engem kausalen und finalem Zusammenhang. Leiden und Sündenbekenntnis werden Ea und Marduk unter Opferungen vorgetragen, um [Er-]Lösung zu erlangen. Die Erwähnung von Schmerzempfindung wirkt wie ein Nachweis ehrlicher Bußabsicht.

> Häufiger entstammen Klage- und Bußgedanken im Babylonischen den Unterwerfungskulten in Hymnen an Götter, insbesondere an den Sonnengott Schamasch, der ein Schützer von Recht ist und der neben Marduk und Ischtar am meisten angebetet wird. Das entnehmen wir einem langen Preislied, in dessen Mitte wenige solcher Verse inseriert sind: „Du hörst, Schamasch, das Gebet,| Beten und Segnen, Niederwerfung, Knien,| Gebetsflüstern und Prostration. Aus tiefster Kehle| ruft der Schwächling dich an; der Kümmerling, der Schwache,| der Hörige trägt den Klagegesang | ständig dir vor ...."[391] Damit entfernen wir uns weit von der therapeutischen Intention und nähern uns den in **C1** beschriebenen Opferbräuchen.
>
> Ziehen wir auch materielle symbolische Anteile für kathartische Technik heran. Es konnten Pflanzen wie Tamariske und Dattelpalme (➜43) im Vergleich mit einem altenglischen Segen als universale ‚archaische Denkstruktur' dargestellt werden: „In den keilschriftlichen Kräuterrezepten ist [...] die Legitimierung einer quasi göttlichen Natur und Herkunft der verwendeten Kräuter vorausgesetzt. So heißt es z. B. von der Tamariske: ‚Vom Hals an ist sie Vater Enlil'". Durch das Abfallen ihrer Triebe im Herbst sei sie zum „Abwerfen des Bösen" analogisiert. [392] Andererseits werden Olivenzweige in einem frühchristlich-koptischen Text zur Dämonenaustreibung herangezogen (➜61)

C4.3.1.1b Die zwei Typen kathartischer Imagination
(introversiv und extroversiv) in zwei vedischen Heilsprüchen:
Sündenbekenntnis vs. theriomorphe Vertreibung

Mit einem Rückgriff auf zwei zum Thema ausgewählte Sprüche aus den kaschmirischen Veden bieten sich uns konträr arbeitende Strategien bei der Nutzung göttlicher Hilfe.

Die Eine: **Beispieltext** ➜56 erzielt mit Hilfe einer kleinen Litanei die zauberhafte Überkreuzung von Raum und Zeit. Es vereinigen sich Götter in Lieder und Lieder mit Göttern, woraus der Gesang selbst zum Werkzeug des performierenden Aktes wird. Dabei geschieht mehrfaches kathartisches Durchleben von

---

391 Ebd., S. 245.
392 Schulz, Monika, Nigon wyrta galdor. Zur Rationalität der ars magica, in: Zeitschrift für Literaturwissenschaft und Linguistik 130 (2003), S. 8–24.

Schmerz und Sünde. Gesündigt hat der Kranke in drei Ebenen, an Göttern, Verstorbenen und Mitmenschen, nicht anders als in der mesopotamischen Kultur. Dieser vertikalen, zeitlichen Achse der Personen stellt sich eine räumliche Ebene der vier Himmelsrichtungen quer. Sühne wird an das Sühnemittel des raumzeitlich wirksamen Gesanges abgegeben, um von Abzehrung, ein Begriff für viele infektiöse und konsumierende Krankheiten, zu befreien.

Die Andere: Im **Beispieltext** ➜57 ist der Kranke ganz anders beteiligt; er überläßt sich und vertraut völlig der kolossalen Gewalt seines Gottes, dessen Keule, Mühlstein und Mordwaffe. Krankheit ist theriomorph in die Namen, Gestalten, Herrschafts- und Verwandtschaftsgrade des Wurmes übertragen wie in den Beispielen aus anderen Kulturen. Hier bedarf es keiner Katharsis des Mitleidens, keines schmerzlichen Rückzugs des Kranken; die Methode ist die andere Seite des imaginierten Psychodramas, ist imaginierter Kriegszug, delegiert an einen hohen Stellvertreter mit distinkter emotionaler Ausrichtung, eine Methode, die als exorzistisch bezeichnet werden kann. Und noch eindeutiger als im ➜56 ist hier das ‚Wort' selbst zum Akt der Vertreibung der Würmer ernannt.

*C4.3.1.1c Introversiv kathartische Elemente in Texten
des christlichen Mittelalters*

Über die Haltung der Demut hinaus und über den tiefenseelischen Dialog mit den Figuren und Bildern ihres Religionsstifters und der Heiligen hinaus hat die mittelalterliche Gesellschaft mit ihrer ‚Kirche der Sünder' in ihrer Heilspruchkultur Zeugnisse zu einer angeleiteten therapeutischen Versenkung geschaffen.

Sie produzierte eine große Anzahl von nicht kanonischen Texten, eine umfangreiche Palette in weitgehend christozentrisch gepolter Orientierung, wie wir sie in **Kap. B4.1** dargestellt haben. Hatte schon Aristoteles in seiner Katharsislehre auf die Bedeutung der Identifizierung des Zuschauers, des sich Hineinversetzens in Leid und Leidenschaft eines Helden hingewiesen, so wird das Leiden Christi und der Märtyrer für den mittelalterlichen Menschen ganz zum existentiellen Element. Zu Beginn des 14. Jahrhundert fordert Pseudo-Bonaventura, die Passion gefühlsmäßig meditativ zu betrachten, „als wäre man selbst anwesend" und es wird die Heilstat Christi in enge Beziehung zum Individuum gestellt[393] wie in ➜**32**.

– Die Lebensstationen Jesu Christi als Kurzcredo (➜**33**) bilden in Ablösung der heidnisch-römischen Gottheit Asklepius frühzeitig verschiedenste meist außerkirchliche Heilformeln.

---

393 Dinzelbacher, Peter, Religiöses Erleben von bildender Kunst, in: Schreiner, Klaus (Hg.), S. 327.

- Die Heilwunder Christi und die Heiligenlegenden werden in die ebenfalls meist außerkirchlichen therapeutischen Sprüche eingebaut (➔25).
- Die seit der karolingischen Renaissance mit Hrabanus Maurus entwickelte Kreuzesverehrung konnte kontemplativ und therapeutisch, später als Amulettbrief für viele Zwecke verwendet werden (➔34).
- In besonders reizvoller erzählerischer Bildhaftigkeit führen die sog. Begegnungssprüche allegorisch an eine Quelle christlicher Leidensmystik, den Kreuzestod sowie die Seitenwunde durch Longinus mit seinem Speer, die zugleich eine der Quellen für Wundheilung wird (➔35).
- Zunehmend gegen Ende des Mittelalters und bis in die Zeiten und Regionen der Gegenreformation ist es die Passion des Heilands, die als wirkungsmächtiges Erzählelement – immer auch jenseits der biblischen Texte – in hochkonzentrierter Form den Patienten ergreifen soll. Mit der Imagination seines Leidensweges, der Kreuzigung, seiner Schmerzen und seines Todes wird der Gläubige seelisch erfaßt und auf diesen Zustand des Gottessohnes eingestimmt und ihm angenähert. Dieser Vergegenwärtigung dienen appellativ zur Imagination die Verben „betrachte, bemerke, bedenke, stell dir vor".[394] Die um 1555 entstandenen geistlichen Übungen des Ignatius von Loyola geben in ihrer tageszeitlich gebunden hochstrukturierten, alle Sinnesorgane involvierenden Gestaltung ein anschauliches Vorbild für derartige Exerzitienprogramme. Und immer auch iuxtaecclesial wurden Wortfiguren, neuartige Verortungen mit Ergänzungen für alle Krankheiten performativ gestaltet.

Daß die offizielle Kirchenlehre von Paulus und Augustinus bis zu Papst Johannes Paul II (Apostolisches Schreiben Salvifici Doloris vom 11. Februar 1984) immer wieder den Vorrang und hohen Wert der Leidenstheologie betonte, hängt nicht zuletzt mit dem zusammen, was Christgläubige als Auferstehung, Erlösertat und metonym als Ostermorgen benennen und was in säkularer Sprache das Christentum als die ‚*Weltreligion der Therapie*' kennzeichnet.

Der Dichter Werner Bergengruen (1892–1964) sieht es so:

Jeder Schmerz entläßt dich reicher.
Preise die geweihte Not.
Und aus nie geleertem Speicher
nährt sich das geheime Brot.

Ob losgelöst oder ob verflochten mit einer individuellen Schuldzuweisung für seine Krankheit und sein persönliches Leid, wird in den spezifischen Heilspruchtexten der Kranke doch berührt von und geleitet in ein religiöses Mitleiden. Damit erlebt er gleichzeitig die eigenen und die allgemein menschlichen

---

394 Ebd., S. 26.

Grenzen im Schmerz Christi, der als Mensch und Gott zum Sieger und Erlöser wird, und der ihn mitnimmt. Es kann in der Regel ein mehrstufiger Prozess angenommen werden, der von einer schmerzhaft schamvollen und bedrückend erniedrigend wahrgenommenen Selbstverleugnung im Zustand der Not und des Kompetenzverlustes auf den Pfad der Einstimmung in die mittelalterlich anerkannte Bekennens- und Bußkultur führt. Das antike ‚per aspera ad astra' kennzeichnet den Rückweg in gesellschaftliche Anpassung und in die Versicherung von Geborgenheit. Im mittelalterlichen Empfinden dürfte die Furcht, aus der Gesellschaft ausgeschlossen zu werden, von großer Bedeutung gewesen sein (**Kap. B4.1.5**). Die Worte Buße und Heil/ Heilung hängen denn auch etymologisch zusammen (➔**20, 21**).

Eine Fülle von Texten hat ein Muster der Passionslitanei zur Heilung von Wunden, Schmerzen und Gicht. In einem ausgedehnten steiermärkischen Text ist es Vergycht, das beschworen wird. (➔**36**). Hier wird nun schon deutlich, daß Schmerzhaftes mit Erlösungsbitte und Hoffnungsmotiven alterniert, sodaß Katharsis auch den Weg zum Licht, zum Heil und zur Osterauferstehung und Ermutigung vorbereitet.

C4.3.1.2 Die neurobiologischen Effekte von introversiver Katharsis:
Redundante Aktivierung mit Exposition von Schmerz und Leid

Eine gewollte therapeutische Erzeugung von seelischem Schmerz, mit dem eine gründliche innere Reinigung, ein Aufwühlen von Affektstaus immer verbunden ist, scheint zunächst paradox und widerspricht jenen Methoden, die eine ‚Ablenkung' vom Schmerz durch Reappraisal oder Labeling der Gefühle, durch behagliche Regression oder Reframing anzielen. Manche derzeitigen Methoden, wie die Humanistische Psychotherapie, scheuen kathartische Eingriffe. Die in obigen Kapiteln behandelten Methoden bewirken vor allem kognitive Aktivierung mit Schwerpunkt in lateral praefrontalen Gebieten bei modulativem Ausgleich in einer Wippenfunktion zu den mit Emotionsverarbeitung betrauten limbischen Gebieten. Dagegen soll introversive Katharsis als ein Einwirken und als performierender Eingriff in die Regelkreise von Schmerzverarbeitung, in die mittleren Strukturen des Gehirns also, dienen, zu allen Zeiten und in vielen Methoden erfahrungsgemäß als Prozess der ‚Reinigung'. Kann Schmerz, körperlich, seelisch oder psychosomatisch ausgelöst, durch suggerierten Schmerz behandelt werden? Ist eine solche Behandlung teilweise mit der modernen ‚Symptomverschreibung' vergleichbar? Wie ist dies vorzustellen?

Zunächst ist davon auszugehen, daß Schmerz keinesfalls nur an körperliche Unversehrtheit oder körperliche Funktionsstörung gebunden werden kann. So definiert die International Association for the Study of Pain: ‚An unpleasant sensory and emotional experience associated with actual or potential tissue damage, or described in terms of such damage.' Und wenn dabei indirekt auch von bloßer Erfahrung, bloßer Möglichkeit und bloßer Schmerz*beschreibung* die Rede ist, „this tendency in the description of pain involves both metonymy and metaphors."[395] Vergleichbar mit Affekten der Angst und der Depressivität wurde Schmerz wegen der empirischen psychosomatischen Befunde und wegen der neuroanatomischen Überlappungen zwischen schmerz- und emotionsrelevanten peripheren und zentralnervösen Regelkreisen als „homoeostatische Emotion" (Craig) bezeichnet. Vorderes Cingulum (ACC), Insula, ventrolaterale praemotorische Rinde, Sympathicus, kardiovasculäres System, Blutdruck, rechter praefrontaler Cortex und Amygdala gehören zu diesem Kreissystem, das der Lebenserhaltung dient.[396] Zwar wurden differente neuronale Systeme für die Verarbeitung von Körperschmerz und Seelenschmerz entdeckt, wobei ersterer nach PET-Untersuchungen in den seitlichen Strukturen contralateral triggert.[397] Viele Untersucher haben jedoch auf die enge Verwandtschaft von seelischem und physischem Schmerz nicht nur im subjektivem Erleben, sondern auch durch Nachweis mit den neurophysiologischen und bildgebenden Verfahren hingewiesen, und wir beziehen uns hier auch auf den oben (**Kap. B3**) genannten ‚autonomen spezifischen Spiegelmechanismus.'[398] Beobachtetes, Gehörtes, Erlebtes, Erlittenes und Vorgestelltes findet sich in *einer* inneren neuronalen Darstellung.

Die Einstimmung auf einen Schmerzpatienten beim Heilspruch führt umso mehr zu einer neuralen ‚therapeutischen Allianz' (**Kap. B3**), als Schmerz im allgemeinen zu den wichtigsten menschlichen Erfahrungen zählt und vom einfachen Ungenügen und Verlassensein bis zum Organschmerz des Unfalls und zur Kolik reicht und bis zu Angst und Stimmungsverlust. Gleichzeitig gilt er als ein warnender, vorbeugender und heilsamer Mechanismus. Der akute Körperschmerz aktiviert nociceptive afferente Neuronen des Rückenmarks und koordiniert via Thalamus zum limbischen System, zur Insula und zum praefrontalen Cortex; damit werden absteigende schmerzhemmende und kontrollierende Impulse in Gang gesetzt. Auch ist Schmerzperzeption in hohem Maße von psychologischen und kulturellen Faktoren abhängig.[399] Eigentlich wird in den meisten

---

395 Semino, Elena, Descriptions of pain, metaphors and embodied simulation, in: Metaphor and Symbol 25 (2010), issue 4, doi: 10.1080/10926488.2010.510926.
396 Vgl. Henningsen, Peter et al.: Neuropsychosomatik. Grundlagen und Klinik, Stuttgart 2006, S. 18–20.
397 Kulkarni et al. Attention to pain localization and unpleasantness discriminates the functions of the medial and lateral pain system, in: Eur. J. Neurosci. (2005), Jun 21 (11), 3133–42.
398 Rizzolatti/ Sinigaglia: Empathie und Spiegelneurone (2006) /deutsch 2008.
399 Guindon, Josée and A. G. Hohmann, Pain Mechanisms and Measurement, in: Berntson/ Cacioppo S. 638–643.

Heilspruchtexten ausgesprochen oder unausgesprochen dem Schmerz und der Angst Rechnung getragen.

Nach den letzten Forschungen gilt als Hauptumschlagsplatz für Schmerzreize die in der Tiefe der Seitenfurche gelegene, von den Opercula überdeckte **Insula Reili**.[400] Ihre Schmerzwahrnehmung wird nach den fMRI-Untersuchungen weitgehend durch situative Umstände und sehr distinkt geprägt. Eine Anzahl von Untersuchungen überprüfte verschiedenste Einflüsse auf Schmerz und Angst: Erwartungshaltungen, Placebos, Aufmerksamkeit, emotionale Reize durch böse oder gute Gesichter, gute und schlechte Gerüche, Launen und Stimmungen, Stress und Disstress. Sie alle haben übereinstimmend ergeben, daß die Insula inmitten einer Fülle differenter Modulationen steht. Gemeinsam mit ihr werden reizabhängig teils frontoparietale und temporale, teils mehr mit dem Cingulum oder der Amygdala korrelierende neurale Kreisläufe aktiviert. Unter der Beeinflussung von Schmerz durch stimmungsbezogene Modulationen korrelierte zum Beispiel die laterale inferiore Frontalrinde (BA 45,47), bei Aufmerksamkeitstesten die superiore Postparietalrinde (BA7) zusammen mit entorhinalen Netzen.[401] Auch schmerzende Worte allein, unabhängig von Emotionen und Situationen, sind in der Lage, spezifische Aktivierungen, getrennt nach schmerzimaginierenden Adjektiven (quälend, zermürbend) und anderen unangenehmen Wörtern (widerlich, ekelig) auszulösen.[402] Und Worte, die als Metaphern und Metonyme – zumal suggestiv oder unter Trance – Schmerzlokalisationen am Heros des Kreuzes beschreiben – Geißel, Ruten, Lanzen – lösen generell die bekannte verkörperliche Mitempfindung (embodied simulation) aus mit ihren entsprechenden empathischen Antwortpotentialen (**Kap. B3**).

> Hirnlokalisatorisch hat eine Bochumer Arbeitsgruppe bei nach depressiven Psychosen Verstorbenen anatomisch einen Neuronentyp untersucht, der in der Evolution des Menschen eine gewaltige Zunahme aufweist: die spindelförmigen von Economo-Neuronen, die im vorderen Cingulum und in der vorderen Insula liegen. Deren Dichte fand sich bei Suicidopfern signifikant erhöht; die Untersucher vermuten eine spezielle Rolle dieser Zellformation für

---

400 Dieses Gebiet wird nach dem Hallenser Christian Reil benannt, der 1808 den Begriff Psychiatrie einführte und damit die Zuständigkeit der Ärzte für seelisch Kranke in humanistischer Absicht anstrebte.
401 Vgl. z. B. die folg. Arbeiten: Mathieu Roy et al., Cerebral and spinal modulation of pain by emotions, PNAS (2009), doi:10.1073/pnas0904706106; Villemure, Chantal and Bushnell, M.C., Mood influences supraspinal pain processing separately from attention (2009) doi: 10.1523/JNEUROSCI.3822-08.2009; Ploner, Markus et al., Flexible cerebral connectivity patterns subserve contextual modulations of pain, (2011) doi: 10.1093/cercor/bhq146; Atlas, Lauren Y. and Tor D. Wager, How expectations shape pain, in: Neuroscience Letters (2012), 520, 140–48.
402 Richter, M., Miltner, W.H.R. und Weiss, T., Schmerzwörter aktivieren schmerzverarbeitende Hirnareale, in: Der Schmerz, 25 (2011), S. 322–324.

Selbsteinschätzung, einschließlich einem negativen self-appraisal.[403] Auch todesbezogene Worte allein vermögen schon bei gesunden Menschen die Insula zu stimulieren und ihr Bewußtsein dahin zu lenken.[404] Bei krankhaften Läsionen der Insula führten Experimente, die die Perzeption von Schmerzen anderer Menschen prüften, im Vergleich zu Läsionen des vorderen Cingulum zu Anzeichen verminderter Mitempfindung. Damit wird die Wichtigkeit einer intakten Insula für Schmerzverarbeitung unterstrichen.[405] Die Wiedererinnerung an soziale Zurückweisungen und Psychotraumen (Bilddarbietung schwerwiegender intimer persönlicher Verluste und Kränkungen in der Vergangenheit) führten zu vergleichbaren, sich überlappenden lokalen Aktivierungen in der Insula ebenso wie die Zufügung körperlicher Schmerzen durch Hitzereize.[406] Einen Beleg für die enge Verbindung von gastrointestinalen Störungen mit kognitiven und motivationalen Einflüssen ergaben Studien, die eine bidirektionale komplexe Modulation, einen balancierenden Mechanismus mit hemmenden und fördernden Aktivitätsmustern darstellten.[407] Dabei zeigte sich im Blick auf unsere sowohl spinal als auch vagal vorzustellende Innervationsbeteiligung bei Koliken und coenästhetisch wandernden ‚Bärmutter'-Schmerzen (→13, 25, 60), daß therapeutische Beeinflussung auch ohne Pharmaka möglich ist. Schließlich werden durch Schmerzreize nicht nur umschriebene Gebiete, sondern stets Netzwerke aktiviert oder gehemmt, und unter Beachtung dieser Funktionsbasis ist es der Schmerz, der auch Aufmerksamkeit und kognitive Fähigkeiten zur Remobilisierung von Ressourcen auf wichtige eintreffende Informationen hin zu fokussieren vermag.[408] Daß ‚eingebildeter' Schmerz, sog. Allodynie, auf schmerzlose Berührungen ebenfalls in den medianen Hirngebieten nachweisbar ist und zwar lokalisatorisch unterschiedlich ausgeprägt, je nachdem ob eine Vorerfahrung gegeben war oder nicht, ist denn nach allem bereits Erforschten selbstverständlich; es scheint – funktionell nur konträr – einer positiven weltweiten therapeutischen Erfahrung bei Übungsbehandlungen wie dem Autogenen Training zu entsprechen.

Diese breiten neuromodulatorischen Verbindungen, deren genaue Funktionsvielfalt und Kooperation auch mit den hormonellen Interaktionen, u. a. über Opioidrezeptoren und den Cortisol-Cytokine-Mechanismus sowie der im **Kap. C4.1** genannten Dopaminregulierung, noch weiterer Forschungen bedarf, geben eine Ahnung davon, wie gleichermaßen verwandt und doch differenziert auch die gewählte Thematik jeder Erzähltherapie die Verarbeitungsweisen und damit Cluster- und Hormonauswahl des Gehirns tangieren kann. Von der Schmerz-

---

403 Brüne, Martin et al., (2011) Neuroanatomical correlates of suicide in psychosis: the possible role of von Economo neurons, PLoS One 6(6): e20936. doi:10.1371/j.pone.0020936.
404 Han, S., Qin, J. and Ma,Y, Neurocognitive processes of linguistic cues related to death, in: Neuropsychologia 48 (2010), S. 3436–3442.
405 Xiaosi Gu et al., Anterior insular cortex is necessary for empathic pain perception, Brain 135 (2012), 2726–2735, doi: 10.1093/brain/aws199.
406 Kross, Ethan et al. EurekAlert 28–Mar–2011.
407 Mayer, Emeran A., Gut feelings: The emerging biology of gut-brain communication, in: Nature Reviews Neuroscience 12 (2011), 453–466; Wilder-Smith, Clive H., The balancing act: endogenous modulation of pain in functional gastrointestinal disorders, in: GUT 60 (2011), S. 1589–1599.
408 Seminowicz, D.A. and K.D. Davis, Pain enhances functional connectivity of a brain network evoked by performance of a cognitive task, in: Journal of Neurophysiology 97, 5 (2007), S. 3651–9.

regulierung über Infektionsanfälligkeit, Wundheilungsprozessen bis zur Stressverarbeitung entfalten psychologische Einflüsse und Meditation via Immunsystem oft nachhaltige Wirkung.[409] Damit ist aber deutlich, daß eine Verbaltherapie mit Spruchtexten, die erlebtes Ungemach und durchlittenen Schmerz imaginierend einbezieht, auch differenziert wirken kann.

In der modernen Hypnotherapie ist die Beeinflussung von Schmerzen ein wichtiges Anwendungsgebiet. Dabei werden u. a. sogenannte assoziative Techniken angewendet, die dem Patienten zunächst eine Zuwendung zum Schmerz zumuten, um anschließend mittels Metapher und Symbol dessen Verarbeitung zu imaginieren. Wir haben schon erwähnt, wie Hypnotherapeuten mit der Metapher ‚Hausgast' arbeiten und diesen Gast oder Eindringling exorzistisch ausfragen. Das heißt eben auch, daß Bedrückung und Schmerz gefühlt werden dürfen und sollen, um sie dann angreifen zu können. In der Einleitung von Trance können etwa Suggestionen wie „Sie spüren, wie erschöpft sie sind, wie schwer ihnen alles fällt, wie weh ihnen das tut" und dergleichen gegeben werden; es sind Bildgebungen, die an die Schicksalssynchronisierung mit Christus (➔2, 32, 34, 36) erinnern. Es ist auch in der modernen christlich spirituellen Therapie die Begegnung mit dem Gottessohn entscheidendes Element, weil Christus nach den Evangelien den Kranken als Depressiven und Schmerzgeplagten „annimmt, so wie er ist."[410]

> Gelegentlich haben in unserer Zeit praktische Übungen in Bibliodrama gezeigt, daß ein biblisches Wunder emotional nachgespielt und nacherlebt werden kann und zu Stimmungsveränderungen mit anfänglich gedrückter Gespanntheit bei den Teilnehmern führte. Erst angesichts der Machtausübung Jesu bei der Verwandlung der Bosheit des Besessenen von Gerasa in eine Herde Schweine und deren exorzistischer Vertreibung zum Ertrinken weicht die depressive Gespanntheit einer bewegt-aufgelockerten Szene. „Das bibliodramatische Gruppenspiel hat die Spannungen gut/böse bzw. hell/dunkel, von denen die Perikope lebt, verlebendigt."[411]

Nach allen diesen Befunden können wir sicher sein, daß Worte, die Trauer beschreiben und damit gleichzeitig die Stimmung verdüstern, eine erhöhte Aktivierung der inferioren Frontalgebiete, der Insula und ihrer jeweiligen Verbindungen und der Amygdala mit Zunahme des Schmerzes auszulösen[412] vermochten und in der Lage waren, eine gespeicherte Schmerzerinnerung durch schmerzverwandte

---

409 Christian, Lisa M. et al. Psychological influences on neuroendocrine and immune outcomes, in: Berntson/Cacioppo (Hg.), Handbook, S. 1260–1279.
410 Grün, Anselmus, Spiritualität und Depression, in: Schweizer Archiv für Neurologie und Psychiatrie 160 (2009), S. 182–187.
411 Frick, E. Der Besessene von Gerasa. Ein Bibliodrama zu Mk. 5,1–20, in: Geist und Leben 64 (1991), S. 385.
412 Berna, C. et al, Induction of depressed mood disrupts emotion regulation neurocircuitry, in: Biol. Psychiatry 67 (2010), S. 1083–1090.

Worte geradezu redundant zu remobilisieren.[413] Man weiß, welch fatale Rolle die ständige Wiederholung der Klagen als eine Art Teufelskreis bei somatoformen Depressionen und Hypochondrie spielen kann. Das heißt aber: Nach modernen Vorstellungen sind diese Therapiemethoden nur dann günstig, wenn sie im Rahmen eines zeitlich abgestimmten sequenziellen Programms geboten werden und das Ziel verfolgen, durch Übertreibung Gegenkräfte gegen Schmerz und Leid zu aktivieren. Denn die kathartische Versenkung muß genauso wie die Regression um der Progression willen (**vgl. Kap. C4.1**) als Therapie*abschnitt* begrenzt organisiert sein. Sie muß von einem regenerierenden Abschnitt abgelöst werden. Das ist im Hinblick auf autoregulative cerebrale Modulationen, die nun nach forcierter Belastung geradezu eine überschießende Gegenreaktion von Selbstheilungskräften auslösen können, zwar theoretisch nicht unbedingt erforderlich. Aber es dürfte keinen Heilspruch geben, der in inversiver schmerzaktivierender Strategie verharrt. Mißbräuchliche Textanwendungen ließen an die in Geschichte und Gegenwart bekannte Gefahr zügelloser Voodookulte, Geißlerheere und Sekten denken.

Moderne Therapieformen, die eine inversive Katharsis einbeziehen, integrieren diese Funktion in einen differenzierteren Zyklus, der anschließende Reintegration betreibt, wie z. B. die Bioenergetik oder die Funktionale Psychotherapie.[414] Die konsequente Einfügung und Abschlußimagination mit der Erlösungstat Christi, die soteriologische Therapie durch den Opfertod, oder die Vorwegnahme bzw. die Erinnerung an den guten Ausgang von Hiobs Schicksalsschlag (→3, 29, 30) durfte für den Kranken – im Gegensatz zum Nachtgebet des Gesunden (→2) – keinem dieser Texte fehlen und ist unabdingbarer Bestandteil der introversiven Katharsis im europäischen Mittelalter. Als entscheidendes therapeutisches Prinzip ist dabei nach heutigen Vorstellungen der Rückgriff auf die Ressourcen (**vgl. bes. Kap. B4.1.2, C1.2**) des mittelalterlichen christozentrischen Weltbildes als salutogenes Wertesystem zu betrachten, das im Trancezustand suggestiv cerebrale, hormonelle und Immuninduktionen vermittelt. Wie dabei durch die Induktion introversiv kathartischer affektiver Schmerzzustände jene neurochemischen Prozesse gefördert werden, die einer Beeinflussung des parasympathischen Systems durch Sympathikolytika entsprechen, muß die weitere Forschung ergeben. Wahrscheinlich fördert die Methode der psychischen Mobilisierung von initial zündenden zielkonträren Kräften auch die Bildung

---

413 Weiß, Thomas, Worte fügen Schmerzen zu, EurekAlert 30–Mar–2010.
414 Vgl. Corsini, Handbuch, S. 94,102,265f.

biochemischer Antagonisten und die Modulationen in Insula, Nucleus accumbens und Locus caeruleus.[415]

C5 Strategien der extroversiven Katharsis: Direkte Bannung des Übels durch Befehl und mittels überirdischer Helfer: Die Expulsion

Psychotherapeutische Strategie
moderne Begriffe:
Expulsive Technik, kreative bzw. kanalisierte Aggression,
Konfrontative Therapie, Impactmethode

historische/ alltägliche/ volkstümliche Begriffe:
(wie Tab.1:) Wortkanonade, Verbaler Befreiungsschlag, Vertreibung der Geister, Verfluchung, Verbannung, exorzistische Techniken; – Anstiftung und Hilfe zur Gegenwehr, Kampf oder Flucht

Anwendungsfelder
Panik und andere Angstformen, seelische Verletzung, Hilflosigkeit,
(wie Tabelle 1)

Neurophysiologie
Sensomotorische Aktivierung durch aktivierende Imperative,
Sympathikotone Aktivierung zur Equilibrierung hormoneller Systeme
(Homöostase)

Tabelle 5: Die unmittelbare beschwörende Aktion als Aktivierung sensomotorischer Cluster und sympathikomimetischer Hormonsysteme

Während eine Vielzahl der Krisen- Notfall- Situationen per se eine Stimmungsabsenkung, Angst, Panik oder Schmerzreaktion mit sich bringt oder auslöst und u. U. damit die innere Reinigung, die introversive Katharsis, schon vorlaufend angebahnt werden kann, weil sie stimmungskonform ist, setzt die extroversive Katharsis, der wir hier nachgehen, einen anderen, unmittelbaren Kontrapunkt gegen defiziente Modi emotionaler und kognitiver Verfasstheit. Es geht um Wortgewalt, Beschwörung und exorzistische Maßnahmen, um eine *unmittelbare* Vertreibung von Dämonen, Krankheitsgeistern, bösen Gestalten,

---

415 Soeter, M. and Kindt, M., Stimulation of the noradrenergic system during memory formation impairs extinction learning but not the disruption of reconsolidation [in anxiety disorders], Neuropsychopharmacology 2012 Apr., 37(5), 1204–15. doi:101038/npp.2011.307.

die als Verursacher von oder selbst als die Krankheit oder Schädigung angesehen werden. Hier scheint zunächst die Mitwirkung des Heilers besonders wichtig. Die Strategie war uns bereits in Zusammenhang mit anderen Methoden der psychoperformativen Therapie begleitend begegnet, bei der Emotionsverarbeitung per Etikettierung (**Kap.C1.2; C1.3**) und bei der Anwendung von Wiederholungsfiguren (**Kap. C4.2**). Dort hatten wir diese anderen Wirkanteile der neurophysiologischen Mechanismen der Dämonenvertreibung in den Vordergrund gestellt. Zur vollen Effektivität benutzte der Zauberspruch vielfach die Einbeziehung der expulsiven Methode, ob als direktes Vorgehen oder mit Hilfe göttlicher oder sonstiger hilfreicher überirdischer Personen.

C5.1 Die Anwendung der expulsiven Methode im Kulturvergleich

Der Begriff der extroversiven Katharsis, wie ich ihn hier verstehe, geht also von einer Sichtweise aus, die einen individuellen emotionalen Status als sympathikotone Spannung voraussetzt, ein Zustand, der den Gegenangriff vorträgt. Von wo und von wem diese Spannung innerhalb der Triade ausgeht, muß jeweils gefragt werden. Es ist ein Sektor der Spruchkultur, der von Anfang an in der germanistischen Forschung anhand der Schriften Aufmerksamkeit gefunden hat. Der Typ mit einem Befehl gegen den Wurm: „Geh heraus, Nesso ... aus dem Mark ... bis in die Fußsohle!" oder gegen den Wundfieberkönig: „Fahr du nun ... Thursenkönig ... Thor schlage dich!" (➜27, 37) oder mit einstreuenden Befehlen an den Alp (➜1) wurde als subjektive Form (eines wollenden Subjektes) des Zauberspruchs beschrieben, zu dem als Grundmerkmal ein „mehr oder weniger anschauliches Verb der Bewegung" im Imperativ gehört.[416] Oder es wurde ihm einmal einleitend „nichts anderes als der exorzistische Imperativ"[417] zuerkannt, der aber sogleich Fragen nach der Identität von res und signum und der Erschaffung des Wortes für den archaischen Menschen aufwerfe. Wir hatten im **Kapitel A1** zu Magietheorien darauf hingewiesen.

Allerdings besitzen wir für diesen Spruchtyp im Mittelalter bis zum Beginn der Neuzeit nur wenige Schriftzeugen, eben die genannten Reste vorchristlicher germanischer Urformen. Der Befehl an diese Dämonen oder an eine personalisierte Krankheit ist für diesen Zeitabschnitt relativ selten, weil die Übel verchristlicht und ihrer heidnischen Götter in Missionszeiten beraubt wurden. Erst mit dem 14./15. Jahrhundert, in der Zeit einer sich allmählich entwickelnden Individualisierung, als viele Texte die Klostermauern verlassen hatten, wurden

---

416 Hampp, Irmgard, Beschwörung Segen Gebet, S. 115,.117.
417 Schulz, Monika (2001), Beschwörungen im Mittelalter, S. 13.

Ich-Autoritäten in die Texte installiert. Damit entwickelte sich aber auch eine engere therapeutische Symbiose zwischen Heiler und schädigender Kraft, die einmal als ‚Ich-Du-Formel'[418] benannt wurde und die die archaischen Formen nachbildet. Meistens jedoch wurden indirekte Befehle erteilt, die bittend mit Gott oder heiligen Gestalten operierten, „auf daß ihr – alp und elbynnen – müsset alle tot sein, das gebiet euch Gott und der liebe Herr Job!" (→**3, vgl. 4, 7, 13, 17, 18 u. a.**). Vielfach wurde bei Gott beschworen (→**22, 25, 39 u. a.**) oder es wurden rituelle Ergänzungen hinzugefügt wie beim Heilspruch gegen Überbein und Geschwulst (→**39**), wo das Holz des Siegeszeichens Christi als Leidenswerkzeug und als Therapieinstrument kontaminiert. Die Vertreibung des Bösen und des Satans blieb dem kirchlichen Exorzismus weitgehend vorbehalten; in langen immer wieder einstreuenden Passagen werden als satanisch imaginierte Krankheitsverursacher kontaktiert, exploriert und verbannt.

Ihre epochale Bedeutung hatten direkte und indirekte Expulsionstechniken in Riten und Sprüchen der mesopotamischen Spruchkultur. In der babylonisch-assyrischen Maqlû-Keilschrift der Beschwörungstafel I (→**43**) folgt auf Klage- Trauer- und Enttäuschungsdramatik der Zeilen 1–16 die Beschuldigung des Anderen, das entschlossene „und erhebe nun Anklage in meiner Sache". Und es bleibt nicht bei Anklage, die eigentlich schon vollzogen wurde, eingestreut in die Larmoyanz des Vorgebrachten, sondern es folgen Angriff und Verurteilung: Hexer und Hexe, dargestellt in Figuren, mögen sterben (Zeile 19), die Götter mögen sie schlagen (Zeile 29), die bösen Zungen sollen sich wie Salz auflösen (Zeile 33). Ein Rachegedanke taucht auf und der Umgang mit Überirdischen wird sehr ‚menschlich' (**Kap. B4.2**). Am Schluß wird der Vollzug dieser Wünsche festgestellt. Aber nicht eigene Kraft der Worte, sondern die Befehle der Götter haben es bewirkt (Zeile 36). In **Beispieltext** →**44** erfolgt schon die Vollstreckung des Urteils an der Figur des Übels ohne direkten Einsatz der Götter, es werden die bösen Nachtgeister-Figuren verbrannt, sollen zergehen, zerfliegen, zertropfen, verrauchen. In **Beispieltext** →**45** werden Ea und Marduk und ihr Held Gira als Retter genannt, erhofft und gerufen. Zumeist folgt extroversive Aktion auf introversive Klage, eine Reinigung von innen nach außen. Die Götter werden nicht immer konsequent bemüht und Helden werden eingesetzt.

Wir erinnern uns an die mesopotamischen Besonderheiten der Vermeidung von gerichtlichen Anklagen wegen Schadenszauber. Vielleicht machte sich dies ersatzweise in der Strategie der Sprüche Luft, weil die eigentliche Anklage gefährlich war. Wir erinnern ferner an die alltägliche Nähe der Menschen zu anthropomorphen Dämonen, Geistern und Göttern, selbst Totengeistern, mit all denen man oft wie mit seinesgleichen umging, selbst wenn man zornig war. Es

---

418 Hampp, ebd., S. 121.

besteht kein Zweifel, daß das gesamte kulturelle und gesellschaftliche Umfeld einen adäquaten Nährboden besonders für die Anwendung expulsiver Spruchtexte bot, zumal deshalb, weil der Kranke selbst als dämonengepackt unter der Vorstellung leiden mußte, einer anderen Welt anzugehören (**Kap. B4.2.3**).

Auch in den altindischen Veden wird jene zupackende Triebkraft gelegentlich deutlich, die den selbstbewußten Heilkundigen auszeichnet, dessen Autorität im Stakkato verkündet: „Aus jedem Glied ... treib ich die Krankheit aus" und „O Schwinden flieg mit dem blauen Häher fort, schwinde dahin!" (➔ **54, 55**). Teilweise wird, wie gesagt, die ganze Wucht göttlicher Kraft bemüht, wenn Indras Mühlstein und Mordwaffen zum Einsatz gegen die Hörner der Krankheitswürmer kommen sollen (➔**57**). In einem frühchristlich koptischen Expulsionstext (➔**61**) wird der Dämon des Besessenen exorzistisch vertrieben, in ein dunkles Chaos der Verlorenheit.

C5.2 Der Befehlsstand der Expulsion: Die Bedeutung des Helfers

Die Betrachtung der Dämonenbannung durch Etikettierung mit emotionsverarbeitenden Metaphern und anderen Sprachfiguren hatte die therapeutischen Wirkkräfte innerhalb der Triade Helfer – Hilfsbedürftiger – Umwelt bisher in unseren Ausführungen eher auf die Verarbeitung durch den Empfänger und sein Gehirn gewichtet. Diese Gewichtung darf nicht darüber hinwegtäuschen, daß dem Heilkundigen Funktionen zukommen, die über die emotionale Konspiration durch Metaphern-Induktion (**Kap. C1.2**) hinausgehen. Am deutlichsten wird dies bei der Betrachtung der Expulsion. Zwar war als Vorform der extroversiven Katharsis schon das imaginative Ansammeln von Waffenarsenalen in der Stresssituation genannt worden, das Aufbäumen gegen Schicksalsschläge, was der Gesunde noch vermag. Wir können es auch als Fluchen und Verdammen begreifen, als Entschlossenheit zur Aggression. In vielen Notfall- und Krisensituationen ist aber ein kalkuliertes sinnvolles Aufbäumen nicht mehr möglich, sodaß es der Unterstützung durch den Helfer bedarf, um Wut und Spannung zu kanalisieren, den oft chaotischen Gefühlen eine Richtung zu geben, sie gegebenenfalls auch mehr oder weniger erst zu mobilisieren. Er ist der Vermittler derjenigen Maßnahmen, die kulturadäquat sind. Das wird ihm zugetraut. Er kennt nicht nur das Instrument der Texte und Riten, sondern auch deren Angriffsflächen und Wirkorte, wo der Schädiger angegangen werden kann. Nur er als Fachmann vermag direkt jenen Ungeist zu packen, der den Kranken der mesopotamischen Kultur gepackt hat. Er kann sozusagen dem Dämonen die Schlüssel zu Handschellen am gepackten Kranken abhandeln. Und nur er vermag in allen Kulturen helfenden Gottheiten besonders nahe zu treten. Er tut dies scheinbar ausschließlich für den Kranken; er vertritt ihn im Kampf aber nur initial oder intermittierend im Verlauf des Heilspruches. Und er sorgt gewollt oder ungewollt in der neuroperformativen Allianz dafür, daß die energetischen Gewichte zwischen ihm und dem Kranken ausgeglichen werden. Der Kranke wird zu Widerstand und Aufruhr angestiftet oder gezielt geführt; der Helfer vermag eine aktive intervenierende Rolle zu spielen, schon weil allein der Anblick einer ranghöheren Person und ihrer Rangabzeichen Hirnregionen aktiviert, die an der Herstellung von Aufmerksamkeit beteiligt sind.[419]

Wer einen Bogen zu heutigen Therapieformen schlagen will, wird besonders die ‚Kreative Aggression' und manche Techniken des Psychodramas heranziehen. In der Kreativen Aggression nach George R. Bach werden sämtliche Formen von Aggressivität, von einer offen-brutalen bis zur hinterhältigen, von sozialer bis zu privater Form einbezogen. Im Bestreben, die jedem Menschen eigenen auch latenten Kampfhaltungen vor den Fluchthandlungen derartig zu organisieren, daß sie nichts und niemanden zerstören, sondern kathartisch Wut und Spannung ablassen können, werden die Kranken verbal und rituell in Feindseligkeitsverhalten trainiert. Die populären Vergleiche mit Dampfablassen und Furunkelaufschneiden sollten nicht vergessen lassen, daß die tatsächliche Lokalisierung des Bedrückenden meist unbekannt und variabel ist. Dabei werden heute nicht anders als beim Maqlûritual auch Puppen und Figuren zur Verwünschung eingesetzt.[420] Vor allem aber übernimmt der Therapeut ähnlich wie der Beschwörer eine sehr aktive Rolle im Sinne einer Beeinflussung bzw. einer Einwirkung (impact) auf den Patienten. Diese Therapie geht theoretisch davon aus, daß nicht die natürliche Aggression sondern nur deren Fehlleitung neu zu programmieren ist. Eine mit pazifistischer und auf Harmonie angelegter Aggressionsphobie argumentierende ideologische Gegenposition gegen dieses Konzept steht denjenigen nicht fern, die am Modell geheimnisvoller magischer Ur-Zauberkräfte für Schamanen festhalten.

---

419 Meyer-Lindenberg, Andreas, Tief verwurzeltes Statusdenken, in: Spektrum der Wissensch., März 2009,.S. 19.
420 Bach, George R., Kreative Aggression, in: Corsini, Handbuch I, S. 571–586.

C5.3 Neurobiologische Effizienz durch extroversive Katharsis: Die Sensomotorik aktivierender Imperative und das homöostatische Gleichgewicht

Wie wir wissen, bewertet das bedeutungsermittelnde Perzeptionssystem des Empfängers die gegen Krankheit und Bedrohung gerichteten Worte. Was an der Botschaft also nutzbar ist, interpretiert das Empfängerhirn in situativ gegebener und systemischer Abstimmung; ein zentrales Modul für Sinn und Bedeutung gibt es nicht. Die Verben dienen der verkörperlichenden Simulation (**Kap. B3; C1.3.4**) mit somatischer Parallelität; sie bedienen die Sensorimotorik, gehen ‚in Fleisch und Blut' über, wie wir dies am **Beispieltext** ➔1 darstellen konnten. Hier wie dort wird der okkupierende Ungeist angesprochen und beschworen wie es in einer modernen Hypnotherapie dem Fremdkörper mittels moderner Begriffe und Namen geschieht. Das bewegende Wort des Heilkundigen aktiviert als Katalysator prämotorische und parietale Schaltstellen. Durch die kunstvolle Simulation des Erlebten per Spruchtext vermag der ‚professionelle Abreagierer' (Lévi-Strauss) das Grauen in eine neue verarbeitbare Zone zu verlegen.

Dabei muß auffallen und muß herausgestellt werden, daß bei den ausführlichen Spruchtexten immer wieder intermittierende Perduktionen und Wiederholungen der Befehlsverben samt ihrer Befehlsobjekte und Verschickungsorte vorkommen. Es begegnet ja immer wieder der Wechsel zwischen intro- und extroversiven Textpassagen, zum Beispiel im mesopotamischen Maqlû-Ritus. Oder wir treffen auf eine Art Einstreutechnik wie im Münchner Nachtsegen mit seiner anfangs fast nur einstreuenden Beiläufigkeit des indirekten Befehls, die sich erst im Verlaufe steigert wie in modernen Hypnotherapiemethoden zum Unterlaufen von ‚Widerständen';[421] dies läßt sich in den ziemlich kurzen alten Schriftscherben des Mittelalters nicht mehr feststellen.

Die durch kombinierte Stimulierung bewirkte Aktivierung sensomotorischer Schaltkreise mit kognitiven und emotionalen Folgen entspricht letztlich einer Art intermittierendem Priming, das mit aktionsbezogener Rede betrieben wird. Darüberhinaus können auch in dieser Methode Verschaltungen zur bewußten Selbstreflexion mit solchen hochaversiven situationsgeprägten Stimuli in der präfrontalen Rinde rechts und unter Minderung der Amygdalaaktivität gebahnt werden.

> Aus der Sicht einer auf akute Traumen gerichteten modernen Hypnotherapie ist die perduzierte Imagination bewegender Verben in Befehlsform besonders effektiv, zumal dann, wenn in die Imagination hilfreiche Personen einbezogen werden. Ein solches Vorgehen entspricht der

---

421 Vgl. Revenstorf und U. Freund, nach M.H. Erickson: sog. Einstreutechnik, in: Revenstorf/ Peter (2009) S. 210.

modernen dissoziativ-symbolischen Methode[422], die den Schmerz und die Angst durch Aktivierung nichtschmerzhafter Teile des Körpers von Teilen des Bewußtseins, das den Schmerz empfindet, abtrennt. Dies geschieht durch Fokussierung auf angenehme Vorstellungen, in diesem Fall auf endgültige Vertreibung des Übels. Es ist eine Vorwärtsverteidigung, die mit Aktivierung sympathischer Systeme betrieben wird.

In der Wachtraumtherapie ist schon von Robert Desoille der Wert der Bewegung nicht nur bei der Wortwahl, sondern bei der Wahl der Bilder und Haltungen im Imaginieren hervorgehoben worden. Er hat seinen Patienten schon initial herausfordernde Bildphantasien empfohlen oder zugelassen, die weder Regression noch Introversion anzielen sollten. Er ließ einen Tiefseegrund einstellen, ließ Riesentintenfische, Haifische und Monster imaginieren und stellte Patienten damit vor Situationen des inneren Kampfes mit dem Unheil, das in der Folge personalisiert werden kann. Kraken und Quallen waren von Angstkranken beschreibend zu attackieren, und in einer quasi magischen Ausdrucksform zu besiegen. Der Therapeut sollte weniger analysieren, sondern «il doit porter surtout sur la modification de l'attitude du sujet dans son rêve.» Desoille hat in dieser aversiven Methode selbst auch archaische Elemente gesehen, besonders wenn er Gallier-Riesen als Väter oder Eulen als Großmütter einführte, oder wenn er Himalaya- und Himmelsaufstiege imaginieren ließ.[423] Diese Therapie produziert eine Dynamik, die im flexiblen und direkt gezielten Angriff durch Metonyme eine Chance erkennt, Krisensituationen zu meistern. Sie arbeitet mit scheinbar milde gelenktem, aber doch konsequentem Vorgehen nach Art einer getarnten und säkularisierten Beschwörung.

All diese alten und neuen Therapieansätze korrespondieren durch ihre Verbal- und Ritualaktionen immer auch mit einem biophysiologischen Gleichgewichtssystem, das in vegetative Zentren für die Körperfunktionen eingreift. Je nach der Art des Krisen-Notfalls herrscht einer der beiden Zügel dieses Systems vor, Sympathikus oder Parasympathikus, in Funktion als Sympathikotonie oder Vagotonie. Extreme Formen ethnischer Rituale konnten eine Beeinflussung bis zum Tabu-Tod über die Drosselung von Gefäßpermeabilität, Herz- und Kreislauffunktionen bewirken. Während sich bei der introversiven Katharsis eine nach innen gerichtete Aufmerksamkeit sammelt und der Blutdruck sich erniedrigen kann, ganz im Sinne der Vagotonie, kann extroversive Katharsis mit aggressiven Durchbrüchen und Enthemmtheit einhergehen, körperlich mit erhöhter Muskelspannung und erhöhtem Blutdruck. Die nach außen gerichtete Aufmerksamkeit wird sympathikoton gesteigert. Wir werden nicht erkunden können, in welchem Zustand jeweils welcher Zauberspruch von einem Heilkundigen gesprochen wurde, inwieweit also eine dem vegetativen System immer gerecht werdende Differentialtherapie geleistet wurde. Es scheint aber von entscheidender Bedeutung zu sein, daß die Verbaltherapie auf eine Balance angelegt war. Und in der Tat kann an manchen Texten dieses Hin und Her, dieses Wechselbad sympathikotoner und vagotoner Reize und Reizzustände gefunden oder vermutet werden (**➔36, 38, 50**).

---

422 Vgl. Halsband, in: Revenstorf/ Peter, S. 817.
423 Desoille, Robert: Théorie et pratique du rêve éveillé dirigé, Genf 1961 (passim).

Man kann auch von einer nicht näher bestimmbaren akuten Stresssituation ausgehen. Der Organismus zeigt gewöhnlich eine Reihe biologischer Reaktionen, die die Homöostase, also das selbst-regulative lebenserhaltende Stoffwechselgleichgewicht angemessen stabilisiert. Stress ist eine Hirn-Körper-Reaktion durch äußere oder innere Impulse, die eine Unterbrechung der Homöostase bewirken.[424] Es sind komplexe neuroautonome und neuroendokrine Systeme, die sowohl auf psychische als auch auf physische Stressoren ansprechen müssen.

> Unter diesen Systemen fungieren vor allem das Sympathico-adrenomedulläre System (SAM), das Hypothalamus-Hypophysen-Nebennierenrinden-System (HHNA) und das Argenin-Vasopressin-System auf seelischen, emotionalen, Kälte- und Schmerz-Stress. Das SAM verbessert Atmung, allgemeine Durchblutung und Energiefreisetzung und es erhöht die Aufmerksamkeit nach außen. Es bildet die sekundenschnelle Initialreaktion auf Bedrohungen und wurde von seinem Erstbeschreiber Cannon (1929) mit der ‚fight-and flight'-Reaktion zusammengesehen. Stress also aktiviert in diesem System die Freisetzung von Adrenalin und Noradrenalin, wobei die noradrenergen Zellen des Locus caeruleus direkt aus dem zentralen Kern der Amygdala innerviert werden, von deren Kompetenz bei der Kampf-Flucht-Entscheidung schon die Rede war. Und es bestehen reziproke Verbindungen zum Hypothalamus. So kann unter Stress sowohl der Hypothalamus die Noradrenalinausschüttung über aufsteigende Bahnen des Locus caeruleus zum Cortex verstärken als auch der L.C. Einfluß auf die Aktivität des Sympathicus nehmen.[425] Der aktivierte Hypothalamus bewirkt über die Nebennierenrinde auch die Freisetzung des ‚Stresshormons' Cortisol, das sehr rasch die genannte Aufmerksamkeit über den Thalamus verbessert.[426] Auch die Bedeutung der homöostatischen Funktion des Oxytocins, eines weiterhin bei Notfällen aktivierten Hormons, dessen Wirkung durch die soziale Zuwendung zum Kranken und unspezifisch einfach ‚fesselnde' Worte nachgewiesen ist,[427] war uns bei der Annäherung an die mittelalterliche Alptraumbeschwörung begegnet (**Kap. C1.3.4**).

Die Forschungsergebnisse zur Wirkung von akutem Stress, der einen Teil der Krisen-Notfallsituationen beherrscht oder begleitet, auf psychische Funktionen, sind aufgrund der verschiedenen Forschungsansätze nicht eindeutig. Übereinstimmend wird jedoch festgestellt, daß die selektive Aufmerksamkeit gesteigert wird. Das heißt aber, Stress blockiert das Langzeitgedächtnis des Hippocampussystems, während das Arbeitsgedächtnis präsent bleibt. Dies trägt zur erhöhten Suggestibilität für die Heilspruchtexte und zur Wirkung der Imaginationen bei.

---

424 Mora, Francisco et al., Stress, neurotransmitters, corticosterone and body-brain integration, in: Brain Research 1476 (2012), 71–85.
425 Petzold, Antje: Auswirkungen ak. psychosoz. Stresses auf Feedback-basiertes Lernen, Dresden 2010, S. 14–17.
426 Strelzyk, Florian et al., Tune it down to live it up? Rapid, nongenomic effects of cortisol in human brain, in: Journal of Neuroscience 32 (2012), S. 616.
427 Olff, Miranda, Bonding after trauma: on the role of social support and the oxytocin system in traumatic stress, in: European Journal of Psychotraumatology 3 (2012) 18597–doi.org/10.3402/ejpt.v3io.

Insofern besteht kein Zweifel, daß die im Notfall zugesprochenen oder gesungenen ergotropen, sympathikotonen aktiven Verben wirkungsvoll und oft auch zielgerichtet ‚ankommen' können. Kraftworte wie das ‚Geh heraus!', das ‚Schwinde dahin!', das ‚Beschwöre dich!' und ‚Zerschmettere dich!' aktivieren nicht nur die Hörgebiete, sondern über die Spiegelneurone auch die jeweils dazugehörigen motorischen Areale, die in Bereitschaft gestellt werden. Armbezogene Bewegungswörter wie das Zerschmettern und Beschwören (Beschweren), beinbezogene Wörter wie das Geh-heraus und Schwinde mobilisieren den extroversiv gerichteten Verteidigungsapparat und unterlaufen damit jene situationsbedingten Störfaktoren, die eine Heilung behindern. Sie fördern Lösungen, die einer natürlichen Regulierung und Ausbalancierung dienen. Modern gesagt, einem ‚Zurück zur Natur nach deinen besten Möglichkeiten'. In der Koaktivierung der Hirnareale vollzieht sich eine zellulär-materielle Einheit zwischen Wort und Aktion.

D Nachbemerkungen

1. Mit der vorliegenden Studie kann zunächst einem verbreiteten intrikaten Widerspruch entgegengetreten werden. Er ist beispielsweise für das 14. Jahrhundert von Umberto Eco seinem Romanhelden, dem Ex-Inquisitor William von Baskerville, Verehrer des Roger Bacon, in den Mund gelegt worden: „Ich habe sehr tüchtige Ärzte gekannt, die hervorragende Medikamente zu mischen wußten, […] aber den Laien verabreichten sie ihre Salben und Säfte nur unter Rezitation von heiligen Worten und Sprüchen, die wie Gebete klangen. Nicht weil diese Sprüche irgendeine heilende Kraft gehabt hätten, sondern damit die Patienten glaubten, daß die Heilung durch die Gebete käme, sodaß sie gesund wurden, ohne allzu sehr auf die Medikamente zu achten. Außerdem hat der Körper, wenn die Seele auf rechte Weise zum Vertrauen in die fromme Formel gebracht wird, mehr Aufnahmebereitschaft für die heilende Wirkung der Medikamente."

Waren also all die Heilsprüche, jene des Mittelalters und jene fremder Kulturen lediglich illusionistisches Therapiebeiwerk, bei den einen zur religiösen Erziehung, bei den anderen zur Festigung der Priesterarzt-Herrschaft oder der Schamanenurkräfte oder wurden sie gar einer Art Placebo verglichen?

Hier irrt Umberto Ecos Held. Nicht daß es allein um historische Feststellungen ginge, also um das fast obligate Nebeneinander pragmatischer und verbaler therapeutischer Textanteile in den Schriften im Sinne einer – modern gesprochen – wenigstens literarisch nachweisbaren Ganzheitsmedizin. Auch nicht um die Vorstellung einer psycho-physischen Einheit in den konvergierenden Theorien und Praktiken von Theologie und Medizin unter Mönchsärzten oder um die antiken griechischen und arabischen Einflüsse nördlich der Alpen seit dem 12. Jahrhundert. Für das europäische Mittelalter haben wir viele solcher Befunde, die entgegen einer gelegentlich vermuteten Ambivalenz irdischer und religiöser Heilsversprechen bei Volk und Klerus viel eher für ein harmonisches, komplementäres Verhältnis beider Therapieansätze sprechen.

Gewichtiger als diese Argumente sind die hier im Kapitel C modellhaft nachgestellten Wirkungsmechanismen für die psychoperformative Spruchtherapie in alten Kulturen, auch wenn sie einer historisch schlüssigen praktischen Beweisführung nicht unterliegen können. Es gibt keine nach modernen wissenschaftlichen Grundlagen aus früheren Zeiten verwertbaren Krankheitskatamnesen und es wird keine reproduzierenden Experimente geben. Tempi passati. Von einer möglichen neurophysiologischen Relevanz haben ehemalige Heilkundige nichts geahnt; jedoch ist die weitgehend feste kulturelle Verankerung dieser Verbalmedizin und ihre stete Weitergabe von Generation zu Generation ein wichtiges Indiz für die allgemeine Anerkennung eines Erfahrungsschatzes.

Aber es gibt auch und vor allem keinen Anlass, unseren und anderer Völker Vorfahren völlig abweichende zum Beispiel rein spirituelle oder dämonengeprägte Hirnstrukturen zu unterstellen. Manche Historiker sagen den Menschen der Antike nach, sie hätten doch auch gewußt, daß es Geister nur im Traum gibt. Bis auf gesellschaftlich bedingte Nuancen wichen die Hirne der Vorfahren nicht von den unseren ab, weil sie wie die unseren ihre prüfende Instanz hatten, die ähnlich wie heute die Bedeutung alles sinnlich Wahrnehmbaren im Notfall auf Überlebenschancen testen, egal wie die Schriften der Mächtigen es wollten. Die Hardware der Gehirne arbeitete und arbeitet für sich und für ihren Leib.

2. Mit den einleitenden Kapiteln zur Geschichte der Deutung der Heilspruchkultur soll keinesfalls eine Liquidierung der Megafiktion Magie gewagt sein, obwohl manche ihr schon ein Ende voraussagten. Mit Magie konstruiert der Mensch eine seiner wichtigsten Tröstungen. Wer sie angreift, spürt umgehend den Bumerang. Magievorstellungen werden ein zähes Leben haben, weil unsere Hirne nicht gegenwärtig und nicht zukünftig in absehbarer Zeit an sich selbst herankommen, nicht sich selbst greifen und begreifen, wie sie es nicht mit dem Himmelsblau vermögen – Urknall und Homunculus sind nur Metaphern. Metaphern gesellen sich janusköpfig um unser Verständnis von den neuronalen Vorgängen: Wir benennen Kehrtwenden im Leben und Rekurrenzen der Hirnaktivität. Wir vermuten Oszillieren zwischen Hirngebieten und imaginieren vielleicht dazu das Schwingen im Tanzen und im Pendeln unseres Verhaltens. Wir erleben Feedback-Schleifen und Balancen wie im Alltagsleben so im Stoffwechsel und seiner Homöostasis. Ist alles nur Rhetorik, modisches Vokabular?

Und doch läßt sich das scheinbar Magische um den Zauberspruch besser eingrenzen, wenn wir versuchen, seinen *natürlichen* Wurzeln und Blüten näher zu kommen, auch wenn die bildgebenden Verfahren nur Anzeichen von frischer Saftverteilung, Füllung und Knospung, sprich Sauerstoff- und Durchblutungsverhältnissen abbilden. Immerhin sind Zauberworte nach Zeitablauf, Verbindung und Verortung messbar und experimentell nachprüfbar. Wie etwa das Fressen von Kieselsteinen in einem alten psychoperformativen Text wirkte, wissen wir nicht, aber daß die gleiche unsinnige Unmöglichkeit experimentell bestimmte Regionen aktiviert, können wir heute beweisen. Der Textvortrag war für manches Gehirn einfach relevant.

Die hier vorgestellten Modelle im Rahmen einer gleichsam kulturneurologischen Theorie psychoperformativer Funktionalität des Zauberspruches verstehen sich als ergänzende Beiträge zu einer Klärung des Verhältnisses von Kultur und Natur auf einem scheinbar engen interdisziplinären Spezialgebiet. Sprechakttheorie, Soziologie und Ethnologie bieten dazu entscheidende Voraussetzungen, besonders auch die Arbeiten von Claude Lévi-Strauss. Dabei vertrete ich nun dezidiert die Ansicht, daß das Bild vom Bau einer Brücke von Kultur zu Natur diesem Thema nicht gerecht wird. Es geht nicht um etwas Künstliches, Prothetisches; es gibt keinen Tiefgrund, kein Gefälle und kein Tal zwischen gesprochenem und imaginiertem Wort, zwischen Zauberspruch und Neuronenaktivitäten. Vielmehr sollten wir uns Flüsse vorstellen, die von Kultur zu Natur und vice versa und in Kreisläufen fließen, wobei ihre Fließeigenschaften und Flußformen variabel sind. Zentrale Stromzungen und marginale Umkehrströme, Strömungswalzen und Neuronencluster treiben ihr Spiel immer wieder in Turbulenzen mit vertrackten Hindernissen und vermeiden damit kurzschlüssige Katastrophen.

3. Es muß hier nicht daran erinnert werden, daß inter- und multidisziplinäre Studien keine Darstellung oder gar Vertiefung der am Thema beteiligten Einzeldisziplinen geben. Für diese gibt es eine Fülle von Veröffentlichungen, auf die ich verweise. Vielmehr geht es um das ergänzend Zusammenführende und vor allem um einen Versuch, das Lebenswirkliche historischer Notfall-Psychotherapie abzubilden. Sinnvoll verbindende Funktionen stehen im Mittelpunkt.

Obwohl die Gehirnforschung noch am Anfang steht, haben sich eben für diese natürlichen Verbindungen doch beträchtliche Erkenntnisse – gerade auch für die Geheimnisse magischer Texte – ergeben. Einige Schwerpunkte seien gerafft herausgegriffen:

Seit den ältesten Schriften der Menschheit sind Verfolgungsvorstellungen vor bösen Augen und bösem Blick nachweisbar. Sie wurden immer wieder bedichtet, mit Gegenzaubern beschworen und gaben Anlaß, Angst, Furcht, Panik und vermeintliche Nachstellungen auf fremde Gestalten, Nachbarn und Dämonen zu projizieren. Mit dem Nachweis weitgehend spezifischer Veränderungen von Hirnfunktionen als Ursache für derartige wahnähnliche, aber kulturell fest verankerte und weithin ausgestaltete Denk- und Gefühlsstörungen, die auf erhöhte Sensitivität des visuellen Systems zurückgehen, ist ein wichtiger Schritt zum Verständnis dieser menschlichen Eigenart gewonnen. Es geht um das Verstehen einer der Ursachen von Vorurteil und Feindschaft.

Mit der Unterscheidung von introversiver und extroversiver Katharsis kann die Aktivierung verschiedener Hirnsysteme postuliert werden. Es gibt zwei Seiten der per Zauberspruch aktivierten neuronalen Systeme. Einerseits wird das mitfühlende, nach- und mitleidende, selbstkritisch reflektierende, aber auch dabei Schmerz empfindende Mittelliniensystem des Gehirns mit seiner sehr menschlichen, sensiblen und vielfältig verflochtenen Insula-Amygdala-Interaktion ‚angesprochen'. Andererseits wird mit der expulsiven extroversiven Dämonenevokation der Gegenangriff vorgenommen, der eher in lateralen und frontalen Hirngebieten ‚Anklang' findet. Damit kann die philologische Sortierung in Erzählformen und Befehlsformen der Texte aufgegriffen, aber im Hinblick auf Effizienz wesentlich ergänzt werden: Die meisten Texte variieren in ihren Strategien und mögen damit auf eine Art Wechselbad abzielen. Sie triggern abtastend Anregungen in das sympathisch/ parasympathische System des vegetativen Stoffwechsels, um zur Regulierung des homöostatischen Gleichgewichts beizutragen.

Die in verschiedenen Kulturen zu findende regressionsfördernde, den Weg seit der Empfängnis symbolisch nachvollziehende Erzähltechnik zur Geburtserleichterung kann in ihrer neurophysiologischen Dynamik nachgezeichnet werden. Ebenso ergeben sich Hinweise für eine Relevanz der Regression in die legendären Verhältnisse des Mutterschoßes bei den Sieben Schläfern von Ephesus.

Eine durch die philologische Forschung weit vorangetriebene Analyse der Sprechakte ist im Hinblick auf die Verwendung von Sprachfiguren als Eingriff in lexikalische Bestände und sensomotorische Simulationen wirkungspotente Maßnahme beschreibbar. Insbesondere lassen sich erste Erklärungsmodelle für die Verarbeitung emotionalisierender Sprachbilder in frontalen Gebieten nachweisen, das Etikettieren des Unheimlichen, das Benennen des Horrors. Dafür wie auch für die eben genannte Balancierung introversiver und extroversiver Stimuli fungiert ein in der neurobiologischen Forschung immer wieder bestätigtes ausgleichendes Zusammenspiel mit den Schwerpunkten Amygdala und Präfrontalhirn.

Besonders zwingend als Befundbestätigung für die Perzeption von Sprachfiguren hat sich auch das mittels modernen EEG-Ableitungen gefundene N400-Potential erwiesen, das die Herausforderungen dokumentiert, die dem Denken durch Inkongruenzen und Widersprüche, den unmöglichen Aufgaben in vielen Zaubertexten, den Adynata, ähnlich wie in Rätseln geboten wird.

Als elementare Voraussetzung für die Möglichkeit einer therapeutischen Einflußnahme durch Verbaltherapie konnten verschiedene Ergebnisse der bildgebenden Verfahren am Gehirn skizziert werden. Sie wollen der hinsichtlich einer Wirkungsfrage an ihrem Ende angelangten konventionellen Sprechakttheorie eine fast erlösende Botschaft geben. Die Entdeckung der Spiegelneurone, deren Bearbeiter nach Jahren der Forschung auch deren Grenzen erkannt haben, bot einen wichtigen Nachweis für die Natur menschlichen Mitfühlens und für die Möglichkeit des *gelingenden* Wortes. Mehr noch ist von den seit wenigen Jahren erfolgreichen Simultanableitungen an synchron korrespondierenden Gehirnen, der sog. Hyperscanning-Forschung zu erwarten, die uns lehrt, daß das sprichwörtliche ‚auf einer Wellenlänge mit jemandem sein' nicht Poesie und Trope ist, sondern flüchtige materielle Wirklichkeit eines natürlichen Gemeinsinns.

4. Die hier nachgewiesenen Wirkungsmodelle geben Möglichkeiten der Bereitstellung primärer Verbalmaßnahmen an, sie beschränken sich auf eine akute Erstversorgung im Notfall. Obwohl wir eine relevante ‚Ansprache' des Gehirns durch diese Texte weitgehend sichern können, wäre die Annahme einer Nachhaltigkeit und Sicherheit der Spruchtherapie als Bewirkung struktureller Veränderungen spekulativ: Wir können *bleibende* Verbesserungen der limbisch-präfrontalen Emotionskontrolle oder die Lösung krankhafter neuronaler Verknotungen oder bleibende Verbesserungen des homöostatischen Balance-Vermögens im historischen Rückblick weder beweisen noch ausschließen.

5. Vernachlässigt sind hier Themen, die man in Zukunft wohl ebenfalls von zentraler Bedeutung für die Wirkung von Zauberspruchpartikeln erkennen wird, die aber noch nicht ohne allzu hohe spekulative Risiken aufgegriffen werden können. Ich meine zum Beispiel die Koaktivierung von kinästhetischen Systemen des Gehirns mit den 3D-Kompass-Formeln in manchen Zaubertexten, schon in den altindischen Veden wie in **Beispieltext** ➔**56** so in christlichen Formeln wie **Beispieltext** ➔**34** angelegt. Sie weisen auf Himmelsrichtungen oder auf individuelle Körperumgebung. Diese Spruchanteile bieten ein oft ausgeprägtes Koordinatensystem und erinnern an primäre Aufmerksamkeits- und Orientierungsprobleme, an Behandlung von Raumangst und auch von geistig-seelischer Desorientiertheit. Sie mögen, wenn sie jemals auch rituell und anhaltend gewichtet wurden, zur Wiedergewinnung von Umgebungs- und Weltbewußtsein beigetragen haben.

E Die Beispieltexte

E1 Europäisches Mittelalter bis Neuzeit; und Marcellus (4./5. Jh.)

Beispieltext 1
Der Münchner Nachtsegen (BSB München, Clm 615, 127r, 14.Jh.)
Erwähnung/Erläuterung siehe B4.1.2; C1.3; C1.3.1; C1.3.4; C4.2.2; C5.3

Der Segenstext in historisch-kritischer Edition nach T. von Grienberger	Erläuterungen und Anmerkungen in Stichworten
Daz saltir ‚deus virtutum' daz hohiste ‚numen divinum' daz heilige ‚sancte spiritus' daz saltir ‚sanctus dominus' 5 daz muze mich noch hint bewarn vor den bosen nahtvarn und muze mich bikrizen vor den swarzen unde wizen di di guten sint genant 10 unde zu dem Brockelsberge sint vor den bilewizen, [gerant, vor den manezzen, vor den wegeschriten vor den zunriten, 15 vor den klingenden golden, vor den unholden Glozan unde Lodevan, Trutan unde Wutan, Wutanes her und alle sine man, 20 di di reder und die wit tragen geradebreht und irhangin, ir sult von hinnen gangin! Alb unde elbelin ir sult nicht lenger bliben hinn, 25 albes swestir unde vatir, ir sult zu varen obir den gatir;	1–4 Als Schutzschild die Kennungen dreier Psalmen (saltir) und der Pfingstsequenz ‚Veni creator spiritus' [?] 5–7 mögen in kommender Nacht (hint= hi-naht) bewahren und umringen (bikrizen) vor dem wilden Heer streunender, plündernder Nachtfahrer (ein Sammelbegriff) ab 8 Aufzählung der einzelnen Geister. Auch die schwarz-weißen, d. h. hier Teils-teils-Hausgeister (Penaten?) sind Blocksberg-entrückt. 11 = Bilwiz, Pilmaz, Pulwechs u. a., Krankheit anschießend, Kornfeld schneidend, 12 Tote Seelen, röm. ‚Manes' 14 zaunreitende Hexen 15 Unklare Klanggeister (Sirenen?) 16 nachtfahrende ursprüngl. Hexengestalten 17 meist als bedrohliche Slaven gedeutet 19 Wotans Heer, Totengott 20/21 Hingerichtete (durch Räder und Strang =Wied) als Mitläufer im wilden Heer 22 erste Forderung an genannte Gestalten 23–38 Familie der Alben mit Schwester, Vater, Mutter, Trud und Mar, Kindern (Wichtel) ihre Methoden: Drücken, Kneifen, Aufreiten und Beschreiten, Anblasen, Räuchern (ruchen)

albes mutir, trute unde marn,
ir sult uz zu dem virste varn !
Noch mich di mare drucke,
30 noch mich di trute zucke,
noch mich di mare rite,
noch mich di mare beschrite !
Alb mit diner krummen nasen,
ich vorbite dir aneblasen;

35 ich vorbite dir, alb ruchen,
kruchen unde anehuchen.
albes kinder, ir wihtelin,
lazet uwer tastin nach mir sin !
Und du klagemutir
40 gedenke min zu gute !
Herbrot unde herebrant
vart zu in ein andir ant !
du ungetruwe molkenstelen
du salt minir tur vorvelen;
45 daz biver unde daz vuzspor,
daz blibe mit dir da vor !
Du salt mich niht beruren,
du salt mich niht zuvuren,
du salt mich niht enschechen,
50 den lebenden vuz abemehen,
daz herze niht uz sugen,
einen strowisch darin schuben !
Ich vorspige dich hute und alle tage,
Ich trete dich baz, wan ich dich trage;
55 nun hin balde, du unreiniz getwas,
wan du wesens hi nicht has !
Ich beswere dich ungehure
bi dem wazzer und bi dem vure,
und alle dine genozen
60 bi dem namen grozen
des visches, der da zelebrant
in der messe wirt genant.
Ich beswere dich vil sere
bi dem miserere,
65 bi dem laudem deo,
bi dem voce mea
bi dem de profundis
bi dem salm coheuntes,

kriechendes Anschleichen (kruchen), vergiften-
des Anhauchen (anehuchen; vgl. Pesthauch),
Betasten
24 sollen nicht länger bleiben
26 sollen ober den Zaun (gatir, Gatter) und
28 zum Dachfirst ausfahren
34,35,38 wiederholte nun ausdrückliche,
verschärft formulierte Verbote

39–40 Klageweib, Winselmutter, Holzweib,
(vgl. Nachtkauz als Vorboten und Verkünder)
41 Personifikation des abgegifteten Augenübels
(~Hordeolum, Gerstenkorn, ~Wurm?)
43 treulose Milchhexe, die den Nutzen stiehlt,
44 sollst meine Tür verfehlen (ver-vaelen)
45 Fieber und schmerzh. Fußkrampf, Fußsparr
mögen fern bleiben.
47–52 Spezifizierung der Verbote: das Tasten,
48 das Verwirren (zevüeren, mnd. tovoren)
49 das Verführen (zur Brockenausfahrt?)
50 das lebendig den Fuß Rauben [?]
51/52 das Herz-Aussaugen und mit Strohbü-
schel (Herrschaftssymbol) verstopfen [?]
53–56 Heute und immer will der Beschwörende
das unreine Gespenst (getwas) mit Ausspeien
(verspigen) verachtend abwenden und es treten,
mehr als er selbst, ihn tragend, getreten wird.
57–75 Direkte autoritäre Beschwörung des
einzig
als Ungeheuer benannten Geistes, wahrschein.
wieder des Albs mit seinen Genossen, als ob er
anwesend ist. Zunächst bei den Elementen Feuer
und Wasser (sich aufhebende Gegensätze), dann
mit christlichen Motiven.
60–62 Zelebrant ist der Priester, der als Christi
Stellvertreter (Fischsymbol) die Messe liest.
64–73 Aufreihung von Psalmeninitien,
Evangelien (oder z. T. Messformeln)
(nach v. Grienberger:) 64 Psalm 4,2;

bi dem dimittis, 65 Lucas 18,43; 66 Psalm 3,5
70 bi dem benedictus 67 Psalm 129,1 (130,1) „De profundis"
bi dem magnificat 68 AT Makkabäer 6,11; 69 Lucas 2,29;
bi der alten trinitat 70 Lucas 1,68; 71 Lucas 1,46
bi den salmen also her, 72 „alt" = ewig
daz du varez obir mer
75 und mich gerures nummermer! Amen

Beispieltext 2
Elsässisches Nachtgebet (Universitätsbibliothek Straßburg L. germ 645, S. 15f, 15.Jh.)
siehe B4.1.2; C13; C4.3.1.2; C5.1

Eyn andre schöne nacht gebett
O here jhesu christe jch befille mich diße nacht jnn dine heilgen
    funf wunden auch jnn die beschirmung
O selige jungfraw maria vnd jnn die beschirmung aller Engel vnd aller heilgen
    vnd hiet nacht behuete mich vor ghachen (jähen) vnd vnversehenen dott,
    vor beßen dreumen (Träumen) vnd jnbiltungen (Einbildungen) vor
    beßer anfechtung mynes fleisches vnd des beßen geistes, vor schrecken
    vnd vor allem schaden des liabs vnd der selen, vmb der angst willen
    die du guetiger herr an dem olberg hast gelieten Amen

Beispieltext 3
Wider die Elben (BSB München Clm 849, 131v, 15. Jh.), ed. Schönbach, Anal. graec. S. 43
siehe B4.1.2; C1.3; C3.1.2; C4.3.1.2

Wider Elbe
Item eyn mentsche daz besessen ist mit den elben, den sal man also peboren: hait daz mentsche eyne müter addir vater (eyn elich man mag es syme wibe thun, ein wib siner husere), daz mentsche daz mit der krangheide befallen ist, daz sall dem jhenen mit syme nagkenden libe uff syme nagketen beyne siczin eyne gude wile; wan dan daz geschehin ist, so sall der gsonde mentsche dem krangkin mentschin mit synere czongen fharen ubir sin naßen: smagkit dan dy naße gesalczen, so sint es dy elbe. so ist auch eyn ander czeichen: dem krangken mentschen zwiddern sine ougen unde syne adern dorch sim lib. –
  Item man saill es also besweren: In dem namen des Vaters, des sones unde des heilgen geistes. Amen.
    By dem heilgen Pater Noster, by dem heilgen Ave Maria unde by dem heilgen glouben, den dy heilgen XII appostln machten, da midde unde da by **beswere ich uch, alp unde elbynnen, unde mit allen uwern**

nachkomelingen, ir syhit wiß addir roit, brün, swarcz, gell, addir yn wilcherley wiß ir syhit, daz ir alle mußit sin tot an dem dritten tage. daz gebudit uch got unde der liebe herre sente Job. fort mehe so wiel ich uch gebieden, daz ir sollit kommen yn eyne widen, dy sollit ir schudden und ryden also lange, daz ditte mentsche nach uch begynnet czu vorlangen. **aber woldit y widder kommen, ir brengit daz heilge frone crucze yn uwerin henden.** daz gebudit uch der man, der sin ende an dem heilgen fronen crucze nam, unser lieben frauwen kuscheit, allir propheten wißheit, alle heilgen mertelere unde alle glaubigen sele, unde III meße, dy dissem mentschen sollen gehalten werden czü helffe unde czü troste; unde dy III naill, dy got dem herren dorch hende unde fuße worden geslain, unde dy crone, dy got dem almechtigen uff sin gebenediet houbit wart gedrocht. dy minsten wonden, dy got der herre enphangen hat, dy mupin myr helffen an disem mentschen, daz em dissir suchte werde buß und baß; daz helffe myr got unde daz heilige grab! Alßo vor war als daz ist, alßo rümet, ir elbe unde elbynnen, mit allen uwerin kommelingen! Amen.
 Disse besweronge sal man thun V morgen nach eyn .1. Pater Noster und Ave Maria.

Beispieltext 4
Aus dem 12(13)-bändigen Buch der Medizin des Pfalzgrafen Ludwig V. bei Rhein (Universitätsbibliothek Heidelberg CPG 271, S. 229, 16. Jh.)
siehe B4.1.2; C1.3; C5.1

   Ein segen fur den alp
+ Amara + Tanta + Cyri + Sicaliri
+ Adjuro vos Elphos + Et Elphos   Ich beschwöre euch Elfen
+ Et omne Jenus demoniorum   und alle Geschlechter der Dämonen
+ per patrem + filium + spiritum s.   durch Vater, Sohn und Heiligen Geist,
+ Ut per virtutem + Domini nostri   daß ihr durch die Kraft unseres Herrn
Jhesu Cristi + Et per omnes sanctos   Jesus Christus und aller Heiligen
et sanctas + Et per omnes electos dei   und aller Auserwählten Gottes
+ ut recedatis ab hoc famulo dei N.   zurückweicht vom Diener Gottes N.
+ Et amplius + eo nichil faciatis   und ihm weiter nicht schadet.
+ Jn nomine dei patris + Et filii …   Im Namen Gott Vaters und Sohnes …

Beispieltext 5
Aus den Gebeten und Segen von Muri (Codex membr. 69 des Benediktinerkollegiums Sarnen), ed. Wilhelm, F.: Denkmäler deutscher Prosa des 11. und 12. Jh., München 1960
siehe C2.1

Oratio bona ad deum	Kräftiges Gebet zu Gott.
Herre almehtige got	Herr Allmächtiger Gott
ich bite dich dur din heiligis hŏbit	ich bitte dich durch dein heiliges Haupt
unde dur allv dinu heiligin werch	und durch alle deine heiligen Werke
uñ dur allv div heiligin wort […]	und durch all deine heiligen Worte […]
bitwinc hute an disime tage	bezwinge heute an diesem Tage
alle die zungin. di minin scadin	all die Zungen, die meinen Schaden
sprechin wellen.	sprechen wollen.
alde die mich hute ansehn suln	**all die, die mich heute ansehen**
odir diheinen giwalt ubir mich	oder die keine Gewalt über mich
habin suln […]	haben sollen […]

Beispieltext 6
Grundtyp aus mehreren Sprüchen des Urteilsbuches des Rostocker Niedergerichts von 1539–1586 (modifiziert nach Spamer/Nickel: Romanusbüchlein, S. 112)
siehe C2.1

Zwei böse Augen haben dich angesehen.
Drei gute sehen dich wieder an,
der eine ist der Vater, der andere ist der Sohn,
der dritte ist der heilige Geist.
Christus Jhesus helfe ihm allermeist. Amen

Beispieltext 7
(München BSB Cgm 54, fol. 106v, 14.Jh.)
siehe C2.1; C5.1

O jesu christ marien sun, dein marter sey heut mein frum [Nutzen].
daz all mein feind vorcheren sich.dez pit ich lieber herre dich.
daz si erstarren und erstummen an mund und an zungen,
an augen und an henden, daz sy ymmer vollenden …

Beispieltext 8
Vor das Berufen oder Verschreien (Wunsiedel, Fichtelgebirgsmuseum, Handschrift C213/2823 des Bauern Ächtner 1769, Seite 31); ed. Ernst, W.: Beschwörungen und Segen
siehe C2.1

Falsche Augen, falsche Sinne, falsches Herz und falsch Beginnen
ist gegangen über mich; Gott wird mein erbarmen sich.
Hats gethan ein Mann aus Neide, kommts ihm in die Brust und Seite,
Hats gethan eine Frau aus Hass, kommts ihr ins Haupt und Hals und Nas',
Hats gethan ein Knecht aus Eifer, kommts ihm in Mund und Zungen Geifer,
Hats gethan eine Magd aus Zorn, kommts ihr in Hand Fuß und Korn
Gott der wird es ihnen senden, alles Böse von mir wenden.

Beispieltext 9
(Marcellus Burdigalensis, De medicamentis 28,74) ed. Heim, Ricardus: Incantamenta magica, Leipzig 1892, S. 496, Nr. 107
siehe C3

Ad rosus tam hominibus quam iumentorum praecantatio sic:
pollice sinistro et duobus minimis digitis ventrem confricans dices:
(Reibe den Bauch mit dem linken Daumen und mit beiden kleinen Fingern und sprich:)

„Stabat arbor in medio mare	Es steht ein Baum in Meeresmitte
et ibi pendebat situla plena	und da hängt ein Gefäß voll
intestinorum humanorum,	menschlicher Gedärme
tres virgines circumibant,	drei Jungfrauen gehen herum,
duae alligabant, una resolvebat"	zwei binden an, eine löst auf

Beispieltext 10
(Marcellus von Bordeaux, De medicamentis 21,3) ed. Niedermann, M. und E. Liechtenhan mit deutscher Übers. J. Kollesch und D. Nickel, Berlin 1968; zit. nach Önnerfors, Alf: Antike Zaubersprüche, Stuttgart 1991
siehe C3.1.1; C4

(Ad corcum) „Corce corcedo stagne:	Gegen Kollern im Leib „Corce c. stagne:
[var.: Stolpus a caelo cecidit	Die Krankheit Stolpus fiel vom Himmel]
Pastores te invenerunt,	Hirten fanden dich,
sine manibus colligerunt,	lasen dich ohne Hände auf,
sine foco coxerunt,	kochten dich ohne Feuer,
sine dentibus comederunt	aßen dich ohne Zähne.

Tres virgines in medio mare mensam marmoream positam habebant; duae torquebant et una retorquebat. Quomoda hoc numquam factum est, sic numquam sciat illa [Hs.: gaioseia] Gaia Seia corci dolorem!"

    Mitten im Meer hatten drei Jungfrauen einen Marmortisch aufgestellt: zwei drehten ihn, eine drehte ihn zurück. Wie dies niemals geschehen ist, so soll diese N.N. niemals Schmerz vom Kollern im Leib spüren."

Beispieltext 11
Der Vogel federlos (BLB Karlsruhe, Cod. Aug. CCV), ed. Eis, Gerhard: Altdeutsche Zaubersprüche, Berlin 1964
siehe C3.1.1

Volauit uolucer sine plumis.	Es flog ein Vogel ohne Federn
sedit in arbore sine foliis.	Er saß auf einem Baume ohne Blätter
venit homo absque manibus.	Es kam ein Mensch ohne Hände
conscendit illum sine pedibus.	Er bestieg ihn ohne Füße
assauit eum sine igne.	Er briet ihn ohne Feuer
comedit eum sine ore	Er verzehrte ihn ohne Mund.
nxtz a titane – :Rätsellösung:	(nvtz statt ‚Nix': Schnee in der Sonne)

Beispieltext 12
Neuzeitlicher Fieberspruch (aus: Bartsch, Karl: Sagen, Märchen und Gebräuche aus Mecklenburg, gesammelt und herausgegeben Wien 1879 Band 2, S. 396, 19. Jh.)
siehe C3.1.1

Für das Fieber
Ein Vogel ohne Lung, ein Storch ohne Zung,
eine Taube ohne Gall, so vertreibe ich die Fieber all. Im Namen Gottes ...

Beispieltext 13
Hebräische Bärmutterbeschwörung (Sefer Hasofoth (hebräischer Codex), Bl. 88v, 14. Jh.)
ed. Holzmann, Verena: „Ich beswer dich, wurm vnd wyrmin ...", Bern u. a. 2001
siehe C3.1.1; C4.3.1.2; C5.1

Bärmutter leg dich, bist so alt als ich.
Bringst du mich zu der erden, du musst mit mir begraben werden.
Ein buch heisst die bibel, bärmutter leg dich nieder.
Du sollst dich legen nieder an deine rechte stätte,
das gebeut dir die heiligen-gotts-kraft:
wermut und hegemut und lieggemut und das biege.
Und filia terrae! das (leiden) fahre unter meine sohle.

Darunter kann ichs wol erdulden.
Darunter fliesst ein bodenloser see,
da geht drinnen ein gratloser fisch,
den sollst essen und sollst menschliches gar vergessen. [...]

Beispieltext 14
Gegen die Ruhr (aus: Das Pfuhler Hausbuch, um 1800) ed. Kropp, A., in: Ulmer Kulturanthropol. Schriften, Band 10, 1998, S. 134
siehe C3.1.2

Vor das roth
stehe unter den freyen Him[m]ele u: heb die
3 Eides finger in die höhe und SPrich 3.M:[al]
(„)N.N. roth roth roth ich stelle dir die ruhr
u:daß roth ich stelle dir die ruhr u: daß
roth **wie gott der herr die Him[m]els Tühr**
Ver schlossen hatt Vor einen Man im
Mehr X X X („) warte 2:biß 3.stund brauchs wieder
in 3.stund wieder so ist geholfen

Beispieltext 15
Blutungsbeschwörung (aus: Kunstbuch des Johannes Zahn von Dürnberg (vor 1691), S. 103), ed. Ernst, Wolfgang, in: Arch. f. Geschichte von Oberfranken 76 (1996), S. 187
siehe C3.1.2

Ein anderes [für Wunden]
Steh bludt wie deß mill [Müllers] seel duth
die vohr die Himmels thur kompt
wan ße den Metzen mit Vn Recht nimptt.
Jn Namen + vatters des Sohnes + des
heiligen geistes + Amen
benenne des Menschen Nahmen dar bey

Beispieltext 16
Gegen Unholde (Württembergische Landesbibliothek Stuttgart Cod. med. et phys. 4°, Nr. 29, Bl. 111a, 15. Jh.), ed. Pfeiffer, Anz. f. d. Kunde d. deutschen Vorzeit NF 1 (1853/54), S. 36
siehe C3.1.2

Von unholden
Item wan dir die milch von unholden wirt genumen dinen Kieen, sprich drimal also, wan du wilt melcken, iij pater noster und iij ave Maria;

sprich, wil man uch uber (1.uwer) milch nemen, sprich: nematz,
ich verputt uch unholden mein milch bi der hailigen gottes krafft und
wil ich si uch nitt laussen, **ir bringt mir dan des vass, da gott
selber in lag, die windlen und die war, da gott selber in gewunden
und gewicklet wart.** Des helf mir got der vatter und der sun und Gott
der hailig gaist. in gottes namen. amen.

Beispieltext 17
Gegen Milzbrand u. a. (aus Ammann, J. J., Volkssegen aus dem Böhmerwald I,
19. Jh., in: Zs. des Vereins für Volkskunde 1(1891), S. 208)
siehe C3.1.2; C4; C5.1

Gegen den kalten Brand
Christus der Herr, ging über Land.Es begegnete ihm ein kaltes
Gesicht. Christus der Herr sprach: wo willst du hin, kaltes Gesicht?
Das kalte Gesicht sprach: ich will in den Menschen fahren. –
Christus der Herr sprach: was willst du in dem Menschen tun?
– sein Bein fressen, sein Fleisch essen, sein Blut trinken.
Christus der Herr sprach: **kaltes Gesicht, das sollst du nicht tun!
Erbsen mußt du brechen, Kieselsteine mußt du essen,
aus einem Brunnen mußt du trinken, darin mußt du versinken!**

Beispieltext 18
Beschwörung der Mora (aus: Krauß, Friedrich: Slavische Volksforschungen.
Abhandl. über Glauben, Gewohnheiten, Sitten, Bräuche und die Guslarenlieder
der Südslaven, Leipzig 1908, S. 150, 19. Jahrhundert, aus Grbalj in Dalmatien)
siehe C3.1.2; C5.1

Wer von einer Mora geplagt wird, pflegt vor dem Schlafengehen
folgendes Gebet aufzusagen:
Mora (bora) überschreit' nicht dieses weissen Hofes Schwelle; denn an
ihm sind feste Schlüssel von unserem Siodorus, Siodorus, Theodorus
und Maria und Matthias und der Schwester Levantija, der allda kein
Eintritt zusteht über dieses Hofes Schwelle; keiner Steinhex, sie
versteiner', keiner Windhex, sie verwehe, keiner Plaghex, sie geplagt
sie, keiner Witwe, zweimal Witwe, keiner magischen Magierin, **eh sie
nicht zu End gezählt hätt: am Himmel alle Sterne, im Hochgebirg die
Blätter, die Sandkörner am Meer, auf der Hindin das Haar, auf der
Ziege die Zoten, auf dem Schaf die Wolle, auf der Wolle die Haare.**
Und sollt' sie dies zu Ende zählen, gürt' sie sich mit einem Webbaum,
Webstuhlnagel sei ihr Stecken, sie fahre in eine Eierschale hinein,
sie soll in der Meerflut ersaufen; ihr Eingeweide dem Bandwurm, ihr

Kopf falle dem Teufel zu, der Teufel' hol ihr alle Ziegen; ihre Milch soll sich nicht verkäsen; sie soll vielmehr schreien: o weh! o weh! bescher' ihr Lena plena und Maria Magdalena! Amen!

Beispieltext 19
Gegen die Würmer (Landesbibliothek Luxemburg, Cod. 109, f.ult. (gem. Katalog 239), aus Echternach) ed. Jacoby, Adolf: Segenstexte aus Luxemburger Handschriften (ca. 1923), S. 17
siehe C3.1.2

Martha super pontem maris stabat,	Marta stand auf der Meeresbrücke,
harenam maris numerare	**den Meeressand zu zählen,**
tantum quantum possit.	so viel sie könne.
Vermes in isto caballo nec vivere	Die Würmer in diesem Pferd mögen
nec crescere nec multiplicare	nicht leben, noch wachsen, noch
nec nocere nec malefacere. Pater n.,	sich vermehren oder schaden ...

Beispieltext 20
Gegen die Wirwelein (aus Kunstbuch des Johannes Zahn, S. 92/93), ed. Ernst, W. S. 180
siehe C3.1.2; C4.3.1.1c

gott vatter + gott Sohn + gott heiliger Geist +
Jhr wirwelein und ihr Medlein
laß(t) euch **das Fleisch zu wider sein**
den(n) ich verbiete eüchs bey gottes Krafft
weder zu nagen noch zu hauen oder zu stechen,
den(n) Christus der herr spricht,
der auff seinen gerechten gericht Stul sietz,
der **kein unrecht urdel (Urteil)** spricht
das sey dir zu Buß gezehlt. ...

Beispieltext 21
Bamberger Blutungssegen (Staatsbibliothek Bamberg Msc. Med. 6 fol. 139rb, 12./13. Jh.)
siehe B4.1.4; C4; C4.3.1.1c

Crist unte iudas spîliten mit spîeza.	Christ und Judas spielten mit Spießen.
do wart der heiligo crist wund in	Da ward der heilige Christ wund an
sine sîton. do nâm er den dvmen.	seiner Seite. Da nahm er den Daumen
unte uor duhta se uorna. So uerstant	und presste ihn vorne darauf. So stehe
du bluod. so se iordanis aha uerstunt.	du Blut wie das Jordanwasser stand,

do der heiligo iohanes den heilanden crist in iro toufta. daz dir zo bvza.	als der hl. Johannes den Heiland Christ in ihm taufte. Dies dir zur Heilung.
Crist wart hi erden wunt. daz wart da ze himele chunt. iz ne blŏtete. noch ne svar. noch nechein eiter ne bar. taz was ein file gŏte stunte heil sis tu wŭnte. In nomine ihesu christi. daz dir ze bvza […]	Christ ward auf Erden wund. Das wurde im Himmel kund. Es blutete nicht, es schwärte nicht. Nicht einmal Eiter bildete sich. Das war eine sehr gute Stunde. Heil sei dir Wunde. Im Namen Jesu Christi. Dies dir zur Heilung.

Beispieltext 22
Fiebersegen mit den Siebenschläfern (BSB München, Clm 14179, f.1', 9. Jh., aus St. Emmeram Regensburg) ed. Franz, Ad.: Die kirchl. Benediktionen II, 480
siehe C4.1.1; C5.1

In nomine domini nostri Iesu Christi. In ephasa ciuitate in monte Celio ibi requiescunt VII dormientes:	Im Namen unseres Herrn Jesus Christus. in der Stadt Ephesus im Celioberg dort ruhten die sieben Schläfer:
Maximinianus, Malchus, Martinianus, Constantinus, Dionisius, Iohannes, Serapion.	
In nomine patris et filii et spiritus sancti. Per merita et intercessionem eorum omnipotens deus dignetur saluare istum famulum illum ab omni infirmitate febris uel frigoris. **Et sicut illi more infantium in utero quiescentium non sentientes laborem neque dolorem neque mortem,** in isto famulo dei illo [fac, ut] neque dormiendo neque uigilando sentiat infirmitatem febris uel frigoris. In nomine patris et filii et spiritus sancti.	Im N. d. Vaters, des Sohnes u. d. hl. Geistes. Durch ihre Verdienste und Fürsprache möge der allmächtige Gott seinen Diener vor aller Krankheit von Fieber und Kälte bewahren. **Und wie jene nach der Kinder Art im Mutterschoß ruhend weder Mühe noch Schmerz oder Tod spürten,** lass deinen Diener weder im Schlaf noch im Wachen an Fieber oder Kälte leiden. Im Namen des Vaters und des Sohnes und des heiligen Geistes.
Eugenius, Stephanus, Protasius, Sambatius, Dionisius, Chelisius, Quirianus	
adiuuando famulo dei ill[um] liberate a febris uel frigore. Adiuro uos [febres], per patrem et filium et spiritum sanctum, de **quacunque natione estis**, atque exorzizo uos per sanctam Mariam, matrem domini nostri Iesu Christi, […] ut non habeatis locum neque potestatem in isto famulo dei illo, sed redeatis, unde uenistis. […]	Befreit den bedürftigen Diener Gottes von Fieber oder Kälte [Schüttelfrost] Ich beschwöre euch Fieber, bei Vater und Sohn und heiligem Geist, von **welcher Herkunft/ Gattung ihr auch seid**, und exorziere euch durch Sancta Maria, die Mutter unseres Herr Jesus Christus, […] daß ihr keinen Ort an und keine Gewalt über diesem(n) Gottesdiener habt, sondern weichet, woher ihr gekommen seid. […]

Beispieltext 23
Der erste Merseburger Zauberspruch (Handschrift: z. Z. Domstiftsbibliothek Naumburg, ehemals Merseburg Hs 136, fol. 85R, 9./10. Jh.)
a) Übertragung Beck, Wolfgang: Die Merseburger Zaubersprüche, Wiesbaden 2003
b) nach Riesel, Elise, Der erste Merseburger Zauberspruch, in: Deutsches Jahrbuch für Volkskunde 4 (1958), S. 53–58, sinngem. Übertragung nach Konzeption Riesel durch Verf.
siehe B2; B4.1.2; C4.1.2; C4.1.2c

Eiris sazun idisi, sazun hera duoder,
Suma hapt heptidun, suma heri lezidun,
suma clubodun umbi cuonio uuidi.
Insprinc haptbandun, inuar uigandun!

a) Einst saßen Idisi (Schlachtjungfrauen), saßen auf den Kriegerscharen.
   Einige fesselten einen Gefangenen, einige hemmten die Heere.
   Einige zerstrennten ringsherum die scharfen Fesseln.
   ‚Entspringe den Fesseln, entfahre den Feinden!'

b) Einst saßen Idisi (weise Helferinnen), da über die Erde dahin.
   Einige knüpften Knoten, einige hielten die verknoteten Fäden zurück.
   Einige entwirrten die verknüpften Fäden.
   ‚Entspringe den Haftbanden ! Entfahre den feindlichen (Krankheits-) Dämonen!'

Beispieltext 24
Drei-Engel-Segen (Zentralbibliothek Zürich, Ms Rheinauer Codex 67, fol. 47, 12. Jh.), ed. Steinmeyer, Elias v., Zeitschr. für deutsches Altertum 21 (1877), S. 209
siehe C4; C4.2.2

In nomine domini tres angeli ambulauerunt super montem synay et obuiauerunt illis. Nessia. nagido. crampho. tropho. Stechido paralisis. Gegihte. Quibus angeli dixerunt ‚vnde venitis. aut quo pergitis?' – [...] ‚imus ad famulam dei N. Caput eius conterere. collum. humeros. brachia. scapulas, dorsum. latera. ventrem. vmbilicum [...] calces. plantas. pedes. debilitare. et medullas omnium memrorum suorum euacuare'

In Gottes Namen wanderten drei Engel über den Berg Sinai und begegneten den Nage- Krampf- Tropf- Stech- Lähm- und Gichtwürmern.* Sprachen die Engel: ‚Woher kommt ihr ? Was habt ihr vor?' ‚Wir gehen zum Gottesdiener N., seinen Kopf verstören, den Hals, die Arme, die Schultern, Seiten, Bauch, Nabel [...] Fersen, Sohlen, Füße lähmen und das Mark aus allen seinen Gliedern saugen.'

* Es handelt sich um verschiedene Krankheiten und Symptome: Atrophien, Schmerzen, Schlaganfall, Gicht, Lähmungen usw.

Beispieltext 25
Bärmutterbeschwörung (Stiftsbibl. St. Gallen, Codex 752, fol. 159f, 10. Jh.), ed. Bernfeld, Werner, in: Kyklos. Jb. d. Instituts für Gesch. der Medizin an der Univ. Leipzig 2 (1929)
siehe C4.2.2; C4.3.1.1c; C4.3.1.2; C5.1

In nomine dei patris ... [...]
Coniuro te, matrix
– per sanctam trinitatem, ut sine
aliqua molestia redeas ad locum tuum [...]
– per novem ordines angelorum ...
ut cum omni mansuetudine ... revertaris [..]
– per patriarchas et prophetas ... , ut
non noceas huic famulae dei N.

Coniuro te, matrix, per dominum nostrum
Jesum Christum, qui siccis pedibus mare
ambulavit, infirmos curavit, demones
effugavit, mortuos suscitavit [...], ut non
noceas huic famulam dei N.

Ut non caput eius teneas, non collum,
non guttur, non pectus, non aures, non
dentes, non oculos, non nares, non scapulas, non brachia, non cor, non stoma-
chum, non epar, non splen, non renes,
non dorsum, non latus, non artus, non
umbilicum, non viscera, non vessicam,
non femora, non tibias, non talones, non
pedes, non ungues teneas, sed quieta
pausas in loco, quem tibi deus delegit,
ut sana sit haec ancilla dei N. [...]

Ich beschwöre dich Bärmutter
– durch die heilige Dreifaltigkeit, daß du
ohne Verdruß an deinen Ort zurückkehrst
– durch die neun Ordnungen der Engel ...
daß du in aller Sanftmut ... umkehrst ...
– durch Patriarchen und Propheten ...
daß du diesem Gottesdiener nicht schadest

Ich beschwöre dich, Bärmutter, durch
unseren Herrn Jesus Christus, der
trockenen Fußes übers Meer ging, der
Kranke heilte, Dämonen vertrieb und Tote
erweckte ..., daß du diesem Gottesdiener
nicht schadest.

Daß du seinen Kopf nicht besetzt, nicht
Hals, nicht Gurgel, nicht Brust, nicht
Ohren, nicht Zähne, Augen, Nase und
Schulterblätter, nicht Arme, nicht Herz,
nicht Magen, nicht Leber, nicht Milz, nicht
Nieren, nicht Rücken nicht Lunge, nicht
die Gliedmaßen, nicht Nabel, nicht Eingeweide, nicht die Harnblase, nicht Ober-
schenkel, nicht Unterschenkel, nicht die
Fersen und nicht die Füße und die Finger-
spitzen, sondern daß du ruhig schläfst an
dem Ort, an den dich Gott gesetzt hat,
damit diese Dienerin Gottes gesundet.

Beispieltext 26
Der zweite Merseburger Zauberspruch (Naumburg, Domstiftsbibliothek, ehemals Merseburg, Domstiftsbibliothek, Handschrift 136, fol.85R), ed. Beck,
Wolfgang, Wiesbaden 2003
siehe B4.1.2; C4.2.2

Phol ende uuodan uuorun zi holza;
Du uuart demo balderes uolon sin uuoz
birenkict.

Phol und Wodan ritten in den Wald;
da ward Balders' Ross der Fuß verrenkt.

Thu biguol'en sinthgunt, sunna, era suister; thu biguol'en friia, uolla, era suister; thu biguol'en uuodan, so he uuola conda: „sose benrenki, sose bluotrenki, sose lidirenki: ben zi bena! bluot zi bluoda! lid zi geliden, sose gelimida sin!"	Da besprach ihn Sinthgunt u. Sunna, ihre Schwester; da besprach ihn Freja und Volla, ihre Schwester; da besprach ihn Wodan, so gut (nur) er es vermochte: „Sei es Knochenverrenkung, sei es Blutverrenkung, sei es Gliedverrenkung: Knochen zu Knochen! Blut zu Blut! Glied zu Gliedern, daß sie gelenkig sind!"

Beispieltext 27
Tegernseer Wurmbeschwörung (München BSB Clm 18524b, 203v, Digitalisat 417, 10.Jh.)
siehe C4.2.2; C5.1

Pro nessia: Gang uz Nesso, mit niun nessinchilinon, uz fonna marge In deo adra, vonna den adrun In daz fleisk, fonna demu fleiske In daz fel, vonn demu velle In diz tulli. Ter Pater noster.	Für Würmer Geh heraus, Nesso, mit neun Nesslein! heraus aus dem Mark in die Ader, von der Ader in das Fleisch, vom Fleisch in das Fell, vom Fell in die Tülle (Hornsohle)! Drei Vater unser.

Beispieltext 28
Rosenminzen-Ansprechformel gegen Hexerei (BSB München, Clm 7021, 167d), ed. Schönbach, Anton E.: Studien zur Gesch. der altdt. Predigt, Hildesheim 1968
siehe C4.2.2

Benedictio ad ḡrosam frangendam. Primo, cum inveneris herbam, dic: ‚In nomine patris quero te, in nomine filii invenio te, et in nomine spiritus te carpo, ut sis mihi et omnibus te portantibus obstaculum contra omnia seva jacula omnium inimicorum nostrorum, incantationes repellas, incarceratos absolvas, dampnatos liberes, gratiam omnium hominum mihi conserves. In nomine patris …	Ein Segen zum Pflücken der Rosenminze. Zuerst, wenn du das Kraut gefunden, sprich: ‚Im Namen des Vaters suche ich dich, im Namen des Sohnes finde ich dich, und im Namen des heiligen Geistes pflücke ich dich, daß du mir und allen, die dich tragen, Hindernis seist gegen alle wütigen Anwürfe all unserer Feinde, daß du Zaubersprüche zurückstößt, Gefangene losmachst, Verdammte freisprichst, und gnädig mich behütest unter allen Menschen. Im Namen des Vaters …'

Anschließend ist das Evangelium ‚Im Anfang war das Wort' und sind fünf ‚Ave Maria' und fünf ‚Credo' zu sprechen; und mit der rechten Hand ist das Kraut herauszuziehen und muß an Mariae Himmelfahrt auf den Altar gelegt werden. Nach der heiligen Messe mit Weihwasser besprengt, gilt es als geweiht.

Beispieltext 29
Trierer Wurmsegen (Stadtbibliothek Trier, Hs40/1018 8°, fol. 41v–43r, 10./11. Jh.), ed. Embach, M., Trierer Zauber- und Segensspr. in: Kurtrierisches Jahrbuch 44 (2004), S. 49
siehe C4; C4.3.1.2

Ad uermem qui dicit[ur] talpam tollendam	Zur Entfernung des Wurmes, gen. Talpa.
Si quis homo et equus vel aliud pecus habet illum uermem […], accipe illum ac conuerte in orientalem plagam, in decrescente luna tuumque dextrum pedem pone super illius et dic in eius aurem subscriptam sentenciam cum dominica oratione. Et post semel dictum gira eum per dextram partem ac dic iterum sicut prius feceras. iterumque gira sicque facias tercia uice prius duabus vicibus. non dicas in dominica oratione, ‚sed libera nos a malo', nisi tercio. hec est sentencia. Piupi et uripi inopia est arapere est. Beatus Hiob tenebatur uermibus modo non habetur. Sic non habeat iste homo et equus albus aut niger. ita uelit dominus ac sancta maria ac bonus iob.	Wenn ein Mensch, ein Pferd oder ein anderes Vieh jenen Wurm in sich trägt, so nimm ihn/es und wende ihn/es n. Osten. Dann stelle bei aufg. Mond deinen rechten auf seinen re. Fuß und sprich ihm den Spruch mit einem Vaterunser ins Ohr. Und nach einmaligem Besprechen umschreite ihn von der rechten Seite her und sprich noch einmal, so wie du es zuvor getan hast. Umschreite ihn wiederum und tue dies ein drittes mal […]. doch sprich nur beim dritten Mal im Vater unser den Satz ‚Sondern befreie uns von dem Bösen'. Dies ist der Spruch: ‚Piupi und Uripi, es ist zwecklos, daß ihr euch heranschleicht. Der hl. Hiob wurde von den Würmern vorübergehend befallen. Deshalb möge auch jener Mensch und jenes Pferd, weiß oder schwarz, die Würmer nicht behalten. So will es der Herr, die heilige Maria und der gute Hiob.'

Beispieltext 30
Regensburger Hiobsegen (BSB München, Clm 536, fol. 84a, 12. Jahrh.); ed. Müllenhoff/ Scherer: Denkmäler I, S. 181
siehe A2; C4.3.1.2

Jôb lag in dem miste. / er rief ze Criste,
    er chot du gnâdige Crist, / du der in demo himile bist,
    du buoze demo mennisken des wrmis. N.
Durch die Jôbes bete / die er zuo dir tete,
doer in demo miste lag, / doer in demo miste rief
zuo demo heiligin Crist.
der wrm ist tôt, / tôt ist der wrm.

Beispieltext 31
Fiebersegen mit dem Haselzweig (Handschrift DON 792, fol. 143v–144r, 15. [?] Jh.) der Badischen Landesbibl. Karlsruhe, ed. Ernst, W.: Beschwörungen und Segen, S. 57f
siehe B2; C4

>Einen Haselzweig zu segnen. Das ist gut für den Frörer [Fieber] und wenn sich jemand etwas gebrochen hat.

So nimm einen einjährigen Haselzweig und spalte ihn mittig der Länge nach entzwei und gieb ihn zwei gegenüber stehenden Männern in die Hände, die also die ‚Schienen' mit ihren Fingern seitlich an ihren Hüften halten [...] Und wenn du die Ruten so zusammengesegnet hast, so fasse sie in der Mitte, wo sie zusammen kommen und binde sie mit einem Faden zusammen. Und dies sind die Worte:

+ Arcus superius stabat	Ein gewölbtes Schutzdach darüber
virgo maria notabat	die Jungfrau Maria nahm es wahr;
+ virga viridis lux	da glänzte die frisch-grüne Gerte
et aurora fulgebant +	im Lichte der Morgenröte.

Beispieltext 32
Passionsmeditation des Mittelalters (aus D. Pezzini, ed., Il sogno della croce e liriche del Duecento inglese sulla Passione, Parma 1992); zit. nach Dinzelbacher, Peter, in: Schreiner, Klaus (Hg.): Frömmigkeit im Mittelalter, München 2002, S. 327
siehe C4.3.1.1c; C4.3.1.2

Blicke auf das Haupt, geneigt, dich zu grüßen,
den Mund, geschlossen, dich zu küssen,
die Arme, zerspannt, dich zu umfangen,
die Seite, geöffnet, dich zu lieben,
die Füße, durchbohrt mit Nägeln, bei dir zu weilen,
den ganzen Leib, am Kreuz ausgespannt, sich dir ganz zu schenken.

Beispieltext 33
Millstätter Blutstellung (Wien, ÖNB, Codex 1705, fol.32r, 12. Jh.) ed. Müllenhoff/Scherer, Denkmäler, Berlin 1892, I, S. 180
siehe B4.1.4; C4.3.1.1c

Der hêligo Christ wart geboren ce Bethlehem,	Der hl. Christ ward in Bethlehem geboren
dannen quam er widere ce Jerusalem.	von dannen kam er wieder nach Jerusalem.
dâ ward er getoufet vone Jôhanne in demo Jordâne.	Da wurde er von Johannes getauft im Jordan.
Duo verstuont der Jordânis fluz	Da stand der Jordanfluß still

unt der sîn runst.	in seinem Fluten/ Rinnen.
Also verstant dû, bluotrinna,	Also stehe auch du still, Blutfluß,
durh des heiligen Christes minna:	durch des heiligen Christs Minne.
Du verstant an der nôte,	Du steh still aus Zwang
alsô der Jordân tâte,	wie es der Jordan getan,
duo der guote sancte Jôhannes	als der gute St. Johannes
den heiligen Christ toufta.	den heiligen Christ taufte.
verstant dû, bluotrinna,	Steh still, Blutrinnen,
durch des hêligen Cristes minna.	durch des heiligen Christs Minne.

Beispieltext 34
Tiroler Kreuzanbetung (Innsbrucker Statthalterei-Archiv A 7. 29., 1400), ed. Prem, S.M.: Tirolischer Glaube und Abergl. des 15. Jh., in: Z. für dt. Altertum 36 (1892), S. 51f
siehe C4.3.1.1c; C4.3.1.2

+ Christus chreutz daz ich zu allen tzeiten anpete
+ Christus chreutz sey mit mir
+ Christus chreucz ist ain warez hail
+ Christus chreucz vberwindet die pant dez todez
+ Christus vberwindet fewer
+ Christus chreucz ist ain schirm für allez waffen
+ Christus chreucz ist ain vngemailtez zaichen
+ Christus chreucz sey myt mir in allen meinen leben an wegen an stegen
   dew ere deu chrauft dez heiligen chreuczez sey mit mir
   mit disem chreucz vber winde ich alle schedleicheiu ding
+ Christus chreucz öffen mir allez gut
+ Christus chreucz enphür mir allez vbel
+ Christus chreucz enphür mir die weiczen dez todez
   daz götlich chreucz hail mich zu allen zeiten hinder mich für mich vnter mich
   dan der alte und laidig tiefel zu allen zeiten fleucht dich wan er waiz dich. a.m.e.n.

Beispieltext 35
Drei-Brüder-Segen (München, BSB, Clm 23374 fol. 16b, 12. Jh.), ed. Müllenhoff, K., in: Zeitschrift für dt. Altertum 15 (1872), S. 454
siehe C4; C4.3.1.1c

Drî guot pruoder giengen ainen wech: dâ bechom in unser hêrre Jhêsus Christus und sprach „wanne vart ir drî guot pruoder?" – „Hêrre wir varn zæinem perge und suochen æin chrût des gewaltes daz iz guot sî zaller slath wnden, si sî geslagen oder gestochen oder swâ von si sî." –
dô sprach unser hêrre Jhesus Christ „chomet zuo mir, ir drî guot pruoder, […] und vart hinz zuo dem mont Olivêt [Ölberg] und nemt ole des olepoumes und scâphwolle, und

leget die uber die wndin und sprechet [:] ‚also de Jud Longinus der unsern hêrren Jhesus Christum stæch in di sîten mit dem sper, – daz eneitert nith, noch gewan hitze, noch enswar, noch enbluotet zevil, noch enfuelt: alsô tuo disiu wnde, diu enbluot nith zevil noch engewinne hitze, noch enswær [schwärte] , noch enhatter [eneiter], noch enfuoel, di ich gesent [gesegnet] hab. In dem namen des vaters und des suns und des hæiligen gaist. Âmen.' Sprich den segen drîstunt und alsô manigen pâternoster, und tuo nith mêr, wan als hie gescriben sî.

Beispieltext 36
Grazer Gichtbeschwörung (Graz, Handschrift 476, Arznei- und Alchemiebuch des Matheus S. von 1587, S. 176–178)
siehe B.1.4; C4.3.1.1c; C4.3.1.2; C5.3

Ain Bewarter Vnd Schenner Segen für das Vergücht Zu sprechen.
Jesus + von Nasareth + Ein Khinig der Juden + Jech Beschöer dich Vergücht Vnd 77 Vergicht pey der sonnen Vnd pey dem Mann Vnnd pey der Heilligen wandtlung Vnsere Lieben herren Jesus Cristus + Jch beschwör dich Vergücht Vnnd die 77 Vergücht pey den Heilligen 5 wunden Vnsers Lieben herren Jesus Cristus
[folgen ebenso Beschwörungen bei heiligem Leichnam, beim ersten erschaffenen Menschen, beim Jüngsten Gericht, beim Urteil, das über „mich und alle Menschen" gesprochen wird, beim heiligen Abendmal, beim Gebet am Ölberg, bei den heiligen Worten am Kreuz, beim blutigen Angstschweiß, bei den Nägeln, Blut aus Hand-Fuß- und Brustwunde, bei den Rutenschlägen und der Geiselung, beim rosenfarbenen Blut und beim heiligen Grab. Dann heißt es:]
vnd pey der Heilligen Vrstenndt [Auferstehung] Vnd Himelfardt Vnsers Lieben Herren Jesus Cristus + Jech Beschwör dich Vergycht Vnd die 77: Vergücht pey dem Heilligen wunder Zaichen das Gott jm Himmel Vnd Auf erdten gethann Hatt pey Himel Vnnd erden pey alles daz gott Erschaffen Hatt. [Es folgen Beschwörungen bei Gottes Worten, bei Maria und Elisabeth und sodann eine Erlösungsbitte:]
Nun bith ich Dich Lieber herr Jesu Crist Du wollest mich N. Lassen geniessen die selbige Liebe die du vnnd sant Johannes Zu ain ander Hettest, das du mich N. wellest Erlösen vnnd belädtigen von der grossen Kranckheit die ich N. Hie Leidten mueß
[Und es erfolgt nochmal eine Aufzählung einiger Geißelwerkzeuge Christi, des Leides und der Tränen seiner Mutter und wieder eine Einbindung der eigenen Glieder, schießlich mit dem bekannten Schlußreim:]
wer da von dem Vergücht Helffe khann, dann der mann, der den Todt vmb der sinder willen an den stamen des Heilligen Creitz Namb. Jch gebeut dier das du dich Legest, Vnnd dich ... Mich N nimer regest pey der Crafft Vnnd macht Gottes Himelischen Vatters ...

Beispieltext 37
Beschwörung des Wundfiebererzeugers (altnordische Runen-Handschrift von Canterbury), ed. Genzmer, Felix, in: Germanisch-Romanische Monatsschrift NF 1(1950), S. 24
siehe C5.1

    Gyrill* des Wundfiebers!
Fahr du nun! Gefunden bist du
Thor schlage dich, Thursenkönig!
    Jyrill des Wundfiebers!
* böser Geist, Riese und Thursenkönig, der Erreger des Wundfiebers

Beispieltext 38
Teilstücke aus einem Großen Exorzismus gegen Besessenheit (CKC Coloniensis Capituli: Bibliothek des Domkapitels Köln 15, Bl. 96'–98', mbr., 9. Jh.), ed. Franz: Bened. II, S. 588f
siehe B2; C5.3

Domine sancte, pater omnipotens ...	Heiliger Herr, allmächtiger Vater,
deus Moysen, deus Aaron, deus	Gott Moses', Gott Arons, Gott der
patriarcharum [...].	Patriarchen ...
in nomine tuo digneris mihi auxi-	Du wollest mir zu Hilfe kommen
lium pietatis tue [prestare ...]	in Deiner Güte ...
aduersus hunc nequissimum	gegen diesen unsäglich nutzlosen
[spiritum]. [...]	Geist ...
Vade inde, uade, inimice satanas	Gehe also davon, gehe, du Feind, Satan!
Audi, inimice, quod uobis dixi.	Höre Feind, was ich gesagt habe !
Maledicti satanas [...]	Verfluchter Satan ...
Exorcizo te, inmundue spiritus,	Ich beschwöre dich, unrein sündiger Geist,
auctor scelerum, deceptor ani-	Schöpfer der Verbrechen, Betrüger der Seelen,
marum, inuide, insaciabilis	neidisch unersättlicher Mörder, der du ver-
homicida, qui conatus es	suchst, dem Menschen die unsterbliche Seele
hoccidere immortalem hominum ...	zu vernichten ...

Beispieltext 39
(Kreuz-) Holzritus und Beschwörung gegen das Überbein (Paris, Bibliothèque Nat. de France, Nouv. acquisitions lat. 229, 12. Jh.), ed. MSD Müllenhoff / Scherer Denkmäler II, 1892, 304f
siehe C5.1

Contra uberbein.	Gegen Überbein
Lignum de sepe vel aliunde sumptum	Lege ein Zaunholz oder ein anderes Nutzholz
pone super uberbein faciens crucem	auf das Überbein, mache ein Kreuzzeichen,

et ter dicens pater noster,   sprich drei Vaterunser und füge diese
additis his teutonicis verbis   deutschen Worte an:

,Ih besueren dich, uberbein, | bî dem holze | dâ der almahtigo got | an ersterban wolda | durch meneschon sunda, | daz dû suînest | unde in al suachost.'

E2 Mesopotamische Beispieltexte (Ausschnitte)

Beispieltext 40
Textstelle Zeile 34–45 aus CT (Cuneiform Texts from Babylonian tablets in the British Museum London) 16,27f (Falkenstein S. 22)
siehe B4.2.3; C1.4.1; C4.2.2; C4.3.1.1a

Magst du ein böser Alû sein, der wie eine Höhlen-Fledermaus(?) in der Nacht [fliegt],
Magst du ein böser Alû sein, der wie ein Vogel an dunklen Orten fliegt,
Magst du ein böser Alû sein, der einen Menschen wie ein bedeckendes Netz zudeckt,
Magst du ein böser Alû sein, der einen Menschen wie ein Fangnetz hinwirft,
Magst du ein böser Alû sein, der wie die Nacht nicht gesehen wird,
Magst du ein böser Alû sein, der wie ein Fuchs in einer verödeten Stadt umherläuft

Beispieltext 41
Textstelle Zeile I 42ff aus CT 16,25 (Falkenstein S. 48)
siehe B4.2.3; C1.4.1; C4.2.2; C4.3.1.1a

Der böse Utukku, der böse Alû,
    der dem in der Nacht gehenden Menschen die Straße sperrt,
das böse Gespenst, der böse Gallû,
    der dem in der Nacht gehenden Menschen den Weg versperrt,
ein überwältigender Löwe, der nichts verschont,
der Böse, der dem Menschen ein böses Auge, Schreckensglanz …,
ein Löwe, der das Maul aufsperrt, der Gnade nicht kennt,
der dem Menschen aus Bosheit wie ein Stern funkelt,
der einen gebundenen Menschen nicht freiläßt. –

Beispieltext 42
Textstelle CT 16,1–7 und CT 16,5, 180–193 (Falkenstein S. 29f)
(Tätigkeit des Beschwörungspriesters)
siehe B4.2.3

Wenn ich an den Kranken herantrete, sein Haus betrete,
meine Hand auf seinen Kopf lege,
die „Sehnen" seiner Glieder untersuche,

die Beschwörung von Eridu über ihn spende,
auf den Kranken die Beschwörung lege …

… wenn ich seine Gliedmaßen prüfe,
das Wasser Eas auf den Kranken sprenge
den Kranken erschrecke, auf die Wange des Kranken schlage,
über den Kopf schreie …

Beispieltext 43
Assyrische Beschwörung gegen nächtliche Verzauberung (Großritual Maqlû)
Tafel 1, Zeile 1–36, nach Tzvi Abusch/Daniel Schwemer TUAT NF4 (2008),
S. 136;
(In Winkelklammern einige Übersetzungen nach Gerhard Meier 1937/67, S. 7f)
siehe B2; B4.2.3; C1.4.1; C4.2.2; C4.3.1.1a; C5.1

1 Beschwörung. „Ich rufe euch, Götter der Nacht, […]
4 weil die Hexe <Zauberin> mich behext hat,
die Lügnerische <der Alp> (Elinitu) mich bezichtigt <gebunden> hat.
sie meinen Gott und meine Göttin von mir entfernt hat,
bin ich dem, der mich sieht, widerwärtig geworden,
leide ich an Schlaflosigkeit bei Nacht und Tag.
Fäden <Zauberknoten> füllen meinen Mund immerzu aus,
10 haben das Mehl (die Nahrung) meines Mundes abgeschnitten,
haben mein Trinkwasser vermindert.
Mein Jubel ist Klage geworden, meine Freude Trauer.
Tretet herbei zu mir, große Götter, hört meine Klage,
sprecht mir Recht, erfahrt, wie es mir ergangen ist!
15 Ich habe (je) eine Figur meines Hexers <Z.> und meiner Hexe <Z.in> angefertigt,
meines Zauberers und meiner Bezauberin,
habe (sie) unter euch (Sternen) <euch zu Füßen> hingelegt und erhebe nun Anklage
Weil sie mir Böses tat, nach gar nicht Gutem gegen mich trachtete,   [in meiner Sache:
möge sie sterben, und ich möge leben!
20 Ihre Hexereien, ihre Zaubereien, ihre magischen Manipulationen <Spuk> mögen
gelöst sein!
Die Tamariske, die im Wipfel üppig ist, möge mich reinigen,
die Dattelpalme, die allem Wind standhält, möge mich lösen!
Das Seifenkraut <mastakal>, das die Erde anfüllt, möge mich rein machen!
[…] Ihr Zauberspruch <ihre Beschwörung> ist der einer bösen Hexe,
ihr Wort ist zurückgesandt zu ihrem Mund, ihre Zunge unbeweglich <festgebunden>.
Wegen ihrer Hexereien mögen die Götter der Nacht sie schlagen,
30 die drei Nachtwachen* mögen ihre bösen Zaubereien <Spuk> lösen!
Ihr Mund sei Talg, ihre Zunge sei Salz:

Der (Mund, der) Böses <die böse Zauberformel> gegen mich gesprochen hat, soll wie
Talg zerstopfen
Die (Zunge, die) Hexereien <Zauberei> durchgeführt hat, soll sich wie Salz auflösen!
Ihre Knoten sind gelöst, ihre Schadenzaubereien <Machenschaften> vernichtet,
35 alle ihre Worte füllen das unbebaute Land <die Steppe>
auf den Befehl hin, den die Götter der Nacht sprachen!"
 * Dämmerung, Mitternacht und Morgengrauen sind die drei Nachtwachen, d. h. im Babylonischen
 die drei Zeitabschnitte, in die man die Nacht einteilte.

Beispieltext 44
Assyrische Beschwörung gegen Nachtgeister (Großritual Maqlû)
Tafel 1, Zeile 135–143, nach Tzvi Abusch/Daniel Schwemer TUAT NF4 (2008),
S. 140;
(In Winkelklammern einige Übersetzungen nach Gerhard Meier 1937/67, S. 12)
siehe B2; C1.4.1; C4.3.1.1a; C5.1

Beschwörung. „Ich erhebe die Fackel, ihre Figuren verbrenne ich!
(Die Figuren) des utukku-Dämons, des šedu-Dämons, des Laurer-Dämons
<des Hockers>, des Totengeistes
der Lamaštu, des labasu-Dämons, des Packers-Dämons,
des lilû-Dämons, der lilitu-Dämonin, des lilû-Mädchens
        <des Nachtmännchens, des Nachtweibchens, der Nachtmagd>
und alles Schlechten, das einen Menschen packen kann,
[…] Zergeht, zerfließt und zertropft nach und nach!
Euer Rauch möge stetig in den Himmel aufsteigen,
eure gühende Asche möge die Sonne auslöschen.
Der Sohn des Ea, der Beschwörer, möge den Schrecken <euren Spion>, der von euch
ausgeht, fernhalten."

Beispieltext 45
Assyrische Beschwörung gegen Verzauberung mit Hilfe Ea und Marduk (Groß-
ritual Maqlû)
Tafel 3, Zeile 39–59 nach Tzvi Abusch/Daniel Schwemer TUAT NF4 (2008),
S. 148f;
(In Winkelklammern einige Übersetzungen nach Gerhard Meier 1937/67, S. 23)
siehe B4.2.3; C1.4.1; C4; C4.2.2; C4.3.1.1a; C5.1

Beschwörung: „ Hexe <Zauberin>, Mörderin,
Lügnerische <Alp>, naršindatu-Zauberin,
Beschwörerin, eššebutu-Ekstatikerin <Zauberpriesterin>,
Schlangenbeschwörerin aguiltu-Zauberin,
qadistu-Kultdienerin <Dirne>, nadistu-Priesterin <Hierodule>,
Ištar-Kultdienerin <Ištargeweihte>, kulmasitu-Hierodule <zermasitu>.

die du des Nacht jagst,
den ganzen Tag umherspähst,
die du den Himmel beschmutzst,
die Erde besudelst <anrührst>,
die du den Mund der Götter bindest,
die Beine <Kniee> der Göttinnen fesselst,
die du die Männer tötest,
die Frauen nicht verschonst. […]
Jetzt sahen dich, packten dich
veränderten dich, verkehrten dich <haben dich ins Wanken gebracht>,
vertauschten den Wortlaut deiner Zaubereien
Ea und Marduk, sie übergaben dich Gira*, dem Helden.
Gira, der Held, möge deine Bindung zerschlagen <deinen Knoten zerbrechen>
und er möge alles, was du gezaubert <gehext> hast, dich selbst empfangen lassen!"
* Gira übernimmt als Gott des Lichtes nachts die Rolle anderer Götter

Beispieltext 46
Assyrische Beschwörung gegen Verzauberung der Körperteile (Großritual Maqlû)
Tafel II, Zeile 19–36 nach Tzvi Abusch/Daniel Schwemer TUAT NF4 (2008), S. 141)
siehe B4.2.3; C1.4.1; C4.2.2; C4.3.1.1a

Beschwörung: „Gira, vollkommener Herr, du bist die Leuchte ist dein Name genannt!
Du spendest den Tempeln aller Götter Licht […]
und erbarme dich meiner, o Herr! Die Hexe hat wie eine Kesselpauke gegen mich gebrüllt,
meinen Kopf, meinen Hals und meinen Schädel hat sie gepackt,
meine Augen, mit denen ich sehen konnte, hat sie gepackt,
meine Füße, mit denen ich einhergehen konnte, hat sie gepackt,
meine Knie, mit denen ich über (Hindernisse) steigen konnte, hat sie gepackt,
meine Arme, mit denen ich gewohnt war (Lasten) zu tragen, hat sie gepackt" […]

Beispieltext 47
Aus: Großritual Maqlû III, Zeile 1–13, ed. Schwemer, Daniel: Abwehrzauber und Behexung, Wiesbaden 2007, S. 68
siehe C21; C4.3.1.1a

Beschwörung: „Die Hexe, die unablässig in den Straßen umherstreift,
    die unentwegt in die Häuser eindringt,
    die in den Gassen umherläuft,
    die auf den Hauptstraßen umherspäht:
        immer wieder wendet sie sich nach vorn und nach hinten,

stellt sich auf der Straße hin und blockiert den Zugang,
auf der Hauptstraße hat sie den Verkehr abgeschnitten.
Dem schönen jungen Mann raubte sie seine männliche Kraft
der schönen jungen Frau trug sie ihren Liebreiz davon,
durch ihren übelwollenden Blick nahm sie ihre Attraktivität fort.
Sie sah den jungen Mann an und raubte seine Lebenskraft,
sie sah die junge Frau an und trug ihren Liebreiz davon.
Es sah mich die Hexe, sie kam hinter mir her ..."

Beispieltext 48

(Aus: Nebnadinirbu- [Niburu-] Keilschrift ed. Weinberger, Bernhard Wolf: An Introduction to the History of Dentistry, St.Louis (USA), 1948, Vol. I, S. 23), zit. n. Kobusch, Hellmuth: Der Zahnwurmglaube 1955
siehe B4.2.3; C4

After Anu had created the heavens, the heavens created the earth,
the earth created the rivers, the rivers created the brooks,
the brooks created the swamps, the swamps created the worm,
then came the worm weeping befor Shamash*.
Before Ea** came her tears: „What willst thou give me to destroy it?"
Ripe figs I will give thee and the meat (pomegranates) of large figs.
„What good are ripe figs to me and the meat of large figs?
Take me up and let me reside between the teeth and the gums,
so that I may destroy the blood of the tooth
and ruin their strength, the roots of the tooth I will eat."
Since thou hast said it, Worm, may Ea strike thee with the power of her fist.
This is the magic ritual: Mix together beer, the sa-kil-bir-plant, and oil.
Then repeat the magic formula thrice and place it on the tooth.
* Sonnengott **Göttin der Tiefe

Beispieltext 49

(aus: Incantations in the neo-Sumerian period, in: Graham Cunningham: ‚Deliver me from evil': Mesopotamian incantations 2500–1500 BC, Roma 1997, S. 71–72)
siehe B4.2.3; C4; C4.1.2; C4.1.2b

1 Incantation:
The just breeding bull has mounted this woman in the pen, the pure fold,
has poured the just seed of mankind into the womb.
The semen poured into the womb having taken form, having given the man an offspring,
5 the woman ate the honey-plant
(and) became full on it.
She ate the honey-plant, her beloved food,

(and) became full on it.
It was indeed its time of birth: she crouched down.
10 The cry approached heaven, the cry approached the underworld,
the whole cry covered the horizon like a garment.
She unfurled the sail like the en-priest's ship,
she filled the king's ship with goods,
she filled the šu-lú ship with carnelian and lapis lazuli.
15 From the vagina of the troubled woman a cord hung down.
   Asalluhi took notice,
   approached his father Enki in the temple,
   spoke to him:
   ‚My father, the just breeding bull has mounted in the pen, the pure fold …
   [wie oben, bis … a cord hung down.]
35 I do not know what to say about this. What will soothe her?'
Enki answers his son Asalluhi:
‚My son, what do jou not know, what shall I add for jou;
Asalluhi, what do jou not know, what shall I add for jou?
Whatever I know jou now too.
After jou the fat from a pure cow, the cream from a mother-cow
have received in the erected dwelling of the agrun,
after jou over the vagina of the troubled woman
from which a cord hangs
have cast the incantation of Eridu,
may it be released like rain from haven,
may it run away like drain-pipe water from a high roof,
may it [x-x] like a river pouring into a lagoon,
may it smashed like a smashed pot.
If it is a male may it weapon (and) axe, its strength of heroism,
seize in the hand;
if it is a female may spindle and hair-clasp be in its hand.
May Gula, the just administrator with meticulous hands,
once she has cut the umbilical cord, determine the destiny.'
Incantation formula

Beispieltext 50
Beschwörung gegen Nachtgeister (Großritual Maqlû Tafel VII, Z. 114–144), ed. Abusch/ Schwemer TUAT NF4 (2008), S. 174f; Ritustafel VII, Zeile 138–156, 185f.
siehe B2; C4.3.1.1a; C5.3

  114 Beschwörung. „Gewaschen habe ich meine Hände, gereinigt meinen Leib
      mit dem reinen Wasser der Grundwassertiefe, das in Eridu erschaffen wurde.
      Alles Übel (und) alles gar nicht Gute,

das in meinem Leib, meinem Fleisch (und) meinen Sehnen vorhanden ist,
das Übel von schlechten (und) gar nicht guten Träumen, Anzeichen (und) Vorzeichen,
[…]
124 Krankheit, Kopfschmerz, Schlaflosigkeit,
Sprachlosigkeit, Benommenheit, Jammer, Kummer, Verluste, Leid,
Weh (und) Ach, Traurigkeit,
Angst, Schrecken, Furcht,
der Fluch der Götter, die Klage der Götter (gegen mich), die Beschwerde [der Götter (gegen mich)], der gebrochene Eid beim Gott, der (verletzte) Schwur durch Erheben der Hände, der (vom gebrochenen Eid hervorgerufene) >Bann<,
das Übel der Hexereien, Zaubereien, magischen Manipulationen (und) bösen Machenschaften der Menschen:
130 mit dem von meinem Körper (abfließenden Wasch)wasser und dem Dreckwasser von meinen Händen werde es (von mir)
abgestreift un(d) gehe auf die Ersatzfigur über:
Die Ersatzfigur trage meine Sünde (, diene) als mein Stellvertreter:
Straße und Gasse mögen meine Sünden lösen:*
Irgendeine möge mich ablösen, irgendeine möge von mir empfangen!
135 Schlimmes habe ich empfangen: sie mögen es von mir empfangen!
Der Tag bringe Heil, der Monat Freude, das Jahr seinen Überfluß!
Ea, Šamaš und Marduk, kommt mir zu Hilfe, auf daß
die Hexereien, Zaubereien, magischen Manipulationen
(und) bösen Machenschaften der Menschen gelöst werden
140 und der >Bann< ausziehe aus meinem Leib!"

Beschwörung: „Erhebe dich, Morgen, wasche meine Hände!
142 Mach auf, Erde, empfange meine Sünden! […]
* D. h. irgendein Passant soll da Übel vom Patienten übernehmen und wegtragen.

Ritustafel zu Beschwörungstafel VII:
[…]
141 Ein Bild der Hexe aus Mehl zeichnest du in eine Waschschüssel hinein.
Du legst eine Lehmfigur der Hexe darauf. Er [der Beschwörer] wäscht seine Hände darüber. Zum Ha[us kehrst du zurück] […]
148 Beschwörung: „Deine Zaubereien, deine Untaten": Handwaschung
Beschwörung: „Deine fest geknoteten Knoten": Handwaschung; du schüttest Erde in die Waschschüssel.
[Beschwörung]: „Gewaschen habe [ich] meine Hände, gereinigt meinen Leib": Er wäscht [übe]r einer Ersatzfigur seine Hände.
[Beschwörung: „Er]hebe dich, Morgen(, wasche meine Hände!)": Handwaschung.
Beschwörung: „Hell erstrahlt ist der Morgen": Handwaschung.
Beschwörung: „Morgen ist es, Morgen": Handwaschung.
Beschwörung: „Am Morgen sind meine Hände gewaschen": Handwaschung.

Beschwörung: „Ich habe meine Hände gewaschen, nochmals meine Hände gewaschen": Handwaschung.

Beispieltext 51
(aus: Ünal, Ahmet, Hethitische Hymnen und Gebete, Hymnus und Gebet an den Sonnengott, CTH 372, in: Kaiser, Otto (Hg.:) Texte aus der Umwelt des Alten Testamentes (TUAT), Band II, Gütersloh 1986–1991, S. 796/798/799
siehe C4.3.1.1a

Vs.I (Zeile 1–13): O Sonnengott, mein Herr, gerechter Herr des Gerichtes, o König des Himmels und der Erde! Du begnadigst das Land, die Macht erteilst du, Du, als gerechter Gott hegst du stets gütige Gesinnung. Du bist erbarmender Gott […] Dein Bart ist aus Lapislazuli; siehe dieser Sterbliche, dein Diener, hat sich vor dir niedergeworfen und spricht (nun) zu dir.

Vs. II (Zeile 24–26/29–34/ 51–54) Da ich seit meiner Kindheit das Mitleid meines Gottes nicht erfahre, werde ich nun danach suchen. […] Auf den Namen meines Gottes habe ich niemals einen Eid geleistet, den Eid habe ich niemals verletzt. Was meinem Gott heilig und daher nicht Rechtens ist zu essen, das habe ich niemals gegessen. Ich habe meinen Leib niemals besudelt. Ich habe niemals eine Kuh aus dem Stall entwendet […] Wenn ich nun genesen bin, bin ich nicht deinem, des Gottes Willen gemäß genesen? […] Nun möge doch mein Gott mir sein Herz und seinen Willen gänzlich kundtun und mir meine Verbrechen bekanntgeben, auf daß ich mich dazu bekenne. Zu mir möge mein Gott durch einen Traum sprechen...

Beispieltext 52
(aus: Bußliturgie mit Fürbitte des Priesters (B 18, 1–21;), Hymnen, Buß- und Klagepsalmen), ed. A. Falkenstein und W. von Soden: Sumerische und Akkadische Hymnen, Zürich und Stuttgart 1953, S. 270, Nr. 18
siehe B4.2.3; C4.3.1.1a

(Der Priester spricht:)
Krankheit, Kopfkrankheit, [Fluch] (und) Schlaflosigkeit
    gossen sie immer wieder auf ihn aus, | (dazu) Verarmung, Seufzen.
............., Widerstand, | Schrecken, Schrecknis
    sind ihm geschickt, | und haben entfernt seine Klage.
Er schaute immer wieder auf sein Leiden, | weint zu dir;
    sein Inneres glüht, | er ver[zehrt] sich nach dir.
In Tränen gebadet ist er, | läßt (sie) träufeln wie einen Nebel;
    bedrückt ist er | und läßt weinen (sogar), die [ihn] nicht gebar.
Wie ein Klagepriester | preßt er Klagen heraus,
    erzählt im flehentlichen Gebet | seine Ruhelosigkeit.
Was hat mein Herr gesagt | und gegen seinen Knecht geplant?
    Bringen will ich | [das Wort] seines kraftlosen Mundes:

(Der Büßer:) Viel sind meiner Sünden, | und allenthalben verfehlte ich mich;
  mag ich (aber) auch diese ..... überschritten haben, |
  möge ich (doch) aus der Not herauskommen !
[Marduk,] viel sind meiner Sünden,| und allenthalben verfehlte ich mich;
  mag ich (aber) auch diese .... überschritten haben, |
  möge ich (doch) aus der Not herauskommen !
(Der Priester:) [B 18, 23–33 beschreibt die Strafen und Bußen des Patienten,
        die ihn, den finster Blickenden, angesichts des Gericht ereilt haben,
        die ihn gebunden haben, und bittet nun auch Ea um Erbarmen:]
In eine Katastrophe geworfen | ist dein Knecht;
  nimm fort deine Strafe | aus dem Morast zieh ihn heraus !
[Zerbrich] seine Kette | löse seine Bande;
  helle auf [seine Wirrnisse], übergib ihn dem Gott, der ihn schuf !
Schenke deinem Knecht Leben, […]

E3 Beispieltexte aus den altindischen Atharvaveden

Beispieltext 53
(aus: Atharvaveda IV, 12, historische Übertragung durch Friedrich Rückert, aus seinem ungedruckten Coburger Nachlass hsgg. von Kreyenborg, Hermann, Hannover 1923, S. 35; weitere, abweichende Übersetzungen finden sich z. B. bei Kuhn, Adalbert, Indische und germanische Segenssprüche, in: Zeitschrift für vergleichende Sprachforschung 13 (1864), S. 58f sowie in: Schulbuchschriften a. d. Königreich Württemberg, (Hundert Lieder des Atharva-Veda) Tübingen 1879, Nachtrag 1869–80, S. 14–15 Online www.zeno.org/nid/20009126139. siehe C4.2.2

  Wachskraut, Wunden verwachsen zu machen.
1. Wachskraut bist du, Wachskraut, zerbrochnen Knochens Wachsekraut;
  Mach du wachsen, Arundhati !
2. Was dir verletzt, was dir verbrannt, was dir zerschossen ist am Leib;
  Der Füger* mit dem Segenskraut füg' es zusammen, Glied an Glied.
3. Es gehe mit dem Marke Mark zusammen dir und Glied mit Glied,
  Zerrissnes Fleisch zusammen wachs' und zusammen der Knochen selbst.
4. Zusammen gehe Mark mit Mark, zusammen wachse Haut mit Haut. –
  Dein Blut, Kraut, wachs' als Knochen, und aus dem Fleische wachse Fleisch.
5. Leibhaar bilde mit deinem Haar, bilde die Haut mit deiner Haut;
  Dein Blut, Kraut, wachs' als Knochen; des Zerschnitte verbinde du ! –
6. Du steh nun auf und geh und lauf,
  Ein Wagen gut an Rädern, Achsen, Naben;
  Steh aufgerichtet !

7. Sei es entzwei durch einen Fall gebrochen,
   Hab' ein Stein, ein geworfner, es geschlagen;
   Ribhu** des Wagens Stücken gleich füg' es zusammen, Glied an Glied !
* Füger, Schöpfergott Dhatar   ** Ribhu, göttliche Werkmeister

Beispieltext 54
(aus: Atharvaveda XX,96.17–22, ed. a) Kuhn, Adalbert, Indische und germanische Segenssprüche, in: Zeitschrift für vergleichende Sprachforschung 13 (1864), S. 66f); ed. b) Online ancient voice. wikidot. com, Hymn XCVI (96), 2009617–19 […] 22.
siehe C4.2.2; C5.1

[Gegen Lungen- und Hauptschwindsucht]
Aus den Augen, aus der Nase, aus den Ohren und aus dem Kinn,
   das Schwinden, das im Kopf, vertreib ich dir aus der Zunge, dem Hirn heraus.
Aus dem Nacken, aus dem Genick, aus dem Brustbein, dem Rückgrat auch,
   treib ich das Schwinden, das im Arme, dir aus den Schultern, Armen, aus.
Aus Eingeweiden und Gedärmen, aus dem Herzen und dem Enddarm,
   aus den Nieren, aus der Leber und der Milz, treib ich die Krankheit aus.
[….] Aus jedem Glied, aus jedem Haar, aus jedem Gelenk,
   aus deinem ganzen Leib, von Kopf bis Fuß treib ich die Krankheit aus.

Beispieltext 55
(aus Rig-Veda X, 97, ed. Kuhn, Adalbert, Indische und germanische Segenssprüche, in: Zeitschrift für vergleichende Sprachforschung 13 (1864), S. 69f); siehe C5.1

[Heilkraft der Pflanzen]
Wenn ich die Kräuter mit Verehrung in die Hand aufnehme,
   dann flieht wie vor dem Totfeind das Schwinden eilig fort.
Wem ihr, o Kräuter, Glied für Glied durchdringt und jegliches Gelenk,
   aus dem nehmt ihr heraus das Schwinden.
O Schwinden, flieg dahin, flieg mit dem blauen Häher fort,
geselle dich dem wilden Zuge und dem Sturm, so schwinde hin!

Beispieltext 56
(aus: Atharvaveda-Paippalāda 2.49, aus einer Sammlung altindischer Zaubersprüche in einer Handschrift auf Birkenrinde vom Beginn des 1. Jahrtausends vor Christus, herausgegeben von Thomas Zehnder, Idstein 1999, S. 116, Prosaformel gegen Schmerzen zur Entsühnung:)
siehe C4.3.1.1a; C4.3.1.1b

2.49,1 Östlich ist die Richtung, das Gāyatra*-Lied die Gottheit.
Die Sünde, welche er hier an den Göttern,
an den Vätern,
an den Menschen begangen hat,
deren Sühnemittel bist du.
Befreie ihn von dieser Auszehrung hier,
von dem, was da weh tut; „svāha".

2.49,2–5
Südlich ist die Richtung, das Rathantara*-Lied die Gottheit. Die Sünde ...
[wie bei 1]
Westlich ist die Richtung, das Vāmadevya-Lied die Gottheit. Die Sünde ...
[wie bei 1]
Nördlich ist die Richtung, das Yajñāyajñiya-Lied die Gottheit, Die Sünde ...
Senkrecht ist die Richtung, das Bṛhant-Lied die Gottheit. Die Sünde ...

* Vedische Versformbezeichnungen, sie werden oft als Gottheiten personifiziert.

Beispieltext 57
(aus: Atharvaveda-Paippalāda 2.14, aus einer Sammlung altindischer Zaubersprüche vom Beginn des 1. Jahrtausends vor Christus, hg. von Thomas Zehnder, Idstein 1999, S. 51–55)
siehe C4.3.1.1b; C5.1

2.14,1 Aufgehend soll der Āditya die Würmer schlagen: Sūriya; untergehend soll er mit seinen Strahlen die Würmer schlagen.

2.14,2 Der, welcher der allgestaltige, vieräugige, der scheckige, weissliche Wurm ist: geschlagen ist der Wurm mit geschlagenem Bruder, mit geschlagener Mutter, mit geschlagener Schwester.

2.14,3 Geschlagen ist der König der Würmer und ihr Oberhaupt ist geschlagen; geschlagen sind seine Untergebenen, geschlagen die Umwohner.

2.14,4 Ich zerschmettere dir die Hörner, mit denen du zustichst; und dann spalte ich da Gefäß, in dem sich dein Gift befindet.

2.14,5 Wie (einst) Atri ich dich, Wurm, wie Kaṇva, wie Jamadagni; mit dem Zauberspruch des Agastiya sind all diese Würmer geschlagen

2.15,1 Indras großer Mühlstein, Zerschmetterer jeden Wurms, mit dem Mühlstein zerstampfe ich die Würmer wie Khava-Körner.

2.15,2 Den sichtbaren, den unsichtbaren habe ich zerschmettert, und dann den Kurūru habe ich zerschmettert; die Algaṇdus, alle Śalūlas (Würmer) zermalmen wir mit einem Spruch.

2.15,3 Die Algaṇdus schlage ich mit der großen Mordwaffe: die getroffenen (und auch) die nicht getroffenen sind wirkungslos geworden; die entwischten (u.) die nicht entwischten werfe ich mit der Stimme nieder, damit auch nicht einer der Würmer übrigbleibt.

2.15,4 Den in den Eingeweiden, den am Kopf und auch den Wurm in den Rippen, den Avaskava (Herabstocherer?) (u.) den Viyadvara (Zernager) zermalmen wir mit dem Spruch.

2.15,5 Die Würmer, welche auf den Bergen, welche in den Wäldern, welche in den Pflanzen, in den Tieren, im Wasser drinnen sind, welche sich unseren Körper zum Standort gemacht haben, die soll Indra mit der großen Keule schlagen.

E4 Texte aus der Tradition der Maya-Kulturen

Beispieltext 58
(aus: a) Mu-Igala or the Way of Muu, a medicine song from the Cuña-Indians of Panama with translation and comments by Nils M. Holmer and Henry Wassen, Göteborg 1947;
b) zum Teil verwendete deutsche Übersetzung der Bearbeitung von Lévi-Strauss (Strukturale Anthropologie I, S. 211–219) durch Hans Naumann)
siehe B2; C4.1.2; C4.1.2a; C4.2.2

Die Kranke sagt zur Hebamme: ‚Wirklich, ich bin bekleidet mit dem heißen Gewand der Krankheit' [= Fieber]. – Die Hebamme antwortet ihr: ‚Wirklich, du bist mit dem heißen Gewand der Krankheit bekleidet, so habe ich dich auch verstanden.' – Die Hebamme fragt nach: ‚Seit wieviel Tagen widerstehst du schon dem üblen Gewand der Krankheit?' – Die Frau sagt: ‚Seit zwei Tagen widerstehe ich dem heißen Gewand der Krankheit.' [Zeile 1–4] – Die Hebamme: ‚Wir [sie meint sich selbst] müssen zum Medizinmann gehen, mit ihm reden wegen dir, du weißt.' [Z. 6] –
Die Hebamme geht in der Hütte umher; die Hebamme sucht Perlen; die Hebamme geht umher, die Hebamme setzt einen Fuß vor den anderen; berührt mit dem einen Fuß den Boden; setzt den anderen Fuß vor. Die Hebamme stößt die Hüttentür auf, die Hüttentür knarrt. Die Hebamme geht hinaus und schaut zögernd und bestürzt um sich. [Zeile 7–14; dann geht sie zur Hütte des Medizinmannes und berichtet ihm.] –
Der Medizinmann sagt ihr: ‚Wenn ich auch zunächst kein Licht, keinen Durchblick habe, ich werde für dich ins dunkle Innere steigen' [Zeile 32, und er verläßt seine Hängematte, geht zur Hütte der Kreißenden, wo er sich unter der Hängematte der Frau auf einem kleinen ‚goldenen' Sitz niederläßt und ‚heilsame' Cocoa-Bohnen in einer Kupferpfanne anbrennt, bis die Hütte vom Rauch erfüllt ist. [Zeile 33–64] – [Nach Vorstellung der Cuñas hat der Foetusbildner Muu des Uterus seine Macht überschritten, hat sich des Purba, d. h. der Seele der Frau bemächtigt und muß nun aufgesucht werden.]
Nele Tionuchunana und ihre Kinder [= geschnitzte Schutzgeisterfiguren] kräftigen die Macht des Medizinmannes gegenüber Muu. Der Medizinmann singt: ‚Unter der Hängematte der kranken Frau sammle ich dich, nelegan [= die Lebenskräfte der Menschen].

Unter der Hängematte der kranken Frau stelle ich dich auf, nelegan.' Und sie steigen empor auf dem Wege des Muu. [Zeile 65–72, 80, 83 Die so personalisierte Widerstandskraft verfolgt den Weg des Muu in die Vagina hinauf.]
Der Medizinmann sagt: ‚Ihr weißes Gewebe [= Vulva] liegt im Schoße und bewegt sich zärtlich. Der Körper der kranken Frau liegt matt. Wenn Muu's Weg erleuchtet wird [durch die glänzenden penisähnlichen Kappen der nelegan, dann ist zu sehen:], läuft Flüssigkeit wie Blut über. Ihr Fluß tropft unter die Hängematte, blutig rot. Das innere weiße Gewebe erstreckt sich bis ins Innere der Erde [kosmische Sicht]; in der Mitte des weißen Gewebes steigt ein menschliches Wesen herab. Die Frau sitzt nach Luft ringend gegen Osten; sitzt mit gespreizten Knien [übliche Gebärposition]; ihr Schweiß bildet eine Lache. Im Schoße der Erde sammelt sich ihr Fließen, blutrot. Neletionuchunana [s. o.] und ihre Kinder schicken sich an: ‚Wir müssen nun die kranke Frau über die Kraft ihrer niga purbalele befragen.' [Zeile 85–93]
Die Frau sagt: ‚Meine Sehkraft ist zerstreut und eingeschlafen auf Muu's Pfad.' [Zeile 97] ‚Muu Puklip ... Muu Nauryaiti ... Muu Kepuniti ist (sind) gekommen und will (wollen) meine niga purbalele für immer behalten.' [Zeile 98–100]
[Der Medizinmann rüstet nun die einzelnen Lebensgeister der Frau magisch aus, mit Perlen vieler Farben, Knochen verschiedener Tiere und Halsbändern. Er mobilisiert die nelegan durch vielfach aufreihende Aufrufe [Zeile 104–229] unter anderem auch unter Beteiligung einer vermuteten magisch-kosmischen Manneskraft:] ‚Ibelele Nuchulele, die große Nele, geschaffen gemeinsam mit der Erde; ich rufe dich zu Hilfe ... ; Nele, die Freunde macht, die mit geweint hat ...' [Zeile 138–140]
[Danach folgt die erzählerische Reproduktion des von der Schwangeren Erlebten, um ihr die Augen wieder zu öffnen, die Wege zu erhellen, auch für den einstigen Liebesakt, indem die nelegan die Gestalt und die Bewegung eines belichtenden Penis annehmen:]
Die Kappen der nelegan glänzen, die Kappen der nelegan werden weiß; die nelegan werden flach und und niedrig (?), ganz wie Spitzen, ganz aufgerichtet; die nelegan werden furchtbar (?), werden alle furchtbar (?), zum Heil der nigapurbalele der Kranken. Sie gehen auf die Suche nach ihrer nigapurbalele zu Muu Puklip's Festung [= Uterus]. Sie wollen ihre niga purbalele wieder aufrichten. [Zeile 230–233]
Der Medizinmann weist die nelegan an: ‚Muu Puklip's Festung ist nun unschwer zu sehen, seit ihr wisst, wie man unsichtbar wird, ihr werdet neles, ihr werdet Dinge verändern durch Sehen'. ‚Der Medizinmann gibt euch eine lebendige Seele ...' – Die nelegan öffnen nun der kranken Frau die Augen. [Zeile 235–238, 240; im folgenden nach Lévi-Strauss die ‚mythische Anatomie' mit ‚einer Art affektiver Geografie' unter Beschreibung von Schmerz- und Widerstandspunkten, nachdem das erhellende Licht gewirkt hat:]
Die nelegan machen sich auf den Weg, die nelegan gehen einer nach dem anderen den Weg des Muu bis zum niedrigen Gebirge; die nelegan gehen ... bis zum kurzen Gebirge; die nelegan gehen ... bis zum langen Gebirge ... bis zum Yale Pokuna Yale ... bis zum Mittelpunkt des flachen Gebirges; die nelegan setzen sich in Bewegung, die nelegan gehen einer nach dem anderen den Weg des Muu. [Zeile 241–248]
[Weitere Widerstände in Form der zu Schmerzen personalisierten Tiere und Untiere sind zu überwinden:] In Muu Puklip's Strudel sind Wohnungen der Alligatoren. Onkel Alliga-

tor Towili bewegt sich mit hervorquellenden Augen, ... mit gewundenem gefleckten Leibe hin und her ... krümmt sich zusammen und schlägt mit dem Schwanze ...; der Tintenfisch streckt seine schleimigen Flossen aus und zieht sie wieder ein; ... Tiger, schwarze Tiere mit heraushängenden Zungen und gefletschten Zähnen ... [Zeile 253–298; schließlich müssen die nelegan auch materielle Hindernisse überwinden:] Es flattern die regenbogenfarbigen Vorhänge; es wogen die regenbogenfarbigen Vorhänge; über Muu Puklip's Haus flattern seidene Vorhänge ... rote ... schwarze ... weiße ... dunkle ... gelbe ... Vorhänge; ... Fasern ... Stricke ... Fäden. [Zeile 308–330; es folgt die Einnahme des Muu Puklip-Hauses und Dialoge und weitere Kämpfe:] Die nelegan treten durch Muu Puklip's Tor wie eine Menge Männer. Sie erfüllen (besetzen) Muu Puklip's Haus. Tatsächlich will Muu Puklip die niga purbalele der kranken Frau besitzen. [Zeile 355–357]... Die nelegan streiten mit Muu Puklip: ‚Lass uns einmal richtig mit unseren goldenen Kappen kämpfen, damit wir sehen, wer die stärkeren hat!' ... [Zeile 386] Die nelegan marschieren alle, die Kappen steigen auf, die Kappen rauchen, als ob ein Mann steht wie ein brennendes Maisstroh. Sie lassen Muu's Haus total bluten. [Zeile 390–391; Muu Puklip verliert den Kampf.] Die nelegan suchen die niga purbale der kranken Frau im Hause, ohne auch nur einen Winkel zu unterlassen. Wenn sie die niga purbalele der Frau gerettet haben ... ihre Herzseele ... ihre Knochenseele ... ihre Zahnseele ... ihre Haarseele ... ihre Nagelseele ... ihre Fußseele, (dann) ruft Nele Upina Wakasip stehend aus: ‚Ich verspreche dir, in acht Tagen wird purbalele wiederhergestellt sein' [Zeile 400–409; aber noch einmal muß die niga ermutigt werden. Die Dorfbewoher müssen zum Krautsammeln auf Geheiß des Medizinmanns in die Berge gehen. Und nach der Entbindung dringt der Medizinmann mit Adstringentien in den Weg des Muu ein, vielleicht zur Lösung der Nachgeburt:]
Der Medizinmann geht zur Öffnung von Muu; wie Nusupane bewegt er sich voran; er geht das Innere zu wischen, um alles komplett zu trocknen. [Er hat sein eingangs gegebenes Versprechen, ins Innere zu steigen, erfüllt.]

Beispieltext 59
(aus: Christian Rätsch: Zaubersprüche bei den Maya und Lakandonen, München 1985, S. 135f)
siehe C4.2.2

Ein kleiner Nikte-Ast wird an einem Ende mit etwas Weihrauch bestrichen und angezündet. Das glimmende Hölzchen wird kreisend über Weihrauchkugeln und Kautschukfiguren auf Opferbrettern bewegt und dazu wird geflüstert:
    Zerplatzende Blüten, erwachet!
Die Streben deines Hauses sind die des Herrn.
Sie sind aus dem Holz des rote Anona-Baumes;
Das sind die Streben des Daches deines Hauses.
Das sind die Streben des Daches deines Hauses,
sie gehören dem Herrn.
Aus Fächerpalmenwedeln ist dein Haus,

sie gehören dem Herrn.
>Ich setze die Knochen ein,
>ich setze das Herz ein,
>ich setze die Lunge ein.
>Ich erwecke die am Fuße des Opferbrettes Sitzenden.

Sie gehören unserem Herrn,
Sie gehören unserem Herrn,
Sie sind deine Opfer.
Erwachet, schlaft nicht, zerplatzende Blüten!

E5 Beispieltexte aus Papyri graecae magicae

Beispieltext 60
(aus: Papyri graecae magicae (PGM) VII. 260–71), ed. Betz, Hans Dieter: The greek magical papyri, Chicago 1986, S. 123f
siehe B2; C4; C4.2.1; C4.2.2; C4.3.1.2

„I conjure you, O Womb, [by the] one established over the Abyss, before heaven, earth, sea, light, or darkness came to be; [you?] who created the angels, being foremost, AMICHAM-CHOU and CHOUCHAŌ CHĒRŌEI OUEIACHŌ ODOU PROSEIOGGĒS, and who sit over the cherubim, who bear / jour (?) own throne, that you retourn again to jour seat, and that you do not turn [to one side?] into the right part of the ribs, or into the left part of the ribs, and that you do not gnaw into the heart like a dog, but remain indeed in your own intended and proper place, not chewing [as long as] I conjure by the one who, / in the beginning, made the heaven and earth and all that is therein. Hallelujah! Amen!" – Write this on a tin tablet and „clothe" it in 7 colors.

Beispieltext 61
(aus PGM IV. 1227–64, ed. Betz, H.D., S. 62
siehe B2; C4.3.1.1b; C5.1

Formula to be spoken over his head: Place olive branches before him, / and stand behind him and say: „Hail. God of Abraham; hail God of Isaac; hail God of Jacob; Jesus Chrestos, the Holy Spirit, the Son of the Father, who is above the Seven,/ who is within the Seven. Bring Iao Sabaoth; may jour power issue forth from him, NN, until jou drive away this unclean daimon Satan, who is in him. I conjure you, daimon, / whoever you are, by this god, SABARBARBATHIOTH SABARBARBATHIOUTH SABARBAR-BATHIONETH SABARBARBAPHAI. Come out, daimon, whoever you are, and stay away from him, NN, / now, now; immediately [...] I deliver you into black chaos in perdition"
Take 7 olive branches; for six of them / tie together the two ends of each one, but for the remaining one use it like a whip as you utter the conjuration. Keep it secret; it is proven.

After driving out the daimon, hang around him, NN, a phylactery, which the patient puts on after the expulsion of the daimon – a phylactery with these things [written] on / a tin metal leaf: BOR PHOR PHORBA […]

Beispieltext 62
(aus PGM CXXII. 51–55, ed. Betz, H.D., S. 317
siehe B4.2.4

„Osiris has headache; Ammon has headache at the temples of his head; [Isis-]Nephthys has a headache all over her head. May Osiris' headache not stop, may Ammon's headache at the temples of his head not stop, until / first he, NN, stops all …"

F Register

F1 Literaturverzeichnis

Abdullaev, Y. and M. I. Posner, Attentional mechanisms, in: Berntson/ Cacioppo, Handbook, I, 370–382 : 54
Abusch, Tzvi und Schwemer, Daniel: Das Abwehrzauber-Ritual Maqlû, in: TUAT NF4 (2008) : 112, 207
Abusch, Tzvi, Einleitung zu Schwemer, Daniel: Abwehrzauber und Behexung, Wiesbaden 2007 : 87
Adolphs, Ralph, Cognitive neuroscience of human social behavior, in: Nat. Rev. Neurosc. 4 (2003), S. 165–178 : 19, 74
Adolphs, R. und Michael Spezio: Social cognition, in: Berntson/ Cacioppo, Handbook II, Kap.47 : 74
Agrippa von Nettesheim: Der Geheimen Philosophie od. Magie drittes Buch, in: Magische Werke, Berlin 1916 : 26
Ahrens, Ullrich: Fremde Träume, Berlin 1996 : 109
Alter, Kai. et al. (Hg.) Brain Talk: Discourse with and in brain, Lund 2009 : 155
Amanzio M, Benedetti F (1999), Neuropharmacol. dissection of placebo analgesia: Expectation-activated opioid system vs. conditioning-activated spec. subsystems, in: J. of neuroscience 19(1), 484–494 : 50
Anders, Silke et al., Flow of affective information between communic. brains, Neuroimage 54 (2011), 439–446 : 68
Angst, Beatrice E.: Magische Praktiken des Menschen unserer Zeit, Bern u. a. 1972 : 24
Arbib, Michael A.: From grasp to language: Embodied concepts and the challenge of abstraction, Journal of physiology (Paris) 102(2008), 4–20 : 66
Atlas, Lauren Y. and Tor D. Wager, How expectations shape pain, in: Neuroscience Letters (2012), 520, 140–48 : 165
Atmaca, Silke et al.: Action co-representation: The joint SNARC effect, in: Social Neuroscience 2008 3(3–4), 410–420 : 71
Aziz-Zadeh, Lisa et al., Common premotor regions for the perception and production of prosody, (2009) PLoS One 5(1): e8759. doi: 10.1371/j.pone.0008759 : 140

Bach, George R., Kreative Aggression, in: Corsini, Handbuch I, S. 571–586 : 106, 173
Baeken, Chris et al., The influence of emotional priming on the neural substrates of memory. A prospective fMRI study using portrait art stimuli, in: Neuroimage 61 (2012), 876–883, doi.org/10.1016/j. : 78
Bagliani, Agostini P. und Geogio Stabile: Träume im Mittelalter: Ikonolog. Studien, Stuttgart Zürich 1989 : 103
Balint, Michael: Therapeutische Aspekte der Regression, Stuttgart 1997² : 59
Bamert, Martin: Magia – Maga – Makha, Zwischenprüfungsarbeit Leipzig WS 06/07 : 25
Bandler, Richard und J. Grinder: Neue Wege der Kurzzeit-Therapie, Paderborn 1994 :63

Baron, Katja: Der Zusammenhang zwischen kognitiver Emotionsregulierung und der Fähigkeit, negative Emotionen zu beeinflussen, eine fMRI-Studie, Diss. Ulm 2011 : 95
Bässler, Andreas, Die Funktion des Rätsels im Lalebuch, in: Daphnis 26 (1997), S. 53–84 : 124
Bauer, Roman, Marburg, Universitas-online, o. J. : 22
Bayle, Dimitri J. et al. Unconsciously perceived fear in peripheral vision alerts the limbic system, (2009) PLos One 4(12) e8207. doi:10.1371/j.pone.0008207 : 119
Bender, Wolfram und Christian Stadler: Psychodrama-Therapie, Stuttgart 2012 : 122
Benedetti F Pollo A et al. Conscious expectation an unconscious conditioning in analgesic, motor, and hormonal placebo/nocebo responses, in: The journal of neuroscience 2003, 23(10): 4315–4323 : 50
Benedetti, Gaetano: Die Botschaft der Träume, Göttingen 1998 : 64, 109, 114
Benedikt XVI, Papst, Generalaudienz 30. Mai 2007 : 77
Benelli, E. et al., Emotional and cognitive processing of narratives and individual appraisal styles, Front. Hum. Neurosciences 2012; 6:239 : 96
Berna, C. et al, Induction of depressed mood disrupts emotion regulation neurocircuitry, in: Biol. Psychiatry 67 (2010), S. 1083–1090 : 167
Berntson, G.G. & Cacioppo, J.T. (Hg.): Handbook of neuroscience for the behavioral sciences, NY 2009
Betz, Hans Dieter (Ed.): The greek magical papyri, Chicago & London 1986 : 58, 218, 219
Bihler, S.M.E. et al., Auditory repetition of words and pseudowords: An fMRI study, in: Alter, K. et al. (Hg.) : 155
Birkhan, Helmut: Magie im Mittelalter, München 2010 : 26
Boesch, E.E.: Ritual und Psychotherapie, in: Z. für kl. Psychologie und Psychotherapie 30 (1982), S. 214–234 : 35
Borgolte, Michael: Sozialgeschichte des Mittelalters, München 1996 : 76
Bouchard, Jr, Thomas J et al., Intrinsic and extrinsic religiousness: genetic and environmental influences and personality correlates, in: Twin Research 2(1999), S. 88–98, doi.org/10.1375 : 84
Bradshaw, Matt et al.: Do genetic factors influence religious life? in: Journal for the scientific study of religion 47 (2008), S. 529–544 doi org/10.1111/j.1468–5906 : 84
Breithaupt, Fritz: Kulturen der Empathie, Frankfurt/M., 2009 : 60, 70
Brenner, Peter J., Hörkulturen. Stimme und Schrift, Hören und Lesen in der abendländischen Kultur, in: Universitas 61 (2006), 224–235 : 78
Brisch, Karl Heinz, Online, Universitätskinderklinik München 2012 : 146
Browne, Edward G.: Arabian Medicine, Cambridge 1921 : 17
Brüne, Martin et al., (2011) Neuroanatomical correlates of suicide in psychosis: the possible role of von Economo neurons, PLoS One 6(6): e20936. doi:10.1371/j.pone.0020936 : 166
Brüne, Martin et al., An fMRI study of ‚theory of mind' in at-risk states of psychosis, in: Neuroimage 55(2011) S. 329–337, doi.org/j. 2010.12.018 : 121
Budygin, E.A. et al.: Aversive stimulus differentially triggers subsecond dopamin release in reward regions, in: Neuroscience 201 (2012), 331–337 : 106
Butler, S.A.L. Mesopotamian conceptions of dreams and dream rituals, Münster 1998 : 110

Cacciari, C. et al., Literal, fictive and metaphorical motion sentences preserve the motion component of the verb: A TMS study, in: Brain & Language, 119 (2011), S. 149–157 : 139
Canetti, Elias: Nachträge aus Hampstead. Aufzeichnungen 1957–1959, München 1994 : 27
Capitão, Liliane et al., MRI amygdala volume in Williams Syndrome, in: Research in Developmental Disabilities 32(2011), 2767–2772; doi:10.1016/j.ridd.2011.05.033 : 75

Chiao, Joan Y. and Genna M. Bebko, Cultural Neuroscience of Social Cognition, in: Han, Shihui/ Pöppel (Hg.): Culture and Neural Frames (2011) : 74

Christian, Lisa M. et al. Psychological influences on neuroendocrine and immune outcomes, in: Berntson/ Cacioppo, Handbook S. 1260–1279 : 167

Churchland, Paul M.: Die Seelenmaschine, Heidelberg u. a. 1997 : 54

Cierpka, Manfred (Hg.:) Frühe Kindheit, Heidelberg 2012

Corsini, Raymond J. (Hg.:) Handbuch der Psychotherapie (deutsche Ausgabe, bearb. von Gerd Wenninger), München 1983 : 168

Coull J T (2009) Neural Substrates of mounting temporal expectation, PLoS Biol 7(8): e1000166. : 51

CSB = Corpus der deutschen Segen- und Beschwörungsformeln am Institut für Sächsische Geschichte und Volkskunde Dresden : 37, 118

Cunningham, Graham, ‚Deliver me from evil', Roma 1997 : 149

David, Nicole et al., Differential involvement of the posterior temporal cortex in mentalizing but not perspective taking, in: Soc.Cogn.Affect. Neuroscience (2008) 3(3):279–289, doi:10.1093/scan/nsn023 : 71

De la Fuente-Fernandez R et al., Expectation and dopamin release: mechanism of the placebo effect in Parkinson's disease, in: Science 2001, 293 (5532), 1164–1166 : 50

De Zubicaray, G. et al., Mirror neurons, the representation of word meaning, and the foot of the third left frontal convolution, in: Brain & Language 112,1 (2010), S. 77–84 : 70

Decety, J. und Lamm,C., The role of the right temporoparietal junction in social interaction, in: Neuroscientist. (2007) 13(6): 580–93 : 84

Decety, Jean and C. Lamm: Empathy and Intersubjectivity, in: Berntson/ Cacioppo, Handbook II, 940–957 : 67

Demare-Lafont, Sophie, Judicial decision-making: Jugdes and arbitrators, in: Radner/Robson, S. 335 : 90

Deremble-Mannes, Colette, Die Traumwelt der Legenden in den Glasmalereien von Chartres, in: Bagliani/ Stabile 1989 : 103

Desoille, Robert: Théorie et pratique du rêve éveillé dirigé, Genf 1961 : 175

Dichgans, Johannes, Mimik, Gesten und Sprachmelodie. Medien sozialer Kommunikation und ihre neuronalen Grundlagen, in: Frühwald, Wolfgang et al. (Hg.): Das Design des Menschen, Köln 2004, S. 217–231 : 22, 116, 141

Dilts, Robert: Sleight of mouth. The magic of conversational belief change (1999/2006), deutsch u. d. T. Die Magie der Sprache, Paderborn 2008² : 125

Dinzelbacher, Peter (Hg.:) Europäische Mentalitätsgeschichte, Stuttgart 1993 : 48

Dinzelbacher, Peter, Religiöses Erleben von bildender Kunst, in: Schreiner, Klaus (Hg.) Frömmigkeit im Mittelalter, München 2002 : 78, 79, 161

Dinzelbacher, Peter: Unglaube im „Zeitalter des Glaubens", Badenweiler 2009, S. 151 : 81

Dinzelbacher, Peter und Dieter R. Bauer (Hg.:) Volksreligion im hohen und späten Mittelalter, Paderborn 1990 : 24, 26, 103

Dirks, Burkhard: Die Notfallmedizin Heidelberg 2007 : 98

Drenhaus, H. und b.Graben, P, nach Nieuwland, M.S. und van Berkum, J.J.A., When peanuts fall in love: N400 evidence for the power of discourse, in: Journal of cognitive neuroscience 18 (2006) : 130

Drenhaus, Heiner und Peter beim Graben, Ereigniskorrelierte Potenziale (EKPs), in: Zeitschrift für germanistische Linguistik 40 (2012), S. 68–96 : 129

Dumas, Guillaume et al.: Inter-brain synchronisation during social interaction, PLoSONE 5(8):(2010) e12166.doi:10.1371/j.pone : 68

Dux, Günter: Historisch-genetische Theorie der Kultur, Weilerswist 2000 : 21

Dzhelyova, M. et al., Temporal dynamics of trustworthiness perception, in: Brain Research 1435 (2012), 81 : 51

Ebeling, Erich, Artikel Dämonen im Reallexikon der Assyrologie, Berlin und Leipzig 1938 II : 88

Ehring, Thomas et al., Emotion and vulnerability to depression: Spontaneous versus instructed use of emotion suppression and reappraisal, in: Emotion 10(2010), 563–572 : 98

Eibl-Eibesfeldt, Irenäus und Christa Sütterlin: Im Banne der Angst. München 1992: 117

Eis, Gerhard: Altdeutsche Zaubersprüche, Berlin 1964 : 124, 191

Englund, Robert K., Accounting in Proto-cuneiform, in: Radner/Robson, Handbook: 86

Enzensberger, Hans Magnus: Neuronales Netz, in: Die Elexiere der Wissenschaft, Frankfurt 2002 : 21

Eply, Nicholas et al., Believer's estimates of God's beliefs are more egocentric than estimates of other people's beliefs, in: PNAS (2009), doi: 10.1073/pnas.0908374106 : 85

Ernst, Wolfgang, Das Kunstbuch des Johannes Zahn von Dürnberg bei Wunsiedel, in: Archiv für Geschichte von Oberfranken 76 (Bayreuth 1996), S. 169–195 : 127

Ernst, Wolfgang: Beschwörungen u. Segen – Angew. Psychotherapie im Mittelalter, Köln Weimar Wien 2011 : 17, 83, 87, 99, 149, 154

Ernst, Wolfgang: Oberpfälzischer Heilzauber, Pressath 2011^2 : 127

Etkin, Amit et al. (2011) Emotional processing in anterior cingulate and medial prefrontal cortex, in: Trends in cognitive sciences, doi:10.1016 /j.tics.2010.11.004 : 95

Falkenstein, A. und W. von Soden (Hg.:) Sumerische und akkadische Hymnen und Gebete, Stuttgart 1953 : 159, 211

Falkenstein, Adam: Die Haupttypen der sumerischen Beschwörung, Diss. Leipzig 1931 : 88, 89, 204

Faust, Miriam (Hg.) Handbook of the Neuropsychology of language, 2012 (Einleitung) : 16

Faust, Miriam, Thinking outside the left box: The role of the right hemisphere in novel metaphor comprehension, in: dies. (Hg.:) Handbook, S. 425–448 : 139

Faust, Volker und Günter Hole: Der gestörte Schlaf, Ulm 1991 : 106

Faust, Volker: Von Amok bis Zwang, Bd. II, Heidelberg u. a. 2013 : 119

Feldes, Roderich: Das Wort als Werkzeug, Göttingen 1976 : 40

Fenichel, Otto: The collected papers, New York 1953 : 54

Fertuck E.A. et al., Enhanced ‚Reading the Mind in the Eyes' in borderline personality disorder, in: Psychological Medicine 39(2009), doi: 10.1017/S003329170900600X : 120

Fiske, Susan T., Cultural Processes, in: Berntson/ Cacioppo, Handbook II, Kap. 51 : 75

Fonagy, Peter und Mary Target, Neubewertung der Entwicklung der Affektregulation vor dem Hintergrund von Winnicotts Konzept des ‚falschen Selbst', in: Psyche 56 (2002), 839–862 : 63

Fossier, Robert: Das Leben im Mittelalter, München 2008 : 83

Foster, Benjamin, R., The person in Mesopotamian thought, in: Radner/ Robson (Hg.:) Handbook S. 117–139 : 86, 90, 108

Franz, Adolph: Die kirchlichen Benediktionen im Mittelalter Freiburg/Br. 1909 : 27, 78, 100

Freud, Sigmund: Der Dichter und das Phantasieren (1908) : 146

Freud, Sigmund: Totem und Tabu, Frankfurt/M. und Hamburg 1964 : 30

Frick, Carina et al., Hypersensitivity in Borderline Personality Disorder during Mindreading, PLoS One 2012, doi 0.1371/journal.pone0041650) : 120

Frick, E. Der Besessene von Gerasa. Ein Bibliodrama zu Mk. 5,1–20, in: Geist und Leben 64 (1991), S. 385 : 167

Friederici, Angela D., Menschliche Sprachverarbeitung und ihre neuronalen Grundlagen, in: Meier/ Ploog, (Hg.) Der Mensch und sein Gehirn, München u. a. 1998², S. 137–156 : 129

Friederici, Angela, in: Spektrum der Wissenschaft, Januar 2010, S. 66–71 : 73

Frühwald, Wolfgang et al. (Hg.): Das Design des Menschen, Köln 2004 : 19, 22, 116, 141

Fuchs, Thomas: Das Gehirn – ein Beziehungsorgan. Stuttgart 2008/ 2013 : 19

Gallese, Vittorio und George Lakoff: The brain's concepts: The role of the sensory-motor system in conceptual knowledge, in: Cognitive Neuropsychology, 22 (2005), 455–479 : 66

Gardner, Julian, Päpstliche Träume, in: Bagliani/ Stabile (Hg.): Träume im Mittelalter 1989 : 103

Geier, Manfred: Die magische Kraft der Poesie. Zur Geschichte, Struktur und Funktion des Zauberspruchs, in: Deutsche Vierteljschr. f. Literaturwissensch. und Geistesgesch. Halle / sp. Stuttgart 56 (1982), S. 359–385 : 18

Gerlitz, Peter, in: Theologische Realenzyklopädie (TRE) Bd. XIV (1985) : 45

Glucklich, Ariel: The end of magic, New York u. a. 1997 : 28, 36, 117

Godet, Alain: „Nun was ist die Imagination anderst als ein Sonn im Menschen", Zürich 1982 : 27, 35

Goethe, Maximen und Reflexionen, Nr. 188 : 15

Gross, James, J.(Hg.): Handbook of Emotions³, New York London 2008 : 97

Grün, Anselmus, Spiritualität und Depression, in: Schweizer Archiv für Neurologie und Psychiatrie 160 (2009), S. 182–187 : 167

Grunewald, Eckard: „Der túfel in der helle ist úwer schlaf geselle", in Dinzelbacher/ Bauer (Hg.): Volksreligion, S. 130–143 : 103

Guindon, Josée and Hohmann, A.G. Pain Mechanisms and Measurement, in: Berntson/ Cacioppo II, S. 638–643 : 164

Guionnet, Sophie et al.: Reciprocal imitation: Toward a neural basis of social interaction, in: Cereb. Cortex (2011) doi: 10.1093/bhr177 : 68

Haeseli, C.M.: Sprachmagische Texte des Clm 536, in: Herberichs, Cornelia und Christian Kiening (Hg.): Literarische Performativität, Zürich 2008, S. 63–81 : 41

Haeseli, Christa M.: Magische Performativität, Würzburg 2011 : 18, 41, 61

Halsband, Ulrike, Neurobiologie der Hypnose, in: Revenstorf / Peter, Hypnose, S. 809f : 136, 142, 175

Hampp, Irmgard: Beschwörung Segen Gebet, Stuttgart 1961 : 93, 128, 133, 170, 171

Han, Shihui, Qin, J. and Ma,Y. „ Neurocognitive processes of linguistic cues related to death, in: Neuropsychologia 48 (2010), S. 3436–3442 : 166

Han, Shihui and E. Pöppel (Hg.:) Culture and Neural Frames of Cognition and Communication, Heidelbg. 2011 : 74

Handbuch Religionswissenschaftl. Grundbegriffe 1998 : 18

Happich, Carl, Das Bildbewußtsein als Ansatzstelle psychischer Behandlung, in: Zentralblatt für Psychotherapie und ihre Grenzgebiete 5 (1932), S. 663–677 : 34

Hauner, K.K. et al., Exposure therapy triggers lasting reorgaization of neural fear processing, in: Proc. Natl. Acad. Sci USA 2012 Jun 5; 109(23): :9203–8. Epub 2012 May 23 : 156

Hauschild, Thomas: Der böse Blick, Berlin 1982² : 23, 119

Heinzle, Joachim (Hg.): Literarische Interessenbildung im Mittelalter, Stuttgart/ Weimar 1993 : 79

Hempfer, Klaus, in: Hempfer, K.W. und J. Volbers (Hg.:) Theorien des Performativen, Bielefeld 2011 : 40

Henningsen, Peter et al.: Neuropsychosomatik. Grundlagen und Klinik neurologischer Psychosomatik, Stuttgart 2006 : 164

Hermann, Fritz, Double-bind und Linguistik. Zur quasi-magischen Zerstörung von Persönlichkeit durch kommunikative Paradoxien, in: Lange-Seidl, Annemarie (Hg.): Zeichen und Magie, Tübingen 1988, S. 57–69 : 129

Hinz, Eike, Cognitive Structures and Processes in Verbal Magic, in: Sociologus 28 (1978) : 39

Hirsch, Joy et al., Fleeting images of fearful faces reveal neurocircuitry of unconscious anxiety, EurekAlert, 17–Dec–2004 : 119

Hofstätter, Peter Robert: Einführung in die Sozialpsychologie, Stuttgart 1963 : 60

Hole, Günter: Direkte Induktionen, in: Revenstorf/ Peter: Hypnose 2009, S. 182 : 58

Holland, Alisha C. and Elizabeth A. Kensinger, The neural correlates of cognition reappraisal during emotional autobiographical memory recall, in: Journal of Cognitive Neuroscience 25 (2013), S. 87–108 : 96

Holzmann, Verena: „Ich beschwör dich wurm vnd wyrmin …". Bern u. a. 2001 : 93

Hüther, Gerald, Pränatale Einflüsse auf die Hirnentwicklung, in: Krens, Inge und Hans (Hg.): Grundlagen einer vorgeburtlichen Psychologie, Göttingen 2005 : 145

Ito, A. et al., The contribution of the dorsolateral prefrontal cortex to the preparation for deception and truth-telling, in: Brain Research 1464 (2012), S. 43–52 : 51

Janota, Johannes et al. (Hg.): FS Haug/ Wachinger, Tübingen 1992 : 103

Jaramillo S, Zador A M The auditory cortex mediates the perceptual effects of acoustic temporal expectation, Nature Neuroscience (2010), doi: 10.1038/nn.2688 : 50

Jaspers, Karl: Allgemeine Psychopathologie, Berlin u. a. 1948 : 85

Jeannerod, Marc and Thierry Anquetil: Putting oneself in the perspective of the other: A framework for self-other differentiation, in: Neuroscience 2008, 3(3–4), 356–367 : 70

Jimenez-Ortega, Laura et al., How the content of discourse affects language comprehension (2012) PLoS ONE 7(3): e33718. doi:101371/j. pone.0033718 : 131

Jinkyung Na and Shinobu Kitayama, Spontaneous trait inference is culture-specific, in: Psychological Science 22,8 (2011), S. 1025–1032 : 74

Josse, Goulven et al., The brain's dorsal route for speech represents word meaning: Evidence from gesture (2012), PLoS One 7(9): e46108. doi: 10.1371/j.pone.0046108 : 140, 156

Jung, Richard: Neurophysiologie und Psychiatrie, in: Psychiatrie der Gegenwart, Bd. I/Ia, Hg. von H.W.Gruhle et al., Berlin Heidelberg New York 1967 : 35

Kaiser, Otto (Hg.) TUAT (Texte aus der Umwelt des Alten Testaments), II, Gütersloh 1986–1991 : 159, 211

Kandyba, Kristina: The Modification and Creation of Memories in Regression to early Childhood and Uterus. A Thesis, Montreal 1996 : 145

Kapogiannis, Dimitrios et al., Neuroanatomical variability of religiosity (2009), PLoSOne 4(9): e7180. : 84

Kaptchuk, Ted and Jensen, Karin, Physicians' brain scans indicate doctors can feel their patients' pain – and their relief, EurekAlert 29–Jan–2013: 62

Kawabata H, Zeki S (2008) The neural Correlates of Desire. PLoS One 3(8): e3027 doi 10.1371/j : 49

Khaleb et al.: On the origin of the N400 Effects: An ERP waveform and source localization analysis in three matching tasks, in: Brain topography 23 (2010), 311–320 : 130

Kieckenhefer, Richard: Forbidden Rites, A Necromancer's Manual, Pennsylvania State Univ. Press 1979 : 100

Kiening, Christian, Mediologie – Christologie. Konturen einer Grundfigur mittelalterlicher Medialität, in: Das Mittelalter 15 (2010), S. 16–32 : 81

Kippenberg, Hans G. und Brigitte Lucchesi (Hg.:) Magie. Die sozialwissenschaftliche Kontroverse über das Verstehen fremden Denkens, Frankfurt/M. 1978 : 39

Klucharev, Vasily, in: EurekAlert 14–Jan–2009 : 79

Knape, Joachim, Rez. zu Heimann-Seelbach, S., Ars und scientia. Genese, Überlieferung und Funktionen der mnemotechnischen Traktatliteratur im 15. Jh., in: Zeitschrift f. dt. Altertum und dt. Lit., 134 (2005), S. 123–128 : 79

Könneker, Carsten: (Hg.): Wer erklärt den Menschen? Frankfurt/M, 2006 : 71

Kosslyn, S M et al., Mental Imagery, in: Berntson/ Cacioppo Handbook, S. 383–394 : 138

Krendl, Anne C. und Heatherton Todd F., Self versus Others/Self-Regulation, in: Berntson/ Cacioppo, Handbook II, Kap. 44, S. 859–875 : 75

Kringelbach, L. Morton, Neural basis of mental representations of motivation, emotion and pleasure, in: Berntson/ Cacioppo, Handbook II, Kap. 41, 807–828 : 47

Krohn, A.J., Borderline ‚empathy' and differentiation of object representation: a contribution to the psychology of object relations, in: Intern. J. of Psychoanal. Psychoth. 3 (1974), 142–165 : 119

Kropp, Amina: Sprachliche Betrachtungen zu den lateinischen Defixionum tabellae, in: Acta antiqua Academiae Scientiarum Hungaricae 49 (2009), S. 77–93 : 18, 41

Kross, Ethan et al. EurekAlert 28–Mar–2011 : 166

Krüger, Klaus, Bilder als Medien der Kommunikation, in: Spieß, Karl Heinz (Hg.) (2003), S. 155–204 : 77

Kuhl, Brice A. and Wagner, Anthony D., Forgetting and Retrieval, in: Berntson/ Cacioppo I, S. 586–605 : 146

Kühn, Simone et al.(2011), „Keep calm and carry on", in: PLoS One 6(1): e16569.doi:10.1371/j.pone : 45

Kulkarni et al. Attention to pain localization and unpleasantness discriminates the functions of the medial and lateral pain system, in: Eur. J. Neurosci. (2005), Jun 21 (11), 3133–42 : 164

Kumaran, Dharshan und Eleanor A. Maguire, A unexpected sequence of events: mismatch detection in the human hippocampus (2006) doi: 10/1371/j.pbio. 0040424 : 131

Lacey, Simon et al., Metaphorically feeling: Comprehending textural metaphors activates somatosensory cortex, in: Brain & Language 120,3 (2012), S. 416–421 : 139

Lange-Seidl, Annemarie (Hg.:) Zeichen und Magie. Kolloquium der Bereiche Kultur und Recht der deutschen Gesellschaft für Semiotik, 5.9.1986 München, Tübingen 1988 : 129

LeDoux, Joseph, Rethinking the emotional brain, in: Neuron 73 (2012), 653–676 : 52

Ledoux, Kerry, (2008) Online studies of repeated name and pronominal coreference : 155

Lee, H. et al., Amygdala-prefrontal coupling underlies individual differences in emotion regulation, in: Neuroimage 62 (2012), 1575–1581 : 95

Leipoldt, Joh. und Siegfried Morenz: Heilige Schriften, Leipzig 1953 : 124

Lentes, Thomas: Inneres Auge, äußerer Blick und heilige Schau, in: Schreiner, K. (Hg.): Frömmigk. S. 185f : 79

Leuzinger-Bohleber, M., Roth, G. und A. Buchheim: Psychoanalyse, Neurobiologie, Trauma. Stuttgart u. a. 2008 : 145

Lévi-Strauss, Claude: Marcel Mauss – Sociologie et Anthropologie, Introduction, Paris 1968 : 34

Lévi-Strauss, Claude: Strukturale Anthropologie, I, Paris 1958, (dt.:) Frankfurt/M. 1981² : 24, 35, 85, 148, 153

Lévi-Strauss, Claude: Das wilde Denken. Frankfurt/M. 1986 [Paris 1962] : 32, 33, 133

Lévy-Bruhl, Lucien: Les fonctions mentales dans les sociétés inférieures, Paris 1928 : 30

Lieberman, Matthew D. et al., Putting feelings into words: Affect labeling disrupts amygdala activity to affective stimuli, in : Psychological Science 18 (2007), 421–428 : 94, 95, 97

Lieberman, M.D. et al., (2011) Subjective responses to emotional stimuli during labeling, reappraisal and distraction, in : Emotion 11, 468–480 : 94, 95

Lohse, Martin J., Notrufe und andere Signale. Die Sprache der Zellen, in: Universitas 55 (2000), S. 718f : 47

Lubrich, Oliver, Figuralität und Persuasion, in: Paragrana 20 (2011), S. 248–265 : 150

Lupyan, Gary and Michael J. Spivey, (2010) Making the invisible visible: Verbal but not visual cues enhance visuel detection, PLoS One 5(7), doi: 10.1371/j.pone.0011452 : 50

Mano Y et al., Perspective-taking as part of narrative comprehension: a functional MRI study, Neurophysiologia 2009 Feb;47(3): 813–24 : 143

Martinez-Jauand, M. et al., Somatosensory activity modulation during observation of other's pain and touch, in: Brain Research 1467 (2012), 48–55 : 67

Mason, R.A. und J.A. Just, The Role of the Theory-of-Mind cortical network in the comprehension of narratives, in: Lang. Linguist. Compass (2009) Jan 1;3(1):157–174; Abstract PMID: 19809575 : 134

Mathieu Roy et al., Cerebral and spinal modulation of pain by emotions, PNAS (2009), doi:10.1073/pnas0904706106 : 165

Maturana, Humberto R., in: Riegas, Volker und C. Vetter: Zur Biologie der Kognition, Frankfurt/M, 1993³ : 45

Mayer, Emeran A., Gut feelings: The emerging biology of gut-brain communication, in: Nature Reviews Neuroscience 12 (2011), 453–466 : 166

Meier, Gerhard: Die assyrische Beschwörungssammlung Maqlû, Berlin 1937, in: Archiv für Orientforschung, Beiheft 2, Neudruck Osnabrück 1967 : 112

Meier, Heinrich und Detlev Ploog (Hg.:) Der Mensch und sein Gehirn, München u. a. 1998² : 129

Meisen, Karl, Der böse Blick und anderer Schadenzauber, in: Rhein. Jahrb. f. Volkskunde 1(1950), 3(1952) : 117

Meiss, Ortwin, Hypnotherapeutische Verfahren in der Arbeit mit depressiven Patienten, in: Revenstorf/ Peter: Hypnose, S. 502: 103

Meyer-Lindenberg, Andreas, Tief verwurzeltes Statusdenken, in: Spektrum der Wissenschaft März 2009, S. 19 : 173

Mocaiber, T.A. et al. Antecedent descriptions change brain reactivity, (2011).doi.org/10.1016/j.neuroscience : 98

Mongardini, Carlo, Über die soziologische Bedeutung des magischen Denkens, in: Zingerle, Arnold und Mongardini, Carlo (Hg.:) Magie und Moderne, Berlin 1987 : 24

Monti, Martin M. and D.N. Osherson, Logic, language and the brain, in: Brain research 1428 (2012), S. 33–42 : 20

Mora, Francisco et al., Stress, neurotransmitters, corticosterone and body-brain integration, in: Brain Research 1476 (2012),71–85 : 176

Müller, Notger et al., Repetition Suppression versus Enhancement – It's Quantity that matters, in: Cerebral Cortex 23 (2013), S. 315–322 : 155

Nahab, Fatta B. et al., The neural processes underlying self-agency, in: Cerebr. Cortex (2011) 21(1): 48–55 : 70

Norenzayan, Ara et al., Memory and mystery: The cultural selection of minimally counterintuitive narratives, in: Cognitive Science 30 (2006), S. 531–553 : 130

Nikki L. Pratt and Spencer D. Kelly, Emotional states influence the neural processing of affectiv language, in: Social Neuroscience 3 (3–4), (2008), S. 434–442 : 129

Ochsner K N, Bunge S A, Gross J J, Gabrieli J D E, Rethinking feelings: An fMRI study of the regulation of emotion, Journal of cognitive neuroscience 14 (2002), 1215–1220 : 95
Ohrt, Ferdinand, Ungerechter Mann (Segen) in: Handbuch des deutschen Aberglaubens VIII,1416 : 127
Olff, Miranda, Bonding after trauma: on the role of social support and the oxytocin system in traumatic stress, in: European Journal of Psychotraumatology 3(2012) 18597–doi.org/10.3402/ejpt.v3io. : 176
Önnerfors, Alf: Antike Zaubersprüche, Stuttgart 1991 : 42, 128
Oppenheim, A., Leo: The interpretation of dreams in the ancient Near East, Philadelphia 1956 : 108

Palmer, Nigel F., Von den naturlichen troymen, in: Janota, Johannes et al. (Hg.): FS Haug/ Wachinger : 103
Peter, Burkhard, Therapeutisches Tertium und hypnotische Rituale, in: Revenstorf/ Peter: Hypnose, S. 70–77 : 59
Petrovic P, Kalso, E, Petersohn K M und Ingvar M (2002), Placebo and opioid analgesia – Imaging a shared neuronal network, in: Science 295 (5560), 1737–1740 : 51
Pettersson, Olof, Magie – Religion. Some marginal notes to an old problem (1957), deutsche Übersetzung in: Petzoldt, L. (1978) : 26
Petzold, Antje: Auswirkungen ak. psychosoz. Stresses auf Feedback-basiertes Lernen, Dresden 2010, S. 14–17 : 176
Petzoldt, Leander (Hg.:) Magie und Religion, Darmstadt 1978 : 24, 26, 30
Petzoldt, Leander, Magie und Religion, in: Dinzelbacher/ Bauer (Hg.) Volksreligion (1990)
Platon, Charmides 157b; Phaidon75e : 17, 143
Plett, F. Heinrich: Einführung in die rhetorische Textanalyse, Hamburg 1991 : 139
Ploner, Markus et al., Flexible cerebral connectivity patterns subserve contextual modulations of pain, (2011) doi: 10.1093/cercor/bhq146 : 165
Pollo A, Amancio M et al. Response expectancies in placebo analgesia and their clinical relevance, in: Pain 2001 Jul; 93(1):77–84 : 50
Posner, Michael I. et al., Attention and the detection of signals. In: Journal of experimental psychology, general. 109 (1980), 160–174 : 53

Radner, Karen and Eleanor Robson (Hg.:) The Oxford Handbook of cuneiform culture, Oxford 2011 : 90
Raposo, A. et al., Repetition suppression and semantic enhancement: An investigation of the neural correlats of priming, in: Neuropsychologia 44 (2006), S. 2284–2295 : 155
Rapp, A. et al., Neural correlates of metonymy resolution, in: Brain & Language 119 (2011), S. 196–205 : 138
Rätsch, Christian: Bilder aus der unsichtbaren Welt, München 1985 : 153, 218
Raz, Amir, Varieties of attention, in: Berntson/ Cacioppo Handbook I, 361–369 : 54
Redcay, Elizabeth et al., Live face-to-face interaction during fMRI: A new tool for social cognitive neuroscience, Neuroimage 50 (2010), S. 1639–47 : 68
Reichertz, Jo: Kommunikationsmacht, VS Verlag für Sozialwissenschaften, 2009 : 18, 43, 46, 57, 74, 82
Revenstorf, Dirk und Peter, Burkhard: Hypnose, Heidelberg 2009 : 58, 59, 103, 136, 145, 174
Revenstorf, D. und U. Freund, in: Revenstorf/ Peter, S. 211 : 145, 174

Richter, Gisela, „Auf dem Birnbaum ohne Blätter ...", in: Innsbrucker Beiträge zur Kulturwissenschaft, Germanistische Reihe Bd. 57 (1997), S. 375–378 : 125

Richter, M., Miltner, W.H.R. und Weiss, T., Schmerzwörter aktivieren schmerzverarbeitende Hirnareale, in: Der Schmerz, 25 (2011), S. 322–324 : 165

Ripper, Kathrin: Hypnose und Geburt, Diss. Tübingen 2003 : 147

Rizzolatti, Giacomo und Corrado Sinigaglia: Empathie und Spiegelneurone. Die biologische Basis des Mitgefühls, Frankfurt/ M. 2008 : 65, 67, 69, 164

Röckelein, Hedwig, Nonverbale Kommunikationsformen- und -medien beim Transfer von Heiligen im Frühmittelalter, in: Spieß, Karl-Heinz (Hg.) S. 83–104 : 78

Röhrich, Lutz: Lexikon der sprichwörtlichen Redensarten, Freiburg 2001 : 65

Roth, Gerhard und Nicole Strüber, Pränatale Entwicklung, in: Cierpka, Manfred (Hg.:) Frühe Kindheit, S. 4ff : 145

Roth, Gerhard: Das Gehirn und seine Wirklichkeit, Frankfurt/ M., 1998^2 : 19, 54, 71, 84, 142

Roth, Gerhard: Fühlen, Denken, Handeln. Frankfurt/ M. 2001 : 47

Rugg, M.D., Event-related potentials in phonol. matching tasks, in: Brain & Language 23 (1984), 225–240 : 131

Sander, Kerstin et al.: fMRI Activations of Amygdala, Cingulate Cortex and Auditory Cortex by Infant Laughing and Crying, in: Human Brain Mapping 28 (2007), 1007–1022 : 48

Schippers, Marleen B. et al. Mapping the information flow from one brain to another during gestural communication, doi 10.1073/pnas.1001791107 (2010) : 68, 140

Schmidbauer, Wolfgang: Helfen als Beruf. Hamburg 1983 : 58

Schmidbauer, Wolfgang: Vom Umgang mit der Seele, München 1998 : 31

Schott, Heinz und Rainer Tölle: Geschichte der Psychiatrie, München 2006 : 45

Schramm, Wolfgang: Ein Compendium sumerisch-akkadischer Beschwörungen, Göttingen 2008 : 88

Schreiner, Klaus (Hg.) Frömmigkeit im Mittelalter, München 2002 : 77, 78, 79, 161, 200

Schreiner, K.: Soziale, visuelle und körperliche Dimensionen mittelalterl. Frömmigkeit, in: Schreiner (2002) : 77

Schröder, Franz Rolf, in: Germanisch-Romanische Monatsschrift 4 (1954), S. 179–185 : 150

Schrott, Raoul und Arthur Jacobs: Gehirn und Gedicht, München 2011 : 15, 53, 64, 83, 106, 129, 140

Schulz, Monika, Nigon wyrta galdor. Zur Rationalität der ars magica, in: Zeitschrift für Literaturwissenschaft und Linguistik 130 (2003), S. 8–24 : 160

Schulz, Monika: Beschwörungen im Mittelalter, Heidelberg 2003 : 128, 170

Schulz, Monika: Magie oder Die Wiederherstellung der Ordnung, Frankfurt u. a. 2000 : 37, 104, 115, 133

Schulze, Lars et al. Neuronal correlates of cognitive reappraisal in Borderline patients, Biolog. Psychiatry (2010), doi:10.1016/j.biopsych. : 121

Schupp, Volker (Hg.) Deutsches Rätselbuch, Stuttgart 1972 : 124

Schwemer, Daniel, Magic rituals: conceptualisation and performance, in: Radner/Robson: Handbook S. 418–442 : 57, 86, 112

Schwemer, Daniel: Abwehrzauber und Behexung, Wiesbaden 2007 : 57, 118, 207

Schwemer, Daniel: Rituale und Beschwörungen gegen Schadenzauber Wiesbaden 2007 : 89, 108

Scott D J, Zubieta JK et al., (2007) Individual differences in reward responding explain placebo-induced expectations and effects, in: Neuron 55,325–336 doi 10.1016/j : 50

Scott, L.N. et al., (2011) Mental state decoding abilities in young adults with borderline personality disorder traits. Personality Disorders: Theory, Res Treatment 2:98–112 : 121

Seligmann, Siegfried: Der böse Blick, Berlin 1910 : 117

Seligmann, Siegfried: Die Zauberkaft des Auges und das Berufen, Hamburg 1922 : 117

Semino, Elena, Descriptions of pain, metaphors and embodied simulation, in: Metaphor and Symbol 25 (2010), issue 4, doi: 10.1080/10926488.2010.510926 : 164
Seminowicz, D.A. and K.D. Davis, Pain enhances functional connectivityof a brain network evoked by performance of a cognitive task, in: Journal of Neurophysiology 97, 5 (2007), S. 3651–9 : 166
Sigerist, Henry E.: Der Arzt in der Mesopotamischen Gesellschaft, Zürich 1963 : 87
Simor, Péter et al., Impaired executive functions in subjects with frequent nightmares, in: Brain and Cognition 78 (2012), S. 274–283 : 99
Singer, Wolf, FAZ 25.11.2004 : 19
Singer, Wolf, Unser Menschenbild im Spannungsfeld zwischen Selbsterfahrung und neurobiologischer Fremdbeschreibung, in: Frühwald, Wolfgang u. a.: Das Design des Menschen, Köln 2004, S. 182–215 : 19
Singer, Wolf, Was kann ein Mensch wann lernen?, in: Universitas 56 (2001), S. 882 : 20
Singer, Wolf, wie oben, Universitas 56 (2001), S. 888; Online: Vortrag 1. Werkstattgespräch Initiative McKinsey der Deutschen Bibliothek Frankfurt/M am 12. 6. 2001 : 73
Singer, Wolf: Der Beobachter im Gehirn, Frankfurt 2002 : 22
Slaby, Jan: Die Objektivitätsmaschine, in: Mertens, K. und I. Günzler (Hg.): Wahrnehmen, Fühlen, Handeln, Paderborn 2012 (Onlineankündigung) : 21
Soeter, M. and Kindt, M., Stimulation of the noradrenergic system during memory formation impairs extinction learning but not the disruption of reconsolidation, Neuropsychopharmacology 2012, 37(5), 1204–15 : 169
Sørensen, Jesper: A cognitive Theory of Magic, Lanham et al. 2007 (Ph. D. Diss. Aarhus 2000/01) : 28, 29
Spaemann, Robert: „Ich und Gott ein Hirngespinst" Kooperationsveranstaltung 13. und 14. Februar 2009, Bistum Augsburg : 21
Spamer, Adolf: Romanusbüchlein. Aus seinem Nachlaß bearbeitet von Johanna Nickel Berlin 1958 : 118
Spieß, Karl-Heinz (Hg.): Medien der Kommunikation im Mittelalter, Wiesbaden/ Stuttgart 2003 "77, 78
Spitzer, Manfred: Lernen. Heidelberg 2002 : 144
Staffelt, Sven: Einführung in die Sprechakttheorie. Tübingen 2008 : 41
Steer, Georg, Bettelorden-Predigt als ‚Massenmedium', in: Heinzle, Joachim (Hg.) (1993) S. 314–336 : 79
Stein, Claudius: Spannungsfelder der Krisenintervention, Stuttgart 2009 : 145
Stephens, Greg J. et al., Speaker-listener neural coupling underlies successful communication, PNAS 107 (2010), 14425–14430; doi:10.1073/pnas.1008662107 : 69
Stewart-Williams S, Podd J (2004) The placebo effect: dissolving the expectancy versus conditioning debate, in: Psychol Bull 130(2):341–3 : 51
Stewart-Williams, Steve, Innate ideas as a naturalistic source of metaphysical knowledge, Biology and Philosophy 20 (2005), S. 791–814 : 30
Stoffers et al., Psychological therapies for people with borderline personality disorder, Cochrane Database Syst Rev. 2012 Aug 15;8:CD005652 : 122
Strassberg, Daniel, Scham als Problem der psychoanalytischen Therapie, in: Schweizer Archiv für Neurologie und Psychiatrie 155 (2004), S. 225–228 : 122
Strauß, Botho: Vom Aufenthalt, München 2009 : 21
Strelzyk, Florian et al., Tune it down to live it up? Rapid, nongenomic effects of cortisol in human brain, in: Journal of Neuroscience 32 (2012), S. 616 : 176
Summerfield, C., et al., Neural repetition suppression reflects fulfilled perceptual expectations, in: Nature Neuroscience 11 (2008), doi:10.1038/ nn.2163 : 154

Tambiah S.J., Form und Bedeutung magischer Akte, in: Kippenberg/ Lucchesi (Hg.) Magie (1978) : 39
Tambiah, S.J., Magic, Science, Religion, and the Scope of Rationality, Cambridge 1990 : 39
Tambiah, Stanley Jeyaraja (1968) The magical power of words, Man 3, 175–208 (189) : 39
Telle, Joachim: Petrus Hispanus in der altdeutschen Medizinliteratur, Heidelberg 1972 : 45
Thakkar, N.K. et al., (2009) Exploring empathic space, PLoS ONE 4(6): e5864.doi:10.1371/j : 66
Thomsen, M.-L.: Zauberdiagnose und schwarze Magie in Mesopotamien, Museum Tusculanum Press 1987 : 89
Tolentini, L.C., Are pumpkins better than heaven? An ERP investigation of order effects in the concret-word advantage, in: Brain & Language 110,1 (2009), S. 12–22 : 136
Tomasek, Tomas: Das deutsche Rätsel im Mittelalter, Tübingen 1994 : 125
Tomasello, Michael: Die Ursprünge der menschlichen Kommunikation, (dt.:) Frankfurt/M. 2009 : 47, 49, 140
TUAT siehe Kaiser, Otto (Hg.) Texte
Turella, M. et al., Mirror neurons in humans: consisting or confounding evidence?, in: Brain & Language 108, 1 (2009), S. 10–21 : 70

Ünal, Ahmet, Hethitische Hymnen und Gebete, in: Kaiser, Otto (Hg.) TUAT, S. 791 : 159
Urgesi, Cosimo et al: The spiritual brain: Selective cortical lesions modulate human self-transcendence, in: Neuron 65 (2010), S. 309–319 : 84

Van Lancker Sidtis, Diana, Two-Track Mind: Formulaic and novel language support a dual-process model, in: Faust, Miriam (Hg.:) Handbook, S. 342–367 : 134
Vanderhasselt, Marie-Anne et al., Interindividual differences in the habitual use of cognitive reappraisal, in: Biological Psychology (2012) doi.org/10.1016/j : 96
Vanja, Christa, Krankheit im Mittelalter, in: Dinzelbacher, P. (Hg.) Mentalitätsgeschichte (1993), 195–200 : 48
Villemure, Chantal and Bushnell, M.C., Mood influences supraspinal pain processing separately from attention (2009) doi: 10.1523/JNEUROSCI.3822-08.2009 : 165
Vogeley, Kai und Albert Newen, Ich denke was, was du nicht denkst, in: Könneker, C. (Hg.) (2006), 59–73 : 71

Waerzeggers, Caroline, The Pious King: Patronage of Temples, in: Radner/Robson, Handbook, S. 746 : 86
Wager T D, Nitschke J B Placebo effects in the brain: Linking mental and physiological processes, in: Brain, Behavior and Immunity 19 (2005), 281–282 : 51
Walter, Henrik: Neurophilosophie der Willensfreiheit, Paderborn u. a. 1997 : 46
Watzlawick, Paul und Giorgio Nardone (Hg.:) Kurzzeittherapie und Wirklichkeit, München 2012 : 19
Watzlawik, Paul: Wie wirklich ist die Wirklichkeit?, München 1976 : 125
Weddige, Hilkert: Einführung in die Mediävistik, München 1992² : 25
Wehr, Michael et al., Researchers find how brain hears the sound of silence, EurekAlert 10–Feb–2010 : 151
Weiss, Thomas, Psychophysiologische Aspekte des Placeboeffekts bei Schmerz, in: Zeitschrift für Neuropsychologie 15(2), 2004, S. 99–110 : 51
Weiß, Thomas, Worte fügen Schmerzen zu, EurekAlert 30–Mar–2010 : 168
Wencai (2011) The integrative effect of cognitive reappraisal on negative affect, PLoS One 7(2):e30761 : 96

Wencai Zhang et al., Neural mechanism of placebo effects and reappraisal in emotion regulation, in: Progress in Neuro-Psychopharmacology and Biological Psychiatry 40 (2013), S. 364–379 : 96

Wendt, J. et al. Mechanisms of change: effects of repetitive exposure to feared stimuli in the brain's fear network, in: Psychophysiology (2012) Oct; 49(10): 1319–29 : 156

Wenzel, Horst: Hören und Sehen, Schrift und Bild. Kultur und Gedächtnis im Mittelalter, München 1995 : 77

Werbow, Stanley Newman: Martin von Amberg. Der Gewissensspiegel, Berlin 1958 : 157

West, W.C. Holcomb, P.J., Imaginal, semantic, and surface-level processing of concret and abstract words: an eletrophysiological investigation, in: Journal of Cognitiv Neuroscience 12, 6 (2000), S. 1024–37 : 136

Wilder-Smith, Clive H., The balancing act: endogenous modulation of pain in functional gastrointestinal disorders, in: GUT 60 (2011), S. 1589–1599 : 166

Winckelmann, J., Die Herkunft von Max Webers „Entzauberungs"-Konzept, in: Kölner Zeitschrift für Soziologie und Sozialpsychologie 32 (2005), S. 12–53 : 80

Wolf, Christoph, Digi 20 der Bayerischen Staatsbibliothek : 40

Woods D L et al. (2009) Functional maps of human auditory cortex: effects of acoustic features and attention, doi: 10.1371/journal pone 0005183 : 50

Xiaosi Gu et al., Anterior insular cortex is necessary for empathic pain perception, Brain 135 (2012), 2726–2735, doi: 10.1093/brain/aws199 : 166

Yamamoto, S. Humle, T. and Tanaka, M.: Chimpanzees help each other upon request (2009) PlosONE 4(10): e7416, doi:10.1371 : 49

Yomogida, Yukihito et al.: The neural basis of agency: An fMRI study, in: Neuroimage 2009.12.054, doi:10.1016 : 70

Zaboura, Nadia: Das empathische Gehirn, Vs-Verlag 2009 : 72

Zingerle, Arnold und Carlo Mongardini (Hg.:) Magie und Moderne, Berlin 1987 : 24

Zgoll, Annette: Traum und Welterleben im antiken Mesopotamien, Münster 2006 : 64, 108

F2 Verzeichnis der Beispieltexte

D1 Marcellus, Europäisches Mittelalter, Neuzeit bis 19. Jh.

Beispieltext 1: Der Münchner Nachtsegen (BSB München, Clm 615, 127r, 14.Jh.) Hist.-krit. Edition nach T. von Grienberger, in: Z. f. das dt. Altertum 41 (1897), S. 335–363

Beispieltext 2: Elsässisches Nachtgebet (Univ.bibl. Straßburg L. germ 645, S. 15f, 15.Jh.)

Beispieltext 3: Wider die Elben (BSB München Clm 849, 131v, 15. Jh.), ed. Schönbach, Analecta Graec., S. 43

Beispieltext 4: Ein Segen fur den Alp, aus dem 12(13)-bändigen Buch der Medizin des Pfalzgrafen Ludwig V. bei Rhein (Universitätsbibliothek Heidelberg CPG 271, S. 229, 16. Jh.)

Beispieltext 5: Gebete und Segen aus Muri (Codex membr. 69 des Benediktinerkollegiums Sarnen), ed. Wilhelm, Friedrich: Denkmäler deutscher Prosa des 11. und 12. Jahrhunderts, München 1960, S. 73

Beispieltext 6: Zwei böse Augen, (Urteilsbuch des Rostocker Niedergerichts von 1539–1586) var. nach ed. Spamer/ Nickel: Romanusbüchlein, S. 112

Beispieltext 7: O jesu marien sun ... daz si erstarren, (München BSB Cgm 54, f.106v,14.Jh.)

Beispieltext 8: Falsche Augen (Wunsiedel, Fichtelgebirgsmuseum, Handschrift 213/2823 des Bauern Ächtner 1769, Seite 31), ed. Ernst, Wolfgang: Beschwörungen und Segen, S. 237

Beispieltext 9: Steht ein Baum in Meeresmitte (Marcellus Burdigalensis, De medicamentis 28,74, 4./5.Jh.), ed. Heim, Ricardus: Incantamenta magica, Leipzig 1892, S. 496, Nr. 107

Beispieltext 10: Gegen Kollern (Marcellus Burdig., De medicamentis 21,3) ed. Niedermann, M. und E. Liechtenhan mit deutscher Übers. J. Kollesch und D. Nickel, Berlin 1968; zit. nach Önnerfors, Alf: Antike Zaubersprüche, Stuttgart 1991, S. 23

Beispieltext 11: Vogel ohne Federn (BLB Karlsruhe, Cod. Aug. CCV), ed. Eis, Gerhard: Altdeutsche Zaubersprüche, Berlin 1964, S. 68

Beispieltext 12: Vogel ohne Lung (ed. Bartsch, Karl: Sagen, Märchen und Gebräuche aus Mecklenburg, gesammelt und herausgegeben Wien 1879, Band 2, S. 396, 19. Jh.)

Beispieltext 13: Bärmutter, leg dich (Sefer Hasofoth, hebräischer Codex, Bl. 88v, 14.Jh.) ed. Holzmann, Verena: „Ich beswer dich, wurm ...", Bern 2001, S. 141; vgl. Müller, Alois, Zeitschr. f. dt. Altertum 19 (1976), S. 474f

Beispieltext 14: Vor das roth ... Himmelstür verschlossen (Pfuhler Hausbuch, um 1800) ed. Kopp, Andreas, in: Ulmer Kulturanthropol. Schriften, Band 10, 1998, S. 134

Beispieltext 15: Steh Blut wie des Müllers Seel (Kunstbuch Johannes Zahn von Dürnberg (vor 1691), S. 103), ed. Ernst,W., in: Archiv für Geschichte von Oberfranken 76 (1996), S. 187

Beispieltext 16: Vor unholden (Württemb. Landesbibl. Stuttgart Cod. med. et phys. 4°, Nr. 29, Bl. 111a, 15. Jh.), ed Pfeiffer, Anz. f. d. Kunde d. dt. Vorzeit NF 1 (1853/54), S. 36

Beispieltext 17: Kaltes Gesicht, Kieselsteine fressen, ed. Ammann, J. J., Volkssegen aus dem Böhmerwald I, in: Zs. des Vereins für Volkskunde 1(1891), S. 208, 19. Jh.

Beispieltext 18: Mora ... zähl die Sterne, (Dalmatinisch, 19. Jh.), ed. Krauß, Friedrich: Slavische Volksforschungen. Abh. über Glauben, ... Gewohnheiten und die Guslarenlieder der Südslaven, Leipzig 1908, S. 150

Beispieltext 19: Marta stand auf der Brücke (Landesbibliothek Luxemburg, Cod. 109, f.ult., gem. Katalog 239, aus Echternach, ed. Jacoby, Adolf: Segenstexte aus Luxemburger Handschriften (1923), S. 17

Beispieltext 20: Ihr wirwelein, Kunstbuch des Joh. Zahn, S. 92/93), ed. Ernst, W. ebd., S. 180

Beispieltext 21: Christ und Judas spielten mit Spießen (Staatsbibl. Bamberg Msc. Med. 6 fol. 139rb, 12./13.Jh.)

Beispieltext 22: Die Sieben Schläfer ...wie im Mutterschoß (Bayerische Staatsbibliothek München, Clm 14179, f.1', 9. Jh., aus St. Emmeram Regensburg), ed. Franz, A.: Die kirchlichen Benediktionen im Mittelalter II, 480

Beispieltext 23: Einst saßen Idisi, (Domstiftsbibl. Naumburg, ehemals Merseburg Hs 136, fol. 85R, 9.Jh.) – a) ed. Beck, Wolfgang: Die Merseburger Zaubersprüche, Wiesbaden 2003, b) Riesel, Elise, Der erste Merseburger Zauberspruch, in: Deutsches Jahrb. für Volkskunde 4 (1958), S. 53–58

Beispieltext 24: Drei Engel begegneten ‚Gichtern', (Zentralbibliothek Zürich, Ms Rheinauer Codex 67, fol. 47, 12. Jh.), ed. Steinmeyer, Elias v., Z. für dt. Altertum 22 (1878), S. 246

Beispieltext 25: Ich beschwöre dich Bärmutter, (Stiftsbibl. St. Gallen, Codex 752, fol. 159f, 10. Jahrhundert) ed. Bernfeld, Werner, in: Kyklos. Jb. des Instituts für Geschichte der Medizin Univers. Leipzig 2 (1929), S. 272–274

Beispieltext 26: Wodan: ben zi bena, bluot zi bluoda! (Naumburg, Domstiftsbibliothek, ehemals Merseburg, Domstiftsbibl., HS 136, fol.85R, 9.Jh.), ed. Beck, Wolfgang: Die Merseburger Zaubersprüche, Wiesbaden 2003

Beispieltext 27: Geh heraus, Nesso! (BSB München Clm 18524b, 203v, Digitalisat 417, Wurmbeschwörung aus Kloster Tegernsee, 10. Jh.)

Beispieltext 28: Im Namen des Vaters such' ich dich, (BSB München, Clm 7021, 167d, Rosenminzen-Ansprechformel gegen Hexerei und Spuk), ed. Schönbach, Anton E.: Studien zur Geschichte der altdeutschen Predigt, Hildesheim 1968, S. 146f

Beispieltext 29: Zur Entfernung des Talpa-Wurmes (Stadtbibliothek Trier, Hs40/1018 8°, fol. 41v–43r, 10./11. Jh.), ed. Embach, Michael, Trierer Zauber- und Segenssprüche des Mittelalters, Kurtrier. Jahrb. 44 (2004), S. 48f

Beispieltext 30: Hiob lag im Mist, (BSB München, Clm 536, fol. 84a, 12. Jh., Wurmsegen aus Kloster Prüll, Regensburg) ed. Müllenhoff/ Scherer: Denkmäler I, S. 181

Beispieltext 31: Haselzweig zu segnen (Handschrift DON 792, fol. 143v–144r, 15.[?] Jh. der Badischen Landesbibl. Karlsruhe), ed. Ernst, W. Beschwörungen und Segen, S. 57f

Beispieltext 32: Blicke auf das Haupt!, ed. D. Pezzini, Il sogno della croce e liriche del Duecento inglese sulla Passione, Parma 1992; zit. nach Dinzelbacher, P., in: Schreiner, K. (Hg.): Frömmigkeit, München 2002, S. 327

Beispieltext 33: Der Millstätter Blutsegen (Wien, ÖNB, Cod. 1705, f.32r, aus dem Benediktinerkloster Millstatt, 12. Jh.), ed.Müllenhoff /Scherer, Denkmäler, Berlin 1892, I. S. 180

Beispieltext 34: Kreuzanbetung (Innsbrucker Statthalterei-Archiv A 7. 29), 1400: ed. Prem, S.M.: Tirolischer Glaube und Aberglaube des 15. Jahrhunderts, in: Z. für dt. Altertum 36 (1892), S. 51f; ed. Holzmann, V., S. 263

Beispieltext 35: Drei gute Brüder begegnen Christus, (BSB München, Clm 23374 fol. 16b, 12. Jh.), ed. Müllenhoff, K., in: Zeitschrift für dt. Altertum 15 (1872), S. 454

Beispieltext 36: Für das Vergücht (Steiermärk. Landesarchiv Graz, Handschrift 476, Arznei- und Alchemiebuch des Matheus S. von 1587, S. 176–178)

Beispieltext 37: Thor schlage dich, Wundfieber! (Altnordische Runen-Handschrift von Canterbury), ed. Genzmer, Felix, Germanische Zaubersprüche, in: Germanisch-Romanische Monatsschrift NF 1(1950), S. 24

Beispieltext 38: Fliehe, Feind Satan! (CKC Coloniensis Capituli: Bibliothek des Domkapitels Köln 15, Bl. 96'–98', mbr., 9. Jh.) Großer Exorzismus für Besessene ed. Franz, Benediktionen II, S. 587–596 (wenige Zeilen)

Beispieltext 39: Bei dem Holze gegen Überbein (Paris, Bibliothèque Nationale de France, Nouv. acqu. lat. 229, 12. Jh.), ed. MSD Müllenhoff/ Scherer Denkmäler II, 1892, S. 305

D2 Mesopotamische Beispieltexte (Ausschnitte)

Beispieltext 40: Magst du ein böser Alû sein (CT= Cuneiform Texts from Babylonian tablets in the British Museum London, 16,27f, Zeile 34–45), ed. Falkenstein, Adam: Die Haupttypen der sumerischen Beschwörung, Leipzig 1931, S. 22

Beispieltext 41: Ein Löwe ... ein böses Auge (CT 16,25, Zeile 42ff) ed. Falkenstein, S. 48

Beispieltext 42: Tätigkeit des Beschwörungspriesters (CT 16,1–7 und CT 16,5, 180–193), ed. Falkenstein S. 29f

Beispieltext 43: Elinitu, die Lügnerische/ Abwehr nächtlicher Verzauberung (Großritual Maqlû Tafel 1, Zeile 1–36), ed. Tzvi Abusch/ Daniel Schwemer: Das Abwehrzauber-Ritual Maqlû, in: TUAT NF4 (2008), S. 136

Beispieltext 44: Assyrische Beschwörung gegen Nachtgeister (Großritual Maqlû Tafel 1, Zeile 135–143), ed. Tzvi Abusch/Daniel Schwemer TUAT NF4 (2008), S. 140

Beispieltext 45: Beschwörung gegen Verzauberung mit Hilfe Ea und Marduk (Großritual Maqlû Tafel 3, Zeile 39–59) ed. Abusch/ Schwemer TUAT NF4 (2008), S. 148f

Beispieltext 46: Beschwörung gegen Verzauberung der Körperteile (Großritual Maqlû Tafel II, Zeile 19–36), ed. Abusch/ Schwemer TUAT NF4 (2008), S. 141

Beispieltext 47: Gegen die Hexe, die in den Straßen umherschweift, (Großritual Maqlû III, Zeile 1–13) ed. Schwemer, D.: Abwehrzauber und Behexung, Wiesbaden 2007, S. 68

Beispieltext 48: Von der Erschaffung der Welt zur Zahnwurmbehandlung (Nebnadinirbu- [Niburu-] Keilschrift), ed. Weinberger, B. W.: An Introduction to the History of Dentistry, St.Louis (USA), 1948, I, S. 23, zit. nach Kobusch, Hellmuth: Der Zahnwurmglaube, Frankfurt/M 1955, S. 7 – vgl. ed. Clifford V, Incantations of the Minor Akkadian Cosmogonies 2005

Beispieltext 49: Entbindungszauber (Incantations in the neo-Sumerian period), ed. Graham Cunningham: ‚Deliver me from evil' 2500–1500 BC, Roma 1997, S. 71–72

Beispieltext 50: Beschwörung gegen Nachtgeister (Großritual Maqlû Tafel VII, Z. 114–144), ed. Abusch/ Schwemer TUAT NF4 (2008), S. 174f; Ritustafel VII, Zeile 138–156,185f.

Beispieltext 51: O Gott, ich suche nach deinem Mitleid, (Hethitische Hymnen und Gebete, an den Sonnengott, CTH 372), ed. Ünal, Ahmet (TUAT), Bd. II, S. 796/798/799

Beispieltext 52: Behandlung einer Kopfkrankheit (Bußliturgie mit Fürbitte des Priesters B 18, 1–21, Hymnen, Buß- und Klagepsalmen), ed. Falkenstein/ von Soden: Sumerische und Akkadische Hymnen, Zürich und Stuttgart 1953, S. 270, Nr. 18

D3 Beispieltexte aus den altindischen Atharvaveden

Beispieltext 53: Wachskraut, Wunden verwachsen zu machen. (Atharvaveda IV, 12, historische Übertragung durch Friedrich Rückert, aus seinem ungedruckten Coburger Nachl.), ed. Kreyenborg, H., Hannover 1923, S. 35

Beispieltext 54: Ich treib die Krankheit aus (Atharvaveda XX,96.17–22, ed. a) Kuhn, Adalbert, Indische und germanische Segenssprüche, in: Z. für vergleichende Sprachforschung 13 (1864), S. 66f); ed. b) ancient voice. wikidot.com, Hymn XCVI (96), 2009617–19 [...] 22

Beispieltext 55: Schwinden, flieg dahin! (Rig-Veda X, 97), ed. Kuhn, Adalbert, Indische und germanische Segenssprüche, in: Z. für vergleichende Sprachforschung 13 (1864), S. 69f

Beispieltext 56: Entsühnung gegen Schmerzen mit Windrose (Atharvaveda-Paippalāda 2.49, 1. Jahrtausend vor Christus), ed. Thomas Zehnder, Idstein 1999, S. 116

Beispieltext 57: Wurmbeschwörung mit Indras Mühlstein (Atharvaveda-Paippalāda 2.14), ed. Thomas Zehnder, Idstein 1999, S. 51–55

D 4 Texte aus der Tradition der Maya-Kulturen

Beispieltext 58: Großer Geburtsgesang (a) Mu-Igala or the Way of Muu, a medicine song from the Cuña-Indians of Panama, ed. Nils M. Holmer and Henry Wassen, Göteborg 1947; b) ed. Lévi-Strauss, Str.: Anthropologie I, S. 211–219

Beispieltext 59: Schamanengesang: Ich setze den Knochen ein! (Zaubersprüche bei den Maya und Lakandonen), ed. Christian Rätsch, München 1985, S. 135f

D5 Beispieltexte aus Papyri graecae magicae

Beispieltext 60: Antike Bärmutterbeschwörung (PGM VII. 260–71, London, Br. Mus., P.Lond.121, 3./4.Jh.), ed. Betz, Hans Dieter: The greek magical papyri, Chicago 1986, S. 123f
Beispieltext 61: Antike Dämonenaustreibung gegen ‚Besessenheit' frühchristlich koptisch? (PGM IV. 1227–64, Paris Bibl.Nat. Suppl. gr.no.574, 4.Jh.), ed. Betz, H.D., S. 62
Beispieltext 62: Antike Formel gegen Kopfschmerz, Erpressung der Götter (PGM CXXII, 51–55, aus dem hl. Buch des Trismegistos von Heliopolis, ägyptisch, 1.Jh vor/1.Jh. nach Chr.), ed. Betz, H.D., S. 317

F3 Kurze Erläuterung zu einigen cerebralen Funktionsgebieten

Amygdala, Mandelkern, wichtige Region für Entstehung und Steuerung von Emotionen. Sie besteht aus einer dem Riechhirn verbundenen unbewußt für Sexualkontakte wichtigen Zellgruppe, aus einem Zentralkern für vegetative Reaktionen und zur Stressregulation und aus einer Kerngruppe für Furchtkonditionierung als Abteilung für Entscheidung über Angriff oder Flucht mit emotionaler Erinnerung. Hier liegt eine Zellgruppe, die wie das ‚gebrannte Kind' nichts vergißt, um Schaden vorzubeugen oder Vorteil zu erlangen. Die Verbindungen zu allen Teilen des Gehirns sind sehr eng und nehmen Anteil an allen Geschehen in kognitiver, emotionaler und motorischer Hinsicht. Da die Reifung der A. im Leben früher abgeschlossen ist, als die Reifung des Hippocampus, dem Archivverwalter des bewußtseinsfähigen Gedächtnisses, so geht die Bildung von bewußtseinsfernen Emotionen der Sprachentwicklung voraus. Dies ist die Ursache für die Entstehung des Unbewußten. Für die Perzeption oder Abweisung von verbalen Triggern (psychoperformative Texte) hat die A. eine entscheidende Schlüsselfunktion (siehe B1.1, B4, B4.1.2, C1, C1.1, C2.2, C4, C4.2.3, C4.3.1.2, C5.3).

Cingulum (Gürtel), vorderes Cingulum (ACC, anteriorer cingulärer Cortex) liegt zwischen den beiden Hemisphären über dem diese verbindenden Balken. Er vermittelt zwischen Emotionen, Denken und Verhalten. Als ein Verbindungsglied im System zwischen Stirnhirn und Amygdala ist er an der Verarbeitung von Gefühlen und Schmerz beteiligt. Besonders wichtig soll er auch für die Lenkung von Aufmerksamkeit und die Taxierung von kognitiven Neuigkeiten sein. Deswegen dürfte er eine Relaisstation für außergewöhnliche Sprachgestalten, z. B. für Pseudoworte, Unmöglichkeiten, untriviale Sprach- und Erzählelemente der psychoperformativen Sprüche sein. Störungen von Form und Funktion finden sich bei Geisteskranken, bei denen die Unterscheidung des Absurden vom Gewöhnlichen behindert ist. Von manchen Forschern wird der ACC in einen zentralnervösen Regelkreis für die lebenserhaltende Homöostase, das Gleichgewicht des Stoffwechsels einbezogen. Andere Forscher sehen hier einen Kernpunkt für Konfliktverarbeitung zur Entlastung des limbischen Systems (siehe B1.1, B3, B4, C1, C4.3.1.2).

Frontaler Cortex (Stirnhirnrinde), besonders der praefrontale Cortex wird allgemein als beim Menschen für persönlichkeitsbildend gehalten. Handlungsplanung und Ich-Bewußtsein, Initiative und sozialer Gemeinsinn werden bei Störungen behindert. Dabei darf nicht vergessen werden, daß die Leistungen des Frontalhirns sehr eng mit emotional wichtigen Gebieten kooperieren, sodaß es für die Verarbeitung emotionaler Reize auch bei Therapie mit psychoperformativen Sprüchen hoch bedeutsam ist. Kognition ist nicht ohne Emotion möglich (Gerhard Roth (1998²), S. 211). Der praefrontale Cortex greift bei der Auswahl von Sinneseindrücken auf gespeicherte Erfahrungen bei Hilfsangeboten zurück oder kann sie verwerfen (B1.1, C4.1.1); seine dorsolateralen Anteile dienen der kritischen Entscheidungsfindung. Die linksseitigen Gebiete (B1.2, C4) unter Verbindung zum motorischen Sprachzentrum sind für Assoziationen und zielgerichtete Tropenauswahl wichtig und überlappen sich im inferioren Bereich mit Neuronen für die Perzeption von Prosodie, während rechtsseitig zumeist Konflikt- und Gefühlsbearbeitung zusammen mit Cingulum und Amygdala angenommen wird (B1.2, C1, C1.1, C1.3.2, C1.4.3). In dorsomedialen praefrontalen Gebieten soll es eine Korrelation mit Gefühlsunterdrückung geben (C1.1). Auch bei Schmerzempfindung (C4.3.1.2) und bei hochaversiven Stimuli (C5.3) findet Aktivierung praefrontaler Bereiche statt. Einbezogen in die Aufrechterhaltung der lebenswichtigen Homöostase wird auch der rechte praefrontale Cortex (C4.3.1.2).

Hippocampus, (Seepferd), liegt beidseits am unteren inneren Rande des Schläfenlappens. Er gilt als Archivverwalter des bewußtseinsfähigen Gehirns, der nicht selbst speichert, sondern Wissen und Erfahrungen in die entsprechenden Rindengebiete einordnet bzw. abruft. Bei den sog. Inselbegabungen ist er defekt, sodaß die betroffenen Musik-, Rechen-, Sprach- und Erinnerungsgenies (‚idiot savants') oft gleichsam in ihrem eigenen einzigen Archiv leben. An der Taxierung von Neuigkeiten, wie sie psychoperformative Sprüche bieten, ist er gemeinsam mit dem anterioren Cingulum beteiligt. Es wird vermutet, daß Neuigkeitssignale zur erhöhten Ausschüttung von Dopamin führen, während der Stress einer Notlage durch seine Anregung der Cortisolausschüttung kurzzeitig neue Gedächtnisinhalte abberuft und gleichzeitig das Altgedächtnis verschlechtert, sodaß die Suggestionen der Spruchtexte via imaginativer Produktivitätssteigerung besser wirksam werden können.
(siehe B4, C3.2, C4, C4.1.1, C5.3)

Hypothalamus gehört zum Zwischenhirn. Er ist mit der Hypophyse, der Hirnanhangdrüse durch einen Stiel verbunden (**Abb. 3**). Damit dirigiert er in enger Verbindung mit der Amygdala die wichtigsten hormonellen Funktionen des Körpers und das Vegetativum. Biologische Grundfunktionen wie Atmung, Kreislauf und Schlaf sowie Angriffs-, Sexual- oder Verteidigungsverhalten werden über ihn kontrolliert und modifiziert. Der Hypothalamus wird von Außenreizen nicht direkt erreicht, stets sind Strukturen des dopaminergen Systems, neben Amygdala auch der Nucleus accumbens, die Insula sowie die praefrontale Rinde vorgeschaltet (siehe B1.1, C5.3).

Insula (insulärer Cortex) wird oft als fünfter Gehirnlappen verstanden, ist nur aufgrund der Ontogenese von anderen überdeckt. Er ist ursprünglich eng mit Geschmacks- und Geruchsfunktionen verbunden, hat aber viele Aufgaben, besonders die emotionale Schmerzbewertung durch die viscerosensiblen Bahnen zu inneren Organen (Hunger, Durst, Übelkeit, Ekel, Atemnot usw.). Wegen dieser direkten Verbindung zum Körper gibt es Forscher, die für die Insula sogar eine der Amygdala vorgeschaltete Aktivität bei Emotionen vermuten. Für den psychoperformativen Spruch ist die Insula vielfach involviert: Sie ist aktiviert bei emotionaler Schmerzbewertung, Mitgefühl mit Schmerzen anderer, Bewertung von Gerüchen (Weihrauch!), zur Regulierung der Homöostase (Stoffwechsel) in Schocksituationen, bei Wortwiederholungen u. a., vieles ist noch unklar, insbesondere ob die enge Zusammenarbeit mit der Amygdala synergetisch oder hemmend abläuft und inwieweit die relative Größe beim Menschen mit elementaren Gefühlen und mit Introspektion zusammenhänge (siehe B3, c4, C4.2.3, C4.3.1.2).

Locus caeruleus (coeruleus) liegt in der vorderen Rautengrube der Brückenhaube des Mittelhirns. Er ist vermutlich an der Steuerung von Orientierung und Aufmerksamkeit und am Kurzzeitgedächtnis beteiligt. Er ist Ursprung des zentralnervösen Noradrenalinsystems, das durch Betarezeptoren-Antagonisten wie Propranolol (Betablocker) gehemmt werden kann. Wird er erregt, so kommt es zur Freigabe von Transmittern, die ihrerseits breite Erregung hervorrufen. Opiate und Alkohol dämpfen seine Aktivität. Für die Spruchkultur ist es wahrscheinlich, daß aversive Inhalte das Hormonsystem auch antagonistisch anregen und zur Bereitstellung von Cortisol beitragen. (siehe C4.3.1.2, C5.3).

Nucleus accumbens ist eine Neuronenansammlung an der Basis des Vorderhirns in der Nähe der Basalganglien, die man zum mesolimbischen Grenzgebiet zählt. Funktionell wird er als Belohnungssystem beschrieben und wurde zuerst bei Tieren entdeckt. Die Ausschüttung von Dopamin ist wichtig für sein Funktionieren, das letztlich die erfüllte positive momentane, nicht langfristige Erwartung mit Zufriedenheit belohnt. Bei Stress etwa nach aversiven Stimuli bekommt der Nucleus accumbens Reize von Hippocampus, Amygdala und Praefrontalrinde, verbunden mit Stresshormonausschüttung. Diese Reize können unter veränderten Bedingungen des Erregungsinputs via Plastizität zu belohnungsbezogenem Verhalten mit Sucht und Suchtrückfällen beitragen. Die Forschung versucht vor allem in Tierversuchen den involvierten Dopamin-Oxytocin- und Glutamatstoffwechsel zu ergründen. Für den performativen Text relevant sind sowohl die Erwartungshaltung, die die Aufmerksamkeit erhöht, als auch die Suche nach befriedigenden, Angst und Schmerz lösenden Ressourcen sowie die Befreiung von jedem Druck, z. B. der Beendigung introversiver Katharsis als ‚Belohnung' und ‚Motivation' (siehe B1.1, C1.3.5, C4.3.1.2).

Temporoparietaler Übergang (temporoparietal junction): Während die temporale Rinde besonders mit Raumordnungsaufgaben betraut ist und die parietale Rinde links mit Sprach- und Bildverarbeitung, wird von vielen Autoren ihrer Übergangsregion rechts eine besondere Rolle in integrierenden Netzwerken zugeschrieben.
Einerseits soll diese Verbindungsstelle an der Konstanz von Ich-Identität und Unterscheidung vom Verhalten anderer beteiligt sein. Zusammen mit dem dorsomedialen praefrontalen Cortex soll das Verstehen anderer im Rahmen eines Mentalisierungsnetzwerks ermöglicht werden, einschließlich der Mitempfindung der Schmerzen anderer (**Abb. 1**). Für die Verarbeitung von regressionsfördernden Erzählungen und Metaphern mag dieses Gebiet zusammen mit dem hinteren Cingulum, dem Hippocampus und mit speichernden Arealen von Bedeutung sein (siehe B3, C4. C4.1.1).

F4 Nachweis der Abbildungen

Abb. 00 (Titelbild) Zeichnung Fritz Klier, Vornbach am Inn, 2010
Abb. 1   Gerrans, Philip, Imitation and Theory of Mind, in: Berntson / Cacioppo, Handbook of Neuroscience, II, Fig. 43, S. 910
Abb. 2   Decety, Jean and Claus Lamm, Empathy and Intersubjectivity, in: Berntson / Cacioppo II, Fig. 48.2 (color C45), S. 948
Abb. 3   Ernst, Wolfgang, Beschwörungen und Segen, Abb. 88, S. 370
Abb. 4   Gehirn und Geist, Spektrum der Wissenschaft, Nr. 1/2002, S. 83
Abb. 5   Dumas, Guillaume et al., Inter-Brain Synchronisation during social Interaction, in : PLoS ONE 5(8) : e12166. doi:10.1371/j.pone, Seite 8, Fig.3
Abb. 6   Drenhaus, Heiner und Peter beim Graben: Grundlagen der EEG/ EKP-Methodik, in: Zeitschrift für Germanistische Linguistik 40 (2012), S. 73

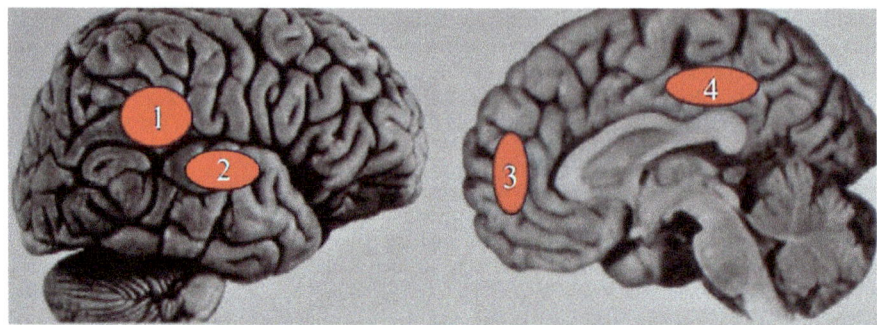

Abb. 1: Seit- und Medialansicht des Gehirns mit den bei durch eine Metaanalyse ermittelten häufigsten Überzeugungs- und Glaubens-Attributionen einschließlich Verstehen von Erzählungen, Erwartungen, Comic-Bilder-Deutungen, fremdem Augenausdruck (RME-Test), Vertrauenszusage und anderen Testergebnissen. (Neurale Substrate für ‚Theory of mind', die Annahmen über Zustand, Verhalten und Denken anderer) 1= Temporoparietaler Übergang, 2= Rechter oberer Temporalsulcus, 3= Praefrontaler Cortex, 4= Posteriores Cingulum.

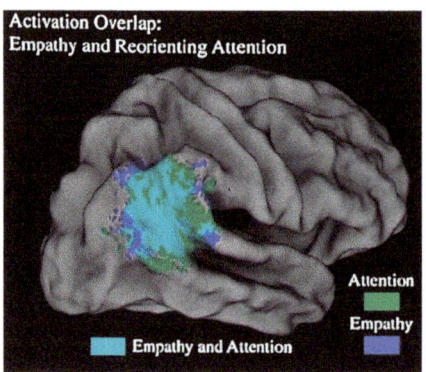

Abb. 2: Befunde sozialer Interaktion im rechten Temporo-parietal-Bereich: Die Überlappung von Schmerzmitgefühl (grün) und Aufmerksamkeitsorientierung (blau)

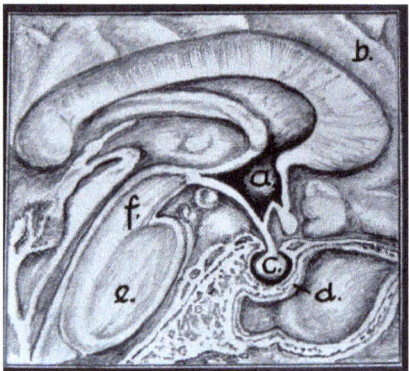

Abb. 3: Schema der anatom. Lage des Hypothalamus a= Hypothalamus, b= Cingulum c= Hypophyse, d= Türkensattel, e= Brücke, f= Mittelhirn.

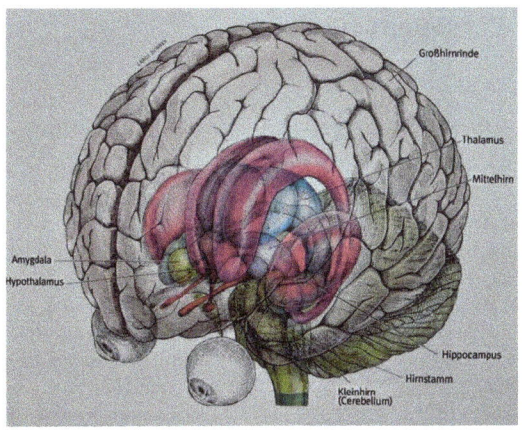

Abb. 4 Schematische Projektion der mittleren tiefen Strukturen in das Gehirn

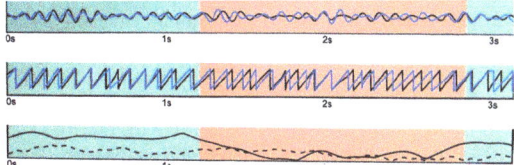

Abb. 5 Hyperscanning: Simultane EEG-Ableitung zweier Subjekte mit intermittierender Phasen-Synchronisation (grüner Bereich) bei imitativem Austausch.

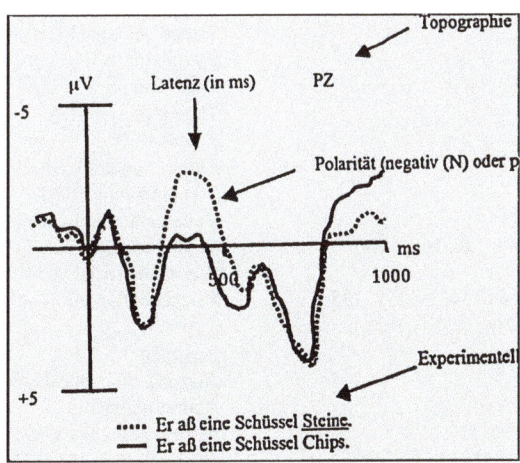

Abb. 6 Schema des N400-Potentials im EEG (punktiert) bei Präsentation eines semantischen Unsinns. Negativität nach oben gepolt.

F5 Sach- und Personenverzeichnis

Sachen
Aberglaube 26, 80
Alptraum-Syndrom 6, 88, 94, 99ff
Anthropomorphismus 31, 59, 108, 113, 140
Arzt-Patient-Beziehung 45f, 73
Atram-hasis-Epos 86
Babyschreien 48
Beschwörungstypen 170f
Besessenheit 87, 100, 167, 172
Borderline-Syndrom 119ff
Buß-und Reue-Übung 162f
Christozentrik 56, 80, 91, 168
Coping (Stress) 97f, 172, 176
Cortisol 106, 146, 166f
Dopamin 50, 54, 84, 106, 166
Embodied Simulation 66, 72, 139, 146, 164f
Endorphine 71
Ephesia grammata 42, 104
Ereignis-Potentiale 129f
Exorzismus 105, 171
Expositionstherapie 106, 156, 163f
Figurinen 113, 149
Freiheitsbegriff 46, 52
Ganzheitsmedizin 45, 179
Geheimwissenschaft 26, 54
Gilgames-Epos 86f, 109
Heilige Hochzeit 149
Helfersyndrom 64
Hyperscanning 68
Hypnotherapie 58f, 103
Idisi (weise Frauen) 149, 196, 167, 174
Isiskult 124
Katathymes Bilderleben 6, 34, 79, 132, 145
Katechese 78, 102, 117
Kindchenschema 47
Klagelieder 93
Klosterkultur 76ff, 81, 91
Kreative Aggression 106, 113, 173
Kultur-Natur-Gegensatz 15, 32, 74, 181
Magia naturalis 25, 27
Maqlû-Beschwörung 158f, 171, 174

Marduk-Ea-Typ 88f, 133, 149, 152, 171
Memoria 79
Mutterschoß-Imagination 56, 141ff, 182
Nacht-Symbolik 101f
Namensgebung 104
Nature and Nurture 74
Neuroling. Programmieren 35, 62f, 125
Oxytocin 106, 147f, 176
Participation mystique 30, 56
Placebobedingung 50f, 96, 165, 179
Polytheismus 86, 92, 113, 142
Pränatalerinnerung 143ff
Primingforschung 154f, 174
Prosodie 73, 116, 140, 153
Psychoanalyse 33f, 54, 61, 64, 109, 146
Psychodrama 100, 122, 167
Reframing 125, 163
Reimverarbeitung 130f
Rituelle Techniken 58, 114, 141, 147
Schadenzauber 89, 108, 117
Schmerzinduktion 157–163
Sensorimotorik 65f, 105f, 174
Sieben Schläfer 141ff, 182
Spiritualität 83ff
Tabu-Tod 35, 175
Theory of mind 134
Totenkult 90
Traum-Symbolik 64, 102ff, 108ff
Traumdeutung 108ff, 114
Unbewußtes, kollektives 30
Vogel federlos 124f, 191
Westöstlicher Diwan 63f
Zeichenhaftigkeit 77, 81f, 101

Personen
Agrippa von Nettesheim 23, 26, 40
Alanus ab Insulis 77
Albertus Magnus 23, 25
Aristoteles 78, 104, 157, 161
Arnald von Villanova 102
Assurbanipal 111, 118

Augustinus 25, 81, 102, 104, 162
Austin, John L. 29, 38f, 42
Bateson, Gregory 129
Bergengruen, Werner 162
Berger, Hans 128
Beth, Karl 26, 37
Boëthius 81
Carossa, Hans 151
Durkheim, Emile 28
Eco, Umberto 179
Erickson, M.H. 62, 125, 146
Foucault, Michael 22, 104
Frazer, James G. 281, 37, 139
Goethe, Joh. Wolfgang 15, 63f
Hrabanus Maurus 162
Hildegard von Bingen 61, 81
Ignatius von Loyola 79, 162
Johannes Nider 79
Johannes Paul II 162
Jung, C.G. 30, 58
Kandel, Eric 36
Malinowski, Bronislaw 29, 31, 36, 39
Marcellus von Bordeaux 123f
Mirandola, G.P. della 179
Paracelsus 27
Petrus Hispanus 45
Plath, Sylvia 107
Pseudo-Bonaventura 161
Rhazes (Al Razi) 16, 103
Saint-Exupéry, A. de 116f
Schiller, Friedrich 143
Searle, John 29, 38, 43
Tertullian 77
Thomas von Aquin 79
Tylor, Edward B. 28, 139

www.ingramcontent.com/pod-product-compliance
Ingram Content Group UK Ltd.
Pitfield, Milton Keynes, MK11 3LW, UK
UKHW021828210426
5322IPUK00004B/78